FORMULAS/EQUATIONS

Distance Formula

If $P_1 = (x_1, y_1)$ and $P_2 = (x_2, y_2)$, the distance from P_1 to P_2 is

$$d(P_1, P_2) = \sqrt{(x_2 - x_1)^2 + (y_2 - y_1)^2}$$

Standard Equation of a Circle

The standard equation of a circle of radius r with center at (h, k) is

$$(x - h)^2 + (y - k)^2 = r^2$$

Slope Formula

The slope m of the line containing the points $P_1 = (x_1, y_1)$ and $P_2 = (x_2, y_2)$ is

$$m = \frac{y_2 - y_1}{x_2 - x_1} \qquad \text{if } x_1 \neq x_2$$

$$m \text{ is undefined} \qquad \text{if } x_1 = x_2$$

Point–Slope Equation of a Line

The equation of a line with slope m containing the point (x_1, y_1) is

$$y - y_1 = m(x - x_1)$$

Slope–Intercept Equation of a Line

The equation of a line with slope m and y-intercept b is

$$y = mx + b$$

Quadratic Formula

The solutions of the equation $ax^2 + bx + c = 0, a \neq 0$, are

$$x = \frac{-b \pm \sqrt{b^2 - 4ac}}{2a}$$

If $b^2 - 4ac > 0$, there are two unequal real solutions.
If $b^2 - 4ac = 0$, there is a repeated real solution.
If $b^2 - 4ac < 0$, there are two complex solutions that are not real.

GEOMETRY FORMULAS

Circle

r = Radius, A = Area, C = Circumference
$A = \pi r^2 \qquad C = 2\pi r$

Triangle

b = Base, h = Altitude (Height), A = area
$A = \frac{1}{2}bh$

Rectangle

l = Length, w = Width, A = area, P = perimeter
$A = lw \qquad P = 2l + 2w$

Rectangular Box

l = Length, w = Width, h = Height, V = Volume
$V = lwh$

Sphere

r = Radius, V = Volume, S = Surface area
$V = \frac{4}{3}\pi r^3 \qquad S = 4\pi r^2$

Right Circular Cylinder

r = Radius, h = Height, V = Volume
$V = \pi r^2 h$

Trigonometry

SIXTH EDITION

Michael Sullivan
Chicago State University

Prentice Hall
Upper Saddle River, New Jersey 07458

Library of Congress Cataloging-in-Publication Data

Sullivan, Michael,
   Trigonometry / Michael Sullivan.—6th ed.
      p. cm.
   ISBN 0-13-041224-4
   1. Trigonometry.   I. Title.
   QA531.S85 2002
   516'.24—dc21   2001040037

Editor-in-Chief/Acquisitions Editor: Sally Yagan
Associate Editor: Dawn Murrin
Vice President/Director of Production and Manufacturing: David W. Riccardi
Executive Managing Editor: Kathleen Schiaparelli
Senior Managing Editor: Linda Mihatov Behrens
Production Editor: Bob Walters
Manufacturing Buyer: Alan Fischer
Manufacturing Manager: Trudy Pisciotti
Marketing Manager: Patrice Lumumba Jones
Assistant Managing Editor, Math Media Production: John Matthews
Art Director: John Christiana
Interior Designer: Judith A. Matz-Coniglio
Cover Designer: Maureen Eide
Creative Director: Carole Anson
Managing Editor Audio/Video Assets: Grace Hazeldine
Director of Creative Services: Paul Belfanti
Photo Editor: Beth Boyd
Cover Photo: George B. Diebold/The Stock Market
Art Studio: Artworks:
   Senior Manager: Patty Burns
   Production Manager: Ronda Whitson
   Manager, Production Technologies: Matt Haas
   Project Coordinator: Jessica Einsig
   Illustrator: Steve McKinley

 © 2002, 1999, 1996, 1993, 1990, 1987 by Prentice-Hall, Inc.
Upper Saddle River, New Jersey 07458

Printed in the United States of America
10 9 8 7 6 5 4 3 2 1

ISBN: 0-13-041224-4

Pearson Education LTD., *London*
Pearson Education Australia PTY, Limited, *Sydney*
Pearson Education Singapore, Pte. Ltd.
Pearson Education North Asia Ltd., *Hong Kong*
Pearson Education Canada Ltd., *Toronto*
Pearson Education de Mexico, S.A. de C.V.
Pearson Education—Japan, *Tokyo*
Pearson Education Malaysia, Pte. Ltd.

To the Memory of Joe and Rita
and my sister Maryrose

Contents

APPENDIX A Review 487

APPENDIX B Graphing Utilities 560

Preface to the Instructor

As a professor at an urban public university for over 30 years, I am aware of the varied needs of trigonometry students who range from having little mathematical background and a fear of mathematics courses to those who have had a strong mathematical education and are highly motivated. For some of your students, this will be their last course in mathematics, while others may decide to further their mathematical education. I have written this text for both groups. As the author of precalculus, engineering calculus, finite mathematics, and business calculus texts, and, as a teacher, I understand what students must know if they are to be focused and successful in upper level mathematics courses. However, as a father of four college graduates, I also understand the realities of college life. I have taken great pains to insure that the text contains solid, student-friendly examples and problems, as well as a clear, seamless, writing style. I encourage you to share with me your experiences teaching from this text.

THE SIXTH EDITION

The Sixth Edition builds upon a solid foundation by integrating new features and techniques that further enhance student interest and involvement. The elements of previous editions that have proved successful remain, while many changes, some obvious, others subtle, have been made. A huge benefit of authoring a successful series is the broad-based feedback upon which improvements and additions are ultimately based. Virtually every change to this edition is the result of thoughtful comments and suggestions made from colleagues and students who have used previous editions. I am sincerely grateful for this feedback and have tried to make changes that improve the flow and usability of the text.

NEW TO THE SIXTH EDITION

Real Mathematics at Motorola

Each chapter begins with ▌ Field Trip to Motorola, a brief description of a current situation at Motorola, followed by ▌ Interview at Motorola, a biographical sketch of a Motorola employee. At the end of each chapter is ▌ Project at Motorola, written by the Motorola employee, that contains a description, with exercises, of a problem at Motorola that relates to the mathematics found in the chapter. It doesn't get more REAL than this.

Preparing for This Section

Most sections now open with a referenced list (by section and page number) of key items to review in preparation for the section ahead. This provides a just-in-time review for students.

Appendix A Review

This Appendix has been renamed to more accurately reflect its content.

The content here consists of a more detailed version of the first section of the old Chapter 1, an expansion of the material found in the old Appendix A, and Complex Numbers. Although it could be used as the first part of a course in Trigonometry, its real value lies in its use as a just-in-time review of material. Specific references to Appendix A occur throughout the text to assist in the review process. Appropriate use of this appendix will allow students to review when they need to and will allow the instructor more time to cover the course content.

Content

- Appendix B, Graphing Utilities, has been updated and expanded to include the latest features of the graphing calculator. While the graphing calculator remains an option, identified by a graphing icon 📷, references to Appendix B occur at appropriate places in the text for those inclined to use the graphing calculator features of the text.
- The Cross Product is a new section in Chapter 5, Polar Coordinates and Vectors.
- Area of a Sector is a new sub-section in Chapter 2, Section 2.1, Angles and Their Measure.
- Combining Waves is a new sub-section in Chapter 4. Applications of Trigonometry, Section 4.5.

Organization

- The chapter on Trigonometric Functions now has a single section devoted to the graphs of the sine and cosine functions, including a discussion of sinusoidal graphs. Separate sections follow on Graphs of the Remaining Trigonometric Functions and Phase Shift; Sinusoidal Curve Fitting. These changes will allow the material of each section to be taught in a single period and provide flexibility in choice of content.
- The chapter on Analytic Trigonometry now begins with two sections that discuss the inverse trigonometric functions. The chapter concludes with two sections devoted to Trigonometric Equations. These changes will allow each section to be taught in a single period.

FEATURES IN THE 6TH EDITION

- Section **OBJECTIVES** appear in a numbered list to begin each section.
- ✏️ **NOW WORK PROBLEM XX** appears after a concept has been introduced. This directs the student to a problem in the exercises that tests the concept, insuring that the concept has been mastered before moving on. The Now Work problems are identified in the exercises using yellow numbers and a ✏️ pencil icon.
- Optional Comments, Explorations, Seeing the Concept, Examples, and Exercises that utilize the graphing calculator are clearly marked with a calculator icon. Calculator exercises are also identified by the 📷 icon and green numbers.
- Discussion, Writing, and Research problems appear in most exercise sets, identified by an 🖉 icon and red numbers. These provide the basis for class discussion, writing projects, and collaborative learning experiences.
- References to Calculus are identified by a △ calculus icon.
- Historical Perspectives, sometimes with exercises, are presented in context and provide interesting anecdotal information.

- Varied applications are abundant both in Examples and in Exercises. Many contain sourced data.
- An extensive Chapter Review provides a list of important formulas, definitions, theorems, and objectives, as well as a complete set of Review Exercises, with sample test questions identified by blue numbers.

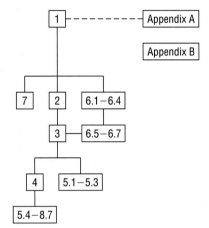

USING THE 6TH EDITION EFFECTIVELY AND EFFICIENTLY WITH YOUR SYLLABUS

To meet the varied needs of diverse syllabi, this book contains more content than expected in a trigonometry course. The illustration shows the dependencies of chapters on each other.

As the chart indicates, this book has been organized with flexibility of use in mind. Even within a given chapter, certain sections can be skipped without fear of future problems.

Chapter 1 Functions and Their Graphs

This chapter is now more streamlined than before. A quick coverage of this chapter, which is mainly review material, will enable you to get to Chapter 2 Trigonometric Functions earlier.

Chapter 2 Trigonometric Functions

The sections follow in sequence.

Chapter 3 Analytic Trigonometry

The sections follow in sequence. Sections 3.2, 3.6, and 3.8 may be skipped in a brief course.

Chapter 4 Applications of Trigonometric Functions

The sections follow in sequence. Sections 4.4 and 4.5 may be skipped in a brief course.

Chapter 5 Polar Coordinates; Vectors

Sections 5.1–5.3 and Sections 5.4–5.7 are independent and may be covered separately.

Chapter 6 Analytic Geometry

Sections 6.1–6.4 follow in sequence. Sections 6.5, 6.6, and 6.7 are independent of each other, but do depend on Sections 6.1–6.4.

Chapter 7 Exponential and Logarithmic Functions

Sections 7.1–7.5 follow in sequence; Sections 7.6, 7.7, and 7.8 each require Section 7.3.

ACKNOWLEDGMENTS

Textbooks are written by authors, but evolve from an idea into final form through the efforts of many people. Special thanks to Don Dellen, who first suggested this book and the other books in this series. Don's extensive contributions to publishing and mathematics are well known; we all miss him dearly.

I would like to thank Motorola and its people who helped make the projects in this new edition possible. Special thanks to Iwona Turlik, Vice President and Director of the Motorola Advanced Technology Center (MATC), for providing the opportunity to share with students examples of their experience in applying mathematics to engineering tasks.

I would also like to thank the authors of these projects:

- Tomasz Klosowiak, the Automotive and Industrial Electronics Group of Integrated Electronic Systems Segment
- Nick Buris, Brian Classon, Terri Fry, and Margot Karam, Motorola Laboratories, Communications Systems and Technologies Labs (CSTL)
- Bill Oslon, Andrew Skipor, John St. Peter, Tom Tirpak, and George Valliath, Motorola Laboratories, Motorola Advanced Technology Center (MATC)
- Jocelyn Carter-Miller, Corporate Vice President and Chief Marketing Officer
- Sheila MB. Griffen, Vice President and Director, Corporate Strategic Marketing, Chief Marketing Office
- Sue Eddins, Curriculum and Assessment Leader for Mathematics and Chuck Hamberg, Mathematics Faculty of the Illinois Mathematics and Science Academy for their generous help and contributions to Chapters 1 and 2.

Special thanks also go to the following:

- Douglas Fekete, Intellectual Property Department
- Jim Coffiing, Director Communications Future Business and Technology
- Anne Stuessy, Director Communications Future Business and Technology
- Vesna Arsic, Director Corporate Marketing Strategy, Chief Marketing Office
- Rosemarie Broda, Chief Marketing Office
- Rita Browne, CSTL
- David Broth, Vice President and Director of the Communications Systems and Technologies Labs
- Joseph Nowack and Bruce Eastmond, CSTL managers
- Tom Babin, Kevin Jelley and Bill Olson, MATC managers

Last, but not least, for his dedication to this project and the daunting task of managing it, I thank Andrew Skipor.

There are many colleagues I would like to thank for their input, encouragement, patience, and support. They have my deepest thanks and appreciation. I apologize for any omissions…

James Africh, *College of DuPage*
Steve Agronsky, *Cal Poly State University*
Grant Alexander, *Joliet Junior College*
Dave Anderson, *South Suburban College*
Joby Milo Anthony, *University of Central Florida*
James E. Arnold, *University of Wisconsin-Milwaukee*
Carolyn Autray, *University of West Georgia*
Agnes Azzolino, *Middlesex County College*
Wilson P Banks, *Illinois State University*
Sudeshna Basu, *Howard University*
Dale R. Bedgood, *East Texas State University*
Beth Beno, *South Suburban College*
Carolyn Bernath, *Tallahassee Community College*

William H. Beyer, *University of Akron*
Annette Blackwelder, *Florida State University*
Richelle Blair, *Lakeland Community College*
Trudy Bratten, *Grossmont College*
Joanne Brunner, *Joliet Junior College*
Warren Burch, *Brevard Community College*
Mary Butler, *Lincoln Public Schools*
William J. Cable, *University of Wisconsin-Stevens Point*
Lois Calamia, *Brookdale Community College*
Jim Campbell, *Lincoln Public Schools*
Roger Carlsen, *Moraine Valley Community College*

Elena Catoiu, *Joliet Junior College*
John Collado, *South Suburban College*
Nelson Collins, *Joliet Junior College*
Jim Cooper, *Joliet Junior College*
Denise Corbett, *East Carolina University*
Theodore C. Coskey, *South Seattle Community College*
John Davenport, *East Texas State University*
Faye Dang, *Joliet Junior College*
Antonio David, *Del Mar College*
Duane E. Deal, *Ball State University*
Timothy Deis, *University of Wisconsin-Platteville*
Vivian Dennis, *Eastfield College*
Guesna Dohrman, *Tallahassee Community College*

Karen R. Dougan, *University of Florida*
Louise Dyson, *Clark College*
Paul D. East, *Lexington Community College*
Don Edmondson, *University of Texas-Austin*
Erica Egizio, *Joliet Junior College*
Christopher Ennis, *University of Minnesota*
Ralph Esparza, Jr., *Richland College*
Garret J. Etgen, *University of Houston*
W.A. Ferguson, *University of Illinois-Urbana/Champaign*
Iris B. Fetta, *Clemson University*
Mason Flake, *student at Edison Community College*
Timothy W. Flood, *Pittsburg State University*
Merle Friel, *Humboldt State University*
Richard A. Fritz, *Moraine Valley Community College*
Carolyn Funk, *South Suburban College*
Dewey Furness, *Ricke College*
Dawit Getachew, *Chicago State University*
Wayne Gibson, *Rancho Santiago College*
Robert Gill, *University of Minnesota Duluth*
Sudhir Kumar Goel, *Valdosta State University*
Joan Goliday, *Sante Fe Community College*
Frederic Gooding, *Goucher College*
Sue Graupner, *Lincoln Public Schools*
Jennifer L. Grimsley, *University of Charleston*
Ken Gurganus, *University of North Carolina*
James E. Hall, *University of Wisconsin-Madison*
Judy Hall, *West Virginia University*
Edward R. Hancock, *DeVry Institute of Technology*
Julia Hassett, *DeVry Institute-Dupage*
Michah Heibel, *Lincoln Public Schools*
LaRae Helliwell, *San Jose City College*
Brother Herron, *Brother Rice High School*
Robert Hoburg, *Western Connecticut State University*
Lee Hruby, *Naperville North High School*
Kim Hughes, *California State College-San Bernardino*
Ron Jamison, *Brigham Young University*
Richard A. Jensen, *Manatee Community College*
Sandra G. Johnson, *St. Cloud State University*
Tuesday Johnson, *New Mexico State University*
Moana H. Karsteter, *Tallahassee Community College*
Arthur Kaufman, *College of Staten Island*
Thomas Kearns, *North Kentucky University*

Shelia Kellenbarger, *Lincoln Public Schools*
Keith Kuchar, *Manatee Community College*
Tor Kwembe, *Chicago State University*
Linda J. Kyle, *Tarrant Country Jr. College*
H.E. Lacey, *Texas A & M University*
Harriet Lamm, *Coastal Bend College*
Matt Larson, *Lincoln Public Schools*
Christopher Lattin, *Oakton Community College*
Adele LeGere, *Oakton Community College*
Kevin Leith, *University of Houston*
Jeff Lewis, *Johnson County Community College*
Stanley Lukawecki, *Clemson University*
Janice C. Lyon, *Tallahassee Community College*
Virginia McCarthy, *Iowa State University*
Jean McArthur, *Joliet Junior College*
Tom McCollow, *DeVry Institute of Technology*
Laurence Maher, *North Texas State University*
Jay A. Malmstrom, *Oklahoma City Community College*
Sherry Martina, *Naperville North High School*
Alec Matheson, *Lamar University*
James Maxwell, *Oklahoma State University-Stillwater*
Judy Meckley, *Joliet Junior College*
David Meel, *Bowling Green State University*
Carolyn Meitler, *Concordia University*
Sarnia Metwali, *Erie Community College*
Rich Meyers, *Joliet Junior College*
Eldon Miller, *University of Mississippi*
James Miller, *West Virginia University*
Michael Miller, *Iowa State University*
Kathleen Miranda, *SUNY at Old Westbury*
Thomas Monaghan, *Naperville North High School*
Craig Morse, *Naperville North High School*
Samad Mortabit, *Metropolitan State University*
A. Muhundan, *Manatee Community College*
Jane Murphy, *Middlesex Community College*
Richard Nadel, *Florida International University*
Gabriel Nagy, *Kansas State University*
Bill Naegele, *South Suburban College*
Lawrence E. Newman, *Holyoke Community College*
James Nymann, *University of Texas-El Paso*
Sharon O'Donnell, *Chicago State University*

Seth F. Oppenheimer, *Mississippi State University*
Linda Padilla, *Joliet Junior College*
E. James Peake, *Iowa State University*
Thomas Radin, *San Joaquin Delta College*
Ken A. Rager, *Metropolitan State College*
Kenneth D. Reeves, *San Antonio College*
Elsi Reinhardt, *Truckee Meadows Community College*
Jane Ringwald, *Iowa State University*
Stephen Rodi, *Austin Community College*
Bill Rogge, *Lincoln Public Schools*
Howard L. Rolf, *Baylor University*
Phoebe Rouse, *Lousiana State University*
Edward Rozema, *University of Tennessee at Chattanooga*
Dennis C. Runde, *Manatee Community College*
John Sanders, *Chicago State University*
Susan Sandmeyer, *Jamestown Community College*
A.K. Shamma, *University of West Florida*
Martin Sherry, *Lower Columbia College*
Tatrana Shubin, *San Jose State University*
Anita Sikes, *Delgado Community College*
Timothy Sipka, *Alma College*
Lori Smellegar, *Manatee Community College*
John Spellman, *Southwest Texas State University*
Becky Stamper, *Western Kentucky University*
Judy Staver, *Florida Community College-South*
Neil Stephens, *Hinsdale South High School*
Christopher Terry, *Augusta State University*
Diane Tesar, *South Suburban College*
Tommy Thompson, *Brookhaven College*
Richard J. Tondra, *Iowa State University*
Marvel Townsend, *University of Florida*
Jim Trudnowski, *Carroll College*
Robert Tuskey, *Joliet Junior College*
Richard G. Vinson, *University of South Alabama*
Mary Voxman, *University of Idaho*
Jennifer Walsh, *Daytona Beach Community College*
Donna Wandke, *Naperville North High School*
Darlene Whitkenack, *Northern Illinois University*
Christine Wilson, *West Virginia University*
Brad Wind, *Florida International University*
Canton Woods, *Auburn University*
George Zazi, *Chicago State University*

Recognition and thanks are due particularly to the following individuals for their valuable assistance in the preparation of this edition: Sally Yagan, for her continued support and genuine interest; Patrice Jones, for his innovative marketing efforts; Bob Walters, for his organizational skills as production supervisor; Phoebe Rouse, for her specific suggestions for this edition and careful proofreading of page proof; Teri Lovelace of Laurel Technical Services for her proofreading skill and checking of my answers; and to the entire Prentice-Hall sales staff for their continuing confidence in this book.

Michael Sullivan

Preface to the Student

As you begin your study of Trigonometry you may feel overwhelmed by the number of theorems, definitions, procedures, and equations that confront you. You may even wonder whether or not you can learn all of this material in the time allotted. These concerns are normal. Keep in mind that many elements of Trigonometry are all around us as we go through our daily routines. Many of the concepts you will learn to express mathematically, you already know intuitively. For many of you, this may be your last math course, while for others, just the first in a series of many. Either way, this text was written with you in mind. I have taught trigonometry courses for over thirty years. I am also the father of four college graduates who called home from time to time, frustrated and with questions. I know what you're going through. So I have written a text that doesn't overwhelm, or unnecessarily complicate Trigonometry, while at the same time providing you the skills and practice you need to be successful.

This text is designed to help you, the student, master the terminology and basic concepts of Trigonometry. These aims have helped to shape every aspect of the book. Many learning aids are built into the format of the text to make your study of the material easier and more rewarding. This book is meant to be a "machine for learning," one that can help you focus your efforts and get the most from the time and energy you invest.

HOW TO USE THIS BOOK EFFECTIVELY
AND EFFICIENTLY

First, and most important, this book is meant to be read—so please, begin by reading the material assigned. You will find that the text has additional explanations and examples that will help you. Also, it is best to read the section before the lecture, so you can ask questions right away about anything you didn't understand.

Many sections begin with "Preparing for This Section," a list of concepts that will be used in the section. Take the short amount of time required to refresh your memory. This will make the section easier to understand and will actually save you time and effort.

A list of *OBJECTIVES* is provided at the beginning of each section. Read them. They will help you recognize the important ideas and skills developed in the section.

After a concept has been introduced and an example given, you will see NOW WORK PROBLEM XX. Go to the exercises at the end of the section, work the problem cited, and check your answer in the back of the book. If you get it right, you can be confident in continuing on in the section. If you don't get it right, go back over the explanations and examples to see what you might have missed. Then rework the problem. Ask for help if you miss it again.

If you follow these practices throughout the section, you will find that you have probably done many of your homework problems. In the exercises, every "Now Work Problem" number is in yellow with a pencil icon . All the odd-numbered problems have answers in the back of the book and worked-out solutions in the Student Solutions Manual supplement. Be sure you have made an honest effort before looking at a worked-out solution.

At the end of each chapter is a Chapter Review. Use it to be sure you are completely familiar with the equations and formulas listed under "Things to Know." If you are unsure of an item here, use the page reference to go back and review it. Go through the Objectives and be sure you can answer "Yes" to the question "I should be able to. . . ." If you are uncertain, a page reference to the objective is provided.

Spend the few minutes necessary to answer the "Fill-in-the-Blank" items and the "True/False" items. These are quick and valuable questions to answer.

Lastly, do the problems identified with blue numbers in the Review Exercises. These are my suggestions for a Practice Test. Do some of the other problems in the review for more practice to prepare for your exam.

Please do not hesitate to contact me, through Prentice Hall, with any suggestions or comments that would improve this text. I look forward to hearing from you.

Best Wishes!

Michael Sullivan

Trigonometric Functions

Field Trip to Motorola

For many centuries, trigonometry has been the basis for science, astronomy, and navigation. Today trigonometry is employed in various disciplines of common life, such as civil engineering, acoustics, analog and digital wireline and wireless communications, and many other applications related to wave theory.

77

Page 77

Interview at Motorola

Margot Karam is a senior staff engineer working as a researcher for Motorola Labs in Schaumburg, Illinois. She is part of a dynamic team designing a digital cellular phone able to operate in every country of the world. Her work consists of providing a solution to process fast data at low cost. Before joining Motorola Labs in the United States she worked in a product group of Motorola in France as a designer with a focus on delivering high-quality, te... ed digital circuits. She is pro... such an international compan... top provider for the best qua...

Page 78

Project at Motorola

Digital Transmission over the Air

Digital communications is a revolutionary technology of the century. For many years, Motorola has been one of the leading companies to employ digital communication in wireless devices, such as cell phones.

Figure 1 shows a simplified overview of a digital communication transmission over the air. The information source to be transmitted can be audio, video, or data. The information source may be formatted into a digital sequence of symbols from a finite set $\{\alpha_n\} = \{0, 1\}$. So 0110100 is an example of a digital sequence. The period of the symbols is denoted by T.

The principle of digital communication systems is that, during the finite interval of time T, the information symbol is represented by one digital waveform from a finite set of digital waveforms before it is sent. This technique is called **modulation.**

3. Evaluate $s(t)$ for $t = 0, \dfrac{1}{4f_0}, \dfrac{1}{2f_0}, \dfrac{3}{4f_0},$ and $\dfrac{1}{f_0}$.

4. Graph $s(t)$ for $0 \le t \le 12T_0$. That is, graph 12 cycles of the function.

5. For what values of t does the function reach its maximum value?
   [**Hint:** Express t in terms of f_0].

Three modulation techniques are used for transmission over the air: amplitude modulation, frequency modulation and phase modulation. In this project, we are interested in phase modulation. Figure 2 illustrates this process. An information symbol is mapped onto a phase that modulates the carrier. The modulated carrier is expressed by $S_i(t) = \sin(2\pi f_0 t + \psi_i)$.

Let's assume the following mapping scheme:

$$\{\alpha_n\} \;\rightarrow\; \{\psi_n\}$$

Page 160

CLEAR WRITING STYLE

Sullivan's **accessible writing style** is apparent throughout, often utilizing various approaches to the same concept. An author who writes clearly makes potentially difficult concepts intuitive, making class time more productive.

Sometimes it is helpful to think of a function f as a machine that receives as input a number from the domain, manipulates it, and outputs the value. See Figure 6.

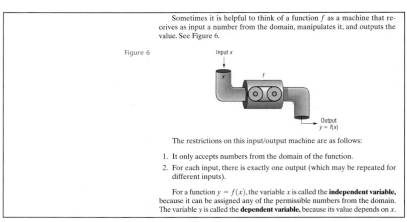

Figure 6

The restrictions on this input/output machine are as follows:

1. It only accepts numbers from the domain of the function.
2. For each input, there is exactly one output (which may be repeated for different inputs).

For a function $y = f(x)$, the variable x is called the **independent variable,** because it can be assigned any of the permissible numbers from the domain. The variable y is called the **dependent variable,** because its value depends on x.

Pages 25-26

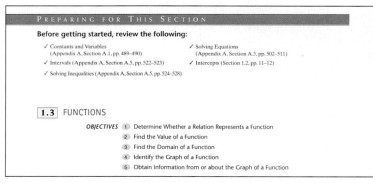

PREPARING FOR THIS SECTION

Before getting started, review the following:

✓ Constants and Variables (Appendix A, Section A.1, pp. 489–490)

✓ Intervals (Appendix A, Section A.5, pp. 522–523)

✓ Solving Inequalities (Appendix A, Section A.5, pp. 524–528)

✓ Solving Equations (Appendix A, Section A.3, pp. 502–511)

✓ Intercepts (Section 1.2, pp. 11–12)

1.3 FUNCTIONS

OBJECTIVES
1. Determine Whether a Relation Represents a Function
2. Find the Value of a Function
3. Find the Domain of a Function
4. Identify the Graph of a Function
5. Obtain Information from or about the Graph of a Function

Page 21

PREPARING FOR THIS SECTION

The **"Preparing for this Section"** feature provides you and your instructor with a list of skills and concepts needed to approach the section, along with page references. You can use the feature to determine what you should review before tackling each section.

STEP-BY-STEP EXAMPLES

Step-by-step examples ensure that you follow the entire solution process and give you an opportunity to check your understanding of each step.

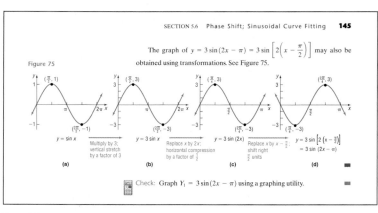

SECTION 5.6   Phase Shift; Sinusoidal Curve Fitting   **145**

The graph of $y = 3 \sin(2x - \pi) = 3 \sin\left[2\left(x - \dfrac{\pi}{2}\right)\right]$ may also be obtained using transformations. See Figure 75.

Figure 75

Check: Graph $Y_1 = 3 \sin(2x - \pi)$ using a graphing utility.

Page 145

TABLE 1	
Date	**Closing Price ($)**
8/31/99	41.09
9/30/99	37.16
10/31/99	38.72
11/30/99	38.34
12/31/99	41.16
1/31/00	49.47
2/29/00	56.50
3/31/00	65.97
4/30/00	63.41
5/31/00	62.34
6/30/00	66.84
7/31/00	66.75
8/31/00	74.88

Courtesy of A.G. Edwards & Sons, Inc.

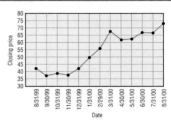

Figure 34
Monthly closing prices
of Intel stock 8/31/99
through 8/31/00

We can see from the graph that the price of the stock was rising rapidly from 11/30/99 through 3/31/00 and was falling slightly from 3/31/00 through 5/31/00. The graph also shows that the lowest price occurred at the end of September, 1999, whereas the highest occurred at the end of August, 2000. Equations and tables, on the other hand, usually require some calculations and interpretation before this kind of information can be "seen."

Look again at Figure 8. The graph shows that for each date on the horizontal axis there is only one price on the vertical axis. Thus, the graph represents a function, although the exact rule for getting from date to price is not

---

Finding the Distance between Two Cities

See Figure 13(a). The latitude L of a location is the angle formed by a ray drawn from the center of Earth to the Equator and a ray drawn from the center of Earth to L. See Figure 13(b). Glasgow, Montana, is due north of Albuquerque, New Mexico. Find the distance between Glasgow (48°9′ north latitude) and Albuquerque (35°5′ north latitude). Assume that the radius of Earth is 3960 miles.

Figure 13

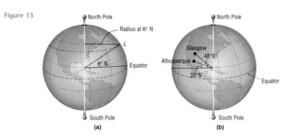

(a)   (b)

Solution   The measure of the central angle between the two cities is $48°9′ - 35°5′ = 13°4′$. We use equation (4), $s = r\theta$, but first we must convert the angle of 13°4′ to radians.

$$\theta = 13°4′ \approx 13.0667° = 13.0667 \cdot \frac{\pi}{180}\text{ radian} \approx 0.228\text{ radian}$$

We use $\theta = 0.228$ radian and $r = 3960$ miles in equation (4). The distance between the two cities is

$$s = r\theta = 3960 \cdot 0.228 \approx 903\text{ miles}\quad\blacksquare$$

When an angle is measured in degrees, the degree symbol will always be shown. However, when an angle is measured in radians, we will follow the usual practice and omit the word *radians*. So, if the measure of an angle is given as $\frac{\pi}{6}$, it is understood to mean $\frac{\pi}{6}$ radian.

  NOW WORK PROBLEM 91.

Figure 14

AREA OF A SECTOR

5   Consider a circle of radius r. Suppose that θ, measured in radians, is a central angle of this circle. See Figure 14. We seek a formula for the area A of the sector formed by the angle θ (shown in blue).

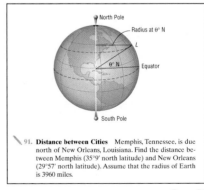

91. **Distance between Cities**   Memphis, Tennessee, is due north of New Orleans, Louisiana. Find the distance between Memphis (35°9′ north latitude) and New Orleans (29°57′ north latitude). Assume that the radius of Earth is 3960 miles.

OVERVIEW

SOLUTIONS

Solutions, both algebraic and graphical, are clearly expressed throughout the text.

EXAMPLE 1 **Graphing a Quadratic Function Using Trasformations**

Graph the function $f(x) = 2x^2 + 8x + 5$. Find the vertex and axis of symmetry.

Solution We begin by completing the square on the right side.

$$f(x) = 2x^2 + 8x + 5$$
$$= 2(x^2 + 4x) + 5 \qquad \text{Factor out the 2 from } 2x^2 + 8x.$$
$$= 2(x^2 + 4x + 4) + 5 - 8 \qquad \begin{array}{l}\text{Complete the square of } 2(x^2 + 4x).\\ \text{Notice that the factor of 2 requires}\end{array}$$
$$= 2(x + 2)^2 - 3 \qquad \text{that 8 be added and subtracted.} \qquad \textbf{(2)}$$

The graph of f can be obtained in three stages, as shown in Figure 6. Now compare this graph to the graph in Figure 5(a). The graph of $f(x) = 2x^2 + 8x + 5$ is a parabola that opens up and has its vertex (lowest point) at $(-2, -3)$. Its axis of symmetry is the line $x = -2$.

Figure 6

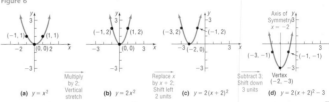

(a) $y = x^2$ Multiply by 2; Vertical stretch

(b) $y = 2x^2$ Replace x by $x + 2$; Shift left 2 units

(c) $y = 2(x + 2)^2$ Subtract 3; Shift down 3 units

(d) $y = 2(x + 2)^2 - 3$ Vertex $(-2, -3)$

 Check: Graph $f(x) = 2x^2 + 8x + 5$ and use the MINIMUM command to locate its vertex.

NOW WORK PROBLEM 17.

The method used in Example 1 can be used to graph any quadratic function $f(x) = ax^2 + bx + c, a \neq 0$, as follows:

$$f(x) = ax^2 + bx + c$$
$$= a\left(x^2 + \frac{b}{a}x\right) + c \qquad \text{Factor out } a \text{ from } ax^2 + bx.$$
$$= a\left(x^2 + \frac{b}{a}x + \frac{b^2}{4a^2}\right) + c - a\left(\frac{b^2}{4a^2}\right) \qquad \begin{array}{l}\text{Complete the square by adding}\\ \text{and subtracting } a(b^2/4a^2).\\ \text{Look closely at this step!}\end{array}$$
$$= a\left(x + \frac{b}{2a}\right)^2 + c - \frac{b^2}{4a}$$
$$= a\left(x + \frac{b}{2a}\right)^2 + \frac{4ac - b^2}{4a} \qquad c - \frac{b^2}{4a} = c \cdot \frac{4a}{4a} - \frac{b^2}{4a} = \frac{4ac - b^2}{4a}$$

Based on these results, we conclude the following:

GRAPHING UTILITIES

Graphing utilities are optional in this text and their use is clearly identified by the use of the graphing utility icon.

Pages 4-5

In Problems 1 and 2, plot each point in the xy-plane. Tell in which quadrant or on what coordinate axis each point lies.

1. (a) $A = (-3, 2)$
   (b) $B = (6, 0)$
   (c) $C = (-2, -2)$
   (d) $D = (6, 5)$
   (e) $E = (0, -3)$
   (f) $F = (6, -3)$

2. (a) $A = (1, 4)$
   (b) $B = (-3, -4)$
   (c) $C = (-3, 4)$
   (d) $D = (4, 1)$
   (e) $E = (0, 1)$
   (f) $F = (-3, 0)$

3. Plot the points $(2, 0), (2, -3), (2, 4), (2, 1),$ and $(2, -1)$. Describe the set of all points of the form $(2, y)$, where y is a real number.

4. Plot the points $(0, 3), (1, 3), (-2, 3), (5, 3),$ and $(-4, 3)$. Describe the set of all points of the form $(x, 3)$, where x is a real number.

In Problems 5–18, find the distance $d(P_1, P_2)$ between the points P_1 and P_2.

5.

6.

7.

8.

9. $P_1 = (3, -4); \quad P_2 = (5, 4)$
10. $P_1 = (-1, 0); \quad P_2 = (2, 4)$
11. $P_1 = (-3, 2); \quad P_2 = (6, 0)$
12. $P_1 = (2, -3); \quad P_2 = (4, 2)$

END-OF-SECTION EXERCISES

Sullivan's exercises are unparalleled in terms of thorough coverage and accuracy. Each **end-of-section exercise** set begins with visual- and concept-based problems, starting you out with the basics of the section. Well-thought-out exercises better prepare you for exams.

Page 7

MODELING

Many examples and exercises connect real-world situations to mathematical concepts. Learning to work with **models** is a skill that transfers to many disciplines.

(a) Draw a scatter diagram of the data for one period.
(b) Find a sinusoidal function of the form
$$y = A \sin(\omega x - \phi) + B$$ that fits the data.
(c) Draw the sinusoidal function found in part (b) on the scatter diagram.
(d) Use a graphing utility to find the sinusoidal function of best fit.
(e) Graph the sinusoidal function of best fit on the scatter diagram.

21. Monthly Temperature The following data represent the average monthly temperatures for Indianapolis, Indiana.

Month, x	Average Monthly Temperature, °F
January, 1	25.5
February, 2	29.6
March, 3	41.4
April, 4	52.4
May, 5	62.8
June, 6	71.9
July, 7	75.4
August, 8	73.2
September, 9	66.6
October, 10	54.7
November, 11	43.0
December, 12	30.9

Source: U.S. National Oceanic and Atmospheric Administration.

Page 214

Month, x	Average Monthly Temperature, °F
January, 1	31.8
February, 2	34.8
March, 3	44.1
April, 4	53.4
May, 5	63.4
June, 6	72.5
July, 7	77.0
August, 8	75.6
September, 9	68.5
October, 10	56.6
November, 11	46.8
December, 12	36.7

Source: U.S. National Oceanic and Atmospheric Administration.

23. Tides Suppose that the length of time between consecutive high tides is approximately 12.5 hours. According to the National Oceanic and Atmospheric Administration, on Saturday, June 28, 1997, in Savannah, Georgia, high tide occurred at 3:38 AM (3.6333 hours) and low tide occurred at 10:08 AM (10.1333 hours). Water heights are measured as the amounts above or below the mean lower low water. The height of the water at high tide was 8.2 feet and the height of the water at low tide was −0.6 foot.
(a) Approximately when will the next high tide occur?
(b) Find a sinusoidal function of the form
$$y = A \sin(\omega x - \phi) + B$$ that fits the data.
(c) Draw a graph of the function found in part (b).
(d) Use the function found in part (b) to predict the height of the water at the next high tide.

Page 153

4 ◢ GRAPHING UTILITY SOLUTIONS

The techniques introduced in this section apply only to certain types of trigonometric equations. Solutions for other types are usually studied in calculus, using numerical methods. In the next example, we show how a graphing utility may be used to obtain solutions.

Solving Trigonometric Equations Using a Graphing Utility

Solve: $5 \sin x + x = 3$
Express the solution(s) rounded to two decimal places.

Solution This type of trigonometric equation cannot be solved by previous methods. A graphing utility, though, can be used here. The solution(s) of this equation is the same as the points of intersection of the graphs of $Y_1 = 5 \sin x + x$ and $Y_2 = 3$. See Figure 34.

Figure 34

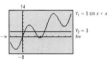

$Y_1 = 5 \sin x + x$
$Y_2 = 3$

There are three points of intersection; the x-coordinates are the solutions that we seek. Using INTERSECT, we find

$$x = 0.52, \qquad x = 3.18, \qquad x = 5.71$$

rounded to two decimal places.

GRAPHING UTILITIES AND TECHNIQUES

Increase your understanding, visualize, discover, explore, and solve problems using a **graphing utility.** Sullivan uses the graphing utility to further your understanding of concepts not to circumvent essential math skills.

DISCUSSION WRITING AND READING EXERCISES

These exercises, clearly identified by the notebook icon and/or red numeration, are designed to get you to "think outside the box." These exercises seek to foster an intuitive understanding of key mathematical concepts. It is easy to find these exercises as they are highlighted by the book icon and red exercise number.

Graph this function for $0 \le x \le 4$ and compare the result to the graphs obtained in part (a) and (b).
(d) What do you think the next approximation to the sawtooth curve is?

32. Graph the sound emitted by the * key on a Touch-Tone phone. See Problem 31.
33. Graph the function $f(x) = \dfrac{\sin x}{x}$, $x > 0$. Based on the graph, what do you conjecture about the value of $\dfrac{\sin x}{x}$ for x close to 0?
34. Graph $y = x \sin x$, $y = x^2 \sin x$, and $y = x^3 \sin x$ for $x > 0$. What patterns do you observe?
35. Graph $y = \dfrac{1}{x} \sin x$, $y = \dfrac{1}{x^2} \sin x$, and $y = \dfrac{1}{x^3} \sin x$ for $x > 0$. What patterns do you observe?
36. How would you explain to a friend what simple harmonic motion is? How would you explain damped motion?

Page 266

LINKS TO CALCULUS

This icon draws attention to the underpinnings of calculus.

4 ◢ GRAPHING UTILITY SOLUTIONS

The techniques introduced in this section apply only to certain types of trigonometric equations. Solutions for other types are usually studied in calculus, using numerical methods. In the next example, we show how a graphing utility may be used to obtain solutions.

Page 214

OVERVIEW

CHAPTER REVIEW

The Chapter Review helps check your understanding of the chapter materials in several ways. **"Things to Know"** gives a general overview of review topics. The **"How To"** section provides a concept-by-concept listing of operations you are expected to perform. The **"Review Exercises"** then serve as a chance to practice the concepts presented within the chapter. Several of the Review Exercises are numbered in blue. These exercises can be combined to create the Chapter Test. Since these problems are odd numbered, you can check your answers in the back of the book. The review materials are designed to make you, the student, confident in knowing the chapter material.

CHAPTER REVIEW

Library of Functions

Linear function (p. 235)
$f(x) = mx + b$
Graph is a line with slope m and y-intercept b.

Constant function (p. 235)
$f(x) = b$
Graph is a horizontal line with y-intercept b.
See Figure 22.

Identity function (p. 235)
$f(x) = x$
Graph is a line with slope 1 and y-intercept 0.
See Figure 23.

Cube function (p. 236)
$f(x) = x^3$
See Figure 25.

Reciprocal function (p. 236)
$f(x) = 1/x$
See Figure 27.

Square function (p. 235)
$f(x) = x^2$
Graph is a parabola with intercept at $(0, 0)$.
See Figure 24.

Square root function (p. 236)
$f(x) = \sqrt{x}$
See Figure 26.

Absolute value function (p. 236)
$f(x) = |x|$
See Figure 28.

Greatest integer function (p. 237)
$f(x) = \text{int}(x)$
See Figure 29.

Things To Know

Function (p. 206)

A relation between two sets of real numbers so that each number x in the first set, the domain, has corresponding to it exactly one number y in the second set. The range is the set of y values of the function for the x values in the domain.

x is the independent variable; y is the dependent variable.

A function f may be defined implicitly by an equation involving x and y or explicitly by writing $y = f(x)$.

Objectives

You should be able to:

Determine whether a relation represents a function (p. 205)

Find the value of a function (p. 209)

Find the domain of a function (p. 211)

Identify the graph of a function (p. 212)

Obtain information from or about the graph of a function (p. 213)

Find the average rate of a change of a function (p. 224)

Use a graph to determine where a function is increasing, is decreasing , or is constant (p. 227)

Use a graph to locate local maxima and minima (p. 228)

Determine even or odd functions from a graph (p. 229)

Identify even or odd functions from the equation (p. 230)

Graph the functions listed in the library of functions (p. 235)

Graph piecewise-defined functions (p. 238)

Graph functions using horizontal and vertical shifts (p. 242)

Graph functions using compressions and stretches (p. 246)

Graph functions using reflections

Fill-in-the-Blank Items

1. If f is a function defined by the equation $y = f(x)$, then x is called _____ variable.

2. A set of points in the xy-plane is the graph of a function if and only _____ graph in at most one point.

3. The average rate of change of a function equals the _____ graph.

4. A(n) _____ function f is one for which $f(-x) = f(x)$ for ev _____ function f is one for which $f(-x) = -f(x)$ for every x in the domain _____

5. Suppose that the graph of a function f is known. Then the graph of $v =$ _____

True/False Items

T F **1.** Every relation is a function.

T F **2.** Vertical lines intersect the graph of a function in no more than one point.

T F **3.** The y-intercept of the graph of the function $y = f(x)$, whose domain is all real numbers, is $f(0)$.

T F **4.** A function f is decreasing on an open interval I if, for any choice of x_1 and x_2 in I, with $x_1 < x_2$, we have $f(x_1) < f(x_2)$.

T F **5.** Even functions have graphs that are symmetric with respect to the origin.

T F **6.** The graph of $y = f(-x)$ is the reflection about the y-axis of the graph of $y = f(x)$.

T F **7.** $f(g(x)) = f(x) \cdot g(x)$.

T F **8.** The domain of the composite function $(f \circ g)(x)$ is the same as that of $g(x)$.

Review Exercises

Blue problem numbers indicate the author's suggestions for use in a Practice Test.

1. Given that f is a linear function, $f(4) = -5$ and $f(0) = 3$, write the equation that defines f.

2. Given that g is a linear function with slope $= -4$ and $g(-2) = 2$, write the equation that defines g.

3. A function f is defined by $f(x) = \dfrac{Ax + 5}{6x - 2}$. If $f(1) = 4$, find A.

4. A function g is defined by $g(x) = \dfrac{A}{x} + \dfrac{8}{x^2}$. If $g(-1) = 0$, find A.

5. Tell which of the following graphs are graphs of functions.

(a)

(b)

(c)

(d)

Sullivan M@thPak
An Integrated Learning Environment

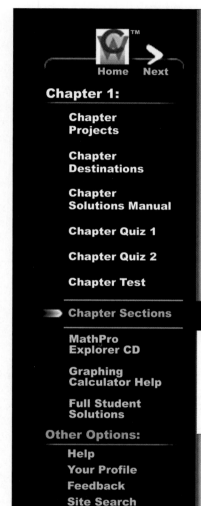

Home **Next**

Chapter 1:

Chapter Projects

Chapter Destinations

Chapter Solutions Manual

Chapter Quiz 1

Chapter Quiz 2

Chapter Test

Chapter Sections

MathPro Explorer CD

Graphing Calculator Help

Full Student Solutions

Other Options:

Help

Your Profile

Feedback

Site Search

➡ Syllabus

GET IT TOGETHER!

M@THP@K integrates and organizes all major student supplements into an easy-to-use format at a price that can't be beat!

Here's just a sample of what you'll find in MathPak:

1.1 Real Numbers

Preparing for this Section

Reading Quiz

MathPro Objectives

PowerPoints

System Requirements:
- 32MB of random access memory (RAM); 64MB or more recommended
- 200MB free hard disk space
- CD-ROM drive
- QuickTime™ 4.0 or better
- Internet Browser 4.5 or higher
- Internet Access 28.8k or better

- **MultiMedia MathPro 4.0**
  This interactive tutorial program offers unlimited practice on College Algebra content. Watch the author work the problems via videos, view other examples, and see a fully worked out solution to the problem you are working on.

- **Graphing Calculator Manuals**
  Includes step by step procedures and screen shots for working with your TI-82, TI-83, TI-85, TI-86, TI-89, TI-92, HP48G, CFX-9850 GaPlus, and SharpE 9600c.

- **Full Student Solutions Manual**
  Includes step by step solutions for all the odd numbered exercises in the text.

Sullivan M@thPak
Helping Students
Get it Together

Additional Media

Sullivan Companion Website

www.prenhall.com/sullivan
This text-specific website beautifully complements the text. Here students can find chapter tests, section-specific links, and PowerPoint downloads in addition to other helpful features.

MathPro 4.0 (network version)

This networkable version of MathPro is *free* to adopters. Contact your Prentice Hall Sales Representative.

Test Gen-EQ

CD-ROM (Windows/Macintosh)

- Algorithmically driven, text-specific testing program
- Networkable for administering tests and capturing grades
- Edit existing test items or add your own questions to create a nearly unlimited number of tests and drill worksheets

ISBN: 0-13-041229-5

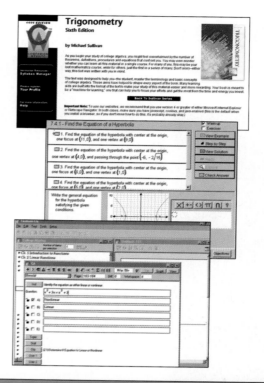

Student Supplements

Student Solutions Manual

Worked solutions to all odd-numbered exercises from the text and complete solutions for chapter review problems and chapter tests. ISBN: 0-13-041227-9

Lecture Videos

The instructional tapes, in a lecture format, feature worked-out examples and exercises taken from each section of the text. ISBN: 0-13-041220-1

New York Times
Themes of the Times

A *free* newspaper from Prentice Hall and *The New York Times.* Interesting and current articles on mathematics which invite discussion and writing about mathematics.

Mathematics on the Internet

Free guide providing a brief history of the Internet, discussing the use of the World Wide Web, and describing how to find your way within the Internet and how to find others on it.

Instructor Supplements

Instructor's Resource Manual

Contains complete step-by-step worked-out solutions to all even-numbered exercises in the textbook. ISBN: 0-13-041226-0

Test Item File

Hard copy of the algorithmic computerized testing materials. ISBN: 0-13-041231-7

List of Applications

Photo and Illustration Credits

Functions and Their Graphs

PREPARING FOR THIS BOOK

Before getting started, read Preface to the Student, p. xv

Field Trip to Motorola

During the past decade the availability and usage of wireless Internet services has increased manyfold. The industry has developed a number of pricing proposals for such services. Marketing data have indicated that subscribers of wireless Internet services have tended to desire flat rate fee structures as compared with rates based totally on usage.

PREPARING FOR THIS SECTION

Before getting started, review the following:

✓ Algebra Review (Appendix A, Section A.1, pp. 487–495)   ✓ Geometry Review (Appendix A, Section A.2, pp. 497–500)

1.1   RECTANGULAR COORDINATES

OBJECTIVES   1   Use the Distance Formula
2   Use the Midpoint Formula

We locate a point on the real number line by assigning it a single real number, called the *coordinate of the point.* For work in a two-dimensional plane, we locate points by using two numbers.

We begin with two real number lines located in the same plane: one horizontal and the other vertical. We call the horizontal line the **x-axis;** the vertical line, the **y-axis;** and the point of intersection, the **origin O.** We assign

Figure 1

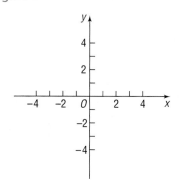

Figure 2

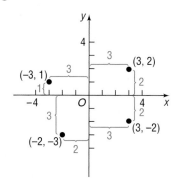

Figure 3

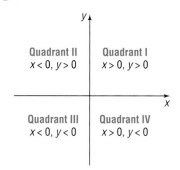

Figure 4

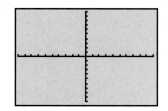

coordinates to every point on these number lines as shown in Figure 1, using a convenient scale. In mathematics, we usually use the same scale on each axis; in applications, a different scale is often used on each axis.

The origin O has a value of 0 on both the x-axis and the y-axis. We follow the usual convention that points on the x-axis to the right of O are associated with positive real numbers, and those to the left of O are associated with negative real numbers. Points on the y-axis above O are associated with positive real numbers, and those below O are associated with negative real numbers. In Figure 1, the x-axis and y-axis are labeled as x and y, respectively, and we have used an arrow at the end of each axis to denote the positive direction.

The coordinate system described here is called a **rectangular** or **Cartesian*** **coordinate system.** The plane formed by the x-axis and y-axis is sometimes called the **xy-plane,** and the x-axis and y-axis are referred to as the **coordinate axes.**

Any point P in the xy-plane can then be located by using an **ordered pair** (x, y) of real numbers. Let x denote the signed distance of P from the y-axis (*signed* in the sense that, if P is to the right of the y-axis, then $x > 0$, and if P is to the left of the y-axis, then $x < 0$); and let y denote the signed distance of P from the x-axis. The ordered pair (x, y), also called the **coordinates** of P, then gives us enough information to locate the point P in the plane.

For example, to locate the point whose coordinates are $(-3, 1)$, go 3 units along the x-axis to the left of O and then go straight up 1 unit. We **plot** this point by placing a dot at this location. See Figure 2, in which the points with coordinates $(-3, 1)$, $(-2, -3)$, $(3, -2)$, and $(3, 2)$ are plotted.

The origin has coordinates $(0, 0)$. Any point on the x-axis has coordinates of the form $(x, 0)$, and any point on the y-axis has coordinates of the form $(0, y)$.

If (x, y) are the coordinates of a point P, then x is called the **x-coordinate,** or **abscissa,** of P and y is the **y-coordinate,** or **ordinate,** of P. We identify the point P by its coordinates (x, y) by writing $P = (x, y)$, referring to it as "the point (x, y)," rather than "the point whose coordinates are (x, y)."

The coordinate axes divide the xy-plane into four sections, called **quadrants,** as shown in Figure 3. In quadrant I, both the x-coordinate and the y-coordinate of all points are positive; in quadrant II, x is negative and y is positive; in quadrant III, both x and y are negative; and in quadrant IV, x is positive and y is negative. Points on the coordinate axes belong to no quadrant.

NOW WORK PROBLEM **1.**

 COMMENT: On a graphing calculator, you can set the scale on each axis. Once this has been done, you obtain the **viewing rectangle.** See Figure 4 for a typical viewing rectangle. You should now read Section B.1, The Viewing Rectangle, in Appendix B, pages 561–562.

DISTANCE BETWEEN POINTS

① If the same unit of measurement, such as inches or centimeters, is used for both the x-axis and the y-axis, then all distances in the xy-plane can be measured using this unit of measurement.

* Named after René Descartes (1596–1650), a French mathematician, philosopher, and theologian.

The **distance formula** provides a straightforward method for computing the distance between two points.

Theorem   Distance Formula

The distance between two points $P_1 = (x_1, y_1)$ and $P_2 = (x_2, y_2)$, denoted by $d(P_1, P_2)$, is

$$d(P_1, P_2) = \sqrt{(x_2 - x_1)^2 + (y_2 - y_1)^2} \qquad \textbf{(1)}$$

EXAMPLE 1   Finding the Distance between Two Points

Find the distance d between the points $(-4, 5)$ and $(3, 2)$.

Solution   Using the distance formula (1), the solution is obtained as follows:

$$d = \sqrt{[3 - (-4)]^2 + (2 - 5)^2} = \sqrt{7^2 + (-3)^2}$$

$$= \sqrt{49 + 9} = \sqrt{58} \approx 7.62$$

NOW WORK PROBLEMS **5** AND **9**.

Figure 5

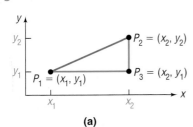

(a)

(b)

Proof of the Distance Formula   Let (x_1, y_1) denote the coordinates of point P_1, and let (x_2, y_2) denote the coordinates of point P_2. Assume that the line joining P_1 and P_2 is neither horizontal nor vertical. Refer to Figure 5(a). The coordinates of P_3 are (x_2, y_1). The horizontal distance from P_1 to P_3 is the absolute value of the difference of the x-coordinates, $|x_2 - x_1|$. The vertical distance from P_3 to P_2 is the absolute value of the difference of the y-coordinates, $|y_2 - y_1|$. See Figure 5(b). The distance $d(P_1, P_2)$ that we seek is the length of the hypotenuse of the right triangle, so, by the Pythagorean Theorem, it follows that

$$[d(P_1, P_2)]^2 = |x_2 - x_1|^2 + |y_2 - y_1|^2$$

$$= (x_2 - x_1)^2 + (y_2 - y_1)^2$$

$$d(P_1, P_2) = \sqrt{(x_2 - x_1)^2 + (y_2 - y_1)^2}$$

Now, if the line joining P_1 and P_2 is horizontal, then the y-coordinate of P_1 equals the y-coordinate of P_2; that is, $y_1 = y_2$. Refer to Figure 6(a). In this case, the distance formula (1) still works, because, for $y_1 = y_2$, it reduces to

$$d(P_1, P_2) = \sqrt{(x_2 - x_1)^2 + 0^2} = \sqrt{(x_2 - x_1)^2} = |x_2 - x_1|$$

A similar argument holds if the line joining P_1 and P_2 is vertical. See Figure 6(b).

Figure 6

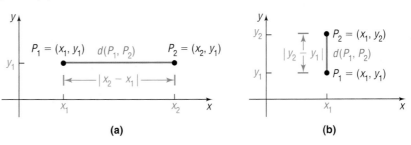

(a)   **(b)**

The distance formula is valid in all cases.

The distance between two points $P_1 = (x_1, y_1)$ and $P_2 = (x_2, y_2)$ is never a negative number. Furthermore, the distance between two points is 0 only when the points are identical, that is, when $x_1 = x_2$ and $y_1 = y_2$. Also, because $(x_2 - x_1)^2 = (x_1 - x_2)^2$ and $(y_2 - y_1)^2 = (y_1 - y_2)^2$, it makes no difference whether the distance is computed from P_1 to P_2 or from P_2 to P_1; that is, $d(P_1, P_2) = d(P_2, P_1)$.

Rectangular coordinates enable us to translate geometry problems into algebra problems, and vice versa. The next example shows how algebra (the distance formula) can be used to solve geometry problems.

| **EXAMPLE 2** | **Using Algebra to Solve Geometry Problems** |

Consider the three points $A = (-2, 1)$, $B = (2, 3)$, and $C = (3, 1)$.

(a) Plot each point and form the triangle ABC.

(b) Find the length of each side of the triangle.

(c) Verify that the triangle is a right triangle.

(d) Find the area of the triangle.

Solution   (a) Points A, B, and C and triangle ABC are plotted in Figure 7.

Figure 7

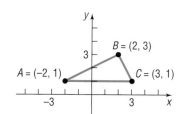

(b) $d(A, B) = \sqrt{[2 - (-2)]^2 + (3 - 1)^2} = \sqrt{16 + 4} = \sqrt{20} = 2\sqrt{5}$

$d(B, C) = \sqrt{(3 - 2)^2 + (1 - 3)^2} = \sqrt{1 + 4} = \sqrt{5}$

$d(A, C) = \sqrt{[3 - (-2)]^2 + (1 - 1)^2} = \sqrt{25 + 0} = 5$

(c) To show that the triangle is a right triangle, we need to show that the sum of the squares of the lengths of two of the sides equals the square of the length of the third side. (Why is this sufficient?) Looking at Figure 7, it seems reasonable to conjecture that the right angle is at vertex B. To verify, we check to see whether

$$[d(A, B)]^2 + [d(B, C)]^2 = [d(A, C)]^2$$

We find that

$$[d(A, B)]^2 + [d(B, C)]^2 = (2\sqrt{5})^2 + (\sqrt{5})^2$$
$$= 20 + 5 = 25 = [d(A, C)]^2$$

so it follows from the converse of the Pythagorean Theorem that triangle ABC is a right triangle.

(d) Because the right angle is at B, the sides AB and BC form the base and altitude of the triangle. Its area is therefore

$$\text{Area} = \frac{1}{2}(\text{Base})(\text{Altitude}) = \frac{1}{2}(2\sqrt{5})(\sqrt{5}) = 5 \text{ square units}$$

━━━━━━━ NOW WORK PROBLEM **19**.

MIDPOINT FORMULA

②   We now derive a formula for the coordinates of the **midpoint of a line segment.** Let $P_1 = (x_1, y_1)$ and $P_2 = (x_2, y_2)$ be the endpoints of a line segment, and let $M = (x, y)$ be the point on the line segment that is the same distance from P_1 as

Figure 8

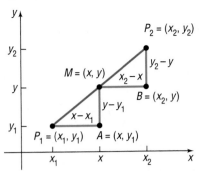

it is from P_2. See Figure 8. The triangles P_1AM and MBP_2 are congruent.* [Do you see why? Angle AP_1M = Angle BMP_2.† Angle P_1MA = Angle MP_2B, and $d(P_1, M) = d(M, P_2)$ is given. We have Angle–Side–Angle.] As a result, corresponding sides are equal in length. That is,

$$x - x_1 = x_2 - x \quad \text{and} \quad y - y_1 = y_2 - y$$
$$2x = x_1 + x_2 \qquad\qquad 2y = y_1 + y_2$$
$$x = \frac{x_1 + x_2}{2} \qquad\qquad y = \frac{y_1 + y_2}{2}$$

Theorem

Midpoint Formula

The midpoint $M = (x, y)$ of the line segment from $P_1 = (x_1, y_1)$ to $P_2 = (x_2, y_2)$ is

$$M = (x, y) = \left(\frac{x_1 + x_2}{2}, \frac{y_1 + y_2}{2} \right) \qquad \textbf{(2)}$$

To find the midpoint of a line segment, we average the x-coordinates and the y-coordinates of the endpoints.

EXAMPLE 3   Finding the Midpoint of a Line Segment

Find the midpoint of the line segment from $P_1 = (-5, 3)$ to $P_2 = (3, 1)$. Plot the points P_1 and P_2 and their midpoint. Check your answer.

Solution   We apply the midpoint formula (2) using $x_1 = -5$, $y_1 = 3$, $x_2 = 3$, and $y_2 = 1$. Then the coordinates (x, y) of the midpoint M are

$$x = \frac{x_1 + x_2}{2} = \frac{-5 + 3}{2} = -1 \quad \text{and} \quad y = \frac{y_1 + y_2}{2} = \frac{3 + 1}{2} = 2$$

Figure 9

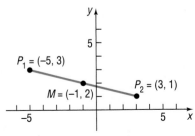

That is, $M = (-1, 2)$. See Figure 9.

Check: Because M is the midpoint, we check the answer by verifying that $d(P_1, M) = d(M, P_2)$.

$$d(P_1, M) = \sqrt{[-1 - (-5)]^2 + (2 - 3)^2} = \sqrt{16 + 1} = \sqrt{17}$$
$$d(M, P_2) = \sqrt{[3 - (-1)]^2 + (1 - 2)^2} = \sqrt{16 + 1} = \sqrt{17}$$

NOW WORK PROBLEM 29.

---

*The following statement is a postulate from geometry. Two triangles are congruent if their sides are the same length (SSS), or if two sides and the included angle are the same (SAS), or if two angles and the included side are the same (ASA).
†Another postulate from geometry states that the transversal $\overline{P_1P_2}$ forms equal corresponding angles with the parallel lines $\overline{P_1A}$ and $\overline{MB}$.

1.1   EXERCISES

In Problems 1 and 2, plot each point in the xy-plane. Tell in which quadrant or on what coordinate axis each point lies.

1. (a) $A = (-3, 2)$
   (b) $B = (6, 0)$
   (c) $C = (-2, -2)$
   (d) $D = (6, 5)$
   (e) $E = (0, -3)$
   (f) $F = (6, -3)$

2. (a) $A = (1, 4)$
   (b) $B = (-3, -4)$
   (c) $C = (-3, 4)$
   (d) $D = (4, 1)$
   (e) $E = (0, 1)$
   (f) $F = (-3, 0)$

3. Plot the points $(2, 0), (2, -3), (2, 4), (2, 1)$, and $(2, -1)$. Describe the set of all points of the form $(2, y)$, where y is a real number.

4. Plot the points $(0, 3), (1, 3), (-2, 3), (5, 3)$, and $(-4, 3)$. Describe the set of all points of the form $(x, 3)$, where x is a real number.

In Problems 5–18, find the distance $d(P_1, P_2)$ between the points P_1 and P_2.

5.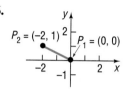

6.

7.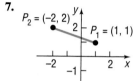

8.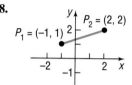

9. $P_1 = (3, -4)$; $P_2 = (5, 4)$

10. $P_1 = (-1, 0)$; $P_2 = (2, 4)$

11. $P_1 = (-3, 2)$; $P_2 = (6, 0)$

12. $P_1 = (2, -3)$; $P_2 = (4, 2)$

13. $P_1 = (4, -3)$; $P_2 = (6, 4)$

14. $P_1 = (-4, -3)$; $P_2 = (6, 2)$

15. $P_1 = (-0.2, 0.3)$; $P_2 = (2.3, 1.1)$

16. $P_1 = (1.2, 2.3)$; $P_2 = (-0.3, 1.1)$

17. $P_1 = (a, b)$; $P_2 = (0, 0)$

18. $P_1 = (a, a)$; $P_2 = (0, 0)$

In Problems 19–24, plot each point and form the triangle ABC. Verify that the triangle is a right triangle. Find its area.

19. $A = (-2, 5)$; $B = (1, 3)$; $C = (-1, 0)$

20. $A = (-2, 5)$; $B = (12, 3)$; $C = (10, -11)$

21. $A = (-5, 3)$; $B = (6, 0)$; $C = (5, 5)$

22. $A = (-6, 3)$; $B = (3, -5)$; $C = (-1, 5)$

23. $A = (4, -3)$; $B = (0, -3)$; $C = (4, 2)$

24. $A = (4, -3)$; $B = (4, 1)$; $C = (2, 1)$

25. Find all points having an x-coordinate of 2 whose distance from the point $(-2, -1)$ is 5.

26. Find all points having a y-coordinate of -3 whose distance from the point $(1, 2)$ is 13.

27. Find all points on the x-axis that are 5 units from the point $(4, -3)$.

28. Find all points on the y-axis that are 5 units from the point $(4, 4)$.

In Problems 29–38, find the midpoint of the line segment from P_1 to P_2.

29. $P_1 = (5, -4)$; $P_2 = (3, 2)$

30. $P_1 = (-1, 0)$; $P_2 = (2, 4)$

31. $P_1 = (-3, 2)$; $P_2 = (6, 0)$

32. $P_1 = (2, -3)$; $P_2 = (4, 2)$

33. $P_1 = (4, -3)$; $P_2 = (6, 1)$

34. $P_1 = (-4, -3)$; $P_2 = (2, 2)$

35. $P_1 = (-0.2, 0.3)$; $P_2 = (2.3, 1.1)$

36. $P_1 = (1.2, 2.3)$; $P_2 = (-0.3, 1.1)$

37. $P_1 = (a, b)$; $P_2 = (0, 0)$

38. $P_1 = (a, a)$; $P_2 = (0, 0)$

39. The **medians** of a triangle are the line segments from each vertex to the midpoint of the opposite side (see the figure). Find the lengths of the medians of the triangle with vertices at $A = (0, 0), B = (0, 6)$, and $C = (4, 4)$.

40. An **equilateral triangle** is one in which all three sides are of equal length. If two vertices of an equilateral triangle are $(0, 4)$ and $(0, 0)$, find the third vertex. How many of these triangles are possible?

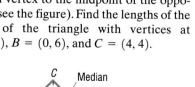

*In Problems 41–44, find the length of each side of the triangle determined by the three points P_1, P_2, and P_3. State whether the triangle is an isosceles triangle, a right triangle, neither of these, or both. (An **isosceles triangle** is one in which at least two of the sides are of equal length.)*

41. $P_1 = (2, 1)$; $P_2 = (-4, 1)$; $P_3 = (-4, -3)$

42. $P_1 = (-1, 4)$; $P_2 = (6, 2)$; $P_3 = (4, -5)$

43. $P_1 = (-2, -1)$; $P_2 = (0, 7)$; $P_3 = (3, 2)$

44. $P_1 = (7, 2)$; $P_2 = (-4, 0)$; $P_3 = (4, 6)$

In Problems 45–48, find the length of the line segment. Assume that the endpoints of each line segment have integer coordinates.

45.
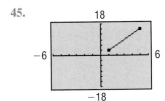

46.
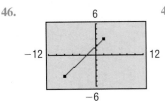

47.
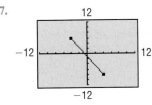

48.
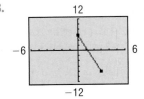

---

49. Geometry   Find the midpoint of each diagonal of a square with side of length s. Draw the conclusion that the diagonals of a square intersect at their midpoints.
[**Hint:** Use $(0, 0)$, $(0, s)$, $(s, 0)$, and (s, s) as the vertices of the square.]

50. Geometry   Verify that the points $(0, 0)$, $(a, 0)$, and $\left(\dfrac{a}{2}, \dfrac{\sqrt{3}a}{2} \right)$ are the vertices of an equilateral triangle.

Then show that the midpoints of the three sides are the vertices of a second equilateral triangle (refer to Problem 40.)

51. Baseball   A major league baseball "diamond" is actually a square, 90 feet on a side (see the figure). What is the distance directly from home plate to second base (the diagonal of the square)?

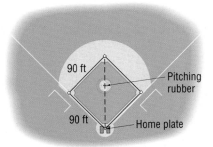

52. Little League Baseball   The layout of a Little League playing field is a square, 60 feet on a side.* How far is it directly from home plate to second base (the diagonal of the square)?

53. Baseball   Refer to Problem 51. Overlay a rectangular coordinate system on a major league baseball diamond so that the origin is at home plate, the positive x-axis lies in the direction from home plate to first base, and the positive y-axis lies in the direction from home plate to third base.
(a) What are the coordinates of first base, second base, and third base? Use feet as the unit of measurement.

(b) If the right fielder is located at $(310, 15)$, how far is it from the right fielder to second base?
(c) If the center fielder is located at $(300, 300)$, how far is it from the center fielder to third base?

54. Little League Baseball   Refer to Problem 52. Overlay a rectangular coordinate system on a Little League baseball diamond so that the origin is at home plate, the positive x-axis lies in the direction from home plate to first base, and the positive y-axis lies in the direction from home plate to third base.
(a) What are the coordinates of first base, second base, and third base? Use feet as the unit of measurement.
(b) If the right fielder is located at $(180, 20)$, how far is it from the right fielder to second base?
(c) If the center fielder is located at $(220, 220)$, how far is it from the center fielder to third base?

55. A Dodge Intrepid and a Mack truck leave an intersection at the same time. The Intrepid heads east at an average speed of 30 miles per hour, while the truck heads south at an average speed of 40 miles per hour. Find an expression for their distance d apart (in miles) at the end of t hours.

56. A hot-air balloon, headed due east at an average speed of 15 miles per hour and at a constant altitude of 100 feet, passes over an intersection (see the figure). Find an expression for its distance d (measured in feet) from the intersection t seconds later.

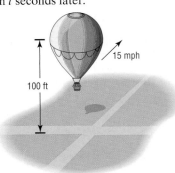

* *Source: Little League Baseball, Official Regulations and Playing Rules, 2000.*

Before getting started, review the following:

✓ Solving Equations (Appendix A, Section A.3, pp. 502–505)

✓ Completing the Square (Appendix A, Section A.3, pp. 507–508)

1.2 GRAPHS OF EQUATIONS; CIRCLES

OBJECTIVES

① Graph Equations by Plotting Points
② Find Intercepts from a Graph
③ Find Intercepts from an Equation
④ Test an Equation for Symmetry with Respect to the *x*-axis, the *y*-axis, and the origin
⑤ Write the Standard Form of the Equation of a Circle
⑥ Graph a Circle
⑦ Find the Center and Radius of a Circle in General Form and Graph It

An **equation in two variables,** say x and y, is a statement in which two expressions involving x and y are equal. The expressions are called the **sides** of the equation. Since an equation is a statement, it may be true or false, depending on the value of the variables. Any values of x and y that result in a true statement are said to **satisfy** the equation.

For example, the following are all equations in two variables x and y:

$$x^2 + y^2 = 5 \qquad 2x - y = 6 \qquad y = 2x + 5 \qquad x^2 = y$$

The first of these, $x^2 + y^2 = 5$, is satisfied for $x = 1$, $y = 2$, since $1^2 + 2^2 = 1 + 4 = 5$. Other choices of x and y also satisfy this equation. It is not satisfied for $x = 2$ and $y = 3$, since $2^2 + 3^2 = 4 + 9 = 13 \neq 5$.

① The **graph of an equation in two variables** x and y consists of the set of points in the xy-plane whose coordinates (x, y) satisfy the equation.

EXAMPLE 1 **Determining Whether a Point Is on the Graph of an Equation**

Determine if the following points are on the graph of the equation $2x - y = 6$.

(a) $(2, 3)$ (b) $(2, -2)$

Solution (a) For the point $(2, 3)$, we check to see if $x = 2$, $y = 3$ satisfies the equation $2x - y = 6$.

$$2x - y = 2(2) - 3 = 4 - 3 = 1 \neq 6$$

The equation is not satisfied, so the point $(2, 3)$ is not on the graph.

(b) For the point $(2, -2)$, we have

$$2x - y = 2(2) - (-2) = 4 + 2 = 6$$

The equation is satisfied, so the point $(2, -2)$ is on the graph. ■

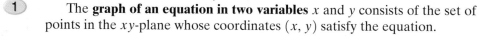 NOW WORK PROBLEM 23.

EXAMPLE 2	Graphing an Equation by Plotting Points

Graph the equation: $y = 2x + 5$

Solution   We want to find all points (x, y) that satisfy the equation. To locate some of these points (and thus get an idea of the pattern of the graph), we assign some numbers to x and find corresponding values for y.

Figure 10
$y = 2x + 5$

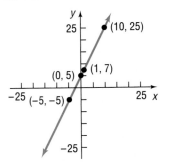

If	Then	Point on Graph
$x = 0$	$y = 2(0) + 5 = 5$	$(0, 5)$
$x = 1$	$y = 2(1) + 5 = 7$	$(1, 7)$
$x = -5$	$y = 2(-5) + 5 = -5$	$(-5, -5)$
$x = 10$	$y = 2(10) + 5 = 25$	$(10, 25)$

By plotting these points and then connecting them, we obtain the graph of the equation (a *line*), as shown in Figure 10. ■

EXAMPLE 3	Graphing an Equation by Plotting Points

Graph the equation: $y = x^2$

Solution   Table 1 provides several points on the graph. In Figure 11 we plot these points and connect them with a smooth curve to obtain the graph (a *parabola*).

Figure 11
$y = x^2$

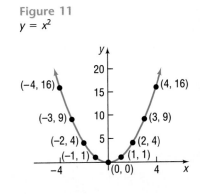

TABLE 1		
x	$y = x^2$	(x, y)
-4	16	$(-4, 16)$
-3	9	$(-3, 9)$
-2	4	$(-2, 4)$
-1	1	$(-1, 1)$
0	0	$(0, 0)$
1	1	$(1, 1)$
2	4	$(2, 4)$
3	9	$(3, 9)$
4	16	$(4, 16)$

■

The graphs of the equations shown in Figures 10 and 11 do not show all points. For example, in Figure 10, the point $(20, 45)$ is a part of the graph of $y = 2x + 5$, but it is not shown. Since the graph of $y = 2x + 5$ could be extended out as far as we please, we use arrows to indicate that the pattern shown continues. It is important when illustrating a graph to present enough of the graph so that any viewer of the illustration will "see" the rest of it as an obvious continuation of what is actually there. This is referred to as a **complete graph.**

So, one way to obtain a complete graph of an equation is to plot a sufficient number of points on the graph until a pattern becomes evident. Then these points are connected with a smooth curve following the suggested pattern. But how many points are sufficient? Sometimes knowledge about the equation tells us. For example, if an equation is of the form $y = mx + b$, then its graph is a line. In this case, only two points are needed to obtain the graph. Refer to Appendix A, Section A.7 for a discussion of lines.

One purpose of this book is to investigate the properties of equations in order to decide whether a graph is complete. At first we shall graph equations by plotting a sufficient number of points. Shortly, we shall investigate various techniques that will enable us to graph an equation without plotting so many points.

 COMMENT: Another way to obtain the graph of an equation is to use a graphing utility. Read Section B.2, Using a Graphing Utility to Graph Equations, in Appendix B, page 563-566.   ∎

Two techniques that reduce the number of points required to graph an equation involve finding *intercepts* and checking for *symmetry*.

INTERCEPTS

② The points, if any, at which a graph crosses or touches the coordinate axes are called the **intercepts.** See Figure 12. The x-coordinate of a point at which the graph crosses or touches the x-axis is an ***x*-intercept,** and the y-coordinate of a point at which the graph crosses or touches the y-axis is a ***y*-intercept.**

Figure 12

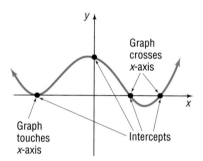

| EXAMPLE 4 | **Finding Intercepts from a Graph** |

Figure 13

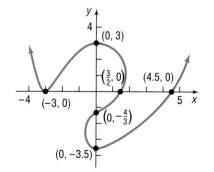

Find the intercepts of the graph in Figure 13. What are its x-intercepts? What are its y-intercepts?

Solution   The intercepts of the graph are the points

$$(-3, 0), \quad (0, 3), \quad \left(\frac{3}{2}, 0\right), \quad \left(0, -\frac{4}{3}\right), \quad (0, -3.5), \quad (4.5, 0)$$

The x-intercepts are $-3, \dfrac{3}{2}$, and 4.5; the y-intercepts are $-3.5, -\dfrac{4}{3}$, and 3.   ∎

✎ **NOW WORK PROBLEM 11(a).**

③   The intercepts of the graph of an equation can be found by using the fact that points on the x-axis have y-coordinates equal to 0, and points on the y-axis have x-coordinates equal to 0.

> **PROCEDURE FOR FINDING INTERCEPTS**
>
> 1. To find the *x*-intercept(s), if any, of the graph of an equation, let $y = 0$ in the equation and solve for *x*.
> 2. To find the *y*-intercept(s), if any, of the graph of an equation, let $x = 0$ in the equation and solve for *y*.

EXAMPLE 5	**Finding Intercepts from an Equation**

Find the *x*-intercept(s) and the *y*-intercept(s) of the graph of $y = x^2 - 4$.

Solution   To find the *x*-intercept(s), we let $y = 0$ and obtain the equation

$$x^2 - 4 = 0$$

$$(x + 2)(x - 2) = 0 \qquad \text{Factor.}$$

$$x + 2 = 0 \quad \text{or} \quad x - 2 = 0 \qquad \text{Zero-Product Property.}$$

$$x = -2 \quad \text{or} \qquad x = 2$$

The equation has the solution set $\{-2, 2\}$. The *x*-intercepts are -2 and 2.
To find the *y*-intercept(s), we let $x = 0$ and obtain the equation

$$y = -4$$

The *y*-intercept is -4.   ■

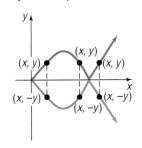 NOW WORK PROBLEM **33 (list the intercepts).**

 COMMENT: For many equations, finding intercepts may not be so easy. In such cases, a graphing utility can be used. Read Section B.3, Using a Graphing Utility to Locate Intercepts and Check for Symmetry in Appendix B, (pp. 567–568), to find out how a graphing utility locates intercepts.   ■

SYMMETRY

We have just seen the role that intercepts play in obtaining key points on the graph of an equation. Another helpful tool for graphing equations involves *symmetry,* particularly symmetry with respect to the *x*-axis, the *y*-axis, and the origin.

Figure 14
Symmetry with respect to the *x*-axis

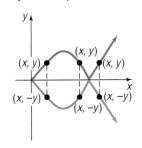

> A graph is said to be **symmetric with respect to the x-axis** if, for every point (x, y) on the graph, the point $(x, -y)$ is also on the graph.

Figure 14 illustrates the definition. Notice that, when a graph is symmetric with respect to the *x*-axis, the part of the graph above the *x*-axis is a reflection or mirror image of the part below it, and vice versa.

EXAMPLE 6	**Points Symmetric with Respect to the *x*-Axis**

If a graph is symmetric with respect to the *x*-axis and the point $(3, 2)$ is on the graph, then the point $(3, -2)$ is also on the graph.   ■

Figure 15
Symmetry with respect to the *y*-axis

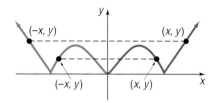

A graph is said to be **symmetric with respect to the y-axis** if, for every point (x, y) on the graph, the point $(-x, y)$ is also on the graph.

Figure 15 illustrates the definition. Notice that, when a graph is symmetric with respect to the *y*-axis, the part of the graph to the right of the *y*-axis is a reflection of the part to the left of it, and vice versa.

EXAMPLE 7   **Points Symmetric with Respect to the *y*-Axis**

If a graph is symmetric with respect to the *y*-axis and the point $(5, 8)$ is on the graph, then the point $(-5, 8)$ is also on the graph. ■

Figure 16
Symmetry with respect to the origin

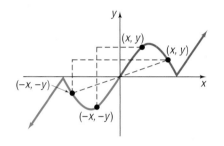

A graph is said to be **symmetric with respect to the origin** if, for every point (x, y) on the graph, the point $(-x, -y)$ is also on the graph.

Figure 16 illustrates the definition. Notice that symmetry with respect to the origin may be viewed in two ways:

1. As a reflection about the *y*-axis, followed by a reflection about the *x*-axis.
2. As a projection along a line through the origin so that the distances from the origin are equal.

EXAMPLE 8   **Points Symmetric with Respect to the Origin**

If a graph is symmetric with respect to the origin and the point $(4, 2)$ is on the graph, then the point $(-4, -2)$ is also on the graph. ■

NOW WORK PROBLEMS **1** AND **11(b)**.

④   When the graph of an equation is symmetric with respect to a coordinate axis or the origin, the number of points that you need to plot in order to see the pattern is reduced. For example, if the graph of an equation is symmetric with respect to the *y*-axis, then, once points to the right of the *y*-axis are plotted, an equal number of points on the graph can be obtained by reflecting them about the *y*-axis. Thus, before we graph an equation, we first want to determine whether it has any symmetry. The following tests are used for this purpose.

TESTS FOR SYMMETRY

To test the graph of an equation for symmetry with respect to the

x-Axis   Replace *y* by $-y$ in the equation. If an equivalent equation results, the graph of the equation is symmetric with respect to the *x*-axis.

y-Axis   Replace *x* by $-x$ in the equation. If an equivalent equation results, the graph of the equation is symmetric with respect to the *y*-axis.

Origin   Replace *x* by $-x$ and *y* by $-y$ in the equation. If an equivalent equation results, the graph of the equation is symmetric with respect to the origin.

EXAMPLE 9   Graphing an Equation by Finding Intercepts and Checking for Symmetry

Graph the equation $y = x^3$. Find any intercepts and check for symmetry first.

Solution   First, we seek the intercepts. When $x = 0$, then $y = 0$; and when $y = 0$, then $x = 0$. The origin $(0, 0)$ is the only intercept. Now we test for symmetry.

x-Axis   Replace y by $-y$. Since the result, $-y = x^3$, is not equivalent to $y = x^3$, the graph is not symmetric with respect to the x-axis.

y-Axis   Replace x by $-x$. Since the result, $y = (-x)^3 = -x^3$, is not equivalent to $y = x^3$, the graph is not symmetric with respect to the x-axis.

Origin   Replace x by $-x$ and y by $-y$. Since the result, $-y = -x^3$, is equivalent to $y = x^3$, the graph is symmetric with respect to the origin.

To graph $y = x^3$, we use the equation to obtain several points on the graph. Because of the symmetry, we only need to locate points on the graph for which $x \geq 0$. See Table 2. Figure 17 shows the graph.

TABLE 2

x	$y = x^3$	(x, y)
0	0	$(0, 0)$
1	1	$(1, 1)$
2	8	$(2, 8)$
3	27	$(3, 27)$

Figure 17
$y = x^3$

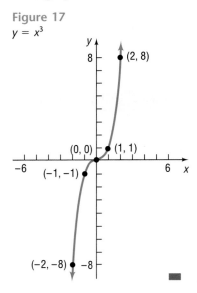

NOW WORK PROBLEM **33** (TEST FOR SYMMETRY).

EXAMPLE 10   Graphing an Equation

Graph the equation $x = y^2$. Find any intercepts and check for symmetry first.

Figure 18
$x = y^2$

Solution   The lone intercept is $(0, 0)$. The graph is symmetric with respect to the x-axis. (Do you see why? Replace y by $-y$.) Figure 18 shows the graph.

If we restrict y so that $y \geq 0$, the equation $x = y^2$, $y \geq 0$, may be written equivalently as $y = \sqrt{x}$. The portion of the graph of $x = y^2$ in quadrant I is therefore the graph of $y = \sqrt{x}$. See Figure 19.

COMMENT: To see the graph of the equation $x = y^2$ on a graphing calculator, you will need to graph two equations: $Y_1 = \sqrt{x}$ and $Y_2 = -\sqrt{x}$. We discuss why in the next section. See Figure 20.

Figure 19

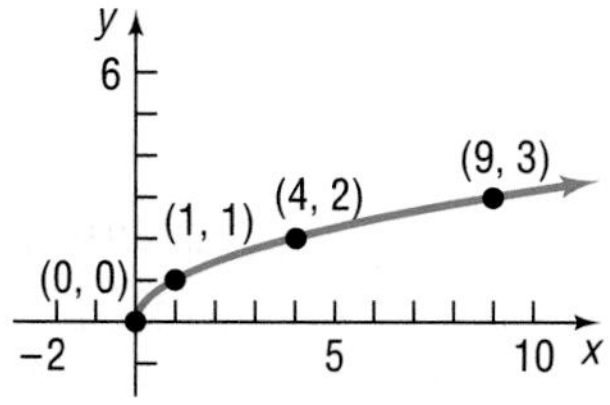

Figure 20

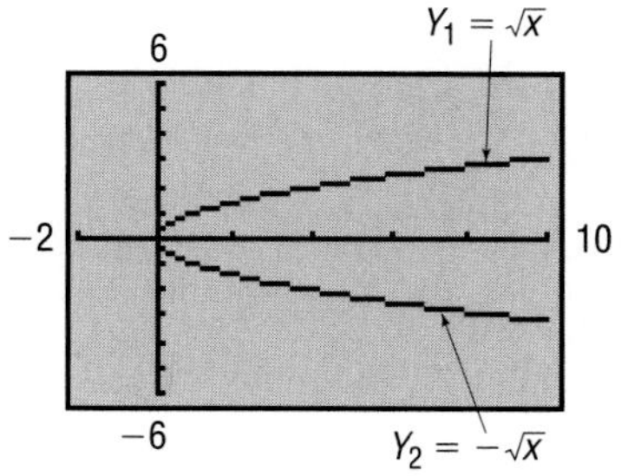

EXAMPLE 11

Graphing the Equation $y = \dfrac{1}{x}$

Graph the equation: $y = \dfrac{1}{x}$

Find any intercepts and check for symmetry first.

Solution   We check for intercepts first. If we let $x = 0$, we obtain a 0 denominator, which is not defined. We conclude that there is no y-intercept. If we let $y = 0$, we get the equation $\dfrac{1}{x} = 0$, which has no solution. We conclude that there is no x-intercept. The graph of $y = \dfrac{1}{x}$ does not cross or touch the coordinate axes.

Next we check for symmetry:

x-Axis   Replacing y by $-y$ yields $-y = \dfrac{1}{x}$, which is not equivalent to $y = \dfrac{1}{x}$.

y-Axis   Replacing x by $-x$ yields $y = \dfrac{1}{-x}$, which is not equivalent to $y = \dfrac{1}{x}$.

Origin   Replacing x by $-x$ and y by $-y$ yields $-y = \dfrac{1}{-x}$, which is equivalent to $y = \dfrac{1}{x}$.

The graph is symmetric with respect to the origin.

Finally, we set up Table 3, listing several points on the graph. Because of the symmetry with respect to the origin, we use only positive values of x. From Table 3 we infer that if x is a large and positive number then $y = \dfrac{1}{x}$ is a positive number close to 0. We also infer that if x is a positive number close to 0 then $y = \dfrac{1}{x}$ is a large and positive number. Armed with this information, we can graph the equation. Figure 21 illustrates some of these points and the graph of $y = \dfrac{1}{x}$. Observe how the absence of intercepts and the existence of symmetry with respect to the origin were utilized.

Figure 21

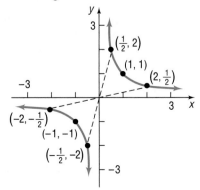

TABLE 3

x	$y = \dfrac{1}{x}$	(x, y)
$\frac{1}{10}$	10	$\left(\frac{1}{10}, 10\right)$
$\frac{1}{3}$	3	$\left(\frac{1}{3}, 3\right)$
$\frac{1}{2}$	2	$\left(\frac{1}{2}, 2\right)$
1	1	$(1, 1)$
2	$\frac{1}{2}$	$\left(2, \frac{1}{2}\right)$
3	$\frac{1}{3}$	$\left(3, \frac{1}{3}\right)$
10	$\frac{1}{10}$	$\left(10, \frac{1}{10}\right)$

COMMENT: Refer to Example 3 in Appendix B, Section B.3, for the graph of $y = \dfrac{1}{x}$ using a graphing utility.   ▬

CIRCLES

⑤   One advantage of a coordinate system is that it enables us to translate a geometric statement into an algebraic statement, and vice versa. Consider, for example, the following geometric statement that defines a circle.

> A **circle** is a set of points in the xy-plane that are a fixed distance r from a fixed point (h, k). The fixed distance r is called the **radius,** and the fixed point (h, k) is called the **center** of the circle.

Figure 22

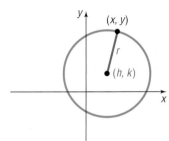

Figure 22 shows the graph of a circle. Is there an equation having this graph? If so, what is the equation? To find the equation, we let (x, y) represent the coordinates of any point on a circle with radius r and center (h, k). Then the distance between the points (x, y) and (h, k) must always equal r. That is, by the distance formula

$$\sqrt{(x - h)^2 + (y - k)^2} = r$$

or, equivalently,

$$(x - h)^2 + (y - k)^2 = r^2$$

> The **standard form of an equation of a circle** with radius r and center (h, k) is
>
> $$(x - h)^2 + (y - k)^2 = r^2 \qquad \textbf{(1)}$$

The standard form of an equation of a circle of radius r with center at the origin $(0, 0)$ is

> $$x^2 + y^2 = r^2$$

Figure 23
Unit circle $x^2 + y^2 = 1$

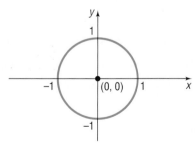

If the radius $r = 1$, the circle whose center is at the origin is called the **unit circle** and has the equation

> $$x^2 + y^2 = 1$$

See Figure 23.

EXAMPLE 12   **Writing the Standard Form of the Equation of a Circle**

Write the standard form of the equation of the circle with radius 5 and center $(-3, 6)$.

Solution   Using the form of equation (1) and substituting the values $r = 5$, $h = -3$, and $k = 6$, we have

$$(x - h)^2 + (y - k)^2 = r^2$$
$$(x + 3)^2 + (y - 6)^2 = 25$$

▬

✏   NOW   WORK   PROBLEM   **53.**

6　The graph of any equation of the form of equation (1) is that of a circle with radius r and center (h, k).

| EXAMPLE 13 | **Graphing a Circle** |

Graph the equation: $(x + 3)^2 + (y - 2)^2 = 16$

Solution　The equation is of the form of equation (1), so its graph is a circle. To graph the equation, we first compare the given equation to the standard form of the equation of a circle. The comparison yields information about the circle.

$$(x + 3)^2 + (y - 2)^2 = 16$$
$$(x - (-3))^2 + (y - 2)^2 = 4^2$$
$$\underset{\uparrow}{(x - h)^2} + \underset{\uparrow}{(y - k)^2} = \underset{\uparrow}{r^2}$$

Figure 24

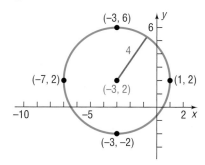

We see that $h = -3$, $k = 2$, and $r = 4$. The circle has center $(-3, 2)$ and a radius of 4 units. To graph this circle, we first plot the center $(-3, 2)$. Since the radius is 4, we can locate four points on the circle by plotting points 4 units to the left, to the right, up, and down from the center. These four points can then be used as guides to obtain the graph. See Figure 24. ∎

NOW WORK PROBLEM **67**.

If we eliminate the parentheses from the standard form of the equation of the circle given in Example 13, we get

$$(x + 3)^2 + (y - 2)^2 = 16$$
$$x^2 + 6x + 9 + y^2 - 4y + 4 = 16$$

which we find, upon simplifying, is equivalent to

$$x^2 + y^2 + 6x - 4y - 3 = 0$$

It can be shown that any equation of the form

$$x^2 + y^2 + ax + by + c = 0$$

has a graph that is a circle or a point, or it has no graph at all. For example, the graph of the equation $x^2 + y^2 = 0$ is the single point $(0, 0)$. The equation $x^2 + y^2 + 5 = 0$, or $x^2 + y^2 = -5$, has no graph, because sums of squares of real numbers are never negative.

When its graph is a circle, the equation

$$\boxed{x^2 + y^2 + ax + by + c = 0}$$

is referred to as the **general form of the equation of a circle.**

NOW WORK PROBLEM **59**.

7　If an equation of a circle is in the general form, we use the method of completing the square to put the equation in standard form so that we can identify its center and radius.

EXAMPLE 14   **Graphing a Circle Whose Equation Is in General Form**

Graph the equation: $x^2 + y^2 + 4x - 6y + 12 = 0$

Solution   We complete the square in both x and y to put the equation in standard form. Group the expressions involving x, group the expressions involving y, and put the constant on the right side of the equation. The result is

$$(x^2 + 4x) + (y^2 - 6y) = -12$$

Next, complete the square of each expression in parentheses. Remember that any number added on the left side of the equation must be added on the right.

Figure 25

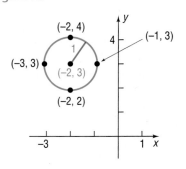

$$(x^2 + 4x + 4) + (y^2 - 6y + 9) = -12 + 4 + 9$$

$$\left(\tfrac{4}{2}\right)^2 = 4 \qquad \left(\tfrac{-6}{2}\right)^2 = 9$$

$$(x + 2)^2 + (y - 3)^2 = 1 \qquad \text{Factor.}$$

We recognize this equation as the standard form of the equation of a circle with radius 1 and center $(-2, 3)$.

To graph the equation, we use the center $(-2, 3)$ and the radius 1. See Figure 25.   ■

 NOW WORK PROBLEM 69.

COMMENT: Now read Section B.5, Square Screens, in Appendix B, page 572.   ■

EXAMPLE 15   **Using a Graphing Utility to Graph a Circle**

Graph the equation: $x^2 + y^2 = 4$

Solution   This is the equation of a circle with center at the origin and radius 2. To graph this equation, we must first solve for y.

Figure 26

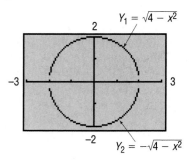

$$x^2 + y^2 = 4$$
$$y^2 = 4 - x^2 \qquad \text{Subtract } x^2 \text{ from each side}$$
$$y = \pm \sqrt{4 - x^2} \qquad \begin{array}{l}\text{Apply the Square Root}\\ \text{Method to solve for } y\end{array}$$

There are two equations to graph: first, we graph $Y_1 = \sqrt{4 - x^2}$ and then $Y_2 = -\sqrt{4 - x^2}$ on the same square screen. (Your circle will appear oval if you do not use a square screen.) See Figure 26.   ■

1.2 EXERCISES

In Problems 1–10, plot each point. Then plot the point that is symmetric to it with respect to (a) the x-axis; (b) the y-axis; (c) the origin.

1. $(3, 4)$ **2.** $(5, 3)$ **3.** $(-2, 1)$ **4.** $(4, -2)$ **5.** $(1, 1)$

6. $(-1, -1)$ **7.** $(-3, -4)$ **8.** $(4, 0)$ **9.** $(0, -3)$ **10.** $(-3, 0)$

In Problems 11–22, the graph of an equation is given.
  (a) *List the intercepts of the graph.*
  (b) *Based on the graph, tell whether the graph is symmetric with respect to the x-axis, the y-axis, and/or the origin.*

11.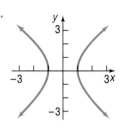

12.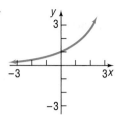

13.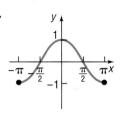

14.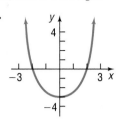

15.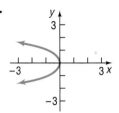

16.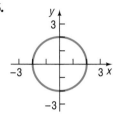

17.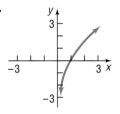

18.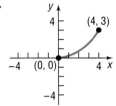

19.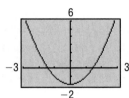

20.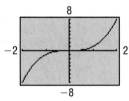

21.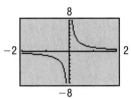

22.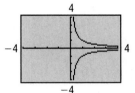

In Problems 23–28, determine whether the given points are on the graph of the equation.

23. Equation: $y = x^4 - \sqrt{x}$
    Points: $(0,0); (1,1); (-1,0)$

24. Equation: $y = x^3 - 2\sqrt{x}$
    Points: $(0,0); (1,1); (1,-1)$

25. Equation: $y^2 = x^2 + 9$
    Points: $(0,3); (3,0); (-3,0)$

26. Equation: $y^3 = x + 1$
    Points: $(1,2); (0,1); (-1,0)$

27. Equation: $x^2 + y^2 = 4$
    Points: $(0,2); (-2,2); (\sqrt{2}, \sqrt{2})$

28. Equation: $x^2 + 4y^2 = 4$
    Points: $(0,1); (2,0); (2, \frac{1}{2})$

In Problems 29–44, list the intercepts and test for symmetry.

29. $x^2 = y$

30. $y^2 = x$

31. $y = 3x$

32. $y = -5x$

33. $x^2 + y - 9 = 0$

34. $y^2 - x - 4 = 0$

35. $9x^2 + 4y^2 = 36$

36. $4x^2 + y^2 = 4$

37. $y = x^3 - 27$

38. $y = x^4 - 1$

39. $y = x^2 - 3x - 4$

40. $y = x^2 + 4$

41. $y = \dfrac{3x}{x^2 + 9}$

42. $y = \dfrac{x^2 - 4}{2x}$

43. $y = \dfrac{-x^3}{x^2 - 9}$

44. $y = \dfrac{x^4 + 1}{2x^5}$

In Problems 45–48, draw a quick sketch of each equation.

45. $y = x^3$

46. $x = y^2$

47. $y = \sqrt{x}$

48. $y = \dfrac{1}{x}$

49. If $(a, 2)$ is a point on the graph of $y = 3x + 5$, what is a?

50. If $(2, b)$ is a point on the graph of $y = x^2 + 4x$, what is b?

51. If (a, b) is a point on the graph of $2x + 3y = 6$, write an equation that relates a to b.

52. If $(2, 0)$ and $(0, 5)$ are points on the graph of $y = mx + b$, what are m and b?

In Problems 53–56, find the center and radius of each circle. Write the standard form of the equation.

53.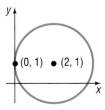

54.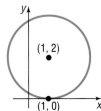

55.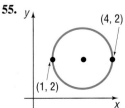

56.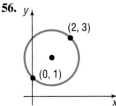

In Problems 57–64, write the standard form of the equation and the general form of the equation of each circle of radius r and center (h, k). Graph each circle.

57. $r = 2$; $(h, k) = (0, 0)$

58. $r = 3$; $(h, k) = (0, 0)$

59. $r = 1$; $(h, k) = (1, -1)$

60. $r = 2$; $(h, k) = (-2, 1)$

61. $r = 2$; $(h, k) = (0, 2)$

62. $r = 3$; $(h, k) = (1, 0)$

63. $r = 5$; $(h, k) = (4, -3)$

64. $r = 4$; $(h, k) = (2, -3)$

In Problems 65–74, find the center (h, k) and radius r of each circle. Graph each circle.

65. $x^2 + y^2 = 4$

66. $x^2 + (y - 1)^2 = 1$

67. $2(x - 3)^2 + 2y^2 = 8$

68. $3(x + 1)^2 + 3(y - 1)^2 = 6$

69. $x^2 + y^2 + 4x - 4y - 1 = 0$

70. $x^2 + y^2 - 6x + 2y + 9 = 0$

71. $x^2 + y^2 - x + 2y + 1 = 0$

72. $x^2 + y^2 + x + y - \dfrac{1}{2} = 0$

73. $2x^2 + 2y^2 - 12x + 8y - 24 = 0$

74. $2x^2 + 2y^2 + 8x + 7 = 0$

In Problems 75–80, find the general form of the equation of each circle.

75. Center at the origin and containing the point $(-3, 2)$

76. Center at the point $(1, 0)$ and containing the point $(-2, 3)$

77. Center at the point $(2, 3)$ and tangent to the x-axis

78. Center at the point $(-3, 1)$ and tangent to the y-axis

79. With endpoints of a diameter at the points $(1, 4)$ and $(-3, 2)$

80. With endpoints of a diameter at the points $(4, 3)$ and $(0, 1)$

In Problems 81–84, match each graph with the correct equation.

(a) $(x - 3)^2 + (y + 3)^2 = 9$

(c) $(x - 1)^2 + (y + 2)^2 = 4$

(b) $(x + 1)^2 + (y - 2)^2 = 4$

(d) $(x + 3)^2 + (y - 3)^2 = 9$

81.
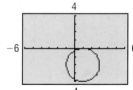

82.
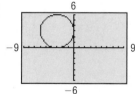

83.
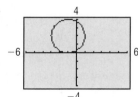

84.
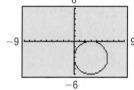

85. Which of the following equations might have the graph shown below? (More than one answer is possible.)

(a) $(x - 2)^2 + (y + 3)^2 = 13$

(b) $(x - 2)^2 + (y - 2)^2 = 8$

(c) $(x - 2)^2 + (y - 3)^2 = 13$

(d) $(x + 2) + (y - 2)^2 = 8$

(e) $x^2 + y^2 - 4x - 9y = 0$

(f) $x^2 + y^2 + 4x - 2y = 0$

(g) $x^2 + y^2 - 9x - 4y = 0$

(h) $x^2 + y^2 - 4x - 4y = 4$

86. Which of the following equations might have the graph shown below? (More than one answer is possible.)

(a) $(x - 2)^2 + y^2 = 3$

(b) $(x + 2)^2 + y^2 = 3$

(c) $x^2 + (y - 2)^2 = 3$

(d) $(x + 2)^2 + y^2 = 4$

(e) $x^2 + y^2 + 10x + 16 = 0$

(f) $x^2 + y^2 + 10x - 2y = 1$

(g) $x^2 + y^2 + 9x + 10 = 0$

(h) $x^2 + y^2 - 9x - 10 = 0$

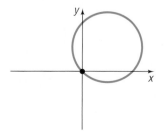

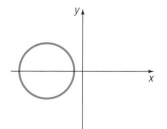

87. Weather Satellites   Earth is represented on a map of a portion of the solar system so that its surface is the circle with equation $x^2 + y^2 + 2x + 4y - 4091 = 0$. A weather satellite circles 0.6 unit above Earth with the center of its circular orbit at the center of Earth. Find the equation for the orbit of the satellite on this map.

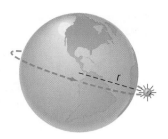

In Problem 88, you may use a graphing utility, but it is not required.

88. (a) Graph $y = \sqrt{x^2}$, $y = x$, $y = |x|$, and $y = \left(\sqrt{x}\right)^2$, noting which graphs are the same.
   (b) Explain why the graphs of $y = \sqrt{x^2}$ and $y = |x|$ are the same.
   (c) Explain why the graphs of $y = x$ and $y = \left(\sqrt{x}\right)^2$ are not the same.
   (d) Explain why the graphs of $y = \sqrt{x^2}$ and $y = x$ are not the same.

89. Make up an equation with the intercepts $(2, 0)$, $(4, 0)$, and $(0, 1)$. Compare your equation with a friend's equation. Comment on any similarities.

90. An equation is being tested for symmetry with respect to the x-axis, the y-axis, and the origin. Explain why, if two of these symmetries are present, the remaining one must also be present.

91. Draw a graph that contains the points $(-2, -1)$, $(0, 1)$, $(1, 3)$, and $(3, 5)$. Compare your graph with those of other students. Are most of the graphs almost straight lines? How many are "curved"? Discuss the various ways that these points might be connected.

P R E P A R I N G   F O R   T H I S   S E C T I O N

Before getting started, review the following:

✓ Constants and Variables (Appendix A, Section A.1, pp. 489–490)

✓ Intervals (Appendix A, Section A.5, pp. 522–523)

✓ Solving Inequalities (Appendix A, Section A.5, pp. 524–528)

✓ Solving Equations (Appendix A, Section A.3, pp. 502–511)

✓ Intercepts (Section 1.2, pp. 11–12)

1.3 FUNCTIONS

OBJECTIVES
1 Determine Whether a Relation Represents a Function
2 Find the Value of a Function
3 Find the Domain of a Function
4 Identify the Graph of a Function
5 Obtain Information from or about the Graph of a Function

1   A **relation** is a correspondence between two sets. If x and y are two elements in these sets and if a relation exists between x and y, then we say that x **corresponds** to y or that y **depends on** x, and we write $x \rightarrow y$. We may also write $x \rightarrow y$ as the ordered pair (x, y).

| EXAMPLE 1 | An Example of a Relation |

Figure 27 depicts a relation between four individuals and their birthdays. The relation might be named "was born on." Then Katy corresponds to June 20, Dan corresponds to Sept 4, and so on. Using ordered pairs, this relation would be expressed as

$$\{(\text{Katy, June 20}), (\text{Dan, Sept 4}), (\text{Patrick, Dec 31}), (\text{Phoebe, Dec 31})\}$$

Figure 27

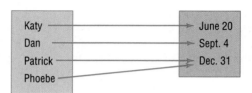

Often, we are interested in specifying the type of relation (such as an equation) that exists between the two variables. For example, the relation between the revenue R resulting from the sale of x items selling for \$10 each may be expressed by the equation $R = 10x$. If we know how many items have been sold, then we can calculate the revenue by using the equation $R = 10x$. This equation is an example of a *function*.

As another example, suppose that an icicle falls off a building from a height of 64 feet above the ground. According to a law of physics, the distance s (in feet) of the icicle from the ground after t seconds is given (approximately) by the formula $s = 64 - 16t^2$. When $t = 0$ seconds, the icicle is $s = 64$ feet above the ground. After 1 second, the icicle is $s = 64 - 16(1)^2 = 48$ feet above the ground. After 2 seconds, the icicle strikes the ground. The formula $s = 64 - 16t^2$ provides a way of finding the distance s for any time t ($0 \le t \le 2$). There is a correspondence between each time t in the interval $0 \le t \le 2$ and the distance s. We say that the distance s is a function of the time t because:

1. There is a correspondence between the set of times and the set of distances.

2. There is exactly one distance s obtained for any time t in the interval $0 \le t \le 2$.

Let's now look at the definition of a function.

Definition of Function

Let X and Y be two nonempty sets.* A **function** from X into Y is a relation that associates with each element of X exactly one element of Y.

---

*The sets X and Y will usually be sets of real numbers. The sets X and Y can also be sets of complex numbers (discussed in Appendix A, Section A.4), and then we have defined a complex function. In the broad definition (due to Lejeune Dirichlet), X and Y can be any two sets.

The set X is called the **domain** of the function. For each element x in X, the corresponding element y in Y is called the **value** of the function at x, or the image of x. The set of all images of the elements of the domain is called the **range** of the function. See Figure 28.

Since there may be some elements in Y that are not the image of any x in X, it follows that the range of a function is a subset of Y, as shown in Figure 28.

Figure 28

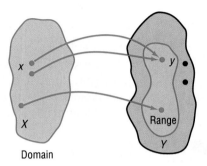

Not all relations between two sets are functions. The next example shows how to determine whether a relation is a function or not.

EXAMPLE 2

Determining Whether a Relation Represents a Function

Determine whether the following relations represent functions.

(a) See Figure 29. For this relation, the domain represents four individuals and the range represents their birthdays.

Figure 29

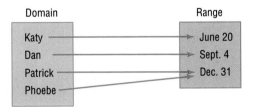

(b) See Figure 30. For this relation, the domain represents the employees of Sara's Pre-Owned Car Mart and the range represents their phone number(s).

Figure 30

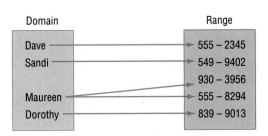

Solution (a) The relation is a function because each element in the domain corresponds to a unique element in the range. Notice that more than one element in the domain can correspond to the same element in the range (Phoebe and Patrick were born on the same day of the year).

(b) The relation is not a function because each element in the domain does not correspond to a unique element in the range. Maureen has two telephone numbers; therefore, if Maureen is chosen from the domain, a unique telephone number cannot be assigned to her. ∎

 NOW WORK PROBLEM **1.**

We may think of a function as a set of ordered pairs (x, y) in which no two distinct pairs have the same first element. The set of all first elements x is the domain of the function, and the set of all second elements y is its range. Associated with each element x in the domain, there is a unique element y in the range.

EXAMPLE 3 | **Determining Whether a Relation Represents a Function**

Determine whether each relation represents a function. For those that are functions, state the domain and range.

(a) $\{(1, 4), (2, 5), (3, 6), (4, 7)\}$
(b) $\{(1, 4), (2, 4), (3, 5), (6, 10)\}$
(c) $\{(-3, 9), (-2, 4), (0, 0), (1, 1), (-3, 8)\}$

Solution (a) This relation is a function because there are no ordered pairs with the same first element and different second elements. The domain of this function is $\{1, 2, 3, 4\}$ and its range is $\{4, 5, 6, 7\}$.

(b) This relation is a function because there are no ordered pairs with the same first element and different second elements. The domain of this function is $\{1, 2, 3, 6\}$ and its range is $\{4, 5, 10\}$.

(c) This relation is not a function because there are two ordered pairs $(-3, 9)$ and $(-3, 8)$ that have the same first element, but different second elements. ∎

In Example 3(b), notice that 1 and 2 in the domain each have the same image in the range. This does not violate the definition of a function; two different first elements can have the same second element. A violation of the definition occurs when two ordered pairs have the same first element and different second elements, as in Example 3(c).

NOW WORK PROBLEM **5.**

Example 2(a) demonstrates that a function may be defined by some correspondence between two sets. Examples 3(a) and 3(b) demonstrate that a function may be defined by a set of ordered pairs. A function may also be defined by an equation in two variables, usually denoted x and y.

EXAMPLE 4 | **Example of a Function**

Consider the function defined by the equation

$$y = 2x - 5, \qquad 1 \le x \le 6$$

The domain $1 \leq x \leq 6$ specifies that the number x is restricted to the real numbers from 1 to 6, inclusive. The equation $y = 2x - 5$ specifies that the number x is to be multiplied by 2 and then 5 is to be subtracted from the result to get y. For example, if $x = \dfrac{3}{2}$, then $y = 2 \cdot \dfrac{3}{2} - 5 = -2.$ ■

FUNCTION NOTATION

Functions are often denoted by letters such as f, F, g, G, and others. If f is a function, then for each number x in its domain the corresponding image in the range is designated by the symbol $f(x)$, read as "f of x" or as "f at x." We refer to $f(x)$ as the **value of f at the number x;** $f(x)$ is the number that results when x is given and the function f is applied; $f(x)$ does *not* mean "f times x." For example, the function given in Example 4 may be written as

$$y = f(x) = 2x - 5, 1 \leq x \leq 6. \text{ Then } f\left(\dfrac{3}{2}\right) = -2.$$

Figure 31 illustrates some other functions. Note that in every function illustrated, for each x in the domain there is one value in the range.

Figure 31

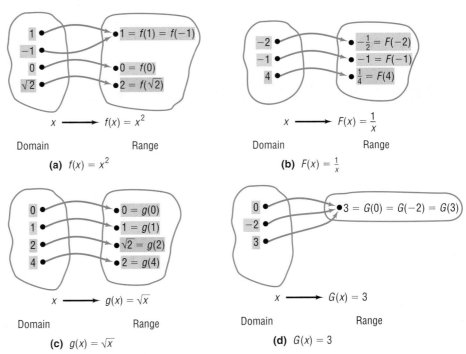

(a) $f(x) = x^2$

(b) $F(x) = \dfrac{1}{x}$

(c) $g(x) = \sqrt{x}$

(d) $G(x) = 3$

Sometimes it is helpful to think of a function f as a machine that receives as input a number from the domain, manipulates it, and outputs the value. See Figure 32.

Figure 32

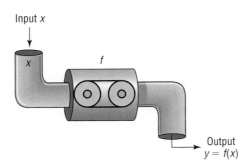

The restrictions on this input/output machine are as follows:

1. It only accepts numbers from the domain of the function.
2. For each input, there is exactly one output (which may be repeated for different inputs).

For a function $y = f(x)$, the variable x is called the **independent variable,** because it can be assigned any of the permissible numbers from the domain. The variable y is called the **dependent variable,** because its value depends on x.

Any symbol can be used to represent the independent and dependent variables. For example, if f is the *cube function*, then f can be given by $f(x) = x^3$ or $f(t) = t^3$ or $f(z) = z^3$. All three functions are the same. Each tells us to cube the independent variable. In practice, the symbols used for the independent and dependent variables are based on common usage, such as using C for cost in business.

② The independent variable is also called the **argument** of the function. Thinking of the independent variable as an argument can sometimes make it easier to find the value of a function. For example, if f is the function defined by $f(x) = x^3$, then f tells us to cube the argument. Thus, $f(2)$ means to cube 2, $f(a)$ means to cube the number a, and $f(x + h)$ means to cube the quantity $x + h$.

EXAMPLE 5   **Finding Values of a Function**

For the function f defined by $f(x) = 2x^2 - 3x$, evaluate

(a) $f(3)$ (b) $f(x) + f(3)$ (c) $f(-x)$

(d) $-f(x)$ (e) $f(x + 3)$

Solution   (a) We substitute 3 for x in the equation for f to get

$$f(3) = 2(3)^2 - 3(3) = 18 - 9 = 9$$

(b) $f(x) + f(3) = (2x^2 - 3x) + (9) = 2x^2 - 3x + 9$

(c) We substitute $-x$ for x in the equation for f.

$$f(-x) = 2(-x)^2 - 3(-x) = 2x^2 + 3x$$

(d) $-f(x) = -(2x^2 - 3x) = -2x^2 + 3x$

(e) $f(x + 3) = 2(x + 3)^2 - 3(x + 3)$ Notice the use of parentheses here.

$$= 2(x^2 + 6x + 9) - 3x - 9$$
$$= 2x^2 + 12x + 18 - 3x - 9$$
$$= 2x^2 + 9x + 9$$ ■

Notice in this example that $f(x + 3) \neq f(x) + f(3)$ and $f(-x) \neq -f(x)$.

 ✏ NOW WORK PROBLEM **13**.

Most calculators have special keys that enable you to find the value of certain commonly used functions. For example, you should be able to find the square function $f(x) = x^2$, the square root function $f(x) = \sqrt{x}$, the reciprocal function $f(x) = \dfrac{1}{x} = x^{-1}$, and many others that will be discussed later in this book (such as $\ln x$ and $\log x$). Verify the results of Example 6, which follows, on your calculator.

| EXAMPLE 6 | Finding Values of a Function on a Calculator |

(a) $f(x) = x^2$; $f(1.234) = 1.522756$

(b) $F(x) = \dfrac{1}{x}$; $F(1.234) = 0.8103727715$

(c) $g(x) = \sqrt{x}$; $g(1.234) = 1.110855526$ ∎

COMMENT   Graphing calculators can be used to evaluate any function that you wish. Figure 33 shows the result obtained in Example 5(a) on a TI-83 graphing calculator with the function to be evaluated, $f(x) = 2x^2 - 3x$, in Y_1.*

Figure 33

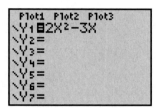   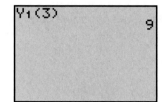

IMPLICIT FORM OF A FUNCTION

In general, when a function f is defined by an equation in x and y, we say that the function f is given **implicitly.** If it is possible to solve the equation for y in terms of x, then we write $y = f(x)$ and say that the function is given **explicitly.** For example,

Implicit Form	**Explicit Form**
$3x + y = 5$	$y = f(x) = -3x + 5$
$x^2 - y = 6$	$y = f(x) = x^2 - 6$
$xy = 4$	$y = f(x) = \dfrac{4}{x}$

Not all equations in x and y define a function $y = f(x)$. If an equation is solved for y and two or more values of y can be obtained for a given x, then the equation does not define a function.

| EXAMPLE 7 | Determining Whether an Equation Is a Function |

Determine if the equation $x^2 + y^2 = 1$ is a function.

Solution   To determine whether the equation $x^2 + y^2 = 1$, which defines the unit circle, is a function, we need to solve the equation for y.

$$x^2 + y^2 = 1$$
$$y^2 = 1 - x^2$$
$$y = \pm\sqrt{1 - x^2}$$

For values of x between -1 and 1, two values of y result. This means that the equation $x^2 + y^2 = 1$ does not define a function.   ∎

NOW WORK PROBLEM **2 7 .**

* Consult your owner's manual for the required keystrokes.

COMMENT The explicit form of a function is the form required by a graphing calculator. Now do you see why it is necessary to graph a circle in two "pieces"? ■

We list next a summary of some important facts to remember about a function f.

SUMMARY OF IMPORTANT FACTS ABOUT FUNCTIONS

(a) To each x in the domain of f, there is exactly one image $f(x)$ in the range; however, an element in the range can result from more than one x in the domain.

(b) f is the symbol that we use to denote the function. It is symbolic of the equation that we use to get from an x in the domain to $f(x)$ in the range.

(c) If $y = f(x)$, then x is called the independent variable or argument of f, and y is called the dependent variable or the value of f at x.

DOMAIN OF A FUNCTION

③ Often the domain of a function f is not specified; instead, only the equation defining the function is given. In such cases, we agree that the domain of f is the largest set of real numbers for which the value $f(x)$ is a real number. The domain of a function f is the same as the domain of the variable x in the expression $f(x)$.

EXAMPLE 8

Finding the Domain of a Function

Find the domain of each of the following functions:

(a) $f(x) = x^2 + 5x$ (b) $g(x) = \dfrac{3x}{x^2 - 4}$ (c) $h(t) = \sqrt{4 - 3t}$

Solution

(a) The function tells us to square a number and then add five times the number. Since these operations can be performed on any real number, we conclude that the domain of f is all real numbers.

(b) The function g tells us to divide $3x$ by $x^2 - 4$. Since division by 0 is not defined, the denominator $x^2 - 4$ can never be 0 so x can never equal -2 or 2. The domain of the function g is $\{x \mid x \neq -2, x \neq 2\}$.

(c) The function h tells us to take the square root of $4 - 3t$. But only non-negative numbers have real square roots so the expression under the square root must be nonnegative. This requires that

$$4 - 3t \geq 0$$
$$-3t \geq -4$$
$$t \leq \frac{4}{3}$$

The domain of h is $\left\{ t \mid t \leq \dfrac{4}{3} \right\}$ or the interval $\left(-\infty, \dfrac{4}{3} \right]$. ■

NOW WORK PROBLEM **37**.

If x is in the domain of a function f, we shall say that **f is defined at x,** or **$f(x)$ exists.** If x is not in the domain of f, we say that **f is not defined at x,** or **$f(x)$ does not exist.** For example, if $f(x) = \dfrac{x}{x^2 - 1}$, then $f(0)$ exists, but $f(1)$ and $f(-1)$ do not exist. (Do you see why?)

We have not said much about finding the range of a function. The reason is that when a function is defined by an equation it is often difficult to find the range.* Therefore, we shall usually be content to find just the domain of a function when only the rule for the function is given. We shall express the domain of a function using inequalities, interval notation, set notation, or words, whichever is most convenient.

THE GRAPH OF A FUNCTION

In applications, a graph often demonstrates more clearly the relationship between two variables than, say, an equation or table would. For example, Table 4 shows the price per share of Intel stock at the end of each month from 8/31/99 through 8/31/00. If we plot these data using the date as the x-coordinate and the price as the y-coordinate and then connect the points, we obtain Figure 34.

T A B L E 4	
Date	**Closing Price ($)**
8/31/99	41.09
9/30/99	37.16
10/31/99	38.72
11/30/99	38.34
12/31/99	41.16
1/31/00	49.47
2/29/00	56.50
3/31/00	65.97
4/30/00	63.41
5/31/00	62.34
6/30/00	66.84
7/31/00	66.75
8/31/00	74.88
Courtesy of A.G. Edwards & Sons, Inc.	

Figure 34
Monthly closing prices
of Intel stock 8/31/99
through 8/31/00

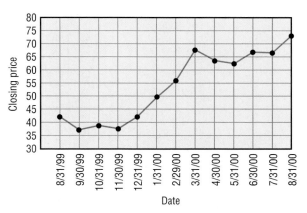

We can see from the graph that the price of the stock was rising rapidly from 11/30/99 through 3/31/00 and was falling slightly from 3/31/00 through 5/31/00. The graph also shows that the lowest price occurred at the end of September, 1999, whereas the highest occurred at the end of August, 2000. Equations and tables, on the other hand, usually require some calculations and interpretation before this kind of information can be "seen."

Look again at Figure 34. The graph shows that for each date on the horizontal axis there is only one price on the vertical axis. Thus, the graph represents a function, although the exact rule for getting from date to price is not given.

When a function is defined by an equation in x and y, the **graph** of the function is the graph of the equation, that is, the set of points (x, y) in the xy-plane that satisfies the equation.

 COMMENT   When we select a viewing rectangle to graph a function, the values of Xmin, Xmax give the domain that we wish to view, while Ymin, Ymax give the range that we wish to view. These settings usually do not represent the actual domain and range of the function.   ▬

 Not every collection of points in the xy-plane represents the graph of a function. Remember, for a function, each number x in the domain has exactly one image y in the range. This means that the graph of a function cannot contain two points with the same x-coordinate and different y-coordinates. Therefore, the graph of a function must satisfy the following **vertical-line test.**

*In Section 1.6, we discuss a way to find the range of a certain class of functions.

Theorem   Vertical-line Test

A set of points in the xy-plane is the graph of a function if and only if every vertical line intersects the graph in at most one point.

◼

It follows that, if any vertical line intersects a graph at more than one point, the graph is not the graph of a function.

EXAMPLE 9   Identifying the Graph of a Function

Which of the graphs in Figure 35 are graphs of functions?

Figure 35

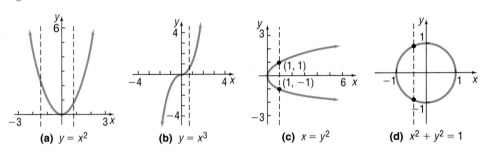

(a) $y = x^2$ **(b)** $y = x^3$ **(c)** $x = y^2$ **(d)** $x^2 + y^2 = 1$

Solution   The graphs in Figures 35(a) and 35(b) are graphs of functions, because every vertical line intersects each graph in at most one point. The graphs in Figures 35(c) and 35(d) are not graphs of functions, because there is a vertical line that intersects each graph in more than one point.

◼

✏ NOW WORK PROBLEM **53.**

⑤   If (x, y) is a point on the graph of a function f, then y is the value of f at x; that is, $y = f(x)$. The next example illustrates how to obtain information about a function if its graph is given.

EXAMPLE 10   Obtaining Information from the Graph of a Function

Figure 36

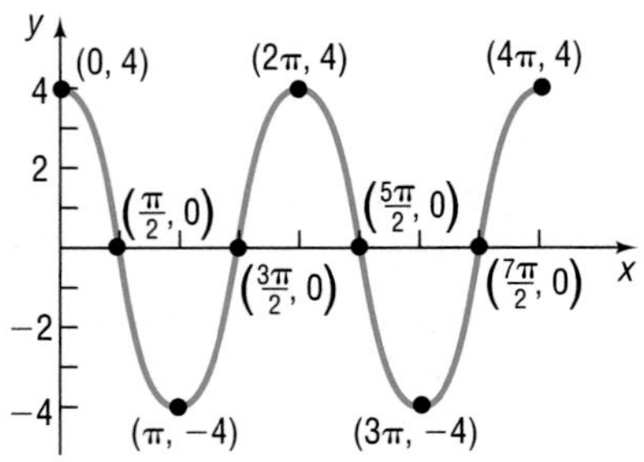

Let f be the function whose graph is given in Figure 36. (The graph of f might represent the distance that the bob of a pendulum is from its *at-rest* position. Negative values of y mean that the pendulum is to the left of the at-rest position, and positive values of y mean that the pendulum is to the right of the at-rest position.)

(0, 4)   (2π, 4)   (4π, 4)

$\left(\frac{\pi}{2}, 0\right)$ $\left(\frac{5\pi}{2}, 0\right)$

$\left(\frac{3\pi}{2}, 0\right)$ $\left(\frac{7\pi}{2}, 0\right)$

(π, −4)   (3π, −4)

(a) What is $f(0)$, $f\left(\dfrac{3\pi}{2}\right)$, and $f(3\pi)$?

(b) What is the domain of f?

(c) What is the range of f?

(d) List the intercepts. (Recall that these are the points, if any, where the graph crosses or touches the coordinate axes.)

(e) How often does the line $y = 2$ intersect the graph?

(f) For what values of x does $f(x) = -4$?

Solution   (a) Since $(0, 4)$ is on the graph of f, the y-coordinate 4 is the value of f at the x-coordinate 0; that is, $f(0) = 4$. In a similar way, we find that when $x = \dfrac{3\pi}{2}$ then $y = 0$, so $f\left(\dfrac{3\pi}{2}\right) = 0$. When $x = 3\pi$, then $y = -4$, so $f(3\pi) = -4$.

(b) To determine the domain of f, we notice that the points on the graph of f will have x-coordinates between 0 and 4π, inclusive; and for each number x between 0 and 4π there is a point $(x, f(x))$ on the graph. The domain of f is $\{x | 0 \le x \le 4\pi\}$ or the interval $[0, 4\pi]$.

(c) The points on the graph all have y-coordinates between -4 and 4, inclusive; and for each such number y there is at least one number x in the domain. The range of f is $\{y | -4 \le y \le 4\}$ or the interval $[-4, 4]$.

(d) The intercepts are $(0, 4)$, $\left(\dfrac{\pi}{2}, 0\right)$, $\left(\dfrac{3\pi}{2}, 0\right)$, $\left(\dfrac{5\pi}{2}, 0\right)$, and $\left(\dfrac{7\pi}{2}, 0\right)$.

(e) Draw the horizontal line $y = 2$ on the graph in Figure 10. Then we find that it intersects the graph four times.

(f) Since $(\pi, -4)$ and $(3\pi, -4)$ are the only points on the graph for which $y = f(x) = -4$, we have $f(x) = -4$ when $x = \pi$ and $x = 3\pi$. ∎

When the graph of a function is given, its domain may be viewed as the shadow created by the graph on the x-axis by vertical beams of light. Its range can be viewed as the shadow created by the graph on the y-axis by horizontal beams of light. Try this technique with the graph given in Figure 36.

✏️ ─── NOW WORK PROBLEMS **47** AND **51**.

EXAMPLE 11   **Obtaining Information about the Graph of a Function**

Consider the function: $f(x) = \dfrac{x}{x + 2}$

(a) Is the point $\left(1, \dfrac{1}{2}\right)$ on the graph of f?

(b) If $x = -1$, what is $f(x)$? What point is on the graph of f?

(c) If $f(x) = 2$, what is x? What point is on the graph of f?

Solution   (a) When $x = 1$, then

$$f(x) = \frac{x}{x + 2}$$

$$f(1) = \frac{1}{1 + 2} = \frac{1}{3}$$

The point $\left(1, \dfrac{1}{3}\right)$ is on the graph of f; the point $\left(1, \dfrac{1}{2}\right)$ is not.

(b) If $x = -1$, then

$$f(x) = \frac{x}{x + 2}$$

$$f(-1) = \frac{-1}{-1 + 2} = -1$$

The point $(-1, -1)$ is on the graph of f.

(c) If $f(x) = 2$, then

$$f(x) = 2$$

$$\frac{x}{x + 2} = 2$$

$$x = 2(x + 2) \qquad \text{Multiply both sides by } x + 2$$

$$x = 2x + 4 \qquad \text{Remove parentheses}$$

$$x = -4 \qquad \text{Solve for } x$$

If $f(x) = 2$, then $x = -4$. The point $(-4, 2)$ is on the graph of f. ∎

NOW WORK PROBLEM **63**.

When we use functions in applications, the domain may be restricted by physical or geometric considerations. For example, the domain of the function f defined by $f(x) = x^2$ is the set of all real numbers. However, if f is used to obtain the area of a square when the length x of a side is known, then we must restrict the domain of f to the positive real numbers, since the length of a side can never be 0 or negative.

EXAMPLE 12

Area of a Circle

Express the area of a circle as a function of its radius.

Figure 37

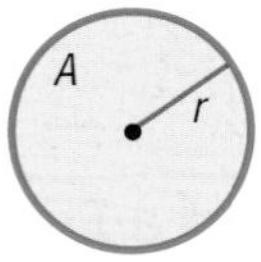

Solution   See Figure 37. We know that the formula for the area A of a circle of radius r is $A = \pi r^2$. If we use r to represent the independent variable and A to represent the dependent variable, the function expressing this relationship is

$$A(r) = \pi r^2$$

In this setting, the domain is $\{r \mid r > 0\}$. (Do you see why?) ∎

SUMMARY

We list here some of the important vocabulary introduced in this section, with a brief description of each term.

Function

A relation between two sets of real numbers so that each number x in the first set, the domain, has corresponding to it exactly one number y in the second set.

A set of ordered pairs (x, y) or $(x, f(x))$ in which no first element is paired with two different second elements.

The range is the set of y values of the function for the x values in the domain.

A function f may be defined implicitly by an equation involving x and y or explicitly by writing $y = f(x)$.

Unspecified domain

If a function f is defined by an equation and no domain is specified, then the domain will be taken to be the largest set of real numbers for which the equation defines a real number.

Function notation

$y = f(x)$

f is a symbol for the function.

x is the independent variable or argument.

y is the dependent variable.

$f(x)$ is the value of the function at x, or the image of x.

Graph of a function

The collection of points (x, y) that satisfies the equation $y = f(x)$.

A collection of points is the graph of a function provided that every vertical line intersects the graph in at most one point (vertical-line test).

1.3 EXERCISES

In Problems 1–12, determine whether each relation represents a function. For each function, state the domain and range.

1.

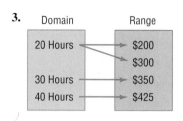

2.

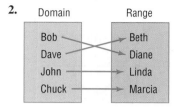

3.

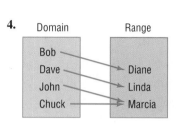

4.

Domain   Range

Bob
Dave ──────► Diane
John ──────► Linda
Chuck ─────► Marcia

5. $\{(2, 6), (-3, 6), (4, 9), (2, 10)\}$

6. $\{(-2, 5), (-1, 3), (3, 7), (4, 12)\}$

7. $\{(1, 3), (2, 3), (3, 3), (4, 3)\}$

8. $\{(0, -2), (1, 3), (2, 3), (3, 7)\}$

9. $\{(-2, 4), (-2, 6), (0, 3), (3, 7)\}$

10. $\{(-4, 4), (-3, 3), (-2, 2), (-1, 1), (-4, 0)\}$

11. $\{(-2, 4), (-1, 1), (0, 0), (1, 1)\}$

12. $\{(-2, 16), (-1, 4), (0, 3), (1, 4)\}$

In Problems 13–20, find the following values for each function:
    (a) $f(0)$ (b) $f(1)$ (c) $f(-1)$ (d) $f(-x)$ (e) $-f(x)$ (f) $f(x + 1)$ (g) $f(2x)$ (h) $f(x + h)$

13. $f(x) = 3x^2 + 2x - 4$ **14.** $f(x) = -2x^2 + x - 1$ **15.** $f(x) = \dfrac{x}{x^2 + 1}$ **16.** $f(x) = \dfrac{x^2 - 1}{x + 4}$

17. $f(x) = |x| + 4$ **18.** $f(x) = \sqrt{x^2 + x}$ **19.** $f(x) = \dfrac{2x + 1}{3x - 5}$ **20.** $f(x) = 1 - \dfrac{1}{(x + 2)^2}$

In Problems 21–32, determine whether the equation is a function.

21. $y = x^2$ **22.** $y = x^3$ **23.** $y = \dfrac{1}{x}$ **24.** $y = |x|$

25. $y^2 = 4 - x^2$ **26.** $y = \pm\sqrt{1 - 2x}$ **27.** $x = y^2$ **28.** $x + y^2 = 1$

29. $y = 2x^2 - 3x + 4$ **30.** $y = \dfrac{3x - 1}{x + 2}$ **31.** $2x^2 + 3y^2 = 1$ **32.** $x^2 - 4y^2 = 1$

In Problems 33–46, find the domain of each function.

33. $f(x) = -5x + 4$ **34.** $f(x) = x^2 + 2$ **35.** $f(x) = \dfrac{x}{x^2 + 1}$ **36.** $f(x) = \dfrac{x^2}{x^2 + 1}$

37. $g(x) = \dfrac{x}{x^2 - 16}$ **38.** $h(x) = \dfrac{2x}{x^2 - 4}$ **39.** $F(x) = \dfrac{x - 2}{x^3 + x}$ **40.** $G(x) = \dfrac{x + 4}{x^3 - 4x}$

41. $h(x) = \sqrt{3x - 12}$ **42.** $G(x) = \sqrt{1 - x}$ **43.** $f(x) = \dfrac{4}{\sqrt{x - 9}}$ **44.** $f(x) = \dfrac{x}{\sqrt{x - 4}}$

45. $p(x) = \sqrt{\dfrac{2}{x - 1}}$ **46.** $q(x) = \sqrt{-x - 2}$

47. Use the graph of the function f given below to answer parts (a)–(n).

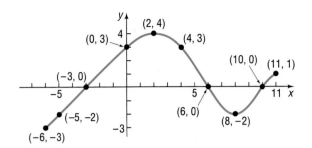

(a) Find $f(0)$ and $f(-6)$.

(b) Find $f(6)$ and $f(11)$.

(c) Is $f(3)$ positive or negative?

(d) Is $f(-4)$ positive or negative?

(e) For what numbers x is $f(x) = 0$?

(f) For what numbers x is $f(x) > 0$?

(g) What is the domain of f?

(h) What is the range of f?

(i) What are the x-intercepts?

(j) What is the y-intercept?

(k) How often does the line $y = \dfrac{1}{2}$ intersect the graph?

(l) How often does the line $x = 5$ intersect the graph?

(m) For what values of x does $f(x) = 3$?

(n) For what values of x does $f(x) = -2$?

48. Use the graph of the function f given below to answer parts (a)–(n).

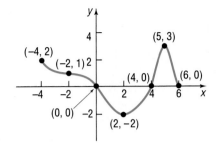

(a) Find $f(0)$ and $f(6)$.

(b) Find $f(2)$ and $f(-2)$.

(c) Is $f(3)$ positive or negative?

(d) Is $f(-1)$ positive or negative?

(e) For what numbers is $f(x) = 0$?

(f) For what numbers is $f(x) < 0$?

(g) What is the domain of f?

(h) What is the range of f?

(i) What are the x-intercepts?

(j) What is the y-intercept?

(k) How often does the line $y = -1$ intersect the graph?

(l) How often does the line $x = 1$ intersect the graph?

(m) For what value of x does $f(x) = 3$?

(n) For what value of x does $f(x) = -2$?

In Problems 49–60, determine whether the graph is that of a function by using the vertical-line test. If it is, use the graph to find:
(a) *Its domain and range* (b) *The intercepts, if any* (c) *Any symmetry with respect to the x-axis, the y-axis, or the origin*

49.

50.

51.

52.

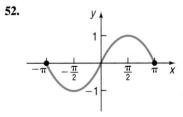

53.

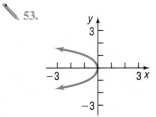

54.

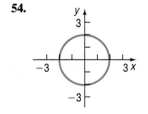

55.

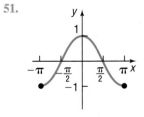

56.

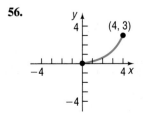

57.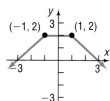

58.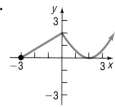

59.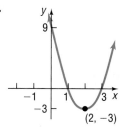

60.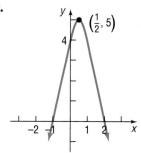

In Problems 61–66, answer the questions about the given function.

61. $f(x) = 2x^2 - x - 1$
(a) Is the point $(-1, 2)$ on the graph of f?
(b) If $x = -2$, what is $f(x)$? What point is on the graph of f?
(c) If $f(x) = -1$, what is x? What point(s) are on the graph of f?
(d) What is the domain of f?
(e) List the x-intercepts, if any, of the graph of f.
(f) List the y-intercept, if there is one, of the graph of f.

62. $f(x) = -3x^2 + 5x$
(a) Is the point $(-1, 2)$ on the graph of f?
(b) If $x = -2$, what is $f(x)$? What point is on the graph of f?
(c) If $f(x) = -2$, what is x? What point(s) are on the graph of f?
(d) What is the domain of f?
(e) List the x-intercepts, if any, of the graph of f.
(f) List the y-intercept, if there is one, of the graph of f.

63. $f(x) = \dfrac{x + 2}{x - 6}$
(a) Is the point $(3, 14)$ on the graph of f?
(b) If $x = 4$, what is $f(x)$? What point is on the graph of f?
(c) If $f(x) = 2$, what is x? What point(s) are on the graph of f?
(d) What is the domain of f?
(e) List the x-intercepts, if any, of the graph of f.
(f) List the y-intercept, if there is one, of the graph of f.

64. $f(x) = \dfrac{x^2 + 2}{x + 4}$
(a) Is the point $\left(1, \dfrac{3}{5}\right)$ on the graph of f?
(b) If $x = 0$, what is $f(x)$? What point is on the graph of f?
(c) If $f(x) = \dfrac{1}{2}$, what is x? What point(s) are on the graph of f?
(d) What is the domain of f?
(e) List the x-intercepts, if any, of the graph of f.
(f) List the y-intercept, if there is one, of the graph of f.

65. $f(x) = \dfrac{2x^2}{x^4 + 1}$
(a) Is the point $(-1, 1)$ on the graph of f?
(b) If $x = 2$, what is $f(x)$? What point is on the graph of f?
(c) If $f(x) = 1$, what is x? What point(s) are on the graph of f?
(d) What is the domain of f?
(e) List the x-intercepts, if any, of the graph of f.
(f) List the y-intercept, if there is one, of the graph of f.

66. $f(x) = \dfrac{2x}{x - 2}$
(a) Is the point $\left(\dfrac{1}{2}, -\dfrac{2}{3}\right)$ on the graph of f?
(b) If $x = 4$, what is $f(x)$? What point is on the graph of f?
(c) If $f(x) = 1$, what is x? What point(s) are on the graph of f?
(d) What is the domain of f?
(e) List the x-intercepts, if any, of the graph of f.
(f) List the y-intercept, if there is one, of the graph of f.

67. If $f(x) = 2x^3 + Ax^2 + 4x - 5$ and $f(2) = 5$, what is the value of A?

68. If $f(x) = 3x^2 - Bx + 4$ and $f(-1) = 12$, what is the value of B?

69. If $f(x) = \dfrac{3x + 8}{2x - A}$ and $f(0) = 2$, what is the value of A?

70. If $f(x) = \dfrac{2x - B}{3x + 4}$ and $f(2) = \dfrac{1}{2}$, what is the value of B?

71. If $f(x) = \dfrac{2x - A}{x - 3}$ and $f(4) = 0$, what is the value of A? Where is f not defined?

72. If $f(x) = \dfrac{x - B}{x - A}$, $f(2) = 0$, and $f(1)$ is undefined, what are the values of A and B?

73. Match each function with the graph that best describes the situation.
   (a) The cost of building a house as a function of its square footage
   (b) The height of an egg dropped from a 300-foot building as a function of time
   (c) The height of a human as a function of time
   (d) The demand for Big Macs as a function of price
   (e) The height of a child on a swing as a function of time

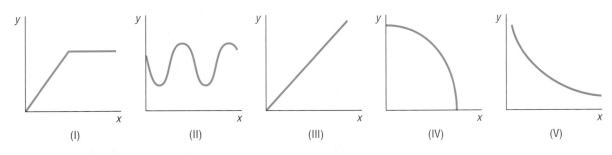

74. Match each function with the graph that best describes the situation.
   (a) The temperature of a bowl of soup as a function of time
   (b) The number of hours of daylight per day over a two-year period
   (c) The population of Florida as a function of time
   (d) The distance of a car traveling at a constant velocity as a function of time
   (e) The height of a golf ball hit with a 7-iron as a function of time

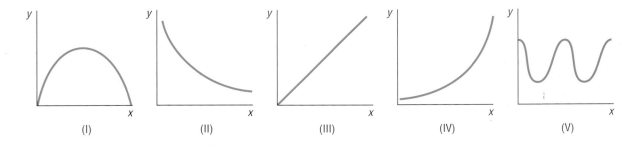

75. Consider the following scenario: Barbara decides to take a walk. She leaves home, walks 2 blocks in 5 minutes at a constant speed, and realizes that she forgot to lock the door. So Barbara runs home in 1 minute. While at her doorstep, it takes her 1 minute to find her keys and lock the door. Barbara walks 5 blocks in 15 minutes and then decides to jog home. It takes her 7 minutes to get home. Draw a graph of Barbara's distance from home (in blocks) as a function of time.

76. Consider the following scenario: Jayne enjoys riding her bicycle through the woods. At the forest preserve, she gets on her bicycle and rides up a 2000-foot incline in 10 minutes. She then travels down the incline in 3 minutes. The next 5000 feet is level terrain and she covers the distance in 20 minutes. She rests for 15 minutes. Jayne then travels 10,000 feet in 30 minutes. Draw a graph of Jayne's distance traveled (in feet) as a function of time.

77. The following sketch represents the distance d (in miles) that Kevin is from home as a function of time t (in hours). Answer the questions based on the graph. In parts (a)–(g),

how many hours elapsed and how far was Kevin from home during this time?
   (a) From $t = 0$ to $t = 2$
   (b) From $t = 2$ to $t = 2.5$
   (c) From $t = 2.5$ to $t = 2.8$
   (d) From $t = 2.8$ to $t = 3$
   (e) From $t = 3$ to $t = 3.9$
   (f) From $t = 3.9$ to $t = 4.2$
   (g) From $t = 4.2$ to $t = 5.3$
   (h) What is the farthest distance that Kevin is from home?
   (i) How many times did Kevin return home?

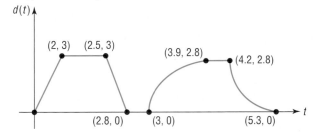

78. The following sketch represents the speed v (in miles per hour) of Michael's car as a function of time t (in minutes).

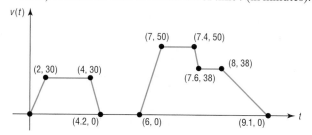

(a) Over what interval of time is Michael traveling fastest?
(b) Over what interval(s) of time is Michael's speed zero?
(c) What is Michael's speed between 0 and 2 minutes?
(d) What is Michael's speed between 4.2 and 6 minutes?
(e) What is Michael's speed between 7 and 7.4 minutes?
(f) When is Michael's speed constant?

79. Effect of Gravity on Earth   If a rock falls from a height of 20 meters on Earth, the height H (in meters) after x seconds is approximately

$$H(x) = 20 - 4.9x^2$$

(a) What is the height of the rock when $x = 1$ second? $x = 1.1$ seconds? $x = 1.2$ seconds? $x = 1.3$ seconds?
(b) When is the height of the rock 15 meters? When is it 10 meters? When is it 5 meters?
(c) When does the rock strike the ground?

80. Effect of Gravity on Jupiter   If a rock falls from a height of 20 meters on the planet Jupiter, its height H (in meters) after x seconds is approximately

$$H(x) = 20 - 13x^2$$

(a) What is the height of the rock when $x = 1$ second? $x = 1.1$ seconds? $x = 1.2$ seconds?
(b) When is the height of the rock 15 meters? When is it 10 meters? When is it 5 meters?
(c) When does the rock strike the ground?

81. Motion of a Golf Ball   A golf ball is hit with an initial velocity of 130 feet per second at an inclination of 30° to the horizontal. In physics, it is established that the height h of the golf ball is given by the function

$$h(x) = \frac{-32x^2}{130^2} + x$$

where x is the horizontal distance that the golf ball has traveled.

(a) Determine the height of the golf ball after it has traveled 100 feet.
(b) 300 feet
(c) 500 feet
(d) How far was the golf ball hit?

82. Cross-sectional Area   The cross-sectional area of a beam cut from a log with radius 1 foot is given by the function $A(x) = 4x\sqrt{1 - x^2}$, where x represents the length of half the base of the beam. See the figure. Determine the cross-sectional area of the beam if the length of half the base of the beam is as follows:
(a) One-third of a foot
(b) One-half of a foot
(c) Two-thirds of a foot

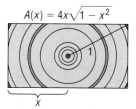

83. Cost of Trans-Atlantic Travel   A Boeing 747 crosses the Atlantic Ocean (3000 miles) with an airspeed of 500 miles per hour. The cost C (in dollars) per passenger is given by

$$C(x) = 100 + \frac{x}{10} + \frac{36{,}000}{x}$$

where x is the ground speed (airspeed ± wind).
(a) What is the cost per passenger for quiescent (no wind) conditions?
(b) What is the cost per passenger with a head wind of 50 miles per hour?
(c) What is the cost per passenger with a tail wind of 100 miles per hour?
(d) What is the cost per passenger with a head wind of 100 miles per hour?
(e) Graph the function $C = C(x)$.
(f) As x varies from 400 to 600 miles per hour, how does the cost vary?

84. Effect of Elevation on Weight   If an object weighs m pounds at sea level, then its weight W (in pounds) at a height of h miles above sea level is given approximately by

$$W(h) = m\left(\frac{4000}{4000 + h}\right)^2$$

(a) If Amy weighs 120 pounds at sea level, how much will she weigh on Pike's Peak, which is 14,110 feet above sea level?
(b) Use a graphing utility to graph the function $W = W(h)$. Use $m = 120$ pounds.
(c) Use the TRACE function to see how weight W varies as h changes from 0 to 5 miles.
(d) At what height will Amy weigh 119.5 pounds?
(e) Does your answer to part (d) seem reasonable?

85. Geometry   Express the area A of a rectangle as a function of the length x if the length is twice the width of the rectangle.

86. **Geometry** Express the area A of an isosceles right triangle as a function of the length x of one of the two equal sides.

87. Express the gross salary G of a person who earns $10 per hour as a function of the number x of hours worked.

88. Tiffany, a commissioned salesperson, earns $100 base pay plus $10 per item sold. Express her gross salary G as a function of the number x of items sold.

89. Some functions f have the property that $f(a + b) = f(a) + f(b)$ for all real numbers a and b. Which of the following functions have this property?
    (a) $h(x) = 2x$ (b) $g(x) = x^2$
    (c) $F(x) = 5x - 2$ (d) $G(x) = 1/x$

90. Draw the graph of a function whose domain is $\{x \mid -3 \le x \le 8, \ x \ne 5\}$ and whose range is $\{y \mid -1 \le y \le 2, y \ne 0\}$. What point(s) in the rectangle $-3 \le x \le 8, -1 \le y \le 2$ cannot be on the graph? Compare your graph with those of other students. What differences do you see?

91. Are the functions $f(x) = x - 1$ and $g(x) = \dfrac{x^2 - 1}{x + 1}$ the same? Explain.

92. Describe how you would proceed to find the domain and range of a function if you were given its graph. How would your strategy change if, instead, you were given the equation defining the function?

93. How many x-intercepts can the graph of a function have? How many y-intercepts can it have?

94. Is a graph that consists of a single point the graph of a function? If so, can you write the equation of such a function?

95. Is there a function whose graph is symmetric with respect to the x-axis? Explain.

96. Investigate when, historically, the use of the function notation $y = f(x)$ first appeared.

P R E P A R I N G   F O R   T H I S   S E C T I O N

Before getting started, review the following:

✓ Intervals (Appendix A, Section A.5, pp. 522–523)

✓ Point–slope Form of a Line
  (Appendix A, Section A.7, p. 542)

✓ Slope of a Line (Appendix A, Section A.7, pp. 537–541)

✓ Tests for Symmetry of an Equation
  (Section 1.2, p. 13)

✓ Graphs of Certain Equations (Section 1.2, Example 3, p. 10, Example 9, p. 14, Example 10, p. 14, Example 11, p. 15)

1.4 PROPERTIES OF FUNCTIONS; LIBRARY OF FUNCTIONS

OBJECTIVES
1. Use a Graph to Determine Where a Function Is Increasing, Is Decreasing, or Is Constant
2. Use a Graph to Locate Local Maxima and Minima
3. Determine Even and Odd Functions from a Graph
4. Identify Even and Odd Functions from the Equation
5. Graph the Functions Listed in the Library of Functions

INCREASING AND DECREASING FUNCTIONS

1. Consider the graph given in Figure 38. If you look from left to right along the graph of the function, you will notice that parts of the graph are rising, parts are falling, and parts are horizontal. In such cases, the function is described as *increasing*, *decreasing*, or *constant*, respectively.

Figure 38

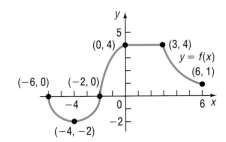

| EXAMPLE 1 | Determining Where a Function Is Increasing, Decreasing, or Constant from Its Graph |

Where is the function in Figure 38 increasing? Where is it decreasing? Where is it constant?

Solution   To answer the question of where a function is increasing, where it is decreasing, and where it is constant, we use nonstrict inequalities involving the independent variable x, or we use open intervals* of x-coordinates. The graph in Figure 38 is rising (increasing) from the point $(-4, -2)$ to the point $(0, 4)$, so we conclude that it is increasing on the open interval $(-4, 0)$ (or for $-4 < x < 0$). The graph is falling (decreasing) from the point $(-6, 0)$ to the point $(-4, -2)$ and from the point $(3, 4)$ to the point $(6, 1)$. We conclude that the graph is decreasing on the open intervals $(-6, -4)$ and $(3, 6)$ (or for $-6 < x < -4$ and $3 < x < 6$). The graph is constant on the open interval $(0, 3)$ (or $0 < x < 3$). ∎

More precise definitions follow:

> A function f is **increasing** on an open interval I if, for any choice of x_1 and x_2 in I, with $x_1 < x_2$, we have $f(x_1) < f(x_2)$.

> A function f is **decreasing** on an open interval I if, for any choice of x_1 and x_2 in I, with $x_1 < x_2$, we have $f(x_1) > f(x_2)$.

> A function f is **constant** on an open interval I if, for all choices of x in I, the values $f(x)$ are equal.

Figure 39 illustrates the definitions. The graph of an increasing function goes up from left to right, the graph of a decreasing function goes down from left to right, and the graph of a constant function remains at a fixed height.

Figure 39

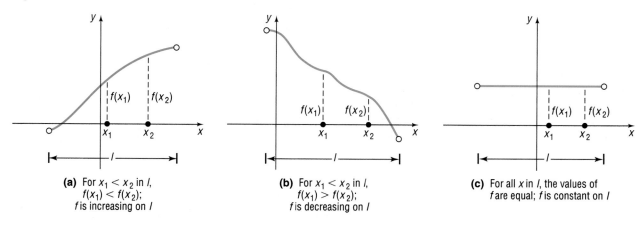

(a) For $x_1 < x_2$ in I, $f(x_1) < f(x_2)$; f is increasing on I

(b) For $x_1 < x_2$ in I, $f(x_1) > f(x_2)$; f is decreasing on I

(c) For all x in I, the values of f are equal; f is constant on I

NOW WORK PROBLEMS **15, 17,** AND **19.**

*The open interval (a, b) consists of all real numbers x for which $a < x < b$. Refer to Appendix A, Section A.5, if necessary.

LOCAL MAXIMUM; LOCAL MINIMUM

△ ② Suppose a function f is defined at the number c.

When the graph of a function is increasing to the left of $x = c$ and decreasing to the right of $x = c$, then at c the value of f is largest. This value is called a *local maximum* of f.

When the graph of a function is decreasing to the left of $x = c$ and is increasing to the right of $x = c$, then at c the value of f is the smallest. This value is called a *local minimum* of f.

If f has a local maximum at c, then the value of f at c is greater than the values of f near c. If f has a local minimum at c, then the value of f at c is less than the values of f near c. The word *local* is used to suggest that it is only near c that the value $f(c)$ is largest or smallest. See Figure 40.

Figure 40
f has a local maximum at x_1 and x_3;
f has a local minimum at x_2

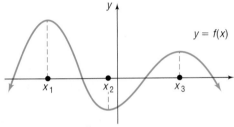

More precise definitions follow:

A function f has a **local maximum at c** if there is an open interval I containing c so that, for all $x \neq c$ in I, $f(x) < f(c)$. We call $f(c)$ a **local maximum of f.**

A function f has a **local minimum at c** if there is an open interval I containing c so that, for all $x \neq c$ in I, $f(x) > f(c)$. We call $f(c)$ a **local minimum of f.**

EXAMPLE 2 | **Finding Local Maxima and Local Minima from a Graph**

Figure 41

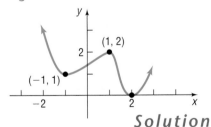

Figure 41 shows the graph of a function f.

(a) At what number(s), if any, does f have a local maximum?
(b) What are the local maxima?
(c) At what number(s), if any, does f have a local minimum?
(d) What are the local minima?

Solution   The domain of f is the set of real numbers.

(a) f has a local maximum at 1, since for all x close to 1, $x \neq 1$, we have $f(x) < f(1)$.
(b) The local maximum is $f(1) = 2$.
(c) f has a local minimum at -1 and at 2.
(d) The local minima are $f(-1) = 1$ and $f(2) = 0$.   ■

△ If the graph of a function f is not given, then to locate the exact value at which f has a local maximum or a local minimum usually requires calculus.

 NOW WORK PROBLEMS **21** AND **23**.

EVEN AND ODD FUNCTIONS

③ A function f is even if and only if whenever the point (x, y) is on the graph of f then the point $(-x, y)$ is also on the graph. Algebraically, we define an even function as follows:

> A function f is **even** if for every number x in its domain the number $-x$ is also in the domain and
>
> $$f(-x) = f(x)$$

A function f is odd if and only if whenever the point (x, y) is on the graph of f then the point $(-x, -y)$ is also on the graph. Algebraically, we define an odd function as follows:

> A function f is **odd** if for every number x in its domain the number $-x$ is also in the domain and
>
> $$f(-x) = -f(x)$$

Refer to Section 1.2, where the tests for symmetry are listed. The following results are then evident.

Theorem   A function is even if and only if its graph is symmetric with respect to the y-axis. A function is odd if and only if its graph is symmetric with respect to the origin.

| EXAMPLE 3 | **Determining Even and Odd Functions from the Graph** |

Determine whether each graph given in Figure 42 is the graph of an even function, an odd function, or a function that is neither even nor odd.

Figure 42

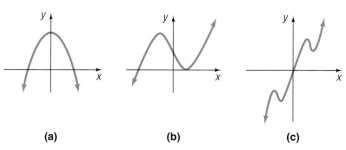

(a)   (b)   (c)

Solution   The graph in Figure 42(a) is that of an even function, because the graph is symmetric with respect to the y-axis. The function whose graph is given in Figure 42(b) is neither even nor odd, because the graph is neither symmetric with respect to the y-axis nor symmetric with respect to the origin. The function whose graph is given in Figure 42(c) is odd, because its graph is symmetric with respect to the origin. ■

NOW WORK PROBLEM **25**.

④   In the next example, we use algebraic techniques to verify whether a given function is even, odd, or neither.

| EXAMPLE 4 | **Identifying Even and Odd Functions Algebraically** |

Determine whether each of the following functions is even, odd, or neither. Then determine whether the graph is symmetric with respect to the y-axis or with respect to the origin.

(a) $f(x) = x^2 - 5$ (b) $g(x) = x^3 - 1$

(c) $h(x) = 5x^3 - x$ (d) $F(x) = |x|$

Solution

(a) To determine whether f is even, odd, or neither, we replace x by $-x$ in $f(x) = x^2 - 5$. Then

$$f(-x) = (-x)^2 - 5 = x^2 - 5 = f(x)$$

Since $f(-x) = f(x)$, we conclude that f is an even function, and the graph is symmetric with respect to the y-axis.

(b) We replace x by $-x$ in $g(x) = x^3 - 1$. Then

$$g(-x) = (-x)^3 - 1 = -x^3 - 1$$

Since $g(-x) \neq g(x)$ and $g(-x) \neq -g(x) = -(x^3 - 1) = -x^3 + 1$, we conclude that g is neither even nor odd. The graph is not symmetric with respect to the y-axis nor is it symmetric with respect to the origin.

(c) We replace x by $-x$ in $h(x) = 5x^3 - x$. Then

$$h(-x) = 5(-x)^3 - (-x) = -5x^3 + x = -(5x^3 - x) = -h(x)$$

Since $h(-x) = -h(x)$, h is an odd function, and the graph of h is symmetric with respect to the origin.

(d) We replace x by $-x$ in $F(x) = |x|$. Then

$$F(-x) = |-x| = |-1| \cdot |x| = |x| = F(x)$$

Since $F(-x) = F(x)$, F is an even function, and the graph of F is symmetric with respect to the y-axis. ∎

 NOW WORK PROBLEM 37.

LIBRARY OF FUNCTIONS

⑤ We now give names to some of the functions that we have encountered. In going through this list, pay special attention to the properties of each function, particularly to the shape of each graph. Knowing these graphs will lay the foundation for later graphing techniques.

Linear Functions

$$f(x) = mx + b \qquad m \text{ and } b \text{ are real numbers}$$

The domain of a **linear function** f consists of all real numbers. The graph of this function is a nonvertical line with slope m and y-intercept b. A linear function is increasing if $m > 0$, decreasing if $m < 0$, and constant if $m = 0$.

Figure 43
Constant Function

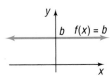

Constant Function

$$f(x) = b \qquad b \text{ is a real number}$$

See Figure 43.

A **constant function** is a special linear function ($m = 0$). Its domain is the set of all real numbers; its range is the set consisting of a single number b. Its graph is a horizontal line whose y-intercept is b. The constant function is an even function whose graph is constant over its domain.

Identity Function

Figure 44
Identity Function

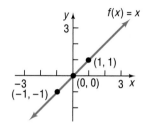

$$f(x) = x$$

See Figure 44.

The **identity function** is also a special linear function. Its domain and range are the set of all real numbers. Its graph is a line whose slope is $m = 1$ and whose y-intercept is 0. The line consists of all points for which the x-coordinate equals the y-coordinate. The identity function is an odd function that is increasing over its domain. Note that the graph bisects quadrants I and III.

Square Function

Figure 45
Square Function

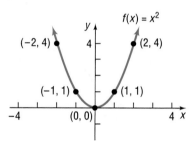

$$f(x) = x^2$$

See Figure 45.

The domain of the **square function** f is the set of all real numbers; its range is the set of nonnegative real numbers. The graph of this function is a parabola whose intercept is at $(0, 0)$. The square function is an even function that is decreasing on the interval $(-\infty, 0)$ and increasing on the interval $(0, \infty)$.

Cube Function

Figure 46
Cube Function

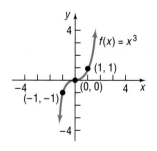

$$f(x) = x^3$$

See Figure 46.

The domain and range of the **cube function** are the set of all real numbers. The intercept of the graph is at $(0, 0)$. The cube function is odd and is increasing on the interval $(-\infty, \infty)$.

Square Root Function

Figure 47
Square Root Function

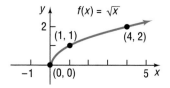

$$f(x) = \sqrt{x}$$

See Figure 47.

The domain and range of the **square root function** are the set of non-negative real numbers. The intercept of the graph is at $(0, 0)$. The square root function is neither even nor odd and is increasing on the interval $(0, \infty)$.

Figure 48
Reciprocal Function

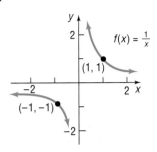

Reciprocal Function

$$f(x) = \frac{1}{x}$$

Refer to Example 11, p. 15, for a discussion of the equation $y = \dfrac{1}{x}$. See Figure 48.

The domain and range of the **reciprocal function** are the set of all nonzero real numbers. The graph has no intercepts. The reciprocal function is decreasing on the intervals $(-\infty, 0)$ and $(0, \infty)$ and is an odd function.

Absolute Value Function

$$f(x) = |x|$$

Figure 49
Absolute Value Function

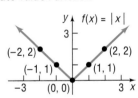

See Figure 49.

The domain of the **absolute value function** is the set of all real numbers; its range is the set of nonnegative real numbers. The intercept of the graph is at $(0, 0)$. If $x \geq 0$, then $f(x) = x$, and the graph of f is part of the line $y = x$; if $x < 0$, then $f(x) = -x$, and the graph of f is part of the line $y = -x$. The absolute value function is an even function; it is decreasing on the interval $(-\infty, 0)$ and increasing on the interval $(0, \infty)$.

SEEING THE CONCEPT: Graph $y = |x|$ on a square screen and compare what you see with Figure 49. Note that some graphing calculators use the symbols abs (x) for absolute value. If your utility has no built-in absolute value function, you can still graph $y = |x|$ by using the fact that $|x| = \sqrt{x^2}$.

 NOW WORK PROBLEMS 1–7.

1.4 EXERCISES

In Problems 1–7, match each graph to the function listed whose graph most resembles the one given.

   A. *Constant function*   B. *Linear function*   C. *Square function*

   D. *Cube function*   E. *Square root function*   F. *Reciprocal function*

   G. *Absolute value function*

1.

2.

3.

4.

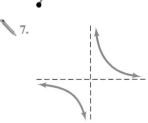

5.

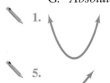

6.

7.

In Problems 8–14, sketch the graph of each function. Be sure to label three points on the graph.

8. $f(x) = 2$ **9.** $f(x) = x$ **10.** $f(x) = x^2$ **11.** $f(x) = x^3$

12. $f(x) = \sqrt{x}$ **13.** $f(x) = \dfrac{1}{x}$ **14.** $f(x) = |x|$

In Problems 15–24, use the graph of the function f given below.

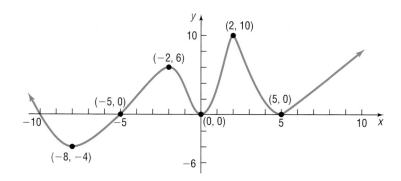

15. Is f increasing on the interval $(-8, -2)$?

16. Is f decreasing on the interval $(-8, -4)$?

17. Is f increasing on the interval $(2, 10)$?

18. Is f decreasing on the interval $(2, 5)$?

19. List the interval(s) on which f is increasing.

20. List the interval(s) on which f is decreasing.

21. Is there a local maximum at 2? If yes, what is it?

22. Is there a local maximum at 5? If yes, what is it?

23. List the numbers at which f has a local maximum. What are these local maxima?

24. List the numbers at which f has a local minimum. What are these local minima?

In Problems 25–32, the graph of a function is given. Use the graph to find:

  (a) *The intercepts, if any*
  (b) *Its domain and range*
  (c) *The intervals on which it is increasing, decreasing, or constant*
  (d) *Whether it is even, odd, or neither*

25.
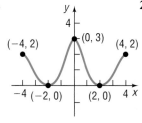

26.
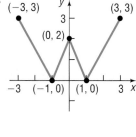

27.
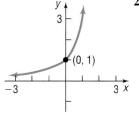

28.
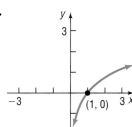

29.
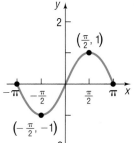

30.
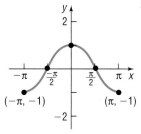

31.
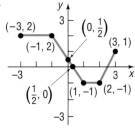

32.
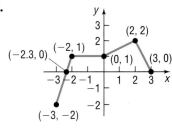

In Problems 33–36, the graph of a function f is given. Use the graph to find:
  (a) *The numbers, if any, at which f has a local maximum. What are these local maxima?*
  (b) *The numbers, if any, at which f has a local minimum. What are these local minima?*

33.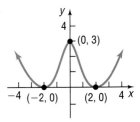

34.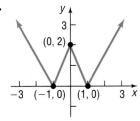

35.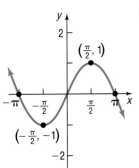

36.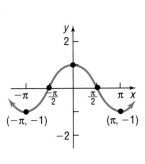

In Problems 37–48, determine algebraically whether each function is even, odd, or neither.

37. $f(x) = 4x^3$

38. $f(x) = 2x^4 - x^2$

39. $g(x) = -3x^2 - 5$

40. $h(x) = 3x^3 + 5$

41. $F(x) = \sqrt[3]{x}$

42. $G(x) = \sqrt{x}$

43. $f(x) = x + |x|$

44. $f(x) = \sqrt[3]{2x^2 + 1}$

45. $g(x) = \dfrac{1}{x^2}$

46. $h(x) = \dfrac{x}{x^2 - 1}$

47. $h(x) = \dfrac{-x^3}{3x^2 - 9}$

48. $F(x) = \dfrac{2x}{|x|}$

---

49. Exploration Graph $y = x^2$. Then on the same screen graph $y = x^2 + 2$, followed by $y = x^2 + 4$, followed by $y = x^2 - 2$. What pattern do you observe? Can you predict the graph of $y = x^2 - 4$? Of $y = x^2 + 5$?

50. Exploration Graph $y = x^2$. Then on the same screen graph $y = (x - 2)^2$, followed by $y = (x - 4)^2$, followed by $y = (x + 2)^2$. What pattern do you observe? Can you predict the graph of $y = (x + 4)^2$? Of $y = (x - 5)^2$?

51. Exploration Graph $y = |x|$. Then on the same screen graph $y = 2|x|$, followed by $y = 4|x|$, followed by $y = \frac{1}{2}|x|$. What pattern do you observe? Can you predict the graph of $y = \frac{1}{4}|x|$? Of $y = 5|x|$?

52. Exploration Graph $y = x^2$. Then on the same screen graph $y = -x^2$. What pattern do you observe? Now try $y = |x|$ and $y = -|x|$. What do you conclude?

53. Exploration Graph $y = \sqrt{x}$. Then on the same screen graph $y = \sqrt{-x}$. What pattern do you observe? Now try $y = 2x + 1$ and $y = 2(-x) + 1$. What do you conclude?

54. Exploration Graph $y = x^3$. Then on the same screen graph $y = (x - 1)^3 + 2$. Could you have predicted the result?

55. Exploration Graph $y = x^2$, $y = x^4$, and $y = x^6$ on the same screen. What do you notice is the same about each graph? What do you notice that is different?

56. Exploration Graph $y = x^3$, $y = x^5$, and $y = x^7$ on the same screen. What do you notice is the same about each graph? What do you notice that is different?

57. Consider the equation

$$y = \begin{cases} 1 & \text{if } x \text{ is rational} \\ 0 & \text{if } x \text{ is irrational} \end{cases}$$

Is this a function? What is its domain? What is its range? What is its y-intercept, if any? What are its x-intercepts, if any? Is it even, odd, or neither? How would you describe its graph?

58. Define some functions that pass through $(0, 0)$ and $(1, 1)$ and are increasing for $x \geq 0$. Begin your list with $y = \sqrt{x}$, $y = x$, and $y = x^2$. Can you propose a general result about such functions?

59. How many x-intercepts can a function defined on an interval have if it is increasing on that interval? Explain.

60. Can you think of a function that is both even and odd?

1.5   GRAPHING TECHNIQUES: TRANSFORMATIONS

OBJECTIVES   **1**   Graph Functions Using Horizontal and Vertical Shifts
   2   Graph Functions Using Compressions and Stretches
   3   Graph Functions Using Reflections about the x-Axis or y-Axis

At this stage, if you were asked to graph any of the functions defined by

$y = x, y = x^2, y = x^3, y = \sqrt{x}, y = |x|,$ or $y = \dfrac{1}{x}$, your response should be,

"Yes, I recognize these functions and know the general shapes of their graphs." (If this is not your answer, review the previous section, Figures 43 through 49.)

Sometimes we are asked to graph a function that is "almost" like one that we already know how to graph. In this section, we look at some of these functions and develop techniques for graphing them. Collectively, these techniques are referred to as **transformations.**

1   **VERTICAL SHIFTS**

EXAMPLE 1   **Vertical Shift Up**

Use the graph of $f(x) = x^2$ to obtain the graph of $g(x) = x^2 + 3$.

Solution   We begin by obtaining some points on the graphs of f and g. For example, when $x = 0$, then $y = f(0) = 0$ and $y = g(0) = 3$. When $x = 1$, then $y = f(1) = 1$ and $y = g(1) = 4$. Table 5 lists these and a few other points on each graph. We conclude that the graph of g is identical to that of f, except that it is shifted vertically up 3 units. See Figure 50.

	T A B L E 5	
x	$y = f(x)$ $= x^2$	$y = g(x)$ $= x^2 + 3$
-2	4	7
-1	1	4
0	0	3
1	1	4
2	4	7

Figure 50

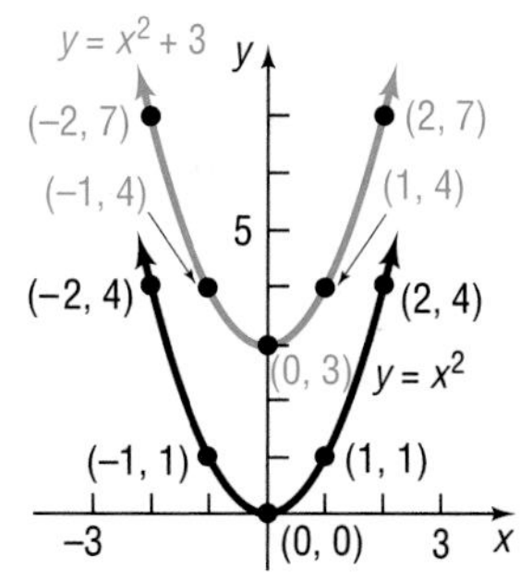

Figure 51

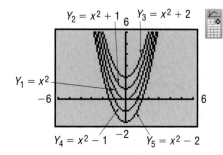

SEEING THE CONCEPT: On the same screen, graph each of the following functions:

$$Y_1 = x^2$$
$$Y_2 = x^2 + 1$$
$$Y_3 = x^2 + 2$$
$$Y_4 = x^2 - 1$$
$$Y_5 = x^2 - 2$$

Figure 51 illustrates the graphs. You should have observed a general pattern. With $Y_1 = x^2$ on the screen, the graph of $Y_2 = x^2 + 1$ is identical to that of $Y_1 = x^2$, except that it is shifted vertically up 1 unit. Similarly, $Y_3 = x^2 + 2$ is identical to that of $Y_1 = x^2$, except that it is shifted vertically up 2 units. The graph of $Y_4 = x^2 - 1$ is identical to that of $Y_1 = x^2$, except that it is shifted vertically down 1 unit. ■

We are led to the following conclusion:

> If a real number k is added to the right side of a function $y = f(x)$, the graph of the new function $y = f(x) + k$ is the graph of f **shifted vertically up** (if $k > 0$) or **down** (if $k < 0$).

Let's look at another example.

EXAMPLE 2 | **Vertical Shift Down**

Use the graph of $f(x) = x^2$ to obtain the graph of $g(x) = x^2 - 4$.

Solution Table 6 lists some points on the graphs of f and g. Notice that each y-coordinate of g is 4 units less than the corresponding y-coordinate of f. The graph of g is identical to that of f, except that it is shifted down 4 units. See Figure 52.

	TABLE 6	
x	$y = f(x)$ $= x^2$	$y = g(x)$ $= x^2 - 4$
-2	4	0
-1	1	-3
0	0	-4
1	1	-3
2	4	0

Figure 52

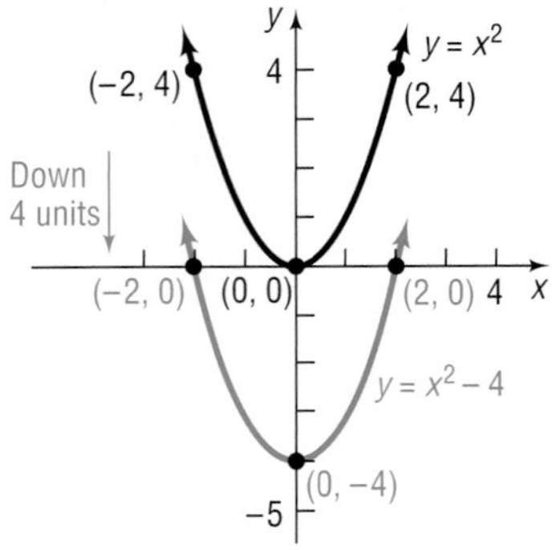

■

NOW WORK PROBLEM **29**.

HORIZONTAL SHIFTS

EXAMPLE 3 | **Horizontal Shift to the Right**

Use the graph of $f(x) = x^2$ to obtain the graph of $g(x) = (x - 2)^2$.

Solution The function $g(x) = (x - 2)^2$ is basically a square function. Table 7 lists some points on the graphs of f and g. Note that when $f(x) = 0$ then $x = 0$, and when $g(x) = 0$, then $x = 2$. Also, when $f(x) = 4$, then $x = -2$ or 2, and when $g(x) = 4$, then $x = 0$ or 4. We conclude that the graph of g is identical to that of f, except that it is shifted 2 units to the right. See Figure 53.

TABLE 7

x	$y = f(x)$ $= x^2$	$y = g(x)$ $= (x - 2)^2$
-2	4	16
0	0	4
2	4	0
4	16	4

Figure 53

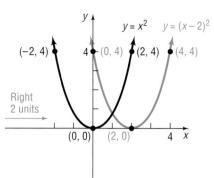

SEEING THE CONCEPT: On the same screen, graph each of the following functions:

$$Y_1 = x^2$$
$$Y_2 = (x - 1)^2$$
$$Y_3 = (x - 3)^2$$
$$Y_4 = (x + 2)^2$$

Figure 54

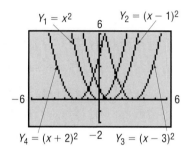

Figure 54 illustrates the graphs.

You should have observed the following pattern. With the graph of $Y_1 = x^2$ on the screen, the graph of $Y_2 = (x - 1)^2$ is identical to that of $Y = x^2$, except that it is shifted horizontally to the right 1 unit. Similarly, the graph of $Y_3 = (x - 3)^2$ is identical to that of $Y_1 = x^2$, except that it is shifted horizontally to the right 3 units. Finally, the graph of $Y_4 = (x + 2)^2$ is identical to that of $Y_1 = x^2$, except that it is shifted horizontally to the left 2 units.

We are led to the following conclusion.

> If the argument x of a function f is replaced by $x - h$, h a real number, the graph of the new function $y = f(x - h)$ is the graph of f **shifted horizontally left** (if $h < 0$) or **right** (if $h > 0$).

EXAMPLE 4   Horizontal Shift to the Left

Use the graph of $f(x) = x^2$ to obtain the graph of $g(x) = (x + 4)^2$.

Solution   Again, the function $g(x) = (x + 4)^2$ is basically a square function. Its graph is the same as that of f, except that it is shifted horizontally 4 units to the left. (Do you see why? $(x + 4)^2 = [x - (-4)]^2$) See Figure 55.

Figure 55

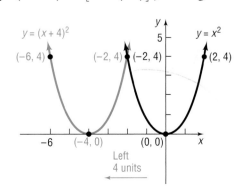

NOW WORK PROBLEM **33**.

Vertical and horizontal shifts are sometimes combined.

EXAMPLE 5 | **Combining Vertical and Horizontal Shifts**

Graph the function: $f(x) = (x + 3)^2 - 5$

Solution We graph f in steps. First, we note that the rule for f is basically a square function, so we begin with the graph of $y = x^2$ as shown in Figure 56(a). Next, to get the graph of $y = (x + 3)^2$, we shift the graph of $y = x^2$ horizontally 3 units to the left. See Figure 56(b). Finally, to get the graph of $y = (x + 3)^2 - 5$, we shift the graph of $y = (x + 3)^2$ vertically down 5 units. See Figure 56(c). Note the points plotted on each graph. Using key points can be helpful in keeping track of the transformation that has taken place.

Figure 56

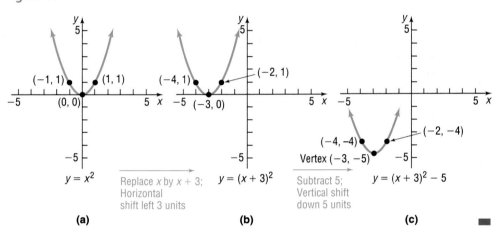

$y = x^2$ Replace x by $x + 3$; Horizontal shift left 3 units $y = (x + 3)^2$ Subtract 5; Vertical shift down 5 units $y = (x + 3)^2 - 5$

(a)   (b)   (c)

Check: Graph $Y_1 = f(x) = (x + 3)^2 - 5$ and compare the graph to Figure 56(c).

In Example 5, if the vertical shift had been done first, followed by the horizontal shift, the final graph would have been the same. Try it for yourself.

NOW WORK PROBLEM **35**.

2 **COMPRESSIONS AND STRETCHES**

EXAMPLE 6 | **Vertical Stretch**

Use the graph of $f(x) = |x|$ to obtain the graph of $g(x) = 2|x|$.

Solution To see the relationship between the graphs of f and g, we form Table 8, listing points on each graph. For each x, the y-coordinate of a point on the graph of g is 2 times as large as the corresponding y-coordinate on the graph of f. The graph of $f(x) = |x|$ is vertically stretched by a factor of 2 [for example, from $(1, 1)$ to $(1, 2)$] to obtain the graph of $g(x) = 2|x|$. See Figure 57.

TABLE 8		
x	$y = f(x)$ $= \|x\|$	$y = g(x)$ $= 2\|x\|$
-2	2	4
-1	1	2
0	0	0
1	1	2
2	2	4

Figure 57

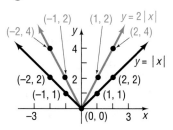

EXAMPLE 7	**Vertical Compression**

Use the graph of $f(x) = |x|$ to obtain the graph of $g(x) = \frac{1}{2}|x|$.

Solution For each x, the y-coordinate of a point on the graph of g is $\frac{1}{2}$ as large as the corresponding y-coordinate on the graph of f. The graph of $f(x) = |x|$ is vertically compressed by a factor of $\frac{1}{2}$ [for example, from $(2, 2)$ to $(2, 1)$] to obtain the graph of $g(x) = \frac{1}{2}|x|$. See Table 9 and Figure 58.

TABLE 9		
x	$y = f(x)$ $= \|x\|$	$y = g(x)$ $= \frac{1}{2}\|x\|$
-2	2	1
-1	1	$\frac{1}{2}$
0	0	0
1	1	$\frac{1}{2}$
2	2	1

Figure 58

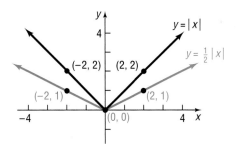

> When the right side of a function $y = f(x)$ is multiplied by a positive number a, the graph of the new function $y = af(x)$ is obtained by multiplying each y-coordinate of $y = f(x)$ by a. A **vertical compression** results if $0 < a < 1$ and a **vertical stretch** occurs if $a > 1$.

 NOW WORK PROBLEM **37.**

What happens if the argument x of a function $y = f(x)$ is multiplied by a positive number a, creating a new function $y = f(ax)$? To find the answer, we look at the following Exploration.

 EXPLORATION On the same screen, graph each of the following functions:

$$Y_1 = f(x) = x^2$$
$$Y_2 = f(2x) = (2x)^2$$
$$Y_3 = f\left(\frac{1}{2}x\right) = \left(\frac{1}{2}x\right)^2$$

RESULT You should have obtained the graphs shown in Figure 59. The graph of $Y_2 = (2x)^2$ is the graph of $Y_1 = x^2$ compressed horizontally. Look at Table 10(a).

Figure 59

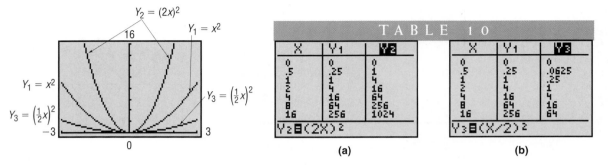

Notice that $(1, 1), (2, 4), (4, 16),$ and $(16, 256)$ are points on the graph of $Y_1 = x^2$. Also, $(0.5, 1), (1, 4), (2, 16),$ and $(8, 256)$ are points on the graph of $Y_2 = (2x)^2$. For each y-coordinate, the x-coordinate on the graph of Y_2 is $\dfrac{1}{2}$ the x-coordinate on Y_1. The graph of $Y_2 = (2x)^2$ is obtained by multiplying the x-coordinate of each point on the graph of $Y_1 = x^2$ by $\dfrac{1}{2}$. The graph of $Y_3 = \left(\dfrac{1}{2}x\right)^2$ is the graph of $Y_1 = x^2$ stretched horizontally. Look at Table 10(b). Notice that $(0.5, 0.25), (1, 1), (2, 4),$ and $(4, 16)$ are points on the graph of $Y_1 = x^2$. Also, $(1, 0.25), (2, 1), (4, 4),$ and $(8, 16)$ are points on the graph of $Y_3 = \left(\dfrac{1}{2}x\right)^2$. For each y-coordinate, the x-coordinate on the graph of Y_3 is 2 times the x-coordinate on Y_1. The graph of $Y_3 = \left(\dfrac{1}{2}x\right)^2$ is obtained by multiplying the x-coordinate of each point on the graph of $Y_1 = x^2$ by a factor of 2. ∎

> If the argument x of a function $y = f(x)$ is multiplied by a positive number a, the graph of the new function $y = f(ax)$ is obtained by multiplying each x-coordinate of $y = f(x)$ by $\dfrac{1}{a}$. A **horizontal compression** results if $a > 1$, and a **horizontal stretch** occurs if $0 < a < 1$.

Let's look at an example.

EXAMPLE 8

Graphing Using Stretches and Compressions

The graph of $y = f(x)$ is given in Figure 60. Use this graph to find the graphs of:

(a) $y = 3f(x)$ (b) $y = f(3x)$

Figure 60

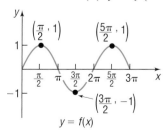

$y = f(x)$

Solution   (a) The graph of $y = 3f(x)$ is obtained by multiplying each y-coordinate of $y = f(x)$ by a factor of 3. See Figure 61(a).

(b) The graph of $y = f(3x)$ is obtained from the graph of $y = f(x)$ by multiplying each x-coordinate of $y = f(x)$ by a factor of $\dfrac{1}{3}$. See Figure 61(b).

Figure 61

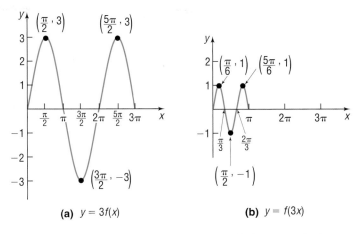

(a) $y = 3f(x)$

(b) $y = f(3x)$

NOW WORK PROBLEMS **57(e)** AND **(g)**.

③   REFLECTIONS ABOUT THE x-AXIS AND THE y-AXIS

EXAMPLE 9

Reflection about the x-Axis

Graph the function: $f(x) = -x^2$

Solution   We begin with the graph of $y = x^2$, as shown in Figure 45. For each point (x, y) on the graph of $y = x^2$, the point $(x, -y)$ is on the graph of $y = -x^2$, as indicated in Table 11. We can draw the graph of $y = -x^2$ by reflecting the graph of $y = x^2$ about the x-axis. See Figure 62.

TABLE 11		
x	$y = x^2$	$y = -x^2$
-2	4	-4
-1	1	-1
0	0	0
1	1	-1
2	4	-4

Figure 62

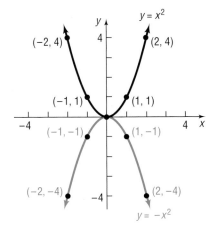

When the right side of the function $y = f(x)$ is multiplied by -1, the graph of the new function $y = -f(x)$ is the **reflection about the x-axis** of the graph of the function $y = f(x)$.

NOW WORK PROBLEM **41.**

EXAMPLE 10	Reflection about the *y*-Axis

Graph the function: $f(x) = \sqrt{-x}$

Solution   First, notice that the domain of f consists of all real numbers x for which $-x \geq 0$ or, equivalently, $x \leq 0$. To get the graph of $f(x) = \sqrt{-x}$, we begin with the graph of $y = \sqrt{x}$, as shown in black in Figure 63. For each point (x, y) on the graph of $y = \sqrt{x}$, the point $(-x, y)$ is on the graph of $y = \sqrt{-x}$. We obtain the graph of $y = \sqrt{-x}$ by reflecting the graph of $y = \sqrt{x}$ about the *y*-axis. See the blue graph in Figure 63.

Figure 63

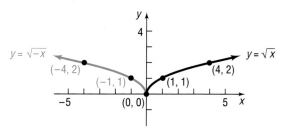

> When the graph of the function $y = f(x)$ is known, the graph of the new function $y = f(-x)$ is the **reflection about the *y*-axis** of the graph of the function $y = f(x)$.

SUMMARY

Summary of Graphing Techniques

Table 12 summarizes the graphing procedures that we have just discussed.

TABLE 12		
To Graph:	**Draw the Graph of *f* and:**	**Functional Change to *f(x)***

Vertical shifts
  $y = f(x) + k, \quad k > 0$
  $y = f(x) - k, \quad k > 0$
	Raise the graph of f by k units.
	Lower the graph of f by k units.
		Add k to $f(x)$.
		Subtract k from $f(x)$.

Horizontal shifts
  $y = f(x + h), \quad h > 0$
  $y = f(x - h), \quad h > 0$
	Shift the graph of f to the left h units.
	Shift the graph of f to the right h units.
		Replace x by $x + h$.
		Replace x by $x - h$.

Compressing or stretching $y = af(x), \quad a > 0$	Multiply each y-coordinate of $y = f(x)$ by a. Stretch the graph of f vertically if $a > 1$. Compress the graph of f vertically if $0 < a < 1$.	Multiply $f(x)$ by a.
$y = f(ax), \quad a > 0$	Multiply each x-coordinate of $y = f(x)$ by $\dfrac{1}{a}$. Stretch the graph of f horizontally if $0 < a < 1$. Compress the graph of f horizontally if $a > 1$.	Replace x by ax.
Reflection about the x-axis $y = -f(x)$	Reflect the graph of f about the x-axis.	Multiply $f(x)$ by -1.
Reflection about the y-axis $y = f(-x)$	Reflect the graph of f about the y-axis.	Replace x by $-x$.

The examples that follow combine some of the procedures outlined in this section to get the required graph.

EXAMPLE 11

Determining the Function Obtained from a Series of Transformations

Find the function that is finally graphed after the following three transformations are applied to the graph of $y = |x|$.

1. Shift left 2 units.   2. Shift up 3 units.   3. Reflect about the y-axis.

Solution   1. Shift left 2 units: Replace x by $x + 2$. $y = |x + 2|$

2. Shift up 3 units: Add 3. $y = |x + 2| + 3$

3. Reflect about the y-axis: Replace x by $-x$. $y = |-x + 2| + 3$ ■

   NOW WORK PROBLEM **21.**

EXAMPLE 12

Combining Graphing Procedures

Graph the function: $f(x) = \dfrac{3}{x - 2} + 1$

Solution   We use the following steps to obtain the graph of f:

STEP 1: $y = \dfrac{1}{x}$ Reciprocal function.

STEP 2: $y = \dfrac{3}{x}$ Multiply by 3; vertical stretch of the graph of $y = \dfrac{1}{x}$ by a factor of 3.

STEP 3: $y = \dfrac{3}{x - 2}$ Replace x by $x - 2$; horizontal shift to the right 2 units.

STEP 4: $y = \dfrac{3}{x - 2} + 1$ Add 1; vertical shift up 1 unit.

See Figure 64.

Figure 64

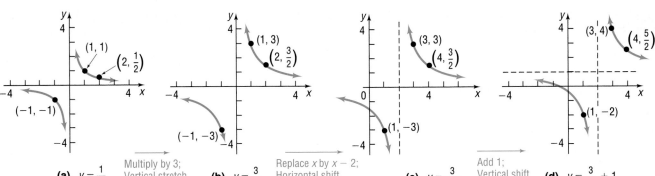

(a) $y = \dfrac{1}{x}$ Multiply by 3; Vertical stretch

(b) $y = \dfrac{3}{x}$ Replace x by $x - 2$; Horizontal shift right 2 units

(c) $y = \dfrac{3}{x-2}$ Add 1; Vertical shift up 1 unit

(d) $y = \dfrac{3}{x-2} + 1$ ■

Other orderings of the steps shown in Example 12 would also result in the graph of f. For example, try this one:

STEP 1: $y = \dfrac{1}{x}$ Reciprocal function

STEP 2: $y = \dfrac{1}{x - 2}$ Replace x by $x - 2$; horizontal shift to the right 2 units.

STEP 3: $y = \dfrac{3}{x - 2}$ Multiply by 3; vertical stretch of the graph of $y = \dfrac{1}{x - 2}$ by factor of 3.

STEP 4: $y = \dfrac{3}{x - 2} + 1$ Add 1; vertical shift up 1 unit.

◢ NOW WORK PROBLEM **47.**

EXAMPLE 13   **Combining Graphing Procedures**

Graph the function: $f(x) = \sqrt{1 - x} + 2$

Solution   We use the following steps to get the graph of $y = \sqrt{1 - x} + 2$:

STEP 1: $y = \sqrt{x}$ Square root function

STEP 2: $y = \sqrt{x + 1}$ Replace x by $x + 1$; horizontal shift left 1 unit.

STEP 3: $y = \sqrt{-x + 1} = \sqrt{1 - x}$ Replace x by $-x$; reflect about y-axis.

STEP 4: $y = \sqrt{1 - x} + 2$ Add 2; vertical shift up 2 units.

See Figure 65.

Figure 65

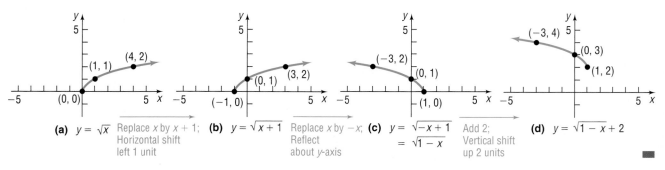

(a) $y = \sqrt{x}$ Replace x by $x + 1$; Horizontal shift left 1 unit   **(b)** $y = \sqrt{x + 1}$ Replace x by $-x$; Reflect about y-axis   **(c)** $y = \sqrt{-x + 1} = \sqrt{1 - x}$ Add 2; Vertical shift up 2 units   **(d)** $y = \sqrt{1 - x} + 2$

1.5 EXERCISES

In Problems 1–12, match each graph to one of the following functions.

A. $y = x^2 + 2$ B. $y = -x^2 + 2$ C. $y = |x| + 2$ D. $y = -|x| + 2$

E. $y = (x - 2)^2$ F. $y = -(x + 2)^2$ G. $y = |x - 2|$ H. $y = -|x + 2|$

I. $y = 2x^2$ J. $y = -2x^2$ K. $y = 2|x|$ L. $y = -2|x|$

1.

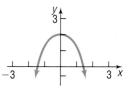

2.

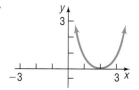

3.

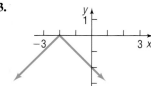

4.

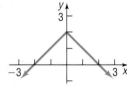

5.

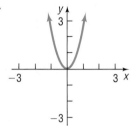

6.

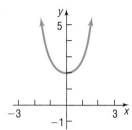

7.

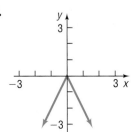

8.

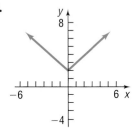

9.

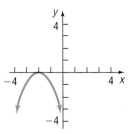

10.

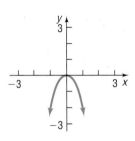

11.

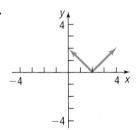

12.

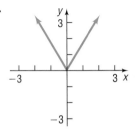

In Problems 13–20, write the function whose graph is the graph of $y = x^3$, but is:

13. Shifted to the right 4 units

14. Shifted to the left 4 units

15. Shifted up 4 units

16. Shifted down 4 units

17. Reflected about the y-axis

18. Reflected about the x-axis

19. Vertically stretched by a factor of 4

20. Horizontally stretched by a factor of 4

In Problems 21–24, find the function that is finally graphed after the following transformations are applied to the graph of $y = \sqrt{x}$.

21. (1) Shift up 2 units
    (2) Reflect about the x-axis
    (3) Reflect about the y-axis

22. (1) Reflect about the x-axis
    (2) Shift right 3 units
    (3) Shift down 2 units

23. (1) Reflect about the x-axis
    (2) Shift up 2 units
    (3) Shift left 3 units

24. (1) Shift up 2 units
    (2) Reflect about the y-axis
    (3) Shift left 3 units

25. If $(3, 0)$ is a point on the graph of $y = f(x)$, which of the following must be on the graph of $y = -f(x)$?
    (a) $(0, 3)$ (b) $(0, -3)$
    (c) $(3, 0)$ (d) $(-3, 0)$

26. If $(3, 0)$ is a point on the graph of $y = f(x)$, which of the following must be on the graph of $y = f(-x)$?
    (a) $(0, 3)$ (b) $(0, -3)$
    (c) $(3, 0)$ (d) $(-3, 0)$

27. If $(0, 3)$ is a point on the graph of $y = f(x)$, which of the following must be on the graph of $y = 2f(x)$?

(a) $(0, 3)$ (b) $(0, 2)$

(c) $(0, 6)$ (d) $(6, 0)$

28. If $(3, 0)$ is a point on the graph of $y = f(x)$, which of the following must be on the graph of $y = \dfrac{1}{2}f(x)$?

(a) $(3, 0)$ (b) $\left(\dfrac{3}{2}, 0\right)$

(c) $\left(0, \dfrac{3}{2}\right)$ (d) $\left(\dfrac{1}{2}, 0\right)$

In Problems 29–56, graph each function using the techniques of shifting, compressing, stretching, and/or reflecting. Start with the graph of the basic function (for example, $y = x^2$) and show all stages.

29. $f(x) = x^2 - 1$

30. $f(x) = x^2 + 4$

31. $g(x) = x^3 + 1$

32. $g(x) = x^3 - 1$

33. $h(x) = \sqrt{x - 2}$

34. $h(x) = \sqrt{x + 1}$

35. $f(x) = (x - 1)^3 + 2$

36. $f(x) = (x + 2)^3 - 3$

37. $g(x) = 4\sqrt{x}$

38. $g(x) = \dfrac{1}{2}\sqrt{x}$

39. $h(x) = \dfrac{1}{2x}$

40. $h(x) = \dfrac{4}{x}$

41. $f(x) = -|x|$

42. $f(x) = -\sqrt{x}$

43. $g(x) = |-x|$

44. $g(x) = (-x)^3$

45. $h(x) = -x^3 + 2$

46. $h(x) = \dfrac{1}{-x}$

47. $f(x) = 2(x + 1)^2 - 3$

48. $f(x) = 3(x - 2)^2 + 1$

49. $g(x) = \sqrt{x - 2} + 1$

50. $g(x) = |x + 1| - 3$

51. $h(x) = \sqrt{-x} - 2$

52. $h(x) = \dfrac{4}{x} + 2$

53. $f(x) = -(x + 1)^3 - 1$

54. $f(x) = -4\sqrt{x - 1}$

55. $g(x) = 2|1 - x|$

56. $g(x) = 4\sqrt{2 - x}$

In Problems 57–60, the graph of a function f is illustrated. Use the graph of f as the first step toward graphing each of the following functions:

(a) $F(x) = f(x) + 3$ (b) $G(x) = f(x + 2)$ (c) $P(x) = -f(x)$ (d) $H(x) = f(x + 1) - 2$

(e) $Q(x) = \dfrac{1}{2}f(x)$ (f) $g(x) = f(-x)$ (g) $h(x) = f(2x)$

57.

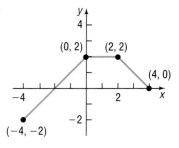

58.

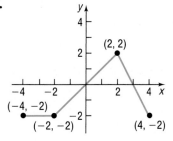

59.

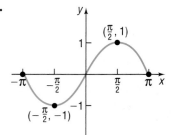

60.

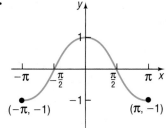

61. Exploration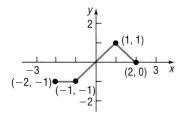
   (a) Use a graphing utility to graph $y = x + 1$ and $y = |x + 1|$.
   (b) Graph $y = 4 - x^2$ and $y = |4 - x^2|$.
   (c) Graph $y = x^3 + x$ and $y = |x^3 + x|$.
   (d) What do you conclude about the relationship between the graphs of $y = f(x)$ and $y = |f(x)|$?

63. The graph of a function f is illustrated in the figure.
   (a) Draw the graph of $y = |f(x)|$.
   (b) Draw the graph of $y = f(|x|)$.

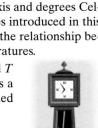

62. Exploration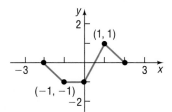
   (a) Use a graphing utility to graph $y = x + 1$ and $y = |x| + 1$.
   (b) Graph $y = 4 - x^2$ and $y = 4 - |x|^2$.
   (c) Graph $y = x^3 + x$ and $y = |x|^3 + |x|$.
   (d) What do you conclude about the relationship between the graphs of $y = f(x)$ and $y = f(|x|)$?

64. The graph of a function f is illustrated in the figure.
   (a) Draw the graph of $y = |f(x)|$.
   (b) Draw the graph of $y = f(|x|)$.

In Problems 65–70, complete the square of each quadratic expression. Then graph each function using the technique of shifting. (If necessary, refer to Appendix A, Section A.3 to review completing the square.)

65. $f(x) = x^2 + 2x$

66. $f(x) = x^2 - 6x$

67. $f(x) = x^2 - 8x + 1$

68. $f(x) = x^2 + 4x + 2$

69. $f(x) = x^2 + x + 1$

70. $f(x) = x^2 - x + 1$

---

71. The equation $y = (x - c)^2$ defines a *family of parabolas*, one parabola for each value of c. On one set of coordinate axes, graph the members of the family for $c = 0$, $c = 3$, and $c = -2$.

72. Repeat Problem 71 for the family of parabolas $y = x^2 + c$.

73. Temperature Measurements   The relationship between the Celsius (°C) and Fahrenheit (°F) scales for measuring temperature is given by the equation

$$F = \frac{9}{5}C + 32$$

The relationship between the Celsius (°C) and Kelvin (K) scales is $K = C + 273$. Graph the equation $F = \frac{9}{5}C + 32$ using degrees Fahrenheit on the y-axis and degrees Celsius on the x-axis. Use the techniques introduced in this section to obtain the graph showing the relationship between Kelvin and Fahrenheit temperatures.

74. Period of a Pendulum   The period T (in seconds) of a simple pendulum is a function of its length l (in feet) defined by the equation

$$T = 2\pi\sqrt{\frac{l}{g}}$$

where $g \approx 32.2$ feet per second per second is the acceleration of gravity.
   (a) Use a graphing utility to graph the function $T = T(l)$.

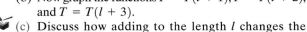

   (b) Now graph the functions $T = T(l + 1)$, $T = T(l + 2)$, and $T = T(l + 3)$.
   (c) Discuss how adding to the length l changes the period T.
   (d) Now graph the functions $T = T(2l)$, $T = T(3l)$, and $T = T(4l)$.
   (e) Discuss how multiplying the length l by factors of 2, 3, and 4 changes the period T.

75. Cigar Company Profits   The daily profits of a cigar company from selling x cigars are given by

$$p(x) = -0.05x^2 + 100x - 2000$$

The government wishes to impose a tax on cigars (sometimes called a *sin tax*) that gives the company the option of either paying a flat tax of \$10,000 per day or a tax of 10% on profits. As chief financial officer of the company, you need to decide which tax is the better option for the company.
   (a) On the same screen, graph $Y_1 = p(x) - 10{,}000$ and $Y_2 = (1 - 0.10)p(x)$.
   (b) Based on the graph, which option would you select? Why?
   (c) Using the terminology learned in this section, describe each graph in terms of the graph of $p(x)$.
   (d) Suppose that the government offered the options of a flat tax of \$4800 or a tax of 10% on profits. Which would you select? Why?

Before getting started, review the following:

✓ Functions (Section 1.3, pp. 21–32)

✓ Increasing/Decreasing Functions (Section 1.4, pp. 38–39)

1.6 ONE-TO-ONE FUNCTIONS; INVERSE FUNCTIONS

OBJECTIVES   ① Determine the Inverse of a Function
② Obtain the Graph of the Inverse Function from the Graph of the Function
③ Find the Inverse Function f^{-1}

①   In Section 1.3 we said that a function f can be thought of as a machine that receives as input a number, say x, from the domain, manipulates it, and outputs the value $f(x)$. The **inverse of f** receives as input a number $f(x)$, manipulates it, and outputs the value x.

EXAMPLE 1   **Finding the Inverse of a Function**

Find the inverse of the following functions.

(a) Let the domain of the function represent the employees of Yolanda's Preowned Car Mart and let the range represent their base salaries.

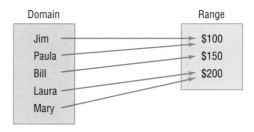

(b) Let the domain of the function represent the employees of Yolanda's Preowned Car Mart and let the range represent their spouse's names.

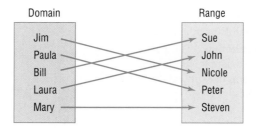

Solution   (a) The elements in the domain represent inputs to the function, and the elements in the range represent the outputs. To find the inverse, interchange the elements in the domain with the elements in the range. For example, the function receives as input Bill and outputs $150. So the inverse receives an input $150 and outputs Bill. The inverse of the given function takes the form

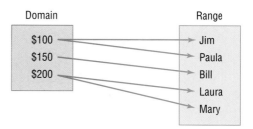

(b) The inverse of the given function is

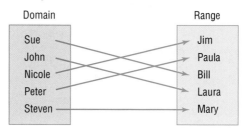

Notice that the inverse found in Example 1(b) is a function, since each element in the domain corresponds to a unique element in the range. The inverse found in Example 1(a) is not a function, since each element in the domain does not correspond to a unique element in the range.

If the function f is a set of ordered pairs (x, y), then the inverse of f is the set of ordered pairs (y, x).

EXAMPLE 2 **Finding the Inverse of a Function**

Find the inverse of the following functions:

(a) $\{(-3, -27), (-2, -8), (-1, -1), (0, 0), (1, 1), (2, 8), (3, 27)\}$
(b) $\{(-3, 9), (-2, 4), (-1, 1), (0, 0), (1, 1), (2, 4), (3, 9)\}$

Solution (a) The inverse of the given function is found by interchanging the entries in each ordered pair and so is given by

$$\{(-27, -3), (-8, -2), (-1, -1), (0, 0), (1, 1), (8, 2), (27, 3)\}$$

(b) The inverse of the given function is
$$\{(9, -3), (4, -2), (1, -1), (0, 0), (1, 1), (4, 2), (9, 3)\}$$

The inverse obtained in the solution to Example 2(a) is a function, but the inverse obtained in Example 2(b) is not a function. So, sometimes the inverse of a function is a function, and sometimes it is not.

When the inverse of a function f is itself a function, then f is said to be a **one-to-one function.** That is, f is **one-to-one** if, for any choice of elements x_1 and x_2 in the domain of f, $x_1 \neq x_2$, then the corresponding values $f(x_1)$ and $f(x_2)$ are unequal, $f(x_1) \neq f(x_2)$.

In other words, if a function f is one-to-one, then for each x in the domain of f there is exactly one y in the range, and no y in the range is the image of more than one x in the domain. See Figure 66.

Figure 66

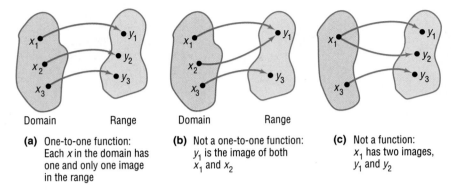

(a) One-to-one function:
Each x in the domain has
one and only one image
in the range

(b) Not a one-to-one function:
y_1 is the image of both
x_1 and x_2

(c) Not a function:
x_1 has two images,
y_1 and y_2

NOW WORK PROBLEMS **1** AND **5**.

If the graph of a function f is known, there is a simple test, called the **horizontal-line test,** to determine whether f is one-to-one.

Theorem   **Horizontal-line Test**

If every horizontal line intersects the graph of a function f in at most one point, then f is one-to-one.

The reason that this test works can be seen in Figure 67, where the horizontal line $y = h$ intersects the graph at two distinct points, (x_1, h) and (x_2, h). Since h is the image of both x_1 and $x_2, x_1 \neq x_2, f$ is not one-to-one.

Figure 67
$f(x_1) = f(x_2) = h,$ and $x_1 \neq x_2;$
f is not a one-to-one function.

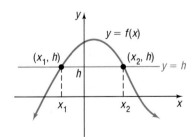

EXAMPLE 3   **Using the Horizontal-line Test**

For each function, use the graph to determine whether the function is one-to-one.

(a) $f(x) = x^2$ (b) $g(x) = x^3$

Solution   (a) Figure 68(a) illustrates the horizontal-line test for $f(x) = x^2$. The horizontal line $y = 1$ intersects the graph of f twice, at $(1, 1)$ and at $(-1, 1)$, so f is not one-to-one.

(b) Figure 68(b) illustrates the horizontal-line test for $g(x) = x^3$. Because every horizontal line will intersect the graph of g exactly once, it follows that g is one-to-one.

Figure 68

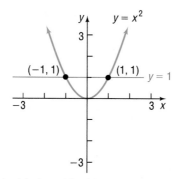

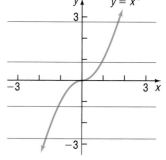

(a) A horizontal line intersects the graph twice; thus, f is not one-to-one

(b) Every horizontal line intersects the graph exactly once; thus, g is one-to-one

NOW WORK PROBLEM **9.**

Let's look more closely at the one-to-one function $g(x) = x^3$. This function is an increasing function. Because an increasing (or decreasing) function will always have different y values for unequal x values, it follows that a function that is increasing (decreasing) over its domain is also a one-to-one function.

Theorem

A function that is increasing over its domain is a one-to-one function.
A function that is decreasing over its domain is a one-to-one function.

INVERSE FUNCTION OF $y = f(x)$

If f is a one-to-one function, its inverse is a function. Then, to each x in the domain of f, there is exactly one y in the range (because f is a function); and to each y in the range of f, there is exactly one x in the domain (because f is one-to-one). The correspondence from the range of f back to the domain of f is called the **inverse function of f** and is denoted by the symbol f^{-1}. Figure 69 illustrates this definition.

Figure 69

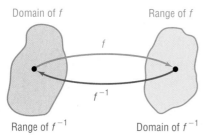

WARNING: Be careful! f^{-1} is a symbol for the inverse function of f. The -1 used in f^{-1} is not an exponent. That is, f^{-1} does *not* mean the reciprocal of f; $f^{-1}(x)$ is not equal to $\dfrac{1}{f(x)}$.

Two facts are now apparent about a function f and its inverse f^{-1}.

Domain of f = Range of f^{-1} Range of f = Domain of f^{-1}

Look again at Figure 69 to visualize the relationship. If we start with x, apply f, and then apply f^{-1}, we get x back again. If we start with x, apply f^{-1}, and then apply f, we get the number x back again. To put it simply, what f does, f^{-1} undoes, and vice versa.

$$\boxed{\text{Input } x} \xrightarrow{\text{Apply } f} \boxed{f(x)} \xrightarrow{\text{Apply } f^{-1}} \boxed{f^{-1}(f(x)) = x}$$

$$\boxed{\text{Input } x} \xrightarrow{\text{Apply } f^{-1}} \boxed{f^{-1}(x)} \xrightarrow{\text{Apply } f} \boxed{f(f^{-1}(x)) = x}$$

In other words,

$$f^{-1}(f(x)) = x \quad \text{and} \quad f(f^{-1}(x)) = x$$

For example, the function $f(x) = 2x$ multiplies the argument x by 2. The inverse function f^{-1} undoes whatever f does. So the inverse function of f is $f^{-1}(x) = \dfrac{1}{2}x$, which divides the argument by 2. We can verify this by showing that

Figure 70

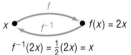

$$f^{-1}(f(x)) = f^{-1}(2x) = \frac{1}{2}(2x) = x \quad \text{and} \quad f(f^{-1}(x)) = f\left(\frac{1}{2}x\right) = 2\left(\frac{1}{2}x\right) = x$$

See Figure 70.

EXAMPLE 4 **Verifying Inverse Functions**

(a) We verify that the inverse of $g(x) = x^3$ is $g^{-1}(x) = \sqrt[3]{x}$ by showing that

$$g^{-1}(g(x)) = g^{-1}(x^3) = \sqrt[3]{x^3} = x$$

and

$$g(g^{-1}(x)) = g(\sqrt[3]{x}) = (\sqrt[3]{x})^3 = x$$

(b) We verify that the inverse of $h(x) = 3x$ is $h^{-1}(x) = \dfrac{1}{3}x$ by showing that

$$h^{-1}(h(x)) = h^{-1}(3x) = \frac{1}{3}(3x) = x$$

and

$$h(h^{-1}(x)) = h\left(\frac{1}{3}x\right) = 3\left(\frac{1}{3}x\right) = x$$

(c) We verify that the inverse of $f(x) = 2x + 3$ is $f^{-1}(x) = \dfrac{1}{2}(x - 3)$ by showing that

$$f^{-1}(f(x)) = f^{-1}(2x + 3) = \frac{1}{2}\left[(2x + 3) - 3\right] = \frac{1}{2}(2x) = x$$

and

$$f(f^{-1}(x)) = f\left(\frac{1}{2}(x - 3)\right) = 2\left[\frac{1}{2}(x - 3)\right] + 3 = (x - 3) + 3 = x$$

 NOW WORK PROBLEM 21.

EXPLORATION: Simultaneously graph $Y_1 = x$, $Y_2 = x^3$, and $Y_3 = \sqrt[3]{x}$ on a square screen, using the viewing rectangle $-3 \le x \le 3$, $-2 \le y \le 2$. What do you observe about the graphs of $Y_2 = x^3$, its inverse $Y_3 = \sqrt[3]{x}$, and the line $Y_1 = x$?

Repeat this experiment by simultaneously graphing $Y_1 = x$, $Y_2 = 2x + 3$, and $Y_3 = \dfrac{1}{2}(x - 3)$, using the viewing rectangle $-6 \le x \le 3$, $-8 \le y \le 4$. Do you see the symmetry of the graph of Y_2 and its inverse Y_3 with respect to the line $Y_1 = x$?

② **GEOMETRIC INTERPRETATION**

Figure 71

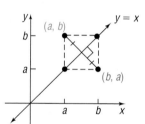

Suppose that (a, b) is a point on the graph of the one-to-one function f defined by $y = f(x)$. Then $b = f(a)$. This means that $a = f^{-1}(b)$, so (b, a) is a point on the graph of the inverse function f^{-1}. The relationship between the point (a, b) on f and the point (b, a) on f^{-1} is shown in Figure 71. The line segment containing (a, b) and (b, a) is perpendicular to the line $y = x$ and is bisected by the line $y = x$. (Do you see why?) It follows that the point (b, a) on f^{-1} is the reflection about the line $y = x$ of the point (a, b) on f.

Theorem   The graph of a function f and the graph of its inverse f^{-1} are symmetric with respect to the line $y = x$.

Figure 72 illustrates this result. Notice that, once the graph of f is known, the graph of f^{-1} may be obtained by reflecting the graph of f about the line $y = x$.

Figure 72

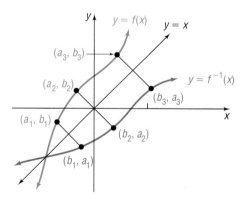

EXAMPLE 5   **Graphing the Inverse Function**

The graph in Figure 73(a) is that of a one-to-one function $y = f(x)$. Draw the graph of its inverse.

Figure 73

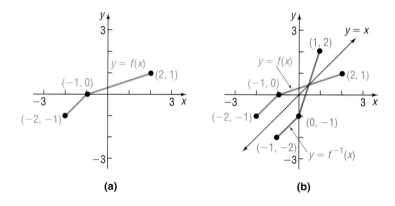

(a)   (b)

Solution We begin by adding the graph of $y = x$ to Figure 73(a). Since the points $(-2, -1), (-1, 0)$, and $(2, 1)$ are on the graph of f, we know that the points $(-1, -2), (0, -1)$, and $(1, 2)$ must be on the graph of f^{-1}. Keeping in mind that the graph of f^{-1} is the reflection about the line $y = x$ of the graph of f, we can draw f^{-1}. See Figure 73(b). ▬

NOW WORK PROBLEM **15**.

③ FINDING THE INVERSE FUNCTION f^{-1}

The fact that the graph of a one-to-one function f and its inverse function f^{-1} are symmetric with respect to the line $y = x$ tells us more. It says that we can obtain f^{-1} by interchanging the roles of x and y in f. Look again at Figure 72. If f is defined by the equation

$$y = f(x)$$

then f^{-1} is defined by the equation

$$x = f(y)$$

The equation $x = f(y)$ defines f^{-1} *implicitly*. If we can solve this equation for y, we will have the *explicit* form of f^{-1}, that is,

$$y = f^{-1}(x)$$

Let's use this procedure to find the inverse of $f(x) = 2x + 3$. (Since f is a linear function and is increasing, we know that f is one-to-one and so has an inverse function.)

EXAMPLE 6 **Finding the Inverse Function f^{-1}**

Find the inverse of $f(x) = 2x + 3$. Also find the domain and range of f and f^{-1}. Graph f and f^{-1} on the same coordinate axes.

Solution In the equation $y = 2x + 3$, interchange the variables x and y. The result,

$$x = 2y + 3$$

is an equation that defines the inverse f^{-1} implicitly. To find the explicit form, we solve for y.

$$2y + 3 = x$$
$$2y = x - 3$$
$$y = \frac{1}{2}(x - 3)$$

The explicit form of the inverse f^{-1} is therefore

$$f^{-1}(x) = \frac{1}{2}(x - 3)$$

which we verified in Example 4(c).

Next we find

$$\text{Domain of } f = \text{Range of } f^{-1} = (-\infty, \infty)$$
$$\text{Range of } f = \text{Domain of } f^{-1} = (-\infty, \infty)$$

The graphs of $f(x) = 2x + 3$ and its inverse $f^{-1}(x) = \frac{1}{2}(x - 3)$ are shown in Figure 74. Note the symmetry of the graphs with respect to the line $y = x$. ▬

Figure 74

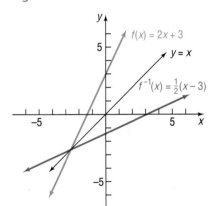

We outline next the steps to follow for finding the inverse of a one-to-one function.

> **PROCEDURE FOR FINDING THE INVERSE OF A ONE-TO-ONE FUNCTION**
>
> **STEP 1:** In $y = f(x)$, interchange the variables x and y to obtain
> $$x = f(y)$$
> This equation defines the inverse function f^{-1} implicitly.
>
> **STEP 2:** If possible, solve the implicit equation for y in terms of x to obtain the explicit form of f^{-1}.
> $$y = f^{-1}(x)$$
>
> **STEP 3:** Check the result by showing that
> $$f^{-1}(f(x)) = x \quad \text{and} \quad f(f^{-1}(x)) = x$$

EXAMPLE 7   **Finding the Inverse Function**

The function

$$f(x) = \frac{2x + 1}{x - 1}, \qquad x \neq 1$$

is one-to-one. Find its inverse and check the result.

Solution   **STEP 1:** Interchange the variables x and y in

$$y = \frac{2x + 1}{x - 1}$$

to obtain

$$x = \frac{2y + 1}{y - 1}$$

STEP 2: Solve for y.

$$x = \frac{2y + 1}{y - 1}$$

$$x(y - 1) = 2y + 1 \qquad \text{Multiply both sides by } y - 1.$$

$$xy - x = 2y + 1 \qquad \text{Apply the distributive property.}$$

$$xy - 2y = x + 1 \qquad \text{Subtract } 2y \text{ from both sides; add } x \text{ to both sides.}$$

$$(x - 2)y = x + 1 \qquad \text{Factor the left side.}$$

$$y = \frac{x + 1}{x - 2} \qquad \text{Divide by } x - 2.$$

The inverse is

$$f^{-1}(x) = \frac{x + 1}{x - 2}, \qquad x \neq 2 \qquad \text{Replace } y \text{ by } f^{-1}(x).$$

STEP 3: Check:

$$f^{-1}(f(x)) = f^{-1}\left(\frac{2x+1}{x-1}\right) = \frac{\dfrac{2x+1}{x-1}+1}{\dfrac{2x+1}{x-1}-2} = \frac{2x+1+x-1}{2x+1-2(x-1)} = \frac{3x}{3} = x$$

$$f(f^{-1}(x)) = f\left(\frac{x+1}{x-2}\right) = \frac{2\left(\dfrac{x+1}{x-2}\right)+1}{\dfrac{x+1}{x-2}-1} = \frac{2(x+1)+x-2}{x+1-(x-2)} = \frac{3x}{3} = x$$

EXPLORATION In Example 7, we found that, if $f(x) = \dfrac{2x+1}{x-1}$, then $f^{-1}(x) = \dfrac{x+1}{x-2}$. Compare the vertical and horizontal asymptotes of f and f^{-1}. What did you find? Are you surprised?

NOW WORK PROBLEM 33.

We said earlier that finding the range of a function f is not easy. However, if f is one-to-one, we can find its range by finding the domain of the inverse function f^{-1}.

EXAMPLE 8   **Finding the Range of a Function**

Find the domain and range of

$$f(x) = \frac{2x+1}{x-1}$$

Solution   The domain of f is $\{x\,|\,x \neq 1\}$. To find the range of f, we first find the inverse f^{-1}. Based on Example 7, we have

$$f^{-1}(x) = \frac{x+1}{x-2}$$

The domain of f^{-1} is $\{x\,|\,x \neq 2\}$, so the range of f is $\{y\,|\,y \neq 2\}$.

NOW WORK PROBLEM 47.

If a function is not one-to-one, then its inverse is not a function. Sometimes, though, an appropriate restriction on the domain of such a function will yield a new function that is one-to-one. Then its inverse is a function. Let's look at an example of this common practice.

EXAMPLE 9   **Finding the Inverse of a Domain-restricted Function**

Find the inverse of $y = f(x) = x^2$ if $x \geq 0$.

Solution   The function $y = x^2$ is not one-to-one. [Refer to Example 3(a).] However, if we restrict this function to only that part of its domain for which $x \geq 0$, as indicated, we have a new function that is increasing and therefore is one-to-one. As a result, the function defined by $y = f(x) = x^2, x \geq 0$, has an inverse function, f^{-1}.

We follow the steps given previously to find f^{-1}.

STEP 1: In the equation $y = x^2$, $x \geq 0$, interchange the variables x and y. The result is

$$x = y^2, \quad y \geq 0$$

This equation defines (implicitly) the inverse function.

STEP 2: We solve for y to get the explicit form of the inverse. Since $y \geq 0$, only one solution for y is obtained.

$$y = \sqrt{x}$$

So

$$f^{-1}(x) = \sqrt{x}$$

STEP 3: Check: $f^{-1}(f(x)) = f^{-1}(x^2) = \sqrt{x^2} = |x| = x$, since $x \geq 0$
$f(f^{-1}(x)) = f(\sqrt{x}) = (\sqrt{x})^2 = x$.

Figure 75 illustrates the graphs of $f(x) = x^2$, $x \geq 0$, and $f^{-1}(x) = \sqrt{x}$.

Figure 75

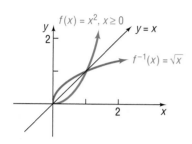

SUMMARY

1. If a function f is one-to-one, then it has an inverse function f^{-1}.
2. Domain of f = Range of f^{-1}; Range of f = Domain of f^{-1}.
3. To verify that f^{-1} is the inverse of f, show that $f^{-1}(f(x)) = x$ and $f(f^{-1}(x)) = x$.
4. The graphs of f and f^{-1} are symmetric with respect to the line $y = x$.
5. To find the range of a one-to-one function f, find the domain of the inverse function f^{-1}.

1.6 EXERCISES

In Problems 1–8, (a) find the inverse and (b) determine whether the inverse is a function.

1.

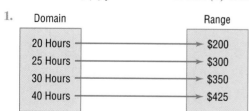

2.

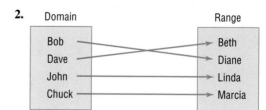

3.

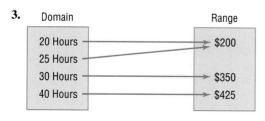

4.

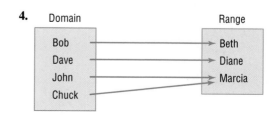

5. $\{(2, 6), (-3, 6), (4, 9), (1, 10)\}$

6. $\{(-2, 5), (-1, 3), (3, 7), (4, 12)\}$

7. $\{(0, 0), (1, 1), (2, 16), (3, 81)\}$

8. $\{(1, 2), (2, 8), (3, 18), (4, 32)\}$

In Problems 9–14, the graph of a function f is given. Use the horizontal-line test to determine whether f is one-to-one.

9.

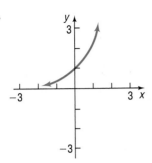

10.

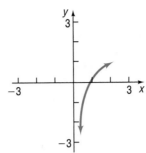

11.

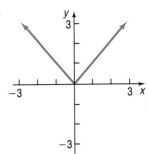

12.

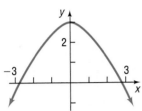

13.

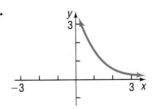

14.

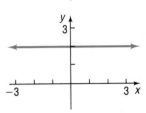

In Problems 15–20, the graph of a one-to-one function f is given. Draw the graph of the inverse function f^{-1}. For convenience (and as a hint), the graph of y = x is also given.

15.

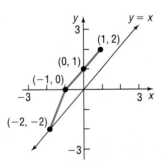

16.

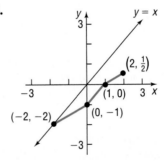

17.

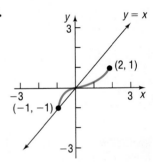

18.

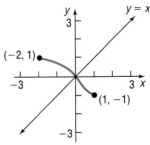

19.

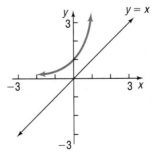

20.

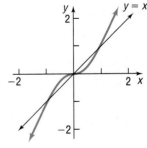

In Problems 21–30, verify that the functions f and g are inverses of each other by showing that $f(g(x)) = x$ and $g(f(x)) = x$.

21. $f(x) = 3x + 4$; $\quad g(x) = \dfrac{1}{3}(x - 4)$

22. $f(x) = 3 - 2x$; $\quad g(x) = -\dfrac{1}{2}(x - 3)$

23. $f(x) = 4x - 8$; $\quad g(x) = \dfrac{x}{4} + 2$

24. $f(x) = 2x + 6$; $\quad g(x) = \dfrac{1}{2}x - 3$

25. $f(x) = x^3 - 8$; $\quad g(x) = \sqrt[3]{x + 8}$

26. $f(x) = (x - 2)^2, x \geq 2$; $\quad g(x) = \sqrt{x} + 2$

27. $f(x) = \dfrac{1}{x}$; $\quad g(x) = \dfrac{1}{x}$

28. $f(x) = x$; $\quad g(x) = x$

29. $f(x) = \dfrac{2x + 3}{x + 4}$; $\quad g(x) = \dfrac{4x - 3}{2 - x}$

30. $f(x) = \dfrac{x - 5}{2x + 3}$; $\quad g(x) = \dfrac{3x + 5}{1 - 2x}$

In Problems 31–42, the function f is one-to-one. Find its inverse and check your answer. State the domain and range of f and f^{-1}. Graph f, f^{-1}, and y = x on the same coordinate axes.

31. $f(x) = 3x$

32. $f(x) = -4x$

33. $f(x) = 4x + 2$

34. $f(x) = 1 - 3x$

35. $f(x) = x^3 - 1$

36. $f(x) = x^3 + 1$

37. $f(x) = x^2 + 4, \quad x \geq 0$

38. $f(x) = x^2 + 9, \quad x \geq 0$

39. $f(x) = \dfrac{4}{x}$

40. $f(x) = -\dfrac{3}{x}$

41. $f(x) = \dfrac{1}{x - 2}$

42. $f(x) = \dfrac{4}{x + 2}$

In Problems 43–54, the function f is one-to-one. Find its inverse and check your answer. State the domain of f and find its range using f^{-1}.

43. $f(x) = \dfrac{2}{3 + x}$

44. $f(x) = \dfrac{4}{2 - x}$

45. $f(x) = \dfrac{3x}{x + 2}$

46. $f(x) = \dfrac{-2x}{x - 1}$

47. $f(x) = \dfrac{2x}{3x - 1}$

48. $f(x) = \dfrac{3x + 1}{-x}$

49. $f(x) = \dfrac{3x + 4}{2x - 3}$

50. $f(x) = \dfrac{2x - 3}{x + 4}$

51. $f(x) = \dfrac{2x + 3}{x + 2}$

52. $f(x) = \dfrac{-3x - 4}{x - 2}$

53. $f(x) = \dfrac{x^2 - 4}{2x^2}, \quad x > 0$

54. $f(x) = \dfrac{x^2 + 3}{3x^2}, \quad x > 0$

---

55. Find the inverse of the linear function
$$f(x) = mx + b, \quad m \neq 0$$

56. Find the inverse of the function
$$f(x) = \sqrt{r^2 - x^2}, \quad 0 \leq x \leq r$$

57. A function f has an inverse function. If the graph of f lies in quadrant I, in which quadrant does the graph of f^{-1} lie?

58. A function f has an inverse function. If the graph of f lies in quadrant II, in which quadrant does the graph of f^{-1} lie?

59. The function $f(x) = |x|$ is not one-to-one. Find a suitable restriction on the domain of f so that the new function that results is one-to-one. Then find the inverse of f.

60. The function $f(x) = x^4$ is not one-to-one. Find a suitable restriction on the domain of f so that the new function that results is one-to-one. Then find the inverse of f.

61. Temperature Conversion   To convert from x degrees Celsius to y degrees Fahrenheit, we use the formula $y = f(x) = \dfrac{9}{5}x + 32$. To convert from x degrees Fahrenheit to y degrees Celsius, we use the formula $y = g(x) = \dfrac{5}{9}(x - 32)$. Show that f and g are inverse functions.

62. Demand for Corn   The demand for corn obeys the equation $p(x) = 300 - 50x$, where p is the price per bushel (in dollars) and x is the number of bushels produced, in millions. Express the production amount x as a function of the price p.

63. Period of a Pendulum   The period T (in seconds) of a simple pendulum is a function of its length l (in feet), given by $T(l) = 2\pi \sqrt{l/g}$, where $g \approx 32.2$ feet per second per second is the acceleration of gravity. Express the length l as a function of the period T.

64. The given function f is one-to-one.

$$f(x) = \dfrac{ax + b}{cx + d}$$

(a) Find the domain of f.
(b) Find f^{-1}.
(c) Find the range of f.
(d) If $c \neq 0$, under what conditions on a, b, c, and d is $f = f^{-1}$?

65. Can an even function be one-to-one? Explain.

66. Is every odd function one-to-one? Explain.

67. If the graph of a function and its inverse intersect, where must this necessarily occur? Can they intersect anywhere else? Must they intersect?

68. Can a one-to-one function and its inverse be equal? What must be true about the graph of f for this to happen? Give some examples to support your conclusion.

69. Draw the graph of a one-to-one function that contains the points $(-2, -3)$, $(0, 0)$, and $(1, 5)$. Now draw the graph of its inverse. Compare your graph to those of other students. Discuss any similarities. What differences do you see?

CHAPTER REVIEW

Things To Know

Formulas/Equations

Distance formula (p. 4)	$d = \sqrt{(x_2 - x_1)^2 + (y_2 - y_1)^2}$
Midpoint formula (p. 6)	$M = (x, y) = \left(\dfrac{x_1 + x_2}{2}, \dfrac{y_1 + y_2}{2} \right)$
Standard form of the equation of a circle (p. 16)	$(x - h)^2 + (y - k)^2 = r^2$; r is the radius of the circle, (h, k) is the center of the circle
Equation of the unit circle (p. 16)	$x^2 + y^2 = 1$
General form of the equation of a circle (p. 17)	$x^2 + y^2 + ax + by + c = 0$
Function (pp. 21–25)	A relation between two sets of real numbers so that each number x in the first set, the domain, has corresponding to it exactly one number y in the second set. The range is the set of y values of the function for the x values in the domain.
	x is the independent variable; y is the dependent variable.
	A function can also be characterized as a set of ordered pairs (x, y) or $(x, f(x))$ in which no first element is paired with two different second elements.
Function notation (pp. 25–28)	A function f may be defined implicitly by an equation involving x and y or explicitly by writing $y = f(x)$.
	f is a symbol for the function.
	x is the argument, or independent variable.
	y is the dependent variable.
	$f(x)$ is the value of the function at x, or the image of x.
Domain (p. 28)	If unspecified, the domain of a function f is the largest set of real numbers for which $f(x)$ is a real number.
Vertical-line test (p. 30)	A set of points in the plane is the graph of a function if and only if every vertical line intersects the graph in at most one point.
Increasing function (pp. 38–39)	A function f is increasing on an open interval I if, for any choice of x_1 and x_2 in I, with $x_1 < x_2$, we have $f(x_1) < f(x_2)$.
Decreasing function (pp. 38–39)	A function f is decreasing on an open interval I if, for any choice of x_1 and x_2 in I, with $x_1 < x_2$, we have $f(x_1) > f(x_2)$.
Constant function (pp. 38–39)	A function f is constant on an interval I if, for all choices of x in I, the values of $f(x)$ are equal.
Local maximum (p. 40)	A function f has a local maximum at c if there is an open interval I containing c so that, for all $x \neq c$ in $I, f(x) < f(c)$.
Local minimum (p. 40)	A function f has a local minimum at c if there is an open interval I containing c so that, for all $x \neq c$ in $I, f(x) > f(c)$.
Even function f (p. 41)	$f(-x) = f(x)$ for every x in the domain ($-x$ must also be in the domain).
Odd function f (p. 41)	$f(-x) = -f(x)$ for every x in the domain ($-x$ must also be in the domain).
One-to-one function f (p. 61)	A function whose inverse is itself a function. For any choice of elements x_1, x_2 in the domain of f, if $x_1 \neq x_2$, then $f(x_1) \neq f(x_2)$.
Horizontal-line test (p. 62)	If every horizontal line intersects the graph of a function f in at most one point, then f is one-to-one.
Inverse function f^{-1} of f (pp. 63–66)	Domain of f = Range of f^{-1}; Range of f = Domain of f^{-1}. $f^{-1}(f(x)) = x$ and $f(f^{-1}(x)) = x$. Graphs of f and f^{-1} are symmetric with respect to the line $y = x$.

Library of Functions

Linear function (p. 42)

$f(x) = mx + b$
Graph is a line with slope m and y-intercept b.

Constant function (p. 42)

$f(x) = b$
Graph is a horizontal line with y-intercept b.
See Figure 43.

Identity function (p. 42)

$f(x) = x$

Graph is a line with slope 1 and y-intercept 0.
See Figure 44.

Cube function (p. 43)

$f(x) = x^3$

See Figure 46.

Reciprocal function (p. 44)

$f(x) = \dfrac{1}{x}$

See Figure 48.

Square function (p. 43)

$f(x) = x^2$
Graph is a parabola with intercept at $(0, 0)$.
See Figure 45.

Square root function (p. 43)

$f(x) = \sqrt{x}$
See Figure 47.

Absolute value function (p. 44)

$f(x) = |x|$
See Figure 49.

Objectives

You should be able to:

Use the distance formula (p. 3)

Use the midpoint formula (p. 5)

Graph equations by plotting points (p. 9)

Find intercepts from a graph (p. 11)

Find intercepts from an equation (p. 11)

Test an equation for symmetry with respect to the x-axis, the y-axis, and the origin (p. 13)

Write the standard form of the equation of a circle (p. 16)

Graph a circle (p. 17)

Find the center and radius of a circle from an equation in general form and graph it (p. 17)

Determine whether a relation represents a function (p. 21)

Find the value of a function (p. 26)

Find the domain of a function (p. 28)

Identify the graph of a function (p. 29)

Obtain information from or about the graph of a function (p. 30)

Use a graph to determine where a function is increasing, is decreasing, or is constant (p. 38)

Use a graph to locate local maxima and minima (p. 40)

Determine even and odd functions from a graph (p. 41)

Identify even or odd functions from the equation (p. 41)

Graph the functions listed in the library of functions (p. 42)

Graph functions using horizontal and vertical shifts (p. 47)

Graph functions using compressions and stretches (p. 50)

Graph functions using reflections about the x-axis or y-axis (p. 53)

Determine the inverse of a function (p. 60)

Obtain the graph of the inverse function from the graph of a function (p. 65)

Find the inverse function f^{-1} (p. 66)

Fill-in-the-Blank Items

1. If (x, y) are the coordinates of a point P in the xy-plane, then x is called the _____ of P and y is the _____ of P.

2. If three distinct points $P, Q,$ and R all lie on a line and if $d(P, Q) = d(Q, R)$, then Q is called the _____ of the line segment from P to R.

3. If for every point (x, y) on a graph the point $(-x, y)$ is also on the graph, then the graph is symmetric with respect to the _____.

4. The set of points in the xy-plane that are a fixed distance from a fixed point is called a(n) _____. The fixed distance is called the _____; the fixed point is called the _____.

5. If f is a function defined by the equation $y = f(x)$, then x is called the _____ variable and y is the _____ variable.

6. A set of points in the xy-plane is the graph of a function if and only if every _____ line intersects the graph in at most one point.

7. A(n) _____ function f is one for which $f(-x) = f(x)$ for every x in the domain of f; a(n) _____ function f is one for which $f(-x) = -f(x)$ for every x in the domain of f.

8. Suppose that the graph of a function f is known. Then the graph of $y = f(x - 2)$ may be obtained by a(n) _____ shift of the graph of f to the _____ a distance of 2 units.

9. If every horizontal line intersects the graph of a function f at no more than one point, then f is a(n) _____ function.

10. If f^{-1} denotes the inverse of a function f, then the graphs of f and f^{-1} are symmetric with respect to the line _____.

True/False Items

T F   **1.** The distance between two points is sometimes a negative number.

T F   **2.** The graph of the equation $y = x^4 + x^2 + 1$ is symmetric with respect to the y-axis.

T F   **3.** Every relation is a function.

T F   **4.** Vertical lines intersect the graph of a function in no more than one point.

T F   **5.** The y-intercept of the graph of the function $y = f(x)$, whose domain is all real numbers, is $f(0)$.

T F   **6.** A function f is decreasing on an open interval I if, for any choice of x_1 and x_2 in I, with $x_1 < x_2$, we have $f(x_1) < f(x_2)$.

T F   **7.** Even functions have graphs that are symmetric with respect to the origin.

T F   **8.** The graph of $y = f(-x)$ is the reflection about the y-axis of the graph of $y = f(x)$.

T F   **9.** If f and g are inverse functions, then the domain of f is the same as the domain of g.

T F   **10.** If f and g are inverse functions, then their graphs are symmetric with respect to the line $y = x$.

Review Exercises

Blue problem numbers indicate the author's suggestions for use in a Practice Test.

In Problems 1–8, find the intercepts and test each equation for symmetry with respect to the x-axis, the y-axis, and the origin.

1. $2x = 3y^2$
2. $y = 5x$
3. $x^2 + 4y^2 = 16$
4. $9x^2 - y^2 = 9$
5. $y = x^4 + 2x^2 + 1$
6. $y = x^3 - x$
7. $x^2 + x + y^2 + 2y = 0$
8. $x^2 + 4x + y^2 - 2y = 0$

In Problems 9–14, find the center and radius of each circle. Graph each circle.

9. $x^2 + (y - 1)^2 = 4$
10. $(x + 2)^2 + y^2 = 9$
11. $x^2 + y^2 - 2x + 4y - 4 = 0$
12. $x^2 + y^2 + 4x - 4y - 1 = 0$
13. $3x^2 + 3y^2 - 6x + 12y = 0$
14. $2x^2 + 2y^2 - 4x = 0$

15. A function f is defined by $f(x) = \dfrac{Ax + 5}{6x - 2}$. If $f(1) = 4$, find A.

16. A function g is defined by $g(x) = \dfrac{A}{x} + \dfrac{8}{x^2}$. If $g(-1) = 0$, find A.

17. Tell which of the following graphs are graphs of functions.

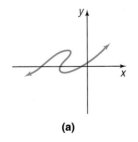

(a)

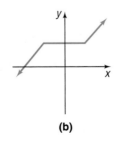

(b)

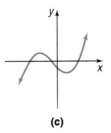

(c)

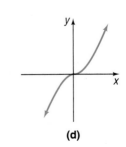

(d)

18. Use the graph of the function f shown to find:

(a) The domain and range of f

(b) $f(-1)$

(c) The intercepts of f

(d) The intervals on which f is increasing, decreasing, or constant

(e) Whether the function is even, odd, or neither

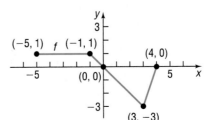

In Problems 19–24, find the following for each function:

(a) $f(-x)$ (b) $-f(x)$ (c) $f(x+2)$ (d) $f(x-2)$ (e) $f(2x)$

19. $f(x) = \dfrac{3x}{x^2 - 4}$

20. $f(x) = \dfrac{x^2}{x + 2}$

21. $f(x) = \sqrt{x^2 - 4}$

22. $f(x) = |x^2 - 4|$

23. $f(x) = \dfrac{x^2 - 4}{x^2}$

24. $f(x) = \dfrac{x^3}{x^2 - 4}$

In Problems 25–32, find the domain of each function.

25. $f(x) = \dfrac{x}{x^2 - 9}$

26. $f(x) = \dfrac{3x^2}{x - 2}$

27. $f(x) = \sqrt{2 - x}$

28. $f(x) = \sqrt{x + 2}$

29. $h(x) = \dfrac{\sqrt{x}}{|x|}$

30. $g(x) = \dfrac{|x|}{x}$

31. $f(x) = \dfrac{x}{x^2 + 2x - 3}$

32. $F(x) = \dfrac{1}{x^2 - 3x - 4}$

In Problems 33–40, determine (algebraically) whether the given function is even, odd, or neither.

33. $f(x) = x^3 - 4x$

34. $g(x) = \dfrac{4 + x^2}{1 + x^4}$

35. $h(x) = \dfrac{1}{x^4} + \dfrac{1}{x^2} + 1$

36. $F(x) = \sqrt[3]{1 - x^3}$

37. $G(x) = 1 - x + x^3$

38. $H(x) = 1 + x + x^2$

39. $f(x) = \dfrac{x}{1 + x^2}$

40. $g(x) = \dfrac{1 + x^2}{x^3}$

In Problems 41–54, graph each function using the techniques of shifting, compressing or stretching, and reflections. Identify any intercepts on the graph. State the domain and, based on the graph, find the range.

41. $F(x) = |x| - 4$

42. $f(x) = |x| + 4$

43. $g(x) = -2|x|$

44. $g(x) = \dfrac{1}{2}|x|$

45. $h(x) = \sqrt{x - 1}$

46. $h(x) = \sqrt{x} - 1$

47. $f(x) = \sqrt{1 - x}$

48. $f(x) = -\sqrt{x + 3}$

49. $h(x) = (x - 1)^2 + 2$

50. $h(x) = (x + 2)^2 - 3$

51. $g(x) = 3(x - 1)^3 + 1$

52. $g(x) = -2(x + 2)^3 - 8$

53. For the graph of the function f shown below, draw the graph of:

(a) $y = f(-x)$ (b) $y = -f(x)$
(c) $y = f(x + 2)$ (d) $y = f(x) + 2$
(e) $y = 2f(x)$ (f) $y = f(3x)$

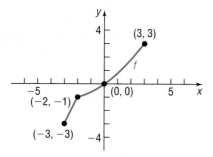

54. For the graph of the function g shown below, draw the graph of:

(a) $y = g(-x)$ (b) $y = -g(x)$
(c) $y = g(x + 2)$ (d) $y = g(x) + 2$
(e) $y = 2g(x)$ (f) $y = g(3x)$

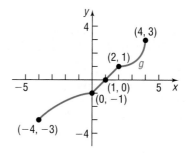

In Problems 55–60, the function f is one-to-one. Find the inverse of each function and check your answer. Find the domain and range of f and f^{-1}.

55. $f(x) = \dfrac{2x + 3}{5x - 2}$

56. $f(x) = \dfrac{2 - x}{3 + x}$

57. $f(x) = \dfrac{1}{x - 1}$

58. $f(x) = \sqrt{x - 2}$

59. $f(x) = \dfrac{3}{x^{1/3}}$

60. $f(x) = x^{1/3} + 1$

---

61. The endpoints of the diameter of a circle are $(-3, 2)$ and $(5, -6)$. Find the center and radius of the circle. Write the general equation of this circle.

62. Show that the points $A = (1, 5)$, $B = (2, 4)$, and $C = (-3, 5)$ lie on a circle with center $(-1, 2)$. What is the radius of this circle?

63. Temperature Conversion The temperature T of the air is approximately a linear function of the altitude h for altitudes within 10,000 meters of the surface of Earth. If the surface temperature is 30°C and the temperature at 10,000 meters is 5°C, find the function $T = T(h)$.

64. Material Needed to Make a Drum A steel drum in the shape of a right circular cylinder is required to have a volume of 100 cubic feet.

(a) Express the amount A of material required to make the drum as a function of the radius r of the cylinder.

(b) How much material is required if the drum is of radius 3 feet?

(c) Of radius 4 feet?

(d) Of radius 5 feet?

(e) Graph $A = A(r)$. For what value of r is A smallest?

65. Cost of a Drum A drum in the shape of a right circular cylinder is required to have a volume of 500 cubic centimeters. The top and bottom are made of material that costs 6¢ per square centimeter; the sides are made of material that costs 4¢ per square centimeter.

500 cc

(a) Express the total cost C of the material as a function of the radius r of the cylinder.

(b) What is the cost if the radius is 4 cm?

(c) What is the cost if the radius is 8 cm?

(d) Graph $C = C(r)$. For what value of r is the cost C least?

66. Page Design A page with dimensions of $8\frac{1}{2}$ inches by 11 inches has a border of uniform width x surrounding the printed matter of the page, as shown in the figure.

$8\frac{1}{2}$ in.

x

And the Rock 'n Roll Beat is On

x | | x | 11 in.

x

(a) Write a formula for the area A of the printed part of the page as a function of the width x of the border.

(b) Give the domain and range of A.

(c) Find the area of the printed page for borders of widths 1 inch, 1.2 inches, and 1.5 inches.

(d) Graph the function $A = A(x)$.

(e) Use TRACE to determine what margin should be used to obtain an area of 70 square inches and of 50 square inches.

67. Describe each of the following graphs in the xy-plane. Give justification.

(a) $x = 0$ (d) $xy = 0$

(b) $y = 0$ (e) $x^2 + y^2 = 0$

(c) $x + y = 0$

Project at Motorola

During the past decade the availability and usage of wireless Internet services have increased. The industry has developed a number of pricing proposals for such services. Marketing data have indicated that subscribers of wireless Internet services have tended to desire flat fee rate structures as compared with rates based totally on usage. The Computer Resource Department of Indigo Media (hypothetical) has entered into a contractual agreement for wireless Internet services. As a part of the contractual agreement, employees are able to sign up for their own wireless services. Three pricing options are available:

Silver Plan: $20/month for up to 200 K-bytes of service plus $0.16 for each additional K-byte of service

Gold Plan: $50/month for up to 1000 K-bytes of service plus $0.08 for each additional K-byte of service

Platinum Plan: $100/month for up to 3000 K-bytes of service plus $0.04 for each additional K-byte of service

You have been requested to write a report that answers the following questions in order to aid employees in choosing the appropriate pricing plan.

(a) If C is the monthly charge for x K-bytes of service, express C as a function of x for each of the three plans.

(b) Graph each of the three functions found in part (a).

(c) For how many K-bytes of service is the Silver Plan the best pricing option? When is the Gold Plan best? When is the Platinum Plan best? Explain your reasoning.

(d) Write a report that summarizes your findings.

Trigonometric Functions

Field Trip to Motorola

For many centuries, trigonometry has been the basis for science, astronomy, and navigation. Today trigonometry is employed in various disciplines of common life, such as civil engineering, acoustics, analog and digital wireline and wireless communications, and many other applications related to wave theory.

P R E P A R I N G F O R T H I S S E C T I O N

Before getting started, review the following:

✓ Circumference and Area of a Circle (Appendix A, Section A.2, pp. 497–502)

2.1 ANGLES AND THEIR MEASURE

OBJECTIVES
1. Convert between Degrees, Minutes, Seconds, and Decimal Forms for Angles
2. Find the Arc Length of a Circle
3. Convert from Degrees to Radians
4. Convert from Radians to Degrees
5. Find the Area of a Sector of a Circle
6. Find the Linear Speed of an Object Traveling in Circular Motion

Figure 1

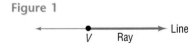

A **ray,** or **half-line,** is that portion of a line that starts at a point V on the line and extends indefinitely in one direction. The starting point V of a ray is called its **vertex.** See Figure 1.

If two rays are drawn with a common vertex, they form an **angle.** We call one of the rays of an angle the **initial side** and the other the **terminal side.** The angle that is formed is identified by showing the direction and amount of rotation from the initial side to the terminal side. If the rotation is in the counterclockwise direction, the angle is **positive;** if the rotation is clockwise, the angle is **negative.**

See Figure 2. Lowercase Greek letters, such as α (alpha), β (beta), γ (gamma), and θ (theta), will be used to denote angles. Notice in Figure 2(a) that the angle α is positive because the direction of the rotation from the initial side to the terminal side is counterclockwise. The angle β in Figure 2(b) is negative because the rotation is clockwise. The angle γ in Figure 2(c) is positive. Notice that the angle α in Figure 2(a) and the angle γ in Figure 2(c) have the same initial side and the same terminal side. However, α and γ are unequal, because the amount of rotation required to go from the initial side to the terminal side is greater for angle γ than for angle α.

Figure 2

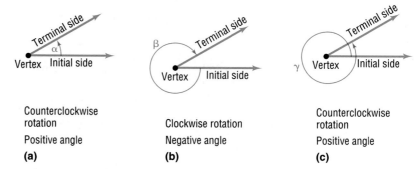

(a) Counterclockwise rotation / Positive angle

(b) Clockwise rotation / Negative angle

(c) Counterclockwise rotation / Positive angle

An angle θ is said to be in **standard position** if its vertex is at the origin of a rectangular coordinate system and its initial side coincides with the positive x-axis. See Figure 3.

Figure 3

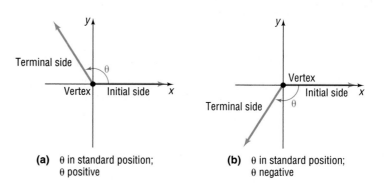

(a) θ in standard position; θ positive

(b) θ in standard position; θ negative

When an angle θ is in standard position, the terminal side will lie either in a quadrant, in which case we say that θ **lies in that quadrant,** or on the x-axis or the y-axis, in which case we say that θ is a **quadrantal angle.** For example, the angle θ in Figure 4(a) lies in quadrant II, the angle θ in Figure 4(b) lies in quadrant IV, and the angle θ in Figure 4(c) is a quadrantal angle.

Figure 4

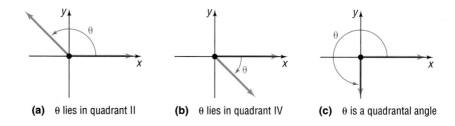

(a) θ lies in quadrant II

(b) θ lies in quadrant IV

(c) θ is a quadrantal angle

We measure angles by determining the amount of rotation needed for the initial side to become coincident with the terminal side. The two commonly used measures for angles are *degrees* and *radians*.

DEGREES

The angle formed by rotating the initial side exactly once in the counterclockwise direction until it coincides with itself (1 revolution) is said to measure 360 degrees, abbreviated 360°. **One degree, 1°,** is $\frac{1}{360}$ revolution. A **right angle** is an angle that measures 90°, or $\frac{1}{4}$ revolution; a **straight angle** is an angle that measures 180°, or $\frac{1}{2}$ revolution. See Figure 5. As Figure 5(b) shows, it is customary to indicate a right angle by using the symbol ∟.

Figure 5

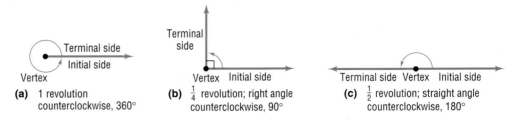

(a) 1 revolution counterclockwise, 360°

(b) $\frac{1}{4}$ revolution; right angle counterclockwise, 90°

(c) $\frac{1}{2}$ revolution; straight angle counterclockwise, 180°

It is also customary to refer to an angle that measures θ degrees as an angle of θ degrees.

EXAMPLE 1

Drawing an Angle

Draw each angle.

(a) 45°   (b) −90°   (c) 225°   (d) 405°

Solution (a) An angle of 45° is $\frac{1}{2}$ of a right angle. See Figure 6.

(b) An angle of −90° is $\frac{1}{4}$ revolution in the clockwise direction. See Figure 7.

Figure 6

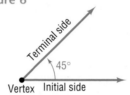

Figure 7

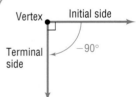

(c) An angle of 225° consists of a rotation through 180° followed by a rotation through 45°. See Figure 8.

(d) An angle of 405° consists of 1 revolution (360°) followed by a rotation through 45°. See Figure 9.

Figure 8

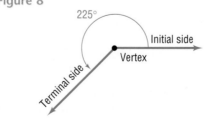

Figure 9

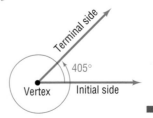

NOW WORK PROBLEM **1**.

① Although subdivisions of a degree may be obtained by using decimals, we also may use the notion of *minutes* and *seconds*. **One minute,** denoted by **1′,** is defined as $\frac{1}{60}$ degree. **One second,** denoted by **1″,** is defined as $\frac{1}{60}$ minute or, equivalently, $\frac{1}{3600}$ degree. An angle of, say, 30 degrees, 40 minutes, 10 seconds is written compactly as $30°40'10''$. To summarize:

$$1 \text{ counterclockwise revolution} = 360°$$

$$1° = 60' \qquad 1' = 60''$$

(1)

It is sometimes necessary to convert from the degree, minute, second notation ($D°M'S''$) to a decimal form, and vice versa. Check your calculator; it should be capable of doing the conversion for you.

Before getting started, though, you must set the mode to degrees, because there are two common ways to measure angles: degree mode and radian mode. (We will define radians shortly.) Usually, a menu is used to change from one mode to another. Check your owner's manual to find out how your particular calculator works.

Now let's see how to convert by hand from the degree, minute, second notation ($D°M'S''$) to a decimal form, and vice versa, by looking at some examples: $15°30' = 15.5°$, because $30' = \frac{1}{2}° = 0.5°$, and $32.25° = 32°15'$, because $0.25° = \frac{1}{4}° = \frac{1}{4}(60') = 15'$.

EXAMPLE 2   Converting between Degrees, Minutes, Seconds, and Decimal Forms by Hand

(a) Convert $50°6'21''$ to a decimal in degrees.

(b) Convert $21.256°$ to the $D°M'S''$ form.

Solution   (a) Because $1' = \frac{1}{60}°$ and $1'' = \frac{1}{60}' = \left(\frac{1}{60} \cdot \frac{1}{60}\right)°$, we convert as follows:

$$50°6'21'' = 50° + 6' + 21'' = 50° + 6 \cdot \frac{1}{60}° + 21 \cdot \frac{1}{60} \cdot \frac{1}{60}°$$

$$\approx 50° + 0.1° + 0.005833°$$

$$= 50.105833°$$

(b) We start with the decimal part of $21.256°$, that is, $0.256°$.

$$0.256° = (0.256)(1°) = (0.256)(60') = 15.36'$$
$$\underset{\substack{\uparrow \\ 1° = 60'}}{}$$

Now we work with the decimal part of $15.36'$, that is, $0.36'$.

$$0.36' = (0.36)(1') = (0.36)(60'') = 21.6'' \approx 22''$$
$$\underset{\substack{\uparrow \\ 1' = 60''}}{}$$

Thus,

$$21.256° = 21° + 0.256° = 21° + 15.36' = 21° + 15' + 0.36'$$

$$= 21° + 15' + 21.6'' \approx 21°15'22''$$

■

   NOW WORK PROBLEMS **69** AND **75.**

In many applications, such as describing the exact location of a star or the precise position of a boat at sea, angles measured in degrees, minutes, and even seconds are used. For calculation purposes, these are transformed to decimal form. In other applications, especially those in calculus, angles are measured using *radians*.

RADIANS

A **central angle** is an angle whose vertex is at the center of a circle. The rays of a central angle subtend (intersect) an arc on the circle. If the radius of the circle is r and the length of the arc subtended by the central angle is also r, then the measure of the angle is **1 radian.** See Figure 10(a).

For a circle of radius 1, the rays of a central angle with measure 1 radian would subtend an arc of length 1. For a circle of radius 3, the rays of a central angle with measure 1 radian would subtend an arc of length 3. See Figure 10(b).

Figure 10

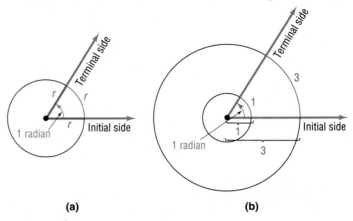

(a)   (b)

②   Now consider a circle of radius r and two central angles, θ and θ_1, measured in radians. Suppose that these central angles subtend arcs of lengths s and s_1, respectively, as shown in Figure 11. From geometry, we know that the ratio of the measures of the angles equals the ratio of the corresponding lengths of the arcs subtended by these angles; that is,

Figure 11
$$\frac{\theta}{\theta_1} = \frac{s}{s_1}$$

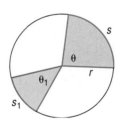

$$\frac{\theta}{\theta_1} = \frac{s}{s_1} \tag{2}$$

Suppose that $\theta_1 = 1$ radian. Refer again to Figure 10(a). The amount of arc s_1 subtended by the central angle $\theta_1 = 1$ radian equals the radius r of the circle. Then $s_1 = r$, so formula (2) reduces to

$$\frac{\theta}{1} = \frac{s}{r} \quad \text{or} \quad s = r\theta \tag{3}$$

Theorem   **Arc Length**

For a circle of radius r, a central angle of θ radians subtends an arc whose length s is

$$s = r\theta \tag{4}$$

NOTE: Formulas must be consistent with regard to the units used. In equation (4), we write
$$s = r\theta$$

To see the units, however, we must go back to equation (3) and write
$$\frac{\theta \text{ radians}}{1 \text{ radian}} = \frac{s \text{ length units}}{r \text{ length units}}$$

$$s \text{ length units} = r \text{ length units} \frac{\theta \text{ radians}}{1 \text{ radian}}$$

Since the radians cancel, we are left with

$$s \text{ length units} = (r \text{ length units})\theta \quad s = r\theta$$

where θ appears to be "dimensionless" but, in fact, is measured in radians. So, in using the formula $s = r\theta$, the dimension for θ is radians, and any convenient unit of length (such as inches or meters) may be used for s and r. ■

EXAMPLE 3

Finding the Length of an Arc of a Circle

Find the length of the arc of a circle of radius 2 meters subtended by a central angle of 0.25 radian.

Solution   We use equation (4) with $r = 2$ meters and $\theta = 0.25$. The length s of the arc is

$$s = r\theta = 2(0.25) = 0.5 \text{ meter} \quad ■$$

▸ **NOW WORK PROBLEM 37.**

RELATIONSHIP BETWEEN DEGREES AND RADIANS

Consider a circle of radius r. A central angle of 1 revolution will subtend an arc equal to the circumference of the circle (Figure 12). Because the circumference of a circle equals $2\pi r$, we use $s = 2\pi r$ in equation (4) to find that, for an angle θ of 1 revolution,

Figure 12
1 revolution = 2π radians

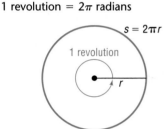

$$s = r\theta$$
$$2\pi r = r\theta \qquad \theta = 1 \text{ revolution}; s = 2\pi r$$
$$\theta = 2\pi \text{ radians} \qquad \text{Solve for } \theta.$$

From this we have,

$$1 \text{ revolution} = 2\pi \text{ radians} \qquad \textbf{(5)}$$

so

$$360° = 2\pi \text{ radians}$$

or

$$180° = \pi \text{ radians} \qquad \textbf{(6)}$$

Divide both sides of equation (6) by 180. Then

$$1 \text{ degree} = \frac{\pi}{180} \text{ radian}$$

Divide both sides of (6) by π. Then

$$\frac{180}{\pi} \text{ degrees} = 1 \text{ radian}$$

We have the following two conversion formulas:

$$1 \text{ degree} = \frac{\pi}{180} \text{ radian} \qquad 1 \text{ radian} = \frac{180}{\pi} \text{ degrees} \qquad \textbf{(7)}$$

EXAMPLE 4   **Converting from Degrees to Radians**

③   Convert each angle in degrees to radians.

(a) 60°   (b) 150°   (c) −45°   (d) 90°

Solution   (a) $60° = 60 \cdot 1 \text{ degree} = 60 \cdot \dfrac{\pi}{180} \text{ radian} = \dfrac{\pi}{3} \text{ radians}$

(b) $150° = 150 \cdot \dfrac{\pi}{180} \text{ radian} = \dfrac{5\pi}{6} \text{ radians}$

(c) $-45° = -45 \cdot \dfrac{\pi}{180} \text{ radian} = -\dfrac{\pi}{4} \text{ radian}$

(d) $90° = 90 \cdot \dfrac{\pi}{180} \text{ radian} = \dfrac{\pi}{2} \text{ radians}$ ∎

Example 4 illustrates that angles that are fractions of a revolution are expressed in radian measure as fractional multiples of π, rather than as decimals. For example, a right angle, as in Example 4(d), is left in the form $\dfrac{\pi}{2}$ radians, which is exact, rather than using the approximation $\dfrac{\pi}{2} \approx \dfrac{3.1416}{2} = 1.5708$ radians.

✏️———▶ NOW WORK PROBLEM 13.

EXAMPLE 5   **Converting Radians to Degrees**

④   Convert each angle in radians to degrees.

(a) $\dfrac{\pi}{6}$ radian   (b) $\dfrac{3\pi}{2}$ radians   (c) $-\dfrac{3\pi}{4}$ radians   (d) $\dfrac{7\pi}{3}$ radians

Solution   (a) $\dfrac{\pi}{6} \text{ radian} = \dfrac{\pi}{6} \cdot 1 \text{ radian} = \dfrac{\pi}{6} \cdot \dfrac{180}{\pi} \text{ degrees} = 30°$

(b) $\dfrac{3\pi}{2} \text{ radians} = \dfrac{3\pi}{2} \cdot \dfrac{180}{\pi} \text{ degrees} = 270°$

(c) $-\dfrac{3\pi}{4} \text{ radians} = -\dfrac{3\pi}{4} \cdot \dfrac{180}{\pi} \text{ degrees} = -135°$

(d) $\dfrac{7\pi}{3} \text{ radians} = \dfrac{7\pi}{3} \cdot \dfrac{180}{\pi} \text{ degrees} = 420°$ ∎

✏️———▶ NOW WORK PROBLEM 25.

Table 1 lists the degree and radian measures of some commonly encountered angles. You should learn to feel equally comfortable using degree or radian measure for these angles.

TABLE 1								
Degrees	0°	30°	45°	60°	90°	120°	135°	150°	180°
Radians	0	$\dfrac{\pi}{6}$	$\dfrac{\pi}{4}$	$\dfrac{\pi}{3}$	$\dfrac{\pi}{2}$	$\dfrac{2\pi}{3}$	$\dfrac{3\pi}{4}$	$\dfrac{5\pi}{6}$	π
Degrees		210°	225°	240°	270°	300°	315°	330°	360°
Radians		$\dfrac{7\pi}{6}$	$\dfrac{5\pi}{4}$	$\dfrac{4\pi}{3}$	$\dfrac{3\pi}{2}$	$\dfrac{5\pi}{3}$	$\dfrac{7\pi}{4}$	$\dfrac{11\pi}{6}$	2π

| EXAMPLE 6 | **Finding the Distance between Two Cities** |

See Figure 13(a). The latitude L of a location is the angle formed by a ray drawn from the center of Earth to the Equator and a ray drawn from the center of Earth to L. See Figure 13(b). Glasgow, Montana, is due north of Albuquerque, New Mexico. Find the distance between Glasgow (48°9′ north latitude) and Albuquerque (35°5′ north latitude). Assume that the radius of Earth is 3960 miles.

Figure 13

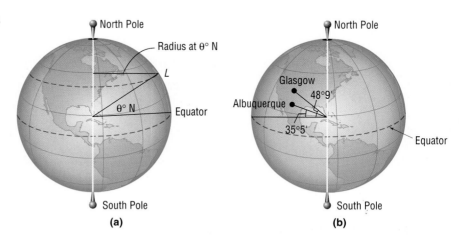

(a)

(b)

Solution   The measure of the central angle between the two cities is $48°9′ - 35°5′ = 13°4′$. We use equation (4), $s = r\theta$, but first we must convert the angle of 13°4′ to radians.

$$\theta = 13°4′ \approx 13.0667° = 13.0667 \cdot \frac{\pi}{180} \text{ radian} \approx 0.228 \text{ radian}$$

We use $\theta = 0.228$ radian and $r = 3960$ miles in equation (4). The distance between the two cities is

$$s = r\theta = 3960 \cdot 0.228 \approx 903 \text{ miles} \qquad \blacksquare$$

When an angle is measured in degrees, the degree symbol will always be shown. However, when an angle is measured in radians, we will follow the usual practice and omit the word *radians*. So, if the measure of an angle is given as $\frac{\pi}{6}$, it is understood to mean $\frac{\pi}{6}$ radian.

Figure 14

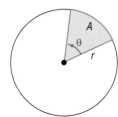

NOW WORK PROBLEM **91.**

AREA OF A SECTOR

⑤   Consider a circle of radius r. Suppose that θ, measured in radians, is a central angle of this circle. See Figure 14. We seek a formula for the area A of the sector formed by the angle θ (shown in blue).

Figure 15

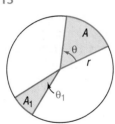

Now consider a circle of radius r and two central angles θ and θ_1, both measured in radians. See Figure 15. From geometry, we know the ratio of the measures of the angles equals the ratio of the corresponding areas of the sectors formed by these angles. That is,

$$\frac{\theta}{\theta_1} = \frac{A}{A_1}$$

Suppose that $\theta_1 = 2\pi$ radians. Then $A_1 = $ area of the circle $= \pi r^2$. Solving for A, we find

$$A = A_1 \frac{\theta}{\theta_1} = \pi r^2 \frac{\theta}{2\pi} = \frac{1}{2} r^2 \theta$$

Theorem

Area of a Sector

The area A of the sector of a circle of radius r formed by a central angle of θ radians is

$$A = \frac{1}{2} r^2 \theta \tag{8}$$

EXAMPLE 7   Finding the Area of a Sector of a Circle

Find the area of the sector of a circle of radius 2 feet formed by an angle of 30°. Round the answer to two decimal places.

Solution   We use equation (8) with $r = 2$ feet and $\theta = 30° = \dfrac{\pi}{6}$ radians. [Remember, in Equation (8), θ must be in radians.] The area A of the sector is

$$A = \frac{1}{2} r^2 \theta = \frac{1}{2} (2)^2 \frac{\pi}{6} = \frac{\pi}{3} \text{ square feet} \approx 1.05 \text{ square feet}$$

rounded to two decimal places.

 NOW WORK PROBLEM 45.

CIRCULAR MOTION

6   We have already defined the average speed of an object as the distance traveled divided by the elapsed time. Suppose that an object moves around a circle of radius r at a constant speed. If s is the distance traveled in time t around this circle, then the **linear speed** v of the object is defined as

Figure 16
$$v = \frac{s}{t} \quad \omega = \frac{\theta}{t}$$

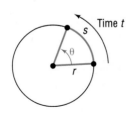

$$v = \frac{s}{t} \tag{9}$$

As this object travels around the circle, suppose that θ (measured in radians) is the central angle swept out in time t. See Figure 16. Then the **angular speed** ω (the Greek letter omega) of this object is the angle (measured in radians) swept out divided by the elapsed time; that is,

$$\omega = \frac{\theta}{t} \qquad \textbf{(10)}$$

Angular speed is the way the turning rate of an engine is described. For example, an engine idling at 900 rpm (revolutions per minute) is one that rotates at an angular speed of

$$900 \, \frac{\text{revolutions}}{\text{minute}} = 900 \, \frac{\cancel{\text{revolutions}}}{\text{minute}} \cdot 2\pi \, \frac{\text{radians}}{\cancel{\text{revolution}}} = 1800\pi \, \frac{\text{radians}}{\text{minute}}$$

There is an important relationship between linear speed and angular speed:

$$\text{linear speed} = v = \underset{\underset{(9)}{\uparrow} \quad \underset{s \, = \, r\theta}{\uparrow}}{\frac{s}{t} = \frac{r\theta}{t}} = r\left(\frac{\theta}{t}\right)$$

Then, using equation (10), we obtain

$$v = r\omega \qquad \textbf{(11)}$$

where ω is measured in radians per unit time.

When using equation (11), remember that $v = \dfrac{s}{t}$ (the linear speed) has the dimensions of length per unit of time (such as feet per second or miles per hour), r (the radius of the circular motion) has the same length dimension as s, and ω (the angular speed) has the dimensions of radians per unit of time. If the angular speed is given in terms of *revolutions* per unit of time (as is often the case), be sure to convert it to *radians* per unit of time before attempting to use equation (11).

EXAMPLE 8   **Finding Linear Speed**

A child is spinning a rock at the end of a 2-foot rope at the rate of 180 revolutions per minute (rpm). Find the linear speed of the rock when it is released.

Solution   Look at Figure 17. The rock is moving around a circle of radius $r = 2$ feet. The angular speed ω of the rock is

Figure 17

$$\omega = 180 \, \frac{\text{revolutions}}{\text{minute}} = 180 \, \frac{\cancel{\text{revolutions}}}{\text{minute}} \cdot 2\pi \, \frac{\text{radians}}{\cancel{\text{revolution}}} = 360\pi \, \frac{\text{radians}}{\text{minute}}$$

From equation (11), the linear speed v of the rock is

$$v = r\omega = 2 \text{ feet} \cdot 360\pi \, \frac{\text{radians}}{\text{minute}} = 720\pi \, \frac{\text{feet}}{\text{minute}} \approx 2262 \, \frac{\text{feet}}{\text{minute}}$$

The linear speed of the rock when it is released is 2262 ft/min $\approx$ 25.7 mi/hr.

NOW WORK PROBLEM **87.**

HISTORICAL FEATURE

Trigonometry was developed by Greek astronomers, who regarded the sky as the inside of a sphere, so it was natural that triangles on a sphere were investigated early (by Menelaus of Alexandria about AD 100) and that triangles in the plane were studied much later. The first book containing a systematic treatment of plane and spherical trigonometry was written by the Persian astronomer Nasîr Eddîn (about AD 1250).

Regiomontanus (1436–1476) is the person most responsible for moving trigonometry from astronomy into mathematics. His work was improved by Copernicus (1473–1543) and Copernicus's student Rhaeticus

(1514–1576). Rhaeticus's book was the first to define the six trigonometric functions as ratios of sides of triangles, although he did not give the functions their present names. Credit for this is due to Thomas Finck (1583), but Finck's notation was by no means universally accepted at the time. The notation was finally stabilized by the textbooks of Leonhard Euler (1707–1783).

Trigonometry has since evolved from its use by surveyors, navigators, and engineers to present applications involving ocean tides, the rise and fall of food supplies in certain ecologies, brain wave patterns, and many other phenomena.

2.1 EXERCISES

In Problems 1–12, draw each angle.

1. $30°$
2. $60°$
3. $135°$
4. $-120°$
5. $450°$
6. $540°$
7. $\dfrac{3\pi}{4}$
8. $\dfrac{4\pi}{3}$
9. $-\dfrac{\pi}{6}$
10. $-\dfrac{2\pi}{3}$
11. $\dfrac{16\pi}{3}$
12. $\dfrac{21\pi}{4}$

In Problems 13–24, convert each angle in degrees to radians. Express your answer as a multiple of π.

13. $30°$
14. $120°$
15. $240°$
16. $330°$
17. $-60°$
18. $-30°$
19. $180°$
20. $270°$
21. $-135°$
22. $-225°$
23. $-90°$
24. $-180°$

In Problems 25–36, convert each angle in radians to degrees.

25. $\dfrac{\pi}{3}$
26. $\dfrac{5\pi}{6}$
27. $-\dfrac{5\pi}{4}$
28. $-\dfrac{2\pi}{3}$
29. $\dfrac{\pi}{2}$
30. 4π
31. $\dfrac{\pi}{12}$
32. $\dfrac{5\pi}{12}$
33. $-\dfrac{\pi}{2}$
34. $-\pi$
35. $-\dfrac{\pi}{6}$
36. $-\dfrac{3\pi}{4}$

In Problems 37–44, s denotes the length of the arc of a circle of radius r subtended by the central angle θ. Find the missing quantity. Round answers to three decimal places.

37. $r = 10$ meters, $\theta = \dfrac{1}{2}$ radian, $s = ?$

38. $r = 6$ feet, $\theta = 2$ radians, $s = ?$

39. $\theta = \dfrac{1}{3}$ radian, $s = 2$ feet, $r = ?$

40. $\theta = \dfrac{1}{4}$ radian, $s = 6$ centimeters, $r = ?$

41. $r = 5$ miles, $s = 3$ miles, $\theta = ?$

42. $r = 6$ meters, $s = 8$ meters, $\theta = ?$

43. $r = 2$ inches, $\theta = 30°$, $s = ?$

44. $r = 3$ meters, $\theta = 120°$, $s = ?$

In Problems 45–52, A denotes the area of the sector of a circle of radius r formed by the central angle θ. Find the missing quantity. Round answers to three decimal places.

45. $r = 10$ meters, $\theta = \dfrac{1}{2}$ radian, $A = ?$

46. $r = 6$ feet, $\theta = 2$ radians, $A = ?$

47. $\theta = \dfrac{1}{3}$ radian, $A = 2$ square feet, $r = ?$

48. $\theta = \dfrac{1}{4}$ radian, $A = 6$ square centimeters, $r = ?$

49. $r = 5$ miles, $A = 3$ square miles, $\theta = ?$

50. $r = 6$ meters, $A = 8$ square meters, $\theta = ?$

51. $r = 2$ inches, $\theta = 30°$, $A = ?$

52. $r = 3$ meters, $\theta = 120°$, $A = ?$

In Problems 53–56, find the length s and area A. Round answers to three decimal places.

53.

54.

55.

56.

In Problems 57–62, convert each angle in degrees to radians. Express your answer in decimal form, rounded to two decimal places.
57. $17°$ **58.** $73°$ **59.** $-40°$ **60.** $-51°$ **61.** $125°$ **62.** $350°$

In Problems 63–68, convert each angle in radians to degrees. Express your answer in decimal form, rounded to two decimal places.
63. 3.14 **64.** 0.75 **65.** 2 **66.** 3 **67.** 6.32 **68.** $\sqrt{2}$

In Problems 69–74, convert each angle to a decimal in degrees. Round your answer to two decimal places.
69. $40°10'25''$ **70.** $61°42'21''$ **71.** $1°2'3''$ **72.** $73°40'40''$ **73.** $9°9'9''$ **74.** $98°22'45''$

In Problems 75–80, convert each angle to D°M'S'' form. Round your answer to the nearest second.
75. $40.32°$ **76.** $61.24°$ **77.** $18.255°$ **78.** $29.411°$ **79.** $19.99°$ **80.** $44.01°$

---

81. Minute Hand of a Clock The minute hand of a clock is 6 inches long. How far does the tip of the minute hand move in 15 minutes? How far does it move in 25 minutes?

82. Movement of a Pendulum A pendulum swings through an angle of $20°$ each second. If the pendulum is 40 inches long, how far does its tip move each second?

83. Area of a Sector Find the area of the sector of a circle of radius 4 meters formed by an angle of $45°$. Round the answer to two decimal places.

84. Area of a Sector Find the area of the sector of a circle of radius 3 centimeters formed by an angle of $60°$. Round the answer to two decimal places.

85. Watering a Lawn A water sprinkler sprays water over a distance of 30 feet while rotating through an angle of $135°$. What area of lawn receives water?

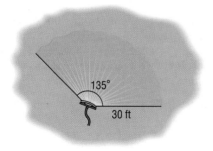

86. Designing a Water Sprinkler An engineer is asked to design a water sprinkler that will cover a field of 100 square yards that is in the shape of a sector of a circle of radius 50 yards. Through what angle should the sprinkler rotate?

87. Motion on a Circle An object is traveling around a circle with a radius of 5 centimeters. If in 20 seconds a central angle of $\frac{1}{3}$ radian is swept out, what is the angular speed of the object? What is its linear speed?

88. Motion on a Circle An object is traveling around a circle with a radius of 2 meters. If in 20 seconds the object travels 5 meters, what is its angular speed? What is its linear speed?

89. Bicycle Wheels The diameter of each wheel of a bicycle is 26 inches. If you are traveling at a speed of 35 miles per hour on this bicycle, through how many revolutions per minute are the wheels turning?

90. Car Wheels The radius of each wheel of a car is 15 inches. If the wheels are turning at the rate of 3 revolutions per second, how fast is the car moving? Express your answer in inches per second and in miles per hour.

In Problems 91–94, the latitude L of a location is the angle formed by a ray drawn from the center of Earth to the Equator and a ray drawn from the center of Earth to L. See the figure.

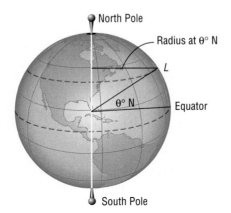

91. **Distance between Cities**   Memphis, Tennessee, is due north of New Orleans, Louisiana. Find the distance between Memphis (35°9′ north latitude) and New Orleans (29°57′ north latitude). Assume that the radius of Earth is 3960 miles.

92. **Distance between Cities**   Charleston, West Virginia, is due north of Jacksonville, Florida. Find the distance between Charleston (38°21′ north latitude) and Jacksonville (30°20′ north latitude). Assume that the radius of Earth is 3960 miles.

93. **Linear Speed on Earth**   Earth rotates on an axis through its poles. The distance from the axis to a location on Earth 30° north latitude is about 3429.5 miles. Therefore, a location on Earth at 30° north latitude is spinning on a circle of radius 3429.5 miles. Compute the linear speed on the surface of Earth at 30° north latitude.

94. **Linear Speed on Earth**   Earth rotates on an axis through its poles. The distance from the axis to a location on Earth 40° north latitude is about 3033.5 miles. Therefore, a location on Earth at 40° north latitude is spinning on a circle of radius 3033.5 miles. Compute the linear speed on the surface of Earth at 40° north latitude.

95. **Speed of the Moon**   The mean distance of the Moon from Earth is 2.39×10^5 miles. Assuming that the orbit of the Moon around Earth is circular and that 1 revolution takes 27.3 days, find the linear speed of the Moon. Express your answer in miles per hour.

96. **Speed of Earth**   The mean distance of Earth from the Sun is 9.29×10^7 miles. Assuming that the orbit of Earth around the Sun is circular and that 1 revolution takes 365 days, find the linear speed of Earth. Express your answer in miles per hour.

97. **Pulleys**   Two pulleys, one with radius 2 inches and the other with radius 8 inches, are connected by a belt.

(See the figure.) If the 2-inch pulley is caused to rotate at 3 revolutions per minute, determine the revolutions per minute of the 8-inch pulley.
[**Hint:** The linear speeds of the pulleys, that is, the speed of the belt, are the same.]

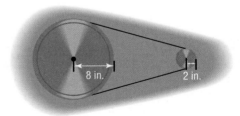

98. **Ferris Wheels**   A neighborhood carnival has a Ferris wheel whose radius is 30 feet. You measure the time it takes for one revolution to be 70 seconds. What is the linear speed (in feet per second) of this Ferris wheel? What is the angular speed in radians per second?

99. **Computing the Speed of a River Current**   To approximate the speed of the current of a river, a circular paddle wheel with radius 4 feet is lowered into the water. If the current causes the wheel to rotate at a speed of 10 revolutions per minute, what is the speed of the current? Express your answer in miles per hour.

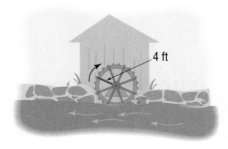

100. **Spin Balancing Tires**   A spin balancer rotates the wheel of a car at 480 revolutions per minute. If the diameter of the wheel is 26 inches, what road speed is being tested? Express your answer in miles per hour. At how many revolutions per minute should the balancer be set to test a road speed of 80 miles per hour?

101. **The Cable Cars of San Francisco**   At the Cable Car Museum you can see the four cable lines that are used to pull cable cars up and down the hills of San Francisco. Each cable travels at a speed of 9.55 miles per hour, caused by a rotating wheel whose diameter is 8.5 feet. How fast is the wheel rotating? Express your answer in revolutions per minute.

102. **Difference in Time of Sunrise**   Naples, Florida, is approximately 90 miles due west of Ft. Lauderdale. How much sooner would a person in Ft. Lauderdale first see the rising Sun than a person in Naples?

[**Hint:** Consult the figure. When a person at Q sees the first rays of the Sun, a person at P is still in the dark. The person at P sees the first rays after Earth has rotated so that P is at the location Q. Now use the fact that at the latitude of Ft. Lauderdale in 24 hours a length of arc of $2\pi(3559)$ miles is subtended.]

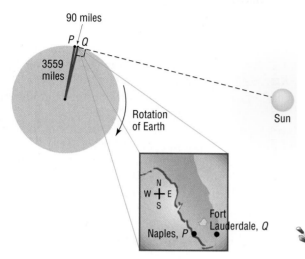

103. Keeping Up with the Sun   How fast would you have to travel on the surface of Earth at the equator to keep up with the Sun (that is, so that the Sun would appear to remain in the same position in the sky)?

104. Nautical Miles   A **nautical mile** equals the length of arc subtended by a central angle of 1 minute on a great circle* on the surface of Earth. (See the figure.) If the ra-

dius of Earth is taken as 3960 miles, express 1 nautical mile in terms of ordinary, or **statute,** miles.

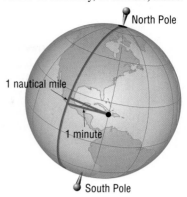

105. Pulleys   Two pulleys, one with radius r_1 and the other with radius r_2, are connected by a belt. The pulley with radius r_1 rotates at ω_1 revolutions per minute, whereas the pulley with radius r_2 rotates at ω_2 revolutions per minute. Show that $r_1/r_2 = \omega_2/\omega_1$.

106. Do you prefer to measure angles using degrees or radians? Provide justification and a rationale for your choice.

107. Discuss why ships and airplanes use nautical miles to measure distance. Explain the difference between a nautical mile and a statute mile.

108. Investigate the way that speed bicycles work. In particular, explain the differences and similarities between 5-speed and 9-speed derailleurs. Be sure to include a discussion of linear speed and angular speed.

*Any circle drawn on the surface of Earth that divides Earth into two equal hemispheres.

PREPARING FOR THIS SECTION

Before getting started, review the following:

✓ Pythagorean Theorem (Appendix A, Section A.2, pp. 497–502)   ✓ Symmetry (Section 1.2, pp. 12–13)

✓ Unit Circle (Section 1.2, p. 16)   ✓ Function Notation (Section 1.3, pp. 25–27)

2.2   TRIGONOMETRIC FUNCTIONS: UNIT CIRCLE APPROACH

OBJECTIVES   **1**   Find the Exact Values of the Trigonometric Functions Using a Point on the Unit Circle

2   Find the Exact Values of the Trigonometric Functions of Quadrantal Angles

3   Find the Exact Values of the Trigonometric Functions of $\dfrac{\pi}{4} = 45°$

4   Find the Exact Values of the Trigonometric Functions of $\dfrac{\pi}{6} = 30°$ and $\dfrac{\pi}{3} = 60°$

5   Find Exact Values for Certain Integral Multiples of $\dfrac{\pi}{6} = 30°$, $\dfrac{\pi}{4} = 45°$, and $\dfrac{\pi}{3} = 60°$

6   Use a Calculator to Approximate the Value of a Trigonometric Function

We are now ready to introduce trigonometric functions. The approach that we take uses the unit circle.

THE UNIT CIRCLE

Recall that the unit circle is a circle whose radius is 1 and whose center is at the origin of a rectangular coordinate system. Also recall that any circle of radius r has circumference of length $2\pi r$. Therefore, the unit circle (radius $= 1$) has a circumference of length 2π. In other words, for 1 revolution around the unit circle the length of the arc is 2π units.

The following discussion sets the stage for defining the trigonometric functions.

Let $t \geq 0$ be any real number and let s be the distance from the origin to t on the real number line. See the red portion of Figure 18(a). Now look at the unit circle in Figure 18(a). Beginning at the the point $(1, 0)$ on the unit circle, travel $s = t$ units in the counterclockwise direction along the circle, to arrive at the point $P = (x, y)$. In this sense, the length $s = t$ units is being **wrapped** around the unit circle.

If $t < 0$, we begin at the point $(1, 0)$ on the unit circle and travel $s = |t|$ units in the clockwise direction to arrive at the point $P = (x, y)$. See Figure 18(b).

Figure 18

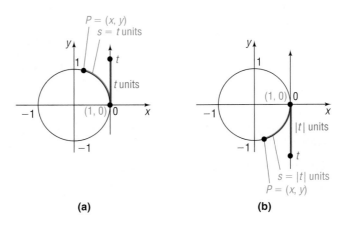

(a)   (b)

If $t > 2\pi$ or if $t < -2\pi$, it will be necessary to travel around the unit circle more than once before arriving at point P. Do you see why?

Let's describe this process another way. Picture a string of length $s = |t|$ units being wrapped around a circle of radius 1 unit. We start wrapping the string around the circle at the point $(1, 0)$. If $t \geq 0$, we wrap the string in the counterclockwise direction; if $t < 0$, we wrap the string in the clockwise direction. The point $P = (x, y)$ is the point where the string ends.

This discussion tells us that, for any real number t, we can locate a unique point $P = (x, y)$ on the unit circle. We call P **the point on the unit circle that corresponds to t.** This is the important idea here. No matter what real number t is chosen, there is a unique point P on the unit circle corresponding to it. We use the coordinates of the point $P = (x, y)$ on the unit circle corresponding to the real number t to define the **six trigonometric functions of t.**

Let t be a real number and let $P = (x, y)$ be the point on the unit circle that corresponds to t.

The **sine function** associates with t the y-coordinate of P and is denoted by

$$\sin t = y$$

The **cosine function** associates with t the x-coordinate of P and is denoted by

$$\cos t = x$$

If $x \neq 0$, the **tangent function** is defined as

$$\tan t = \frac{y}{x}$$

If $y \neq 0$, the **cosecant function** is defined as

$$\csc t = \frac{1}{y}$$

If $x \neq 0$, the **secant function** is defined as

$$\sec t = \frac{1}{x}$$

If $y \neq 0$, the **cotangent function** is defined as

$$\cot t = \frac{x}{y}$$

Notice in these definitions that if $x = 0$, that is, if the point $P = (0, y)$ is on the y-axis, then the tangent function and the secant function are undefined. Also, if $y = 0$, that is, if the point $P = (x, 0)$ is on the x-axis, then the cosecant function and the cotangent function are undefined.

Because we use the unit circle in these definitions of the trigonometric functions, they are also sometimes referred to as **circular functions.**

EXAMPLE 1	Finding the Values of the Six Trigonometric Functions Using a Point on the Unit Circle

① Let t be a real number and let $P = \left(-\dfrac{1}{2}, \dfrac{\sqrt{3}}{2} \right)$ be the point on the unit circle that corresponds to t. Find the values of $\sin t, \cos t, \tan t, \csc t, \sec t,$ and $\cot t$.

Solution See Figure 19. We follow the definition of the six trigonometric functions, using $P = \left(-\dfrac{1}{2}, \dfrac{\sqrt{3}}{2} \right) = (x, y)$. Then, with $x = -\dfrac{1}{2}, y = \dfrac{\sqrt{3}}{2}$, we have

Figure 19

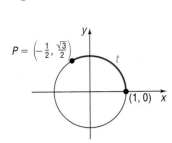

$$\sin t = y = \frac{\sqrt{3}}{2} \qquad \cos t = x = -\frac{1}{2} \qquad \tan t = \frac{y}{x} = \frac{\dfrac{\sqrt{3}}{2}}{-\dfrac{1}{2}} = -\sqrt{3}$$

$$\csc t = \frac{1}{y} = \frac{1}{\dfrac{\sqrt{3}}{2}} = \frac{2\sqrt{3}}{3} \qquad \sec t = \frac{1}{x} = \frac{1}{-\dfrac{1}{2}} = -2 \qquad \cot t = \frac{x}{y} = \frac{-\dfrac{1}{2}}{\dfrac{\sqrt{3}}{2}} = -\frac{\sqrt{3}}{3}$$ ∎

NOW WORK PROBLEM 1.

TRIGONOMETRIC FUNCTIONS OF ANGLES

Let $P = (x, y)$ be the point on the unit circle corresponding to the real number t. See Figure 20(a). Let θ be the angle in standard position, measured in radians, whose terminal side is the ray from the origin through P. See Figure 20(b). Since the unit circle has radius 1 unit, from the formula for arc length, $s = r\theta$, we find that

$$s = r\theta = \theta$$
$$\uparrow$$
$$r = 1$$

Figure 20   So, if $s = |t|$ units, then $\theta = t$ radians. See Figures 20(c) and (d).

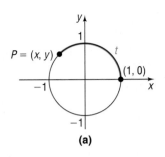

(a)

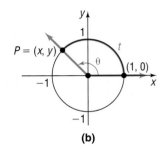

(b)

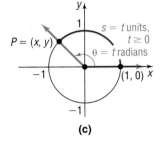

(c)

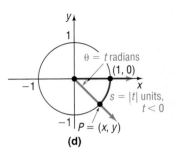

(d)

The point $P = (x, y)$ on the unit circle that corresponds to the real number t is the point P on the terminal side of the angle $\theta = t$ radians. As a result, we can say that

$$\sin t = \sin \theta$$
$$\uparrow \qquad \uparrow$$
Real number $\theta = t$ radians

and so on. We can now define the trigonometric functions of the angle θ.

If $\theta = t$ radians, the **six trigonometric functions of the angle θ** are defined as

$$\sin \theta = \sin t \qquad \cos \theta = \cos t \qquad \tan \theta = \tan t$$
$$\csc \theta = \csc t \qquad \sec \theta = \sec t \qquad \cot \theta = \cot t$$

Even though the distinction between trigonometric functions of real numbers and trigonometric functions of angles is important, it is customary to refer to trigonometric functions of real numbers and trigonometric functions of angles collectively as the *trigonometric functions*. We shall follow this practice from now on.

If an angle θ is measured in degrees, we shall use the degree symbol when writing a trigonometric function of θ, as, for example, in $\sin 30°$ and $\tan 45°$. If an angle θ is measured in radians, then no symbol is used when writing a trigonometric function of θ, as, for example, in $\cos \pi$ and $\sec \dfrac{\pi}{3}$.

Finally, since the values of the trigonometric functions of an angle θ are determined by the coordinates of the point $P = (x, y)$ on the unit circle corresponding to θ, the units used to measure the angle θ are irrelevant. For example, it does not matter whether we write $\theta = \dfrac{\pi}{2}$ radians or $\theta = 90°$. The point on the unit circle corresponding to this angle is $P = (0, 1)$. Hence,

$$\sin \frac{\pi}{2} = \sin 90° = 1 \quad \text{and} \quad \cos \frac{\pi}{2} = \cos 90° = 0$$

EVALUATING THE TRIGONOMETRIC FUNCTIONS

To find the exact value of a trigonometric function of an angle θ or a real number t requires that we locate the point $P = (x, y)$ on the unit circle that corresponds to t. This is not always easy to do. In the examples that follow, we will evaluate the trigonometric functions of certain angles or real numbers for which this process is relatively easy. A calculator will be used to evaluate the trigonometric functions of most other angles.

EXAMPLE 2

Finding the Exact Values of the Six Trigonometric Functions of Quadrantal Angles

Find the exact values of the six trigonometric functions of:

(a) $\theta = 0 = 0°$

(b) $\theta = \dfrac{\pi}{2} = 90°$

(c) $\theta = \pi = 180°$

(d) $\theta = \dfrac{3\pi}{2} = 270°$

Figure 21(a)

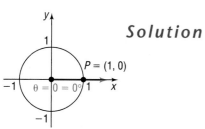

Solution   (a) The point on the unit circle that corresponds to $\theta = 0 = 0°$ is $P = (1, 0)$. See Figure 21(a). Then

$$\sin 0 = \sin 0° = y = 0 \qquad \cos 0 = \cos 0° = x = 1$$
$$\tan 0 = \tan 0° = \frac{y}{x} = 0 \qquad \sec 0 = \sec 0° = \frac{1}{x} = 1$$

Since the y-coordinate of P is 0, $\csc 0$ and $\cot 0$ are not defined.

Figure 21(b)

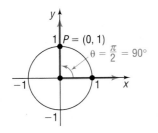

(b) The point on the unit circle that corresponds to $\theta = \dfrac{\pi}{2} = 90°$ is $P = (0, 1)$. See Figure 21(b). Then

$$\sin\frac{\pi}{2} = \sin 90° = y = 1 \qquad \cos\frac{\pi}{2} = \cos 90° = x = 0$$

$$\csc\frac{\pi}{2} = \csc 90° = \frac{1}{y} = 1 \qquad \cot\frac{\pi}{2} = \cot 90° = \frac{x}{y} = 0$$

Since the x-coordinate of P is 0, $\tan\dfrac{\pi}{2}$ and $\sec\dfrac{\pi}{2}$ are not defined.

Figure 21(c)

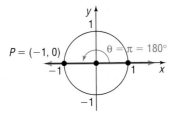

(c) The point on the unit circle that corresponds to $\theta = \pi = 180°$ is $P = (-1, 0)$. See Figure 21(c). Then

$$\sin\pi = \sin 180° = y = 0 \qquad \cos\pi = \cos 180° = x = -1$$

$$\tan\pi = \tan 180° = \frac{y}{x} = 0 \qquad \sec\pi = \sec 180° = \frac{1}{x} = -1$$

Since the y-coordinate of P is 0, $\csc\pi$ and $\cot\pi$ are not defined.

Figure 21(d)

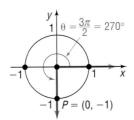

(d) The point on the unit circle that corresponds to $\theta = \dfrac{3\pi}{2} = 270°$ is $P = (0, -1)$. See Figure 21(d). Then

$$\sin\frac{3\pi}{2} = \sin 270° = y = -1 \qquad \cos\frac{3\pi}{2} = \cos 270° = x = 0$$

$$\csc\frac{3\pi}{2} = \csc 270° = \frac{1}{y} = -1 \qquad \cot\frac{3\pi}{2} = \cot 270° = \frac{x}{y} = 0$$

Since the x-coordinate of P is 0, $\tan\dfrac{3\pi}{2}$ and $\sec\dfrac{3\pi}{2}$ are not defined. ∎

Table 2 summarizes the values of the trigonometric functions found in Example 2.

	TABLE 2						
			Quadrantal Angles				
θ (Radians)	θ (Degrees)	$\sin\theta$	$\cos\theta$	$\tan\theta$	$\csc\theta$	$\sec\theta$	$\cot\theta$
0	0°	0	1	0	Not defined	1	Not defined
$\dfrac{\pi}{2}$	90°	1	0	Not defined	1	Not defined	0
π	180°	0	-1	0	Not defined	-1	Not defined
$\dfrac{3\pi}{2}$	270°	-1	0	Not defined	-1	Not defined	0

There is no need to memorize Table 2. To find the value of a trigonometric function of a quadrantal angle, draw the angle and apply the definition, as we did in Example 2.

EXAMPLE 3

Finding Exact Values of the Trigonometric Functions of Angles that Are Integral Multiples of Quadrantal Angles

Find the exact value of:

(a) $\sin(3\pi)$ (b) $\cos(-270°)$

Solution   (a) See Figure 22. The point P on the unit circle that corresponds to $\theta = 3\pi$ is $P = (-1, 0)$ so $\sin(3\pi) = 0$.

(b) See Figure 23. The point P on the unit circle that corresponds to $\theta = -270°$ is $P = (0, 1)$ so $\cos(-270°) = 0$.

Figure 22

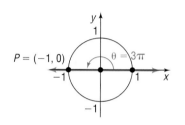

Figure 23

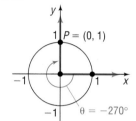

NOW WORK PROBLEMS **9** AND **53**.

③   TRIGONOMETRIC FUNCTIONS OF $\dfrac{\pi}{4} = 45°$

EXAMPLE 4

Finding the Exact Values of the Trigonometric Functions of $\dfrac{\pi}{4} = 45°$

Find the exact values of the six trigonometric functions of $\dfrac{\pi}{4} = 45°$.

Solution   We seek the coordinates of the point $P = (x, y)$ on the unit circle that corresponds to $\theta = \dfrac{\pi}{4} = 45°$. See Figure 24. First, we observe that P lies on the line $y = x$. (Do you see why? Since $\theta = 45° = \dfrac{1}{2} \cdot 90°$, P must lie on the line that bisects quadrant I.) Since $P = (x, y)$ also lies on the unit circle, $x^2 + y^2 = 1$, it follows that

Figure 24

$\theta = \dfrac{\pi}{4} = 45°$

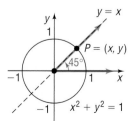

$$x^2 + y^2 = 1 \quad y = x, x > 0, y > 0$$

$$x^2 + x^2 = 1$$

$$2x^2 = 1$$

$$x = \dfrac{1}{\sqrt{2}} = \dfrac{\sqrt{2}}{2}, \qquad y = \dfrac{\sqrt{2}}{2}$$

Then,

$$\sin\frac{\pi}{4} = \sin 45° = \frac{\sqrt{2}}{2} \qquad \cos\frac{\pi}{4} = \cos 45° = \frac{\sqrt{2}}{2} \qquad \tan\frac{\pi}{4} = \tan 45° = \frac{\frac{\sqrt{2}}{2}}{\frac{\sqrt{2}}{2}} = 1$$

$$\csc\frac{\pi}{4} = \csc 45° = \frac{1}{\frac{\sqrt{2}}{2}} = \sqrt{2} \qquad \sec\frac{\pi}{4} = \sec 45° = \frac{1}{\frac{\sqrt{2}}{2}} = \sqrt{2} \qquad \cot\frac{\pi}{4} = \cot 45° = \frac{\frac{\sqrt{2}}{2}}{\frac{\sqrt{2}}{2}} = 1 \quad \blacksquare$$

EXAMPLE 5 **Finding the Exact Value of a Trigonometric Expression**

Find the exact value of each expression.

(a) $\sin 45° \cos 180°$ (b) $\tan\frac{\pi}{4} - \sin\frac{3\pi}{2}$ (c) $\left(\sec\frac{\pi}{4}\right)^2 + \csc\frac{\pi}{2}$

Solution (a) $\sin 45° \cos 180° = \frac{\sqrt{2}}{2} \cdot (-1) = -\frac{\sqrt{2}}{2}$

                      ↑   ↑
          From Example 4   From Table 2

(b) $\tan\frac{\pi}{4} - \sin\frac{3\pi}{2} = 1 - (-1) = 2$

              ↑   ↑
     From Example 4   From Table 2

(c) $\left(\sec\frac{\pi}{4}\right)^2 + \csc\frac{\pi}{2} = (\sqrt{2})^2 + 1 = 2 + 1 = 3$ $\blacksquare$

✏ **NOW WORK PROBLEM 23.**

---

④ **TRIGONOMETRIC FUNCTIONS OF $\frac{\pi}{6} = 30°$ AND $\frac{\pi}{3} = 60°$**

Consider a right triangle in which one of the angles is $\frac{\pi}{6} = 30°$. It then follows that the third angle is $\frac{\pi}{3} = 60°$. Figure 25(a) illustrates such a triangle with hypotenuse of length 1. Our problem is to determine a and b.

We begin by placing next to this triangle another triangle congruent to the first, as shown in Figure 25(b). Notice that we now have a triangle whose angles are each 60°. This triangle is therefore equilateral, so each side is of length 1. In particular, the base is $2a = 1$, and so $a = \frac{1}{2}$. By the Pythagorean Theorem, b satisfies the equation $a^2 + b^2 = c^2$, so we have

Figure 25

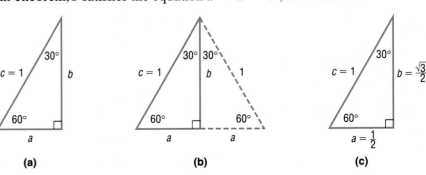

(a)   (b)   (c)

$$a^2 + b^2 = c^2 \qquad a = \tfrac{1}{2}, c = 1$$

$$\frac{1}{4} + b^2 = 1$$

$$b^2 = 1 - \frac{1}{4} = \frac{3}{4}$$

$$b = \frac{\sqrt{3}}{2}$$

This results in Figure 25(c).

EXAMPLE 6	**Finding the Exact Values of the Trigonometric Functions of $\dfrac{\pi}{3} = 60°$**

Find the exact values of the six trigonometric functions of $\dfrac{\pi}{3} = 60°$.

Solution   Position the triangle in Figure 25(c) so that the 60° angle is in the standard position. See Figure 26.

Figure 26

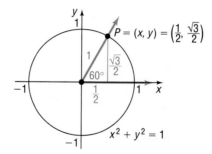

The point on the unit circle that corresponds to $\theta = \dfrac{\pi}{3} = 60°$ is $P = \left(\dfrac{1}{2}, \dfrac{\sqrt{3}}{2}\right)$. Then

$$\sin\frac{\pi}{3} = \sin 60° = \frac{\sqrt{3}}{2} \qquad\qquad \cos\frac{\pi}{3} = \cos 60° = \frac{1}{2}$$

$$\csc\frac{\pi}{3} = \csc 60° = \frac{1}{\dfrac{\sqrt{3}}{2}} = \frac{2}{\sqrt{3}} = \frac{2\sqrt{3}}{3} \qquad\qquad \sec\frac{\pi}{3} = \sec 60° = \frac{1}{\dfrac{1}{2}} = 2$$

$$\tan\frac{\pi}{3} = \tan 60° = \frac{\dfrac{\sqrt{3}}{2}}{\dfrac{1}{2}} = \sqrt{3} \qquad\qquad \cot\frac{\pi}{3} = \cot 60° = \frac{\dfrac{1}{2}}{\dfrac{\sqrt{3}}{2}} = \frac{1}{\sqrt{3}} = \frac{\sqrt{3}}{3}$$ ■

EXAMPLE 7	**Finding the Exact Values of the Trigonometric Functions of $\dfrac{\pi}{6} = 30°$**

Find the exact values of the trigonometric functions of $\dfrac{\pi}{6} = 30°$.

Solution Position the triangle in Figure 25(c) so that the 30° angle is in the standard position. See Figure 27.

The point on the unit circle that corresponds to $\theta = \dfrac{\pi}{6} = 30°$ is $P = \left(\dfrac{\sqrt{3}}{2}, \dfrac{1}{2}\right)$. Then

Figure 27

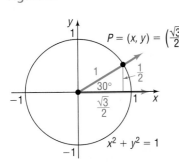

$$\sin\frac{\pi}{6} = \sin 30° = \frac{1}{2} \qquad\qquad \cos\frac{\pi}{6} = \cos 30° = \frac{\sqrt{3}}{2}$$

$$\csc\frac{\pi}{6} = \csc 30° = \frac{1}{\frac{1}{2}} = 2 \qquad\qquad \sec\frac{\pi}{6} = \sec 30° = \frac{1}{\frac{\sqrt{3}}{2}} = \frac{2}{\sqrt{3}} = \frac{2\sqrt{3}}{3}$$

$$\tan\frac{\pi}{6} = \tan 30° = \frac{\frac{1}{2}}{\frac{\sqrt{3}}{2}} = \frac{1}{\sqrt{3}} = \frac{\sqrt{3}}{3} \qquad \cot\frac{\pi}{6} = \cot 30° = \frac{\frac{\sqrt{3}}{2}}{\frac{1}{2}} = \sqrt{3}$$

Table 3 summarizes the information just derived for $\dfrac{\pi}{6} = 30°$, $\dfrac{\pi}{4} = 45°$, and $\dfrac{\pi}{3} = 60°$. Until you memorize the entries in Table 3, you should draw an appropriate diagram to determine the values given in the table.

			T A B L E 3				
θ (Radians)	θ (Degrees)	$\sin\theta$	$\cos\theta$	$\tan\theta$	$\csc\theta$	$\sec\theta$	$\cot\theta$
$\dfrac{\pi}{6}$	30°	$\dfrac{1}{2}$	$\dfrac{\sqrt{3}}{2}$	$\dfrac{\sqrt{3}}{3}$	2	$\dfrac{2\sqrt{3}}{3}$	$\sqrt{3}$
$\dfrac{\pi}{4}$	45°	$\dfrac{\sqrt{2}}{2}$	$\dfrac{\sqrt{2}}{2}$	1	$\sqrt{2}$	$\sqrt{2}$	1
$\dfrac{\pi}{3}$	60°	$\dfrac{\sqrt{3}}{2}$	$\dfrac{1}{2}$	$\sqrt{3}$	$\dfrac{2\sqrt{3}}{3}$	2	$\dfrac{\sqrt{3}}{3}$

NOW WORK PROBLEM **29.**

EXAMPLE 8

Constructing a Rain Gutter

A rain gutter is to be constructed of aluminum sheets 12 inches wide. After marking off a length of 4 inches from each edge, this length is bent up at an angle θ. See Figure 28. The area A of the opening may be expressed as a function of θ as

$$A(\theta) = 16\sin\theta(\cos\theta + 1)$$

Find the area A of the opening for $\theta = 30°$, $\theta = 45°$, and $\theta = 60°$.

Figure 28

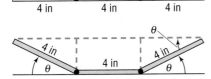

Solution For $\theta = 30°$: $A(30°) = 16\sin 30°\,(\cos 30° + 1)$

$$= 16\left(\frac{1}{2}\right)\left(\frac{\sqrt{3}}{2} + 1\right) = 4\sqrt{3} + 8$$

The area of the opening for $\theta = 30°$ is about 14.9 square inches.

For $\theta = 45°$: $A(45°) = 16 \sin 45° (\cos 45° + 1)$

$$= 16\left(\frac{\sqrt{2}}{2}\right)\left(\frac{\sqrt{2}}{2} + 1\right) = 8 + 8\sqrt{2}$$

The area of the opening for $\theta = 45°$ is about 19.3 square inches.

For $\theta = 60°$: $A(60°) = 16 \sin 60° (\cos 60° + 1)$

$$= 16\left(\frac{\sqrt{3}}{2}\right)\left(\frac{1}{2} + 1\right) = 12\sqrt{3}$$

The area of the opening for $\theta = 60°$ is about 20.8 square inches.   ■

EXACT VALUES FOR CERTAIN INTEGRAL MULTIPLES OF $\frac{\pi}{6} = 30°$, $\frac{\pi}{4} = 45°$, AND $\frac{\pi}{3} = 60°$

(5) We know the exact values of the trigonometric functions of $\frac{\pi}{4} = 45°$. Using symmetry, we can find the exact values of the trigonometric functions of $\frac{3\pi}{4} = 135°$, $\frac{5\pi}{4} = 225°$, and $\frac{7\pi}{4} = 315°$. Figure 29 shows how.

As Figure 29 shows, using symmetry with respect to the y-axis, the point $\left(-\frac{\sqrt{2}}{2}, \frac{\sqrt{2}}{2}\right)$ is the point on the unit circle that corresponds to the angle $\frac{3\pi}{4} = 135°$. Similarly, using symmetry with respect to the origin, the point $\left(-\frac{\sqrt{2}}{2}, -\frac{\sqrt{2}}{2}\right)$ is the point on the unit circle that corresponds to the angle $\frac{5\pi}{4} = 225°$. Finally, using symmetry with respect to the x-axis, the point $\left(\frac{\sqrt{2}}{2}, -\frac{\sqrt{2}}{2}\right)$ is the point on the unit circle that corresponds to the angle $\frac{7\pi}{4} = 315°$.

Figure 29

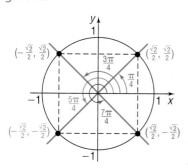

EXAMPLE 9

Finding Exact Values for Multiples of $\frac{\pi}{4} = 45°$

Based on Figure 29, we see that

(a) $\sin 135° = \dfrac{\sqrt{2}}{2}$ (b) $\cos \dfrac{5\pi}{4} = -\dfrac{\sqrt{2}}{2}$ (c) $\tan 315° = \dfrac{-\dfrac{\sqrt{2}}{2}}{\dfrac{\sqrt{2}}{2}} = -1$ ■

Figure 29 can also be used to find exact values for other multiples of $\frac{\pi}{4} = 45°$. For example, the point $\left(\frac{\sqrt{2}}{2}, -\frac{\sqrt{2}}{2}\right)$ is the point on the unit circle that corresponds to the angle $-\frac{\pi}{4} = -45°$; the point $\left(\frac{\sqrt{2}}{2}, \frac{\sqrt{2}}{2}\right)$ is the point on the unit circle that corresponds to the angle $\frac{9\pi}{4} = 405°$.

NOW WORK PROBLEMS **43** AND **47**.

The use of symmetry also provides information about certain integral multiples of the angles $\frac{\pi}{6} = 30°$ and $\frac{\pi}{3} = 60°$. See Figures 30 and 31.

Figure 30

Figure 31

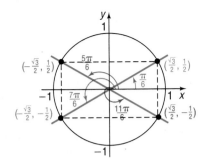

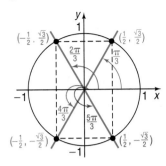

EXAMPLE 10

Finding Exact Values for Multiples of $\dfrac{\pi}{6} = 30°$ and $\dfrac{\pi}{3} = 60°$

Based on Figures 30 and 31, we see that

(a) $\cos 210° = -\dfrac{\sqrt{3}}{2}$ (b) $\sin(-60°) = -\dfrac{\sqrt{3}}{2}$

(c) $\tan \dfrac{5\pi}{3} = \dfrac{-\dfrac{\sqrt{3}}{2}}{\dfrac{1}{2}} = -\sqrt{3}$

■

 NOW WORK PROBLEM **39**.

USING A CALCULATOR TO FIND VALUES OF TRIGONOMETRIC FUNCTIONS

6 Before getting started, you must first decide whether to enter the angle in the calculator using radians or degrees and then set the calculator to the correct MODE.* Check your instruction manual to find out how your calculator handles degrees and radians. Your calculator has the keys marked $\boxed{\sin}$, $\boxed{\cos}$, and $\boxed{\tan}$. To find the values of the remaining three trigonometric functions, secant, cosecant, and cotangent, we use the fact that if $P = (x, y)$ is a point on the unit circle on the terminal side of θ, then

$$\sec\theta = \frac{1}{x} = \frac{1}{\cos\theta} \qquad \csc\theta = \frac{1}{y} = \frac{1}{\sin\theta} \qquad \cot\theta = \frac{x}{y} = \frac{1}{\dfrac{y}{x}} = \frac{1}{\tan\theta}$$

EXAMPLE 11

Using a Calculator to Approximate the Value of a Trigonometric Function

Use a calculator to find the approximate value of:

(a) $\cos 48°$ (b) $\csc 21°$ (c) $\tan \dfrac{\pi}{12}$

Express your answers rounded to two decimal places.

---

* If your calculator does not display the MODE, you can determine the current mode by evaluating $\boxed{\sin}\ \boxed{30}$. If you are in the degree mode, the display will show $\boxed{0.5}$ ($\sin 30° = 0.5$). If you are in the radian mode, the display will show $\boxed{-0.9880316}$.

Solution   (a) Set the mode to receive degrees.

Keystrokes:*

Display:

Then
$$\cos 48° = 0.66991306 = 0.67$$

rounded to two decimal places.

(b) Most calculators do not have a csc key. The manufacturers assume that the user knows some trigonometry. To find the value of csc 21°, we use the fact that $\csc 21° = \dfrac{1}{\sin 21°}$ and proceed as follows:

Keystrokes: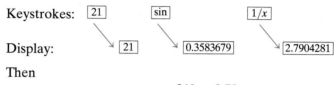

Display:

Then
$$\csc 21° = 2.79$$

rounded to two decimal places.

Figure 32

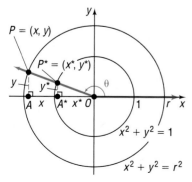

(c) Set the MODE to receive radians. Figure 32 shows the solution using a TI-83 graphing calculator. Then

$$\tan \frac{\pi}{12} = 0.27$$

rounded to two decimal places.

NOW WORK PROBLEM **57.**

USING A CIRCLE OF RADIUS r TO EVALUATE THE TRIGONOMETRIC FUNCTIONS

Until now, to find the exact value of a trigonometric function of an angle θ required that we locate the corresponding point $P = (x, y)$ on the unit circle. In fact, though, any circle whose center is at the origin can be used.

Let θ be any nonquadrantal angle placed in standard position. Let $P = (x, y)$ be the point on the circle $x^2 + y^2 = r^2$ that corresponds to θ. See Figure 33.

Figure 33

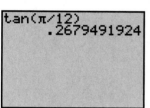

Notice that the triangles $OA*P*$ and OAP are similar; as a result, the ratios of corresponding sides are equal.

$$\frac{y*}{1} = \frac{y}{r} \qquad \frac{x*}{1} = \frac{x}{r} \qquad \frac{y*}{x*} = \frac{y}{x}$$

$$\frac{1}{y*} = \frac{r}{y} \qquad \frac{1}{x*} = \frac{r}{x} \qquad \frac{x*}{y*} = \frac{x}{y}$$

These results lead us to formulate the following theorem:

---

*On some calculators, the function key is pressed first, followed by the angle. Consult your manual.

Theorem   For an angle θ in standard position, let $P = (x, y)$ be the point on the terminal side of θ that is also on the circle $x^2 + y^2 = r^2$. Then

$$\sin\theta = \frac{y}{r} \qquad\qquad \cos\theta = \frac{x}{r} \qquad\qquad \tan\theta = \frac{y}{x}, \quad x \neq 0$$

$$\csc\theta = \frac{r}{y}, \quad y \neq 0 \qquad \sec\theta = \frac{r}{x}, \quad x \neq 0 \qquad \cot\theta = \frac{x}{y}, \quad y \neq 0$$

EXAMPLE 12   **Finding the Exact Values of the Six Trigonometric Functions**

Find the exact values of each of the six trigonometric functions of an angle θ if $(4, -3)$ is a point on its terminal side.

Figure 34

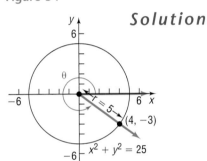

Solution   The point $(4, -3)$ is on a circle of radius $r = \sqrt{4^2 + (-3)^2} = \sqrt{16 + 9} = \sqrt{25} = 5$ with the center at the origin. For the point $(x, y) = (4, -3)$, we have $x = 4$ and $y = -3$. Since $r = 5$, we find

$$\sin\theta = \frac{y}{r} = -\frac{3}{5} \qquad \cos\theta = \frac{x}{r} = \frac{4}{5} \qquad \tan\theta = \frac{y}{x} = -\frac{3}{4}$$

$$\csc\theta = \frac{r}{y} = -\frac{5}{3} \qquad \sec\theta = \frac{r}{x} = \frac{5}{4} \qquad \cot\theta = \frac{x}{y} = -\frac{4}{3}$$

✏️ NOW WORK PROBLEM **73**.

HISTORICAL FEATURE

The name *sine* for the sine function is due to a medieval confusion. The name comes from the Sanskrit word *jīva* (meaning chord), first used in India by Araybhata the Elder (AD 510). He really meant half-chord, but abbreviated it. This was brought into Arabic as *jība*, which was meaningless. Because the proper Arabic word *jaib* would be written the same way (short vowels are not written out in Arabic), *jība* was pronounced as *jaib*, which meant bosom or hollow, and *jaib* remains as the Arabic word for sine to this day. Scholars translating the Arabic works into Latin found that the word *sinus* also meant bosom or hollow, and from *sinus* we get the word *sine*.

The name *tangent*, due to Thomas Finck (1583), can be understood by looking at Figure 35. The line segment $\overline{DC}$ is tangent to the circle at C. If $d(O, B) = d(O, C) = 1$, then the length of the line segment $\overline{DC}$ is

$$d(D, C) = \frac{d(D, C)}{1} = \frac{d(D, C)}{d(O, C)} = \tan\alpha$$

The old name for the tangent is *umbra versa* (meaning turned shadow), referring to the use of the tangent in solving height problems with shadows.

The names of the remaining functions came about as follows. If α and β are complementary angles, then $\cos\alpha = \sin\beta$. Because β is the complement of α, it was natural to write the cosine of α as *sin co α*. Probably for reasons involving ease of pronunciation, the *co* migrated to the front, and then cosine received a three-letter abbreviation to match sin, sec, and tan. The two other cofunctions were similarly treated, except that the long forms *cotan* and *cosec* survive to this day in some countries.

Figure 35

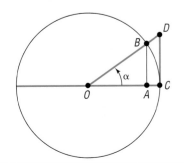

2.2 EXERCISES

In Problems 1–8, t is a real number and P = (x, y) is the point on the unit circle that corresponds to t. Find the exact values of the six trigonometric functions of t.

1. $\left(\dfrac{\sqrt{3}}{2}, \dfrac{1}{2}\right)$

2. $\left(\dfrac{1}{2}, -\dfrac{\sqrt{3}}{2}\right)$

3. $\left(-\dfrac{2}{5}, \dfrac{\sqrt{21}}{5}\right)$

4. $\left(-\dfrac{1}{5}, \dfrac{2\sqrt{6}}{5}\right)$

5. $\left(-\dfrac{\sqrt{2}}{2}, \dfrac{\sqrt{2}}{2}\right)$

6. $\left(\dfrac{\sqrt{2}}{2}, \dfrac{\sqrt{2}}{2}\right)$

7. $\left(\dfrac{2\sqrt{2}}{3}, -\dfrac{1}{3}\right)$

8. $\left(-\dfrac{\sqrt{5}}{3}, -\dfrac{2}{3}\right)$

In Problems 9–18, find the exact value. Do not use a calculator.

9. $\sin \dfrac{11\pi}{2}$

10. $\cos(7\pi)$

11. $\tan(6\pi)$

12. $\cot \dfrac{7\pi}{2}$

13. $\csc \dfrac{11\pi}{2}$

14. $\sec(8\pi)$

15. $\cos\left(-\dfrac{3\pi}{2}\right)$

16. $\sin(-3\pi)$

17. $\sec(-\pi)$

18. $\tan(-3\pi)$

In Problems 19–38, find the exact value of each expression. Do not use a calculator.

19. $\sin 45° + \cos 60°$

20. $\sin 30° - \cos 45°$

21. $\sin 90° + \tan 45°$

22. $\cos 180° - \sin 180°$

23. $\sin 45° \cos 45°$

24. $\tan 45° \cos 30°$

25. $\csc 45° \tan 60°$

26. $\sec 30° \cot 45°$

27. $4 \sin 90° - 3 \tan 180°$

28. $5 \cos 90° - 8 \sin 270°$

29. $2 \sin \dfrac{\pi}{3} - 3 \tan \dfrac{\pi}{6}$

30. $2 \sin \dfrac{\pi}{4} + 3 \tan \dfrac{\pi}{4}$

31. $\sin \dfrac{\pi}{4} - \cos \dfrac{\pi}{4}$

32. $\tan \dfrac{\pi}{3} + \cos \dfrac{\pi}{3}$

33. $2 \sec \dfrac{\pi}{4} + 4 \cot \dfrac{\pi}{3}$

34. $3 \csc \dfrac{\pi}{3} + \cot \dfrac{\pi}{4}$

35. $\tan \pi - \cos 0$

36. $\sin \dfrac{3\pi}{2} + \tan \pi$

37. $\csc \dfrac{\pi}{2} + \cot \dfrac{\pi}{2}$

38. $\sec \pi - \csc \dfrac{\pi}{2}$

In Problems 39–56, find the exact values of the six trigonometric functions of the given angle. If any are not defined, say "not defined." Do not use a calculator.

39. $\dfrac{2\pi}{3}$

40. $\dfrac{5\pi}{6}$

41. $210°$

42. $240°$

43. $\dfrac{3\pi}{4}$

44. $\dfrac{11\pi}{4}$

45. $\dfrac{8\pi}{3}$

46. $\dfrac{13\pi}{6}$

47. $405°$

48. $390°$

49. $-\dfrac{\pi}{6}$

50. $-\dfrac{\pi}{3}$

51. $-45°$

52. $-60°$

53. $\dfrac{5\pi}{2}$

54. 5π

55. $720°$

56. $630°$

In Problems 57–72, use a calculator to find the approximate value of each expression rounded to two decimal places.

57. $\sin 28°$

58. $\cos 14°$

59. $\tan 21°$

60. $\cot 70°$

61. $\sec 41°$

62. $\csc 55°$

63. $\sin \dfrac{\pi}{10}$

64. $\cos \dfrac{\pi}{8}$

65. $\tan \dfrac{5\pi}{12}$

66. $\cot \dfrac{\pi}{18}$

67. $\sec \dfrac{\pi}{12}$

68. $\csc \dfrac{5\pi}{13}$

69. $\sin 1$

70. $\tan 1$

71. $\sin 1°$

72. $\tan 1°$

In Problems 73–82, a point on the terminal side of an angle θ is given. Find the exact values of the six trigonometric functions of θ.

73. $(-3, 4)$

74. $(5, -12)$

75. $(2, -3)$

76. $(-1, -2)$

77. $(-2, -2)$

78. $(1, -1)$

79. $(-3, -2)$

80. $(2, 2)$

81. $\left(\dfrac{1}{3}, -\dfrac{1}{4}\right)$

82. $(-0.3, -0.4)$

83. Find the exact value of
$$\sin 45° + \sin 135° + \sin 225° + \sin 315°.$$

84. Find the exact value of
$$\tan 60° + \tan 150°.$$

85. If $\sin \theta = 0.1$, find $\sin(\theta + \pi)$.

86. If $\cos \theta = 0.3$, find $\cos(\theta + \pi)$.

87. If $\tan \theta = 3$, find $\tan(\theta + \pi)$.

88. If $\cot \theta = -2$, find $\cot(\theta + \pi)$.

89. If $\sin \theta = \dfrac{1}{5}$, find $\csc \theta$.

90. If $\cos \theta = \dfrac{2}{3}$, find $\sec \theta$.

In Problems 91–102, $f(\theta) = \sin \theta$ *and* $g(\theta) = \cos \theta$. *Find the exact value of each function below if* $\theta = 60°$. *Do not use a calculator.*

91. $f(\theta)$

92. $g(\theta)$

73. $f\left(\dfrac{\theta}{2}\right)$

94. $g\left(\dfrac{\theta}{2}\right)$

95. $[f(\theta)]^2$

96. $[g(\theta)]^2$

97. $f(2\theta)$

98. $g(2\theta)$

99. $2f(\theta)$

100. $2g(\theta)$

101. $f(-\theta)$

102. $g(-\theta)$

103. Use a calculator in radian mode to complete the following table.
What can you conclude about the ratio $\dfrac{\sin \theta}{\theta}$ as θ approaches 0?

θ	0.5	0.4	0.2	0.1	0.01	0.001	0.0001	0.00001
$\sin \theta$								
$\dfrac{\sin \theta}{\theta}$								

104. Use a calculator in radian mode to complete the following table.
What can you conclude about the ratio $\dfrac{\cos \theta - 1}{\theta}$ as θ approaches 0?

θ	0.5	0.4	0.2	0.1	0.01	0.001	0.0001	0.00001
$\cos \theta - 1$								
$\dfrac{\cos \theta - 1}{\theta}$								

Projectile Motion   *The path of a projectile fired at an inclination θ to the horizontal with initial speed v_0 is a parabola (see the figure).*

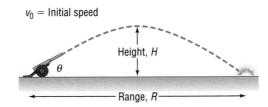

The range R of the projectile, that is, the horizontal distance that the projectile travels, is found by using the formula

$$R = \frac{v_0^2 \sin(2\theta)}{g}$$

where $g \approx 32.2$ feet per second per second ≈ 9.8 meters per second per second is the acceleration due to gravity. The maximum height H of the projectile is

$$H = \frac{v_0^2 \sin^2 \theta}{2g}$$

In Problems 105–108, find the range R and maximum height H.

105. The projectile is fired at an angle of 45° to the horizontal with an initial speed of 100 feet per second.

106. The projectile is fired at an angle of 30° to the horizontal with an initial speed of 150 meters per second.

107. The projectile is fired at an angle of 25° to the horizontal with an initial speed of 500 meters per second.

108. The projectile is fired at an angle of 50° to the horizontal with an initial speed of 200 feet per second.

109. Inclined Plane   If friction is ignored, the time t (in seconds) required for a block to slide down an inclined plane (see the figure) is given by the formula

$$t = \sqrt{\frac{2a}{g \sin \theta \cos \theta}}$$

where a is the length (in feet) of the base and $g \approx 32$ feet per second per second is the acceleration of gravity. How long does it take a block to slide down an inclined plane with base $a = 10$ feet when:
(a) $\theta = 30°$?   (b) $\theta = 45°$?   (c) $\theta = 60°$?

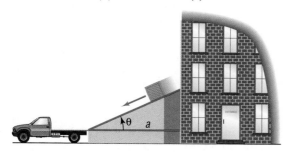

110. Piston Engines   In a certain piston engine, the distance x (in centimeters) from the center of the drive shaft to the head of the piston is given by

$$x = \cos \theta + \sqrt{16 + 0.5 \cos (2\theta)}$$

where θ is the angle between the crank and the path of the piston head (see the figure). Find x when $\theta = 30°$ and when $\theta = 45°$.

111. Calculating the Time of a Trip   Two oceanfront homes are located 8 miles apart on a straight stretch of beach, each a distance of 1 mile from a paved road that parallels the ocean. Sally can jog 8 miles per hour along the paved road, but only 3 miles per hour in the sand on the beach. Because of a river directly between the two hous-

es, it is necessary to jog in the sand to the road, continue on the road, and then jog directly back in the sand to get from one house to the other. See the illustration. The time T to get from one house to the other as a function of the angle θ shown in the illustration is

$$T(\theta) = 1 + \frac{2}{3 \sin \theta} - \frac{1}{4 \tan \theta}, \qquad 0° < \theta < 90°$$

(a) Calculate the time T for $\theta = 30°$. How long is Sally on the paved road?
(b) Calculate the time T for $\theta = 45°$. How long is Sally on the paved road?
(c) Calculate the time T for $\theta = 60°$. How long is Sally on the paved road?
(d) Calculate the time T for $\theta = 90°$. Describe the path taken. Why can't the formula for T be used?

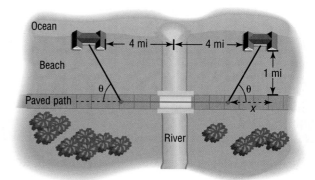

112. Designing Fine Decorative Pieces   A designer of decorative art plans to market solid gold spheres encased in clear crystal cones. Each sphere is of fixed radius R and will be enclosed in a cone of height h and radius r. See the illustration. Many cones can be used to enclose the sphere, each having a different slant angle θ. The volume V of the cone can be expressed as a function of the slant angle θ of the cone as

$$V(\theta) = \frac{1}{3} \pi R^3 \frac{(1 + \sec \theta)^3}{\tan^2 \theta}, \qquad 0° < \theta < 90°$$

What volume V is required to enclose a sphere of radius 2 centimeters in a cone whose slant angle θ is 30°? 45°? 60°?

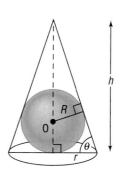

113. Projectile Motion   An object is propelled upward at an angle θ, $45° < \theta < 90°$, to the horizontal with an initial velocity of v_0 feet per second from the base of a plane that makes an angle of $45°$ with the horizontal. See the illustration. If air resistance is ignored, the distance R that it travels up the inclined plane is given by

$$R = \frac{v_0^2 \sqrt{2}}{32}\left[\sin(2\theta) - \cos(2\theta) - 1\right]$$

(a) Find the distance R that the object travels along the inclined plane if the initial velocity is 32 feet per second and $\theta = 60°$.

(b) Graph $R = R(\theta)$ if the initial velocity is 32 feet per second.

(c) What value of θ makes R largest?

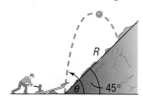

114. If θ $(0 < \theta < \pi)$ is the angle between a horizontal ray directed to the right (say, the positive x-axis) and a nonhorizontal, nonvertical line L, show that the slope m of L equals $\tan\theta$. The angle θ is called the **inclination** of L. [**Hint:** See the illustration, where we have drawn the line L^* parallel to L and passing through the origin. Use the fact that L^* intersects the unit circle at the point $(\cos\theta, \sin\theta)$.]

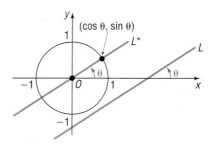

115. Write a brief paragraph that explains how to quickly compute the trigonometric functions of $30°, 45°,$ and $60°$.

116. Write a brief paragraph that explains how to quickly compute the trigonometric functions of $0°, 90°, 180°,$ and $270°$.

119. How would you explain the meaning of the sine function to a fellow student who has just completed college algebra?

PREPARING FOR THIS SECTION

Before getting started, review the following:

✓ Domain and Range of a Function (Section 1.3, pp. 23 and 28–29)

✓ Identity (Appendix A, Section A.3, p. 503)

✓ Even and Odd Functions (Section 1.4, pp. 41–42)

2.3   PROPERTIES OF THE TRIGONOMETRIC FUNCTIONS

OBJECTIVES

1　Determine the Domain and Range of the Trigonometric Functions
2　Determine the Period of the Trigonometric Functions
3　Determine the Signs of the Trigonometric Functions in a Given Quadrant
4　Find the Values of the Trigonometric Functions Utilizing Fundamental Identities
5　Use Even–Odd Properties to Find the Exact Values of the Trigonometric Functions

Figure 36

DOMAIN AND RANGE OF THE TRIGONOMETRIC FUNCTIONS

1　Let θ be an angle in standard position, and let $P = (x, y)$ be the point on the unit circle that corresponds to θ. See Figure 36. Then, by definition,

$$\sin\theta = y \qquad \cos\theta = x \qquad \tan\theta = \frac{y}{x}, \quad x \neq 0$$

$$\csc\theta = \frac{1}{y}, \quad y \neq 0 \qquad \sec\theta = \frac{1}{x}, \quad x \neq 0 \qquad \cot\theta = \frac{x}{y}, \quad y \neq 0$$

For $\sin\theta$ and $\cos\theta$, θ can be any angle, so it follows that the domain of the sine function and cosine function is the set of all real numbers.

> The domain of the sine function is the set of all real numbers.
>
> The domain of the cosine function is the set of all real numbers.

If $x = 0$, then the tangent function and the secant function are not defined. That is, for the tangent function and secant function, the x-coordinate of $P = (x, y)$ cannot be 0. On the unit circle, there are two such points $(0, 1)$ and $(0, -1)$. These two points correspond to the angles $\frac{\pi}{2}$ $(90°)$ and $\frac{3\pi}{2}$ $(270°)$ or, more generally, to any angle that is an odd multiple of $\frac{\pi}{2}$ $(90°)$, such as $\frac{\pi}{2}$ $(90°)$, $\frac{3\pi}{2}$ $(270°)$, $\frac{5\pi}{2}$ $(450°)$, $-\frac{\pi}{2}$ $(-90°)$, $-\frac{3\pi}{2}$ $(-270°)$, and so on. Such angles must therefore be excluded from the domain of the tangent function and secant function.

> The domain of the tangent function is the set of all real numbers, except odd multiples of $\frac{\pi}{2}$ $(90°)$.
>
> The domain of the secant function is the set of all real numbers, except odd multiples of $\frac{\pi}{2}$ $(90°)$.

If $y = 0$, then the cotangent function and the cosecant function are not defined. For the cotangent function and cosecant function, the y-coordinate of $P = (x, y)$ cannot be 0. On the unit circle, there are two such points, $(1, 0)$ and $(-1, 0)$. These two points correspond to the angles $0(0°)$ and $\pi(180°)$ or, more generally, to any angle that is an integral multiple of $\pi(180°)$, such as $0(0°)$, $\pi(180°)$, $2\pi(360°)$, $3\pi(540°)$, $-\pi(-180°)$, and so on. Such angles must therefore be excluded from the domain of the cotangent function and cosecant function.

> The domain of the cotangent function is the set of all real numbers, except integral multiples of $\pi(180°)$.
>
> The domain of the cosecant function is the set of all real numbers, except integral multiples of $\pi(180°)$.

Next, we determine the range of each of the six trigonometric functions. Refer again to Figure 36. Let $P = (x, y)$ be the point on the unit circle that corresponds to the angle θ. It follows that $-1 \le x \le 1$ and $-1 \le y \le 1$. Consequently, since $\sin\theta = y$ and $\cos\theta = x$, we have

> $$-1 \le \sin\theta \le 1 \qquad -1 \le \cos\theta \le 1$$

The range of both the sine function and the cosine function consists of all real numbers between -1 and 1, inclusive. Using absolute value notation, we have $|\sin\theta| \le 1$ and $|\cos\theta| \le 1$.

Similarly, if θ is not a multiple of $\pi(180°)$, then $\csc\theta = \dfrac{1}{y}$. Since $y = \sin\theta$ and $|y| = |\sin\theta| \le 1$, it follows that $|\csc\theta| = \dfrac{1}{|\sin\theta|} = \dfrac{1}{|y|} \ge 1$. The range of the cosecant function consists of all real numbers less than or equal to -1 or greater than or equal to 1. That is,

$$\csc\theta \le -1 \quad \text{or} \quad \csc\theta \ge 1$$

Using absolute value notation, we have $|\csc\theta| \ge 1$.

If θ is not an odd multiple of $\dfrac{\pi}{2}$ (90°), then, by definition, $\sec\theta = \dfrac{1}{x}$. Since $x = \cos\theta$ and $|x| = |\cos\theta| \le 1$, it follows that $|\sec\theta| = \dfrac{1}{|\cos\theta|} = \dfrac{1}{|x|} \ge 1$. The range of the secant function consists of all real numbers less than or equal to -1 or greater than or equal to 1. That is,

$$\sec\theta \le -1 \quad \text{or} \quad \sec\theta \ge 1$$

Using absolute value notation, we have $|\sec\theta| \ge 1$.

The range of both the tangent function and the cotangent function consists of all real numbers. You are asked to prove this in Problems 111 and 112.

$$-\infty < \tan\theta < \infty \qquad -\infty < \cot\theta < \infty$$

Table 4 summarizes these results.

TABLE 4

Function	Symbol	Domain	Range
sine	$f(\theta) = \sin\theta$	All real numbers	All real numbers from -1 to 1, inclusive
cosine	$f(\theta) = \cos\theta$	All real numbers	All real numbers from -1 to 1, inclusive
tangent	$f(\theta) = \tan\theta$	All real numbers, except odd multiples of $\dfrac{\pi}{2}$ (90°)	All real numbers
cosecant	$f(\theta) = \csc\theta$	All real numbers, except integral multiples of π (180°)	All real numbers greater than or equal to 1 or less than or equal to -1
secant	$f(\theta) = \sec\theta$	All real numbers, except odd multiples of $\dfrac{\pi}{2}$ (90°)	All real numbers greater than or equal to 1 or less than or equal to -1
cotangent	$f(\theta) = \cot\theta$	All real numbers, except integral multiples of π (180°)	All real numbers

Figure 37

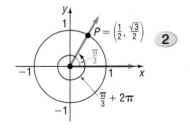

✎ NOW WORK PROBLEM **87.**

PERIOD OF THE TRIGONOMETRIC FUNCTIONS

② Look at Figure 37. This figure shows that for an angle of $\dfrac{\pi}{3}$ radians the corresponding point P on the unit circle is $\left(\dfrac{1}{2}, \dfrac{\sqrt{3}}{2}\right)$. Notice that, for an angle of $\dfrac{\pi}{3} + 2\pi$ radians, the corresponding point P on the unit circle is also $\left(\dfrac{1}{2}, \dfrac{\sqrt{3}}{2}\right)$. Then

$$\sin \frac{\pi}{3} = \frac{\sqrt{3}}{2} \quad \text{and} \quad \sin\left(\frac{\pi}{3} + 2\pi\right) = \frac{\sqrt{3}}{2}$$

$$\cos \frac{\pi}{3} = \frac{1}{2} \quad \text{and} \quad \cos\left(\frac{\pi}{3} + 2\pi\right) = \frac{1}{2}$$

This example illustrates a more general situation. For a given angle θ, measured in radians, suppose that we know the corresponding point $P = (x, y)$ on the unit circle. Now add 2π to θ. The point on the unit circle corresponding to $\theta + 2\pi$ is identical to the point P on the unit circle corresponding to θ. See Figure 38. The values of the trigonometric functions of $\theta + 2\pi$ are equal to the values of the corresponding trigonometric functions of θ.

If we add (or subtract) integral multiples of 2π to θ, the trigonometric values remain unchanged. That is, for all θ,

Figure 38

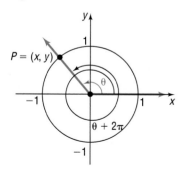

$$\sin(\theta + 2\pi k) = \sin\theta \qquad \cos(\theta + 2\pi k) = \cos\theta \qquad \textbf{(1)}$$

where k is any integer.

Functions that exhibit this kind of behavior are called *periodic functions.*

A function f is called **periodic** if there is a positive number p such that, whenever θ is in the domain of f, so is $\theta + p$, and

$$f(\theta + p) = f(\theta)$$

If there is a smallest such number p, this smallest value is called the **(fundamental) period** of f.

Based on equation (1), the sine and cosine functions are periodic. In fact, the sine and cosine functions have period 2π. You are asked to prove this fact in Problems 113 and 114. The secant and cosecant functions are also periodic with period 2π, and the tangent and cotangent functions are periodic with period π. You are asked to prove these statements in Problems 115 through 118.

These facts are summarized as follows:

Periodic Properties

$$\sin(\theta + 2\pi k) = \sin\theta \quad \cos(\theta + 2\pi k) = \cos\theta \quad \tan(\theta + \pi k) = \tan\theta$$

$$\csc(\theta + 2\pi k) = \csc\theta \quad \sec(\theta + 2\pi k) = \sec\theta \quad \cot(\theta + \pi k) = \cot\theta$$

where k is any integer.

Because the sine, cosine, secant, and cosecant functions have period 2π, once we know their values for $0 \leq \theta < 2\pi$, we know all their values; similarly, since the tangent and cotangent functions have period π, once we know their values for $0 \leq \theta < \pi$, we know all their values.

| EXAMPLE 1 | **Finding Exact Values Using Periodic Properties** |

Find the exact value of:

(a) $\sin \dfrac{17\pi}{4}$ (b) $\cos(5\pi)$ (c) $\tan \dfrac{5\pi}{4}$

Solution   (a) It is best to sketch the angle first, as shown in Figure 39(a). Since the period of the sine function is 2π, each full revolution can be ignored. This leaves the angle $\dfrac{\pi}{4}$. Then,

$$\sin \frac{17\pi}{4} = \sin\left(\frac{\pi}{4} + 4\pi\right) = \sin \frac{\pi}{4} = \frac{\sqrt{2}}{2}$$

(b) See Figure 39(b). Since the period of the cosine function is 2π, each full revolution can be ignored. This leaves the angle π. Then,

$$\cos(5\pi) = \cos(\pi + 4\pi) = \cos \pi = -1$$

(c) See Figure 39(c). Since the period of the tangent function is π, each half-revolution can be ignored. This leaves the angle $\dfrac{\pi}{4}$. Then,

$$\tan \frac{5\pi}{4} = \tan\left(\frac{\pi}{4} + \pi\right) = \tan \frac{\pi}{4} = 1$$

Figure 39

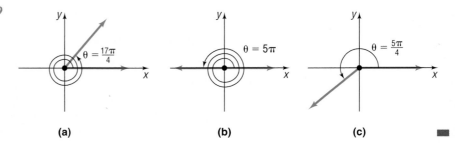

(a)   (b)   (c)

The periodic properties of the trigonometric functions will be very helpful to us when we study their graphs later in the chapter.

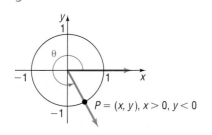 NOW WORK PROBLEM **1**.

THE SIGNS OF THE TRIGONOMETRIC FUNCTIONS

③ Let $P = (x, y)$ be the point on the unit circle that corresponds to the angle θ. If we know in which quadrant the point P lies, then we can determine the signs of the trigonometric functions of θ. For example, if $P = (x, y)$ lies in quadrant IV, as shown in Figure 40, then we know that $x > 0$ and $y < 0$. Consequently,

Figure 40

$P = (x, y)$, $x > 0$, $y < 0$

$$\sin \theta = y < 0 \qquad \cos \theta = x > 0 \qquad \tan \theta = \frac{y}{x} < 0$$

$$\csc \theta = \frac{1}{y} < 0 \qquad \sec \theta = \frac{1}{x} > 0 \qquad \cot \theta = \frac{x}{y} < 0$$

Table 5 lists the signs of the six trigonometric functions for each quadrant. See also Figure 41.

TABLE 5			
Quadrant of P	$\sin\theta$, $\csc\theta$	$\cos\theta$, $\sec\theta$	$\tan\theta$, $\cot\theta$
I	Positive	Positive	Positive
II	Positive	Negative	Negative
III	Negative	Negative	Positive
IV	Negative	Positive	Negative

Figure 41

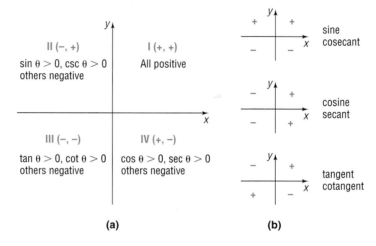

(a)   (b)

EXAMPLE 2

Finding the Quadrant in Which an Angle θ Lies

If $\sin\theta < 0$ and $\cos\theta < 0$, name the quadrant in which the angle θ lies.

Solution   Let $P = (x, y)$ be the point on the unit circle corresponding to θ. Then $\sin\theta = y < 0$ and $\cos\theta = x < 0$. The point $P = (x, y)$ must be in quadrant III, so θ lies in quadrant III. ■

NOW WORK PROBLEM **17**.

FUNDAMENTAL IDENTITIES

If $P = (x, y)$ is the point on the unit circle corresponding to θ, then

$$\sin\theta = y \qquad\qquad \cos\theta = x \qquad\qquad \tan\theta = \frac{y}{x}, \quad \text{if } x \neq 0$$

$$\csc\theta = \frac{1}{y}, \quad \text{if } y \neq 0 \qquad \sec\theta = \frac{1}{x}, \quad \text{if } x \neq 0 \qquad \cot\theta = \frac{x}{y}, \quad \text{if } y \neq 0$$

Based on these definitions, we have the **reciprocal identities:**

Reciprocal Identities

$$\csc\theta = \frac{1}{\sin\theta} \qquad \sec\theta = \frac{1}{\cos\theta} \qquad \cot\theta = \frac{1}{\tan\theta} \qquad \textbf{(2)}$$

Two other fundamental identities are the **quotient identities.**

Quotient Identities

$$\tan \theta = \frac{\sin \theta}{\cos \theta} \qquad \cot \theta = \frac{\cos \theta}{\sin \theta} \qquad\qquad \textbf{(3)}$$

The proofs of formulas (2) and (3) follow from the definitions of the trigonometric functions. (See Problems 119 and 120.)

If $\sin \theta$ and $\cos \theta$ are known, formulas (2) and (3) make it easy to find the values of the remaining trigonometric functions.

EXAMPLE 3

Finding Exact Values Using Identities When Sine and Cosine Are Given

Given $\sin \theta = \dfrac{\sqrt{5}}{5}$ and $\cos \theta = \dfrac{2\sqrt{5}}{5}$, find the exact values of the four remaining trigonometric functions of θ using identities.

Solution   Based on a quotient identity from formula (3), we have

$$\tan \theta = \frac{\sin \theta}{\cos \theta} = \frac{\dfrac{\sqrt{5}}{5}}{\dfrac{2\sqrt{5}}{5}} = \frac{1}{2}$$

Then we use the reciprocal identities from formula (2) to get

$$\csc \theta = \frac{1}{\sin \theta} = \frac{1}{\dfrac{\sqrt{5}}{5}} = \frac{5}{\sqrt{5}} = \sqrt{5} \qquad \sec \theta = \frac{1}{\cos \theta} = \frac{1}{\dfrac{2\sqrt{5}}{5}} = \frac{5}{2\sqrt{5}} = \frac{\sqrt{5}}{2} \qquad \cot \theta = \frac{1}{\tan \theta} = \frac{1}{\dfrac{1}{2}} = 2 \quad \blacksquare$$

NOW WORK PROBLEM **25.**

The equation of the unit circle is $x^2 + y^2 = 1$. If $P = (x, y)$ is the point on the unit circle that corresponds to the angle θ, then

$$y^2 + x^2 = 1$$

But $y = \sin \theta$ and $x = \cos \theta$, so

$$(\sin \theta)^2 + (\cos \theta)^2 = 1 \qquad\qquad \textbf{(4)}$$

It is customary to write $\sin^2 \theta$ instead of $(\sin \theta)^2$, $\cos^2 \theta$ instead of $(\cos \theta)^2$, and so on. With this notation, we can rewrite equation (4) as

$$\sin^2 \theta + \cos^2 \theta = 1 \qquad\qquad \textbf{(5)}$$

If $\cos \theta \neq 0$, we can divide each side of equation (5) by $\cos^2 \theta$.

$$\frac{\sin^2 \theta}{\cos^2 \theta} + 1 = \frac{1}{\cos^2 \theta}$$

$$\left(\frac{\sin \theta}{\cos \theta}\right)^2 + 1 = \left(\frac{1}{\cos \theta}\right)^2$$

Now use formulas (2) and (3) to get

$$\tan^2\theta + 1 = \sec^2\theta \qquad\qquad (6)$$

Similarly, if $\sin\theta \neq 0$, we can divide equation (5) by $\sin^2\theta$ and use formulas (2) and (3) to get the result:

$$1 + \cot^2\theta = \csc^2\theta \qquad\qquad (7)$$

Collectively, the identities in equations (5), (6), and (7) are referred to as the **Pythagorean identities.**

Let's pause here to summarize the fundamental identities.

Fundamental Identities

$$\tan\theta = \frac{\sin\theta}{\cos\theta} \qquad \cot\theta = \frac{\cos\theta}{\sin\theta}$$

$$\csc\theta = \frac{1}{\sin\theta} \qquad \sec\theta = \frac{1}{\cos\theta} \qquad \cot\theta = \frac{1}{\tan\theta}$$

$$\sin^2\theta + \cos^2\theta = 1 \qquad \tan^2\theta + 1 = \sec^2\theta \qquad 1 + \cot^2\theta = \csc^2\theta$$

The Pythagorean identity

$$\sin^2\theta + \cos^2\theta = 1$$

can be solved for $\sin\theta$ in terms of $\cos\theta$ (or vice versa) as follows:

$$\sin^2\theta = 1 - \cos^2\theta$$
$$\sin\theta = \pm\sqrt{1 - \cos^2\theta}$$

where the $+$ sign is used if $\sin\theta > 0$ and the $-$ sign is used if $\sin\theta < 0$. Similarly, in $\tan^2\theta + 1 = \sec^2\theta$, we can solve for $\tan\theta$ (or $\sec\theta$) and in $1 + \cot^2\theta = \csc^2\theta$, we can solve for $\cot\theta$ (or $\csc\theta$).

EXAMPLE 4

Finding the Exact Value of a Trigonometric Expression Using Identities

④ Find the exact value of each expression. Do not use a calculator.

(a) $\tan 20° - \dfrac{\sin 20°}{\cos 20°}$ (b) $\sin^2\dfrac{\pi}{12} + \dfrac{1}{\sec^2\dfrac{\pi}{12}}$

Solution   (a) $\tan 20° - \underset{\substack{\uparrow \\ \frac{\sin\theta}{\cos\theta} = \tan\theta}}{\dfrac{\sin 20°}{\cos 20°}} = \tan 20° - \tan 20° = 0$

(b) $\sin^2\dfrac{\pi}{12} + \underset{\substack{\uparrow \\ \cos\theta = \frac{1}{\sec\theta}}}{\dfrac{1}{\sec^2\dfrac{\pi}{12}}} = \sin^2\dfrac{\pi}{12} + \underset{\substack{\uparrow \\ \sin^2\theta + \cos^2\theta = 1}}{\cos^2\dfrac{\pi}{12}} = 1$ ∎

NOW WORK PROBLEM **69.**

Many problems require finding the exact values of the remaining trigonometric functions when the value of one of them is known and the quadrant θ lies in can be found.

| EXAMPLE 5 | **Finding Exact Values Given One Value and the Sign of Another** |

Given that $\sin \theta = \dfrac{1}{3}$ and $\cos \theta < 0$, find the exact values of each of the remaining five trigonometric functions.

Solution 1
Using the Definition

Suppose that $P = (x, y)$ is the point on the unit circle that corresponds to θ. Since $\sin \theta = \dfrac{1}{3} = y$ and $\cos \theta = x < 0$, the point $P = (x, y) = \left(x, \dfrac{1}{3} \right)$ is in Quadrant II. See Figure 42. Then,

Figure 42

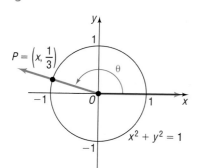

$$x^2 + y^2 = 1 \qquad y = \tfrac{1}{3}, x < 0$$

$$x^2 + \left(\frac{1}{3} \right)^2 = 1$$

$$x^2 = \frac{8}{9}$$

$$x = -\frac{2\sqrt{2}}{3}$$

Since $x = -\dfrac{2\sqrt{2}}{3}$ and $y = \dfrac{1}{3}$, we find that

$$\cos \theta = x = -\frac{2\sqrt{2}}{3} \qquad \tan \theta = \frac{y}{x} = \frac{\dfrac{1}{3}}{-2\sqrt{2}} = -\frac{1}{2\sqrt{2}} = -\frac{\sqrt{2}}{4}$$

$$\csc \theta = \frac{1}{y} = \frac{1}{\dfrac{1}{3}} = 3 \qquad \sec \theta = \frac{1}{x} = \frac{1}{-\dfrac{2\sqrt{2}}{3}} = -\frac{3}{2\sqrt{2}} = -\frac{3\sqrt{2}}{4} \qquad \cot \theta = \frac{x}{y} = \frac{-\dfrac{2\sqrt{2}}{3}}{\dfrac{1}{3}} = -2\sqrt{2}$$

Solution 2
Using Identities

First, we solve equation (5) for $\cos \theta$.

$$\sin^2 \theta + \cos^2 \theta = 1$$

$$\cos^2 \theta = 1 - \sin^2 \theta$$

$$\cos \theta = \pm\sqrt{1 - \sin^2 \theta}$$

Because $\cos \theta < 0$, we choose the minus sign.

$$\cos \theta = -\sqrt{1 - \sin^2 \theta} = -\sqrt{1 - \frac{1}{9}} = -\sqrt{\frac{8}{9}} = -\frac{2\sqrt{2}}{3}$$

$$\uparrow$$
$$\sin \theta = \tfrac{1}{3}$$

Now we know the values of $\sin \theta$ and $\cos \theta$, so we can use formulas (2) and (3) to get

$$\tan \theta = \frac{\sin \theta}{\cos \theta} = \frac{\dfrac{1}{3}}{\dfrac{-2\sqrt{2}}{3}} = \frac{1}{-2\sqrt{2}} = -\frac{\sqrt{2}}{4} \qquad \cot \theta = \frac{1}{\tan \theta} = -2\sqrt{2}$$

$$\sec \theta = \frac{1}{\cos \theta} = \frac{1}{\dfrac{-2\sqrt{2}}{3}} = \frac{-3}{2\sqrt{2}} = -\frac{3\sqrt{2}}{4} \qquad \csc \theta = \frac{1}{\sin \theta} = \frac{1}{\dfrac{1}{3}} = 3 \qquad ■$$

FINDING THE VALUES OF THE TRIGONOMETRIC FUNCTIONS WHEN ONE IS KNOWN

Given the value of one trigonometric function and the quadrant θ lies in, the exact value of each of the remaining five trigonometric functions can be found in either of two ways.

Method 1   Using the Definition

STEP 1: Draw a circle showing the location of the angle θ and the point $P = (x, y)$ that corresponds to θ. The radius of the circle is $r = \sqrt{x^2 + y^2}$.

STEP 2: Assign a value to two of the three variables x, y, r based on the value of the given trigonometric fuction.

STEP 3: Use the fact that P lies on the circle $x^2 + y^2 = r^2$ to find the value of the missing variable.

STEP 4: Apply the Theorem on p. 104 to find the values of the remaining trigonometric functions.

Method 2   Using Identities
Use appropriately selected identities to find the value of each of the remaining trigonometric functions.

EXAMPLE 6	Given One Value of a Trigonometric Function, Find the Remaining Ones

Given that $\tan \theta = \dfrac{1}{2}$ and $\sin \theta < 0$, find the exact value of each of the remaining five trigonometric functions of θ.

Solution 1
Using the Definition

STEP 1: Since $\tan \theta = \dfrac{1}{2} > 0$ and $\sin \theta < 0$, the point $P = (x, y)$ that correspond to θ lies in Quadrant III. See Figure 43.

STEP 2: Since $\tan \theta = \dfrac{1}{2} = \dfrac{y}{x}$ and θ lies in Quadrant III, we let $x = -2$ and $y = -1$.

STEP 3: Then $r = \sqrt{x^2 + y^2} = \sqrt{(-2)^2 + (-1)^2} = \sqrt{5}$ and P lies on the circle $x^2 + y^2 = 5$.

Figure 43

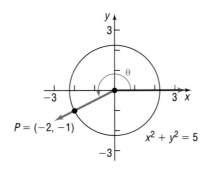

$P = (-2, -1)$

$x^2 + y^2 = 5$

STEP 4:   Now apply the definitions using $x = -2, y = -1, r = \sqrt{5}$.

$$\sin \theta = \frac{y}{r} = \frac{-1}{\sqrt{5}} = -\frac{\sqrt{5}}{5} \qquad \cos \theta = \frac{x}{r} = \frac{-2}{\sqrt{5}} = -\frac{2\sqrt{5}}{5}$$

$$\csc \theta = \frac{r}{y} = \frac{\sqrt{5}}{-1} = -\sqrt{5} \qquad \sec \theta = \frac{r}{x} = \frac{\sqrt{5}}{-2} = -\frac{\sqrt{5}}{2} \qquad \cot \theta = \frac{x}{y} = \frac{-2}{-1} = 2$$

Solution 2
Using Identities

We use the Pythagorean identity that involves $\tan \theta$, namely, $\tan^2 \theta + 1 = \sec^2 \theta$. Since $\tan \theta = \dfrac{1}{2} > 0$ and $\sin \theta < 0$, then θ lies in Quadrant III, where $\sec \theta < 0$. Then

$$\tan^2 \theta + 1 = \sec^2 \theta \qquad \text{Pythagorean Identity}$$

$$\left(\frac{1}{2}\right)^2 + 1 = \sec^2 \theta \qquad \tan \theta = \frac{1}{2}$$

$$\sec^2 \theta = \frac{1}{4} + 1 = \frac{5}{4} \qquad \text{Proceed to solve for } \sec \theta$$

$$\sec \theta = -\frac{\sqrt{5}}{2} \qquad \sec \theta < 0$$

Now,

$$\cos \theta = \frac{1}{\sec \theta} = \frac{1}{-\dfrac{\sqrt{5}}{2}} = -\frac{2}{\sqrt{5}} = -\frac{2\sqrt{5}}{5}$$

$$\tan \theta = \frac{\sin \theta}{\cos \theta}, \text{ so } \sin \theta = \tan \theta \cdot \cos \theta = \left(\frac{1}{2}\right) \cdot \left(-\frac{2\sqrt{5}}{5}\right) = -\frac{\sqrt{5}}{5}$$

$$\csc \theta = \frac{1}{\sin \theta} = \frac{1}{-\dfrac{\sqrt{5}}{5}} = -\frac{5}{\sqrt{5}} = -\sqrt{5}$$

$$\cot \theta = \frac{1}{\tan \theta} = \frac{1}{\dfrac{1}{2}} = 2$$

NOW WORK PROBLEM 33.

EVEN–ODD PROPERTIES

⑤   Recall that a function f is even if $f(-\theta) = f(\theta)$ for all θ in the domain of f; a function f is odd if $f(-\theta) = -f(\theta)$ for all θ in the domain of f. We will now show that the trigonometric functions sine, tangent, cotangent, and cosecant are odd functions, whereas the functions cosine and secant are even functions.

Theorem   **Even–Odd Properties**

$$\sin(-\theta) = -\sin\theta \qquad \cos(-\theta) = \cos\theta \qquad \tan(-\theta) = -\tan\theta$$

$$\csc(-\theta) = -\csc\theta \qquad \sec(-\theta) = \sec\theta \qquad \cot(-\theta) = -\cot\theta$$

Figure 44

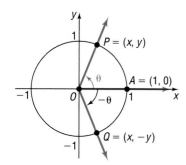

Proof   Let $P = (x, y)$ be the point on the unit circle that corresponds to the angle θ. (See Figure 44.) The point Q on the unit circle that corresponds to the angle $-\theta$ will have coordinates $(x, -y)$. Using the definition for the trigonometric functions, we have

$$\sin\theta = y \qquad \cos\theta = x \qquad \sin(-\theta) = -y \qquad \cos(-\theta) = x$$

so

$$\sin(-\theta) = -y = -\sin\theta \qquad \cos(-\theta) = x = \cos\theta$$

Now, using these results and some of the fundamental identities, we have

$$\tan(-\theta) = \frac{\sin(-\theta)}{\cos(-\theta)} = \frac{-\sin\theta}{\cos\theta} = -\tan\theta \qquad \cot(-\theta) = \frac{1}{\tan(-\theta)} = \frac{1}{-\tan\theta} = -\cot\theta$$

$$\sec(-\theta) = \frac{1}{\cos(-\theta)} = \frac{1}{\cos\theta} = \sec\theta \qquad \csc(-\theta) = \frac{1}{\sin(-\theta)} = \frac{1}{-\sin\theta} = -\csc\theta \quad ■$$

EXAMPLE 7   **Finding Exact Values Using Even–Odd Properties**

Find the exact value of:

(a) $\sin(-45°)$ (b) $\cos(-\pi)$ (c) $\cot\left(-\dfrac{3\pi}{2}\right)$ (d) $\tan\left(-\dfrac{37\pi}{4}\right)$

Solution   (a) $\sin(-45°) = -\sin 45° = -\dfrac{\sqrt{2}}{2}$ (b) $\cos(-\pi) = \cos\pi = -1$
                     ↑   ↑
                 Odd function   Even function

(c) $\cot\left(-\dfrac{3\pi}{2}\right) = -\cot\dfrac{3\pi}{2} = 0$
                        ↑
                  Odd function

(d) $\tan\left(-\dfrac{37\pi}{4}\right) = -\tan\dfrac{37\pi}{4} = -\tan\left(\dfrac{\pi}{4} + 9\pi\right) = -\tan\dfrac{\pi}{4} = -1$
                       ↑   ↑
                 Odd function   Period is π ■

NOW WORK PROBLEM 49.

2.3 EXERCISES

In Problems 1–16, use the fact that the trigonometric functions are periodic to find the exact value of each expression. Do not use a calculator.

1. $\sin 405°$

2. $\cos 420°$

3. $\tan 405°$

4. $\sin 390°$

5. $\csc 450°$

6. $\sec 540°$

7. $\cot 390°$

8. $\sec 420°$

9. $\cos \dfrac{33\pi}{4}$

10. $\sin \dfrac{9\pi}{4}$

11. $\tan (21\pi)$

12. $\csc \dfrac{9\pi}{2}$

13. $\sec \dfrac{17\pi}{4}$

14. $\cot \dfrac{17\pi}{4}$

15. $\tan \dfrac{19\pi}{6}$

16. $\sec \dfrac{25\pi}{6}$

In Problems 17–24, name the quadrant in which the angle θ lies.

17. $\sin \theta > 0, \quad \cos \theta < 0$

18. $\sin \theta < 0, \quad \cos \theta > 0$

19. $\sin \theta < 0, \quad \tan \theta < 0$

20. $\cos \theta > 0, \quad \tan \theta > 0$

21. $\cos \theta > 0, \quad \tan \theta < 0$

22. $\cos \theta < 0, \quad \tan \theta > 0$

23. $\sec \theta < 0, \quad \sin \theta > 0$

24. $\csc \theta > 0, \quad \cos \theta < 0$

In Problems 25–32, $\sin \theta$ and $\cos \theta$ are given. Find the exact value of each of the four remaining trigonometric functions.

25. $\sin \theta = -\dfrac{3}{5}, \quad \cos \theta = \dfrac{4}{5}$

26. $\sin \theta = \dfrac{4}{5}, \quad \cos \theta = -\dfrac{3}{5}$

27. $\sin \theta = \dfrac{2\sqrt{5}}{5}, \quad \cos \theta = \dfrac{\sqrt{5}}{5}$

28. $\sin \theta = -\dfrac{\sqrt{5}}{5}, \quad \cos \theta = -\dfrac{2\sqrt{5}}{5}$

29. $\sin \theta = \dfrac{1}{2}, \quad \cos \theta = \dfrac{\sqrt{3}}{2}$

30. $\sin \theta = \dfrac{\sqrt{3}}{2}, \quad \cos \theta = \dfrac{1}{2}$

31. $\sin \theta = -\dfrac{1}{3}, \quad \cos \theta = \dfrac{2\sqrt{2}}{3}$

32. $\sin \theta = \dfrac{2\sqrt{2}}{3}, \quad \cos \theta = -\dfrac{1}{3}$

In Problems 33–48, find the exact value of each of the remaining trigonometric functions of θ.

33. $\sin \theta = \dfrac{12}{13}, \quad \theta$ in quadrant II

34. $\cos \theta = \dfrac{3}{5}, \quad \theta$ in quadrant IV

35. $\cos \theta = -\dfrac{4}{5}, \quad \theta$ in quadrant III

36. $\sin \theta = -\dfrac{5}{13}, \quad \theta$ in quadrant III

37. $\sin \theta = \dfrac{5}{13}, \quad 90° < \theta < 180°$

38. $\cos \theta = \dfrac{4}{5}, \quad 270° < \theta < 360°$

39. $\cos \theta = -\dfrac{1}{3}, \quad \dfrac{\pi}{2} < \theta < \pi$

40. $\sin \theta = -\dfrac{2}{3}, \quad \pi < \theta < \dfrac{3\pi}{2}$

41. $\sin \theta = \dfrac{2}{3}, \quad \tan \theta < 0$

42. $\cos \theta = -\dfrac{1}{4}, \quad \tan \theta > 0$

43. $\sec \theta = 2, \quad \sin \theta < 0$

44. $\csc \theta = 3, \quad \cot \theta < 0$

45. $\tan \theta = \dfrac{3}{4}, \quad \sin \theta < 0$

46. $\cot \theta = \dfrac{4}{3}, \quad \cos \theta < 0$

47. $\tan \theta = -\dfrac{1}{3}, \quad \sin \theta > 0$

48. $\sec \theta = -2, \quad \tan \theta > 0$

In Problems 49–66, use the even–odd properties to find the exact value of each expression. Do not use a calculator.

49. $\sin(-60°)$

50. $\cos(-30°)$

51. $\tan(-30°)$

52. $\sin(-135°)$

53. $\sec(-60°)$

54. $\csc(-30°)$

55. $\sin(-90°)$

56. $\cos(-270°)$

57. $\tan\left(-\dfrac{\pi}{4}\right)$

58. $\sin(-\pi)$

59. $\cos\left(-\dfrac{\pi}{4}\right)$

60. $\sin\left(-\dfrac{\pi}{3}\right)$

61. $\tan(-\pi)$

62. $\sin\left(-\dfrac{3\pi}{2}\right)$

63. $\csc\left(-\dfrac{\pi}{4}\right)$

64. $\sec(-\pi)$

65. $\sec\left(-\dfrac{\pi}{6}\right)$

66. $\csc\left(-\dfrac{\pi}{3}\right)$

In Problems 67–78, use properties of the trigonometric function to find the exact value of each expression. Do not use a calculator.

67. $\sin^2 40° + \cos^2 40°$

68. $\sec^2 18° - \tan^2 18°$

69. $\sin 80° \csc 80°$

70. $\tan 10° \cot 10°$

71. $\tan 40° - \dfrac{\sin 40°}{\cos 40°}$

72. $\cot 20° - \dfrac{\cos 20°}{\sin 20°}$

73. $\cos 400° \cdot \sec 40°$

74. $\tan 200° \cdot \cot 20°$

75. $\sin\left(-\dfrac{\pi}{12}\right) \csc \dfrac{25\pi}{12}$

76. $\sec\left(-\dfrac{\pi}{18}\right) \cdot \cos \dfrac{37\pi}{18}$

77. $\dfrac{\sin(-20°)}{\cos 380°} + \tan 200°$

78. $\dfrac{\sin 70°}{\cos(-430°)} + \tan(-70°)$

---

79. If $\sin\theta = 0.3$, find the value of:
$\sin\theta + \sin(\theta + 2\pi) + \sin(\theta + 4\pi)$.

80. If $\cos\theta = 0.2$, find the value of:
$\cos\theta + \cos(\theta + 2\pi) + \cos(\theta + 4\pi)$.

81. If $\tan\theta = 3$, find the value of:
$\tan\theta + \tan(\theta + \pi) + \tan(\theta + 2\pi)$.

82. If $\cot\theta = -2$, find the value of:
$\cot\theta + \cot(\theta - \pi) + \cot(\theta - 2\pi)$.

83. Find the exact value of
$\sin 1° + \sin 2° + \sin 3° + \cdots + \sin 358° + \sin 359°$.

84. Find the exact value of
$\cos 1° + \cos 2° + \cos 3° + \cdots + \cos 358° + \cos 359°$.

85. What is the domain of the sine function?

86. What is the domain of the cosine function?

87. For what numbers θ is $f(\theta) = \tan\theta$ not defined?

88. For what numbers θ is $f(\theta) = \cot\theta$ not defined?

89. For what numbers θ is $f(\theta) = \sec\theta$ not defined?

90. For what numbers θ is $f(\theta) = \csc\theta$ not defined?

91. What is the range of the sine function?

92. What is the range of the cosine function?

93. What is the range of the tangent function?

94. What is the range of the cotangent function?

95. What is the range of the secant function?

96. What is the range of the cosecant function?

97. Is the sine function even, odd, or neither? Is its graph symmetric? With respect to what?

98. Is the cosine function even, odd, or neither? Is its graph symmetric? With respect to what?

99. Is the tangent function even, odd, or neither? Is its graph symmetric? With respect to what?

100. Is the cotangent function even, odd, or neither? Is its graph symmetric? With respect to what?

101. Is the secant function even, odd, or neither? Is its graph symmetric? With respect to what?

102. Is the cosecant function even, odd, or neither? Is its graph symmetric? With respect to what?

In Problems 103–108, use the periodic and even–odd properties.

103. If $f(\theta) = \sin\theta$ and $f(a) = \dfrac{1}{3}$, find the exact value of:

(a) $f(-a)$ (b) $f(a) + f(a + 2\pi) + f(a + 4\pi)$

104. If $f(\theta) = \cos\theta$ and $f(a) = \dfrac{1}{4}$, find the exact value of:

(a) $f(-a)$ (b) $f(a) + f(a + 2\pi) + f(a - 2\pi)$

105. If $f(\theta) = \tan\theta$ and $f(a) = 2$, find the exact value of:

(a) $f(-a)$ (b) $f(a) + f(a + \pi) + f(a + 2\pi)$

106. If $f(\theta) = \cot\theta$ and $f(a) = -3$, find the exact value of:

(a) $f(-a)$ (b) $f(a) + f(a + \pi) + f(a + 4\pi)$

107. If $f(\theta) = \sec\theta$ and $f(a) = -4$, find the exact value of:

(a) $f(-a)$ (b) $f(a) + f(a + 2\pi) + f(a + 4\pi)$

108. If $f(\theta) = \csc\theta$ and $f(a) = 2$, find the exact value of:

(a) $f(-a)$ (b) $f(a) + f(a + 2\pi) + f(a + 4\pi)$

109. Calculating the Time of a Trip From a parking lot, you want to walk to a house on the ocean. The house is located 1500 feet down a paved path that parallels the ocean, which is 500 feet away. See the illustration. Along the path you can walk 300 feet per minute, but in the sand on the beach you can only walk 100 feet per minute.

    The time T to get from the parking lot to the beachhouse can be expressed as a function of the angle θ shown in the illustration and is

$$T(\theta) = 5 - \frac{5}{3\tan\theta} + \frac{5}{\sin\theta}, \qquad 0 < \theta < \frac{\pi}{2}$$

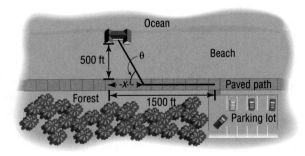

Calculate the time T if you walk directly from the parking lot to the house.

$$\left[\textbf{Hint:}\ \tan\theta = \frac{500}{1500}.\right]$$

110. Calculating the Time of a Trip Two oceanfront homes are located 8 miles apart on a straight stretch of beach, each a distance of 1 mile from a paved road that parallels the ocean. Sally can jog 8 miles per hour along the paved road, but only 3 miles per hour in the sand on the beach. Because of a river directly between the two houses, it is necessary to jog in the sand to the road, continue on the road, and then jog directly back in the sand to get from one house to the other. See the illustration. The time T to get from one house to the other as a function of the angle θ shown in the illustration is

$$T(\theta) = 1 + \frac{2}{3\sin\theta} - \frac{1}{4\tan\theta} \qquad 0 < \theta < \frac{\pi}{2}$$

(a) Calculate the time T for $\tan\theta = \dfrac{1}{4}$.

 (b) Describe the path taken.

(c) Explain why θ must be larger than 14°.

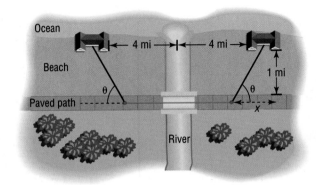

111. Show that the range of the tangent function is the set of all real numbers.

112. Show that the range of the cotangent function is the set of all real numbers.

113. Show that the period of $f(\theta) = \sin\theta$ is 2π.

[**Hint:** Assume that $0 < p < 2\pi$ exists so that $\sin(\theta + p) = \sin\theta$ for all θ. Let $\theta = 0$ to find p. Then let $\theta = \dfrac{\pi}{2}$ to obtain a contradiction.]

114. Show that the period of $f(\theta) = \cos\theta$ is 2π.

115. Show that the period of $f(\theta) = \sec\theta$ is 2π.

116. Show that the period of $f(\theta) = \csc\theta$ is 2π.

117. Show that the period of $f(\theta) = \tan\theta$ is π.

118. Show that the period of $f(\theta) = \cot\theta$ is π.

119. Prove the reciprocal identities given in formula (2).

120. Prove the quotient identities given in formula (3).

121. Establish the identity:

$$(\sin\theta \cos\phi)^2 + (\sin\theta \sin\phi)^2 + \cos^2\theta = 1$$

 122. Write down five properties of the tangent function. Explain the meaning of each.

123. Describe your understanding of the meaning of a periodic function.

PREPARING FOR THIS SECTION

Before getting started, review the following:

✓ Graphing Techniques: Transformations (Section 1.5, pp. 47–56)

2.4 GRAPHS OF THE SINE AND COSINE FUNCTIONS

OBJECTIVES   **1** Graph Transformations of the Sine Function
2 Graph Transformations of the Cosine Function
3 Determine the Amplitude and Period of Sinusoidal Functions
4 Graph Sinusoidal Functions: $y = A \sin(\omega x)$
5 Find an Equation for a Sinusoidal Graph

Since we want to graph the trigonometric functions in the xy-plane, we shall use the traditional symbols x for the independent variable (or argument) and y for the dependent variable (or value at x) for each function. So we write the six trigonometric functions as

$$y = f(x) = \sin x \qquad y = f(x) = \cos x \qquad y = f(x) = \tan x$$
$$y = f(x) = \csc x \qquad y = f(x) = \sec x \qquad y = f(x) = \cot x$$

Here the independent variable x represents an angle, measured in radians. In calculus, x will usually be treated as a real number. As we said earlier, these are equivalent ways of viewing x.

THE GRAPH OF $y = \sin x$

Since the sine function has period 2π, we need to graph $y = \sin x$ only on the interval $[0, 2\pi]$. The remainder of the graph will consist of repetitions of this portion of the graph.

We begin by constructing Table 6, which lists some points on the graph of $y = \sin x$, $0 \le x \le 2\pi$. As the table shows, the graph of $y = \sin x$, $0 \le x \le 2\pi$, begins at the origin. As x increases from 0 to $\dfrac{\pi}{2}$, the value of $y = \sin x$ increases from 0 to 1; as x increases from $\dfrac{\pi}{2}$ to π to $\dfrac{3\pi}{2}$, the value of y decreases from 1 to 0 to -1; as x increases from $\dfrac{3\pi}{2}$ to 2π, the value of y increases from -1 to 0. If we plot the points listed in Table 6 and connect them with a smooth curve, we obtain the graph shown in Figure 45.

TABLE 6		
x	$y = \sin x$	(x, y)
0	0	$(0, 0)$
$\dfrac{\pi}{6}$	$\dfrac{1}{2}$	$\left(\dfrac{\pi}{6}, \dfrac{1}{2}\right)$
$\dfrac{\pi}{2}$	1	$\left(\dfrac{\pi}{2}, 1\right)$
$\dfrac{5\pi}{6}$	$\dfrac{1}{2}$	$\left(\dfrac{5\pi}{6}, \dfrac{1}{2}\right)$
π	0	$(\pi, 0)$
$\dfrac{7\pi}{6}$	$-\dfrac{1}{2}$	$\left(\dfrac{7\pi}{6}, -\dfrac{1}{2}\right)$
$\dfrac{3\pi}{2}$	-1	$\left(\dfrac{3\pi}{2}, -1\right)$
$\dfrac{11\pi}{6}$	$-\dfrac{1}{2}$	$\left(\dfrac{11\pi}{6}, -\dfrac{1}{2}\right)$
2π	0	$(2\pi, 0)$

Figure 45

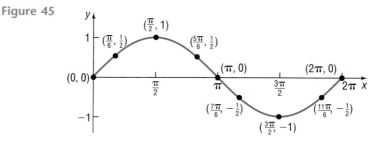

The graph in Figure 45 is one period, or **cycle,** of the graph of $y = \sin x$. To obtain a more complete graph of $y = \sin x$, we repeat this period in each direction, as shown in Figure 46.

Figure 46
$y = \sin x, -\infty < x < \infty$

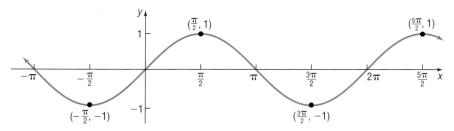

The graph of $y = \sin x$ illustrates some of the facts that we already know about the sine function.

PROPERTIES OF THE SINE FUNCTION

1. The domain is the set of all real numbers.
2. The range consists of all real numbers from -1 to 1, inclusive.
3. The sine function is an odd function, as the symmetry of the graph with respect to the origin indicates.
4. The sine function is periodic, with period 2π.
5. The x-intercepts are $\ldots, -2\pi, -\pi, 0, \pi, 2\pi, 3\pi, \ldots$; the y-intercept is 0.
6. The maximum value is 1 and occurs at $x = \ldots, -\dfrac{3\pi}{2}, \dfrac{\pi}{2}, \dfrac{5\pi}{2}, \dfrac{9\pi}{2}, \ldots$;

 the minimum value is -1 and occurs at $x = \ldots, -\dfrac{\pi}{2}, \dfrac{3\pi}{2}, \dfrac{7\pi}{2}, \dfrac{11\pi}{2}, \ldots$.

NOW WORK PROBLEMS 1, 3, AND 5.

1 The graphing techniques introduced in Chapter 1 may be used to graph functions that are transformations of the sine function (refer to Section 1.5).

EXAMPLE 1

Graphing Variations of $y = \sin x$ Using Transformations

Use the graph of $y = \sin x$ to graph $y = \sin\left(x - \dfrac{\pi}{4}\right)$.

Solution Figure 47 illustrates the steps.

Figure 47

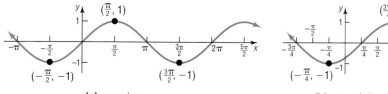

Replace x by $x - \frac{\pi}{4}$; horizontal shift to the right $\frac{\pi}{4}$ units.

Check: Graph $Y_1 = \sin\left(x - \dfrac{\pi}{4}\right)$ and compare the result with Figure 47(b). ∎

| EXAMPLE 2 | **Graphing Variations of $y = \sin x$ Using Transformations** |

Use the graph of $y = \sin x$ to graph $y = -\sin x + 2$.

Figure 48

Solution   Figure 48 illustrates the steps.

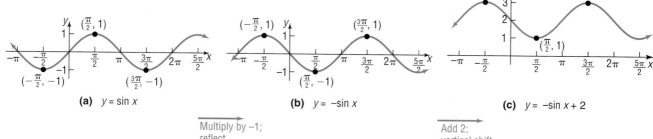

(a) $y = \sin x$ (b) $y = -\sin x$ (c) $y = -\sin x + 2$

Multiply by –1;
reflect
about x-axis.

Add 2;
vertical shift.

∎

 Check: Graph $Y_1 = -\sin x + 2$ and compare the result with Figure 48(c). ∎

✏ NOW WORK PROBLEM 17.

THE GRAPH OF $y = \cos x$

The cosine function also has period 2π. We proceed as we did with the sine function by constructing Table 7, which lists some points on the graph of $y = \cos x$, $0 \leq x \leq 2\pi$. As the table shows, the graph of $y = \cos x$, $0 \leq x \leq 2\pi$, begins at the point $(0, 1)$. As x increases from 0 to $\dfrac{\pi}{2}$ to π, the value of y decreases from 1 to 0 to -1; as x increases from π to $\dfrac{3\pi}{2}$ to 2π, the value of y increases from -1 to 0 to 1. As before, we plot the points in Table 7 to get one period or cycle of the graph. See Figure 49.

TABLE 7

x	$y = \cos x$	(x, y)
0	1	$(0, 1)$
$\dfrac{\pi}{3}$	$\dfrac{1}{2}$	$\left(\dfrac{\pi}{3}, \dfrac{1}{2}\right)$
$\dfrac{\pi}{2}$	0	$\left(\dfrac{\pi}{2}, 0\right)$
$\dfrac{2\pi}{3}$	$-\dfrac{1}{2}$	$\left(\dfrac{2\pi}{3}, -\dfrac{1}{2}\right)$
π	-1	$(\pi, -1)$
$\dfrac{4\pi}{3}$	$-\dfrac{1}{2}$	$\left(\dfrac{4\pi}{3}, -\dfrac{1}{2}\right)$
$\dfrac{3\pi}{2}$	0	$\left(\dfrac{3\pi}{2}, 0\right)$
$\dfrac{5\pi}{3}$	$\dfrac{1}{2}$	$\left(\dfrac{5\pi}{3}, \dfrac{1}{2}\right)$
2π	1	$(2\pi, 1)$

Figure 49
$y = \cos x$, $0 \leq x \leq 2\pi$

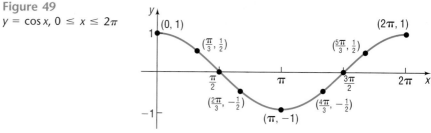

A more complete graph of $y = \cos x$ is obtained by repeating this period in each direction, as shown in Figure 50.

Figure 50
$y = \cos x$,
$-\infty < x < \infty$

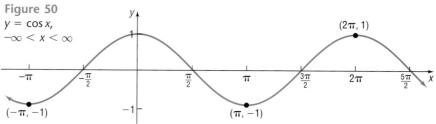

The graph of $y = \cos x$ illustrates some of the facts that we already know about the cosine function.

> ### PROPERTIES OF THE COSINE FUNCTION
>
> 1. The domain is the set of all real numbers.
> 2. The range consists of all real numbers from -1 to 1, inclusive.
> 3. The cosine function is an even function, as the symmetry of the graph with respect to the y-axis indicates.
> 4. The cosine function is periodic, with period 2π.
> 5. The x-intercepts are $\ldots, -\dfrac{3\pi}{2}, -\dfrac{\pi}{2}, \dfrac{\pi}{2}, \dfrac{3\pi}{2}, \dfrac{5\pi}{2}, \ldots$; the y-intercept is 1.
> 6. The maximum value is 1 and occurs at $x = \ldots, -2\pi, 0, 2\pi, 4\pi, 6\pi, \ldots$; the minimum value is -1 and occurs at $x = \ldots, -\pi, \pi, 3\pi, 5\pi, \ldots$.

2   Again, the graphing techniques from Chapter 1 may be used to graph transformations of the cosine function.

EXAMPLE 3   ## Graphing Variations of $y = \cos x$ Using Transformations

Use the graph of $y = \cos x$ to graph $y = 2 \cos x$.

Solution   Figure 51 illustrates the steps.

Figure 51

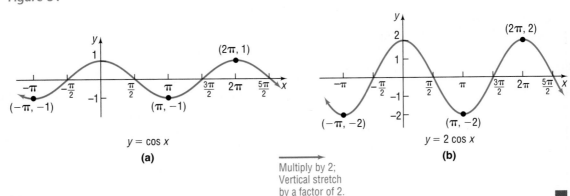

$y = \cos x$
(a)

Multiply by 2;
Vertical stretch
by a factor of 2.

$y = 2 \cos x$
(b)

 Check: Graph $Y_1 = 2 \cos x$ and compare the result with Figure 51.

EXAMPLE 4   ## Graphing Variations of $y = \cos x$ Using Transformations

Use the graph of $y = \cos x$ to graph $y = \cos(3x)$.

Solution   Figure 52 illustrates the graph, which is a horizontal compression of the graph of $y = \cos x.$ $\left(\text{Multiply each } x\text{-coordinate by } \dfrac{1}{3}.\right)$ Notice that, due to this compression, the period of $y = \cos(3x)$ is $\dfrac{2\pi}{3}$, whereas the period of $y = \cos x$ is 2π.

Figure 52

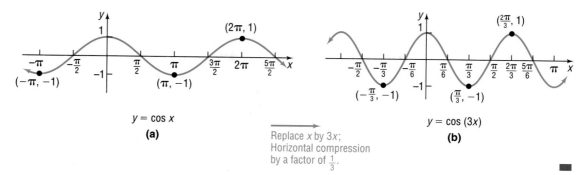

$$y = \cos x$$

(a)

Replace x by $3x$;
Horizontal compression
by a factor of $\frac{1}{3}$.

$$y = \cos (3x)$$

(b)

Check: Graph $Y_1 = \cos (3x)$. Use TRACE to verify that the period is $\frac{2\pi}{3}$.

NOW WORK PROBLEM 25.

SINUSOIDAL GRAPHS

The graph of $y = \cos x$, when compared to the graph of $y = \sin x$, suggests that the graph of $y = \sin x$ is the same as the graph of $y = \cos\left(x - \dfrac{\pi}{2}\right)$. See Figure 53.

Figure 53

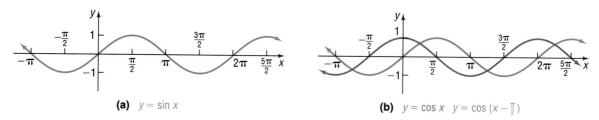

(a) $y = \sin x$

(b) $y = \cos x$ $y = \cos\left(x - \frac{\pi}{2}\right)$

Based on Figure 53, we conjecture that

$$\sin x = \cos\left(x - \frac{\pi}{2}\right)$$

(We shall prove this fact in Chapter 3.) Because of this similarity, the graphs of sine functions and cosine functions are referred to as **sinusoidal graphs.**

 SEEING THE CONCEPT   Graph $Y_1 = \sin x$ and $Y_2 = \cos\left(x - \dfrac{\pi}{2}\right)$. How many graphs do you see?

Let's look at some general properties of sinusoidal graphs.

③ In Example 3 we obtained the graph of $y = 2\cos x$, which we reproduce in Figure 54. Notice that the values of $y = 2\cos x$ lie between -2 and 2, inclusive.

Figure 54
$y = 2\cos x$

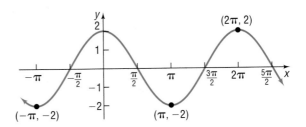

In general, the values of the functions $y = A\sin x$ and $y = A\cos x$, where $A \neq 0$, will always satisfy the inequalities

$$-|A| \leq A\sin x \leq |A| \quad \text{and} \quad -|A| \leq A\cos x \leq |A|$$

respectively. The number $|A|$ is called the **amplitude** of $y = A\sin x$ or $y = A\cos x$. See Figure 55.

Figure 55

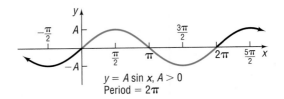

$y = A\sin x,\ A > 0$
Period $= 2\pi$

In Example 4, we obtained the graph of $y = \cos(3x)$, which we reproduce in Figure 56. Notice that the period of this function is $\dfrac{2\pi}{3}$.

Figure 56
$y = \cos(3x)$

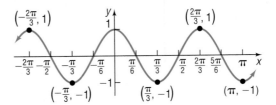

In general, if $\omega > 0$, the functions $y = \sin(\omega x)$ and $y = \cos(\omega x)$ will have period $T = \dfrac{2\pi}{\omega}$. To see why, recall that the graph of $y = \sin(\omega x)$ is obtained from the graph of $y = \sin x$ by performing a horizontal compression or stretch by a factor $\dfrac{1}{\omega}$. This horizontal compression replaces the interval $[0, 2\pi]$, which contains one period of the graph of $y = \sin x$, by the interval $\left[0, \dfrac{2\pi}{\omega}\right]$, which contains one period of the graph of $y = \sin(\omega x)$. The period of the functions $y = \sin(\omega x)$ and $y = \cos(\omega x)$, $\omega > 0$, is $\dfrac{2\pi}{\omega}$.

For example, for the function $y = \cos(3x)$, graphed in Figure 56, $\omega = 3$, so the period is $\dfrac{2\pi}{\omega} = \dfrac{2\pi}{3}$.

One period of the graph of $y = \sin(\omega x)$ or $y = \cos(\omega x)$ is called a **cycle.** Figure 57 illustrates the general situation. The blue portion of the graph is one cycle.

Figure 57

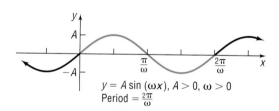

$y = A \sin(\omega x),\ A > 0,\ \omega > 0$
Period $= \frac{2\pi}{\omega}$

If $\omega < 0$ in $y = \sin(\omega x)$ or $y = \cos(\omega x)$, we use the even–odd properties of the sine and cosine functions as follows:

$$-\sin(\omega x) = \sin(-\omega x) \quad \text{and} \quad \cos(\omega x) = \cos(-\omega x)$$

This gives us an equivalent form in which the coefficient of x is positive. For example,

$$\sin(-2x) = -\sin(2x) \quad \text{and} \quad \cos(-\pi x) = \cos(\pi x)$$

Theorem

If $\omega > 0$, the amplitude and period of $y = A \sin(\omega x)$ and $y = A \cos(\omega x)$ are given by

$$\text{Amplitude} = |A| \qquad \text{Period} = T = \frac{2\pi}{\omega} \qquad \textbf{(1)}$$

EXAMPLE 5

Finding the Amplitude and Period of a Sinusoidal Function

Determine the amplitude and period of $y = 3 \sin(4x)$.

Solution   Comparing $y = 3 \sin(4x)$ to $y = A \sin(\omega x)$, we find that $A = 3$ and $\omega = 4$. From equation (1),

$$\text{Amplitude} = |A| = 3 \qquad \text{Period} = T = \frac{2\pi}{\omega} = \frac{2\pi}{4} = \frac{\pi}{2}$$

====🖉 NOW WORK PROBLEM 33.

Earlier, we graphed sine and cosine functions using tranformations. We now introduce another method that can be used to graph these functions.

Figure 58 shows one cycle of the graphs of $y = \sin x$ and $y = \cos x$ on the interval $[0, 2\pi]$. Notice that each graph consists of four parts corresponding to the four subintervals:

$$\left[0, \frac{\pi}{2}\right], \quad \left[\frac{\pi}{2}, \pi\right], \quad \left[\pi, \frac{3\pi}{2}\right], \quad \left[\frac{3\pi}{2}, 2\pi\right]$$

Each of these subintervals is of length $\dfrac{\pi}{2}$ (the period 2π divided by 4) and the endpoints of these intervals give rise to five key points, as shown in Figure 58.

Figure 58

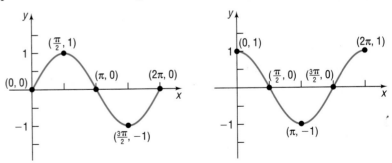

4   When graphing a sinusoidal function of the form $y = A\sin(\omega x)$ or $y = A\cos(\omega x)$, we use the amplitude to determine the maximum and minimum values of the function. The period is used to divide the x-axis into four subintervals. The endpoints of the subintervals give rise to five key points on the graph, which are used to sketch one cycle. Finally, extend the graph in either direction to make it complete. Let's look at an example.

EXAMPLE 6   Graphing a Sinusoidal Function

Graph: $y = 3\sin(4x)$

Solution   From Example 5, the amplitude is 3 and the period is $\dfrac{\pi}{2}$. The graph of $y = 3\sin(4x)$ will lie between -3 and 3 on the y-axis. One cycle will begin at $x = 0$ and end at $x = \dfrac{\pi}{2}$.

We divide the interval $\left[0, \dfrac{\pi}{2}\right]$ into four subintervals, each of length $\dfrac{\pi}{2} \div 4 = \dfrac{\pi}{8}$:

$$\left[0, \frac{\pi}{8}\right], \quad \left[\frac{\pi}{8}, \frac{\pi}{4}\right], \quad \left[\frac{\pi}{4}, \frac{3\pi}{8}\right], \quad \left[\frac{3\pi}{8}, \frac{\pi}{2}\right]$$

The endpoints of these intervals give rise to five key points on the graph:

$$(0, 0), \quad \left(\frac{\pi}{8}, 3\right), \quad \left(\frac{\pi}{4}, 0\right), \quad \left(\frac{3\pi}{8}, -3\right), \quad \left(\frac{\pi}{2}, 0\right)$$

We plot these five points and fill in the graph of the sine curve as shown in Figure 59(a). If we extend the graph in either direction, we obtain the complete graph shown in Figure 59(b).

Figure 59

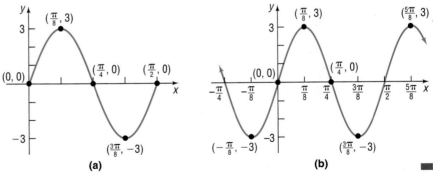

(a)   (b)

Check: Graph $y = 3 \sin(4x)$ using transformations. Which graphing method do you prefer?

 Check: Graph $Y_1 = 3 \sin(4x)$ using a graphing utility.
[**Hint:** Use the amplitude to set Ymin, Ymax. Use the period to set Xmin, Xmax.]

NOW WORK PROBLEM 39.

EXAMPLE 7

Finding the Amplitude and Period of a Sinusoidal Function and Graphing It

Determine the amplitude and period of $y = -4 \cos(\pi x)$, and graph the function.

Solution   Comparing $y = -4 \cos(\pi x)$ with $y = A \cos(\omega x)$, we find that $A = -4$ and $\omega = \pi$. The amplitude is $|A| = |-4| = 4$, and the period is

$$T = \frac{2\pi}{\omega} = \frac{2\pi}{\pi} = 2.$$

The graph of $y = -4 \cos(\pi x)$ will lie between -4 and 4 on the y-axis. One cycle will begin at $x = 0$ and end at $x = 2$. We divide the interval $[0, 2]$ into four subintervals, each of length $2 \div 4 = \frac{1}{2}$:

$$\left[0, \frac{1}{2}\right], \quad \left[\frac{1}{2}, 1\right], \quad \left[1, \frac{3}{2}\right], \quad \left[\frac{3}{2}, 2\right].$$

The five key points on the graph are

$$(0, -4), \quad \left(\frac{1}{2}, 0\right), \quad (1, 4), \quad \left(\frac{3}{2}, 0\right), \quad (2, -4).$$

We plot these five points and fill in the graph of the cosine function as shown in Figure 60(a). Extending the graph in either direction, we obtain Figure 60(b).

Figure 60
$y = -4 \cos(\pi x)$

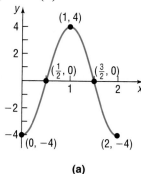
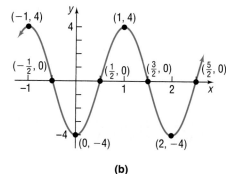

(a)   (b)

 Check: Graph $y = -4 \cos(\pi x)$ using transformations. Which graphing method do you prefer?

 Check: Graph $Y_1 = -4 \cos(\pi x)$ using a graphing utility.

EXAMPLE 8

Finding the Amplitude and Period of a Sinusoidal Function and Graphing It

Determine the amplitude and period of $y = 2 \sin\left(-\frac{\pi}{2} x\right)$, and graph the function.

Solution   Since the sine function is odd, we use the equivalent form:

$$y = -2 \sin\left(\frac{\pi}{2} x\right)$$

Comparing $y = -2 \sin\left(\dfrac{\pi}{2}x\right)$ to $y = A\sin(\omega x)$, we find that $A = -2$ and $\omega = \dfrac{\pi}{2}$. The amplitude is $|A| = 2$, and the period is $T = \dfrac{2\pi}{\omega} = \dfrac{2\pi}{\dfrac{\pi}{2}} = 4$.

The graph of $y = -2\sin\left(\dfrac{\pi}{2}x\right)$ will lie between -2 and 2 on the y-axis. One cycle will begin at $x = 0$ and end at $x = 4$. We divide the interval $[0, 4]$ into four subintervals, each of length $4 \div 4 = 1$:

$$[0, 1], \quad [1, 2], \quad [2, 3], \quad [3, 4]$$

The five key points on the graph are

$$(0, 0), \quad (1, -2), \quad (2, 0), \quad (3, 2), \quad (4, 0)$$

We plot these five points and fill in the graph of the sine function as shown in Figure 61(a). Extending the graph in either direction, we obtain Figure 61(b).

Figure 61

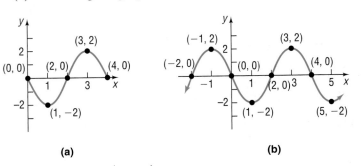

(a)   (b)

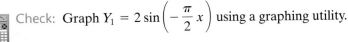

Check: Graph $y = 2\sin\left(-\dfrac{\pi}{2}x\right)$ using transformations. Which graphing method do you prefer?

Check: Graph $Y_1 = 2\sin\left(-\dfrac{\pi}{2}x\right)$ using a graphing utility.

NOW WORK PROBLEM 53.

5 We can also use the ideas of amplitude and period to identify a sinusoidal function when its graph is given.

EXAMPLE 9   **Finding an Equation for a Sinusoidal Graph**

Find an equation for the graph shown in Figure 62.

Figure 62

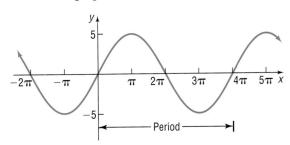

Solution This graph can be viewed as the graph of a sine function* $y = A\sin(\omega x)$ with $A = 5$. The period T is observed to be 4π. By equation (1),

---

*The equation could also be viewed as a cosine function with a horizontal shift, but viewing it as a sine function is easier.

$$T = \frac{2\pi}{\omega}$$

$$4\pi = \frac{2\pi}{\omega}$$

$$\omega = \frac{2\pi}{4\pi} = \frac{1}{2}$$

The sine function whose graph is given in Figure 86 is

$$y = A\sin(\omega x) = 5\sin\left(\frac{1}{2}x\right)$$

 Check: Graph $Y_1 = 5\sin\left(\frac{1}{2}x\right)$ and compare the result with Figure 62. ■

EXAMPLE 10   **Finding an Equation for a Sinusoidal Graph**

Find an equation for the graph shown in Figure 63.

Figure 63

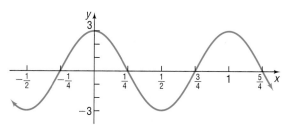

Solution   From the graph we conclude that it is easiest to view the equation as a cosine function $y = A\cos(\omega x)$ with $A = 3$ and period $T = 1$. Then $\dfrac{2\pi}{\omega} = 1$, so $\omega = 2\pi$. The cosine function whose graph is given in Figure 63 is

$$y = A\cos(\omega x) = 3\cos(2\pi x)$$

 Check: Graph $Y_1 = 3\cos(2\pi x)$ and compare the result with Figure 63. ■

✏ NOW WORK PROBLEMS **63** AND **67**.

2.4   EXERCISES

In Problems 1–10, if necessary, refer to the graphs to answer each question.

1. What is the *y*-intercept of $y = \sin x$?

2. What is the *y*-intercept of $y = \cos x$?

3. For what numbers $x, -\pi \le x \le \pi$, is the graph of $y = \sin x$ increasing?

4. For what numbers $x, -\pi \le x \le \pi$, is the graph of $y = \cos x$ decreasing?

5. What is the largest value of $y = \sin x$?

6. What is the smallest value of $y = \cos x$?

7. For what numbers $x, 0 \le x \le 2\pi$, does $\sin x = 0$?

8. For what numbers $x, 0 \le x \le 2\pi$, does $\cos x = 0$?

9. For what numbers $x, -2\pi \le x \le 2\pi$, does $\sin x = 1$? What about $\sin x = -1$?

10. For what numbers $x, -2\pi \le x \le 2\pi$, does $\cos x = 1$? What about $\cos x = -1$?

In Problems 11 and 12, match the graph to a function. Three answers are possible.

A. $y = -\sin x$

B. $y = -\cos x$

C. $y = \sin\left(x - \dfrac{\pi}{2}\right)$

D. $y = -\cos\left(x - \dfrac{\pi}{2}\right)$

E. $y = \sin(x + \pi)$

F. $y = \cos(x + \pi)$

11.

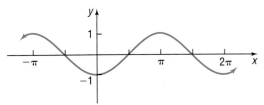

12.

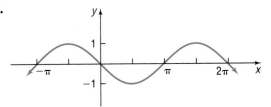

In Problems 13–28, use transformations to graph each function.

13. $y = 3 \sin x$ **14.** $y = 4 \cos x$ **15.** $y = \cos\left(x + \dfrac{\pi}{4}\right)$ **16.** $y = \sin(x - \pi)$

17. $y = \sin x - 1$ **18.** $y = \cos x + 1$ **19.** $y = -2 \sin x$ **20.** $y = -3 \cos x$

21. $y = \sin(\pi x)$ **22.** $y = \cos\left(\dfrac{\pi}{2} x\right)$ **23.** $y = 2 \sin x + 2$ **24.** $y = 3 \cos x + 3$

25. $y = -2 \cos\left(x - \dfrac{\pi}{2}\right)$ **26.** $y = -3 \sin\left(x + \dfrac{\pi}{2}\right)$ **27.** $y = 3 \sin(\pi - x)$ **28.** $y = 2 \cos(\pi - x)$

In Problems 29–38, determine the amplitude and period of each function without graphing.

29. $y = 2 \sin x$ **30.** $y = 3 \cos x$ **31.** $y = -4 \cos(2x)$ **32.** $y = -\sin\left(\dfrac{1}{2} x\right)$

33. $y = 6 \sin(\pi x)$ **34.** $y = -3 \cos(3x)$ **35.** $y = -\dfrac{1}{2} \cos\left(\dfrac{3}{2} x\right)$ **36.** $y = \dfrac{4}{3} \sin\left(\dfrac{2}{3} x\right)$

37. $y = \dfrac{5}{3} \sin\left(-\dfrac{2\pi}{3} x\right)$ **38.** $y = \dfrac{9}{5} \cos\left(-\dfrac{3\pi}{2} x\right)$

In Problems 39–48, match the given function to one of the graphs (A)–(J). (See page 135.)

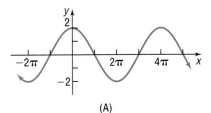

(A)

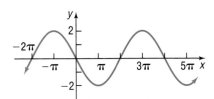

(B)

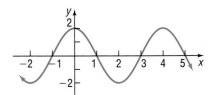

(C)

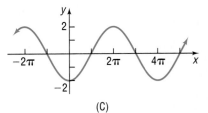

(D)

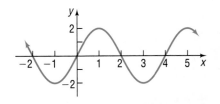

(E)

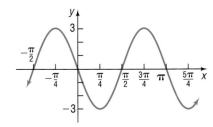

(F)

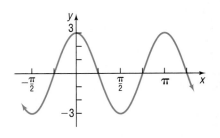

(G)

(H)

(I)

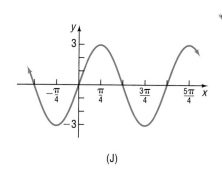

(J)

39. $y = 2 \sin\left(\dfrac{\pi}{2} x\right)$ **40.** $y = 2 \cos\left(\dfrac{\pi}{2} x\right)$ **41.** $y = 2 \cos\left(\dfrac{1}{2} x\right)$

42. $y = 3 \cos(2x)$ **43.** $y = -3 \sin(2x)$ **44.** $y = 2 \sin\left(\dfrac{1}{2} x\right)$

45. $y = -2 \cos\left(\dfrac{1}{2} x\right)$ **46.** $y = -2 \cos\left(\dfrac{\pi}{2} x\right)$ **47.** $y = 3 \sin(2x)$

48 $y = -2 \sin\left(\dfrac{1}{2} x\right)$

In Problems 49–52, match the given function to one of the graphs (A)–(D).

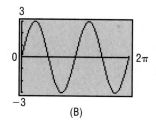
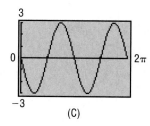
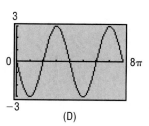
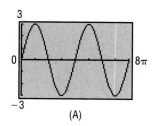

(A)   (B)   (C)   (D)

49. $y = 3 \sin\left(\dfrac{1}{2} x\right)$ **50.** $y = -3 \sin(2x)$ **51.** $y = 3 \sin(2x)$ **52.** $y = -3 \sin\left(\dfrac{1}{2} x\right)$

In Problems 53–62, graph each sinusoidal function.

53. $y = 5 \sin(4x)$ **54.** $y = 4 \cos(6x)$ **55.** $y = 5 \cos(\pi x)$ **56.** $y = 2 \sin(\pi x)$

57. $y = -2 \cos(2\pi x)$ **58.** $y = -5 \cos(2\pi x)$ **59.** $y = -4 \sin\left(\dfrac{1}{2} x\right)$ **60.** $y = -2 \cos\left(\dfrac{1}{2} x\right)$

61. $y = \dfrac{3}{2} \sin\left(-\dfrac{2}{3} x\right)$ **62.** $y = \dfrac{4}{3} \cos\left(-\dfrac{1}{3} x\right)$

In Problems 63–66, write the equation of a sine function that has the given characteristics.

63. Amplitude: 3   **64.** Amplitude: 2   **65.** Amplitude: 3   **66.** Amplitude: 4
  Period: π Period: 4π Period: 2   Period: 1

In Problems 67–80, find an equation for each graph.

67.

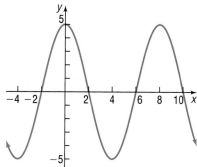

68.

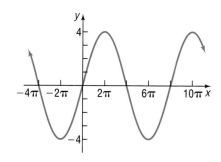

69.

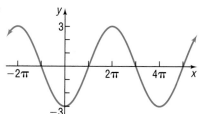

70.

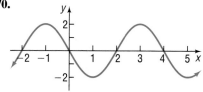

71.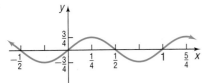

72.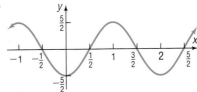

73.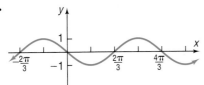

74.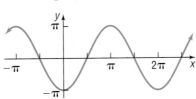

75.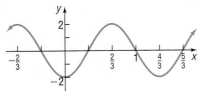

76.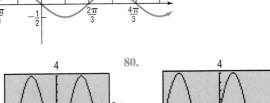

77.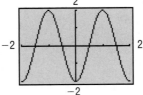

78.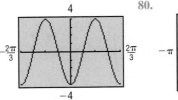

79.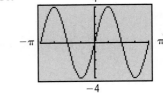

80.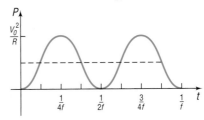

81. Alternating Current (ac) Circuits The current I, in amperes, flowing through an ac (alternating current) circuit at time t is

$$I = 220 \sin(60\pi t), \qquad t \geq 0$$

What is the period? What is the amplitude? Graph this function over two periods.

82. Alternating Current (ac) Circuits The current I, in amperes, flowing through an ac (alternating current) circuit at time t is

$$I = 120 \sin(30\pi t), \qquad t \geq 0$$

What is the period? What is the amplitude? Graph this function over two periods.

83. Alternating Current (ac) Generators The voltage V produced by an ac generator is

$$V = 220 \sin(120\pi t)$$

(a) What is the amplitude? What is the period?
(b) Graph V over two periods, beginning at $t = 0$.
(c) If a resistance of $R = 10$ ohms is present, what is the current I?
  [**Hint:** Use Ohm's Law, $V = IR$.]
(d) What is the amplitude and period of the current I?
(e) Graph I over two periods, beginning at $t = 0$.

84. Alternating Current (ac) Generators The voltage V produced by an ac generator is

$$V = 120 \sin(120\pi t)$$

(a) What is the amplitude? What is the period?
(b) Graph V over two periods, beginning at $t = 0$.
(c) If a resistance of $R = 20$ ohms is present, what is the current I?
  [**Hint:** Use Ohm's Law, $V = IR$.]
(d) What is the amplitude and period of the current I?
(e) Graph I over two periods, beginning at $t = 0$.

85. Alternating Current (ac) Generators The voltage V produced by an ac generator is sinusoidal. As a function of time, the voltage V is

$$V = V_0 \sin(2\pi ft)$$

where f is the **frequency,** the number of complete oscillations (cycles) per second. [In the United States and Canada, f is 60 hertz (Hz).] The **power** P delivered to a resistance R at any time t is defined as

$$P = \frac{V^2}{R}$$

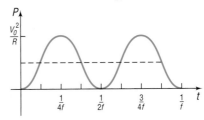

Power in an ac generator

(a) Show that $P = \dfrac{V_0^2}{R} \sin^2(2\pi ft)$.

(b) The graph of P is shown in the figure. Express P as a sinusoidal function.

(c) Deduce that

$$\sin^2(2\pi ft) = \frac{1}{2}\left[1 - \cos(4\pi ft)\right]$$

86. Biorhythms   In the theory of biorhythms, a sine function of the form

$$P = 50\sin(\omega t) + 50$$

is used to measure the percent P of a person's potential at time t, where t is measured in days and $t = 0$ is the person's birthday. Three characteristics are commonly measured:

     Physical potential: period of 23 days
     Emotional potential: period of 28 days
     Intellectual potential: period of 33 days

(a) Find ω for each characteristic.

(b) Graph all three functions.

(c) Is there a time t when all three characteristics have 100% potential? When is it?

(d) Suppose that you are 20 years old today ($t = 7305$ days). Describe your physical, emotional, and intellectual potential for the next 30 days.

87. Graph $y = |\cos x|,\ -2\pi \le x \le 2\pi$.

88. Graph $y = |\sin x|,\ -2\pi \le x \le 2\pi$.

2.5   GRAPHS OF THE TANGENT, COTANGENT, COSECANT, AND SECANT FUNCTIONS

OBJECTIVES   **1**   Graph Transformations of the Tangent Function and Cotangent Function
           2   Graph Transformations of the Cosecant Function and Secant Function

THE GRAPHS OF $y = \tan x$ AND $y = \cot x$

1   Because the tangent function has period π, we only need to determine the graph over some interval of length π. The rest of the graph will consist of repetitions of that graph. Because the tangent function is not defined at ... , $-\dfrac{3\pi}{2}, -\dfrac{\pi}{2}, \dfrac{\pi}{2}, \dfrac{3\pi}{2}, \ldots$, we will concentrate on the interval $\left(-\dfrac{\pi}{2}, \dfrac{\pi}{2}\right)$, of length π, and construct Table 8, which lists some points on the graph of $y = \tan x$, $-\dfrac{\pi}{2} < x < \dfrac{\pi}{2}$. We plot the points in the table and connect them with a smooth curve. See Figure 64 for a partial graph of $y = \tan x$, where $-\dfrac{\pi}{3} \le x \le \dfrac{\pi}{3}$.

x	$y = \tan x$	(x, y)
$-\dfrac{\pi}{3}$	$-\sqrt{3}$	$\left(-\dfrac{\pi}{3}, -\sqrt{3}\right)$
$-\dfrac{\pi}{4}$	-1	$\left(-\dfrac{\pi}{4}, -1\right)$
$-\dfrac{\pi}{6}$	$-\dfrac{\sqrt{3}}{3}$	$\left(-\dfrac{\pi}{6}, -\dfrac{\sqrt{3}}{3}\right)$
0	0	$(0, 0)$
$\dfrac{\pi}{6}$	$\dfrac{\sqrt{3}}{3}$	$\left(\dfrac{\pi}{6}, \dfrac{\sqrt{3}}{3}\right)$
$\dfrac{\pi}{4}$	1	$\left(\dfrac{\pi}{4}, 1\right)$
$\dfrac{\pi}{3}$	$\sqrt{3}$	$\left(\dfrac{\pi}{3}, \sqrt{3}\right)$

TABLE 8

Figure 64

$y = \tan x,\ -\dfrac{\pi}{3} \le x \le \dfrac{\pi}{3}$

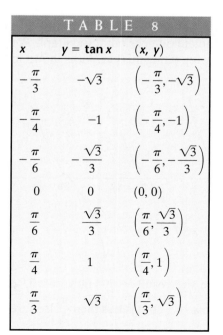

To complete one period of the graph of $y = \tan x$, we need to investigate the behavior of the function as x approaches $-\frac{\pi}{2}$ and $\frac{\pi}{2}$. We must be careful, though, because $y = \tan x$ is not defined at these numbers. To determine this behavior, we use the identity

$$\tan x = \frac{\sin x}{\cos x}$$

See Table 9. If x is close to $\frac{\pi}{2} \approx 1.5708$, but remains less than $\frac{\pi}{2}$, then $\sin x$ will be close to 1 and $\cos x$ will be positive and close to 0. (Refer back to the graphs of the sine function and the cosine function.) Hence, the ratio $\frac{\sin x}{\cos x}$ will be positive and large. In fact, the closer x gets to $\frac{\pi}{2}$, the closer $\sin x$ gets to 1 and $\cos x$ gets to 0, so $\tan x$ approaches ∞ $\left(\lim\limits_{x \to \frac{\pi}{2}} \tan x = \infty \right)$. In other words, the vertical line $x = \frac{\pi}{2}$ is a vertical asymptote to the graph of $y = \tan x$.

TABLE 9			
x	$\sin x$	$\cos x$	$y = \tan x$
$\frac{\pi}{3} \approx 1.05$	$\frac{\sqrt{3}}{2}$	$\frac{1}{2}$	$\sqrt{3} \approx 1.73$
1.5	0.9975	0.0707	14.1
1.57	0.9999	7.96E^{-4}	1255.8
1.5707	0.9999	9.6E^{-5}	10381
$\frac{\pi}{2} \approx 1.5708$	1	0	Undefined

If x is close to $-\frac{\pi}{2}$, but remains greater than $-\frac{\pi}{2}$, then $\sin x$ will be close to -1 and $\cos x$ will be positive and close to 0. Hence, the ratio $\frac{\sin x}{\cos x}$ approaches $-\infty$ $\left(\lim\limits_{x \to -\frac{\pi}{2}^+} \tan x = -\infty \right)$. In other words, the vertical line $x = -\frac{\pi}{2}$ is also a vertical asymptote to the graph.

With these observations, we can complete one period of the graph. We obtain the complete graph of $y = \tan x$ by repeating this period, as shown in Figure 65.

Figure 65

$y = \tan x$, $-\infty < x < \infty$, x not equal to odd multiples of $\frac{\pi}{2}$, $-\infty < y < \infty$

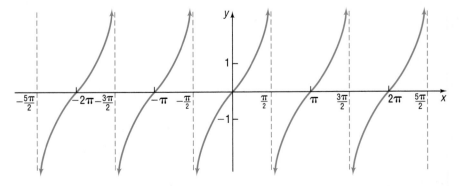

Check: Graph $Y_1 = \tan x$ and compare the result with Figure 65. Use TRACE to see what happens as x gets close to $\frac{\pi}{2}$, but is less than $\frac{\pi}{2}$. Be sure to set the WINDOW accordingly and use DOT mode.

The graph of $y = \tan x$ illustrates some facts that we already know about the tangent function.

> **PROPERTIES OF THE TANGENT FUNCTION**
>
> 1. The domain is the set of all real numbers, except odd multiples of $\dfrac{\pi}{2}$.
> 2. The range consists of all real numbers.
> 3. The tangent function is an odd function, as the symmetry of the graph with respect to the origin indicates.
> 4. The tangent function is periodic, with period π.
> 5. The x-intercepts are $\ldots, -2\pi, -\pi, 0, \pi, 2\pi, 3\pi, \ldots$; the y-intercept is 0.
> 6. Vertical asymptotes occur at $x = \ldots, -\dfrac{3\pi}{2}, -\dfrac{\pi}{2}, \dfrac{\pi}{2}, \dfrac{3\pi}{2}, \ldots$.

NOW WORK PROBLEMS **1** AND **9**.

EXAMPLE 1

Graphing Variations of $y = \tan x$ Using Transformations

Graph: $y = 2 \tan x$

Solution   We start with the graph of $y = \tan x$ and vertically stretch it by a factor of 2. See Figure 66.

Figure 66

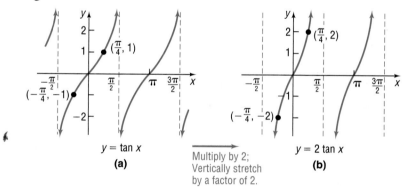

$y = \tan x$
(a)

Multiply by 2;
Vertically stretch
by a factor of 2.

$y = 2 \tan x$
(b)

Check: Graph $Y_1 = 2 \tan x$ and compare the result to Figure 66(b).

EXAMPLE 2

Graphing Variations of $y = \tan x$ Using Transformations

Graph: $y = -\tan\left(x + \dfrac{\pi}{4} \right)$

Solution   We start with the graph of $y = \tan x$. See Figure 67.

Figure 67

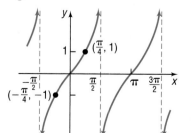

$y = \tan x$

(a)

Replace x by $x + \dfrac{\pi}{4}$;
Shift left $\dfrac{\pi}{4}$ units.

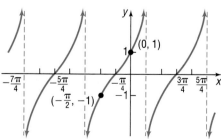

$y = \tan\left(x + \dfrac{\pi}{4}\right)$

(b)

Multiply by -1;
Reflect about
x-axis.

$y = -\tan\left(x + \dfrac{\pi}{4}\right)$

(c)

Check: Graph $Y_1 = -\tan\left(x + \dfrac{\pi}{4}\right)$ and compare the result to Figure 67(c).

NOW WORK PROBLEM 19.

We obtain the graph of $y = \cot x$ as we did the graph of $y = \tan x$. The period of $y = \cot x$ is π. Because the cotangent function is not defined for integral multiples of π, we will concentrate on the interval $(0, \pi)$. Table 10 lists some points on the graph of $y = \cot x$, $0 < x < \pi$. As x approaches 0, but remains greater than 0, the value of $\cos x$ will be close to 1 and the value of $\sin x$ will be positive and close to 0. Hence, the ratio $\dfrac{\cos x}{\sin x} = \cot x$ will be positive and large; so as x approaches 0, with $x > 0$, $\cot x$ approaches ∞ $\left(\lim\limits_{x \to 0^+} \cot x = \infty\right)$. Similarly, as x approaches π, but remains less than π, the value of $\cos x$ will be close to -1, and the value of $\sin x$ will be positive and close to 0. Hence, the ratio $\dfrac{\cos x}{\sin x} = \cot x$ will be negative and will approach $-\infty$ as x approaches π $\left(\lim\limits_{x \to \pi^-} \cot x = -\infty\right)$. Figure 68 shows the graph.

TABLE 10

x	$y = \cot x$	(x, y)
$\dfrac{\pi}{6}$	$\sqrt{3}$	$\left(\dfrac{\pi}{6}, \sqrt{3}\right)$
$\dfrac{\pi}{4}$	1	$\left(\dfrac{\pi}{4}, 1\right)$
$\dfrac{\pi}{3}$	$\dfrac{\sqrt{3}}{3}$	$\left(\dfrac{\pi}{3}, \dfrac{\sqrt{3}}{3}\right)$
$\dfrac{\pi}{2}$	0	$\left(\dfrac{\pi}{2}, 0\right)$
$\dfrac{2\pi}{3}$	$-\dfrac{\sqrt{3}}{3}$	$\left(\dfrac{2\pi}{3}, -\dfrac{\sqrt{3}}{3}\right)$
$\dfrac{3\pi}{4}$	-1	$\left(\dfrac{3\pi}{4}, -1\right)$
$\dfrac{5\pi}{6}$	$-\sqrt{3}$	$\left(\dfrac{5\pi}{6}, -\sqrt{3}\right)$

Figure 68
$y = \cot x$, $-\infty < x < \infty$, x not equal to integral multiples of π, $-\infty < y < \infty$

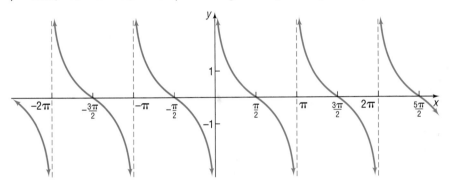

Check: Graph $Y_1 = \cot x$ and compare the result with Figure 68. Use TRACE to see what happens when x is close to 0.

NOW WORK PROBLEM 25.

THE GRAPHS OF $y = \csc x$ AND $y = \sec x$

② The cosecant and secant functions, sometimes referred to as **reciprocal functions,** are graphed by making use of the reciprocal identities

$$\csc x = \frac{1}{\sin x} \quad \text{and} \quad \sec x = \frac{1}{\cos x}$$

For example, the value of the cosecant function $y = \csc x$ at a given number x equals the reciprocal of the corresponding value of the sine function, provided that the value of the sine function is not 0. If the value of $\sin x$ is 0, then, at such numbers x, the cosecant function is not defined. In fact, the graph of the cosecant function has vertical asymptotes at integral multiples of π. Figure 69 shows the graph.

Figure 69
$y = \csc x$, $-\infty < x < \infty$, x not equal to integral multiples of π, $|y| \geq 1$

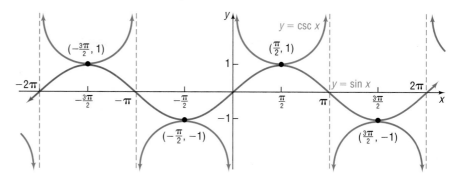

Check: Graph $Y_1 = \csc x$ and compare the result with Figure 69. Use TRACE to see what happens when x is close to 0. ■

EXAMPLE 3   ## Graphing Variations of $y = \csc x$ Using Transformations

Graph: $y = 2 \csc\left(x - \dfrac{\pi}{2}\right)$

Solution   Figure 70 shows the required steps.

Figure 70

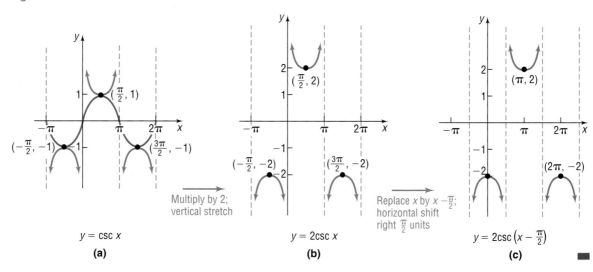

Multiply by 2; vertical stretch

Replace x by $x - \frac{\pi}{2}$; horizontal shift right $\frac{\pi}{2}$ units

$y = \csc x$
(a)

$y = 2\csc x$
(b)

$y = 2\csc\left(x - \frac{\pi}{2}\right)$
(c)   ■

Check: Graph $Y_1 = 2 \csc\left(x - \dfrac{\pi}{2}\right)$ and compare the result with Figure 70. ■

➤ NOW WORK PROBLEM **31.**

Using the idea of reciprocals, we can similarly obtain the graph of $y = \sec x$. See Figure 71.

Figure 71

$y = \sec x$, $-\infty < x < \infty$, x not equal to odd multiples of $\dfrac{\pi}{2}$, $|y| \geq 1$

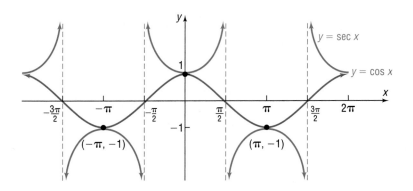

2.5 EXERCISES

In Problems 1–10, if necessary, refer to the graphs to answer each question.

1. What is the y-intercept of $y = \tan x$?

2. What is the y-intercept of $y = \cot x$?

3. What is the y-intercept of $y = \sec x$?

4. What is the y-intercept of $y = \csc x$?

5. For what numbers x, $-2\pi \leq x \leq 2\pi$, does $\sec x = 1$? What about $\sec x = -1$?

6. For what numbers x, $-2\pi \leq x \leq 2\pi$, does $\csc x = 1$? What about $\csc x = -1$?

7. For what numbers x, $-2\pi \leq x \leq 2\pi$, does the graph of $y = \sec x$ have vertical asymptotes?

8. For what numbers x, $-2\pi \leq x \leq 2\pi$, does the graph of $y = \csc x$ have vertical asymptotes?

9. For what numbers x, $-2\pi \leq x \leq 2\pi$, does the graph of $y = \tan x$ have vertical asymptotes?

10. For what numbers x, $-2\pi \leq x \leq 2\pi$, does the graph of $y = \cot x$ have vertical asymptotes?

In Problems 11–14, match each function to its graph.

A. $y = -\tan x$ B. $y = \tan\left(x + \dfrac{\pi}{2}\right)$ C. $y = \tan(x + \pi)$ D. $y = -\tan\left(x - \dfrac{\pi}{2}\right)$

11.
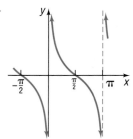

12.
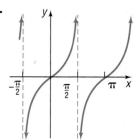

13.
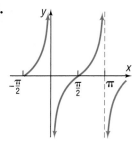

14.
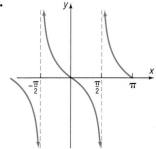

In Problems 15–34, use tranformations to graph each function.

15. $y = -\sec x$

16. $y = -\cot x$

17. $y = \sec\left(x - \dfrac{\pi}{2}\right)$

18. $y = \csc(x - \pi)$

19. $y = \tan(x - \pi)$

20. $y = \cot(x - \pi)$

21. $y = 3\tan(2x)$

22. $y = 4\tan\left(\dfrac{1}{2}x\right)$

23. $y = \sec(2x)$

24. $y = \csc\left(\dfrac{1}{2}x\right)$

25. $y = \cot(\pi x)$

26. $y = \cot(2x)$

27. $y = -3\tan(4x)$

28. $y = -3\tan(2x)$

29. $y = 2\sec\left(\dfrac{1}{2}x\right)$

30. $y = 2\sec(3x)$

31. $y = -3\csc\left(x + \dfrac{\pi}{4}\right)$

32. $y = -2\tan\left(x + \dfrac{\pi}{4}\right)$

33. $y = \dfrac{1}{2}\cot\left(x - \dfrac{\pi}{4}\right)$

34. $y = 3\sec\left(x + \dfrac{\pi}{2}\right)$

35. Carrying a Ladder around a Corner   A ladder of length L is carried horizontally around a corner from a hall 3 feet wide into a hall 4 feet wide. See the illustration.

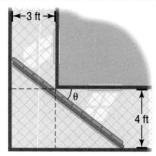

(a) Show that the length L of the ladder as a function of the angle θ is

$$L(\theta) = 3 \sec \theta + 4 \csc \theta$$

(b) Graph $L, 0 < \theta < \dfrac{\pi}{2}$.

(c) For what value of θ is L the least?

(d) What is the length of the longest ladder that can be carried around the corner? Why is this also the least value of L?

36. Exploration   Graph

$$y = \tan x \quad \text{and} \quad y = -\cot\left(x + \frac{\pi}{2}\right)$$

Do you think that $\tan x = -\cot\left(x + \dfrac{\pi}{2}\right)$?

2.6   PHASE SHIFT; SINUSOIDAL CURVE FITTING

OBJECTIVES   **1**   Determine the Phase Shift of a Sinusoidal Function

2   Graph Sinusoidal Functions: $y = A \sin(\omega x - \phi)$

3   Find a Sinusoidal Function from Data

PHASE SHIFT

1

We have seen that the graph of $y = A \sin(\omega x)$, $\omega > 0$, has amplitude $|A|$ and period $T = \dfrac{2\pi}{\omega}$. One cycle can be drawn as x varies from 0 to $\dfrac{2\pi}{\omega}$ or, equivalently, as ωx varies from 0 to 2π. See Figure 72.

We now want to discuss the graph of

$$y = A \sin(\omega x - \phi) = A \sin\left[\omega\left(x - \frac{\phi}{\omega}\right)\right]$$

where $\omega > 0$ and ϕ (the Greek letter phi) are real numbers. The graph will be a sine curve of amplitude $|A|$. As $\omega x - \phi$ varies from 0 to 2π, one period will be traced out. This period will begin when

$$\omega x - \phi = 0 \quad \text{or} \quad x = \frac{\phi}{\omega}$$

and will end when

$$\omega x - \phi = 2\pi \quad \text{or} \quad x = \frac{2\pi}{\omega} + \frac{\phi}{\omega}$$

See Figure 73.

We see that the graph of $y = A \sin(\omega x - \phi) = A \sin\left[\omega\left(x - \dfrac{\phi}{\omega}\right)\right]$ is the same as the graph of $y = A \sin(\omega x)$, except that it has been shifted $\dfrac{\phi}{\omega}$ units (to the right if $\phi > 0$ and to the left if $\phi < 0$). This number $\dfrac{\phi}{\omega}$ is called the **phase shift** of the graph of $y = A \sin(\omega x - \phi)$.

Figure 72
One cycle $y = A \sin(\omega x)$, $A > 0$, $\omega > 0$

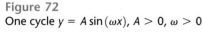

Figure 73
One cycle $y = A \sin(\omega x - \phi)$, $A > 0$, $\omega > 0$, $\phi > 0$

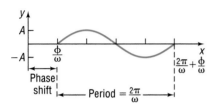

For the graphs of $y = A \sin(\omega x - \phi)$ or $y = A \cos(\omega x - \phi)$, $\omega > 0$,

$$\text{Amplitude} = |A| \qquad \text{Period} = T = \frac{2\pi}{\omega} \qquad \text{Phase shift} = \frac{\phi}{\omega}$$

The phase shift is to the left if $\phi < 0$ and to the right if $\phi > 0$.

EXAMPLE 1

Finding the Amplitude, Period, and Phase Shift of a Sinusoidal Function and Graphing It

2 Find the amplitude, period, and phase shift of $y = 3 \sin(2x - \pi)$, and graph the function.

Solution Comparing

$$y = 3 \sin(2x - \pi) = 3 \sin\left[2\left(x - \frac{\pi}{2}\right)\right]$$

to

$$y = A \sin(\omega x - \phi) = A \sin\left[\omega\left(x - \frac{\phi}{\omega}\right)\right]$$

we find that $A = 3$, $\omega = 2$, and $\phi = \pi$. The graph is a sine curve with amplitude $|A| = 3$, period $T = \dfrac{2\pi}{\omega} = \dfrac{2\pi}{2} = \pi$, and phase shift $= \dfrac{\phi}{\omega} = \dfrac{\pi}{2}$.

The graph of $y = 3 \sin(2x - \pi)$ will lie between -3 and 3 on the y-axis. One cycle will begin at $x = \dfrac{\phi}{\omega} = \dfrac{\pi}{2}$ and end at $x = \dfrac{2\pi}{\omega} + \dfrac{\phi}{\omega} = \pi + \dfrac{\pi}{2} = \dfrac{3\pi}{2}$. We divide the interval $\left[\dfrac{\pi}{2}, \dfrac{3\pi}{2}\right]$ into four subintervals, each of length $\pi \div 4 = \dfrac{\pi}{4}$:

$$\left[\frac{\pi}{2}, \frac{3\pi}{4}\right], \quad \left[\frac{3\pi}{4}, \pi\right], \quad \left[\pi, \frac{5\pi}{4}\right], \quad \left[\frac{5\pi}{4}, \frac{3\pi}{2}\right]$$

The five key points on the graph are

$$\left(\frac{\pi}{2}, 0\right), \quad \left(\frac{3\pi}{4}, 3\right), \quad (\pi, 0), \quad \left(\frac{5\pi}{4}, -3\right), \quad \left(\frac{3\pi}{2}, 0\right)$$

We plot these five points and fill in the graph of the sine function as shown in Figure 74(a). Extending the graph in either direction, we obtain Figure 74(b).

Figure 74

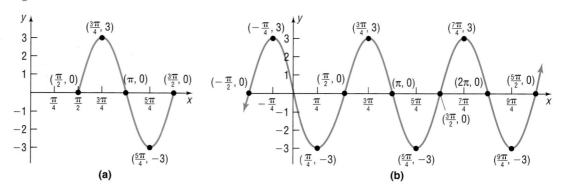

(a)   (b)

The graph of $y = 3 \sin (2x - \pi) = 3 \sin \left[2 \left(x - \dfrac{\pi}{2} \right) \right]$ may also be obtained using transformations. See Figure 75.

Figure 75

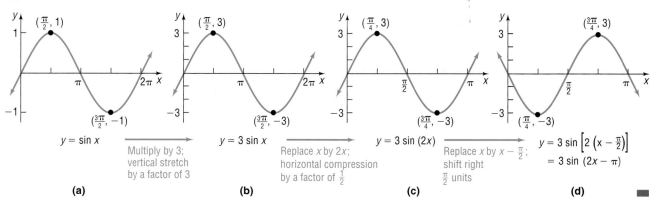

$y = \sin x$ Multiply by 3; vertical stretch by a factor of 3 $y = 3 \sin x$ Replace x by $2x$; horizontal compression by a factor of $\frac{1}{2}$ $y = 3 \sin (2x)$ Replace x by $x - \frac{\pi}{2}$; shift right $\frac{\pi}{2}$ units $y = 3 \sin \left[2 \left(x - \frac{\pi}{2} \right) \right]$ $= 3 \sin (2x - \pi)$

(a)   (b)   (c)   (d)

Check: Graph $Y_1 = 3 \sin (2x - \pi)$ using a graphing utility.

EXAMPLE 2

Finding the Amplitude, Period, and Phase Shift of a Sinusoidal Function and Graphing It

Find the amplitude, period, and phase shift of $y = 2 \cos (4x + 3\pi)$, and graph the function.

Solution   Comparing

$$y = 2 \cos (4x + 3\pi) = 2 \cos \left[4 \left(x + \dfrac{3\pi}{4} \right) \right]$$

to

$$y = A \cos (\omega x - \phi) = A \cos \left[\omega \left(x - \dfrac{\phi}{\omega} \right) \right]$$

we see that $A = 2$, $\omega = 4$, and $\phi = -3\pi$. The graph is a cosine curve with amplitude $|A| = 2$, period $T = \dfrac{2\pi}{\omega} = \dfrac{2\pi}{4} = \dfrac{\pi}{2}$, and phase shift $= \dfrac{\phi}{\omega} = -\dfrac{3\pi}{4}$.

The graph of $y = 2 \cos (4x + 3\pi)$ will lie between -2 and 2 on the y-axis. One cycle will begin at $x = \dfrac{\phi}{\omega} = -\dfrac{3\pi}{4}$ and end at $x = \dfrac{2\pi}{\omega} + \dfrac{\phi}{\omega} = \dfrac{\pi}{2} + \left(-\dfrac{3\pi}{4} \right) = -\dfrac{\pi}{4}$. We divide the interval $\left[-\dfrac{3\pi}{4}, -\dfrac{\pi}{4} \right]$ into four subintervals, each of the length $\dfrac{\pi}{2} \div 4 = \dfrac{\pi}{8}$:

$$\left[-\dfrac{3\pi}{4}, -\dfrac{5\pi}{8} \right], \quad \left[-\dfrac{5\pi}{8}, -\dfrac{\pi}{2} \right], \quad \left[-\dfrac{\pi}{2}, -\dfrac{3\pi}{8} \right], \quad \left[-\dfrac{3\pi}{8}, -\dfrac{\pi}{4} \right]$$

The five key points on the graph are

$$\left(-\dfrac{3\pi}{4}, 2 \right), \quad \left(-\dfrac{5\pi}{8}, 0 \right), \quad \left(-\dfrac{\pi}{2}, -2 \right), \quad \left(-\dfrac{3\pi}{8}, 0 \right), \quad \left(-\dfrac{\pi}{4}, 2 \right)$$

We plot these five points and fill in the graph of the cosine function as shown in Figure 76(a). Extending the graph in either direction, we obtain Figure 76(b).

Figure 76

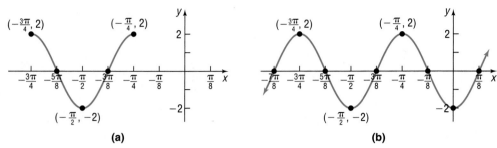

(a)

(b)

The graph of $y = 2\cos(4x + 3\pi) = 2\cos\left[4\left(x + \dfrac{3\pi}{4}\right)\right]$ may also be obtained using transformations. See Figure 77.

Figure 77

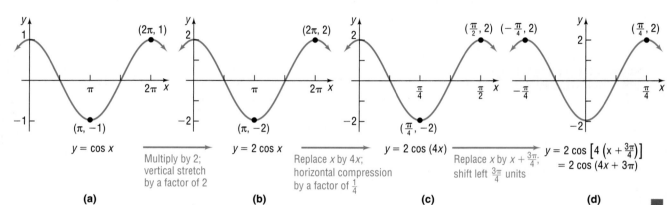

(a) $y = \cos x$

Multiply by 2;
vertical stretch
by a factor of 2

(b) $y = 2\cos x$

Replace x by $4x$;
horizontal compression
by a factor of $\frac{1}{4}$

(c) $y = 2\cos(4x)$

Replace x by $x + \frac{3\pi}{4}$;
shift left $\frac{3\pi}{4}$ units

(d) $y = 2\cos\left[4\left(x + \frac{3\pi}{4}\right)\right]$
        $= 2\cos(4x + 3\pi)$

Check: Graph $Y_1 = 2\cos(4x + 3\pi)$ using a graphing utility.

NOW WORK PROBLEM **1**.

SUMMARY

Steps for Graphing Sinusoidal Functions

To graph sinusoidal functions of the form $y = A\sin(\omega x - \phi)$ or $y = A\cos(\omega x - \phi)$:

STEP 1: Determine the amplitude $|A|$ and period $T = \dfrac{2\pi}{\omega}$.

STEP 2: Determine the starting point of one cycle of the graph, $\dfrac{\phi}{\omega}$.

STEP 3: Determine the ending point of one cycle of the graph, $\dfrac{2\pi}{\omega} + \dfrac{\phi}{\omega}$.

STEP 4: Divide the interval $\left[\dfrac{\phi}{\omega}, \dfrac{2\pi}{\omega} + \dfrac{\phi}{\omega}\right]$ into four subintervals, each of length $\dfrac{2\pi}{\omega} \div 4$.

STEP 5: Use the endpoints of the subintervals to find the five key points on the graph.

STEP 6: Fill in one cycle of the graph.

STEP 7: Extend the graph in each direction to make it complete.

FINDING SINUSOIDAL FUNCTIONS FROM DATA

③ Scatter diagrams of data sometimes take the form of a sinusoidal function. Let's look at an example.

The data given in Table 11 represent the average monthly temperatures in Denver, Colorado. Since the data represent *average* monthly temperatures collected over many years, the data will not vary much from year to year and so will essentially repeat each year. In other words, the data are periodic. Figure 78 shows the scatter diagram of these data repeated over two years, where $x = 1$ represents January, $x = 2$ represents February, and so on.

Figure 78

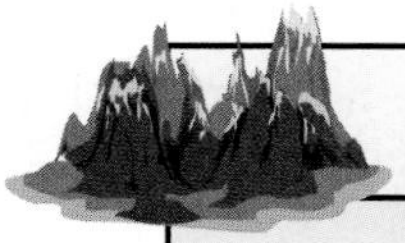

Month, x	Average Monthly Temperature, °F
January, 1	29.7
February, 2	33.4
March, 3	39.0
April, 4	48.2
May, 5	57.2
June, 6	66.9
July, 7	73.5
August, 8	71.4
September, 9	62.3
October, 10	51.4
November, 11	39.0
December, 12	31.0

TABLE 11

Source: U.S. National Oceanic and Atmospheric Administration

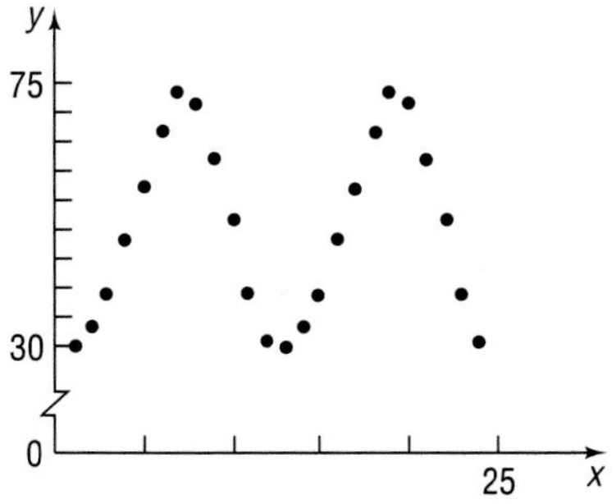

Notice that the scatter diagram looks like the graph of a sinusoidal function. We choose to fit the data to a sine function of the form

$$y = A \sin(\omega x - \phi) + B$$

where A, B, ω, and ϕ are constants.

EXAMPLE 3   Finding a Sinusoidal Function from Temperature Data

Figure 79

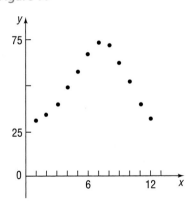

Fit a sine function to the data in Table 11.

Solution   We begin with a scatter diagram of the data for one year. See Figure 79. The data will be fitted to a sine function of the form

$$y = A \sin(\omega x - \phi) + B$$

STEP 1: To find the amplitude A, we compute

$$\text{Amplitude} = \frac{\text{largest data value} - \text{smallest data value}}{2}$$

$$= \frac{73.5 - 29.7}{2} = 21.9$$

Figure 80

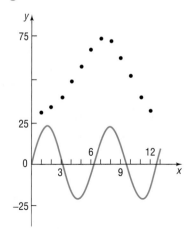

Figure 81

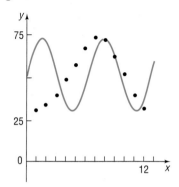

Figure 82

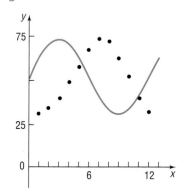

To see the remaining steps in this process, we superimpose the graph of the function $y = 21.9 \sin x$, where x represents months, on the scatter diagram. Figure 80 shows the two graphs.

To fit the data, the graph needs to be shifted vertically, shifted horizontally, and stretched horizontally.

STEP 2: We determine the vertical shift by finding the average of the highest and lowest data value.

$$\text{Vertical shift} = \frac{73.5 + 29.7}{2} = 51.6$$

Now we superimpose the graph of $y = 21.9 \sin x + 51.6$ on the scatter diagram. See Figure 81.

We see that the graph needs to be shifted horizontally and stretched horizontally.

STEP 3: It is easier to find the horizontal stretch factor first. Since the temperatures repeat every 12 months, the period of the function is $T = 12$. Since $T = \dfrac{2\pi}{\omega} = 12$,

$$\omega = \frac{2\pi}{12} = \frac{\pi}{6}$$

Now we superimpose the graph of $y = 21.9 \sin\left(\dfrac{\pi}{6}x\right) + 51.6$ on the scatter diagram. See Figure 82.

We see that the graph still needs to be shifted horizontally.

STEP 4: To determine the horizontal shift, we solve the equation

$$y = 21.9 \sin\left(\frac{\pi}{6}x - \phi\right) + 51.6$$

for ϕ by letting $y = 29.7$ and $x = 1$ (the average temperature in Denver in January).*

$$29.7 = 21.9 \sin\left(\frac{\pi}{6} \cdot 1 - \phi\right) + 51.6$$

$$-21.9 = 21.9 \sin\left(\frac{\pi}{6} - \phi\right) \qquad \text{Subtract 51.6 from both sides of the equation.}$$

$$-1 = \sin\left(\frac{\pi}{6} - \phi\right) \qquad \text{Divide both sides of the equation by 21.9.}$$

$$\frac{\pi}{6} - \phi = -\frac{\pi}{2} \qquad \sin\theta = -1 \text{ when } \theta = -\frac{\pi}{2}.$$

$$\phi = \frac{2\pi}{3} \qquad \text{Solve for } \phi.$$

The sine function that fits the data is

$$y = 21.9 \sin\left(\frac{\pi}{6}x - \frac{2\pi}{3}\right) + 51.6$$

---

*The data point selected here to find ϕ is arbitrary. Selecting a different data point will usually result in a different value for ϕ. To maintain consistency, we will always choose the data point for which y is smallest (in this case, January gives the lowest temperature).

Figure 83

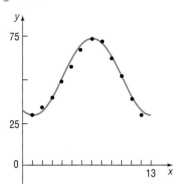

The graph of $y = 21.9 \sin\left(\dfrac{\pi}{6}x - \dfrac{2\pi}{3}\right) + 51.6$ and the scatter diagram of the data are shown in Figure 83.   ∎

The steps to fit a sine function

$$y = A \sin(\omega x - \phi) + B$$

to sinusoidal data follow:

> **STEPS FOR FITTING DATA TO A SINE FUNCTION $Y = A \sin(\omega X - \phi) + B$**
>
> **STEP 1:** Determine A, the amplitude of the function.
>
> $$\text{Amplitude} = \frac{\text{largest data value} - \text{smallest data value}}{2}$$
>
> **STEP 2:** Determine B, the vertical shift of the function.
>
> $$\text{Vertical shift} = \frac{\text{largest data value} + \text{smallest data value}}{2}$$
>
> **STEP 3:** Determine ω. Since the period T, the time it takes for the data to repeat, is $T = \dfrac{2\pi}{\omega}$, we have
>
> $$\omega = \frac{2\pi}{T}$$
>
> **STEP 4:** Determine the horizontal shift of the function by solving the equation
>
> $$y = A \sin(\omega x - \phi) + B$$
>
> for ϕ by choosing an ordered pair (x, y) from the data. Since answers will vary depending on the ordered pair selected, we will always choose the ordered pair for which y is smallest in order to maintain consistency.

 NOW WORK PROBLEM 19(a)–(c).

Certain graphing utilities (such as a TI-83 and TI-86) have the capability of finding the sine function of best fit for sinusoidal data. At least four data points are required for this process.

EXAMPLE 4   Finding the Sine Function of Best Fit

Figure 84

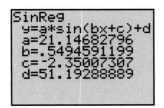

Use a graphing utility to find the sine function of best fit for the data in Table 11. Graph this function with the scatter diagram of the data.

Solution   Enter the data from Table 11 and execute the SINe REGression program. The result is shown in Figure 84.

The output that the utility provides shows us the equation

$$y = a \sin(bx + c) + d$$

Figure 85

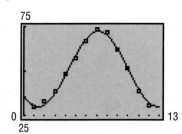

The sinusoidal function of best fit is

$$y = 21.15 \sin(0.55x - 2.35) + 51.19$$

where x represents the month and y represents the average temperature.

Figure 85 shows the graph of the sinusoidal function of best fit on the scatter diagram.   ■

◀━━━━━━ N O W   W O R K   P R O B L E M   **19(d)–(e)**.

Since the number of hours of sunlight in a day cycles annually, the number of hours of sunlight in a day for a given location can be modeled by a sinusoidal function.

The longest day of the year (in terms of hours of sunlight) occurs on the day of the summer solstice. The summer solstice is the time when the sun is farthest north (for locations in the northern hemisphere). In 1997, the summer solstice occurred on June 21 (the 172nd day of the year) at 8:21 AM Greenwich mean time (GMT). The shortest day of the year occurs on the day of the winter solstice. The winter solstice is the time when the sun is farthest south (again, for locations in the northern hemisphere). In 1997, the winter solstice occurred on December 21 (the 355th day of the year) at 8:09 PM (GMT).

| EXAMPLE 5 | **Finding a Sinusoidal Function for Hours of Daylight** |

According to the *Old Farmer's Almanac*, the number of hours of sunlight in Boston on the summer solstice is 15.283 and the number of hours of sunlight on the winter solstice is 9.067.

(a) Find a sinusoidal function of the form $y = A \sin(\omega x - \phi) + B$ that fits the data.*

(b) Use the function found in part (a) to predict the number of hours of sunlight on April 1, the 91st day of the year.

 (c) Draw a graph of the function found in part (a).

 (d) Look up the number of hours of sunlight for April 1 in the *Old Farmer's Almanac* and compare the actual hours of daylight to the results found in part (b).

Solution   (a) STEP 1:   Amplitude $= \dfrac{\text{largest data value} - \text{smallest data value}}{2}$

$$= \dfrac{15.283 - 9.067}{2} = 3.108$$

STEP 2:   Vertical shift $= \dfrac{\text{largest data value} + \text{smallest data value}}{2}$

$$= \dfrac{15.283 + 9.067}{2} = 12.175$$

STEP 3:   The data repeat every 365 days. Since $T = \dfrac{2\pi}{\omega} = 365$, we find

$$\omega = \dfrac{2\pi}{365}$$

So far, we have $y = 3.108 \sin\left(\dfrac{2\pi}{365}x - \phi\right) + 12.175$.

*Notice that only two data points are given, so a graphing utility cannot be used to find the sine function of best fit.

STEP 4:   To determine the horizontal shift, we solve the equation

$$y = 3.108 \sin\left(\frac{2\pi}{365} x - \phi\right) + 12.175$$

for ϕ by letting $y = 9.067$ and $x = 355$ (the number of hours of daylight in Boston on December 21).

$$9.067 = 3.108 \sin\left(\frac{2\pi}{365} \cdot 355 - \phi\right) + 12.175$$

$$-3.108 = 3.108 \sin\left(\frac{2\pi}{365} \cdot 355 - \phi\right) \qquad \text{Subtract 12.175 from both sides of the equation.}$$

$$-1 = \sin\left(\frac{2\pi}{365} \cdot 355 - \phi\right) \qquad \text{Divide both sides of the equation by 3.108.}$$

$$\frac{2\pi}{365} \cdot 355 - \phi = -\frac{\pi}{2} \qquad \text{$\sin\theta = -1$ when $\theta = -\dfrac{\pi}{2}$.}$$

$$\phi = 2.45\pi \qquad \text{Solve for ϕ.}$$

The function that provides the number of hours of daylight in Boston for any day, x, is given by

$$y = 3.108 \sin\left(\frac{2\pi}{365} x - 2.45\pi\right) + 12.175$$

(b) To predict the number of hours of daylight on April 1, we let $x = 91$ in the function found in part (a) and obtain

$$y = 3.108 \sin\left(\frac{2\pi}{365} \cdot 91 - 2.45\pi\right) + 12.175$$

$$\approx 3.108 \sin(-1.95\pi) + 12.175$$

$$\approx 12.65$$

So we predict that there will be about 12.65 hours of sunlight on April 1 in Boston.

Figure 86

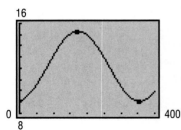

(c) The graph of the function found in part (a) is given in Figure 86.

(d) According to the *Old Farmer's Almanac*, there will be 12 hours 43 minutes of sunlight on April 1 in Boston. Our prediction of 12.65 hours converts to 12 hours 39 minutes.

NOW WORK PROBLEM **25.**

2.6   EXERCISES

In Problems 1–12, find the amplitude, period, and phase shift of each function. Graph each function. Show at least one period.

1. $y = 4 \sin(2x - \pi)$

2. $y = 3 \sin(3x - \pi)$

3. $y = 2 \cos\left(3x + \dfrac{\pi}{2}\right)$

4. $y = 3 \cos(2x + \pi)$

5. $y = -3 \sin\left(2x + \dfrac{\pi}{2}\right)$

6. $y = -2 \cos\left(2x - \dfrac{\pi}{2}\right)$

7. $y = 4 \sin(\pi x + 2)$

8. $y = 2 \cos(2\pi x + 4)$

9. $y = 3 \cos(\pi x - 2)$

10. $y = 2 \cos(2\pi x - 4)$

11. $y = 3 \sin\left(-2x + \dfrac{\pi}{2}\right)$

12. $y = 3 \cos\left(-2x + \dfrac{\pi}{2}\right)$

In Problems 13–16, write the equation of a sine function that has the given characteristics.

13. Amplitude: 2
    Period: π
    Phase shift: $\dfrac{1}{2}$

14. Amplitude: 3
    Period: $\dfrac{\pi}{2}$
    Phase shift: 2

15. Amplitude: 3
    Period: 3π
    Phase shift: $-\dfrac{1}{3}$

16. Amplitude: 2
    Period: π
    Phase shift: -2

17. **Alternating Current (ac) Circuits**   The current I, in amperes, flowing through an ac (alternating current) circuit at time t is

    $$I = 120 \sin\left(30\pi t - \dfrac{\pi}{3}\right), \qquad t \geq 0$$

    What is the period? What is the amplitude? What is the phase shift? Graph this function over two periods.

18. **Alternating Current (ac) Circuits**   The current I, in amperes, flowing through an ac (alternating current) circuit at time t is

    $$I = 220 \sin\left(60\pi t - \dfrac{\pi}{6}\right), \qquad t \geq 0$$

    What is the period? What is the amplitude? What is the phase shift? Graph this function over two periods.

19. **Monthly Temperature**   The following data represent the average monthly temperatures for Juneau, Alaska.

Month, x	Average Monthly Temperature, °F
January, 1	24.2
February, 2	28.4
March, 3	32.7
April, 4	39.7
May, 5	47.0
June, 6	53.0
July, 7	56.0
August, 8	55.0
September, 9	49.4
October, 10	42.2
November, 11	32.0
December, 12	27.1

Source: U.S. National Oceanic and Atmospheric Administration.

(a) Draw a scatter diagram of the data for one period.

(b) Find a sinusoidal function of the form
    $y = A \sin(\omega x - \phi) + B$ that fits the data.

(c) Draw the sinusoidal function found in part (b) on the scatter diagram.

(d) Use a graphing utility to find the sinusoidal function of best fit.

(e) Draw the sinusoidal function of best fit on the scatter diagram.

20. **Monthly Temperature**   The following data represent the average monthly temperatures for Washington, D.C.

Month, x	Average Monthly Temperature, °F
January, 1	34.6
February, 2	37.5
March, 3	47.2
April, 4	56.5
May, 5	66.4
June, 6	75.6
July, 7	80.0
August, 8	78.5
September, 9	71.3
October, 10	59.7
November, 11	49.8
December, 12	39.4

Source: U.S. National Oceanic and Atmospheric Administration.

(a) Draw a scatter diagram of the data for one period.

(b) Find a sinusoidal function of the form
$y = A \sin(\omega x - \phi) + B$ that fits the data.

(c) Draw the sinusoidal function found in part (b) on the scatter diagram.

(d) Use a graphing utility to find the sinusoidal function of best fit.

(e) Graph the sinusoidal function of best fit on the scatter diagram.

21. Monthly Temperature   The following data represent the average monthly temperatures for Indianapolis, Indiana.

Month, x	Average Monthly Temperature, °F
January, 1	25.5
February, 2	29.6
March, 3	41.4
April, 4	52.4
May, 5	62.8
June, 6	71.9
July, 7	75.4
August, 8	73.2
September, 9	66.6
October, 10	54.7
November, 11	43.0
December, 12	30.9

Source: U.S. National Oceanic and Atmospheric Administration.

(a) Draw a scatter diagram of the data for one period.

(b) Find a sinusoidal function of the form
$y = A \sin(\omega x - \phi) + B$ that fits the data.

(c) Draw the sinusoidal function found in part (b) on the scatter diagram.

(d) Use a graphing utility to find the sinusoidal function of best fit.

(e) Graph the sinusoidal function of best fit on the scatter diagram.

22. Monthly Temperature   The data at the top of the right column represent the average monthly temperatures for Baltimore, Maryland.

(a) Draw a scatter diagram of the data for one period.

(b) Find a sinusoidal function of the form
$y = A \sin(\omega x - \phi) + B$ that fits the data.

(c) Draw the sinusoidal function found in part (b) on the scatter diagram.

(d) Use a graphing utility to find the sinusoidal function of best fit.

(e) Graph the sinusoidal function of best fit on the scatter diagram.

Month, x	Average Monthly Temperature, °F
January, 1	31.8
February, 2	34.8
March, 3	44.1
April, 4	53.4
May, 5	63.4
June, 6	72.5
July, 7	77.0
August, 8	75.6
September, 9	68.5
October, 10	56.6
November, 11	46.8
December, 12	36.7

Source: U.S. National Oceanic and Atmospheric Administration.

23. Tides   Suppose that the length of time between consecutive high tides is approximately 12.5 hours. According to the National Oceanic and Atmospheric Administration, on Saturday, June 28, 1997, in Savannah, Georgia, high tide occurred at 3:38 AM (3.6333 hours) and low tide occurred at 10:08 AM (10.1333 hours). Water heights are measured as the amounts above or below the mean lower low water. The height of the water at high tide was 8.2 feet and the height of the water at low tide was −0.6 foot.

(a) Approximately when will the next high tide occur?

(b) Find a sinusoidal function of the form
$y = A \sin(\omega x - \phi) + B$ that fits the data.

(c) Draw a graph of the function found in part (b).

(d) Use the function found in part (b) to predict the height of the water at the next high tide.

24. Tides   Suppose that the length of time between consecutive high tides is approximately 12.5 hours. According to the National Oceanic and Atmospheric Administration, on Saturday, June 28, 1997, in Juneau, Alaska, high tide occurred at 8:11 AM (8.1833 hours) and low tide occurred at 2:14 PM (14.2333 hours). Water heights are measured as the amounts above or below the mean lower low water. The height of the water at high tide was 13.2 feet and the height of the water at low tide was 2.2 feet.

(a) Approximately when will the next high tide occur?

(b) Find a sinusoidal function of the form
$y = A \sin(\omega x - \phi) + B$ that fits the data.

(c) Draw a graph of the function found in part (b).

(d) Use the function found in part (b) to predict the height of the water at the next high tide.

25. Hours of Daylight   According to the *Old Farmer's Almanac*, in Miami, Florida, the number of hours of sunlight on the summer solstice is 12.75 and the number of hours of sunlight on the winter solstice is 10.583.

(a) Find a sinusoidal function of the form
$y = A \sin(\omega x - \phi) + B$ that fits the data.

(b) Draw a graph of the function found in part (a).
(c) Use the function found in part (a) to predict the number of hours of sunlight on April 1, the 91st day of the year

 (d) Look up the number of hours of sunlight for April 1 in the *Old Farmer's Almanac*, and compare the actual hours of daylight to the results found in part (c).

26. **Hours of Daylight** According to the *Old Farmer's Almanac*, in Detroit, Michigan, the number of hours of sunlight on the summer solstice is 13.65 and the number of hours of sunlight on the winter solstice is 9.067.
    (a) Find a sinusoidal function of the form
        $y = A \sin(\omega x - \phi) + B$ that fits the data.
    (b) Draw a graph of the function found in part (a).
    (c) Use the function found in part (a) to predict the number of hours of sunlight on April 1, the 91st day of the year.
    (d) Look up the number of hours of sunlight for April 1 in the *Old Farmer's Almanac*, and compare the actual hours of daylight to the results found in part (c).

27. **Hours of Daylight** According to the *Old Farmer's Almanac*, in Anchorage, Alaska, the number of hours of sunlight on the summer solstice is 16.233 and the number of hours of sunlight on the winter solstice is 5.45.
    (a) Find a sinusoidal function of the form
        $y = A \sin(\omega x - \phi) + B$ that fits the data.

(b) Draw a graph of the function found in part (a).
(c) Use the function found in part (a) to predict the number of hours of sunlight on April 1, the 91st day of the year.

 (d) Look up the number of hours of sunlight for April 1 in the *Old Farmer's Almanac*, and compare the actual hours of daylight to the results found in part (c).

28. **Hours of Daylight** According to the *Old Farmer's Almanac*, in Honolulu, Hawaii, the number of hours of sunlight on the summer solstice is 12.767 and the number of hours of sunlight on the winter solstice is 10.783.
    (a) Find a sinusoidal function of the form
        $y = A \sin(\omega x - \phi) + B$ that fits the data.
    (b) Draw a graph of the function found in part (a).
    (c) Use the function found in part (a) to predict the number of hours of sunlight on April 1, the 91st day of the year.
    (d) Look up the number of hours of sunlight for April 1 in the *Old Farmer's Almanac*, and compare the actual hours of daylight to the results found in part (c).

29. Explain how the amplitude and period of a sinusoidal graph are used to establish the scale on each coordinate axis.

30. Find an application in your major field that leads to a sinusoidal graph. Write a paper about your findings.

CHAPTER REVIEW

Things To Know

Definitions

Angle in standard position (p. 79)	Vertex is at the origin; initial side is along the positive x-axis
Degree (1°) (p. 79)	$1° = \frac{1}{360}$ revolution
Radian (p. 82)	The measure of a central angle of a circle whose rays subtend an arc whose length is the radius of the circle
Trigonometric functions (p. 93)	$P = (x, y)$ is the point on the unit circle corresponding to $\theta = t$ radians:

$$\sin t = \sin\theta = y \qquad\qquad \cos t = \cos\theta = x$$

$$\tan t = \tan\theta = \frac{y}{x}, \quad x \neq 0 \qquad \cot t = \cot\theta = \frac{x}{y}, \quad y \neq 0$$

$$\csc t = \csc\theta = \frac{1}{y}, \quad y \neq 0 \qquad \sec t = \sec\theta = \frac{1}{x}, \quad x \neq 0$$

Periodic function (p. 111)	$f(\theta + p) = f(\theta)$, for all θ, $p > 0$, where the smallest such p is the fundamental period

Formulas

1 revolution = 360°
$\qquad\qquad$ = 2π radians (p. 83)

$s = r\theta$ (p. 82)

$A = \dfrac{1}{2}r^2\theta$ (p. 86)

θ is measured in radians; s is the length of the arc subtended by the central angle θ of the circle of radius r; A is the area of the sector.

$v = r\omega$ (p. 87)

v is the linear speed around the circle of radius r; ω is the angular speed (measured in radians per unit time).

TABLE OF VALUES

θ (Radians)	θ (Degrees)	$\sin\theta$	$\cos\theta$	$\tan\theta$	$\csc\theta$	$\sec\theta$	$\cot\theta$
0	0°	0	1	0	Not defined	1	Not defined
$\dfrac{\pi}{6}$	30°	$\dfrac{1}{2}$	$\dfrac{\sqrt{3}}{2}$	$\dfrac{\sqrt{3}}{3}$	2	$\dfrac{2\sqrt{3}}{3}$	$\sqrt{3}$
$\dfrac{\pi}{4}$	45°	$\dfrac{\sqrt{2}}{2}$	$\dfrac{\sqrt{2}}{2}$	1	$\sqrt{2}$	$\sqrt{2}$	1
$\dfrac{\pi}{3}$	60°	$\dfrac{\sqrt{3}}{2}$	$\dfrac{1}{2}$	$\sqrt{3}$	$\dfrac{2\sqrt{3}}{3}$	2	$\dfrac{\sqrt{3}}{3}$
$\dfrac{\pi}{2}$	90°	1	0	Not defined	1	Not defined	0
π	180°	0	-1	0	Not defined	-1	Not defined
$\dfrac{3\pi}{2}$	270°	-1	0	Not defined	-1	Not defined	0

Identities (p. 115)

$$\tan\theta = \frac{\sin\theta}{\cos\theta}, \quad \cot\theta = \frac{\cos\theta}{\sin\theta}$$

$$\csc\theta = \frac{1}{\sin\theta}, \quad \sec\theta = \frac{1}{\cos\theta}, \quad \cot\theta = \frac{1}{\tan\theta}$$

$$\sin^2\theta + \cos^2\theta = 1, \quad \tan^2\theta + 1 = \sec^2\theta, \quad 1 + \cot^2\theta = \csc^2\theta$$

Properties of the Trigonometric Functions

$y = \sin x$ (p. 124)
Domain: $-\infty < x < \infty$
Range: $-1 \le y \le 1$
Periodic: period $= 2\pi$ (360°)
Odd function

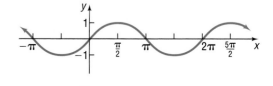

$y = \cos x$ (p. 126)
Domain: $-\infty < x < \infty$
Range: $-1 \le y \le 1$
Periodic: period $= 2\pi$ (360°)
Even function

$y = \tan x$ (p. 139)
Domain: $-\infty < x < \infty$, except odd multiples of $\dfrac{\pi}{2}$ (90°)
Range: $-\infty < y < \infty$
Periodic: period $= \pi$ (180°)
Odd function

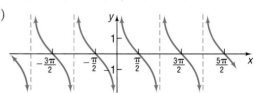

$y = \cot x$ (p. 140)
Domain: $-\infty < x < \infty$, except integral multiples of π (180°)
Range: $-\infty < y < \infty$
Periodic: period $= \pi$ (180°)
Odd function

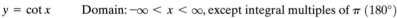

$y = \csc x$ (p. 141)	Domain: $-\infty < x < \infty$, except integral multiples of π (180°) Range: $\lvert y \rvert \geq 1$ Periodic: period $= 2\pi$ (360°) Odd function

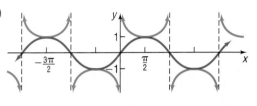

$y = \sec x$ (p. 142)	Domain: $-\infty < x < \infty$, except odd multiples of $\dfrac{\pi}{2}$ (90°) Range: $\lvert y \rvert \geq 1$ Periodic: period $= 2\pi$ (360°) Even function

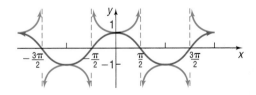

Sinusoidal graphs

$y = A \sin(\omega x), \quad \omega > 0$

$y = A \cos(\omega x), \quad \omega > 0$

$y = A \sin(\omega x - \phi) = A \sin\left[\omega\left(x - \dfrac{\phi}{\omega}\right)\right]$

$y = A \cos(\omega x - \phi) = A \cos\left[\omega\left(x - \dfrac{\phi}{\omega}\right)\right]$

Period $= \dfrac{2\pi}{\omega}$ (p. 129)

Amplitude $= \lvert A \rvert$ (p. 129)

Phase shift $= \dfrac{\phi}{\omega}$ (p. 144)

Objectives

You should be able to:

Convert between degrees, minutes, seconds, and decimal forms for angles (p. 80)

Find the arc length of a circle (p. 82)

Convert from degrees to radians (p. 84)

Convert from radians to degrees (p. 84)

Find the area of a sector of a circle (p. 85)

Find the linear speed of an object traveling in circular motion (p. 86)

Find the exact values of the trigonometric functions using a point on the unit circle (p. 94)

Find the exact values of the trigonometric functions of quadrantal angles (p. 95)

Find the exact values of the trigonometric functions of $\dfrac{\pi}{4} = 45°$ (p. 97)

Find the exact values of the trigonometric functions of $\dfrac{\pi}{6} = 30°$ and $\dfrac{\pi}{3} = 60°$ (p. 98)

Find exact values for certain integral multiples of $\dfrac{\pi}{6} = 30°$, $\dfrac{\pi}{4} = 45°$, and $\dfrac{\pi}{3} = 60°$ (p. 101)

Use a calculator to approximate the value of a trigonometric function (p. 102)

Determine the domain and range of the trigonometric functions (p. 108)

Determine the period of the trigonometric functions (p. 109)

Determine the signs of the trigonometric functions in a given quadrant (p. 112)

Find the values of the trigonometric functions utilizing fundamental identities (p. 115)

Use even-odd properties to find the exact values of the trigonometric functions (p. 119)

Graph transformations of the sine function (p. 124)

Graph transformations of the cosine function (p. 126)

Determine the amplitude and period of sinusoidal functions (p. 128)

Graph sinusoidal functions $y = A \sin(\omega x)$ (p. 130)

Find an equation for a sinusoidal graph (p. 132)

Graph transformations of the tangent and cotangent function (p. 137)

Graph transformations of the secant and cosecant function (p. 140)

Determine the phase shift of a sinusoidal function (p. 143)

Graph sinusoidal functions $y = A \sin(\omega x - \phi)$ (p. 144)

Find a sinusoidal function from data (p. 147)

Fill-in-the-Blank Items

1. Two rays drawn with a common vertex form a(n) _____. One of the rays is called the _____ _____; the other is called the _____ _____.

2. In the formula $s = r\theta$ for measuring the length s of an arc along a circle of radius r, the angle θ must be measured in _____.

3. 180 degrees = _____ radians.

4. An angle is in _____ _____ if its vertex is at the origin and its initial side coincides with the positive x-axis.

5. The sine, cosine, cosecant, and secant functions have period _____; the tangent and cotangent functions have period _____.

6. Which of the trigonometric functions have graphs that are symmetric with respect to the y-axis?

7. Which of the trigonometric functions have graphs that are symmetric with respect to the origin?

8. The following function has amplitude 3 and period 2: $y =$ _____ sin (_____ x).

9. The function $y = 3\sin(6x)$ has amplitude _____ and period _____.

True/False Items

T F   1. In the formula $s = r\theta$, r is the radius of a circle and s is the arc subtended by a central angle θ, where θ is measured in degrees.

T F   2. $|\sin\theta| \leq 1$

T F   3. $1 + \tan^2\theta = \csc^2\theta$

T F   4. The only even trigonometric functions are the cosine and secant functions.

T F   5. The graphs of $y = \tan x$, $y = \cot x$, $y = \sec x$, and $y = \csc x$ each have infinitely many vertical asymptotes.

T F   6. The graphs of $y = \sin x$ and $y = \cos x$ are identical except for a horizontal shift.

T F   7. For $y = 2\sin(\pi x)$, the amplitude is 2 and the period is $\dfrac{\pi}{2}$.

Review Exercises

Blue problem numbers indicate the author's suggestions for use in a Practice Test.

In Problems 1–4, convert each angle in degrees to radians. Express your answer as a multiple of π.

1. 135°   **2.** 210°   **3.** 18°   **4.** 15°

In Problems 5–8, convert each angle in radians to degrees.

5. $\dfrac{3\pi}{4}$ **6.** $\dfrac{2\pi}{3}$ **7.** $-\dfrac{5\pi}{2}$ **8.** $-\dfrac{3\pi}{2}$

In Problems 9–26, find the exact value of each expression. Do not use a calculator.

9. $\tan\dfrac{\pi}{4} - \sin\dfrac{\pi}{6}$ **10.** $\cos\dfrac{\pi}{3} + \sin\dfrac{\pi}{2}$ **11.** $3\sin 45° - 4\tan\dfrac{\pi}{6}$

12. $4\cos 60° + 3\tan\dfrac{\pi}{3}$

13. $6\cos\dfrac{3\pi}{4} + 2\tan\left(-\dfrac{\pi}{3}\right)$

14. $3\sin\dfrac{2\pi}{3} - 4\cos\dfrac{5\pi}{2}$

15. $\sec\left(-\dfrac{\pi}{3}\right) - \cot\left(-\dfrac{5\pi}{4}\right)$

16. $4\csc\dfrac{3\pi}{4} - \cot\left(-\dfrac{\pi}{4}\right)$

17. $\tan\pi + \sin\pi$

18. $\cos\dfrac{\pi}{2} - \csc\left(-\dfrac{\pi}{2}\right)$

19. $\cos 180° - \tan(-45°)$

20. $\sin 270° + \cos(-180°)$

21. $\sin^2 20° + \dfrac{1}{\sec^2 20°}$

22. $\dfrac{1}{\cos^2 40°} - \dfrac{1}{\cot^2 40°}$

23. $\sec 50° \cos 50°$

24. $\tan 10° \cot 10°$

25. $\dfrac{\cos 400°}{\cos(-40°)}$

26. $\dfrac{\tan(-20°)}{\tan 200°}$

In Problems 27–42, find the exact values of each of the remaining trigonometric functions.

27. $\sin\theta = -\dfrac{4}{5}, \quad \cos\theta > 0$

28. $\cos\theta = -\dfrac{3}{5}, \quad \sin\theta < 0$

29. $\tan\theta = \dfrac{12}{5}, \quad \sin\theta < 0$

30. $\cot\theta = \dfrac{12}{5}, \quad \cos\theta < 0$

31. $\sec\theta = -\dfrac{5}{4}, \quad \tan\theta < 0$

32. $\csc\theta = -\dfrac{5}{3}, \quad \cot\theta < 0$

33. $\sin\theta = \dfrac{12}{13}, \quad \theta$ in quadrant II

34. $\cos\theta = -\dfrac{3}{5}, \quad \theta$ in quadrant III

35. $\sin\theta = -\dfrac{5}{13}, \quad \dfrac{3\pi}{2} < \theta < 2\pi$

36. $\cos\theta = \dfrac{12}{13}, \quad \dfrac{3\pi}{2} < \theta < 2\pi$

37. $\tan\theta = \dfrac{1}{3}, \quad 180° < \theta < 270°$

38. $\tan\theta = -\dfrac{2}{3}, \quad 90° < \theta < 180°$

39. $\sec\theta = 3, \quad \dfrac{3\pi}{2} < \theta < 2\pi$

40. $\csc\theta = -4, \quad \pi < \theta < \dfrac{3\pi}{2}$

41. $\cot\theta = -2, \quad \dfrac{\pi}{2} < \theta < \pi$

42. $\tan\theta = -2, \quad \dfrac{3\pi}{2} < \theta < 2\pi$

In Problems 43–54, graph each function. Each graph should contain at least one period.

43. $y = 2\sin(4x)$

44. $y = -3\cos(2x)$

45. $y = -2\cos\left(x + \dfrac{\pi}{2}\right)$

46. $y = 3\sin(x - \pi)$

47. $y = \tan(x + \pi)$

48. $y = -\tan\left(x - \dfrac{\pi}{2}\right)$

49. $y = -2\tan(3x)$

50. $y = 4\tan(2x)$

51. $y = \cot\left(x + \dfrac{\pi}{8}\right)$

52. $y = -4\cot(2x)$

53. $y = \sec\left(x - \dfrac{\pi}{4}\right)$

54. $y = \csc\left(x + \dfrac{\pi}{4}\right)$

In Problems 55–58, determine the amplitude and period of each function without graphing.

55. $y = 4\cos x$

56. $y = \sin(2x)$

57. $y = -8\sin\left(\dfrac{\pi}{2}x\right)$

58. $y = -2\cos(3\pi x)$

In Problems 59–66, find the amplitude, period, and phase shift of each function. Graph each function. Show at least one period.

59. $y = 4\sin(3x)$

60. $2\cos\left(\dfrac{1}{3}x\right)$

61. $y = -2\sin\left(\dfrac{\pi}{2}x + \dfrac{1}{2}\right)$

62. $y = -6\sin(2\pi x - 2)$

63. $y = \dfrac{1}{2}\sin\left(\dfrac{3}{2}x - \pi\right)$

64. $y = \dfrac{3}{2}\cos(6x + 3\pi)$

65. $y = -\dfrac{2}{3}\cos(\pi x - 6) + \dfrac{2}{3}$

66. $y = -7\sin\left(\dfrac{\pi}{3}x + \dfrac{4}{3}\right) - 2$

In Problems 67–70, find a function for the given graph.

67.

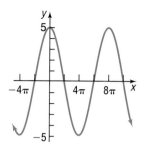

68.

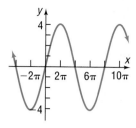

69.

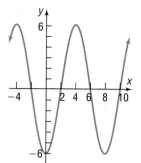

70.

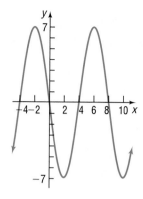

---

71. Find the length of the arc subtended by a central angle of 30° on a circle of radius 2 feet. What is the area of the sector?

72. The minute hand of a clock is 8 inches long. How far does the tip of the minute hand move in 30 minutes? How far does it move in 20 minutes?

73. Angular Speed of a Race Car A race car is driven around a circular track at a constant speed of 180 miles per hour. If the diameter of the track is $\frac{1}{2}$ mile, what is the angular speed of the car? Express your answer in revolutions per hour (which is equivalent to laps per hour).

74. Merry-Go-Rounds A neighborhood carnival has a merry-go-round whose radius is 25 feet. If the time for one revolution is 30 seconds, how fast is the merry-go-round going? Give the linear speed and the angular speed.

75. Lighthouse Beacons The Montauk Point Lighthouse on Long Island has dual beams (two light sources opposite each other). Ships at sea observe a blinking light every 5 seconds. What angular speed is required to do this?

76. Spin Balancing Tires The radius of each wheel of a car is 16 inches. At how many revolutions per minute should a spin balancer be set to balance the tires at a speed of 90 miles per hour? Is the setting different for a wheel of radius 14 inches? What is this setting?

77. Alternating Voltage The electromotive force E, in volts, in a certain ac circuit obeys the equation

$$E(t) = 120 \sin(120\pi t), \qquad t \geq 0$$

where t is measured in seconds.
(a) What is the maximum value of E?
(b) What is the period?
(c) Graph the function over two periods.

78. Alternating Current The current I, in amperes, flowing through an ac (alternating current) circuit at time t is

$$I(t) = 220 \sin\left(30\pi t + \frac{\pi}{6}\right), \qquad t \geq 0$$

(a) What is the period?
(b) What is the amplitude?
(c) What is the phase shift?
(d) Graph this function over two periods.

79. Monthly Temperature The following data represent the average monthly temperatures for Phoenix, Arizona.
(a) Using a graphing utility, draw a scatter diagram of the data for one period.
(b) Find a sinusoidal function of the form $y = A \sin(\omega x - \phi) + B$ that fits the data.
(c) Draw the sinusoidal function found in part (b) on the scatter diagram.
(d) Use a graphing utility to find the sinusoidal function of best fit.
(e) Draw the sinusoidal function of best fit on the scatter diagram.

Month, m	Average Monthly Temperature, T
January, 1	51
February, 2	55
March, 3	63
April, 4	67
May, 5	77
June, 6	86
July, 7	90
August, 8	90
September, 9	84
October, 10	71
November, 11	59
December, 12	52

Source: U.S. National Oceanic and Atmospheric Administration.

Project at Motorola

Digital Transmission over the Air

Digital communications is a revolutionary technology of the century. For many years, Motorola has been one of the leading companies to employ digital communication in wireless devices, such as cell phones.

Figure 1 shows a simplified overview of a digital communication transmission over the air. The information source to be transmitted can be audio, video, or data. The information source may be formatted into a digital sequence of symbols from a finite set $\{\alpha_n\} = \{0, 1\}$. So 0110100 is an example of a digital sequence. The period of the symbols is denoted by T.

The principle of digital communication systems is that, during the finite interval of time T, the information symbol is represented by one digital waveform from a finite set of digital waveforms before it is sent. This technique is called **modulation.**

Modulation techniques use a carrier that is modulated by the information to be transmitted. The modulated carrier is transformed into an electromagnetic field and propagated in the air through an antenna. The unmodulated carrier can be represented in its general form by a sinusoidal function $s(t) = A \sin(\omega_0 t + \phi)$, where A is the amplitude, ω_0 is the radian frequency, and ϕ is the phase.

Let's assume that $A = 1$, $\phi = 0$, and $\omega_0 = 2\pi f_0$ radian, where f_0 is the frequency of the unmodulated carrier.

1. Write $s(t)$ using these assumptions.
2. What is the period, T_0, of the unmodulated carrier?

3. Evaluate $s(t)$ for $t = 0, \dfrac{1}{4f_0}, \dfrac{1}{2f_0}, \dfrac{3}{4f_0},$ and $\dfrac{1}{f_0}$.
4. Graph $s(t)$ for $0 \leq t \leq 12T_0$. That is, graph 12 cycles of the function.
5. For what values of t does the function reach its maximum value?
   [**Hint:** Express t in terms of f_0].

Three modulation techniques are used for transmission over the air: amplitude modulation, frequency modulation and phase modulation. In this project, we are interested in phase modulation. Figure 2 illustrates this process. An information symbol is mapped onto a phase that modulates the carrier. The modulated carrier is expressed by $S_i(t) = \sin(2\pi f_0 t + \psi_i)$.
Let's assume the following mapping scheme:

$$\{\alpha_n\} \rightarrow \{\psi_n\}$$
$$0 \qquad \psi_0 = 0$$
$$1 \qquad \psi_1 = \pi$$

6. Map the binary sequence $M = 010$ into a phase sequence P.
7. What is the expression of the modulated carrier $S_0(t)$ for $\psi_i = \psi_0$ and $S_1(t)$ for $\psi_i = \psi_1$?
8. Let's assume that in the sequence M the period of each symbol is $T = 4T_0$. For each of the three intervals $[0, 4T_0]$, $[4T_0, 8T_0]$, and $[8T_0, 12T_0]$, indicate which of $S_0(t)$ or $S_1(t)$ is the modulated carrier. On the same graph, illustrate $M, P,$ and the modulated carrier for $0 \leq t \leq 12T_0$.

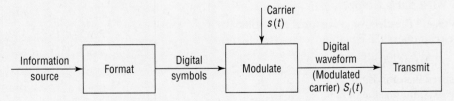

Figure 1   Simplified Overview of a Digital Communication Transmission

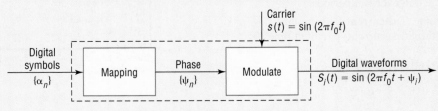

Figure 2   Principle of Phase Modulation

Analytic Trigonometry

Field Trip to Motorola

Trigonometry has a surprisingly wide application. Although it begins as the study of the relation between angles and lines, many functions can be represented as a combination of simple trigonometric functions. This immediately finds application to electric fields, radio waves, light, heat, sound, vibrations in mechanical structures, and other fields. A good understanding of this subject will lay a foundation for a wide range of practical subjects.

PREPARING FOR THIS SECTION

Before getting started, review the following:

✓ One-to-One Functions; Inverse Functions (Section 1.6, pp. 60–69)

✓ Definition of the Trigonometric Functions (Section 2.2, p. 90)

✓ Values of the Trigonometric Functions of Certain Angles (Section 2.2, p. 96 and p. 100)

✓ Domain and Range of the Sine, Cosine, and Tangent Functions (Section 2.3, pp. 108–110)

✓ Graphs of the Sine, Cosine, and Tangent Functions (Section 2.4, pp. 123–127, and Section 2.5, pp. 137–139)

3.1 THE INVERSE SINE, COSINE, AND TANGENT FUNCTIONS

OBJECTIVES
1. Find the Exact Value of the Inverse Sine, Cosine, and Tangent Functions
2. Find an Approximate Value of the Inverse Sine, Cosine, and Tangent Functions

In Section 1.6 we discussed inverse functions, and we noted that if a function is one-to-one it will have an inverse function. We also observed that if a function is not one-to-one it may be possible to restrict its domain in some suitable manner so that the restricted function is one-to-one.

Next, we review some properties of a function f and its inverse function f^{-1}.

1. $f^{-1}(f(x)) = x$ for every x in the domain of f and $f(f^{-1}(x)) = x$ for every x in the domain of f^{-1}.
2. Domain of f = range of f^{-1} and range of f = domain of f^{-1}.
3. The graph of f and the graph of f^{-1} are symmetric with respect to the line $y = x$.
4. If a function $y = f(x)$ has an inverse function, the equation of the inverse function is $x = f(y)$. The solution of this equation is $y = f^{-1}(x)$.

THE INVERSE SINE FUNCTION

In Figure 1, we reproduce the graph of $y = \sin x$. Because every horizontal line $y = b$, where b is between -1 and 1, intersects the graph of $y = \sin x$ infinitely many times, it follows from the horizontal-line test that the function $y = \sin x$ is not one-to-one.

Figure 1

$y = \sin x, -\infty < x < \infty, -1 \leq y \leq 1$

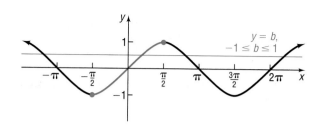

Figure 2

$y = \sin x, -\dfrac{\pi}{2} \leq x \leq \dfrac{\pi}{2}, -1 \leq y \leq 1$

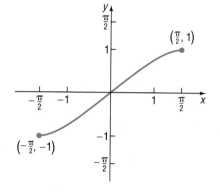

However, if we restrict the domain of $y = \sin x$ to the interval $\left[-\dfrac{\pi}{2}, \dfrac{\pi}{2}\right]$, the restricted function

$$y = \sin x, \qquad -\dfrac{\pi}{2} \leq x \leq \dfrac{\pi}{2}$$

is one-to-one and, hence, will have an inverse function.* See Figure 2.

An equation for the inverse of $y = f(x) = \sin x$ is obtained by interchanging x and y. The implicit form of the inverse function is $x = \sin y$, $-\dfrac{\pi}{2} \leq y \leq \dfrac{\pi}{2}$. The explicit form is called the **inverse sine** of x and is symbolized by $y = f^{-1}(x) = \sin^{-1} x$.

$$y = \sin^{-1} x \quad \text{means} \quad x = \sin y \qquad \textbf{(1)}$$

$$\text{where } -1 \leq x \leq 1 \quad \text{and} \quad -\dfrac{\pi}{2} \leq y \leq \dfrac{\pi}{2}$$

Because $y = \sin^{-1} x$ means $x = \sin y$, we read $y = \sin^{-1} x$ as "y is the angle or real number whose sine equals x." Alternatively, we can say that "y is the inverse sine of x." Be careful about the notation used. The superscript -1 that appears in $y = \sin^{-1} x$ is not an exponent, but is reminiscent of the symbolism f^{-1} used to denote the inverse function of f. [To avoid this notation, some books use the notation $y = \arcsin x$ instead of $y = \sin^{-1} x$.]

The inverse of a function f receives as input an element from the range of f and returns as output an element in the domain of f. The restricted sine function, $y = f(x) = \sin x$, receives as input an angle or real number x in the interval $\left[-\dfrac{\pi}{2}, \dfrac{\pi}{2}\right]$ and outputs a real number in the interval $[-1, 1]$. Therefore, the inverse sine function receives as input a real number in the interval $[-1, 1]$ and outputs an angle or real number in the interval $\left[-\dfrac{\pi}{2}, \dfrac{\pi}{2}\right]$.

Since the domain of f = range of f^{-1} and the range of f = domain of f^{-1}, the domain of the inverse sine function, $y = f^{-1}(x) = \sin^{-1} x$, is $[-1, 1]$ or $-1 \leq x \leq 1$, and the range of the inverse sine function is $\left[-\dfrac{\pi}{2}, \dfrac{\pi}{2}\right]$ or $-\dfrac{\pi}{2} \leq y \leq \dfrac{\pi}{2}$. The graph of the inverse sine function can be obtained by reflecting the restricted portion of the graph of $y = f(x) = \sin x$ about the line $y = x$, as shown in Figure 3.

Figure 3

$y = \sin^{-1} x, -1 \leq x \leq 1, -\dfrac{\pi}{2} \leq y \leq \dfrac{\pi}{2}$

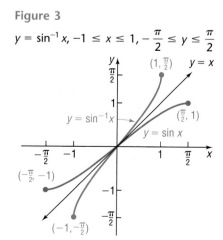

---

*Although there are many other ways to restrict the domain and obtain a one-to-one function, mathematicians have agreed on a consistent use of the interval $\left[-\dfrac{\pi}{2}, \dfrac{\pi}{2}\right]$ in order to define the inverse of $y = \sin x$.

Check: Graph $Y = \sin^{-1} x$ and compare the result with Figure 3.

When we discussed functions and their inverses in Section 1.6, we found that $f^{-1}(f(x)) = x$ and $f(f^{-1}(x)) = x$. In terms of the sine function and its inverse, these properties are of the form

$$f^{-1}(f(x)) = \sin^{-1}(\sin x) = x, \qquad \text{where } -\frac{\pi}{2} \leq x \leq \frac{\pi}{2} \qquad \textbf{(2a)}$$

$$f(f^{-1}(x)) = \sin(\sin^{-1} x) = x, \qquad \text{where } -1 \leq x \leq 1 \qquad \textbf{(2b)}$$

For example,

$$\sin^{-1}\left[\sin\left(\frac{\pi}{8}\right)\right] = \frac{\pi}{8}$$

because $\frac{\pi}{8}$ lies in the interval $\left[-\frac{\pi}{2}, \frac{\pi}{2}\right]$, the restricted domain of the sine function. Also,

$$\sin\left[\sin^{-1}(0.8)\right] = 0.8$$

because 0.8 lies in the interval $[-1, 1]$, the domain of the inverse sine function. See Figure 4 for these calculations on a graphing calculator.

However,

$$\sin^{-1}\left[\sin\left(\frac{5\pi}{8}\right)\right] \neq \frac{5\pi}{8}$$

because $\frac{5\pi}{8}$ is not in the interval $\left[-\frac{\pi}{2}, \frac{\pi}{2}\right]$. See Figure 5. Also,

$$\sin\left[\sin^{-1}(1.8)\right] \neq 1.8$$

because 1.8 is not in the interval $[-1, 1]$. See Figure 6. Can you explain why the error appears?

Figure 4

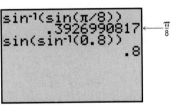

$\leftarrow \frac{\pi}{8}$

Figure 5

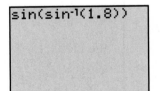

Figure 6

1 For some numbers x it is possible to find the exact value of $y = \sin^{-1} x$.

EXAMPLE 1 **Finding the Exact Value of an Inverse Sine Function**

Find the exact value of: $\sin^{-1} 1$

Solution Let $\theta = \sin^{-1} 1$. We seek the angle θ, $-\frac{\pi}{2} \leq \theta \leq \frac{\pi}{2}$, whose sine equals 1.

$$\theta = \sin^{-1} 1, \qquad -\frac{\pi}{2} \leq \theta \leq \frac{\pi}{2}$$

$$\sin \theta = 1, \qquad -\frac{\pi}{2} \leq \theta \leq \frac{\pi}{2} \qquad \text{By definition of } y = \sin^{-1} x.$$

θ	$\sin\theta$
$-\dfrac{\pi}{2}$	-1
$-\dfrac{\pi}{3}$	$-\dfrac{\sqrt{3}}{2}$
$-\dfrac{\pi}{4}$	$-\dfrac{\sqrt{2}}{2}$
$-\dfrac{\pi}{6}$	$-\dfrac{1}{2}$
0	0
$\dfrac{\pi}{6}$	$\dfrac{1}{2}$
$\dfrac{\pi}{4}$	$\dfrac{\sqrt{2}}{2}$
$\dfrac{\pi}{3}$	$\dfrac{\sqrt{3}}{2}$
$\dfrac{\pi}{2}$	1

TABLE 1

Now look at Table 1 and Figure 7.

Figure 7

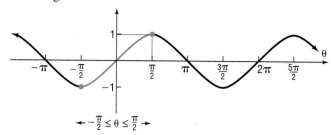

$$-\frac{\pi}{2} \le \theta \le \frac{\pi}{2}$$

We see that the only angle θ within the interval $\left[-\dfrac{\pi}{2}, \dfrac{\pi}{2}\right]$ whose sine is 1 is $\dfrac{\pi}{2}$. [Note that $\sin\dfrac{5\pi}{2}$ also equals 1, but $\dfrac{5\pi}{2}$ lies outside the interval $\left[-\dfrac{\pi}{2}, \dfrac{\pi}{2}\right]$ and hence is not admissible.] So, since $\sin\dfrac{\pi}{2} = 1$ and $\dfrac{\pi}{2}$ is in $\left[-\dfrac{\pi}{2}, \dfrac{\pi}{2}\right]$, we conclude that

$$\sin^{-1} 1 = \frac{\pi}{2}$$

■

NOW WORK PROBLEM **1**.

EXAMPLE 2

Finding the Exact Value of an Inverse Sine Function

Find the exact value of: $\sin^{-1}\left(-\dfrac{1}{2}\right)$

Solution Let $\theta = \sin^{-1}\left(-\dfrac{1}{2}\right)$. We seek the angle θ, $-\dfrac{\pi}{2} \le \theta \le \dfrac{\pi}{2}$, whose sine equals $-\dfrac{1}{2}$.

$$\theta = \sin^{-1}\left(-\frac{1}{2}\right), \qquad -\frac{\pi}{2} \le \theta \le \frac{\pi}{2}$$

$$\sin\theta = -\frac{1}{2}, \qquad\qquad -\frac{\pi}{2} \le \theta \le \frac{\pi}{2}$$

(Refer to Table 1 and Figure 7, if necessary.) The only angle within the interval $\left[-\dfrac{\pi}{2}, \dfrac{\pi}{2}\right]$ whose sine is $-\dfrac{1}{2}$ is $-\dfrac{\pi}{6}$. So, since $\sin\left(-\dfrac{\pi}{6}\right) = -\dfrac{1}{2}$ and $-\dfrac{\pi}{6}$ is in the interval $\left[-\dfrac{\pi}{2}, \dfrac{\pi}{2}\right]$, we conclude that

$$\sin^{-1}\left(-\frac{1}{2}\right) = -\frac{\pi}{6}$$

■

NOW WORK PROBLEM **7**.

2   For most numbers x, the value $y = \sin^{-1} x$ must be approximated.

EXAMPLE 3

Finding an Approximate Value of an Inverse Sine Function

Find an approximate value of:

(a) $\sin^{-1}\dfrac{1}{3}$ (b) $\sin^{-1}\left(-\dfrac{1}{4}\right)$

Express the answer in radians rounded to two decimal places.

Solution Because we want the angle measured in radians, we first set the mode to radians.

(a) Keystrokes:*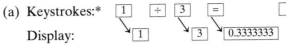

Then, $\sin^{-1}\dfrac{1}{3} = 0.34$, rounded to two decimal places.

(b) Figure 8 shows the solution using a TI-83 graphing calculator. Then

$$\sin^{-1}\left(-\frac{1}{4}\right) = -0.25$$

rounded to two decimal places. ∎

Figure 8

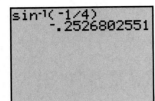

 NOW WORK PROBLEM **13.**

THE INVERSE COSINE FUNCTION

In Figure 9 we reproduce the graph of $y = \cos x$. Because every horizontal line $y = b$, where b is between -1 and 1, intersects the graph of $y = \cos x$ infinitely many times, it follows that the cosine function is not one-to-one.

Figure 9
$y = \cos x, -\infty < x < \infty, -1 \le y \le 1$

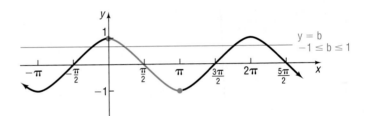

Figure 10
$y = \cos x, 0 \le x \le \pi, -1 \le y \le 1$

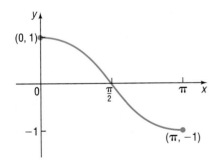

However, if we restrict the domain of $y = \cos x$ to the interval $[0, \pi]$, the restricted function

$$y = \cos x, \qquad 0 \le x \le \pi$$

is one-to-one and hence will have an inverse function.[†] See Figure 10.

An equation for the inverse of $y = f(x) = \cos x$ is obtained by interchanging x and y. The implicit form of the inverse function is $x = \cos y$, $0 \le y \le \pi$. The explicit form is called the **inverse cosine** of x and is symbolized by $y = f^{-1}(x) = \cos^{-1} x$ (or by $y = \arccos x$).

$$y = \cos^{-1} x \quad \text{means} \quad x = \cos y \tag{3}$$

$$\text{where} \quad -1 \le x \le 1 \quad \text{and} \quad 0 \le y \le \pi$$

Here, y is the angle whose cosine is x. The domain of the function $y = \cos^{-1} x$ is $-1 \le x \le 1$, and its range is $0 \le y \le \pi$. (Do you know why?) The graph of $y = \cos^{-1} x$ can be obtained by reflecting the restricted portion of the graph of $y = \cos x$ about the line $y = x$, as shown in Figure 11.

---

*On most calculators, the inverse sine is obtained by pressing SHIFT or 2nd, followed by sin. Also, on some calculators, $\sin^{-1}$ is pressed first; then 1/3 is entered. Consult your owner's manual for the correct sequence.
† This is the generally accepted restriction to define the inverse.

Figure 11
$y = \cos^{-1} x, -1 \le x \le 1, 0 \le y \le \pi$

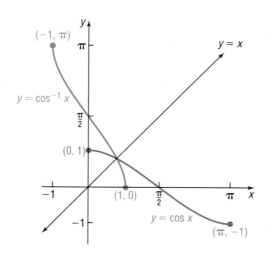

Check: Graph $Y = \cos^{-1} x$ and compare the result with Figure 11.

For the cosine function and its inverse, the following properties hold:

$$f^{-1}(f(x)) = \cos^{-1}(\cos x) = x, \qquad \text{where } 0 \le x \le \pi \qquad \textbf{(4a)}$$
$$f(f^{-1}(x)) = \cos(\cos^{-1} x) = x, \qquad \text{where } -1 \le x \le 1 \qquad \textbf{(4b)}$$

EXAMPLE 4   Finding the Exact Value of an Inverse Cosine Function

Find the exact value of: $\cos^{-1} 0$

Solution Let $\theta = \cos^{-1} 0$. We seek the angle $\theta, 0 \le \theta \le \pi$, whose cosine equals 0.

$$\theta = \cos^{-1} 0, \qquad 0 \le \theta \le \pi$$
$$\cos\theta = 0, \qquad 0 \le \theta \le \pi$$

Look at Table 2 and Figure 12.

Figure 12

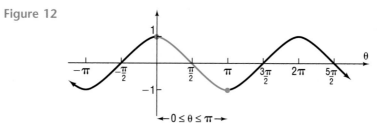

We see that the only angle θ within the interval $[0, \pi]$ whose cosine is 0 is $\dfrac{\pi}{2}$. $\left[\text{Note that } \cos\dfrac{3\pi}{2} \text{ also equals 0, but } \dfrac{3\pi}{2} \text{ lies outside the interval } [0, \pi] \text{ and hence is not admissible.}\right]$ So, since $\cos\dfrac{\pi}{2} = 0$ and $\dfrac{\pi}{2}$ is in the interval $[0, \pi]$, we conclude that

$$\cos^{-1} 0 = \frac{\pi}{2}$$

TABLE	2
θ	$\cos\theta$
0	1
$\dfrac{\pi}{6}$	$\dfrac{\sqrt{3}}{2}$
$\dfrac{\pi}{4}$	$\dfrac{\sqrt{2}}{2}$
$\dfrac{\pi}{3}$	$\dfrac{1}{2}$
$\dfrac{\pi}{2}$	0
$\dfrac{2\pi}{3}$	$-\dfrac{1}{2}$
$\dfrac{3\pi}{4}$	$-\dfrac{\sqrt{2}}{2}$
$\dfrac{5\pi}{6}$	$-\dfrac{\sqrt{3}}{2}$
π	-1

| EXAMPLE 5 | **Finding the Exact Value of an Inverse Cosine Function** |

Find the exact value of: $\cos^{-1}\dfrac{\sqrt{2}}{2}$

Solution Let $\theta = \cos^{-1}\dfrac{\sqrt{2}}{2}$. We seek the angle $\theta, 0 \le \theta \le \pi$, whose cosine equals $\dfrac{\sqrt{2}}{2}$.

$$\theta = \cos^{-1}\dfrac{\sqrt{2}}{2}, \qquad 0 \le \theta \le \pi$$

$$\cos\theta = \dfrac{\sqrt{2}}{2}, \qquad 0 \le \theta \le \pi$$

Look at Table 2 and Figure 13.

Figure 13

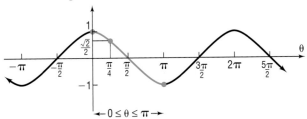

We see that the only angle θ within the interval $[0, \pi]$, whose cosine is $\dfrac{\sqrt{2}}{2}$ is $\dfrac{\pi}{4}$. So, since $\cos\dfrac{\pi}{4} = \dfrac{\sqrt{2}}{2}$ and $\dfrac{\pi}{4}$ is in the interval $[0, \pi]$, we conclude that

$$\cos^{-1}\dfrac{\sqrt{2}}{2} = \dfrac{\pi}{4}$$

NOW WORK PROBLEM **11**.

| EXAMPLE 6 | **Finding the Exact Value of a Composite Function** |

Find the exact value of: (a) $\cos^{-1}\left[\cos\left(\dfrac{\pi}{12}\right)\right]$ (b) $\cos\left[\cos^{-1}(-0.4)\right]$

Solution (a) $\cos^{-1}\left[\cos\left(\dfrac{\pi}{12}\right)\right] = \dfrac{\pi}{12}$ By Property (4a).

(b) $\cos\left[\cos^{-1}(-0.4)\right] = -0.4$ By Property (4b).

NOW WORK PROBLEM **27**.

THE INVERSE TANGENT FUNCTION

In Figure 14 we reproduce the graph of $y = \tan x$. Because every horizontal line intersects the graph infinitely many times, it follows that the tangent function is not one-to-one.

Figure 14

$y = \tan x, -\infty < x < \infty$, x not equal to odd multiples of $\dfrac{\pi}{2}, -\infty < y < \infty$

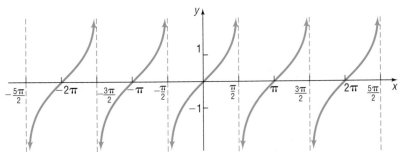

Figure 15

$y = \tan x, -\dfrac{\pi}{2} < x < \dfrac{\pi}{2}, -\infty < y < \infty$

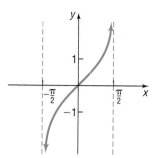

However, if we restrict the domain of $y = \tan x$ to the interval $\left(-\dfrac{\pi}{2}, \dfrac{\pi}{2}\right)$,* the restricted function

$$y = \tan x, \qquad -\dfrac{\pi}{2} < x < \dfrac{\pi}{2}$$

is one-to-one and hence has an inverse function. See Figure 15.

An equation for the inverse of $y = f(x) = \tan x$ is obtained by interchanging x and y. The implicit form of the inverse function is $x = \tan y$, $-\dfrac{\pi}{2} < y < \dfrac{\pi}{2}$. The explicit form is called the **inverse tangent** of x and is symbolized by $y = f^{-1}(x) = \tan^{-1} x$ (or by $y = \arctan x$).

$$y = \tan^{-1} x \quad \text{means} \quad x = \tan y \qquad\qquad (5)$$
$$\text{where} \quad -\infty < x < \infty \quad \text{and} \quad -\dfrac{\pi}{2} < y < \dfrac{\pi}{2}$$

Here, y is the angle whose tangent is x. The domain of the function $y = \tan^{-1} x$ is $-\infty < x < \infty$, and its range is $-\dfrac{\pi}{2} < y < \dfrac{\pi}{2}$. The graph of $y = \tan^{-1} x$ can be obtained by reflecting the restricted portion of the graph of $y = \tan x$ about the line $y = x$, as shown in Figure 16.

Figure 16

$y = \tan^{-1} x, -\infty < x < \infty,$

$-\dfrac{\pi}{2} < y < \dfrac{\pi}{2}$

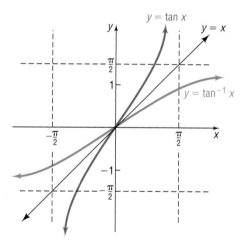

Check: Graph $Y = \tan^{-1} x$ and compare the result with Figure 16.

For the tangent function and its inverse, the following properties hold:

$$f^{-1}\big(f(x)\big) = \tan^{-1}(\tan x) = x, \qquad \text{where } -\dfrac{\pi}{2} < x < \dfrac{\pi}{2}$$
$$f\big(f^{-1}(x)\big) = \tan(\tan^{-1} x) = x, \qquad \text{where } -\infty < x < \infty$$

*This is the generally accepted restriction.

| EXAMPLE 7 | Finding the Exact Value of an Inverse Tangent Function |

Find the exact value of: $\tan^{-1}1$

TABLE 3

θ	$\tan\theta$
$-\dfrac{\pi}{2}$	Undefined
$-\dfrac{\pi}{3}$	$-\sqrt{3}$
$-\dfrac{\pi}{4}$	-1
$-\dfrac{\pi}{6}$	$-\dfrac{\sqrt{3}}{3}$
0	0
$\dfrac{\pi}{6}$	$\dfrac{\sqrt{3}}{3}$
$\dfrac{\pi}{4}$	1
$\dfrac{\pi}{3}$	$\sqrt{3}$
$\dfrac{\pi}{2}$	Undefined

Solution   Let $\theta = \tan^{-1}1$. We seek the angle $\theta, -\dfrac{\pi}{2} < \theta < \dfrac{\pi}{2}$, whose tangent equals 1.

$$\theta = \tan^{-1}1, \qquad -\frac{\pi}{2} < \theta < \frac{\pi}{2}$$

$$\tan\theta = 1, \qquad -\frac{\pi}{2} < \theta < \frac{\pi}{2}$$

Look at Table 3 or Figure 15. The only angle θ within the interval $\left(-\dfrac{\pi}{2}, \dfrac{\pi}{2}\right)$ whose tangent is 1 is $\dfrac{\pi}{4}$. So, since $\tan\dfrac{\pi}{4} = 1$ and $\dfrac{\pi}{4}$ is in the interval $\left(-\dfrac{\pi}{2}, \dfrac{\pi}{2}\right)$, we conclude that

$$\tan^{-1}1 = \frac{\pi}{4}$$

| EXAMPLE 8 | Finding the Exact Value of an Inverse Tangent Function |

Find the exact value of: $\tan^{-1}\left(-\sqrt{3}\right)$

Solution   Let $\theta = \tan^{-1}\left(-\sqrt{3}\right)$. We seek the angle $\theta, -\dfrac{\pi}{2} < \theta < \dfrac{\pi}{2}$, whose tangent equals $-\sqrt{3}$.

$$\theta = \tan^{-1}\left(-\sqrt{3}\right), \qquad -\frac{\pi}{2} < \theta < \frac{\pi}{2}$$

$$\tan\theta = -\sqrt{3}, \qquad -\frac{\pi}{2} < \theta < \frac{\pi}{2}$$

Look at Table 3 or Figure 15 if necessary. The only angle θ within the interval $\left(-\dfrac{\pi}{2}, \dfrac{\pi}{2}\right)$, whose tangent is $-\sqrt{3}$, is $-\dfrac{\pi}{3}$. So, since $\tan\left(-\dfrac{\pi}{3}\right) = -\sqrt{3}$ and $-\dfrac{\pi}{3}$ is in the interval $\left(-\dfrac{\pi}{2}, \dfrac{\pi}{2}\right)$, we conclude that

$$\tan^{-1}\left(-\sqrt{3}\right) = -\frac{\pi}{3}$$

NOW WORK PROBLEM **5**.

3.1 EXERCISES

In Problems 1–12, find the exact value of each expression.

1. $\sin^{-1}0$
2. $\cos^{-1}1$
3. $\sin^{-1}(-1)$
4. $\cos^{-1}(-1)$
5. $\tan^{-1}0$
6. $\tan^{-1}(-1)$
7. $\sin^{-1}\dfrac{\sqrt{2}}{2}$
8. $\tan^{-1}\dfrac{\sqrt{3}}{3}$
9. $\tan^{-1}\sqrt{3}$
10. $\sin^{-1}\left(-\dfrac{\sqrt{3}}{2}\right)$
11. $\cos^{-1}\left(-\dfrac{\sqrt{3}}{2}\right)$
12. $\sin^{-1}\left(-\dfrac{\sqrt{2}}{2}\right)$

In Problems 13–24, use a calculator to find the value of each expression rounded to two decimal places.

13. $\sin^{-1}0.1$
14. $\cos^{-1}0.6$
15. $\tan^{-1}5$
16. $\tan^{-1}0.2$

17. $\cos^{-1}\dfrac{7}{8}$ **18.** $\sin^{-1}\dfrac{1}{8}$ **19.** $\tan^{-1}(-0.4)$ **20.** $\tan^{-1}(-3)$

21. $\sin^{-1}(-0.12)$ **22.** $\cos^{-1}(-0.44)$ **23.** $\cos^{-1}\dfrac{\sqrt{2}}{3}$ **24.** $\sin^{-1}\dfrac{\sqrt{3}}{5}$

In Problems 25–32, find the exact value of the expression. Do not use a calculator.

25. $\sin\left[\sin^{-1}(0.54)\right]$ **26.** $\tan\left[\tan^{-1}(7.4)\right]$ **27.** $\cos^{-1}\left[\cos\left(\dfrac{4\pi}{5}\right)\right]$ **28.** $\sin^{-1}\left[\sin\left(-\dfrac{\pi}{10}\right)\right]$

29. $\tan\left[\tan^{-1}(-3.5)\right]$ **30.** $\cos\left[\cos^{-1}(-0.05)\right]$ **31.** $\sin^{-1}\left[\sin\left(-\dfrac{3\pi}{7}\right)\right]$ **32.** $\tan^{-1}\left[\tan\left(\dfrac{2\pi}{5}\right)\right]$

In Problems 33–44, do not use a calculator.

33. Does $\sin^{-1}\left[\sin\left(-\dfrac{\pi}{6}\right)\right] = -\dfrac{\pi}{6}$? Why or why not?   **34.** Does $\sin^{-1}\left[\sin\left(\dfrac{2\pi}{3}\right)\right] = \dfrac{2\pi}{3}$? Why or why not?

35. Does $\sin\left[\sin^{-1}(2)\right] = 2$? Why or why not?   **36.** Does $\sin\left[\sin^{-1}\left(-\dfrac{1}{2}\right)\right] = -\dfrac{1}{2}$? Why or why not?

37. Does $\cos^{-1}\left[\cos\left(-\dfrac{\pi}{6}\right)\right] = -\dfrac{\pi}{6}$? Why or why not?   **38.** Does $\cos^{-1}\left[\cos\left(\dfrac{2\pi}{3}\right)\right] = \dfrac{2\pi}{3}$? Why or why not?

39. Does $\cos\left[\cos^{-1}\left(-\dfrac{1}{2}\right)\right] = -\dfrac{1}{2}$? Why or why not?   **40.** Does $\cos\left[\cos^{-1}(2)\right] = 2$? Why or why not?

41. Does $\tan^{-1}\left[\tan\left(-\dfrac{\pi}{3}\right)\right] = -\dfrac{\pi}{3}$? Why or why not?   **42.** Does $\tan^{-1}\left[\tan\left(\dfrac{2\pi}{3}\right)\right] = \dfrac{2\pi}{3}$? Why or why not?

43. Does $\tan\left[\tan^{-1}(2)\right] = 2$? Why or why not?   **44.** Does $\tan\left[\tan^{-1}\left(-\dfrac{1}{2}\right)\right] = -\dfrac{1}{2}$? Why or why not?

In Problems 45–50, use the following: The formula

$$D = 24\left[1 - \frac{\cos^{-1}(\tan i \tan\theta)}{\pi}\right]$$

can be used to approximate the number of hours of daylight when the declination of the Sun is $i°$ at a location $\theta°$ north latitude for any date between the vernal equinox and autumnal equinox. The declination of the Sun is defined as the angle i between the equatorial plane and any ray of light from the Sun. The latitude of a location is the angle θ between the Equator and the location on the surface of Earth, with the vertex of the angle located at the center of Earth. See the figure. To use the formula, $\cos^{-1}(\tan i \tan\theta)$ must be expressed in radians.

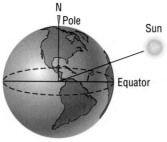
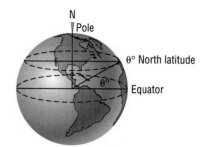

45. Approximate the number of hours of daylight in Houston, Texas (29°45′ north latitude), for the following dates:
(a) Summer solstice ($i = 23.5°$)
(b) Vernal equinox ($i = 0°$)
(c) July 4 ($i = 22°48′$)

46. Approximate the number of hours of daylight in New York, New York (40°45′ north latitude), for the following dates:
(a) Summer solstice ($i = 23.5°$)
(b) Vernal equinox ($i = 0°$)
(c) July 4 ($i = 22°48′$)

47. Approximate the number of hours of daylight in Honolulu, Hawaii (21°18′ north latitude), for the following dates:
(a) Summer solstice ($i = 23.5°$)
(b) Vernal equinox ($i = 0°$)
(c) July 4 ($i = 22°48′$)

48. Approximate the number of hours of daylight in Anchorage, Alaska (61°10′ north latitude), for the following dates:
(a) Summer solstice ($i = 23.5°$)
(b) Vernal equinox ($i = 0°$)
(c) July 4 ($i = 22°48′$)

49. Approximate the number of hours of daylight at the Equator (0° north latitude) for the following dates:
   (a) Summer solstice ($i = 23.5°$)
   (b) Vernal equinox ($i = 0°$)
   (c) July 4 ($i = 22°48'$)
   (d) What do you conclude about the number of hours of daylight throughout the year for a location at the Equator?

50. Approximate the number of hours of daylight for any location that is 66°30' north latitude for the following dates:
   (a) Summer solstice ($i = 23.5°$)
   (b) Vernal equinox ($i = 0°$)
   (c) July 4 ($i = 22°48'$)
   (d) The number of hours of daylight on the winter solstice may be found by computing the number of hours of daylight on the summer solstice and subtracting this result from 24 hours, due to the symmetry of the orbital path of Earth around the Sun. Compute the number of hours of daylight for this location on the winter solstice. What do you conclude about daylight for a location at 66°30' north latitude?

51. Being the First to See the Rising Sun   Cadillac Mountain, elevation 1530 feet, is located in Acadia National Park, Maine, and is the highest peak on the east coast of the United States. It is said that a person standing on the summit will be the first person in the United States to see the rays of the rising Sun. How much sooner would a person atop Cadillac Mountain see the first rays than a person standing below, at sea level?
[**Hint:** Consult the figure. When the person at D sees the first rays of the Sun, the person at P does not. The person at P sees the first rays of the Sun only after Earth has rotated so that P is at location Q. Compute the length of the arc subtended by the central angle θ. Then use the fact that, at the latitude of Cadillac Mountain, in 24 hours a length of 2π (2710) miles is subtended, and find the time it takes to subtend this length.]

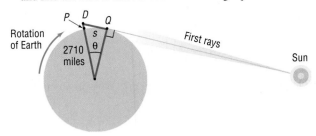

Before getting started, review the following concepts:

✓ Finding Exact Values Given the Value of a Trigonometric Function and the Quadrant of the Angle (Section 2.3, pp. 116–118)

✓ Domain and Range of the Secant, Cosecant, and Cotangent Functions (Section 2.3, pp. 108–110)

✓ Graphs of the Secant, Cosecant, and Cotangent Functions (Section 2.5, pp. 140–142)

3.2   THE INVERSE TRIGONOMETRIC FUNCTIONS (Continued)

OBJECTIVES   ① Find the Exact Value of Expressions Involving the Inverse Sine, Cosine, and Tangent Functions
   ② Know the Definition of the Inverse Secant, Cosecant, and Cotangent Functions
   ③ Use a Calculator to Evaluate $\sec^{-1} x$, $\csc^{-1} x$, and $\cot^{-1} x$

① In this section we continue our discussion of the inverse trigonometric functions.

EXAMPLE 1   **Finding the Exact Value of Expressions Involving Inverse Trigonometric Functions**

Find the exact value of: $\sin^{-1}\left(\sin \dfrac{5\pi}{4} \right)$

Solution $\sin^{-1}\left(\sin \dfrac{5\pi}{4} \right) = \sin^{-1}\left(-\dfrac{\sqrt{2}}{2} \right) = -\dfrac{\pi}{4}$

Notice in the solution to Example 1 that we did not use Property (2a), page 164. This is because the argument of the sine function is not in the interval $\left[-\dfrac{\pi}{2}, \dfrac{\pi}{2}\right]$, as required. If we use the fact that

$$\sin\frac{5\pi}{4} = -\sin\frac{\pi}{4} = \sin\left(-\frac{\pi}{4}\right)$$
$$\underset{\substack{\uparrow \\ y = \sin x \text{ is odd}}}{}$$

then we can use Property (2a):

$$\sin^{-1}\left(\sin\frac{5\pi}{4}\right) = \sin^{-1}\left[\sin\left(-\frac{\pi}{4}\right)\right] = -\frac{\pi}{4}$$
$$\underset{\substack{\uparrow \\ \text{Property (2a)}}}{}$$

NOW WORK PROBLEM **13.**

EXAMPLE 2

Finding the Exact Value of Expressions Involving Inverse Trigonometric Functions

Find the exact value of: $\sin\left(\tan^{-1}\dfrac{1}{2}\right)$

Figure 17

$\tan\theta = \dfrac{1}{2}$

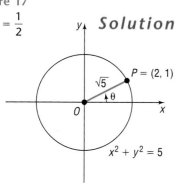

Solution   Let $\theta = \tan^{-1}\dfrac{1}{2}$. Then $\tan\theta = \dfrac{1}{2}$, where $-\dfrac{\pi}{2} < \theta < \dfrac{\pi}{2}$. Because $\tan\theta > 0$, it follows that $0 < \theta < \dfrac{\pi}{2}$, so θ lies in quadrant I. Since $\tan\theta = \dfrac{1}{2} = \dfrac{y}{x}$, we let $x = 2$ and $y = 1$. The point $P = (x, y) = (2, 1)$ is on the circle $x^2 + y^2 = 5$, since $r = d(O, P) = \sqrt{2^2 + 1^2} = \sqrt{5}$. See Figure 17. Then, with $x = 2$, $y = 1$, and $r = \sqrt{5}$, we have

$$\sin\left(\tan^{-1}\frac{1}{2}\right) = \sin\theta = \frac{1}{\sqrt{5}} = \frac{\sqrt{5}}{5}$$
$$\underset{\substack{\uparrow \\ \sin\theta = \frac{y}{r}}}{}$$

EXAMPLE 3

Finding the Exact Value of Expressions Involving Inverse Trigonometric Functions

Find the exact value of: $\cos\left[\sin^{-1}\left(-\dfrac{1}{3}\right)\right]$

Solution   Let $\theta = \sin^{-1}\left(-\dfrac{1}{3}\right)$. Then $\sin\theta = -\dfrac{1}{3}$ and $-\dfrac{\pi}{2} \le \theta \le \dfrac{\pi}{2}$. Because $\sin\theta < 0$, it follows that $-\dfrac{\pi}{2} \le \theta < 0$, so θ lies in quadrant IV. Since $\sin\theta = \dfrac{-1}{3} = \dfrac{y}{r}$, we let $y = -1$ and $r = 3$. The point $P = (x, y) = (x, -1)$,

Figure 18

$\sin\theta = -\dfrac{1}{3}$

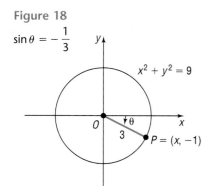

$x > 0$, is on a circle of radius 3, $x^2 + y^2 = 9$. See Figure 18. Then,

$$x^2 + y^2 = 9, \qquad x > 0, y = -1$$
$$x^2 + (-1)^2 = 9$$
$$x^2 = 8$$
$$x = 2\sqrt{2}$$

Then, we have $x = 2\sqrt{2}$, $y = -1$, $r = 3$, so that

$$\cos\left[\sin^{-1}\left(-\frac{1}{3}\right)\right] = \cos\theta = \frac{2\sqrt{2}}{3}$$
$$\underset{\substack{\uparrow \\ \cos\theta = \frac{x}{r}}}{}$$

EXAMPLE 4

Finding the Exact Value of Expressions Involving Inverse Trigonometric Functions

Find the exact value of: $\tan\left[\cos^{-1}\left(-\dfrac{1}{3}\right)\right]$

Solution Let $\theta = \cos^{-1}\left(-\dfrac{1}{3}\right)$. Then $\cos\theta = -\dfrac{1}{3}$ and $0 \le \theta \le \pi$. Because $\cos\theta < 0$, it follows that $\dfrac{\pi}{2} < \theta \le \pi$, so θ lies in quadrant II. Since $\cos\theta = \dfrac{-1}{3} = \dfrac{x}{r}$, we let $x = -1$ and $r = 3$. The point $P = (x, y) = (-1, y)$, $y > 0$, is on a circle of radius $r = 3$, $x^2 + y^2 = 9$. See Figure 19. Then,

$$x^2 + y^2 = 9, \qquad x = -1, y > 0$$
$$(-1)^2 + y^2 = 9$$
$$y^2 = 8$$
$$y = 2\sqrt{2}$$

Then, we have $x = -1$, $y = 2\sqrt{2}$, and $r = 3$, so that

$$\tan\left[\cos^{-1}\left(-\frac{1}{3}\right)\right] = \tan\theta = \frac{2\sqrt{2}}{-1} = -2\sqrt{2}$$

$$\uparrow$$
$$\tan\theta = \frac{y}{x}$$

Figure 19 $\cos\theta = -\dfrac{1}{3}$

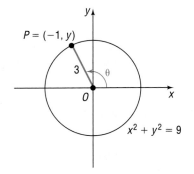

NOW WORK PROBLEMS **1** AND **19**.

THE REMAINING INVERSE TRIGONOMETRIC FUNCTIONS

2 The inverse secant, inverse cosecant, and inverse cotangent functions are defined as follows:

$$y = \sec^{-1}x \quad \text{means} \quad x = \sec y \tag{1}$$
$$\text{where} \quad |x| \ge 1 \quad \text{and} \quad 0 \le y \le \pi, \quad y \ne \frac{\pi}{2} *$$
$$y = \csc^{-1}x \quad \text{means} \quad x = \csc y \tag{2}$$
$$\text{where} \quad |x| \ge 1 \quad \text{and} \quad -\frac{\pi}{2} \le y \le \frac{\pi}{2}, \quad y \ne 0^\dagger$$
$$y = \cot^{-1}x \quad \text{means} \quad x = \cot y \tag{3}$$
$$\text{where} \quad -\infty < x < \infty \quad \text{and} \quad 0 < y < \pi$$

You are encouraged to review the graphs of the secant, cosecant, and cotangent functions in Figures 71 (p. 142), 69 (p. 141), and 68 (p. 140) in Section 2.5 to help you to see the basis for these definitions.

EXAMPLE 5

Finding the Exact Value of an Inverse Cosecant Function

Find the exact value of: $\csc^{-1}2$

Solution Let $\theta = \csc^{-1}2$. We seek the angle θ, $-\dfrac{\pi}{2} \le \theta \le \dfrac{\pi}{2}$, $\theta \ne 0$, whose cosecant equals 2.

$$\theta = \csc^{-1}2, \qquad -\frac{\pi}{2} \le \theta \le \frac{\pi}{2}, \quad \theta \ne 0$$
$$\csc\theta = 2, \qquad -\frac{\pi}{2} \le \theta \le \frac{\pi}{2}, \quad \theta \ne 0$$

---

* Most books use this definition. A few use the restriction $0 \le y < \dfrac{\pi}{2}$, $\pi \le y < \dfrac{3\pi}{2}$.

$\dagger$ Most books use this definition. A few use the restriction $-\pi < y \le -\dfrac{\pi}{2}$, $0 < y \le \dfrac{\pi}{2}$.

The only angle θ in the interval $-\dfrac{\pi}{2} \le \theta \le \dfrac{\pi}{2}$, $\theta \ne 0$, whose cosecant is 2 is $\dfrac{\pi}{6}$, so $\csc^{-1} 2 = \dfrac{\pi}{6}$. ∎

NOW WORK PROBLEM 31.

3   Most calculators do not have keys for evaluating the inverse cotangent, cosecant, and secant functions. The easiest way to evaluate them is to convert to an inverse trigonometric function whose range is the same as the one to be evaluated. In this regard, notice that $y = \cot^{-1} x$ and $y = \sec^{-1} x$ (except where undefined) each have the same range as $y = \cos^{-1} x$; $y = \csc^{-1} x$, except where undefined, has the same range as $y = \sin^{-1} x$.

| EXAMPLE 6 | **Approximating the Value of Inverse Trigonometric Functions** |

Use a calculator to approximate each expression in radians rounded to two decimal places.

(a) $\sec^{-1} 3$ (b) $\csc^{-1}(-4)$ (c) $\cot^{-1} \dfrac{1}{2}$ (d) $\cot^{-1}(-2)$

Solution   First, set your calculator to radian mode.

(a) Let $\theta = \sec^{-1} 3$. Then $\sec \theta = 3$ and $0 \le \theta \le \pi$, $\theta \ne \dfrac{\pi}{2}$. Since $\cos \theta = \dfrac{1}{3}$ and $\theta = \cos^{-1} \dfrac{1}{3}$, we have

$$\sec^{-1} 3 = \theta = \cos^{-1} \dfrac{1}{3} \underset{\underset{\text{Use a calculator.}}{\uparrow}}{\approx} 1.23$$

Figure 20 $\cot \theta = \frac{1}{2}, 0 < \theta < \pi$

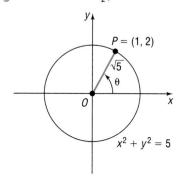

$P = (1, 2)$
$\sqrt{5}$
$x^2 + y^2 = 5$

(b) Let $\theta = \csc^{-1}(-4)$. Then $\csc \theta = -4$, $-\dfrac{\pi}{2} \le \theta \le \dfrac{\pi}{2}$, $\theta \ne 0$. Since $\sin \theta = -\dfrac{1}{4}$ and $\theta = \sin^{-1}\left(-\dfrac{1}{4}\right)$, we have

$$\csc^{-1}(-4) = \theta = \sin^{-1}\left(-\dfrac{1}{4}\right) \approx -0.25$$

(c) Let $\theta = \cot^{-1} \dfrac{1}{2}$. Then $\cot \theta = \dfrac{1}{2}, 0 < \theta < \pi$. From these facts we know that θ lies in quadrant I. Since $\cot \theta = \dfrac{1}{2} = \dfrac{x}{y}$, we let $x = 1$ and $y = 2$. The point $P = (x, y) = (1, 2)$ is on the circle $x^2 + y^2 = 5$, since $r = d(O, P) = \sqrt{1^2 + 2^2} = \sqrt{5}$. See Figure 20. Then $\cos \theta = \dfrac{x}{r} = \dfrac{1}{\sqrt{5}}$,

Figure 21 $\cot \theta = -2, 0 < \theta < \pi$

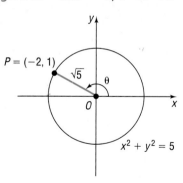

$P = (-2, 1)$
$\sqrt{5}$
$x^2 + y^2 = 5$

so, $\theta = \cos^{-1}\left(\dfrac{1}{\sqrt{5}}\right)$. As a result,

$$\cot^{-1} \dfrac{1}{2} = \theta = \cos^{-1}\left(\dfrac{1}{\sqrt{5}}\right) \approx 1.11$$

(d) Let $\theta = \cot^{-1}(-2)$. Then $\cot \theta = -2, 0 < \theta < \pi$. From these facts we know that θ lies in quadrant II. Since $\cot \theta = -2 = \dfrac{x}{y}$, $x < 0, y > 0$, we let $x = -2$ and $y = 1$. The point $P = (x, y) = (-2, 1)$ is on the circle $x^2 + y^2 = 5$, since $r = d(O, P) = \sqrt{(-2)^2 + 1^2} = \sqrt{5}$. See Figure 21. Then $\cos \theta = \dfrac{x}{r} = \dfrac{-2}{\sqrt{5}}$, so, $\theta = \cos^{-1}\left(\dfrac{-2}{\sqrt{5}}\right)$, and

$$\cot^{-1}(-2) = \theta = \cos^{-1}\left(\frac{-2}{\sqrt{5}}\right) \approx 2.68$$

NOW WORK PROBLEM **37.**

3.2 EXERCISES

In Problems 1–28, find the exact value of each expression.

1. $\cos\left(\sin^{-1}\dfrac{\sqrt{2}}{2}\right)$

2. $\sin\left(\cos^{-1}\dfrac{1}{2}\right)$

3. $\tan\left[\cos^{-1}\left(-\dfrac{\sqrt{3}}{2}\right)\right]$

4. $\tan\left[\sin^{-1}\left(-\dfrac{1}{2}\right)\right]$

5. $\sec\left(\cos^{-1}\dfrac{1}{2}\right)$

6. $\cot\left[\sin^{-1}\left(-\dfrac{1}{2}\right)\right]$

7. $\csc(\tan^{-1}1)$

8. $\sec(\tan^{-1}\sqrt{3})$

9. $\sin[\tan^{-1}(-1)]$

10. $\cos\left[\sin^{-1}\left(-\dfrac{\sqrt{3}}{2}\right)\right]$

11. $\sec\left[\sin^{-1}\left(-\dfrac{1}{2}\right)\right]$

12. $\csc\left[\cos^{-1}\left(-\dfrac{\sqrt{3}}{2}\right)\right]$

13. $\cos^{-1}\left(\cos\dfrac{5\pi}{4}\right)$

14. $\tan^{-1}\left(\tan\dfrac{2\pi}{3}\right)$

15. $\sin^{-1}\left[\sin\left(-\dfrac{7\pi}{6}\right)\right]$

16. $\cos^{-1}\left[\cos\left(-\dfrac{\pi}{3}\right)\right]$

17. $\tan\left(\sin^{-1}\dfrac{1}{3}\right)$

18. $\tan\left(\cos^{-1}\dfrac{1}{3}\right)$

19. $\sec\left(\tan^{-1}\dfrac{1}{2}\right)$

20. $\cos\left(\sin^{-1}\dfrac{\sqrt{2}}{3}\right)$

21. $\cot\left[\sin^{-1}\left(-\dfrac{\sqrt{2}}{3}\right)\right]$

22. $\csc[\tan^{-1}(-2)]$

23. $\sin[\tan^{-1}(-3)]$

24. $\cot\left[\cos^{-1}\left(-\dfrac{\sqrt{3}}{3}\right)\right]$

25. $\sec\left(\sin^{-1}\dfrac{2\sqrt{5}}{5}\right)$

26. $\csc\left(\tan^{-1}\dfrac{1}{2}\right)$

27. $\sin^{-1}\left(\cos\dfrac{3\pi}{4}\right)$

28. $\cos^{-1}\left(\sin\dfrac{7\pi}{6}\right)$

In Problems 29–36, find the exact value of each expression.

29. $\cot^{-1}\sqrt{3}$

30. $\cot^{-1}1$

31. $\csc^{-1}(-1)$

32. $\csc^{-1}\sqrt{2}$

33. $\sec^{-1}\dfrac{2\sqrt{3}}{3}$

34. $\sec^{-1}(-2)$

35. $\cot^{-1}\left(-\dfrac{\sqrt{3}}{3}\right)$

36. $\csc^{-1}\left(-\dfrac{2\sqrt{3}}{3}\right)$

In Problems 37–48, use a calculator to find the value of each expression rounded to two decimal places.

37. $\sec^{-1}4$

38. $\csc^{-1}5$

39. $\cot^{-1}2$

40. $\sec^{-1}(-3)$

41. $\csc^{-1}(-3)$

42. $\cot^{-1}\left(-\dfrac{1}{2}\right)$

43. $\cot^{-1}(-\sqrt{5})$

44. $\cot^{-1}(-8.1)$

45. $\csc^{-1}\left(-\dfrac{3}{2}\right)$

46. $\sec^{-1}\left(-\dfrac{4}{3}\right)$

47. $\cot^{-1}\left(-\dfrac{3}{2}\right)$

48. $\cot^{-1}(-\sqrt{10})$

---

49. Using a graphing utility, graph $y = \cot^{-1}x$.
50. Using a graphing utility, graph $y = \sec^{-1}x$.
51. Using a graphing utility, graph $y = \csc^{-1}x$.

52. Explain in your own words how you would use your calculator to find the value of $\cot^{-1}10$.
53. Consult three books on calculus and write down the definition in each of $y = \sec^{-1}x$ and $y = \csc^{-1}x$. Compare these with the definition given in this book.

PREPARING FOR THIS SECTION

Before getting started, review the following:

✓ Fundamental Identities (Section 2.3, p. 115)

3.3 TRIGONOMETRIC IDENTITIES

OBJECTIVE ① Establish Identities

We saw in the previous chapter that the trigonometric functions lend themselves to a wide variety of identities. Before establishing some additional identities, let's review the definition of an *identity*.

Two functions f and g are said to be **identically equal** if

$$f(x) = g(x)$$

for every value of x for which both functions are defined. Such an equation is referred to as an **identity.** An equation that is not an identity is called a **conditional equation.**

For example, the following are identities:

$$(x + 1)^2 = x^2 + 2x + 1 \qquad \sin^2 x + \cos^2 x = 1 \qquad \csc x = \frac{1}{\sin x}$$

The following are conditional equations:

$$2x + 5 = 0 \qquad \text{True only if } x = -\frac{5}{2}.$$

$$\sin x = 0 \qquad \text{True only if } x = k\pi, \text{ } k \text{ an integer.}$$

$$\sin x = \cos x \qquad \text{True only if } x = \frac{\pi}{4} + 2k\pi \text{ or } x = \frac{5\pi}{4} + 2k\pi, \text{ } k \text{ an integer.}$$

The following boxes summarize the trigonometric identities that we have established thus far.

Quotient Identities

$$\tan\theta = \frac{\sin\theta}{\cos\theta} \qquad \cot\theta = \frac{\cos\theta}{\sin\theta}$$

Reciprocal Identities

$$\csc\theta = \frac{1}{\sin\theta} \qquad \sec\theta = \frac{1}{\cos\theta} \qquad \cot\theta = \frac{1}{\tan\theta}$$

Pythagorean Identities

$$\sin^2\theta + \cos^2\theta = 1 \qquad \tan^2\theta + 1 = \sec^2\theta$$
$$1 + \cot^2\theta = \csc^2\theta$$

Even–Odd Identities

$$\sin(-\theta) = -\sin\theta \qquad \cos(-\theta) = \cos\theta \qquad \tan(-\theta) = -\tan\theta$$
$$\csc(-\theta) = -\csc\theta \qquad \sec(-\theta) = \sec\theta \qquad \cot(-\theta) = -\cot\theta$$

This list of identities comprises what we shall refer to as the **basic trigonometric identities.** These identities should not merely be memorized, but should be *known* (just as you know your name rather than have it memorized). In fact, minor variations of a basic identity are often used. For example, we might want to use $\sin^2\theta = 1 - \cos^2\theta$ or $\cos^2\theta = 1 - \sin^2\theta$ instead of $\sin^2\theta + \cos^2\theta = 1$. For this reason, among others, you need to know these relationships and be quite comfortable with variations of them.

In the examples that follow, the directions will read "Establish the identity…." As you will see, this is accomplished by starting with one side of the given equation (usually the one containing the more complicated expression) and, using appropriate basic identities and algebraic manipulations, arriving at the other side. The selection of appropriate basic identities to obtain the desired result is learned only through experience and lots of practice.

EXAMPLE 1

Establishing an Identity

Establish the identity: $\csc\theta \cdot \tan\theta = \sec\theta$

Solution   We start with the left side, because it contains the more complicated expression, and apply a reciprocal identity and a quotient identity.

$$\csc\theta \cdot \tan\theta = \frac{1}{\sin\theta} \cdot \frac{\sin\theta}{\cos\theta} = \frac{1}{\cos\theta} = \sec\theta$$

Having arrived at the right side, the identity is established.   ■

    COMMENT: A graphing utility can be used to provide evidence of an identity. For example, if we graph $Y_1 = \csc\theta \cdot \tan\theta$ and $Y_2 = \sec\theta$, the graphs appear to be the same. This provides evidence that $Y_1 = Y_2$. However, it does not prove their equality. A graphing utility *cannot be used to establish an identity*—identities must be established algebraically.   ■

── NOW WORK PROBLEM **1.**

EXAMPLE 2

Establishing an Identity

Establish the identity: $\sin^2(-\theta) + \cos^2(-\theta) = 1$

Solution   We begin with the left side and apply even–odd identities.

$$\begin{aligned}
\sin^2(-\theta) + \cos^2(-\theta) &= \left[\sin(-\theta)\right]^2 + \left[\cos(-\theta)\right]^2 \\
&= (-\sin\theta)^2 + (\cos\theta)^2 &&\text{Even–odd identities.} \\
&= (\sin\theta)^2 + (\cos\theta)^2 \\
&= 1 &&\text{Pythagorean Identity.} \quad ■
\end{aligned}$$

EXAMPLE 3

Establishing an Identity

Establish the identity: $\dfrac{\sin^2(-\theta) - \cos^2(-\theta)}{\sin(-\theta) - \cos(-\theta)} = \cos\theta - \sin\theta$

Solution   We begin with two observations: The left side appears to contain the more complicated expression. Also, the left side contains expressions with the argument $-\theta$, whereas the right side contains expressions with the argument θ. We decide, therefore, to start with the left side and apply even–odd identities.

$$\begin{aligned}
\frac{\sin^2(-\theta) - \cos^2(-\theta)}{\sin(-\theta) - \cos(-\theta)} &= \frac{\left[\sin(-\theta)\right]^2 - \left[\cos(-\theta)\right]^2}{\sin(-\theta) - \cos(-\theta)} \\
&= \frac{(-\sin\theta)^2 - (\cos\theta)^2}{-\sin\theta - \cos\theta} &&\text{Even–odd identities.} \\
&= \frac{(\sin\theta)^2 - (\cos\theta)^2}{-\sin\theta - \cos\theta} &&\text{Simplify.} \\
&= \frac{(\sin\theta - \cos\theta)(\sin\theta + \cos\theta)}{-(\sin\theta + \cos\theta)} &&\text{Factor.} \\
&= \cos\theta - \sin\theta &&\text{Cancel and simplify.} \quad ■
\end{aligned}$$

EXAMPLE 4

Establishing an Identity

Establish the identity: $\dfrac{1 + \tan\theta}{1 + \cot\theta} = \tan\theta$

Solution
$$\frac{1 + \tan\theta}{1 + \cot\theta} = \frac{1 + \tan\theta}{1 + \dfrac{1}{\tan\theta}} = \frac{1 + \tan\theta}{\dfrac{\tan\theta + 1}{\tan\theta}} = \frac{\tan\theta\,(1 + \tan\theta)}{\tan\theta + 1} = \tan\theta$$

■

── NOW WORK PROBLEM **9.**

When sums or differences of quotients appear, it is usually best to rewrite them as a single quotient, especially if the other side of the identity consists of only one term.

| EXAMPLE 5 | **Establishing an Identity** |

Establish the identity: $\dfrac{\sin\theta}{1+\cos\theta} + \dfrac{1+\cos\theta}{\sin\theta} = 2\csc\theta$

Solution The left side is more complicated, so we start with it and proceed to add.

$$\frac{\sin\theta}{1+\cos\theta} + \frac{1+\cos\theta}{\sin\theta} = \frac{\sin^2\theta + (1+\cos\theta)^2}{(1+\cos\theta)(\sin\theta)} \qquad \text{Add the quotients.}$$

$$= \frac{\sin^2\theta + 1 + 2\cos\theta + \cos^2\theta}{(1+\cos\theta)(\sin\theta)} \qquad \text{Remove parentheses in numerator.}$$

$$= \frac{(\sin^2\theta + \cos^2\theta) + 1 + 2\cos\theta}{(1+\cos\theta)(\sin\theta)} \qquad \text{Regroup.}$$

$$= \frac{2 + 2\cos\theta}{(1+\cos\theta)(\sin\theta)} \qquad \text{Pythagorean Identity.}$$

$$= \frac{2\cancel{(1+\cos\theta)}}{\cancel{(1+\cos\theta)}(\sin\theta)} \qquad \text{Factor and cancel.}$$

$$= \frac{2}{\sin\theta}$$

$$= 2\csc\theta \qquad \text{Reciprocal Identity.} \blacksquare$$

Sometimes it helps to write one side in terms of sines and cosines only.

| EXAMPLE 6 | **Establishing an Identity** |

Establish the identity: $\dfrac{\tan\theta + \cot\theta}{\sec\theta\csc\theta} = 1$

Solution $\dfrac{\tan\theta + \cot\theta}{\sec\theta\csc\theta} = \dfrac{\dfrac{\sin\theta}{\cos\theta} + \dfrac{\cos\theta}{\sin\theta}}{\dfrac{1}{\cos\theta}\dfrac{1}{\sin\theta}} = \dfrac{\dfrac{\sin^2\theta + \cos^2\theta}{\cos\theta\sin\theta}}{\dfrac{1}{\cos\theta\sin\theta}}$

$\quad\quad\quad\quad\quad\quad\quad\quad\quad$ Change to sines $\quad$ Add the quotients
$\quad\quad\quad\quad\quad\quad\quad\quad\quad$ and cosines. $\quad\quad$ in the numerator.

$$= \frac{1}{\cos\theta\sin\theta} \cdot \frac{\cos\theta\sin\theta}{1} = 1$$

Divide quotient;
$\sin^2\theta + \cos^2\theta = 1$

$\blacksquare$

NOW WORK PROBLEM 51.

Sometimes, multiplying the numerator and denominator by an appropriate factor will result in a simplification.

EXAMPLE 7   **Establishing an Identity**

Establish the identity: $\dfrac{1 - \sin\theta}{\cos\theta} = \dfrac{\cos\theta}{1 + \sin\theta}$

Solution   We start with the left side and multiply the numerator and the denominator by $1 + \sin\theta$. (Alternatively, we could multiply the numerator and denominator of the right side by $1 - \sin\theta$.)

$$\dfrac{1 - \sin\theta}{\cos\theta} = \dfrac{1 - \sin\theta}{\cos\theta} \cdot \dfrac{1 + \sin\theta}{1 + \sin\theta} \quad \text{Multiply numerator and denominator by } 1 + \sin\theta.$$

$$= \dfrac{1 - \sin^2\theta}{\cos\theta(1 + \sin\theta)}$$

$$= \dfrac{\cos^2\theta}{\cos\theta(1 + \sin\theta)} \quad 1 - \sin^2\theta = \cos^2\theta.$$

$$= \dfrac{\cos\theta}{1 + \sin\theta} \quad \text{Cancel.} \quad ■$$

🖉 **NOW WORK PROBLEM 35.**

EXAMPLE 8   **Establishing an Identity Involving Inverse Trigonometric Functions**

Show that $\sin(\tan^{-1}v) = \dfrac{v}{\sqrt{1 + v^2}}$.

Solution   Let $\theta = \tan^{-1}v$ so that $\tan\theta = v$, $-\dfrac{\pi}{2} < \theta < \dfrac{\pi}{2}$. As a result, we know that $\sec\theta > 0$.

$$\sin(\tan^{-1}v) = \sin\theta = \sin\theta \cdot \dfrac{\cos\theta}{\cos\theta} = \tan\theta\cos\theta = \dfrac{\tan\theta}{\sec\theta} = \dfrac{\tan\theta}{\sqrt{1 + \tan^2\theta}} = \dfrac{v}{\sqrt{1 + v^2}}$$

$$\dfrac{\sin\theta}{\cos\theta} = \tan\theta \qquad \sec^2\theta = 1 + \tan^2\theta$$
$$\sec\theta > 0 \qquad ■$$

🖉 **NOW WORK PROBLEM 81.**

Although a lot of practice is the only real way to learn how to establish identities, the following guidelines should prove helpful.

> **GUIDELINES FOR ESTABLISHING IDENTITIES**
>
> 1. It is almost always preferable to start with the side containing the more complicated expression.
> 2. Rewrite sums or differences of quotients as a single quotient.
> 3. Sometimes, rewriting one side in terms of sines and cosines only will help.
> 4. Always keep your goal in mind. As you manipulate one side of the expression, you must keep in mind the form of the expression on the other side.

WARNING: Be careful not to handle identities to be established as if they were conditional equations. You *cannot* establish an identity by such methods as adding the same expression to each side and obtaining a true statement. This practice is not allowed, because the original statement is precisely the one that you are trying to establish. You do not know until it has been established that it is, in fact, true. ∎

3.3 EXERCISES

In Problems 1–80, establish each identity.

1. $\csc\theta \cdot \cos\theta = \cot\theta$

2. $\sec\theta \cdot \sin\theta = \tan\theta$

3. $1 + \tan^2(-\theta) = \sec^2\theta$

4. $1 + \cot^2(-\theta) = \csc^2\theta$

5. $\cos\theta(\tan\theta + \cot\theta) = \csc\theta$

6. $\sin\theta(\cot\theta + \tan\theta) = \sec\theta$

7. $\tan\theta\cot\theta - \cos^2\theta = \sin^2\theta$

8. $\sin\theta\csc\theta - \cos^2\theta = \sin^2\theta$

9. $(\sec\theta - 1)(\sec\theta + 1) = \tan^2\theta$

10. $(\csc\theta - 1)(\csc\theta + 1) = \cot^2\theta$

11. $(\sec\theta + \tan\theta)(\sec\theta - \tan\theta) = 1$

12. $(\csc\theta + \cot\theta)(\csc\theta - \cot\theta) = 1$

13. $\cos^2\theta(1 + \tan^2\theta) = 1$

14. $(1 - \cos^2\theta)(1 + \cot^2\theta) = 1$

15. $(\sin\theta + \cos\theta)^2 + (\sin\theta - \cos\theta)^2 = 2$

16. $\tan^2\theta\cos^2\theta + \cot^2\theta\sin^2\theta = 1$

17. $\sec^4\theta - \sec^2\theta = \tan^4\theta + \tan^2\theta$

18. $\csc^4\theta - \csc^2\theta = \cot^4\theta + \cot^2\theta$

19. $\sec\theta - \tan\theta = \dfrac{\cos\theta}{1 + \sin\theta}$

20. $\csc\theta - \cot\theta = \dfrac{\sin\theta}{1 + \cos\theta}$

21. $3\sin^2\theta + 4\cos^2\theta = 3 + \cos^2\theta$

22. $9\sec^2\theta - 5\tan^2\theta = 5 + 4\sec^2\theta$

23. $1 - \dfrac{\cos^2\theta}{1 + \sin\theta} = \sin\theta$

24. $1 - \dfrac{\sin^2\theta}{1 - \cos\theta} = -\cos\theta$

25. $\dfrac{1 + \tan\theta}{1 - \tan\theta} = \dfrac{\cot\theta + 1}{\cot\theta - 1}$

26. $\dfrac{\csc\theta - 1}{\csc\theta + 1} = \dfrac{1 - \sin\theta}{1 + \sin\theta}$

27. $\dfrac{\sec\theta}{\csc\theta} + \dfrac{\sin\theta}{\cos\theta} = 2\tan\theta$

28. $\dfrac{\csc\theta - 1}{\cot\theta} = \dfrac{\cot\theta}{\csc\theta + 1}$

29. $\dfrac{1 + \sin\theta}{1 - \sin\theta} = \dfrac{\csc\theta + 1}{\csc\theta - 1}$

30. $\dfrac{\cos\theta + 1}{\cos\theta - 1} = \dfrac{1 + \sec\theta}{1 - \sec\theta}$

31. $\dfrac{1 - \sin\theta}{\cos\theta} + \dfrac{\cos\theta}{1 - \sin\theta} = 2\sec\theta$

32. $\dfrac{\cos\theta}{1 + \sin\theta} + \dfrac{1 + \sin\theta}{\cos\theta} = 2\sec\theta$

33. $\dfrac{\sin\theta}{\sin\theta - \cos\theta} = \dfrac{1}{1 - \cot\theta}$

34. $1 - \dfrac{\sin^2\theta}{1 + \cos\theta} = \cos\theta$

35. $\dfrac{1 - \sin\theta}{1 + \sin\theta} = (\sec\theta - \tan\theta)^2$

36. $\dfrac{1 - \cos\theta}{1 + \cos\theta} = (\csc\theta - \cot\theta)^2$

37. $\dfrac{\cos\theta}{1 - \tan\theta} + \dfrac{\sin\theta}{1 - \cot\theta} = \sin\theta + \cos\theta$

38. $\dfrac{\cot\theta}{1 - \tan\theta} + \dfrac{\tan\theta}{1 - \cot\theta} = 1 + \tan\theta + \cot\theta$

39. $\tan\theta + \dfrac{\cos\theta}{1 + \sin\theta} = \sec\theta$

40. $\dfrac{\sin\theta\cos\theta}{\cos^2\theta - \sin^2\theta} = \dfrac{\tan\theta}{1 - \tan^2\theta}$

41. $\dfrac{\tan\theta + \sec\theta - 1}{\tan\theta - \sec\theta + 1} = \tan\theta + \sec\theta$

42. $\dfrac{\sin\theta - \cos\theta + 1}{\sin\theta + \cos\theta - 1} = \dfrac{\sin\theta + 1}{\cos\theta}$

43. $\dfrac{\tan\theta - \cot\theta}{\tan\theta + \cot\theta} = \sin^2\theta - \cos^2\theta$

44. $\dfrac{\sec\theta - \cos\theta}{\sec\theta + \cos\theta} = \dfrac{\sin^2\theta}{1 + \cos^2\theta}$

45. $\dfrac{\tan\theta - \cot\theta}{\tan\theta + \cot\theta} + 1 = 2\sin^2\theta$

46. $\dfrac{\tan\theta - \cot\theta}{\tan\theta + \cot\theta} + 2\cos^2\theta = 1$

47. $\dfrac{\sec\theta + \tan\theta}{\cot\theta + \cos\theta} = \tan\theta\sec\theta$

48. $\dfrac{\sec\theta}{1 + \sec\theta} = \dfrac{1 - \cos\theta}{\sin^2\theta}$

49. $\dfrac{1 - \tan^2\theta}{1 + \tan^2\theta} + 1 = 2\cos^2\theta$

50. $\dfrac{1 - \cot^2\theta}{1 + \cot^2\theta} + 2\cos^2\theta = 1$

51. $\dfrac{\sec\theta - \csc\theta}{\sec\theta\csc\theta} = \sin\theta - \cos\theta$

52. $\dfrac{\sin^2\theta - \tan\theta}{\cos^2\theta - \cot\theta} = \tan^2\theta$

53. $\sec\theta - \cos\theta - \sin\theta\tan\theta = 0$

54. $\tan\theta + \cot\theta - \sec\theta\csc\theta = 0$

55. $\dfrac{1}{1 - \sin\theta} + \dfrac{1}{1 + \sin\theta} = 2\sec^2\theta$

56. $\dfrac{1 + \sin\theta}{1 - \sin\theta} - \dfrac{1 - \sin\theta}{1 + \sin\theta} = 4\tan\theta\sec\theta$

57. $\dfrac{\sec\theta}{1 - \sin\theta} = \dfrac{1 + \sin\theta}{\cos^3\theta}$

58. $\dfrac{1 - \sin\theta}{1 + \sin\theta} = (\sec\theta - \tan\theta)^2$

59. $\dfrac{(\sec\theta - \tan\theta)^2 + 1}{\csc\theta(\sec\theta - \tan\theta)} = 2\tan\theta$

60. $\dfrac{\sec^2\theta - \tan^2\theta + \tan\theta}{\sec\theta} = \sin\theta + \cos\theta$

61. $\dfrac{\sin\theta + \cos\theta}{\cos\theta} - \dfrac{\sin\theta - \cos\theta}{\sin\theta} = \sec\theta\csc\theta$

62. $\dfrac{\sin\theta + \cos\theta}{\sin\theta} - \dfrac{\cos\theta - \sin\theta}{\cos\theta} = \sec\theta\csc\theta$

63. $\dfrac{\sin^3\theta + \cos^3\theta}{\sin\theta + \cos\theta} = 1 - \sin\theta\cos\theta$

64. $\dfrac{\sin^3\theta + \cos^3\theta}{1 - 2\cos^2\theta} = \dfrac{\sec\theta - \sin\theta}{\tan\theta - 1}$

65. $\dfrac{\cos^2\theta - \sin^2\theta}{1 - \tan^2\theta} = \cos^2\theta$

66. $\dfrac{\cos\theta + \sin\theta - \sin^3\theta}{\sin\theta} = \cot\theta + \cos^2\theta$

67. $\dfrac{(2\cos^2\theta - 1)^2}{\cos^4\theta - \sin^4\theta} = 1 - 2\sin^2\theta$

68. $\dfrac{1 - 2\cos^2\theta}{\sin\theta\cos\theta} = \tan\theta - \cot\theta$

69. $\dfrac{1 + \sin\theta + \cos\theta}{1 + \sin\theta - \cos\theta} = \dfrac{1 + \cos\theta}{\sin\theta}$

70. $\dfrac{1 + \cos\theta + \sin\theta}{1 + \cos\theta - \sin\theta} = \sec\theta + \tan\theta$

71. $(a\sin\theta + b\cos\theta)^2 + (a\cos\theta - b\sin\theta)^2 = a^2 + b^2$

72. $(2a\sin\theta\cos\theta)^2 + a^2(\cos^2\theta - \sin^2\theta)^2 = a^2$

73. $\dfrac{\tan\alpha + \tan\beta}{\cot\alpha + \cot\beta} = \tan\alpha\tan\beta$

74. $(\tan\alpha + \tan\beta)(1 - \cot\alpha\cot\beta) + (\cot\alpha + \cot\beta)(1 - \tan\alpha\tan\beta) = 0$

75. $(\sin\alpha + \cos\beta)^2 + (\cos\beta + \sin\alpha)(\cos\beta - \sin\alpha) = 2\cos\beta(\sin\alpha + \cos\beta)$

76. $(\sin\alpha - \cos\beta)^2 + (\cos\beta + \sin\alpha)(\cos\beta - \sin\alpha) = -2\cos\beta(\sin\alpha - \cos\beta)$

77. $\ln|\sec\theta| = -\ln|\cos\theta|$

78. $\ln|\tan\theta| = \ln|\sin\theta| - \ln|\cos\theta|$

79. $\ln|1 + \cos\theta| + \ln|1 - \cos\theta| = 2\ln|\sin\theta|$

80. $\ln|\sec\theta + \tan\theta| + \ln|\sec\theta - \tan\theta| = 0$

81. Show that $\sec(\tan^{-1}v) = \sqrt{1 + v^2}$.

82. Show that $\tan(\sin^{-1}v) = \dfrac{v}{\sqrt{1 - v^2}}$.

83. Show that $\tan(\cos^{-1}v) = \dfrac{\sqrt{1 - v^2}}{v}$.

84. Show that $\sin(\cos^{-1}v) = \sqrt{1 - v^2}$.

85. Show that $\cos(\sin^{-1}v) = \sqrt{1 - v^2}$.

86. Show that $\cos(\tan^{-1}v) = \dfrac{1}{\sqrt{1 + v^2}}$.

87. Write a few paragraphs outlining your strategy for establishing identities.

PREPARING FOR THIS SECTION

Before getting started, review the following:

✓ Distance Formula (Section 1.1, p. 4)

✓ Values of the Trigonometric Functions of Certain Angles (Section 2.2, p. 96 and p. 100)

3.4 SUM AND DIFFERENCE FORMULAS

OBJECTIVES **1** Use Sum and Difference Formulas to Find Exact Values

2 Use Sum and Difference Formulas to Establish Identities

3 Use Sum and Difference Formulas Involving Inverse Trigonometric Functions

In this section, we continue our derivation of trigonometric identities by obtaining formulas that involve the sum or difference of two angles, such as $\cos(\alpha + \beta)$, $\cos(\alpha - \beta)$, or $\sin(\alpha + \beta)$. These formulas are referred to as the **sum and difference formulas.** We begin with the formulas for $\cos(\alpha + \beta)$ and $\cos(\alpha - \beta)$.

Theorem

Sum and Difference Formulas for Cosines

$$\cos(\alpha + \beta) = \cos\alpha\cos\beta - \sin\alpha\sin\beta \qquad (1)$$

$$\cos(\alpha - \beta) = \cos\alpha\cos\beta + \sin\alpha\sin\beta \qquad (2)$$

In words, formula (1) states that the cosine of the sum of two angles equals the cosine of the first angle times the cosine of the second angle minus the sine of the first angle times the sine of the second angle.

Proof   We will prove formula (2) first. Although this formula is true for all numbers α and β, we shall assume in our proof that $0 < \beta < \alpha < 2\pi$. We begin with the unit circle and place the angles α and β in standard position, as shown in Figure 22(a). The point P_1 lies on the terminal side of β, so its coordinates are $(\cos\beta, \sin\beta)$; and the point P_2 lies on the terminal side of α so its coordinates are $(\cos\alpha, \sin\alpha)$.

Now, place the angle $\alpha - \beta$ in standard position, as shown in Figure 22(b). The point A has coordinates $(1, 0)$, and the point P_3 is on the terminal side of the angle $\alpha - \beta$, so its coordinates are $(\cos(\alpha - \beta), \sin(\alpha - \beta))$.

Figure 22

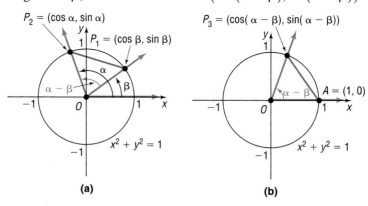

(a)　　　　　　(b)

Looking at triangle OP_1P_2 in Figure 22(a) and triangle OAP_3 in Figure 22(b), we see that these triangles are congruent. (Do you see why? Two sides and the included angle, $\alpha - \beta$, are equal.) As a result, the unknown side of each triangle must be equal; that is,

$$d(A, P_3) = d(P_1, P_2)$$

Using the distance formula, we find that

$\sqrt{[\cos(\alpha - \beta) - 1]^2 + [\sin(\alpha - \beta) - 0]^2} = \sqrt{(\cos\alpha - \cos\beta)^2 + (\sin\alpha - \sin\beta)^2}$ $d(A, P_3) = d(P_1, P_2)$.

$[\cos(\alpha - \beta) - 1]^2 + \sin^2(\alpha - \beta) = (\cos\alpha - \cos\beta)^2 + (\sin\alpha - \sin\beta)^2$ Square both sides.

$\cos^2(\alpha - \beta) - 2\cos(\alpha - \beta) + 1 + \sin^2(\alpha - \beta) = \cos^2\alpha - 2\cos\alpha\cos\beta + \cos^2\beta$ Multiply out the squared terms.
$$+ \sin^2\alpha - 2\sin\alpha\sin\beta + \sin^2\beta$$

$2 - 2\cos(\alpha - \beta) = 2 - 2\cos\alpha\cos\beta - 2\sin\alpha\sin\beta$ Apply a Pythagorean Identity (3 times).

$-2\cos(\alpha - \beta) = -2\cos\alpha\cos\beta - 2\sin\alpha\sin\beta$ Subtract 2 from each side.

$\cos(\alpha - \beta) = \cos\alpha\cos\beta + \sin\alpha\sin\beta$ Divide each side by -2.

which is formula (2).

The proof of formula (1) follows from formula (2) and the Even–Odd Identities. We use the fact that $\alpha + \beta = \alpha - (-\beta)$. Then

$$\cos(\alpha + \beta) = \cos[\alpha - (-\beta)]$$
$$= \cos\alpha\cos(-\beta) + \sin\alpha\sin(-\beta) \quad \text{Use formula (2).}$$
$$= \cos\alpha\cos\beta - \sin\alpha\sin\beta \quad \text{Even–odd Identities.} \quad \blacksquare$$

1 One use of formulas (1) and (2) is to obtain the exact value of the cosine of an angle that can be expressed as the sum or difference of angles whose sine and cosine are known exactly.

EXAMPLE 1 **Using the Sum Formula to Find Exact Values**

Find the exact value of $\cos 75°$.

Solution Since $75° = 45° + 30°$, we use formula (1) to obtain

$$\cos 75° = \cos(45° + 30°) = \cos 45° \cos 30° - \sin 45° \sin 30°$$
$$\uparrow$$
$$\text{Formula (1)}$$

$$= \frac{\sqrt{2}}{2} \cdot \frac{\sqrt{3}}{2} - \frac{\sqrt{2}}{2} \cdot \frac{1}{2} = \frac{1}{4}(\sqrt{6} - \sqrt{2}) \qquad \blacksquare$$

EXAMPLE 2 **Using the Difference Formula to Find Exact Values**

Find the exact value of $\cos \dfrac{\pi}{12}$.

Solution $\cos \dfrac{\pi}{12} = \cos\left(\dfrac{3\pi}{12} - \dfrac{2\pi}{12}\right) = \cos\left(\dfrac{\pi}{4} - \dfrac{\pi}{6}\right)$

$$= \cos \frac{\pi}{4} \cos \frac{\pi}{6} + \sin \frac{\pi}{4} \sin \frac{\pi}{6} \qquad \text{Use formula (2).}$$

$$= \frac{\sqrt{2}}{2} \cdot \frac{\sqrt{3}}{2} + \frac{\sqrt{2}}{2} \cdot \frac{1}{2} = \frac{1}{4}(\sqrt{6} + \sqrt{2}) \qquad \blacksquare$$

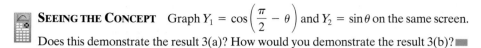 NOW WORK PROBLEM 3.

2 Another use of formulas (1) and (2) is to establish other identities. One important pair of identities is given next.

$$\cos\left(\frac{\pi}{2} - \theta\right) = \sin \theta \qquad \text{(3a)}$$

$$\sin\left(\frac{\pi}{2} - \theta\right) = \cos \theta \qquad \text{(3b)}$$

SEEING THE CONCEPT Graph $Y_1 = \cos\left(\dfrac{\pi}{2} - \theta\right)$ and $Y_2 = \sin \theta$ on the same screen. Does this demonstrate the result 3(a)? How would you demonstrate the result 3(b)? ■

Proof To prove formula (3a), we use the formula for $\cos(\alpha - \beta)$ with $\alpha = \dfrac{\pi}{2}$ and $\beta = \theta$.

$$\cos\left(\frac{\pi}{2} - \theta\right) = \cos \frac{\pi}{2} \cos \theta + \sin \frac{\pi}{2} \sin \theta$$

$$= 0 \cdot \cos \theta + 1 \cdot \sin \theta$$

$$= \sin \theta$$

To prove formula (3b), we make use of the identity (3a) just established.

$$\sin\left(\frac{\pi}{2} - \theta\right) = \cos\left[\frac{\pi}{2} - \left(\frac{\pi}{2} - \theta\right)\right] = \cos \theta \qquad \blacksquare$$
$$\uparrow$$
$$\text{Use (3a).}$$

Since

$$\cos\left(\frac{\pi}{2} - \theta\right) = \cos\left[-\left(\theta - \frac{\pi}{2}\right)\right] = \cos\left(\theta - \frac{\pi}{2}\right)$$

↑
Even Property
of Cosine

and

$$\cos\left(\frac{\pi}{2} - \theta\right) = \sin\theta$$

↑
3(a)

it follows that $\cos\left(\theta - \frac{\pi}{2}\right) = \sin\theta$. The graphs of $y = \cos\left(\theta - \frac{\pi}{2}\right)$ and $y = \sin\theta$ are identical, a fact that we conjectured earlier in Section 2.4.

FORMULAS FOR $\sin(\alpha + \beta)$ AND $\sin(\alpha - \beta)$

Having established the identities in formulas (3a) and (3b), we now can derive the sum and difference formulas for $\sin(\alpha + \beta)$ and $\sin(\alpha - \beta)$.

Proof

$$\sin(\alpha + \beta) = \cos\left[\frac{\pi}{2} - (\alpha + \beta)\right] \qquad \text{Formula (3a).}$$

$$= \cos\left[\left(\frac{\pi}{2} - \alpha\right) - \beta\right]$$

$$= \cos\left(\frac{\pi}{2} - \alpha\right)\cos\beta + \sin\left(\frac{\pi}{2} - \alpha\right)\sin\beta \qquad \text{Formula (2).}$$

$$= \sin\alpha\cos\beta + \cos\alpha\sin\beta \qquad \text{Formulas (3a) and (3b).}$$

$$\sin(\alpha - \beta) = \sin[\alpha + (-\beta)]$$

$$= \sin\alpha\cos(-\beta) + \cos\alpha\sin(-\beta) \qquad \text{Use the sum formula for sine just obtained.}$$

$$= \sin\alpha\cos\beta + \cos\alpha(-\sin\beta) \qquad \text{Even–odd Identities.}$$

$$= \sin\alpha\cos\beta - \cos\alpha\sin\beta \qquad ■$$

Theorem   **Sum and Difference Formulas for Sines**

$$\sin(\alpha + \beta) = \sin\alpha\cos\beta + \cos\alpha\sin\beta \qquad \textbf{(4)}$$

$$\sin(\alpha - \beta) = \sin\alpha\cos\beta - \cos\alpha\sin\beta \qquad \textbf{(5)}$$

In words, formula (4) states that the sine of the sum of two angles equals the sine of the first angle times the cosine of the second angle plus the cosine of the first angle times the sine of the second angle.

EXAMPLE 3

Using the Sum Formula to Find Exact Values

Find the exact value of $\sin \dfrac{7\pi}{12}$.

Solution $\sin \dfrac{7\pi}{12} = \sin\left(\dfrac{3\pi}{12} + \dfrac{4\pi}{12}\right) = \sin\left(\dfrac{\pi}{4} + \dfrac{\pi}{3}\right)$

$$= \sin\dfrac{\pi}{4}\cos\dfrac{\pi}{3} + \cos\dfrac{\pi}{4}\sin\dfrac{\pi}{3} \qquad \text{Formula (4)}$$

$$= \dfrac{\sqrt{2}}{2}\cdot\dfrac{1}{2} + \dfrac{\sqrt{2}}{2}\cdot\dfrac{\sqrt{3}}{2} = \dfrac{1}{4}\left(\sqrt{2} + \sqrt{6}\right) \qquad\blacksquare$$

NOW WORK PROBLEM 9.

EXAMPLE 4

Using the Difference Formula to Find Exact Values

Find the exact value of $\sin 80° \cos 20° - \cos 80° \sin 20°$.

Solution The form of the expression $\sin 80° \cos 20° - \cos 80° \sin 20°$ is that of the right side of the formula (5) for $\sin(\alpha - \beta)$ with $\alpha = 80°$ and $\beta = 20°$. Thus,

$$\sin 80° \cos 20° - \cos 80° \sin 20° = \sin(80° - 20°) = \sin 60° = \dfrac{\sqrt{3}}{2} \qquad\blacksquare$$

NOW WORK PROBLEMS 15 AND 19.

EXAMPLE 5

Finding Exact Values

If it is known that $\sin \alpha = \dfrac{4}{5}, \dfrac{\pi}{2} < \alpha < \pi$, and that $\sin \beta = -\dfrac{2}{\sqrt{5}} = -\dfrac{2\sqrt{5}}{5}$, $\pi < \beta < \dfrac{3\pi}{2}$, find the exact value of

(a) $\cos\alpha$ (b) $\cos\beta$ (c) $\cos(\alpha + \beta)$ (d) $\sin(\alpha + \beta)$

Solution (a) Since $\sin\alpha = \dfrac{4}{5} = \dfrac{y}{r}$ and $\dfrac{\pi}{2} < \alpha < \pi$, we let $y = 4$ and $r = 5$ and place α in quadrant II. The point $P = (x, y) = (x, 4)$, $x < 0$, is on a circle of radius 5, $x^2 + y^2 = 25$. See Figure 23. Then,

Figure 23

Given $\sin\alpha = \dfrac{4}{5}, \dfrac{\pi}{2} < \alpha < \pi$

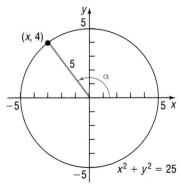

$$x^2 + y^2 = 25, \qquad x < 0, y = 4$$
$$x^2 + 16 = 25$$
$$x^2 = 25 - 16 = 9$$
$$x = -3$$

Then,

$$\cos\alpha = \dfrac{x}{r} = -\dfrac{3}{5}$$

Alternatively, we can find $\cos\alpha$ using identities, as follows:

$$\cos\alpha = -\sqrt{1 - \sin^2\alpha} = -\sqrt{1 - \dfrac{16}{25}} = -\sqrt{\dfrac{9}{25}} = -\dfrac{3}{5}$$

$\uparrow$
α in quadrant II,
$\cos\alpha < 0$

(b) Since $\sin \beta = \dfrac{-2}{\sqrt{5}} = \dfrac{y}{r}$ and $\pi < \beta < \dfrac{3\pi}{2}$, we let $y = -2$ and $r = \sqrt{5}$ and place β in quadrant III. The point $P = (x, y) = (x, -2), x < 0,$ is on a circle of radius $\sqrt{5}, x^2 + y^2 = 5.$ See Figure 24. Then,

$$x^2 + y^2 = 5, \qquad x < 0, y = -2$$
$$x^2 + 4 = 5$$
$$x^2 = 1$$
$$x = -1$$

Then,

$$\cos \beta = \frac{x}{r} = \frac{-1}{\sqrt{5}} = -\frac{\sqrt{5}}{5}$$

Alternatively, we can find $\cos \beta$ using identities, as follows:

$$\cos \beta = -\sqrt{1 - \sin^2 \beta} = -\sqrt{1 - \frac{4}{5}} = -\sqrt{\frac{1}{5}} = -\frac{\sqrt{5}}{5}$$

(c) Using the results found in parts (a) and (b) and formula (1), we have

$$\cos(\alpha + \beta) = \cos \alpha \cos \beta - \sin \alpha \sin \beta$$

$$= -\frac{3}{5}\left(-\frac{\sqrt{5}}{5}\right) - \frac{4}{5}\left(-\frac{2\sqrt{5}}{5}\right) = \frac{11\sqrt{5}}{25}$$

(d) $\sin(\alpha + \beta) = \sin \alpha \cos \beta + \cos \alpha \sin \beta$

$$= \frac{4}{5}\left(-\frac{\sqrt{5}}{5}\right) + \left(-\frac{3}{5}\right)\left(-\frac{2\sqrt{5}}{5}\right) = \frac{2\sqrt{5}}{25}$$

NOW WORK PROBLEM 23(a), (b), AND (c).

Figure 24

Given $\sin \beta = \dfrac{-2}{\sqrt{5}}, \pi < \beta < \dfrac{3\pi}{2}$

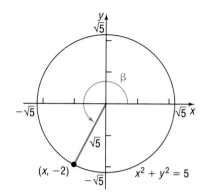

| EXAMPLE 6 | **Establishing an Identity** |

Establish the identity: $\dfrac{\cos(\alpha - \beta)}{\sin \alpha \sin \beta} = \cot \alpha \cot \beta + 1$

Solution

$$\frac{\cos(\alpha - \beta)}{\sin \alpha \sin \beta} = \frac{\cos \alpha \cos \beta + \sin \alpha \sin \beta}{\sin \alpha \sin \beta}$$

$$= \frac{\cos \alpha \cos \beta}{\sin \alpha \sin \beta} + \frac{\sin \alpha \sin \beta}{\sin \alpha \sin \beta}$$

$$= \frac{\cos \alpha}{\sin \alpha} \frac{\cos \beta}{\sin \beta} + 1$$

$$= \cot \alpha \cot \beta + 1$$

NOW WORK PROBLEMS 31 AND 43.

FORMULAS FOR $\tan(\alpha + \beta)$ AND $\tan(\alpha - \beta)$

We use the identity $\tan \theta = \dfrac{\sin \theta}{\cos \theta}$ and the sum formulas for $\sin(\alpha + \beta)$ and $\cos(\alpha + \beta)$ to derive a formula for $\tan(\alpha + \beta)$.

Proof

$$\tan(\alpha + \beta) = \frac{\sin(\alpha + \beta)}{\cos(\alpha + \beta)} = \frac{\sin \alpha \cos \beta + \cos \alpha \sin \beta}{\cos \alpha \cos \beta - \sin \alpha \sin \beta}$$

Now we divide the numerator and denominator by $\cos\alpha\cos\beta$.

$$\tan(\alpha + \beta) = \frac{\dfrac{\sin\alpha\cos\beta + \cos\alpha\sin\beta}{\cos\alpha\cos\beta}}{\dfrac{\cos\alpha\cos\beta - \sin\alpha\sin\beta}{\cos\alpha\cos\beta}} = \frac{\dfrac{\sin\alpha\cos\beta}{\cos\alpha\cos\beta} + \dfrac{\cos\alpha\sin\beta}{\cos\alpha\cos\beta}}{\dfrac{\cos\alpha\cos\beta}{\cos\alpha\cos\beta} - \dfrac{\sin\alpha\sin\beta}{\cos\alpha\cos\beta}}$$

$$= \frac{\dfrac{\sin\alpha}{\cos\alpha} + \dfrac{\sin\beta}{\cos\beta}}{1 - \dfrac{\sin\alpha\sin\beta}{\cos\alpha\cos\beta}} = \frac{\tan\alpha + \tan\beta}{1 - \tan\alpha\tan\beta}$$

We use the sum formula for $\tan(\alpha + \beta)$ and even–odd properties to get the difference formula.

$$\tan(\alpha - \beta) = \tan[\alpha + (-\beta)] = \frac{\tan\alpha + \tan(-\beta)}{1 - \tan\alpha\tan(-\beta)} = \frac{\tan\alpha - \tan\beta}{1 + \tan\alpha\tan\beta} \blacksquare$$

We have proved the following results:

Theorem   **Sum and Difference Formulas for Tangents**

$$\tan(\alpha + \beta) = \frac{\tan\alpha + \tan\beta}{1 - \tan\alpha\tan\beta} \qquad (6)$$

$$\tan(\alpha - \beta) = \frac{\tan\alpha - \tan\beta}{1 + \tan\alpha\tan\beta} \qquad (7)$$

∎

In words, formula (6) states that the tangent of the sum of two angles equals the tangent of the first angle plus the tangent of the second angle, all divided by 1 minus their product.

NOW WORK PROBLEM **23(d)**.

EXAMPLE 7   **Establishing an Identity**

Prove the identity: $\tan(\theta + \pi) = \tan\theta$

Solution $\tan(\theta + \pi) = \dfrac{\tan\theta + \tan\pi}{1 - \tan\theta\tan\pi} = \dfrac{\tan\theta + 0}{1 - \tan\theta\cdot 0} = \tan\theta$ ∎

The result obtained in Example 7 verifies that the tangent function is periodic with period π, a fact that we mentioned earlier.

WARNING: Be careful when using formulas (6) and (7). These formulas can be used only for angles α and β for which $\tan\alpha$ and $\tan\beta$ are defined, that is, all angles except odd multiples of $\dfrac{\pi}{2}$. ∎

| EXAMPLE 8 | Establishing an Identity |

Prove the identity: $\tan\left(\theta + \dfrac{\pi}{2}\right) = -\cot\theta$

Solution   We cannot use formula (6), since $\tan\dfrac{\pi}{2}$ is not defined. Instead, we proceed as follows:

$$\tan\left(\theta + \frac{\pi}{2}\right) = \frac{\sin\left(\theta + \dfrac{\pi}{2}\right)}{\cos\left(\theta + \dfrac{\pi}{2}\right)} = \frac{\sin\theta\cos\dfrac{\pi}{2} + \cos\theta\sin\dfrac{\pi}{2}}{\cos\theta\cos\dfrac{\pi}{2} - \sin\theta\sin\dfrac{\pi}{2}}$$

$$= \frac{(\sin\theta)(0) + (\cos\theta)(1)}{(\cos\theta)(0) - (\sin\theta)(1)} = \frac{\cos\theta}{-\sin\theta} = -\cot\theta \quad\blacksquare$$

| EXAMPLE 9 | Finding the Exact Value of Expressions Involving Inverse Trigonometric Functions |

3   Find the exact value of: $\sin\left(\cos^{-1}\dfrac{1}{2} + \sin^{-1}\dfrac{3}{5}\right)$

Solution   We seek the sine of the sum of two angles, $\alpha = \cos^{-1}\dfrac{1}{2}$ and $\beta = \sin^{-1}\dfrac{3}{5}$. Then

$$\cos\alpha = \frac{1}{2}, \quad 0 \le \alpha \le \pi, \quad \text{and} \quad \sin\beta = \frac{3}{5}, \quad -\frac{\pi}{2} \le \beta \le \frac{\pi}{2}$$

We use Pythagorean Identities to obtain $\sin\alpha$ and $\cos\beta$. Since $\sin\alpha > 0$ and $\cos\beta > 0$ (do you know why?), we find

$$\sin\alpha = \sqrt{1 - \cos^2\alpha} = \sqrt{1 - \frac{1}{4}} = \sqrt{\frac{3}{4}} = \frac{\sqrt{3}}{2}$$

$$\cos\beta = \sqrt{1 - \sin^2\beta} = \sqrt{1 - \frac{9}{25}} = \sqrt{\frac{16}{25}} = \frac{4}{5}$$

As a result,

$$\sin\left(\cos^{-1}\frac{1}{2} + \sin^{-1}\frac{3}{5}\right) = \sin(\alpha + \beta) = \sin\alpha\cos\beta + \cos\alpha\sin\beta$$

$$= \frac{\sqrt{3}}{2} \cdot \frac{4}{5} + \frac{1}{2} \cdot \frac{3}{5} = \frac{4\sqrt{3} + 3}{10} \quad\blacksquare$$

NOW WORK PROBLEM 59.

| EXAMPLE 10 | Writing a Trigonometric Expression as an Algebraic Expression |

Write $\sin(\sin^{-1}u + \cos^{-1}v)$ as an algebraic expression containing u and v (that is, without any trigonometric functions).

Solution   Let $\alpha = \sin^{-1}u$ and $\beta = \cos^{-1}v$. Then

$$\sin\alpha = u, \quad -\frac{\pi}{2} \le \alpha \le \frac{\pi}{2} \quad \text{and} \quad \cos\beta = v, \quad 0 \le \beta \le \pi$$

Since $-\dfrac{\pi}{2} \le \alpha \le \dfrac{\pi}{2}$, we know that $\cos \alpha \ge 0$. As a result,

$$\cos \alpha = \sqrt{1 - \sin^2 \alpha} = \sqrt{1 - u^2}$$

Similarly, since $0 \le \beta \le \pi$, we know that $\sin \beta \ge 0$. Thus,

$$\sin \beta = \sqrt{1 - \cos^2 \beta} = \sqrt{1 - v^2}$$

Now

$$\sin(\sin^{-1} u + \cos^{-1} v) = \sin(\alpha + \beta) = \sin \alpha \cos \beta + \cos \alpha \sin \beta$$
$$= uv + \sqrt{1 - u^2}\sqrt{1 - v^2} \quad \blacksquare$$

NOW WORK PROBLEM **69**.

SUMMARY

The following box summarizes the sum and difference formulas.

Sum and Difference Formulas

$$\cos(\alpha + \beta) = \cos \alpha \cos \beta - \sin \alpha \sin \beta \qquad \cos(\alpha - \beta) = \cos \alpha \cos \beta + \sin \alpha \sin \beta$$

$$\sin(\alpha + \beta) = \sin \alpha \cos \beta + \cos \alpha \sin \beta \qquad \sin(\alpha - \beta) = \sin \alpha \cos \beta - \cos \alpha \sin \beta$$

$$\tan(\alpha + \beta) = \frac{\tan \alpha + \tan \beta}{1 - \tan \alpha \tan \beta} \qquad\qquad \tan(\alpha - \beta) = \frac{\tan \alpha - \tan \beta}{1 + \tan \alpha \tan \beta}$$

3.4   EXERCISES

In Problems 1–12, find the exact value of each trigonometric function.

1. $\sin \dfrac{5\pi}{12}$ 　　**2.** $\sin \dfrac{\pi}{12}$ 　　**3.** $\cos \dfrac{7\pi}{12}$ 　　**4.** $\tan \dfrac{7\pi}{12}$ 　　**5.** $\cos 165°$ 　　**6.** $\sin 105°$

7. $\tan 15°$ 　　**8.** $\tan 195°$ 　　**9.** $\sin \dfrac{17\pi}{12}$ 　　**10.** $\tan \dfrac{19\pi}{12}$ 　　**11.** $\sec\left(-\dfrac{\pi}{12}\right)$ 　　**12.** $\cot\left(-\dfrac{5\pi}{12}\right)$

In Problems 13–22, find the exact value of each expression.

13. $\sin 20° \cos 10° + \cos 20° \sin 10°$

14. $\sin 20° \cos 80° - \cos 20° \sin 80°$

15. $\cos 70° \cos 20° - \sin 70° \sin 20°$

16. $\cos 40° \cos 10° + \sin 40° \sin 10°$

17. $\dfrac{\tan 20° + \tan 25°}{1 - \tan 20° \tan 25°}$

18. $\dfrac{\tan 40° - \tan 10°}{1 + \tan 40° \tan 10°}$

19. $\sin \dfrac{\pi}{12} \cos \dfrac{7\pi}{12} - \cos \dfrac{\pi}{12} \sin \dfrac{7\pi}{12}$

20. $\cos \dfrac{5\pi}{12} \cos \dfrac{7\pi}{12} - \sin \dfrac{5\pi}{12} \sin \dfrac{7\pi}{12}$

21. $\cos \dfrac{\pi}{12} \cos \dfrac{5\pi}{12} + \sin \dfrac{5\pi}{12} \sin \dfrac{\pi}{12}$

22. $\sin \dfrac{\pi}{18} \cos \dfrac{5\pi}{18} + \cos \dfrac{\pi}{18} \sin \dfrac{5\pi}{18}$

In Problems 23–28, find the exact value of each of the following under the given conditions:
(a) $\sin(\alpha + \beta)$ 　　(b) $\cos(\alpha + \beta)$ 　　(c) $\sin(\alpha - \beta)$ 　　(d) $\tan(\alpha - \beta)$

23. $\sin \alpha = \dfrac{3}{5},\quad 0 < \alpha < \dfrac{\pi}{2};\quad \cos \beta = \dfrac{2\sqrt{5}}{5},\quad -\dfrac{\pi}{2} < \beta < 0$

24. $\cos \alpha = \dfrac{\sqrt{5}}{5},\quad 0 < \alpha < \dfrac{\pi}{2};\quad \sin \beta = -\dfrac{4}{5},\quad -\dfrac{\pi}{2} < \beta < 0$

25. $\tan \alpha = -\dfrac{4}{3},\quad \dfrac{\pi}{2} < \alpha < \pi;\quad \cos \beta = \dfrac{1}{2},\quad 0 < \beta < \dfrac{\pi}{2}$

26. $\tan \alpha = \dfrac{5}{12},\quad \pi < \alpha < \dfrac{3\pi}{2};\quad \sin \beta = -\dfrac{1}{2},\quad \pi < \beta < \dfrac{3\pi}{2}$

27. $\sin \alpha = \dfrac{5}{13},\quad -\dfrac{3\pi}{2} < \alpha < -\pi;\quad \tan \beta = -\sqrt{3},\quad \dfrac{\pi}{2} < \beta < \pi$

28. $\cos \alpha = \dfrac{1}{2},\quad -\dfrac{\pi}{2} < \alpha < 0;\quad \sin \beta = \dfrac{1}{3},\quad 0 < \beta < \dfrac{\pi}{2}$

29. If $\sin \theta = \dfrac{1}{3}$, θ in quadrant II, find the exact value of:

(a) $\cos \theta$ (b) $\sin\left(\theta + \dfrac{\pi}{6}\right)$ (c) $\cos\left(\theta - \dfrac{\pi}{3}\right)$ (d) $\tan\left(\theta + \dfrac{\pi}{4}\right)$

30. If $\cos \theta = \dfrac{1}{4}$, θ in quadrant IV, find the exact value of:

(a) $\sin \theta$ (b) $\sin\left(\theta - \dfrac{\pi}{6}\right)$ (c) $\cos\left(\theta + \dfrac{\pi}{3}\right)$ (d) $\tan\left(\theta - \dfrac{\pi}{4}\right)$

In Problems 31–56, establish each identity.

31. $\sin\left(\dfrac{\pi}{2} + \theta\right) = \cos \theta$ 　　　**32.** $\cos\left(\dfrac{\pi}{2} + \theta\right) = -\sin \theta$ 　　　**33.** $\sin(\pi - \theta) = \sin \theta$

34. $\cos(\pi - \theta) = -\cos \theta$ 　　　**35.** $\sin(\pi + \theta) = -\sin \theta$ 　　　**36.** $\cos(\pi + \theta) = -\cos \theta$

37. $\tan(\pi - \theta) = -\tan \theta$ 　　　**38.** $\tan(2\pi - \theta) = -\tan \theta$ 　　　**39.** $\sin\left(\dfrac{3\pi}{2} + \theta\right) = -\cos \theta$

40. $\cos\left(\dfrac{3\pi}{2} + \theta\right) = \sin \theta$ 　　　　　　　**41.** $\sin(\alpha + \beta) + \sin(\alpha - \beta) = 2 \sin \alpha \cos \beta$

42. $\cos(\alpha + \beta) + \cos(\alpha - \beta) = 2 \cos \alpha \cos \beta$ 　　　**43.** $\dfrac{\sin(\alpha + \beta)}{\sin \alpha \cos \beta} = 1 + \cot \alpha \tan \beta$

44. $\dfrac{\sin(\alpha + \beta)}{\cos \alpha \cos \beta} = \tan \alpha + \tan \beta$ 　　　**45.** $\dfrac{\cos(\alpha + \beta)}{\cos \alpha \cos \beta} = 1 - \tan \alpha \tan \beta$

46. $\dfrac{\cos(\alpha - \beta)}{\sin \alpha \cos \beta} = \cot \alpha + \tan \beta$ 　　　**47.** $\dfrac{\sin(\alpha + \beta)}{\sin(\alpha - \beta)} = \dfrac{\tan \alpha + \tan \beta}{\tan \alpha - \tan \beta}$

48. $\dfrac{\cos(\alpha + \beta)}{\cos(\alpha - \beta)} = \dfrac{1 - \tan \alpha \tan \beta}{1 + \tan \alpha \tan \beta}$ 　　　**49.** $\cot(\alpha + \beta) = \dfrac{\cot \alpha \cot \beta - 1}{\cot \beta + \cot \alpha}$

50. $\cot(\alpha - \beta) = \dfrac{\cot \alpha \cot \beta + 1}{\cot \beta - \cot \alpha}$ 　　　**51.** $\sec(\alpha + \beta) = \dfrac{\csc \alpha \csc \beta}{\cot \alpha \cot \beta - 1}$

52. $\sec(\alpha - \beta) = \dfrac{\sec \alpha \sec \beta}{1 + \tan \alpha \tan \beta}$ 　　　**53.** $\sin(\alpha - \beta) \sin(\alpha + \beta) = \sin^2 \alpha - \sin^2 \beta$

54. $\cos(\alpha - \beta) \cos(\alpha + \beta) = \cos^2 \alpha - \sin^2 \beta$ 　　　**55.** $\sin(\theta + k\pi) = (-1)^k \sin \theta$, $\quad k$ any integer

56. $\cos(\theta + k\pi) = (-1)^k \cos \theta$, $\quad k$ any integer

In Problems 57–68, find the exact value of each expression.

57. $\sin\left(\sin^{-1} \dfrac{1}{2} + \cos^{-1} 0\right)$ **58.** $\sin\left(\sin^{-1} \dfrac{\sqrt{3}}{2} + \cos^{-1} 1\right)$ **59.** $\sin\left[\sin^{-1} \dfrac{3}{5} - \cos^{-1}\left(-\dfrac{4}{5}\right)\right]$ **60.** $\sin\left[\sin^{-1}\left(-\dfrac{4}{5}\right) - \tan^{-1} \dfrac{3}{4}\right]$

61. $\cos\left(\tan^{-1} \dfrac{4}{3} + \cos^{-1} \dfrac{5}{13}\right)$ **62.** $\cos\left[\tan^{-1} \dfrac{5}{12} - \sin^{-1}\left(-\dfrac{3}{5}\right)\right]$ **63.** $\cos\left(\sin^{-1} \dfrac{5}{13} - \tan^{-1} \dfrac{3}{4}\right)$ **64.** $\cos\left(\tan^{-1} \dfrac{4}{3} + \cos^{-1} \dfrac{12}{13}\right)$

65. $\tan\left(\sin^{-1} \dfrac{3}{5} + \dfrac{\pi}{6}\right)$ **66.** $\tan\left(\dfrac{\pi}{4} - \cos^{-1} \dfrac{3}{5}\right)$ **67.** $\tan\left(\sin^{-1} \dfrac{4}{5} + \cos^{-1} 1\right)$ **68.** $\tan\left(\cos^{-1} \dfrac{4}{5} + \sin^{-1} 1\right)$

In Problems 69–74, write each trigonometric expression as an algebraic expression containing u and v.

69. $\cos\left(\cos^{-1} u + \sin^{-1} v\right)$ 　　　　**70.** $\sin\left(\sin^{-1} u - \cos^{-1} v\right)$ 　　　　**71.** $\sin\left(\tan^{-1} u - \sin^{-1} v\right)$

72. $\cos\left(\tan^{-1} u + \tan^{-1} v\right)$ 　　　　**73.** $\tan\left(\sin^{-1} u - \cos^{-1} v\right)$ 　　　　**74.** $\sec\left(\tan^{-1} u + \cos^{-1} v\right)$

75. Show that $\sin^{-1} v + \cos^{-1} v = \pi/2$. 　　　　**76.** Show that $\tan^{-1} v + \cot^{-1} v = \pi/2$.

77. Show that $\tan^{-1}(1/v) = \pi/2 - \tan^{-1} v$, if $v > 0$. 　　　　**78.** Show that $\cot^{-1} e^v = \tan^{-1} e^{-v}$.

79. Show that $\sin\left(\sin^{-1} v + \cos^{-1} v\right) = 1$. 　　　　**80.** Show that $\cos\left(\sin^{-1} v + \cos^{-1} v\right) = 0$.

△ **81. Calculus** Show that the difference quotient for $f(x) = \sin x$ is given by

$$\dfrac{f(x + h) - f(x)}{h} = \dfrac{\sin(x + h) - \sin x}{h}$$

$$= \cos x \cdot \dfrac{\sin h}{h} - \sin x \cdot \dfrac{1 - \cos h}{h}$$

△ **82. Calculus** Show that the difference quotient for $f(x) = \cos x$ is given by

$$\dfrac{f(x + h) - f(x)}{h} = \dfrac{\cos(x + h) - \cos x}{h}$$

$$= -\sin x \cdot \dfrac{\sin h}{h} - \cos x \cdot \dfrac{1 - \cos h}{h}$$

83. Explain why formula (7) cannot be used to show that

$$\tan\left(\frac{\pi}{2} - \theta\right) = \cot\theta$$

Establish this identity by using formulas (3a) and (3b).

84. If $\tan\alpha = x + 1$ and $\tan\beta = x - 1$, show that $2\cot(\alpha - \beta) = x^2$.

85. Geometry: Angle between Two Lines Let L_1 and L_2 denote two nonvertical intersecting lines, and let θ denote the acute angle between L_1 and L_2 (see the figure). Show that

$$\tan\theta = \frac{m_2 - m_1}{1 + m_1 m_2}$$

where m_1 and m_2 are the slopes of L_1 and L_2, respectively. [**Hint:** Use the facts that $\tan\theta_1 = m_1$ and $\tan\theta_2 = m_2$.]

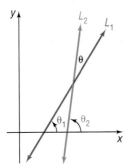

86. If $\alpha + \beta + \gamma = 180°$ and

$$\cot\theta = \cot\alpha + \cot\beta + \cot\gamma, \quad 0 < \theta < 90°$$

show that

$$\sin^3\theta = \sin(\alpha - \theta)\sin(\beta - \theta)\sin(\gamma - \theta)$$

87. Discuss the following derivation:

$$\tan\left(\theta + \frac{\pi}{2}\right) = \frac{\tan\theta + \tan\dfrac{\pi}{2}}{1 - \tan\theta\tan\dfrac{\pi}{2}}$$

$$= \frac{\dfrac{\tan\theta}{\tan\dfrac{\pi}{2}} + 1}{\dfrac{1}{\tan\dfrac{\pi}{2}} - \tan\theta} = \frac{0 + 1}{0 - \tan\theta}$$

$$= \frac{1}{-\tan\theta} = -\cot\theta$$

Can you justify each step?

3.5 DOUBLE-ANGLE AND HALF-ANGLE FORMULAS

OBJECTIVES **1** Use Double-Angle Formulas to Find Exact Values
2 Use Double-Angle and Half-Angle Formulas to Establish Identities
3 Use Half-Angle Formulas to Find Exact Values

In this section we derive formulas for $\sin(2\theta)$, $\cos(2\theta)$, $\sin\left(\frac{1}{2}\theta\right)$, and $\cos\left(\frac{1}{2}\theta\right)$ in terms of $\sin\theta$ and $\cos\theta$. They are easily derived using the sum formulas.

DOUBLE-ANGLE FORMULAS

In the sum formulas for $\sin(\alpha + \beta)$ and $\cos(\alpha + \beta)$, let $\alpha = \beta = \theta$. Then

$$\sin(\alpha + \beta) = \sin\alpha\cos\beta + \cos\alpha\sin\beta$$

$$\sin(\theta + \theta) = \sin\theta\cos\theta + \cos\theta\sin\theta$$

$$\sin(2\theta) = 2\sin\theta\cos\theta \tag{1}$$

and

$$\cos(\alpha + \beta) = \cos\alpha\cos\beta - \sin\alpha\sin\beta$$

$$\cos(\theta + \theta) = \cos\theta\cos\theta - \sin\theta\sin\theta$$

$$\cos(2\theta) = \cos^2\theta - \sin^2\theta \tag{2}$$

An application of the Pythagorean Identity $\sin^2\theta + \cos^2\theta = 1$ results in two other ways to write formula (2) for $\cos(2\theta)$.

$$\cos(2\theta) = \cos^2\theta - \sin^2\theta = (1 - \sin^2\theta) - \sin^2\theta = 1 - 2\sin^2\theta$$

and

$$\cos(2\theta) = \cos^2\theta - \sin^2\theta = \cos^2\theta - (1 - \cos^2\theta) = 2\cos^2\theta - 1$$

We have established the following **double-angle formulas:**

Theorem   **Double-Angle Formulas**

$$\sin(2\theta) = 2\sin\theta\cos\theta \qquad \textbf{(1)}$$

$$\cos(2\theta) = \cos^2\theta - \sin^2\theta \qquad \textbf{(2)}$$

$$\cos(2\theta) = 1 - 2\sin^2\theta \qquad \textbf{(3)}$$

$$\cos(2\theta) = 2\cos^2\theta - 1 \qquad \textbf{(4)}$$

EXAMPLE 1   Finding Exact Values Using the Double-Angle Formula

1   If $\sin\theta = \dfrac{3}{5}, \dfrac{\pi}{2} < \theta < \pi$, find the exact value of:

(a) $\sin(2\theta)$ (b) $\cos(2\theta)$

Solution   (a) Because $\sin(2\theta) = 2\sin\theta\cos\theta$ and we already know that $\sin\theta = \dfrac{3}{5}$, we only need to find $\cos\theta$. Since $\sin\theta = \dfrac{3}{5} = \dfrac{y}{r}, \dfrac{\pi}{2} < \theta < \pi$, we let $y = 3$ and $r = 5$, and place θ in quadrant II. The point $P = (x, y) = (x, 3)$, $x < 0$, is on a circle of radius 5, $x^2 + y^2 = 25$. See Figure 25. Then,

$$x^2 + y^2 = 25, \qquad x < 0, y = 3$$
$$x^2 + 9 = 25$$
$$x^2 = 25 - 9 = 16$$
$$x = -4$$

Figure 25

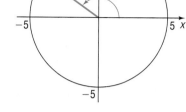

We find that $\cos\theta = \dfrac{x}{r} = \dfrac{-4}{5}$. Now we use formula (1) to obtain

$$\sin(2\theta) = 2\sin\theta\cos\theta = 2\left(\dfrac{3}{5}\right)\left(-\dfrac{4}{5}\right) = -\dfrac{24}{25}$$

(b) Because we are given $\sin\theta = \dfrac{3}{5}$, it is easiest to use formula (1) to get $\cos(2\theta)$.

$$\cos(2\theta) = 1 - 2\sin^2\theta = 1 - 2\left(\dfrac{9}{25}\right) = 1 - \dfrac{18}{25} = \dfrac{7}{25} \qquad \blacksquare$$

WARNING: In finding $\cos(2\theta)$ in Example 1(b), we chose to use a version of the double-angle formula, formula (3). Note that we are unable to use the Pythagorean Identity $\cos(2\theta) = \pm\sqrt{1 - \sin^2(2\theta)}$, with $\sin(2\theta) = -\dfrac{24}{25}$, because we have no way of knowing which sign to choose. $\qquad \blacksquare$

NOW WORK PROBLEMS **1(a)** AND **(b)**.

| EXAMPLE 2 | **Establishing Identities** |

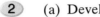   (a) Develop a formula for $\tan(2\theta)$ in terms of $\tan\theta$.

(b) Develop a formula for $\sin(3\theta)$ in terms of $\sin\theta$ and $\cos\theta$.

Solution   (a) In the sum formula for $\tan(\alpha + \beta)$, let $\alpha = \beta = \theta$. Then

$$\tan(\alpha + \beta) = \frac{\tan\alpha + \tan\beta}{1 - \tan\alpha\tan\beta}$$

$$\tan(\theta + \theta) = \frac{\tan\theta + \tan\theta}{1 - \tan\theta\tan\theta}$$

$$\tan(2\theta) = \frac{2\tan\theta}{1 - \tan^2\theta} \tag{5}$$

(b) To get a formula for $\sin(3\theta)$, we use the sum formula and write 3θ as $2\theta + \theta$.

$$\sin(3\theta) = \sin(2\theta + \theta) = \sin(2\theta)\cos\theta + \cos(2\theta)\sin\theta$$

Now use the double-angle formulas to get

$$\sin(3\theta) = (2\sin\theta\cos\theta)(\cos\theta) + (\cos^2\theta - \sin^2\theta)(\sin\theta)$$

$$= 2\sin\theta\cos^2\theta + \sin\theta\cos^2\theta - \sin^3\theta$$

$$= 3\sin\theta\cos^2\theta - \sin^3\theta \qquad\blacksquare$$

The formula obtained in Example 2(b) can also be written as

$$\sin(3\theta) = 3\sin\theta\cos^2\theta - \sin^3\theta = 3\sin\theta(1 - \sin^2\theta) - \sin^3\theta$$

$$= 3\sin\theta - 4\sin^3\theta$$

That is, $\sin(3\theta)$ is a third-degree polynomial in the variable $\sin\theta$. In fact, $\sin(n\theta)$, n a positive odd integer, can always be written as a polynomial of degree n in the variable $\sin\theta$.*

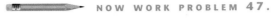   NOW WORK PROBLEM **47.**

OTHER VARIATIONS OF THE DOUBLE-ANGLE FORMULAS

By rearranging the double-angle formulas (3) and (4), we obtain other formulas that we will use later in this section.

We begin with formula (3) and proceed to solve for $\sin^2\theta$.

$$\cos(2\theta) = 1 - 2\sin^2\theta$$

$$2\sin^2\theta = 1 - \cos(2\theta)$$

$$\sin^2\theta = \frac{1 - \cos(2\theta)}{2} \tag{6}$$

---

*Due to the work done by P.L. Chebyshëv, these polynomials are sometimes called *Chebyshëv polynomials*.

Similarly, using formula (4), we proceed to solve for $\cos^2\theta$.

$$\cos(2\theta) = 2\cos^2\theta - 1$$

$$2\cos^2\theta = 1 + \cos(2\theta)$$

$$\cos^2\theta = \frac{1 + \cos(2\theta)}{2} \qquad \textbf{(7)}$$

Formulas (6) and (7) can be used to develop a formula for $\tan^2\theta$.

$$\tan^2\theta = \frac{\sin^2\theta}{\cos^2\theta} = \frac{\dfrac{1 - \cos(2\theta)}{2}}{\dfrac{1 + \cos(2\theta)}{2}}$$

$$\tan^2\theta = \frac{1 - \cos(2\theta)}{1 + \cos(2\theta)} \qquad \textbf{(8)}$$

Formulas (6) through (8) do not have to be memorized since their derivations are so straightforward.

Formulas (6) and (7) are important in calculus. The next example illustrates a problem that arises in calculus requiring the use of formula (7).

EXAMPLE 3 | **Establishing an Identity**

Write an equivalent expression for $\cos^4\theta$ that does not involve any powers of sine or cosine greater than 1.

Solution The idea here is to apply formula (7) twice.

$$\cos^4\theta = \left(\cos^2\theta\right)^2 = \left(\frac{1 + \cos(2\theta)}{2}\right)^2 \qquad \text{Formula (7)}$$

$$= \frac{1}{4}\left[1 + 2\cos(2\theta) + \cos^2(2\theta)\right]$$

$$= \frac{1}{4} + \frac{1}{2}\cos(2\theta) + \frac{1}{4}\cos^2(2\theta)$$

$$= \frac{1}{4} + \frac{1}{2}\cos(2\theta) + \frac{1}{4}\left\{\frac{1 + \cos[2(2\theta)]}{2}\right\} \qquad \text{Formula (7)}$$

$$= \frac{1}{4} + \frac{1}{2}\cos(2\theta) + \frac{1}{8}\left[1 + \cos(4\theta)\right]$$

$$= \frac{3}{8} + \frac{1}{2}\cos(2\theta) + \frac{1}{8}\cos(4\theta) \qquad \blacksquare$$

NOW WORK PROBLEM **23**.

Identities, such as the double-angle formulas, can sometimes be used to rewrite expressions in a more suitable form. Let's look at an example.

| EXAMPLE 4 | **Projectile Motion** |

An object is propelled upward at an angle θ to the horizontal with an initial velocity of v_0 feet per second. See Figure 26. If air resistance is ignored, the **range** R, the horizontal distance that the object travels, is given by

Figure 26

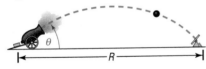

$$R = \frac{1}{16} v_0^2 \sin \theta \cos \theta$$

(a) Show that $R = \frac{1}{32} v_0^2 \sin(2\theta)$.

(b) Find the angle θ for which R is a maximum.

Solution

(a) We rewrite the given expression for the range using the double-angle formula $\sin(2\theta) = 2 \sin \theta \cos \theta$. Then

$$R = \frac{1}{16} v_0^2 \sin \theta \cos \theta = \frac{1}{16} v_0^2 \frac{2 \sin \theta \cos \theta}{2} = \frac{1}{32} v_0^2 \sin(2\theta)$$

(b) In this form, the largest value for the range R can be found. For a fixed initial speed v_0, the angle θ of inclination to the horizontal determines the value of R. Since the largest value of a sine function is 1, occurring when the argument 2θ is $90°$, it follows that for maximum R we must have

$$2\theta = 90°$$
$$\theta = 45°$$

An inclination to the horizontal of $45°$ results in maximum range. ∎

HALF-ANGLE FORMULAS

Another important use of formulas (6) through (8) is to prove the **half-angle formulas.** In formulas (6) through (8), let $\theta = \frac{\alpha}{2}$. Then

$$\sin^2 \frac{\alpha}{2} = \frac{1 - \cos \alpha}{2} \qquad \cos^2 \frac{\alpha}{2} = \frac{1 + \cos \alpha}{2} \qquad \tan^2 \frac{\alpha}{2} = \frac{1 - \cos \alpha}{1 + \cos \alpha} \quad \text{(9)}$$

 NOTE: The identities in box (9) will prove useful in integral calculus.

If we solve for the trigonometric functions on the left sides of equations (9), we obtain the half-angle formulas.

Theorem

Half-Angle Formulas

$$\sin \frac{\alpha}{2} = \pm \sqrt{\frac{1 - \cos \alpha}{2}} \qquad \text{(10a)}$$

$$\cos \frac{\alpha}{2} = \pm \sqrt{\frac{1 + \cos \alpha}{2}} \qquad \text{(10b)}$$

$$\tan \frac{\alpha}{2} = \pm \sqrt{\frac{1 - \cos \alpha}{1 + \cos \alpha}} \qquad \text{(10c)}$$

where the $+$ or $-$ sign is determined by the quadrant of the angle $\frac{\alpha}{2}$.

We use the half-angle formulas in the next example.

| EXAMPLE 5 | **Finding Exact Values Using Half-Angle Formulas** |

(3)   Find the exact value of:

(a) $\cos 15°$ (b) $\sin(-15°)$

Solution   (a) Because $15° = \dfrac{30°}{2}$, we can use the half-angle formula for $\cos\dfrac{\alpha}{2}$ with $\alpha = 30°$. Also, because $15°$ is in quadrant I, $\cos 15° > 0$, so we choose the $+$ sign in using formula (10b):

$$\cos 15° = \cos\frac{30°}{2} = \sqrt{\frac{1 + \cos 30°}{2}}$$

$$= \sqrt{\frac{1 + \sqrt{3}/2}{2}} = \sqrt{\frac{2 + \sqrt{3}}{4}} = \frac{\sqrt{2 + \sqrt{3}}}{2}$$

(b) We use the fact that $\sin(-15°) = -\sin 15°$ and then apply formula (10a).

$$\sin(-15°) = -\sin\frac{30°}{2} = -\sqrt{\frac{1 - \cos 30°}{2}}$$

$$= -\sqrt{\frac{1 - \sqrt{3}/2}{2}} = -\sqrt{\frac{2 - \sqrt{3}}{4}} = -\frac{\sqrt{2 - \sqrt{3}}}{2}$$ ∎

It is interesting to compare the answer found in Example 5(a) with the answer to Example 2 of Section 3.4. There we calculated

$$\cos\frac{\pi}{12} = \cos 15° = \frac{1}{4}\left(\sqrt{6} + \sqrt{2}\right)$$

Based on this and the result of Example 5(a), we conclude that

$$\frac{1}{4}\left(\sqrt{6} + \sqrt{2}\right) \quad \text{and} \quad \frac{\sqrt{2 + \sqrt{3}}}{2}$$

are equal. (Since each expression is positive, you can verify this equality by squaring each expression.) Two very different looking, yet correct, answers can be obtained, depending on the approach taken to solve a problem.

✏ N O W   W O R K   P R O B L E M   **13**.

| EXAMPLE 6 | **Finding Exact Values Using Half-Angle Formulas** |

If $\cos\alpha = -\dfrac{3}{5}$, $\pi < \alpha < \dfrac{3\pi}{2}$, find the exact value of:

(a) $\sin\dfrac{\alpha}{2}$ (b) $\cos\dfrac{\alpha}{2}$ (c) $\tan\dfrac{\alpha}{2}$

Solution   First, we observe that if $\pi < \alpha < \dfrac{3\pi}{2}$, then $\dfrac{\pi}{2} < \dfrac{\alpha}{2} < \dfrac{3\pi}{4}$. As a result, $\dfrac{\alpha}{2}$ lies in quadrant II.

(a) Because $\dfrac{\alpha}{2}$ lies in quadrant II, $\sin\dfrac{\alpha}{2} > 0$, so we use the $+$ sign in formula (10a) to get

$$\sin\frac{\alpha}{2} = \sqrt{\frac{1 - \cos\alpha}{2}} = \sqrt{\frac{1 - \left(-\frac{3}{5}\right)}{2}}$$

$$= \sqrt{\frac{\frac{8}{5}}{2}} = \sqrt{\frac{4}{5}} = \frac{2}{\sqrt{5}} = \frac{2\sqrt{5}}{5}$$

(b) Because $\dfrac{\alpha}{2}$ lies in quadrant II, $\cos\dfrac{\alpha}{2} < 0$, so we use the $-$ sign in formula (10b) to get

$$\cos\frac{\alpha}{2} = -\sqrt{\frac{1 + \cos\alpha}{2}} = -\sqrt{\frac{1 + \left(-\frac{3}{5}\right)}{2}}$$

$$= -\sqrt{\frac{\frac{2}{5}}{2}} = -\frac{1}{\sqrt{5}} = -\frac{\sqrt{5}}{5}$$

(c) Because $\dfrac{\alpha}{2}$ lies in quadrant II, $\tan\dfrac{\alpha}{2} < 0$, so we use the $-$ sign in formula (10c) to get

$$\tan\frac{\alpha}{2} = -\sqrt{\frac{1 - \cos\alpha}{1 + \cos\alpha}} = -\sqrt{\frac{1 - \left(-\frac{3}{5}\right)}{1 + \left(-\frac{3}{5}\right)}} = -\sqrt{\frac{\frac{8}{5}}{\frac{2}{5}}} = -2$$

Another way to solve Example 6(c) is to use the solutions found in parts (a) and (b).

$$\tan\frac{\alpha}{2} = \frac{\sin\dfrac{\alpha}{2}}{\cos\dfrac{\alpha}{2}} = \frac{\dfrac{2\sqrt{5}}{5}}{-\dfrac{\sqrt{5}}{5}} = -2$$

> **NOW WORK PROBLEMS 1(c) AND (d).**

There is a formula for $\tan\dfrac{\alpha}{2}$ that does not contain $+$ and $-$ signs, making it more useful than Formula 10(c). Because

$$1 - \cos\alpha = 2\sin^2\frac{\alpha}{2} \qquad \text{Formula (9)}$$

and

$$\sin\alpha = \sin\left[2\left(\frac{\alpha}{2}\right)\right] = 2\sin\frac{\alpha}{2}\cos\frac{\alpha}{2} \qquad \text{Double-angle formula}$$

we have

$$\frac{1 - \cos\alpha}{\sin\alpha} = \frac{2\sin^2\dfrac{\alpha}{2}}{2\sin\dfrac{\alpha}{2}\cos\dfrac{\alpha}{2}} = \frac{\sin\dfrac{\alpha}{2}}{\cos\dfrac{\alpha}{2}} = \tan\frac{\alpha}{2}$$

Since it also can be shown that

$$\frac{1 - \cos\alpha}{\sin\alpha} = \frac{\sin\alpha}{1 + \cos\alpha}$$

we have the following two half-angle formulas:

Half-Angle Formulas for $\tan\dfrac{\alpha}{2}$

$$\tan\frac{\alpha}{2} = \frac{1 - \cos\alpha}{\sin\alpha} = \frac{\sin\alpha}{1 + \cos\alpha} \qquad \textbf{(11)}$$

With this formula, the solution to Example 6(c) can be given as

$$\cos\alpha = -\frac{3}{5}$$

$$\sin\alpha = -\sqrt{1 - \cos^2\alpha} = -\sqrt{1 - \frac{9}{25}} = -\sqrt{\frac{16}{25}} = -\frac{4}{5}$$

Then, by equation (11),

$$\tan\frac{\alpha}{2} = \frac{1 - \cos\alpha}{\sin\alpha} = \frac{1 - \left(-\dfrac{3}{5}\right)}{-\dfrac{4}{5}} = \frac{\dfrac{8}{5}}{-\dfrac{4}{5}} = -2$$

3.5 EXERCISES

In Problems 1–12, use the information given about the angle θ, $0 \le \theta \le 2\pi$, to find the exact value of

(a) $\sin(2\theta)$ (b) $\cos(2\theta)$ (c) $\sin\dfrac{\theta}{2}$ (d) $\cos\dfrac{\theta}{2}$

1. $\sin\theta = \dfrac{3}{5}$, $0 < \theta < \dfrac{\pi}{2}$

2. $\cos\theta = \dfrac{3}{5}$, $0 < \theta < \dfrac{\pi}{2}$

3. $\tan\theta = \dfrac{4}{3}$, $\pi < \theta < \dfrac{3\pi}{2}$

4. $\tan\theta = \dfrac{1}{2}$, $\pi < \theta < \dfrac{3\pi}{2}$

5. $\cos\theta = -\dfrac{\sqrt{6}}{3}$, $\dfrac{\pi}{2} < \theta < \pi$

6. $\sin\theta = -\dfrac{\sqrt{3}}{3}$, $\dfrac{3\pi}{2} < \theta < 2\pi$

7. $\sec\theta = 3$, $\sin\theta > 0$

8. $\csc\theta = -\sqrt{5}$, $\cos\theta < 0$

9. $\cot\theta = -2$, $\sec\theta < 0$

10. $\sec\theta = 2$, $\csc\theta < 0$

11. $\tan\theta = -3$, $\sin\theta < 0$

12. $\cot\theta = 3$, $\cos\theta < 0$

In Problems 13–22, use the half-angle formulas to find the exact value of each trigonometric function.

13. $\sin 22.5°$

14. $\cos 22.5°$

15. $\tan\dfrac{7\pi}{8}$

16. $\tan\dfrac{9\pi}{8}$

17. $\cos 165°$

18. $\sin 195°$

19. $\sec\dfrac{15\pi}{8}$

20. $\csc\dfrac{7\pi}{8}$

21. $\sin\left(-\dfrac{\pi}{8}\right)$

22. $\cos\left(-\dfrac{3\pi}{8}\right)$

23. Show that $\sin^4\theta = \dfrac{3}{8} - \dfrac{1}{2}\cos(2\theta) + \dfrac{1}{8}\cos(4\theta)$.

24. Develop a formula for $\cos(3\theta)$ as a third-degree polynomial in the variable $\cos\theta$.

25. Show that $\sin(4\theta) = (\cos\theta)(4\sin\theta - 8\sin^3\theta)$.

26. Develop a formula for $\cos(4\theta)$ as a fourth-degree polynomial in the variable $\cos\theta$.

27. Find an expression for $\sin(5\theta)$ as a fifth-degree polynomial in the variable $\sin\theta$.

28. Find an expression for $\cos(5\theta)$ as a fifth-degree polynomial in the variable $\cos\theta$.

In Problems 29–50, establish each identity.

29. $\cos^4\theta - \sin^4\theta = \cos(2\theta)$

30. $\dfrac{\cot\theta - \tan\theta}{\cot\theta + \tan\theta} = \cos(2\theta)$

31. $\cot(2\theta) = \dfrac{\cot^2\theta - 1}{2\cot\theta}$

32. $\cot(2\theta) = \dfrac{1}{2}(\cot\theta - \tan\theta)$

33. $\sec(2\theta) = \dfrac{\sec^2\theta}{2 - \sec^2\theta}$

34. $\csc(2\theta) = \dfrac{1}{2}\sec\theta\csc\theta$

35. $\cos^2(2\theta) - \sin^2(2\theta) = \cos(4\theta)$

36. $(4\sin\theta\cos\theta)(1 - 2\sin^2\theta) = \sin(4\theta)$

37. $\dfrac{\cos(2\theta)}{1 + \sin(2\theta)} = \dfrac{\cot\theta - 1}{\cot\theta + 1}$

38. $\sin^2\theta\cos^2\theta = \dfrac{1}{8}[1 - \cos(4\theta)]$

39. $\sec^2\dfrac{\theta}{2} = \dfrac{2}{1 + \cos\theta}$

40. $\csc^2\dfrac{\theta}{2} = \dfrac{2}{1 - \cos\theta}$

41. $\cot^2\dfrac{\theta}{2} = \dfrac{\sec\theta + 1}{\sec\theta - 1}$

42. $\tan\dfrac{\theta}{2} = \csc\theta - \cot\theta$

43. $\cos\theta = \dfrac{1 - \tan^2\dfrac{\theta}{2}}{1 + \tan^2\dfrac{\theta}{2}}$

44. $1 - \dfrac{1}{2}\sin(2\theta) = \dfrac{\sin^3\theta + \cos^3\theta}{\sin\theta + \cos\theta}$

45. $\dfrac{\sin(3\theta)}{\sin\theta} - \dfrac{\cos(3\theta)}{\cos\theta} = 2$

46. $\dfrac{\cos\theta + \sin\theta}{\cos\theta - \sin\theta} - \dfrac{\cos\theta - \sin\theta}{\cos\theta + \sin\theta} = 2\tan(2\theta)$

47. $\tan(3\theta) = \dfrac{3\tan\theta - \tan^3\theta}{1 - 3\tan^2\theta}$

48. $\tan\theta + \tan(\theta + 120°) + \tan(\theta + 240°) = 3\tan(3\theta)$

49. $\ln|\sin\theta| = \dfrac{1}{2}(\ln|1 - \cos(2\theta)| - \ln 2)$

50. $\ln|\cos\theta| = \dfrac{1}{2}(\ln|1 + \cos(2\theta)| - \ln 2)$

In Problems 51–62, find the exact value of each expression.

51. $\sin\left(2\sin^{-1}\dfrac{1}{2}\right)$

52. $\sin\left[2\sin^{-1}\dfrac{\sqrt{3}}{2}\right]$

53. $\cos\left(2\sin^{-1}\dfrac{3}{5}\right)$

54. $\cos\left(2\cos^{-1}\dfrac{4}{5}\right)$

55. $\tan\left[2\cos^{-1}\left(-\dfrac{3}{5}\right)\right]$

56. $\tan\left(2\tan^{-1}\dfrac{3}{4}\right)$

57. $\sin\left(2\cos^{-1}\dfrac{4}{5}\right)$

58. $\cos\left[2\tan^{-1}\left(-\dfrac{4}{3}\right)\right]$

59. $\sin^2\left(\dfrac{1}{2}\cos^{-1}\dfrac{3}{5}\right)$

60. $\cos^2\left(\dfrac{1}{2}\sin^{-1}\dfrac{3}{5}\right)$

61. $\sec\left(2\tan^{-1}\dfrac{3}{4}\right)$

62. $\csc\left[2\sin^{-1}\left(-\dfrac{3}{5}\right)\right]$

63. If $x = 2\tan\theta$, express $\sin(2\theta)$ as a function of x.

64. If $x = 2\tan\theta$, express $\cos(2\theta)$ as a function of x.

65. Find the value of the number C:
$$\frac{1}{2}\sin^2 x + C = -\frac{1}{4}\cos(2x)$$

66. Find the value of the number C:
$$\frac{1}{2}\cos^2 x + C = \frac{1}{4}\cos(2x)$$

67. If $z = \tan\dfrac{\alpha}{2}$, show that $\sin\alpha = \dfrac{2z}{1 + z^2}$.

68. If $z = \tan\dfrac{\alpha}{2}$, show that $\cos\alpha = \dfrac{1 - z^2}{1 + z^2}$.

---

69. Area of an Isosceles Triangle Show that the area A of an isosceles triangle whose equal sides are of length s and θ is the angle between them is
$$\frac{1}{2}s^2\sin\theta$$

[**Hint:** See the illustration. The height h bisects the angle θ and is the perpendicular bisector of the base.]

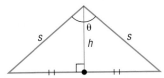

70. Geometry A rectangle is inscribed in a semicircle of radius 1. See the illustration.

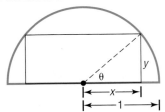

(a) Express the area A of the rectangle as a function of the angle θ shown in the illustration.
(b) Show that $A = \sin(2\theta)$.
(c) Find the angle θ that results in the largest area A.
(d) Find the dimensions of this largest rectangle.

71. Graph $f(x) = \sin^2 x = \dfrac{1 - \cos(2x)}{2}$ for $0 \le x \le 2\pi$ by using transformations.

72. Repeat Problem 71 for $g(x) = \cos^2 x$.

73. Use the fact that
$$\cos\frac{\pi}{12} = \frac{1}{4}\left(\sqrt{6} + \sqrt{2}\right)$$
to find $\sin\dfrac{\pi}{24}$ and $\cos\dfrac{\pi}{24}$.

74. Show that
$$\cos\frac{\pi}{8} = \frac{\sqrt{2 + \sqrt{2}}}{2}$$
and use it to find $\sin\dfrac{\pi}{16}$ and $\cos\dfrac{\pi}{16}$.

75. Show that
$$\sin^3\theta + \sin^3(\theta + 120°) + \sin^3(\theta + 240°) = -\frac{3}{4}\sin(3\theta)$$

76. If $\tan\theta = a\tan\dfrac{\theta}{3}$, express $\tan\dfrac{\theta}{3}$ in terms of a.

77. Projectile Motion An object is propelled upward at an angle θ, $45° < \theta < 90°$, to the horizontal with an initial velocity of v_0 feet per second from the base of a plane that makes an angle of $45°$ with the horizontal. See the illustration. If air resistance is ignored, the distance R that it travels up the inclined plane is given by
$$R = \frac{v_0^2\sqrt{2}}{16}\cos\theta(\sin\theta - \cos\theta)$$

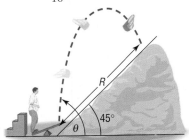

(a) Show that
$$R = \frac{v_0^2\sqrt{2}}{32}\left[\sin(2\theta) - \cos(2\theta) - 1\right]$$

(b) Graph $R = R(\theta)$. (Use $v_0 = 32$ feet per second.)
(c) What value of θ makes R the largest? (Use $v_0 = 32$ feet per second.)

78. Sawtooth Curve An oscilloscope often displays a sawtooth curve. This curve can be approximated by sinusoidal curves of varying periods and amplitudes. A first approximation to the sawtooth curve is given by
$$y = \frac{1}{2}\sin(2\pi x) + \frac{1}{4}\sin(4\pi x)$$
Show that $y = \sin(2\pi x)\cos^2(\pi x)$.

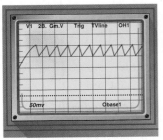

79. Go to the library and research Chebyshëv polynomials. Write a report on your findings.

3.6 PRODUCT-TO-SUM AND SUM-TO-PRODUCT FORMULAS

OBJECTIVES **1** Express Products as Sums
2 Express Sums as Products

1 The sum and difference formulas can be used to derive formulas for writing the products of sines and/or cosines as sums or differences. These identities are usually called the **Product-to-Sum Formulas.**

Theorem **Product-to-Sum Formulas**

$$\sin \alpha \sin \beta = \frac{1}{2}\left[\cos(\alpha - \beta) - \cos(\alpha + \beta)\right] \qquad \textbf{(1)}$$

$$\cos \alpha \cos \beta = \frac{1}{2}\left[\cos(\alpha - \beta) + \cos(\alpha + \beta)\right] \qquad \textbf{(2)}$$

$$\sin \alpha \cos \beta = \frac{1}{2}\left[\sin(\alpha + \beta) + \sin(\alpha - \beta)\right] \qquad \textbf{(3)}$$

These formulas do not have to be memorized. Instead, you should remember how they are derived. Then, when you want to use them, either look them up or derive them, as needed.

To derive formulas (1) and (2), write down the sum and difference formulas for the cosine:

$$\cos(\alpha - \beta) = \cos \alpha \cos \beta + \sin \alpha \sin \beta \qquad \textbf{(4)}$$
$$\cos(\alpha + \beta) = \cos \alpha \cos \beta - \sin \alpha \sin \beta \qquad \textbf{(5)}$$

Subtract equation (5) from equation (4) to get

$$\cos(\alpha - \beta) - \cos(\alpha + \beta) = 2 \sin \alpha \sin \beta$$

from which

$$\sin \alpha \sin \beta = \frac{1}{2}\left[\cos(\alpha - \beta) - \cos(\alpha + \beta)\right]$$

Now, add equations (4) and (5) to get

$$\cos(\alpha - \beta) + \cos(\alpha + \beta) = 2 \cos \alpha \cos \beta$$

from which

$$\cos \alpha \cos \beta = \frac{1}{2}\left[\cos(\alpha - \beta) + \cos(\alpha + \beta)\right]$$

To derive Product-to-Sum Formula (3), use the Sum and Difference Formulas for sine in a similar way. (You are asked to do this in Problem 41.)

EXAMPLE 1   Expressing Products as Sums

Express each of the following products as a sum containing only sines or cosines.

(a) $\sin(6\theta) \sin(4\theta)$ (b) $\cos(3\theta) \cos \theta$ (c) $\sin(3\theta) \cos(5\theta)$

Solution (a) We use formula (1) to get

$$\sin(6\theta) \sin(4\theta) = \frac{1}{2}\left[\cos(6\theta - 4\theta) - \cos(6\theta + 4\theta)\right]$$

$$= \frac{1}{2}\left[\cos(2\theta) - \cos(10\theta)\right]$$

(b) We use formula (2) to get

$$\cos(3\theta)\cos\theta = \frac{1}{2}\left[\cos(3\theta - \theta) + \cos(3\theta + \theta)\right]$$

$$= \frac{1}{2}\left[\cos(2\theta) + \cos(4\theta)\right]$$

(c) We use formula (3) to get

$$\sin(3\theta)\cos(5\theta) = \frac{1}{2}\left[\sin(3\theta + 5\theta) + \sin(3\theta - 5\theta)\right]$$

$$= \frac{1}{2}\left[\sin(8\theta) + \sin(-2\theta)\right] = \frac{1}{2}\left[\sin(8\theta) - \sin(2\theta)\right]$$ ■

NOW WORK PROBLEM 1.

2 The **Sum-to-Product Formulas** are given next.

Theorem **Sum-to-Product Formulas**

$$\sin\alpha + \sin\beta = 2\sin\frac{\alpha + \beta}{2}\cos\frac{\alpha - \beta}{2} \qquad (6)$$

$$\sin\alpha - \sin\beta = 2\sin\frac{\alpha - \beta}{2}\cos\frac{\alpha + \beta}{2} \qquad (7)$$

$$\cos\alpha + \cos\beta = 2\cos\frac{\alpha + \beta}{2}\cos\frac{\alpha - \beta}{2} \qquad (8)$$

$$\cos\alpha - \cos\beta = -2\sin\frac{\alpha + \beta}{2}\sin\frac{\alpha - \beta}{2} \qquad (9)$$

■

We will derive formula (6) and leave the derivations of formulas (7) through (9) as exercises (see Problems 42 through 44).

Proof

$$2\sin\frac{\alpha + \beta}{2}\cos\frac{\alpha - \beta}{2} = 2 \cdot \frac{1}{2}\left[\sin\left(\frac{\alpha + \beta}{2} + \frac{\alpha - \beta}{2}\right) + \sin\left(\frac{\alpha + \beta}{2} - \frac{\alpha - \beta}{2}\right)\right]$$

↑
Product-to-Sum Formula (3)

■

$$= \sin\frac{2\alpha}{2} + \sin\frac{2\beta}{2} = \sin\alpha + \sin\beta$$

EXAMPLE 2 **Expressing Sums (or Differences) as a Product**

Express each sum or difference as a product of sines and/or cosines.

(a) $\sin(5\theta) - \sin(3\theta)$ (b) $\cos(3\theta) + \cos(2\theta)$

Solution (a) We use formula (7) to get

$$\sin(5\theta) - \sin(3\theta) = 2\sin\frac{5\theta - 3\theta}{2}\cos\frac{5\theta + 3\theta}{2}$$

$$= 2\sin\theta\cos(4\theta)$$

(b) $\cos(3\theta) + \cos(2\theta) = 2\cos\dfrac{3\theta + 2\theta}{2}\cos\dfrac{3\theta - 2\theta}{2}$ Formula (8)

$$= 2\cos\frac{5\theta}{2}\cos\frac{\theta}{2}$$

■

NOW WORK PROBLEM **11.**

3.6 EXERCISES

In Problems 1–10, express each product as a sum containing only sines or cosines.

1. $\sin(4\theta)\sin(2\theta)$ **2.** $\cos(4\theta)\cos(2\theta)$ **3.** $\sin(4\theta)\cos(2\theta)$ **4.** $\sin(3\theta)\sin(5\theta)$

5. $\cos(3\theta)\cos(5\theta)$ **6.** $\sin(4\theta)\cos(6\theta)$ **7.** $\sin\theta\sin(2\theta)$ **8.** $\cos(3\theta)\cos(4\theta)$

9. $\sin\dfrac{3\theta}{2}\cos\dfrac{\theta}{2}$ **10.** $\sin\dfrac{\theta}{2}\cos\dfrac{5\theta}{2}$

In Problems 11–18, express each sum or difference as a product of sines and/or cosines.

11. $\sin(4\theta) - \sin(2\theta)$ **12.** $\sin(4\theta) + \sin(2\theta)$ **13.** $\cos(2\theta) + \cos(4\theta)$ **14.** $\cos(5\theta) - \cos(3\theta)$

15. $\sin\theta + \sin(3\theta)$ **16.** $\cos\theta + \cos(3\theta)$ **17.** $\cos\dfrac{\theta}{2} - \cos\dfrac{3\theta}{2}$ **18.** $\sin\dfrac{\theta}{2} - \sin\dfrac{3\theta}{2}$

In Problems 19–36, establish each identity.

19. $\dfrac{\sin\theta + \sin(3\theta)}{2\sin(2\theta)} = \cos\theta$ **20.** $\dfrac{\cos\theta + \cos(3\theta)}{2\cos(2\theta)} = \cos\theta$ **21.** $\dfrac{\sin(4\theta) + \sin(2\theta)}{\cos(4\theta) + \cos(2\theta)} = \tan(3\theta)$

22. $\dfrac{\cos\theta - \cos(3\theta)}{\sin(3\theta) - \sin\theta} = \tan(2\theta)$ **23.** $\dfrac{\cos\theta - \cos(3\theta)}{\sin\theta + \sin(3\theta)} = \tan\theta$ **24.** $\dfrac{\cos\theta - \cos(5\theta)}{\sin\theta + \sin(5\theta)} = \tan(2\theta)$

25. $\sin\theta[\sin\theta + \sin(3\theta)] = \cos\theta[\cos\theta - \cos(3\theta)]$ **26.** $\sin\theta[\sin(3\theta) + \sin(5\theta)] = \cos\theta[\cos(3\theta) - \cos(5\theta)]$

27. $\dfrac{\sin(4\theta) + \sin(8\theta)}{\cos(4\theta) + \cos(8\theta)} = \tan(6\theta)$ **28.** $\dfrac{\sin(4\theta) - \sin(8\theta)}{\cos(4\theta) - \cos(8\theta)} = -\cot(6\theta)$

29. $\dfrac{\sin(4\theta) + \sin(8\theta)}{\sin(4\theta) - \sin(8\theta)} = -\dfrac{\tan(6\theta)}{\tan(2\theta)}$ **30.** $\dfrac{\cos(4\theta) - \cos(8\theta)}{\cos(4\theta) + \cos(8\theta)} = \tan(2\theta)\tan(6\theta)$

31. $\dfrac{\sin\alpha + \sin\beta}{\sin\alpha - \sin\beta} = \tan\dfrac{\alpha + \beta}{2}\cot\dfrac{\alpha - \beta}{2}$ **32.** $\dfrac{\cos\alpha + \cos\beta}{\cos\alpha - \cos\beta} = -\cot\dfrac{\alpha + \beta}{2}\cot\dfrac{\alpha - \beta}{2}$

33. $\dfrac{\sin\alpha + \sin\beta}{\cos\alpha + \cos\beta} = \tan\dfrac{\alpha + \beta}{2}$ **34.** $\dfrac{\sin\alpha - \sin\beta}{\cos\alpha - \cos\beta} = -\cot\dfrac{\alpha + \beta}{2}$

35. $1 + \cos(2\theta) + \cos(4\theta) + \cos(6\theta) = 4\cos\theta\cos(2\theta)\cos(3\theta)$

36. $1 - \cos(2\theta) + \cos(4\theta) - \cos(6\theta) = 4\sin\theta\cos(2\theta)\sin(3\theta)$

37. Touch-Tone Phones On a Touch-Tone phone, each button produces a unique sound. The sound produced is the sum of two tones, given by

$$y = \sin(2\pi l t) \quad \text{and} \quad y = \sin(2\pi h t)$$

where l and h are the low and high frequencies (cycles per second) shown on the illustration. For example, if you touch 7, the low frequency is $l = 852$ cycles per second and the high frequency is $h = 1209$ cycles per second. The sound emitted by touching 7 is

$$y = \sin\left[2\pi(852)t\right] + \sin\left[2\pi(1209)t\right]$$

Touch-Tone phone

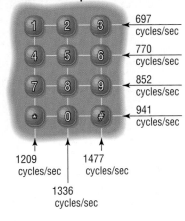

(a) Write this sound as a product of sines and/or cosines.
(b) Determine the maximum value of y.
(c) Graph the sound emitted by touching 7.

38. Touch-Tone Phones
(a) Write the sound emitted by touching the # key as a product of sines and/or cosines.
(b) Determine the maximum value of y.
(c) Graph the sound emitted by touching the # key.

39. If $\alpha + \beta + \gamma = \pi$, show that

$$\sin(2\alpha) + \sin(2\beta) + \sin(2\gamma) = 4\sin\alpha\sin\beta\sin\gamma$$

40. If $\alpha + \beta + \gamma = \pi$, show that

$$\tan\alpha + \tan\beta + \tan\gamma = \tan\alpha\tan\beta\tan\gamma$$

41. Derive formula (3).

42. Derive formula (7).

43. Derive formula (8).

44. Derive formula (9).

PREPARING FOR THIS SECTION

Before getting started, review the following:

✓ Solving Equations (Appendix A, Section A.3, pp. 502–505)   ✓ Values of the Trigonometric Functions of Certain Angles (Section 2.2, p. 99 and p. 100)

3.7 TRIGONOMETRIC EQUATIONS (I)

OBJECTIVE   ① Solve Equations Involving a Single Trigonometric Function

① The previous four sections of this chapter were devoted to trigonometric identities, that is, equations involving trigonometric functions that are satisfied by every value in the domain of the variable. In the remaining two sections, we discuss **trigonometric equations,** that is, equations involving trigonometric functions that are satisfied only by some values of the variable (or, possibly, are not satisfied by any values of the variable). The values that satisfy the equation are called **solutions** of the equation.

EXAMPLE 1   **Checking Whether a Given Number Is a Solution of a Trigonometric Equation**

Determine whether $\theta = \dfrac{\pi}{4}$ is a solution of the equation $\sin\theta = \dfrac{1}{2}$. Is $\theta = \dfrac{\pi}{6}$ a solution?

Solution   Replace θ by $\dfrac{\pi}{4}$ in the given equation. The result is

$$\sin \frac{\pi}{4} = \frac{\sqrt{2}}{2} \neq \frac{1}{2}$$

We conclude that $\dfrac{\pi}{4}$ is not a solution.

Next, replace θ by $\dfrac{\pi}{6}$ in the equation. The result is

$$\sin \frac{\pi}{6} = \frac{1}{2}$$

We conclude that $\dfrac{\pi}{6}$ is a solution of the given equation.   ■

The equation given in Example 1 has other solutions besides $\theta = \dfrac{\pi}{6}$. For example, $\theta = \dfrac{5\pi}{6}$ is also a solution, as is $\theta = \dfrac{13\pi}{6}$. (You should check this for yourself.) In fact, the equation has an infinite number of solutions due to the periodicity of the sine function, as can be seen in Figure 27.

Figure 27

$$\sin x = \frac{1}{2}$$

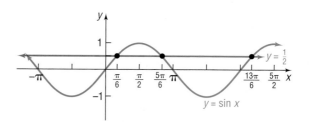

Unless the domain of the variable is restricted, we need to find *all* the solutions of a trigonometric equation. As the next example illustrates, finding all the solutions can be accomplished by first finding solutions over an interval whose length equals the period of the function and then adding multiples of that period to the solutions found. Let's look at some examples.

EXAMPLE 2   **Finding the Solutions of a Trigonometric Equation**

Solve the equation: $\cos \theta = \dfrac{1}{2}$

Give a general formula for all the solutions. List six solutions.

Figure 28
$\cos \theta = \frac{1}{2}$

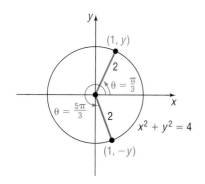

Solution   The period of the cosine function is 2π. In the interval $[0, 2\pi)$, there are two angles θ for which $\cos \theta = \dfrac{1}{2}$: $\theta = \dfrac{\pi}{3}$ and $\theta = \dfrac{5\pi}{3}$. See Figure 28. Because the cosine function has period 2π, all the solutions of $\cos \theta = \dfrac{1}{2}$ may be given by the general formula

$$\theta = \frac{\pi}{3} + 2k\pi \quad \text{or} \quad \theta = \frac{5\pi}{3} + 2k\pi \qquad k \text{ any integer.}$$

Some of the solutions are

$$\underbrace{\frac{\pi}{3}, \ \frac{5\pi}{3}}_{k = 0}, \ \underbrace{\frac{7\pi}{3}, \ \frac{11\pi}{3}}_{k = 1}, \ \underbrace{\frac{13\pi}{3}, \ \frac{17\pi}{3}}_{k = 2}, \quad \text{and so on.}$$ ■

Figure 29

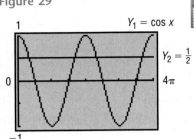

Check: We can verify the solutions by graphing $Y_1 = \cos x$ and $Y_2 = \dfrac{1}{2}$ to determine where the graphs intersect. (Be sure to graph in radian mode.) See Figure 29. The graph of Y_1 intersects the graph of Y_2 at $x = 1.05\left(\approx \dfrac{\pi}{3}\right)$, $5.24\left(\approx \dfrac{5\pi}{3}\right)$, $7.33\left(\approx \dfrac{7\pi}{3}\right)$, and $11.52\left(\approx \dfrac{11\pi}{3}\right)$, rounded to two decimal places. ■

━ NOW WORK PROBLEM **1**.

In most of our work, we shall be interested only in finding solutions of trigonometric equations for $0 \le \theta < 2\pi$.

EXAMPLE 3	**Solving a Linear Trigonometric Equation**

Solve the equation: $2 \sin\theta + \sqrt{3} = 0$, $0 \le \theta < 2\pi$

Solution We solve the equation for $\sin\theta$.

$$2 \sin\theta + \sqrt{3} = 0$$

$$2 \sin\theta = -\sqrt{3} \qquad \text{Subtract } \sqrt{3} \text{ from both sides.}$$

$$\sin\theta = -\frac{\sqrt{3}}{2} \qquad \text{Divide both sides by 2.}$$

The period of the sine function is 2π. In the interval $[0, 2\pi)$, there are two angles θ for which $\sin\theta = -\dfrac{\sqrt{3}}{2}$: $\theta = \dfrac{4\pi}{3}$ and $\theta = \dfrac{5\pi}{3}$. ■

━ NOW WORK PROBLEM **11**.

EXAMPLE 4	**Solving a Trigonometric Equation**

Solve the equation: $\sin(2\theta) = \dfrac{1}{2}$, $0 \le \theta < 2\pi$

Solution The period of the sine function is 2π. In the interval $[0, 2\pi)$, the sine function has a value $\dfrac{1}{2}$ at $\dfrac{\pi}{6}$ and $\dfrac{5\pi}{6}$. See Figure 30. Consequently, because the argument is 2θ in the equation $\sin(2\theta) = \dfrac{1}{2}$, we have

Figure 30
$\sin(2\theta) = \frac{1}{2}, 0 \le \theta < 2\pi$

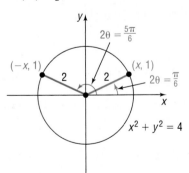

$$2\theta = \frac{\pi}{6} + 2k\pi \quad \text{or} \quad 2\theta = \frac{5\pi}{6} + 2k\pi \quad \text{k any integer.}$$

$$\theta = \frac{\pi}{12} + k\pi \qquad\qquad \theta = \frac{5\pi}{12} + k\pi \quad \text{Divide by 2.}$$

Then

$$\theta = \frac{\pi}{12} + (-1)\pi = \frac{-11\pi}{12} \quad {\scriptstyle k=-1} \quad \theta = \frac{5\pi}{12} + (-1)\pi = \frac{-7\pi}{12}$$

$$\theta = \frac{\pi}{12} + (0)\pi = \frac{\pi}{12} \qquad {\scriptstyle k=0} \quad \theta = \frac{5\pi}{12} + (0)\pi = \frac{5\pi}{12}$$

$$\theta = \frac{\pi}{12} + (1)\pi = \frac{13\pi}{12} \qquad {\scriptstyle k=1} \quad \theta = \frac{5\pi}{12} + (1)\pi = \frac{17\pi}{12}$$

$$\theta = \frac{\pi}{12} + (2)\pi = \frac{25\pi}{12} \qquad {\scriptstyle k=2} \quad \theta = \frac{5\pi}{12} + (2)\pi = \frac{29\pi}{12}$$

In the interval $[0, 2\pi)$, the solutions of $\sin(2\theta) = \dfrac{1}{2}$ are $\theta = \dfrac{\pi}{12}$, $\theta = \dfrac{13\pi}{12}$, $\theta = \dfrac{5\pi}{12}$, and $\theta = \dfrac{17\pi}{12}$. ■

Check: Verify these solutions by graphing $Y_1 = \sin(2x)$ and $Y_2 = \dfrac{1}{2}$ for $0 \le x \le 2\pi$. ▬

WARNING: In solving a trigonometric equation for $\theta, 0 \le \theta < 2\pi$, in which the argument is not θ (as in Example 4), you must write down all the solutions first and then list those that are in the interval $[0, 2\pi)$. Otherwise, solutions may be lost. For example, in solving $\sin(2\theta) = \dfrac{1}{2}$, if you merely write the solutions $2\theta = \dfrac{\pi}{6}$ and $2\theta = \dfrac{5\pi}{6}$, you will find only $\theta = \dfrac{\pi}{12}$ and $\theta = \dfrac{5\pi}{12}$ and miss the other solutions. ▬

✏➤ NOW WORK PROBLEM **17.**

EXAMPLE 5

Solving a Trigonometric Equation

Solve the equation: $\tan\left(\theta - \dfrac{\pi}{2}\right) = 1, \quad 0 \le \theta < 2\pi$

Solution   The period of the tangent function is π. In the interval $[0, \pi)$, the tangent function has the value 1 when the argument is $\dfrac{\pi}{4}$. Because the argument is $\theta - \dfrac{\pi}{2}$ in the given equation, we have

$$\theta - \frac{\pi}{2} = \frac{\pi}{4} + k\pi \qquad \text{\small k any integer.}$$

$$\theta = \frac{3\pi}{4} + k\pi$$

In the interval $[0, 2\pi)$, $\theta = \dfrac{3\pi}{4}$ and $\theta = \dfrac{3\pi}{4} + \pi = \dfrac{7\pi}{4}$ are the only solutions. ▬

 Check: Verify these solutions using a graphing utility. ▬

The next example illustrates how to solve trigonometric equations using a calculator. Remember that the function keys on a calculator will only give values consistent with the definition of the function.

EXAMPLE 6

Solving a Trigonometric Equation with a Calculator

Use a calculator to solve the equation: $\sin\theta = 0.3, \quad 0 \le \theta < 2\pi$
Express any solutions in radians, rounded to two decimal places.

Figure 31
$\sin\theta = 0.3$

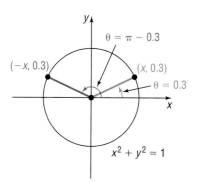

Solution   To solve $\sin\theta = 0.3$ on a calculator, first set the mode to radians. Then use the $\boxed{\sin^{-1}}$ key to obtain

$$\theta = \sin^{-1}(0.3) \approx 0.304692654$$

Rounded to two decimal places, $\theta = \sin^{-1}(0.3) = 0.30$ radian. Because of the definition of $y = \sin^{-1} x$, the angle θ that we obtain is the angle $-\dfrac{\pi}{2} \le \theta \le \dfrac{\pi}{2}$ for which $\sin\theta = 0.3$. Another angle for which $\sin\theta = 0.3$ is $\pi - 0.30$. See Figure 31. The angle $\pi - 0.30$ is the angle in quadrant II, where $\sin\theta = 0.3$. The solutions for $\sin\theta = 0.3, 0 \le \theta < 2\pi$, are

$$\theta = 0.30 \text{ radian} \quad \text{and} \quad \theta = \pi - 0.30 \approx 2.84 \text{ radians} \quad ▬$$

WARNING: Example 6 illustrates that caution must be exercised when solving trigonometric equations on a calculator. Remember that the calculator supplies an angle only within the restrictions of the definition of the inverse trigonometric function. To find the remaining solutions, you must identify other quadrants, if any, in which the angle may be located. ▬

✏➤ NOW WORK PROBLEM **35.**

3.7 EXERCISES

In Problems 1–10, solve each equation. Give a general formula for all the solutions. List six solutions.

1. $\sin \theta = \dfrac{1}{2}$

2. $\tan \theta = 1$

3. $\tan \theta = -\dfrac{\sqrt{3}}{3}$

4. $\cos \theta = -\dfrac{\sqrt{3}}{2}$

5. $\cos \theta = 0$

6. $\sin \theta = \dfrac{\sqrt{2}}{2}$

7. $\cos (2\theta) = -\dfrac{1}{2}$

8. $\sin (2\theta) = -1$

9. $\sin \dfrac{\theta}{2} = -\dfrac{\sqrt{3}}{2}$

10. $\tan \dfrac{\theta}{2} = -1$

In Problems 11–34, solve each equation on the interval $0 \le \theta < 2\pi$.

11. $2 \sin \theta + 3 = 2$

12. $1 - \cos \theta = \frac{1}{2}$

13. $4 \cos^2 \theta = 1$

14. $\tan^2 \theta = \frac{1}{3}$

15. $2 \sin^2 \theta - 1 = 0$

16. $4 \cos^2 \theta - 3 = 0$

17. $\sin (3\theta) = -1$

18. $\tan \dfrac{\theta}{2} = \sqrt{3}$

19. $\cos (2\theta) = -\dfrac{1}{2}$

20. $\tan (2\theta) = -1$

21. $\sec \dfrac{3\theta}{2} = -2$

22. $\cot \dfrac{2\theta}{3} = -\sqrt{3}$

23. $\cos \left(2\theta - \dfrac{\pi}{2} \right) = -1$

24. $\sin \left(3\theta + \dfrac{\pi}{18} \right) = 1$

25. $\tan \left(\dfrac{\theta}{2} + \dfrac{\pi}{3} \right) = 1$

26. $\cos \left(\dfrac{\theta}{3} - \dfrac{\pi}{4} \right) = \dfrac{1}{2}$

27. $2 \sin \theta + 1 = 0$

28. $\cos \theta + 1 = 0$

29. $\tan \theta + 1 = 0$

30. $\sqrt{3} \cot \theta + 1 = 0$

31. $4 \sec \theta + 6 = -2$

32. $5 \csc \theta - 3 = 2$

33. $3 \sqrt{2} \cos \theta + 2 = -1$

34. $4 \sin \theta + 3\sqrt{3} = \sqrt{3}$

In Problems 35–42, use a calculator to solve each equation on the interval $0 \le \theta < 2\pi$. Round answers to two decimal places.

35. $\sin \theta = 0.4$

36. $\cos \theta = 0.6$

37. $\tan \theta = 5$

38. $\cot \theta = 2$

39. $\cos \theta = -0.9$

40. $\sin \theta = -0.2$

41. $\sec \theta = -4$

42. $\csc \theta = -3$

*The following discussion of **Snell's Law of Refraction** (named after Willebrord Snell, 1580–1626) is needed for Problems 43–49. Light, sound, and other waves travel at different speeds, depending on the media (air, water, wood, and so on) through which they pass. Suppose that light travels from a point A in one medium, where its speed is v_1, to a point B in another medium, where its speed is v_2. Refer to the figure, where the angle θ_1 is called the **angle of incidence** and the angle θ_2 is the **angle of refraction**. Snell's Law,* which can be proved using calculus, states that*

$$\frac{\sin \theta_1}{\sin \theta_2} = \frac{v_1}{v_2}$$

*The ratio v_1/v_2 is called the **index of refraction**. Some values are given in the following table.*

SOME INDEXES OF REFRACTION	
Medium	**Index of Refraction***
Water	1.33
Ethyl alcohol	1.36
Carbon bisulfide	1.63
Air (1 atm and 20°C)	1.0003
Methylene iodide	1.74
Fused quartz	1.46
Glass, crown	1.52
Glass, dense flint	1.66
Sodium chloride	1.53

*For light of wavelength 589 nanometers, measured with respect to a vacuum. The index with respect to air is negligibly different in most cases.

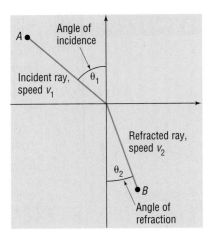

* Because this law was also deduced by René Descartes in France, it is also known as Descartes' Law.

43. The index of refraction of light in passing from a vacuum into water is 1.33. If the angle of incidence is 40°, determine the angle of refraction.

44. The index of refraction of light in passing from a vacuum into dense glass is 1.66. If the angle of incidence is 50°, determine the angle of refraction.

45. Ptolemy, who lived in the city of Alexandria in Egypt during the second century AD, gave the measured values in the table below for the angle of incidence θ_1 and the angle of refraction θ_2 for a light beam passing from air into water. Do these values agree with Snell's Law? If so, what index of refraction results? (These data are interesting as the oldest recorded physical measurements.)*

θ_1	θ_2	θ_1	θ_2
10°	7° 45′	50°	35° 0′
20°	15° 30′	60°	40° 30′
30°	22° 30′	70°	45° 30′
40°	29° 0′	80°	50° 0′

46. The speed of yellow sodium light (wavelength of 589 nanometers) in a certain liquid is measured to be 1.92×10^8 meters per second. What is the index of refraction of this liquid, with respect to air, for sodium light?[†]
[**Hint:** The speed of light in air is approximately 2.99×10^8 meters per second.]

47. A beam of light with a wavelength of 589 nanometers traveling in air makes an angle of incidence of 40° on a slab of transparent material, and the refracted beam makes an angle of refraction of 26°. Find the index of refraction of the material.[†]

48. A light ray with a wavelength of 589 nanometers (produced by a sodium lamp) traveling through air makes an angle of incidence of 30° on a smooth, flat slab of crown glass. Find the angle of refraction.[†]

49. A light beam passes through a thick slab of material whose index of refraction is n_2. Show that the emerging beam is parallel to the incident beam.[†]

 50. Explain in your own words how you would use your calculator to solve the equation $\sin x = 0.3$, $0 \le x < 2\pi$. How would you modify your approach in order to solve the equation $\cot x = 5$, $0 < x < 2\pi$?

* Adapted from Halliday and Resnick, *Physics, Parts 1 & 2*, 3rd ed. New York: Wiley, 1978, p. 953.
[†] Adapted from Serway, *Physics*, 3rd ed. Philadelphia: W. B. Saunders, p. 805.

PREPARING FOR THIS SECTION

Before getting started, review the following:

✓ Solving Quadratic Equations by Factoring (Appendix A, Section A.3, pp. 505–506)

✓ The Quadratic Formula (Appendix A, Section A.3, pp. 508–514)

✓ Using a Graphing Utility to Solve Equations (Appendix B, Section B.4, pp. 569–571)

3.8 TRIGONOMETRIC EQUATIONS (II)

OBJECTIVES
1. Solve Trigonometric Equations That Are Quadratic in Form
2. Solve Trigonometric Equations Using Identities
3. Solve Trigonometric Equations That Are Linear in Sine and Cosine
4. Solve Trigonometric Equations Using a Graphing Utility

1. In this section we continue our study of trigonometric equations. Many trigonometric equations can be solved by applying techniques that we already know, such as applying the quadratic formula (if the equation is a second-degree polynomial) or factoring.

EXAMPLE 1	**Solving a Trigonometric Equation Quadratic in Form**

Solve the equation: $2 \sin^2\theta - 3 \sin\theta + 1 = 0$, $0 \le \theta < 2\pi$

Solution   The equation that we wish to solve is a quadratic equation (in $\sin\theta$) that can be factored.

$$2 \sin^2 \theta - 3 \sin \theta + 1 = 0 \qquad 2x^2 - 3x + 1 = 0, \quad x = \sin \theta$$
$$(2 \sin \theta - 1)(\sin \theta - 1) = 0 \qquad (2x - 1)(x - 1) = 0$$
$$2 \sin \theta - 1 = 0 \quad \text{or} \quad \sin \theta - 1 = 0$$
$$\sin \theta = \frac{1}{2} \qquad\qquad \sin \theta = 1$$

Solving each equation in the interval $[0, 2\pi)$, we obtain

$$\theta = \frac{\pi}{6}, \qquad \theta = \frac{5\pi}{6}, \qquad \theta = \frac{\pi}{2} \qquad\qquad \blacksquare$$

✎ NOW WORK PROBLEM **3**.

②   When a trigonometric equation contains more than one trigonometric function, identities sometimes can be used to obtain an equivalent equation that contains only one trigonometric function.

EXAMPLE 2	Solving a Trigonometric Equation Using Identities

Solve the equation: $3 \cos \theta + 3 = 2 \sin^2 \theta, \quad 0 \le \theta < 2\pi$

Solution   The equation in its present form contains sines and cosines. However, a form of the Pythagorean Identity can be used to transform the equation into an equivalent expression containing only cosines.

$$3 \cos \theta + 3 = 2 \sin^2 \theta$$
$$3 \cos \theta + 3 = 2(1 - \cos^2 \theta) \qquad \sin^2 \theta = 1 - \cos^2 \theta$$
$$3 \cos \theta + 3 = 2 - 2 \cos^2 \theta$$
$$2 \cos^2 \theta + 3 \cos \theta + 1 = 0 \qquad\qquad \text{Quadratic in } \cos \theta$$
$$(2 \cos \theta + 1)(\cos \theta + 1) = 0 \qquad\qquad \text{Factor}$$
$$2 \cos \theta + 1 = 0 \quad \text{or} \quad \cos \theta + 1 = 0$$
$$\cos \theta = -\frac{1}{2} \qquad\qquad \cos \theta = -1$$

Solving each equation in the interval $[0, 2\pi)$, we obtain

$$\theta = \frac{2\pi}{3}, \qquad \theta = \frac{4\pi}{3}, \qquad \theta = \pi \qquad\qquad \blacksquare$$

⌨ Check: Graph $Y_1 = 3 \cos x + 3$ and $Y_2 = 2 \sin^2 x, 0 \le x \le 2\pi$, and find the points of intersection. How close are your approximate solutions to the exact ones found in this example? ∎

EXAMPLE 3	Solving a Trigonometric Equation Using Identities

Solve the equation: $\cos(2\theta) + 3 = 5 \cos \theta, \quad 0 \le \theta < 2\pi$

Solution   First, we observe that the given equation contains two cosine functions, but with different arguments, θ and 2θ. We use the Double-Angle Formula $\cos(2\theta) = 2 \cos^2 \theta - 1$ to obtain an equivalent equation containing only $\cos \theta$.

$$\cos(2\theta) + 3 = 5 \cos \theta$$
$$(2 \cos^2 \theta - 1) + 3 = 5 \cos \theta$$

$$2\cos^2\theta - 5\cos\theta + 2 = 0$$

$$(\cos\theta - 2)(2\cos\theta - 1) = 0$$

$$\cos\theta = 2 \quad \text{or} \quad \cos\theta = \frac{1}{2}$$

For any angle $\theta, -1 \le \cos\theta \le 1$; therefore, the equation $\cos\theta = 2$ has no solution. The solutions of $\cos\theta = \frac{1}{2}, 0 \le \theta < 2\pi$, are

$$\theta = \frac{\pi}{3}, \qquad \theta = \frac{5\pi}{3} \qquad \blacksquare$$

 Check: Graph $Y_1 = \cos(2x) + 3$ and $Y_2 = 5\cos x, 0 \le x \le 2\pi$, and find the points of intersection. Compare your results with those of Example 3.   ■

 N O W   W O R K   P R O B L E M   **19.**

| **EXAMPLE 4** | **Solving a Trigonometric Equation Using Identities** |

Solution

Solve the equation: $\cos^2\theta + \sin\theta = 2, \quad 0 \le \theta < 2\pi$

This equation involves two trigonometric functions, sine and cosine. Since it is easier to work with only one, we use a form of the Pythagorean Identity, $\sin^2\theta + \cos^2\theta = 1$ to rewrite the equation.

$$\cos^2\theta + \sin\theta = 2$$

$$(1 - \sin^2\theta) + \sin\theta = 2 \quad \text{\small $\cos^2\theta = 1 - \sin^2\theta$}$$

$$\sin^2\theta - \sin\theta + 1 = 0$$

This is a quadratic equation in $\sin\theta$. The discriminant is $b^2 - 4ac = 1 - 4 = -3 < 0$. Therefore, the equation has no real solution.   ■

 Check: Graph $Y_1 = \cos^2 x + \sin x$ and $Y_2 = 2$ to see that the two graphs do not intersect anywhere.   ■

| **EXAMPLE 5** | **Solving a Trigonometric Equation Using Identities** |

Solution

Solve the equation: $\sin\theta\cos\theta = -\frac{1}{2}, \quad 0 \le \theta < 2\pi$

The left side of the given equation is in the form of the Double-Angle Formula $2\sin\theta\cos\theta = \sin(2\theta)$, except for a factor of 2. We multiply each side by 2.

$$\sin\theta\cos\theta = -\frac{1}{2}$$

$$2\sin\theta\cos\theta = -1 \quad \text{\small Multiply each side by 2.}$$

$$\sin(2\theta) = -1 \quad \text{\small Double-Angle Formula}$$

The argument here is 2θ. So we need to write all the solutions of this equation and then list those that are in the interval $[0, 2\pi)$.

$$2\theta = \frac{3\pi}{2} + 2k\pi \quad \text{\small k any integer}$$

$$\theta = \frac{3\pi}{4} + k\pi$$

$$\theta = \frac{3\pi}{4} + (-1)\pi = -\frac{\pi}{4}, \quad \theta = \frac{3\pi}{4} + (0)\pi = \frac{3\pi}{4}, \quad \theta = \frac{3\pi}{4} + (1)\pi = \frac{7\pi}{4}, \quad \theta = \frac{3\pi}{4} + (2)\pi = \frac{11\pi}{4}$$

$$\uparrow \qquad\qquad\qquad \uparrow \qquad\qquad\qquad \uparrow \qquad\qquad\qquad \uparrow$$

$$k = -1 \qquad\qquad\quad k = 0 \qquad\qquad\quad k = 1 \qquad\qquad\quad k = 2$$

The solutions in the interval $[0, 2\pi)$ are

$$\theta = \frac{3\pi}{4}, \qquad \theta = \frac{7\pi}{4}$$

■

3 Sometimes it is necessary to square both sides of an equation in order to obtain expressions that allow the use of identities. Remember, however, that when squaring both sides extraneous solutions may be introduced. As a result, apparent solutions must be checked.

| EXAMPLE 6 | **Other Methods for Solving a Trigonometric Equation** |

Solve the equation: $\sin\theta + \cos\theta = 1$, $0 \leq \theta < 2\pi$

Solution A   Attempts to use available identities do not lead to equations that are easy to solve. (Try it yourself.) Given the form of this equation, we decide to square each side.

$$\sin\theta + \cos\theta = 1$$
$$(\sin\theta + \cos\theta)^2 = 1 \qquad \text{Square each side.}$$
$$\sin^2\theta + 2\sin\theta\cos\theta + \cos^2\theta = 1 \qquad \text{Remove parentheses.}$$
$$2\sin\theta\cos\theta = 0 \qquad \sin^2\theta + \cos^2\theta = 1$$
$$\sin\theta\cos\theta = 0$$

Setting each factor equal to zero, we obtain

$$\sin\theta = 0 \quad \text{or} \quad \cos\theta = 0$$

The apparent solutions are

$$\theta = 0, \qquad \theta = \pi, \qquad \theta = \frac{\pi}{2}, \qquad \theta = \frac{3\pi}{2}$$

Because we squared both sides of the original equation, we must check these apparent solutions to see if any are extraneous.

$$\theta = 0: \quad \sin 0 + \cos 0 = 0 + 1 = 1 \qquad \text{A solution.}$$
$$\theta = \pi: \quad \sin\pi + \cos\pi = 0 + (-1) = -1 \qquad \text{Not a solution.}$$
$$\theta = \frac{\pi}{2}: \quad \sin\frac{\pi}{2} + \cos\frac{\pi}{2} = 1 + 0 = 1 \qquad \text{A solution.}$$
$$\theta = \frac{3\pi}{2}: \quad \sin\frac{3\pi}{2} + \cos\frac{3\pi}{2} = -1 + 0 = -1 \qquad \text{Not a solution.}$$

Therefore, $\theta = \frac{3\pi}{2}$ and $\theta = \pi$ are extraneous. The only solutions are $\theta = 0$ and $\theta = \frac{\pi}{2}$. ■

We can solve the equation given in Example 6 in another way.

Solution B   We start with the equation

$$\sin\theta + \cos\theta = 1$$

and divide each side by $\sqrt{2}$. (The reason for this choice will become apparent shortly.) Then

$$\frac{1}{\sqrt{2}} \sin \theta + \frac{1}{\sqrt{2}} \cos \theta = \frac{1}{\sqrt{2}}$$

The left side now resembles the formula for the sine of the sum of two angles, one of which is θ. The other angle is unknown (call it ϕ.) Then

$$\sin(\theta + \phi) = \sin \theta \cos \phi + \cos \theta \sin \phi = \frac{1}{\sqrt{2}} = \frac{\sqrt{2}}{2} \qquad \textbf{(1)}$$

where

$$\cos \phi = \frac{1}{\sqrt{2}} = \frac{\sqrt{2}}{2}, \qquad \sin \phi = \frac{1}{\sqrt{2}} = \frac{\sqrt{2}}{2}, \qquad 0 \le \phi < 2\pi$$

The angle ϕ is therefore $\dfrac{\pi}{4}$. As a result, equation (1) becomes

$$\sin\left(\theta + \frac{\pi}{4}\right) = \frac{\sqrt{2}}{2}$$

These are two angles whose sine is $\dfrac{\sqrt{2}}{2}$: $\dfrac{\pi}{4}$ and $\dfrac{3\pi}{4}$. See Figure 32. As a result,

$$\theta + \frac{\pi}{4} = \frac{\pi}{4} \qquad \text{or} \quad \theta + \frac{\pi}{4} = \frac{3\pi}{4}$$

$$\theta = 0 \qquad\qquad\qquad \theta = \frac{\pi}{2}$$

These solutions agree with the solutions found earlier. ∎

This second method of solution can be used to solve any linear equation in the variables $\sin \theta$ and $\cos \theta$.

Figure 32

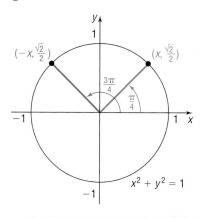

EXAMPLE 7	Solving a Trigonometric Equation

Linear in $\sin \theta$ and $\cos \theta$

Solve:

$$a \sin \theta + b \cos \theta = c, \qquad 0 \le \theta < 2\pi \qquad \textbf{(2)}$$

where a, b, and c are constants and either $a \ne 0$ or $b \ne 0$.

Solution We divide each side of equation (2) by $\sqrt{a^2 + b^2}$. Then

$$\frac{a}{\sqrt{a^2 + b^2}} \sin \theta + \frac{b}{\sqrt{a^2 + b^2}} \cos \theta = \frac{c}{\sqrt{a^2 + b^2}} \qquad \textbf{(3)}$$

There is a unique angle ϕ, $0 \le \phi < 2\pi$, for which

$$\cos \phi = \frac{a}{\sqrt{a^2 + b^2}} \quad \text{and} \quad \sin \phi = \frac{b}{\sqrt{a^2 + b^2}} \qquad \textbf{(4)}$$

(see Figure 33). Equation (3) may be written as

$$\sin \theta \cos \phi + \cos \theta \sin \phi = \frac{c}{\sqrt{a^2 + b^2}}$$

or, equivalently,

$$\sin(\theta + \phi) = \frac{c}{\sqrt{a^2 + b^2}} \qquad \textbf{(5)}$$

where ϕ satisfies equation (4).

Figure 33

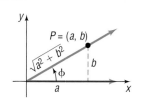

If $|c| > \sqrt{a^2 + b^2}$, then $\sin(\theta + \phi) > 1$ or $\sin(\theta + \phi) < -1$, and equation (5) has no solution.

If $|c| \le \sqrt{a^2 + b^2}$, then the solutions of equation (5) are

$$\theta + \phi = \sin^{-1} \frac{c}{\sqrt{a^2 + b^2}} \quad \text{or} \quad \theta + \phi = \pi - \sin^{-1} \frac{c}{\sqrt{a^2 + b^2}}$$

Because the angle ϕ is determined by equations (4), these are the solutions to equation (2). ■

 NOW WORK PROBLEM 33.

④ ◿ # GRAPHING UTILITY SOLUTIONS

The techniques introduced in this section apply only to certain types of trigonometric equations. Solutions for other types are usually studied in calculus, using numerical methods. In the next example, we show how a graphing utility may be used to obtain solutions.

EXAMPLE 8 **Solving Trigonometric Equations Using a Graphing Utility**

Solve: $5 \sin x + x = 3$
Express the solution(s) rounded to two decimal places.

Solution This type of trigonometric equation cannot be solved by previous methods. A graphing utility, though, can be used here. The solution(s) of this equation is the same as the points of intersection of the graphs of $Y_1 = 5 \sin x + x$ and $Y_2 = 3$. See Figure 34.

Figure 34

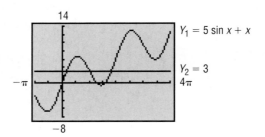

There are three points of intersection; the x-coordinates are the solutions that we seek. Using INTERSECT, we find

$$x = 0.52, \qquad x = 3.18, \qquad x = 5.71$$

rounded to two decimal places. ■

NOW WORK PROBLEM 45.

3.8 EXERCISES

In Problems 1–38, solve each equation on the interval $0 \le \theta < 2\pi$.

1. $2\cos^2\theta + \cos\theta = 0$

2. $\sin^2\theta - 1 = 0$

3. $2\sin^2\theta - \sin\theta - 1 = 0$

4. $2\cos^2\theta + \cos\theta - 1 = 0$

5. $(\tan\theta - 1)(\sec\theta - 1) = 0$

6. $(\cot\theta + 1)\left(\csc\theta - \dfrac{1}{2}\right) = 0$

7. $\sin^2\theta - \cos^2\theta = 1 + \cos\theta$

8. $\cos^2\theta - \sin^2\theta + \sin\theta = 0$

9. $\sin^2\theta = 6(\cos\theta + 1)$

10. $2\sin^2\theta = 3(1 - \cos\theta)$

11. $\cos(2\theta) + 6\sin^2\theta = 4$

12. $\cos(2\theta) = 2 - 2\sin^2\theta$

13. $\cos\theta = \sin\theta$

14. $\cos\theta + \sin\theta = 0$

15. $\tan\theta = 2\sin\theta$

16. $\sin(2\theta) = \cos\theta$ **17.** $\sin\theta = \csc\theta$ **18.** $\tan\theta = \cot\theta$

19. $\cos(2\theta) = \cos\theta$ **20.** $\sin(2\theta)\sin\theta = \cos\theta$ **21.** $\sin(2\theta) + \sin(4\theta) = 0$

22. $\cos(2\theta) + \cos(4\theta) = 0$ **23.** $\cos(4\theta) - \cos(6\theta) = 0$ **24.** $\sin(4\theta) - \sin(6\theta) = 0$

25. $1 + \sin\theta = 2\cos^2\theta$ **26.** $\sin^2\theta = 2\cos\theta + 2$ **27.** $\tan^2\theta = \dfrac{3}{2}\sec\theta$

28. $\csc^2\theta = \cot\theta + 1$ **29.** $3 - \sin\theta = \cos(2\theta)$ **30.** $\cos(2\theta) + 5\cos\theta + 3 = 0$

31. $\sec^2\theta + \tan\theta = 0$ **32.** $\sec\theta = \tan\theta + \cot\theta$ **33.** $\sin\theta - \sqrt{3}\cos\theta = 1$

34. $\sqrt{3}\sin\theta + \cos\theta = 1$ **35.** $\tan(2\theta) + 2\sin\theta = 0$ **36.** $\tan(2\theta) + 2\cos\theta = 0$

37. $\sin\theta + \cos\theta = \sqrt{2}$ **38.** $\sin\theta + \cos\theta = -\sqrt{2}$

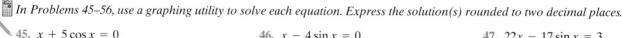

In Problems 39–44, solve each equation for $x, -\pi \le x \le \pi$. *Express the solution(s) rounded to two decimal places.*

39. Solve the equation $\cos x = e^x$ by graphing $Y_1 = \cos x$ and $Y_2 = e^x$ and finding their point(s) of intersection.

40. Solve the equation $\cos x = e^x$ by graphing $Y_1 = \cos x - e^x$ and finding the x-intercept(s).

41. Solve the equation $2\sin x = 0.7x$ by graphing $Y_1 = 2\sin x$ and $Y_2 = 0.7x$ and finding their point(s) of intersection.

42. Solve the equation $2\sin x = 0.7x$ by graphing $Y_1 = 2\sin x - 0.7x$ and finding the x-intercept(s).

43. Solve the equation $\cos x = x^2$ by graphing $Y_1 = \cos x$ and $Y_2 = x^2$ and finding their point(s) of intersection.

43. Solve the equation $\cos x = x^2$ by graphing $Y_1 = \cos x - x^2$ and finding the x-intercept(s).

In Problems 45–56, use a graphing utility to solve each equation. Express the solution(s) rounded to two decimal places.

45. $x + 5\cos x = 0$ **46.** $x - 4\sin x = 0$ **47.** $22x - 17\sin x = 3$

48. $19x + 8\cos x = 2$ **49.** $\sin x + \cos x = x$ **50.** $\sin x - \cos x = x$

51. $x^2 - 2\cos x = 0$ **52.** $x^2 + 3\sin x = 0$ **53.** $x^2 - 2\sin(2x) = 3x$

54. $x^2 = x + 3\cos(2x)$ **55.** $6\sin x - e^x = 2, \quad x > 0$ **56.** $4\cos(3x) - e^x = 1, \quad x > 0$

---

57. Constructing a Rain Gutter A rain gutter is to be constructed of aluminum sheets 12 inches wide. After marking off a length of 4 inches from each edge, this length is bent up at an angle θ. See the illustration. The area A of the opening as a function of θ is given by

$$A(\theta) = 16\sin\theta(\cos\theta + 1), \quad 0° < \theta < 90°$$

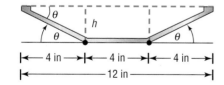

(a) In calculus, you will be asked to find the angle θ that maximizes A by solving the equation

$$\cos(2\theta) + \cos\theta = 0, \quad 0° < \theta < 90°$$

Solve this equation for θ by using the Double-Angle Formula.

(b) Solve the equation for θ by writing the sum of the two cosines as a product.

(c) What is the maximum area A of the opening?

(d) Graph A, $0° \le \theta \le 90°$, and find the angle θ that maximizes the area A. Also find the maximum area. Compare the results to the answers found earlier.

58. Projectile Motion An object is propelled upward at an angle θ, $45° < \theta < 90°$, to the horizontal with an initial velocity of v_0 feet per second from the base of a plane that makes an angle of $45°$ with the horizontal. See the illustration. If air resistance is ignored, the distance R that it travels up the inclined plane is given by

$$R = \frac{v_0^2\sqrt{2}}{32}\left[\sin(2\theta) - \cos(2\theta) - 1\right]$$

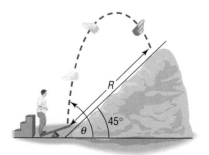

(a) In calculus, you will be asked to find the angle θ that maximizes R by solving the equation

$$\sin(2\theta) + \cos(2\theta) = 0$$

Solve this equation for θ using the method of Example 7.

(b) Solve this equation for θ by dividing each side by $\cos(2\theta)$.

(c) What is the maximum distance R if $v_0 = 32$ feet per second?

(d) Graph R, $45° \le \theta \le 90°$, and find the angle θ that maximizes the distance R. Also find the maximum distance. Use $v_0 = 32$ feet per second. Compare the results with the answers found earlier.

59. **Heat Transfer** In the study of heat transfer, the equation $x + \tan x = 0$ occurs. Graph $Y_1 = -x$ and $Y_2 = \tan x$ for $x \ge 0$. Conclude that there are an infinite number of points of intersection of these two graphs. Now find the first two positive solutions of $x + \tan x = 0$ rounded to two decimal places.

60. **Carrying a Ladder around a Corner** A ladder of length L is carried horizontally around a corner from a hall 3 feet wide into a hall 4 feet wide. See the illustration.

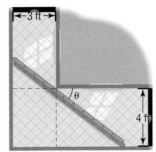

(a) Express L as a function of θ.

(b) In calculus, you will be asked to find the length of the longest ladder that can turn the corner by solving the equation

$$3 \sec \theta \tan \theta - 4 \csc \theta \cot \theta = 0, \quad 0° < \theta < 90°$$

Solve this equation for θ.

(c) What is the length of the longest ladder that can be carried around the corner?

(d) Graph L, $0° \le \theta \le 90°$, and find the angle θ that minimizes the length L.

(e) Compare the result with the one found in part (b). Explain why the two answers are the same.

61. **Projectile Motion** The horizontal distance that a projectile will travel in the air is given by the equation

$$R = \frac{v_0^2 \sin(2\theta)}{g}$$

where v_0 is the initial velocity of the projectile, θ is the angle of elevation, and g is acceleration due to gravity (9.8 meters per second squared).

(a) If you can throw a baseball with an initial speed of 34.8 meters per second, at what angle of elevation θ should you direct the throw so that the ball travels a distance of 107 meters before striking the ground?

(b) Determine the maximum distance that you can throw the ball.

(c) Graph R, with $v_0 = 34.8$ meters per second.

(d) Verify the results obtained in parts (a) and (b) using ZERO or ROOT.

62. **Projectile Motion** Refer to Problem 61.

(a) If you can throw a baseball with an initial speed of 40 meters per second, at what angle of elevation θ should you direct the throw so that the ball travels a distance of 110 meters before striking the ground?

(b) Determine the maximum distance that you can throw the ball.

(c) Graph R, with $v_0 = 40$ meters per second.

(d) Verify the results obtained in parts (a) and (b) using ZERO or ROOT.

CHAPTER REVIEW

Things To Know

Definitions of the six inverse trigonometric functions

$y = \sin^{-1} x$	means	$x = \sin y$	where $-1 \le x \le 1$, $-\dfrac{\pi}{2} \le y \le \dfrac{\pi}{2}$	(p. 163)
$y = \cos^{-1} x$	means	$x = \cos y$	where $-1 \le x \le 1$, $0 \le y \le \pi$	(p. 166)
$y = \tan^{-1} x$	means	$x = \tan y$	where $-\infty < x < \infty$, $-\dfrac{\pi}{2} < y < \dfrac{\pi}{2}$	(p. 169)
$y = \sec^{-1} x$	means	$x = \sec y$	where $|x| \ge 1$, $0 \le y \le \pi$, $y \ne \dfrac{\pi}{2}$	(p. 174)
$y = \csc^{-1} x$	means	$x = \csc y$	where $|x| \ge 1$, $-\dfrac{\pi}{2} \le y \le \dfrac{\pi}{2}$, $y \ne 0$	(p. 174)
$y = \cot^{-1} x$	means	$x = \cot y$	where $-\infty < x < \infty$, $0 < y < \pi$	(p. 174)

Sum and Difference Formulas (pp. 182, 185, and 188)

$$\cos(\alpha + \beta) = \cos \alpha \cos \beta - \sin \alpha \sin \beta \qquad \cos(\alpha - \beta) = \cos \alpha \cos \beta + \sin \alpha \sin \beta$$

$$\sin(\alpha + \beta) = \sin\alpha\cos\beta + \cos\alpha\sin\beta \qquad \sin(\alpha - \beta) = \sin\alpha\cos\beta - \cos\alpha\sin\beta$$

$$\tan(\alpha + \beta) = \frac{\tan\alpha + \tan\beta}{1 - \tan\alpha\tan\beta} \qquad \tan(\alpha - \beta) = \frac{\tan\alpha - \tan\beta}{1 + \tan\alpha\tan\beta}$$

Double-Angle Formulas (pp. 193 and 194)

$$\sin(2\theta) = 2\sin\theta\cos\theta \qquad \cos(2\theta) = \cos^2\theta - \sin^2\theta \qquad \cos(2\theta) = 1 - 2\sin^2\theta$$

$$\cos(2\theta) = 2\cos^2\theta - 1 \qquad \tan(2\theta) = \frac{2\tan\theta}{1 - \tan^2\theta}$$

Half-Angle Formulas (pp. 194, 195, 196, and 198)

$$\sin^2\frac{\alpha}{2} = \frac{1 - \cos\alpha}{2} \qquad \cos^2\frac{\alpha}{2} = \frac{1 + \cos\alpha}{2} \qquad \tan^2\frac{\alpha}{2} = \frac{1 - \cos\alpha}{1 + \cos\alpha}$$

$$\sin\frac{\alpha}{2} = \pm\sqrt{\frac{1 - \cos\alpha}{2}} \qquad \cos\frac{\alpha}{2} = \pm\sqrt{\frac{1 + \cos\alpha}{2}} \qquad \tan\frac{\alpha}{2} = \pm\sqrt{\frac{1 - \cos\alpha}{1 + \cos\alpha}} = \frac{1 - \cos\alpha}{\sin\alpha} = \frac{\sin\alpha}{1 + \cos\alpha}$$

where the $+$ or $-$ sign is determined by the quadrant of $\dfrac{\alpha}{2}$

Product-to-Sum Formulas (p. 201)

$$\sin\alpha\sin\beta = \frac{1}{2}\left[\cos(\alpha - \beta) - \cos(\alpha + \beta)\right] \qquad \cos\alpha\cos\beta = \frac{1}{2}\left[\cos(\alpha - \beta) + \cos(\alpha + \beta)\right]$$

$$\sin\alpha\cos\beta = \frac{1}{2}\left[\sin(\alpha + \beta) + \sin(\alpha - \beta)\right]$$

Sum-to-Product Formulas (p. 202)

$$\sin\alpha + \sin\beta = 2\sin\frac{\alpha + \beta}{2}\cos\frac{\alpha - \beta}{2} \qquad \sin\alpha - \sin\beta = 2\sin\frac{\alpha - \beta}{2}\cos\frac{\alpha + \beta}{2}$$

$$\cos\alpha + \cos\beta = 2\cos\frac{\alpha + \beta}{2}\cos\frac{\alpha - \beta}{2} \qquad \cos\alpha - \cos\beta = -2\sin\frac{\alpha + \beta}{2}\sin\frac{\alpha - \beta}{2}$$

Objectives

You should be able to:

Find the exact value of the inverse sine, cosine, and tangent functions (p. 164)

Find an approximate value of the inverse sine, cosine, and tangent functions (p 165)

Find the exact value of expressions involving the inverse sine, cosine and tangent functions (p. 172)

Know the definition of the inverse secant, cosecant, and cotangent functions (p. 174)

Use a calculator to evaluate $\sec^{-1}x$, $\csc^{-1}x$, and $\cot^{-1}x$ (p. 175)

Establish identities (p. 177)

Use sum and difference formulas to find exact values (p. 184)

Use sum and difference formulas to establish identities (p. 184)

Use sum and difference formulas involving inverse trigonometric functions (p. 189)

Use double-angle formulas to find exact values (p. 193)

Use double-angle and half-angle formulas to establish identities (p. 194)

Use half-angle formulas to find exact values (p. 197)

Express products as sums (p. 201)

Express sums as products (p. 202)

Solve equations involving a single trigonometric function (p. 203)

Solve trigonometric equations that are quadratic in form (p. 209)

Solve trigonometric equations using identities (p. 210)

Solve trigonometric equations that are linear in sine and cosine (p. 212)

Solve trigonometric equations using a graphing utility (p. 214)

Fill-in-the-Blank Items

1. Suppose that f and g are two functions with the same domain. If $f(x) = g(x)$ for every x in the domain, the equation is called a(n) _____. Otherwise, it is called a(n) _____ equation.

2. $\cos(\alpha + \beta) = \cos\alpha\cos\beta$ _____ $\sin\alpha\sin\beta$.

3. $\sin(\alpha + \beta) = \sin\alpha\cos\beta$ _____ $\cos\alpha\sin\beta$.

4. $\cos(2\theta) = \cos^2\theta -$ _____ $=$ _____ $-1 = 1 -$ _____.

5. $\sin^2\dfrac{\alpha}{2} = \dfrac{\rule{1cm}{0.4pt}}{2}$.

6. The function $y = \sin^{-1} x$ has domain _____ and range _____.

7. The value of $\sin^{-1}\left[\cos\left(\dfrac{\pi}{2}\right)\right]$ is _____.

True/False Items

T F **1.** $\sin(-\theta) + \sin\theta = 0$ for all θ.

T F **2.** $\sin(\alpha + \beta) = \sin\alpha + \sin\beta + 2\sin\alpha\sin\beta$.

T F **3.** $\cos(2\theta)$ has three equivalent forms: $\cos^2\theta - \sin^2\theta$, $1 - 2\sin^2\theta$, and $2\cos^2\theta - 1$.

T F **4.** $\cos\dfrac{\alpha}{2} = \pm\dfrac{\sqrt{1 + \cos\alpha}}{2}$, where the $+$ or $-$ sign depends on the angle $\dfrac{\alpha}{2}$.

T F **5.** The domain of $y = \sin^{-1} x$ is $-\dfrac{\pi}{2} \le x \le \dfrac{\pi}{2}$.

T F **6.** $\cos(\sin^{-1} 0) = 1$ and $\sin(\cos^{-1} 0) = 1$.

T F **7.** Most trigonometric equations have unique solutions.

T F **8.** The equation $\tan\theta = \dfrac{\pi}{2}$ has no solution.

Review Exercises

Blue problem numbers indicate the author's suggestions for use in a Practice Test.

In Problems 1–20, find the exact value of each expression. Do not use a calculator.

1. $\sin^{-1} 1$

2. $\cos^{-1} 0$

3. $\tan^{-1} 1$

4. $\sin^{-1}\left(-\dfrac{1}{2}\right)$

5. $\cos^{-1}\left(-\dfrac{\sqrt{3}}{2}\right)$

6. $\tan^{-1}(-\sqrt{3})$

7. $\sin\left(\cos^{-1}\dfrac{\sqrt{2}}{2}\right)$

8. $\cos(\sin^{-1} 0)$

9. $\tan\left[\sin^{-1}\left(-\dfrac{\sqrt{3}}{2}\right)\right]$

10. $\tan\left[\cos^{-1}\left(-\dfrac{1}{2}\right)\right]$

11. $\sec\left(\tan^{-1}\dfrac{\sqrt{3}}{3}\right)$

12. $\csc\left(\sin^{-1}\dfrac{\sqrt{3}}{2}\right)$

13. $\sin\left(\tan^{-1}\dfrac{3}{4}\right)$

14. $\cos\left(\sin^{-1}\dfrac{3}{5}\right)$

15. $\tan\left[\sin^{-1}\left(-\dfrac{4}{5}\right)\right]$

16. $\tan\left[\cos^{-1}\left(-\dfrac{3}{5}\right)\right]$

17. $\sin^{-1}\left(\cos\dfrac{2\pi}{3}\right)$

18. $\cos^{-1}\left(\tan\dfrac{3\pi}{4}\right)$

19. $\tan^{-1}\left(\tan\dfrac{7\pi}{4}\right)$

20. $\cos^{-1}\left(\cos\dfrac{7\pi}{6}\right)$

In Problems 21–52, establish each identity.

21. $\tan\theta\cot\theta - \sin^2\theta = \cos^2\theta$

22. $\sin\theta\csc\theta - \sin^2\theta = \cos^2\theta$

23. $\cos^2\theta(1 + \tan^2\theta) = 1$

24. $(1 - \cos^2\theta)(1 + \cot^2\theta) = 1$

25. $4\cos^2\theta + 3\sin^2\theta = 3 + \cos^2\theta$

26. $4\sin^2\theta + 2\cos^2\theta = 4 - 2\cos^2\theta$

27. $\dfrac{1 - \cos\theta}{\sin\theta} + \dfrac{\sin\theta}{1 - \cos\theta} = 2\csc\theta$

28. $\dfrac{\sin\theta}{1 + \cos\theta} + \dfrac{1 + \cos\theta}{\sin\theta} = 2\csc\theta$

29. $\dfrac{\cos\theta}{\cos\theta - \sin\theta} = \dfrac{1}{1 - \tan\theta}$

30. $1 - \dfrac{\cos^2\theta}{1 + \sin\theta} = \sin\theta$

31. $\dfrac{\csc\theta}{1 + \csc\theta} = \dfrac{1 - \sin\theta}{\cos^2\theta}$

32. $\dfrac{1 + \sec\theta}{\sec\theta} = \dfrac{\sin^2\theta}{1 - \cos\theta}$

33. $\csc\theta - \sin\theta = \cos\theta\cot\theta$

34. $\dfrac{\csc\theta}{1 - \cos\theta} = \dfrac{1 + \cos\theta}{\sin^3\theta}$

35. $\dfrac{1 - \sin\theta}{\sec\theta} = \dfrac{\cos^3\theta}{1 + \sin\theta}$

36. $\dfrac{1 - \cos\theta}{1 + \cos\theta} = (\csc\theta - \cot\theta)^2$

37. $\dfrac{1 - 2\sin^2\theta}{\sin\theta\cos\theta} = \cot\theta - \tan\theta$

38. $\dfrac{(2\sin^2\theta - 1)^2}{\sin^4\theta - \cos^4\theta} = 1 - 2\cos^2\theta$

39. $\dfrac{\cos(\alpha + \beta)}{\cos\alpha\sin\beta} = \cot\beta - \tan\alpha$

40. $\dfrac{\sin(\alpha - \beta)}{\sin\alpha\cos\beta} = 1 - \cot\alpha\tan\beta$

41. $\dfrac{\cos(\alpha - \beta)}{\cos\alpha\cos\beta} = 1 + \tan\alpha\tan\beta$

42. $\dfrac{\cos(\alpha + \beta)}{\sin \alpha \cos \beta} = \cot \alpha - \tan \beta$

43. $(1 + \cos \theta)\left(\tan \dfrac{\theta}{2}\right) = \sin \theta$

44. $\sin \theta \tan \dfrac{\theta}{2} = 1 - \cos \theta$

45. $2 \cot \theta \cot(2\theta) = \cot^2 \theta - 1$

46. $2 \sin(2\theta)(1 - 2 \sin^2 \theta) = \sin(4\theta)$

47. $1 - 8 \sin^2 \theta \cos^2 \theta = \cos(4\theta)$

48. $\dfrac{\sin(3\theta) \cos \theta - \sin \theta \cos(3\theta)}{\sin(2\theta)} = 1$

49. $\dfrac{\sin(2\theta) + \sin(4\theta)}{\cos(2\theta) + \cos(4\theta)} = \tan(3\theta)$

50. $\dfrac{\sin(2\theta) + \sin(4\theta)}{\sin(2\theta) - \sin(4\theta)} + \dfrac{\tan(3\theta)}{\tan \theta} = 0$

51. $\dfrac{\cos(2\theta) - \cos(4\theta)}{\cos(2\theta) + \cos(4\theta)} - \tan \theta \tan(3\theta) = 0$

52. $\cos(2\theta) - \cos(10\theta) = \tan(4\theta)\big[\sin(2\theta) + \sin(10\theta)\big]$

In Problems 53–60, find the exact value of each expression.

53. $\sin 165°$

54. $\tan 105°$

55. $\cos \dfrac{5\pi}{12}$

56. $\sin\left(-\dfrac{\pi}{12}\right)$

57. $\cos 80° \cos 20° + \sin 80° \sin 20°$

58. $\sin 70° \cos 40° - \cos 70° \sin 40°$

59. $\tan \dfrac{\pi}{8}$

60. $\sin \dfrac{5\pi}{8}$

In Problems 61–70, use the information given about the angles α and β to find the exact value of:

(a) $\sin(\alpha + \beta)$ (b) $\cos(\alpha + \beta)$ (c) $\sin(\alpha - \beta)$ (d) $\tan(\alpha + \beta)$

(e) $\sin(2\alpha)$ (f) $\cos(2\beta)$ (g) $\sin \dfrac{\beta}{2}$ (h) $\cos \dfrac{\alpha}{2}$

61. $\sin \alpha = \dfrac{4}{5}, \;\; 0 < \alpha < \dfrac{\pi}{2}; \;\; \sin \beta = \dfrac{5}{13}, \;\; \dfrac{\pi}{2} < \beta < \pi$

62. $\cos \alpha = \dfrac{4}{5}, \;\; 0 < \alpha < \dfrac{\pi}{2}; \;\; \cos \beta = \dfrac{5}{13}, \;\; -\dfrac{\pi}{2} < \beta < 0$

63. $\sin \alpha = -\dfrac{3}{5}, \;\; \pi < \alpha < \dfrac{3\pi}{2}; \;\; \cos \beta = \dfrac{12}{13}, \;\; \dfrac{3\pi}{2} < \beta < 2\pi$

64. $\sin \alpha = -\dfrac{4}{5}, \;\; -\dfrac{\pi}{2} < \alpha < 0; \;\; \cos \beta = -\dfrac{5}{13}, \;\; \dfrac{\pi}{2} < \beta < \pi$

65. $\tan \alpha = \dfrac{3}{4}, \;\; \pi < \alpha < \dfrac{3\pi}{2}; \;\; \tan \beta = \dfrac{12}{5}, \;\; 0 < \beta < \dfrac{\pi}{2}$

66. $\tan \alpha = -\dfrac{4}{3}, \;\; \dfrac{\pi}{2} < \alpha < \pi; \;\; \cot \beta = \dfrac{12}{5}, \;\; \pi < \beta < \dfrac{3\pi}{2}$

67. $\sec \alpha = 2, \;\; -\dfrac{\pi}{2} < \alpha < 0; \;\; \sec \beta = 3, \;\; \dfrac{3\pi}{2} < \beta < 2\pi$

68. $\csc \alpha = 2, \;\; \dfrac{\pi}{2} < \alpha < \pi; \;\; \sec \beta = -3, \;\; \dfrac{\pi}{2} < \beta < \pi$

69. $\sin \alpha = -\dfrac{2}{3}, \;\; \pi < \alpha < \dfrac{3\pi}{2}; \;\; \cos \beta = -\dfrac{2}{3}, \;\; \pi < \beta < \dfrac{3\pi}{2}$

70. $\tan \alpha = -2, \;\; \dfrac{\pi}{2} < \alpha < \pi; \;\; \cot \beta = -2, \;\; \dfrac{\pi}{2} < \beta < \pi$

In Problems 71–76, find the exact value of each expression.

71. $\cos\left(\sin^{-1}\dfrac{3}{5} - \cos^{-1}\dfrac{1}{2}\right)$

72. $\sin\left(\cos^{-1}\dfrac{5}{13} - \cos^{-1}\dfrac{4}{5}\right)$

73. $\tan\left[\sin^{-1}\left(-\dfrac{1}{2}\right) - \tan^{-1}\dfrac{3}{4}\right]$

74. $\cos\left[\tan^{-1}(-1) + \cos^{-1}\left(-\dfrac{4}{5}\right)\right]$

75. $\sin\left[2\cos^{-1}\left(-\dfrac{3}{5}\right)\right]$

76. $\cos\left(2\tan^{-1}\dfrac{4}{3}\right)$

In Problems 77–100, solve each equation on the interval $0 \le \theta < 2\pi$.

77. $\cos \theta = \dfrac{1}{2}$

78. $\sin \theta = -\dfrac{\sqrt{3}}{2}$

79. $2 \cos \theta + \sqrt{2} = 0$

80. $\tan \theta + \sqrt{3} = 0$

81. $\sin(2\theta) + 1 = 0$

82. $\cos(2\theta) = 0$

83. $\tan(2\theta) = 0$

84. $\sin(3\theta) = 1$

85. $\sec^2 \theta = 4$

86. $\csc^2 \theta = 1$

87. $\sin \theta = \tan \theta$

88. $\cos \theta = \sec \theta$

89. $\sin \theta + \sin(2\theta) = 0$

90. $\cos(2\theta) = \sin \theta$

91. $\sin(2\theta) - \cos \theta - 2 \sin \theta + 1 = 0$

92. $\sin(2\theta) - \sin \theta - 2 \cos \theta + 1 = 0$

93. $2 \sin^2 \theta - 3 \sin \theta + 1 = 0$

94. $2 \cos^2 \theta + \cos \theta - 1 = 0$

95. $4 \sin^2 \theta = 1 + 4 \cos \theta$

96. $8 - 12 \sin^2 \theta = 4 \cos^2 \theta$

97. $\sin(2\theta) = \sqrt{2} \cos \theta$

98. $1 + \sqrt{3} \cos \theta + \cos(2\theta) = 0$

99. $\sin \theta - \cos \theta = 1$

100. $\sin \theta - \sqrt{3} \cos \theta = 2$

In Problems 101–106, use a graphing utility to solve each equation on the interval $0 \le x \le 2\pi$. Approximate any solutions rounded to two decimal places.

101. $2x = 5 \cos x$

102. $2x = 5 \sin x$

103. $2 \sin x + 3 \cos x = 4x$

104. $3 \cos x + x = \sin x$

105. $\sin x = \ln x$

106. $\sin x = e^{-x}$

Project at Motorola

Sending Pictures Wirelessly

The picture below on the left shows a special digital camera being developed at Motorola to allow people to take pictures and then send them anywhere in the world right from where they are. This is called an ImagePhone Camera. It attaches to a cellular phone as shown below in the picture on the right. Once attached the images can be sent wirelessly by e-mail.

The images contain a lot of data. A digital image is made up of picture elements, or *pixels* for short. A common type of image is called a VGA image, which is short for video graphic array. The VGA array of pixels has 640 columns and 480 rows for a total of $640 \times 480 = 307,200$ pixels. Each pixel is composed of a red subpixel, a green subpixel, and a blue subpixel, which makes $307,200 \times 3 = 921,600$ subpixels. Each subpixel is represented by 8 bits of data which makes 7,372,800 bits of data for a single VGA picture!

If we want to send this image wirelessly, it will take a long time. Present cellular systems can only send 9600 bits per second (bps) over the air. At this rate the above VGA image would take $7,372,800/9600 = 768$ seconds or approximately 13 minutes.

The ImagePhone camera takes only 1 minute to send a VGA image, instead of the expected 13 minutes. This reduction is made possible by a process called *image compression* that reduces the size of the image data. One common compression method is called JPEG compression, where JPEG stands for Joint Photographic Experts Group. This method can reduce the image data by a factor ranging from 10 to 50. For example, a compression factor of 25 (or 25:1 compression) will reduce the image transmission time from 13 minutes to 30 seconds!

How does compression work? To understand this, consider one row of the VGA image. Each row has 640 pixels corresponding to the 640 columns. If we draw an x-axis through these data points and plot the intensity (brightness) as a function of x across the image, we can obtain an arbitrary function $f(x)$. Using a technique called **Fourier series expansion,** this function can be represented as a series of sines and cosines as follows:

$$f(x) = \frac{a_0}{2} + \sum_{n=1}^{\infty} a_n \cos(nx) + \sum_{n=1}^{\infty} b_n \sin(nx)$$

where a_n and b_n are the **Fourier coefficients.** If the function $f(x)$ were such that an infinite number of terms in the series were not required, but only a few terms (say 10), then all we would have to do is send 10 coefficients and achieve a significant reduction in the amount of data needed to represent the function $f(x)$. Note that JPEG uses a similar type of expression as the Fourier series.

As an example, consider a simple square wave that represents the information that we have to send. Let us call this function $g(x)$ defined as follows:

$$g(x) = 0, \quad -\pi < x < 0, \quad \pi < x < 2\pi, \ldots$$
$$g(x) = h, \quad 0 < x < \pi, \quad 2\pi < x < 3\pi, \ldots$$

where $h > 0$.

Perhaps this represents intensity variation as a row in an image that depicts a solid block. Using a Fourier series expansion, we can represent the square wave as

$$g(x) = \frac{h}{2} + \frac{2h}{\pi} \left[\frac{\sin x}{1} + \frac{\sin(3x)}{3} + \frac{\sin(5x)}{5} + \cdots \right]$$

or

$$g(x) = c_0 + c_1 \sin x + c_3 \sin(3x) + c_5 \sin(5x) + \cdots$$

To completely represent this square wave, we do not need to send all the points on this waveform, but only the coefficients $(c_0, c_1, c_3, c_5, \ldots)$. If we have to send an infinite set of coefficients, then we do not gain much. However, if we find that we only need to send the first N coefficients, where $N = 15$, then we have reduced the data needed to represent the square wave significantly.

The following exercises will help you to understand this process better.

1. Plot the square wave $g(x)$ defined above by hand.
2. Using a graphing utility, plot the Fourier series for $g(x)$ using 5 coefficients with $h = 2$.
3. Plot the Fourier series using 10 coefficients.
4. Plot the Fourier series using 20 coefficients.
5. Plot them all on the same chart and compare the differences in the curves for plots 2, 3 and 4. Which would you use to represent the data?

Applications of Trigonometric Functions

Field Trip to Motorola

In Motorola Automotive we have to test most of our products (automotive electronics, especially engine controllers) in fairly severe vibration conditions in order to ensure a long and healthy life in the field. One tool for analyzing and monitoring any profile of vibrations (whether it is the vibration of a car engine or, for that matter, any acoustic wave) is the Fourier series and Fourier transformations. The use of these transformations is based on the fundamental theorem, stating that any vibration profile (or periodic function) can be expressed as a superposition of elementary sine and cosine profiles. This type of presentation is very conducive to an analysis of a specific frequency content in a complex wave, as well as to a study of the frequency response function of a system.

PREPARING FOR THIS SECTION

Before getting started, review the following:

✓ Pythagorean Theorem
(Appendix A, Section A.2, pp. 497–499)

✓ Trigonometric Equations (I) (Section 3.7, pp. 204–207)

✓ Definition of the Trigonometric Functions
(Section 2.2, pp. 92–95)

✓ Theorem on Trigonometric Functions Using a Circle
of Radius r (Section 2.2, p. 104)

4.1 RIGHT TRIANGLE TRIGONOMETRY

OBJECTIVES
1. Find the Value of Trigonometric Functions of Acute Angles
2. Use the Complementary Angle Theorem
3. Solve Right Triangles
4. Solve Applied Problems

Figure 1

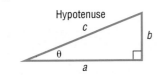

(a)

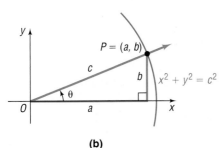

(b)

(1) A triangle in which one angle is a right angle (90°) is called a **right triangle.** Recall that the side opposite the right angle is called the **hypotenuse,** and the remaining two sides are called the **legs** of the triangle. In Figure 1(a), we have labeled the hypotenuse as c to indicate that its length is c units, and, in a like manner, we have labeled the legs as a and b. Because the triangle is a right triangle, the Pythagorean Theorem tells us that

$$a^2 + b^2 = c^2$$

In Figure 1(a), we also show the angle θ. The angle θ is an **acute angle:** that is, $0° < \theta < 90°$, if θ is measured in degrees, or $0 < \theta < \dfrac{\pi}{2}$, if θ is measured in radians. Place θ in standard position, as shown in Figure 1(b). Then the coordinates of the point P are (a, b). Also, P is a point on the terminal side of θ that is also on the circle $x^2 + y^2 = c^2$. (Do you see why?).

Now use the theorem on page 104. By referring to the lengths of the sides of the triangle by the names hypotenuse (c), opposite (b), and adjacent (a), as indicated in Figure 2, we can express the trigonometric functions of θ as ratios of the sides of a right triangle.

Figure 2

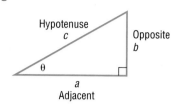

$$\sin\theta = \frac{\text{Opposite}}{\text{Hypotenuse}} = \frac{b}{c} \qquad \cos\theta = \frac{\text{Adjacent}}{\text{Hypotenuse}} = \frac{a}{c}$$

$$\tan\theta = \frac{\text{Opposite}}{\text{Adjacent}} = \frac{b}{a} \qquad \csc\theta = \frac{\text{Hypotenuse}}{\text{Opposite}} = \frac{c}{b} \qquad \textbf{(1)}$$

$$\sec\theta = \frac{\text{Hypotenuse}}{\text{Adjacent}} = \frac{c}{a} \qquad \cot\theta = \frac{\text{Adjacent}}{\text{Opposite}} = \frac{a}{b}$$

Notice that each of the trigonometric functions of the acute angle θ is positive.

EXAMPLE 1   **Finding the Value of Trigonometric Functions from a Right Triangle**

Find the exact value of the six trigonometric functions of the angle θ in Figure 3.

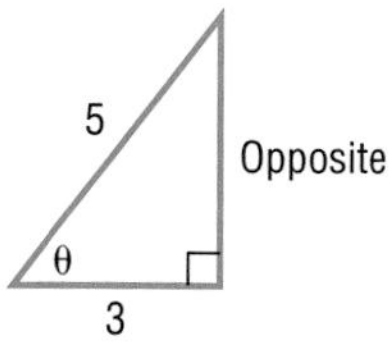

Solution   We see in Figure 3 that the two given sides of the triangle are

Figure 3

$$c = \text{Hypotenuse} = 5 \qquad a = \text{Adjacent} = 3$$

To find the length of the opposite side, we use the Pythagorean Theorem.

$$(\text{Adjacent})^2 + (\text{Opposite})^2 = (\text{Hypotenuse})^2$$

$$3^2 + (\text{Opposite})^2 = 5^2$$

$$(\text{Opposite})^2 = 25 - 9 = 16$$

$$\text{Opposite} = 4$$

Now that we know the lengths of the three sides, we use the ratios in equations (1) to find the value of each of the six trigonometric functions.

$$\sin\theta = \frac{\text{Opposite}}{\text{Hypotenuse}} = \frac{4}{5} \qquad \cos\theta = \frac{\text{Adjacent}}{\text{Hypotenuse}} = \frac{3}{5} \qquad \tan\theta = \frac{\text{Opposite}}{\text{Adjacent}} = \frac{4}{3}$$

$$\csc\theta = \frac{\text{Hypotenuse}}{\text{Opposite}} = \frac{5}{4} \qquad \sec\theta = \frac{\text{Hypotenuse}}{\text{Adjacent}} = \frac{5}{3} \qquad \cot\theta = \frac{\text{Adjacent}}{\text{Opposite}} = \frac{3}{4} \quad ■$$

  NOW WORK PROBLEM 1.

The values of the trigonometric functions of an acute angle are ratios of the lengths of the sides of a right triangle. This way of viewing the trigonometric functions leads to many applications and, in fact, was the point of view used by early mathematicians (before calculus) in studying the subject of trigonometry.

We look at one such application next.

| EXAMPLE 2 | Constructing a Rain Gutter |

Figure 4(a)

A rain gutter is to be constructed of aluminum sheets 12 inches wide. After marking off a length of 4 inches from each edge, this length is bent up at an angle θ. See Figure 4(a).

(a) Express the area A of the opening as a function of θ.

[**Hint:** Let b denote the vertical height of the bend.]

(b) Graph $A = A(\theta)$. Find the angle θ that makes A largest. (This bend will allow the most water to flow through the gutter.)

Solution
(a) Look again at Figure 4(a). The area A of the opening is the sum of the areas of two congruent right triangles and one rectangle. Look at Figure 4(b), showing the triangle in Figure 4(a) redrawn. We see that

Figure 4(b)

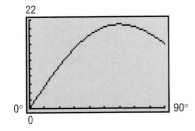

$$\cos\theta = \frac{a}{4} \quad \text{so} \quad a = 4\cos\theta \qquad \sin\theta = \frac{b}{4} \quad \text{so} \quad b = 4\sin\theta$$

The area of the triangle is

$$\text{area} = \tfrac{1}{2}(\text{base})(\text{height}) = \tfrac{1}{2}ab = \tfrac{1}{2}(4\cos\theta)(4\sin\theta) = 8\sin\theta\cos\theta$$

So the area of the two triangles is $16\sin\theta\cos\theta$.
   The rectangle has length 4 and height b, so its area is

$$4b = 4(4\sin\theta) = 16\sin\theta$$

The area A of the opening is

$$A = \text{area of the two triangles} + \text{area of the rectangle}$$
$$A(\theta) = 16\sin\theta\cos\theta + 16\sin\theta = 16\sin\theta(\cos\theta + 1)$$

Figure 5

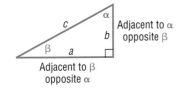

(b) Figure 5 shows the graph of $A = A(\theta)$. Using MAXIMUM, the angle θ that makes A largest is $60°$. ■

COMPLEMENTARY ANGLES: COFUNCTIONS

② Two acute angles are called **complementary** if their sum is a right angle. Because the sum of the angles of any triangle is $180°$, it follows that, for a right triangle, the two acute angles are complementary.
   Refer now to Figure 6; we have labeled the angle opposite side b as β and the angle opposite side a as α. Notice that side b is adjacent to angle α and side a is adjacent to angle β. As a result,

Figure 6

Adjacent to α
opposite β

Adjacent to β
opposite α

$$\sin\beta = \frac{b}{c} = \cos\alpha \qquad \cos\beta = \frac{a}{c} = \sin\alpha \qquad \tan\beta = \frac{b}{a} = \cot\alpha \quad \textbf{(2)}$$

$$\csc\beta = \frac{c}{b} = \sec\alpha \qquad \sec\beta = \frac{c}{a} = \csc\alpha \qquad \cot\beta = \frac{a}{b} = \tan\alpha$$

Because of these relationships, the functions sine and cosine, tangent and cotangent, and secant and cosecant are called **cofunctions** of each other. The identities (2) may be expressed in words as follows:

Complementary Angle Theorem   Cofunctions of complementary angles are equal.

Examples of this theorem are given next:

Complementary angles	Complementary angles	Complementary angles
↓ ↓	↓ ↓	↓ ↓
$\sin 30° = \cos 60°$	$\tan 40° = \cot 50°$	$\sec 80° = \csc 10°$
↑_____↑	↑_____↑	↑_____↑
Cofunctions	Cofunctions	Cofunctions

EXAMPLE 3

Using the Complementary Angle Theorem

(a) $\sin 62° = \cos(90° - 62°) = \cos 28°$

(b) $\tan\dfrac{\pi}{12} = \cot\left(\dfrac{\pi}{2} - \dfrac{\pi}{12}\right) = \cot\dfrac{5\pi}{12}$

(c) $\sin^2 40° + \sin^2 50° = \sin^2 40° + \cos^2 40° = 1$

$$\uparrow$$
$$\sin 50° = \cos 40°$$

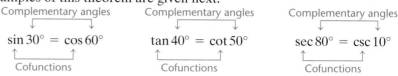 NOW WORK PROBLEM **11.**

SOLVING RIGHT TRIANGLES

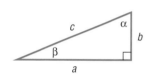 In the discussion that follows, we will always label a right triangle so that side a is opposite angle α, side b is opposite angle β, and side c is the hypotenuse, as shown in Figure 7. **To solve a right triangle** means to find the missing lengths of its sides and the measurements of its angles. We shall follow the practice of expressing the lengths of the sides rounded to two decimal places and of expressing angles in degrees rounded to one decimal place. (Be sure that your calculator is in degree mode.)

Figure 7

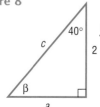

To solve a right triangle, we need to know one of the acute angles α or β and a side, or else two sides. Then we make use of the Pythagorean Theorem and the fact that the sum of the angles of a triangle is 180°. The sum of the angles α and β in a right triangle is therefore 90°.

For the right triangle shown in Figure 7, we have

$$c^2 = a^2 + b^2, \qquad \alpha + \beta = 90°$$

EXAMPLE 4

Solving a Right Triangle

Figure 8

Use Figure 8. If $b = 2$ and $\alpha = 40°$, find $a, c,$ and β.

Solution Since $\alpha = 40°$ and $\alpha + \beta = 90°$, we find that $\beta = 50°$. To find the sides a and c, we use the facts that

$$\tan 40° = \frac{a}{2} \quad \text{and} \quad \cos 40° = \frac{2}{c}$$

Now solve for a and c.

$$a = 2\tan 40° \approx 1.68 \quad \text{and} \quad c = \frac{2}{\cos 40°} \approx 2.61$$

NOW WORK PROBLEM **21.**

| EXAMPLE 5 | Solving a Right Triangle |

Figure 9

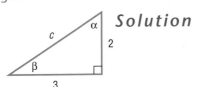

Use Figure 9. If $a = 3$ and $b = 2$, find c, α, and β.

Solution Since $a = 3$ and $b = 2$, then, by the Pythagorean Theorem, we have

$$c^2 = a^2 + b^2 = 3^2 + 2^2 = 9 + 4 = 13$$

$$c = \sqrt{13} \approx 3.61$$

To find angle α, we use the fact that

$$\tan \alpha = \frac{3}{2} \quad \text{so} \quad \alpha = \tan^{-1} \frac{3}{2}$$

Set the mode on your calculator to degrees. Then, rounded to one decimal place, we find that $\alpha = 56.3°$. Since $\alpha + \beta = 90°$, we find that $\beta = 33.7°$. ■

NOTE: To avoid round-off errors when using a calculator, we will store unrounded values in memory for use in subsequent calculations.

NOW WORK PROBLEM **31.**

APPLICATIONS

④ One common use for trigonometry is to measure heights and distances that are either awkward or impossible to measure by ordinary means.

| EXAMPLE 6 | Finding the Width of a River |

A surveyor can measure the width of a river by setting up a transit* at a point C on one side of the river and taking a sighting of a point A on the other side. Refer to Figure 10. After turning through an angle of 90° at C, the surveyor walks a distance of 200 meters to point B. Using the transit at B, the angle β is measured and found to be 20°. What is the width of the river rounded to the nearest meter?

Figure 10

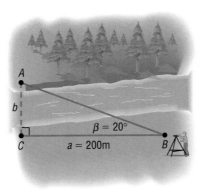

* An instrument used in surveying to measure angles.

Solution   We seek the length of side b. We know a and β, so we use the fact that

$$\tan\beta = \frac{b}{a}$$

to get

$$\tan 20° = \frac{b}{200}$$

$$b = 200\tan 20° \approx 72.79 \text{ meters}$$

The width of the river is 73 meters, rounded to the nearest meter.   ■

 **NOW WORK PROBLEM 41.**

| EXAMPLE 7 | **Finding the Inclination of a Mountain Trail** |

A straight trail leads from the Alpine Hotel, elevation 8000 feet, to a scenic overlook, elevation 11,100 feet. The length of the trail is 14,100 feet. What is the inclination (grade) of the trail? That is, what is the angle β in Figure 11?

Figure 11

Hotel

Overlook
elevation
11,100 ft

Trail
14,100 ft

3100 ft

β

Elevation
8000 ft

Solution   As Figure 11 illustrates, the angle β obeys the equation

$$\sin\beta = \frac{3100}{14,100}$$

Using a calculator,

$$\beta = \sin^{-1}\frac{3100}{14,100} \approx 12.7°$$

The inclination (grade) of the trail is approximately $12.7°$.   ■

Vertical heights can sometimes be measured using either the angle of elevation or the angle of depression. If a person is looking up at an object, the acute angle measured from the horizontal to a line-of-sight observation of the object is called the **angle of elevation.** See Figure 12(a).

If a person is standing on a cliff looking down at an object, the acute angle made by the line-of-sight observation of the object and the horizontal is called the **angle of depression.** See Figure 12(b).

Figure 12

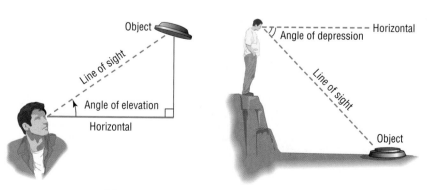

Object

Line of sight

Angle of elevation

Horizontal

Horizontal

Angle of depression

Line of sight

Object

(a)

(b)

| EXAMPLE 8 | Finding the Height of a Cloud |

Meteorologists find the height of a cloud using an instrument called a **ceilometer.** A ceilometer consists of a **light projector** that directs a vertical light beam up to the cloud base and a **light detector** that scans the cloud to detect the light beam. See Figure 13(a). On December 1, 2000, at Midway Airport in Chicago, a ceilometer with a base of 300 feet was employed to find the height of the cloud cover. If the angle of elevation of the light detector is 75°, what is the height of the cloud cover?

Figure 13

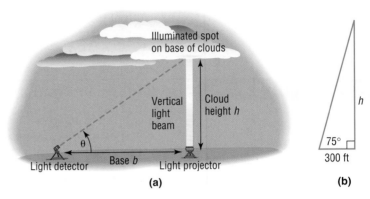

Illuminated spot on base of clouds

Vertical light beam

Cloud height *h*

θ

Light detector   Base *b*   Light projector

(a)

h

75°

300 ft

(b)

Solution   Figure 13(b) illustrates the situation. To find the height h, we use the fact that $\tan 75° = \dfrac{h}{300}$, so

$$h = 300 \tan 75° \approx 1120 \text{ feet}$$

The ceiling (height to the base of the cloud cover) is approximately 1120 feet. ∎

✏ NOW WORK PROBLEM 43.

The idea behind Example 8 can also be used to find the height of an object with a base that is not accessible to the horizontal.

| EXAMPLE 9 | Finding the Height of a Statue on a Building |

Adorning the top of the Board of Trade building in Chicago is a statue of Ceres, the Greek goddess of wheat. From street level, two observations are taken 400 feet from the center of the building. The angle of elevation to the base of the statue is found to be 45°; the angle of elevation to the top of the statue is 47.2°. See Figure 14(a). What is the height of the statue?

Figure 14

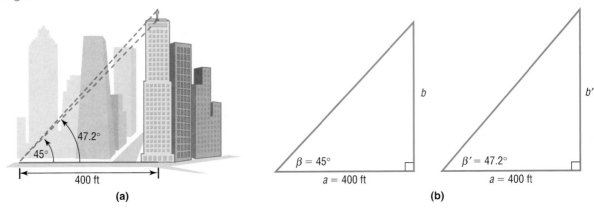

47.2°

45°

400 ft

(a)

b

$\beta = 45°$

$a = 400$ ft

b′

$\beta' = 47.2°$

$a = 400$ ft

(b)

Solution   Figure 14(b) shows two triangles that replicate Figure 14(a). The height of the statue of Ceres will be $b' - b$. To find b and b', we refer to Figure 14(b).

$$\tan 45° = \frac{b}{400} \qquad\qquad \tan 47.2° = \frac{b'}{400}$$

$$b = 400 \tan 45° = 400 \qquad\qquad b' = 400 \tan 47.2° \approx 432$$

The height of the statue is approximately $432 - 400 = 32$ feet.   ■

NOW WORK PROBLEM **51.**

| EXAMPLE 10 | The Gibb's Hill Lighthouse, Southampton, Bermuda |

In operation since 1846, the Gibb's Hill Lighthouse stands 117 feet high on a hill 245 feet high, so its beam of light is 362 feet above sea level. A brochure states that the light can be seen on the horizon about 26 miles distant. Verify the accuracy of this statement.

Solution   Figure 15 illustrates the situation. The central angle θ, positioned at the center of Earth, radius 3960 miles, obeys the equation

Figure 15

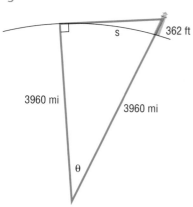

$$\cos \theta = \frac{3960}{3960 + \dfrac{362}{5280}} \approx 0.999982687 \qquad \text{1 mile = 5280 feet}$$

Solving for θ, we find

$$\theta \approx 0.33715° \approx 20.23'$$

The brochure does not indicate whether the distance is measured in nautical miles or statute miles. Let's calculate both distances.

The distance s in nautical miles (refer to Problem 104, p. 91) is the measure of angle θ in minutes, so $s = 20.23$ nautical miles.

The distance s in statute miles is given by the formula $s = r\theta$, where θ is measured in radians. Then, since

$$\theta = 20.23' = 0.33715° = 0.00588 \text{ radian}$$

$$1' = \tfrac{1}{60}° \qquad 1° = \tfrac{\pi}{180} \text{ radian.}$$

Figure 16

we find that

$$s = r\theta = (3960)(0.00588) = 23.3 \text{ miles}$$

In either case, it would seem that the brochure overstated the distance somewhat.   ■

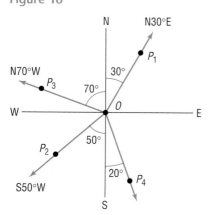

In navigation and surveying, the **direction** or **bearing** from a point O to a point P equals the acute angle θ between the ray OP and the vertical line through O, the north–south line.

Figure 16 illustrates some bearings. Notice that the bearing from O to P_1 is denoted by the symbolism N30°E, indicating that the bearing is 30° east of north. In writing the bearing from O to P, the direction north or south always appears first, followed by an acute angle, followed by east or west. In Figure 16, the bearing from O to P_2 is S50°W, and from O to P_3 it is N70°W.

EXAMPLE 11   Finding the Bearing of an Object

In Figure 16, what is the bearing from O to an object at P_4?

Solution   The acute angle between the ray OP_4 and the north–south line through O is given as 20°. The bearing from O to P_4 is S20°E.   ∎

EXAMPLE 12   Finding the Bearing of an Airplane

A Boeing 777 aircraft takes off from O'Hare Airport on runway 2 LEFT, which has a bearing of N20°E.* After flying for 1 mile, the pilot of the aircraft requests permission to turn 90° and head toward the northwest. The request is granted. After the plane goes 2 miles in this direction, what bearing should the control tower use to locate the aircraft?

Figure 17

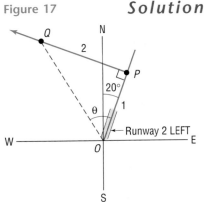

Solution   Figure 17 illustrates the situation. After flying 1 mile from the airport O (the control tower), the aircraft is at P. After turning 90° toward the northwest and flying 2 miles, the aircraft is at the point Q. In triangle OPQ, the angle θ obeys the equation

$$\tan \theta = \frac{2}{1} = 2 \quad \text{so} \quad \theta = \tan^{-1} 2 \approx 63.4°$$

The acute angle between north and the ray OQ is $63.4° - 20° = 43.4°$. The bearing of the aircraft from O to Q is N43.4°W.   ∎

NOW WORK PROBLEM 59.

*In air navigation, the term **azimuth** is employed to denote the positive angle measured clockwise from the north (N) to a ray OP. In Figure 16, the azimuth from O to P_1 is 30°; the azimuth from O to P_2 is 230°; the azimuth from O to P_3 is 290°. In naming runways, the units digit is left off the azimuth. Runway 2 LEFT means the left runway with a direction of azimuth 20° (bearing N20°E). Runway 23 is the runway with azimuth 230° and bearing S50°W.

4.1 EXERCISES

In Problems 1–10, find the exact value of the six trigonometric functions of the angle θ in each figure.

1.

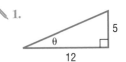

12, 5

2.

4, 3

3.

2, 3

4.

3, 3

5.

2, 4

6.

3, 4

7.

$\sqrt{2}$, 1

8.

$\sqrt{3}$, 2

9.

1, $\sqrt{5}$

10.

2, $\sqrt{5}$

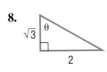

In Problems 11–20, find the exact value of each expression. Do not use a calculator.

11. $\sin 38° - \cos 52°$

12. $\tan 12° - \cot 78°$

13. $\dfrac{\cos 10°}{\sin 80°}$

14. $\dfrac{\cos 40°}{\sin 50°}$

15. $1 - \cos^2 20° - \cos^2 70°$

16. $1 + \tan^2 5° - \csc^2 85°$

17. $\tan 20° - \dfrac{\cos 70°}{\cos 20°}$

18. $\cot 40° - \dfrac{\sin 50°}{\sin 40°}$

19. $\cos 35° \sin 55° + \sin 35° \cos 55°$

20. $\sec 35° \csc 55° - \tan 35° \cot 55°$

In Problems 21–34, use the right triangle shown in the margin. Then, using the given information, solve the triangle.

21. $b = 5$, $\beta = 20°$; find $a, c,$ and α

22. $b = 4$, $\beta = 10°$; find $a, c,$ and α

23. $a = 6$, $\beta = 40°$; find $b, c,$ and α

24. $a = 7$, $\beta = 50°$; find $b, c,$ and α

25. $b = 4$, $\alpha = 10°$; find $a, c,$ and β

26. $b = 6$, $\alpha = 20°$; find $a, c,$ and β

27. $a = 5$, $\alpha = 25°$; find $b, c,$ and β

28. $a = 6$, $\alpha = 40°$; find $b, c,$ and β

29. $c = 9$, $\beta = 20°$; find $b, a,$ and α

30. $c = 10$, $\alpha = 40°$; find $b, a,$ and β

31. $a = 5$, $b = 3$; find $c, \alpha,$ and β

32. $a = 2$, $b = 8$; find $c, \alpha,$ and β

33. $a = 2$, $c = 5$; find $b, \alpha,$ and β

34. $b = 4$, $c = 6$; find $a, \alpha,$ and β

---

35. Geometry A right triangle has a hypotenuse of length 8 inches. If one angle is 35°, find the length of each leg.

36. Geometry A right triangle has a hypotenuse of length 10 centimeters. If one angle is 40°, find the length of each leg.

37. Geometry A right triangle contains a 25° angle. If one leg is of length 5 inches, what is the length of the hypotenuse?
[**Hint:** Two answers are possible.]

38. Geometry A right triangle contains an angle of $\dfrac{\pi}{8}$ radian. If one leg is of length 3 meters, what is the length of the hypotenuse?
[**Hint:** Two answers are possible.]

39. Geometry The hypotenuse of a right triangle is 5 inches. If one leg is 2 inches, find the degree measure of each angle.

40. Geometry The hypotenuse of a right triangle is 3 feet. If one leg is 1 foot, find the degree measure of each angle.

41. Finding the Width of a Gorge Find the distance from A to C across the gorge illustrated in the figure.

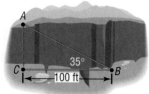

42. Finding the Distance across a Pond Find the distance from A to C across the pond illustrated in the figure.

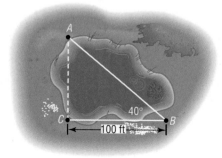

43. The Eiffel Tower The tallest tower built before the era of television masts, the Eiffel Tower was completed on March 31, 1889. Find the height of the Eiffel Tower (before a television mast was added to the top) using the information given in the illustration.

44. Finding the Distance of a Ship from Shore A ship, offshore from a vertical cliff known to be 100 feet in height, takes a sighting of the top of the cliff. If the angle of elevation is found to be 25°, how far offshore is the ship?

45. Finding the Distance to a Plateau Suppose that you are headed toward a plateau 50 meters high. If the angle of elevation to the top of the plateau is 20°, how far are you from the base of the plateau?

46. Statue of Liberty A ship is just offshore of New York City. A sighting is taken of the Statue of Liberty, which is about 305 feet tall. If the angle of elevation to the top of the statue is 20°, how far is the ship from the base of the statue?

47. Finding the Reach of a Ladder A 22-foot extension ladder leaning against a building makes a 70° angle with the ground. How far up the building does the ladder touch?

48. Finding the Height of a Building To measure the height of a building, two sightings are taken a distance of 50 feet apart. If the first angle of elevation is 40° and the second is 32°, what is the height of the building?

49. Directing a Laser Beam A laser beam is to be directed through a small hole in the center of a circle of radius 10 feet. The origin of the beam is 35 feet from the circle (see the figure). At what angle of elevation should the beam be aimed to ensure that it goes through the hole?

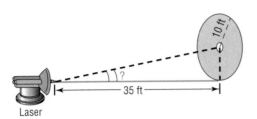

50. Finding the Angle of Elevation of the Sun At 10 AM on April 26, 2000, a building 300 feet high casts a shadow 50 feet long. What is the angle of elevation of the Sun?

51. Mt. Rushmore To measure the height of Lincoln's caricature on Mt. Rushmore, two sightings 800 feet from the base of the mountain are taken. If the angle of elevation to the bottom of Lincoln's face is 32° and the angle of elevation to the top is 35°, what is the height of Lincoln's face?

52. Finding the Distance between Two Objects A blimp, suspended in the air at a height of 500 feet, lies directly over a line from Soldier Field to the Adler Planetarium on Lake Michigan (see the figure). If the angle of depression from the blimp to the stadium is 32° and from the blimp to the planetarium is 23°, find the distance between Soldier Field and the Adler Planetarium.

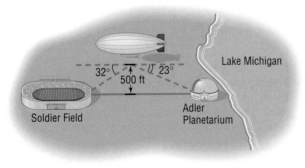

53. Finding the Length of a Guy Wire A radio transmission tower is 200 feet high. How long should a guy wire be if it is to be attached to the tower 10 feet from the top and is to make an angle of 21° with the ground?

54. Finding the Height of a Tower A guy wire 80 feet long is attached to the top of a radio transmission tower, making an angle of 25° with the ground. How high is the tower?

55. Washington Monument The angle of elevation of the Washington Monument is 35.1° at the instant it casts a shadow 789 feet long. Use this information to calculate the height of the monument.

56. Finding the Length of a Mountain Trail A straight trail with an inclination of 17° leads from a hotel at an elevation of 9000 feet to a mountain lake at an elevation of 11,200 feet. What is the length of the trail?

57. Finding the Speed of a Truck A state trooper is hidden 30 feet from a highway. One second after a truck passes, the angle θ between the highway and the line of observation from the patrol car to the truck is measured. See the illustration.

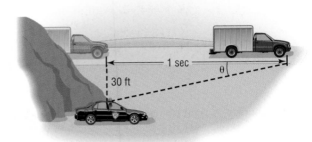

(a) If the angle measures 15°, how fast is the truck traveling? Express the answer in feet per second and in miles per hour.
(b) If the angle measures 20°, how fast is the truck traveling? Express the answer in feet per second and in miles per hour.
(c) If the speed limit is 55 miles per hour and a speeding ticket is issued for speeds of 5 miles per hour or more over the limit, for what angles should the trooper issue a ticket?

58. Security A security camera in a neighborhood bank is mounted on a wall 9 feet above the floor. What angle of depression should be used if the camera is to be directed to a spot 6 feet above the floor and 12 feet from the wall?

59. Finding the Bearing of an Aircraft A DC-9 aircraft leaves Midway Airport from runway 4 RIGHT, whose bearing is N40°E. After flying for 1/2 mile, the pilot requests permission to turn 90° and head toward the southeast. The permission is granted. After the airplane goes 1 mile in this direction, what bearing should the control tower use to locate the aircraft?

60. Finding the Bearing of a Ship A ship leaves the port of Miami with a bearing of S80°E and a speed of 15 knots. After 1 hour, the ship turns 90° toward the south. After 2 hours, maintaining the same speed, what is the bearing to the ship from port?

61. Shooting Free Throws in Basketball The eyes of a basketball player are 6 feet above the floor. The player is at the free-throw line, which is 15 feet from the center of the

basket rim (see the figure). What is the angle of elevation from the player's eyes to the center of the rim?
[**Hint:** The rim is 10 feet above the floor.]

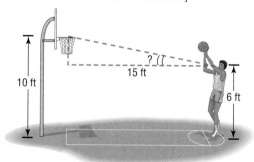

62. **Finding the Pitch of a Roof** A carpenter is preparing to put a roof on a garage that is 20 feet by 40 feet by 20 feet. A steel support beam 46 feet in length is positioned in the center of the garage. To support the roof, another beam will be attached to the top of the center beam (see the figure). At what angle of elevation is the new beam? In other words, what is the pitch of the roof?

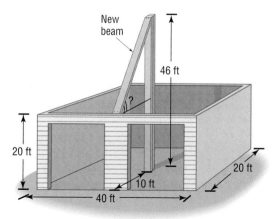

63. **Constructing a Highway** A highway whose primary directions are north–south is being constructed along the west coast of Florida. Near Naples, a bay obstructs the straight path of the road. Since the cost of a bridge is prohibitive, engineers decide to go around the bay. The illustration shows the path that they decide on and the measurements taken. What is the length of highway needed to go around the bay?

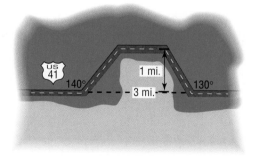

64. **Surveillance Satellites** A surveillance satellite circles Earth at a height of h miles above the surface. Suppose that d is the distance, in miles, on the surface of Earth that can be observed from the satellite. See the illustration.
   (a) Find an equation that relates the central angle θ to the height h.
   (b) Find an equation that relates the observable distance d and θ.
   (c) Find an equation that relates d and h.
   (d) If d is to be 2500 miles, how high must the satellite orbit above Earth?
   (e) If the satellite orbits at a height of 300 miles, what distance d on the surface can be observed?

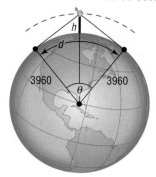

65. **Photography** A camera is mounted on a tripod 4 feet high at a distance of 10 feet from George, who is 6 feet tall. See the illustration. If the camera lens has angles of depression and elevation of 20°, will George's feet and head be seen by the lens? If not, how far back will the camera need to be moved to include George's feet and head?

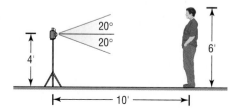

66. **Construction** A ramp for wheelchair accessibility is to be constructed with an angle of elevation of 15° and a final height of 5 feet. How long is the ramp?

67. **Geometry** A rectangle is inscribed in a semicircle of radius 1. See the illustration.

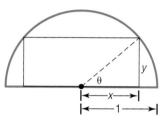

   (a) Express the area A of the rectangle as a function of the angle θ shown in the illustration.
   (b) Show that $A = \sin(2\theta)$.
   (c) Find the angle θ that results in the largest area A.
   (d) Find the dimensions of this largest rectangle.

68. Area of an Isosceles Triangle Show that the area A of an isosceles triangle, whose equal sides are of length s and the angle between them is θ, is

$$A = \tfrac{1}{2}s^2 \sin \theta$$

[**Hint:** See the illustration. The height h bisects the angle θ and is the perpendicular bisector of the base.]

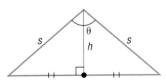

69. Calculating the Time of a Trip Two oceanfront homes are located 8 miles apart on a straight stretch of beach, each a distance of 1 mile from a paved road that parallels the ocean. Sally can jog 8 miles per hour along the paved road, but only 3 miles per hour in the sand on the beach. Because of a river directly between the two houses, it is necessary to jog in the sand to the road, continue on the road, and then jog back in the sand to get from one house to the other. See the illustration.
(a) Express the time T to get from one house to the other as a function of the angle θ shown in the illustration.
(b) Graph $T = T(\theta)$. What angle θ results in the least time? What is the least time? How long is Sally on the paved road?

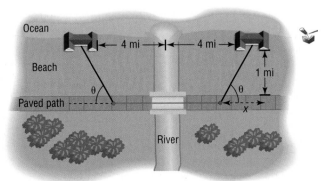

70. Designing Fine Decorative Pieces A designer of decorative art plans to market solid gold spheres enclosed in clear crystal cones. Each sphere is of fixed radius R and will be enclosed in a cone of height h and radius r. See the illustration. Many cones can be used to enclose the sphere, each having a different slant angle θ.
(a) Express the volume V of the cone as a function of the slant angle θ of the cone.
[**Hint:** The volume V of a cone of height h and radius r is $V = \tfrac{1}{3}\pi r^2 h$.]
(b) What slant angle θ should be used for the volume V of the cone to be a minimum? (This choice minimizes the amount of crystal required and gives maximum emphasis to the gold sphere).

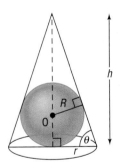

71. Calculating the Time of a Trip From a parking lot, you want to walk to a house on the ocean. The house is located 1500 feet down a paved path that parallels the ocean, which is 500 feet away. See the illustration. Along the path you can walk 300 feet per minute, but in the sand on the beach you can walk only 100 feet per minute.
(a) Calculate the time T if you walk 1500 feet along the paved path and then 500 feet in the sand to the house.
(b) Calculate the time T if you walk in the sand first for 500 feet and then walk along the sand for 1500 feet to the house.
(c) Express the time T to get from the parking lot to the beachhouse as a function of the angle θ shown in the illustration.
(d) Calculate the time T if you walk 1000 feet along the paved path and then walk directly to the house.
(e) Graph $T = T(\theta)$. For what angle θ is T least? What is the least time? What is x for this angle?

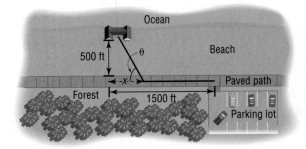

72. Carrying a Ladder around a Corner A ladder of length L is carried horizontally around a corner from a hall 3 feet wide into a hall 4 feet wide. See the illustration. Find the length L of the ladder as a function of the angle θ shown in the illustration.

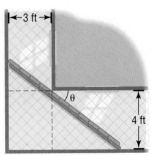

73. Geometry   Show that the area of an isosceles triangle is $A = a^2 \sin\theta \cos\theta$, where a is the length of one of the two equal sides and θ is the measure of one of the two equal angles (see the figure).

74. Let $n > 0$ be any real number, and let θ be any angle for which $0 < \theta < \pi/(1 + n)$. Then we can construct a triangle with the angles θ and $n\theta$ and included side of length 1 (do you see why?) and place it on the unit circle as illustrated. Now, drop the perpendicular from C to $D = (x, 0)$ and show that

$$x = \frac{\tan(n\theta)}{\tan\theta + \tan(n\theta)}$$

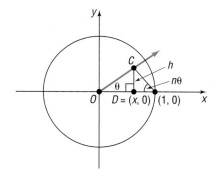

75. Refer to the given figure. The smaller circle, whose radius is a, is tangent to the larger circle, whose radius is b. The ray OA contains a diameter of each circle, and the ray OB is tangent to each circle. Show that

$$\cos\theta = \frac{\sqrt{ab}}{\dfrac{a + b}{2}}$$

(That is, $\cos\theta$ equals the ratio of the geometric mean of a and b to the arithmetic mean of a and b.)

$\left[\textbf{Hint: } \text{First show that } \sin\theta = \dfrac{b - a}{b + a}.\right]$

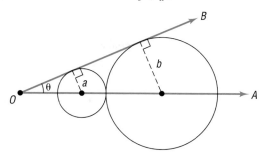

76. Suppose that the angle θ is a central angle of a circle of radius 1 (see the figure). Show that:

(a) Angle $OAC = \dfrac{\theta}{2}$

(b) $|CD| = \sin\theta$ and $|OD| = \cos\theta$

(c) $\tan\dfrac{\theta}{2} = \dfrac{\sin\theta}{1 + \cos\theta}$

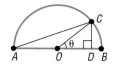

77. If $\cos\alpha = \tan\beta$ and $\cos\beta = \tan\alpha$, where α and β are acute angles, show that

$$\sin\alpha = \sin\beta = \sqrt{\frac{3 - \sqrt{5}}{2}}$$

78. The Gibb's Hill Lighthouse, Southampton, Bermuda   In operation since 1846, the Gibb's Hill Lighthouse stands 117 feet high on a hill 245 feet high, so its beam of light is 362 feet above sea level. A brochure states that ships 40 miles away can see the light and planes flying at 10,000 feet can see it 120 miles away. Verify the accuracy of these statements. What assumption did the brochure make about the height of the ship?

PREPARING FOR THIS SECTION

Before getting started, review the following:

✓ Trigonometric Equations (I) (Section 3.7, pp. 204–207)

4.2   THE LAW OF SINES

OBJECTIVES   **1** Solve SAA or ASA Triangles
   2 Solve SSA Triangles
   3 Solve Applied Problems

If none of the angles of a triangle is a right angle, the triangle is called **oblique.** An oblique triangle will have either three acute angles or two acute angles and one obtuse angle (an angle between 90° and 180°). See Figure 18.

Figure 18

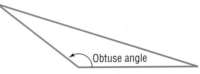

(a) All angles are acute **(b)** Two acute angles and one obtuse angle

In the discussion that follows, we will always label an oblique triangle so that side a is opposite angle α, side b is opposite angle β, and side c is opposite angle γ, as shown in Figure 19.

Figure 19

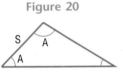

To **solve an oblique triangle** means to find the lengths of its sides and the measurements of its angles. To do this, we shall need to know the length of one side along with two other facts: either two angles, or the other two sides, or one angle and one other side. Knowing three angles of a triangle determines a family of *similar triangles*, that is, triangles that have the same shape but different sizes. There are four possibilities to consider:

CASE 1: One side and two angles are known (ASA or SAA).
CASE 2: Two sides and the angle opposite one of them are known (SSA).
CASE 3: Two sides and the included angle are known (SAS).
CASE 4: Three sides are known (SSS).

Figure 20 illustrates the four cases.

Figure 20

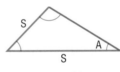
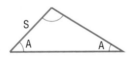
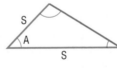
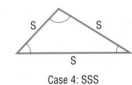

Case 1: ASA   Case 1: SAA   Case 2: SSA   Case 3: SAS   Case 4: SSS

The **Law of Sines** is used to solve triangles for which Case 1 or 2 holds. Cases 3 and 4 are considered when we study the Law of Cosines in Section 4.3.

Theorem   **Law of Sines**

For a triangle with sides a, b, c and opposite angles α, β, γ, respectively,

$$\frac{\sin \alpha}{a} = \frac{\sin \beta}{b} = \frac{\sin \gamma}{c} \qquad (1)$$

Proof   To prove the Law of Sines, we construct an altitude of length h from one of the vertices of such a triangle. Figure 21(a) shows h for a triangle with three acute angles, and Figure 21(b) shows h for a triangle with an obtuse angle. In each case, the altitude is drawn from the vertex at β. Using either illustration, we have

$$\sin \gamma = \frac{h}{a}$$

Figure 21

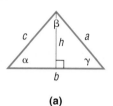

(a)

(b)

Figure 22

(a)

(b)

from which

$$h = a \sin \gamma \qquad \text{(2)}$$

From Figure 21(a), it also follows that

$$\sin \alpha = \frac{h}{c}$$

from which

$$h = c \sin \alpha \qquad \text{(3)}$$

From Figure 21(b), it follows that

$$\sin(180° - \alpha) = \sin \alpha = \frac{h}{c}$$

$\uparrow$
Difference formula

which again gives

$$h = c \sin \alpha$$

So, whether the triangle has three acute angles or has two acute angles and one obtuse angle, equations (2) and (3) hold. As a result, we may equate the expressions for h in equations (2) and (3) to get

$$a \sin \gamma = c \sin \alpha$$

from which

$$\frac{\sin \alpha}{a} = \frac{\sin \gamma}{c} \qquad \text{(4)}$$

In a similar manner, by constructing the altitude h' from the vertex of angle α as shown in Figure 22, we can show that

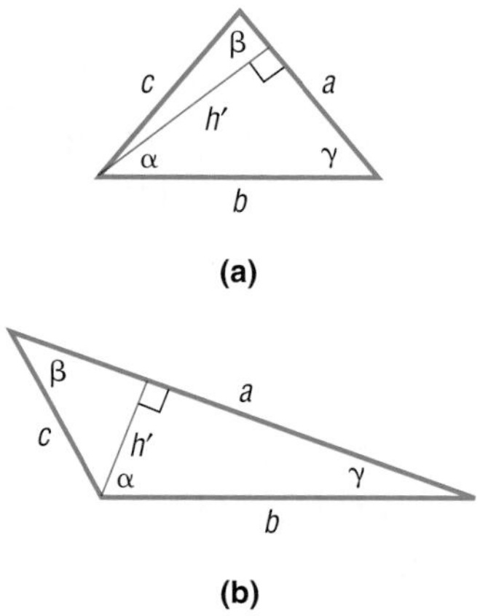

$$\sin \beta = \frac{h'}{c} \quad \text{and} \quad \sin \gamma = \frac{h'}{b}$$

Equating the expressions for h', we find that

$$h' = c \sin \beta = b \sin \gamma$$

from which

$$\frac{\sin \beta}{b} = \frac{\sin \gamma}{c} \qquad \text{(5)}$$

When equations (4) and (5) are combined, we have equation (1), the Law of Sines. ■

In applying the Law of Sines to solve triangles, we use the fact that the sum of the angles of any triangle equals 180°; that is,

$$\boxed{\alpha + \beta + \gamma = 180° \qquad \text{(6)}}$$

①   Our first two examples show how to solve a triangle when one side and two angles are known (Case 1: SAA or ASA).

| EXAMPLE 1 | Using the Law of Sines to Solve a SAA Triangle |

Solve the triangle: $\alpha = 40°, \beta = 60°, a = 4$

Solution　Figure 23 shows the triangle that we want to solve. The third angle γ is found using equation (6).

Figure 23

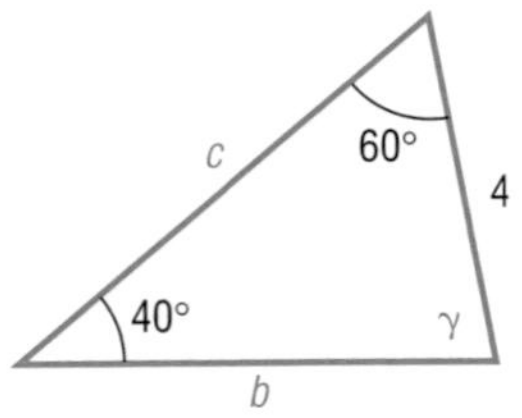

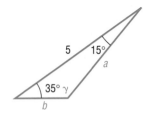

$$\alpha + \beta + \gamma = 180°$$
$$40° + 60° + \gamma = 180°$$
$$\gamma = 80°$$

Now we use the Law of Sines (twice) to find the unknown sides b and c.

$$\frac{\sin \alpha}{a} = \frac{\sin \beta}{b} \qquad \frac{\sin \alpha}{a} = \frac{\sin \gamma}{c}$$

Because $a = 4, \alpha = 40°, \beta = 60°,$ and $\gamma = 80°$, we have

$$\frac{\sin 40°}{4} = \frac{\sin 60°}{b} \qquad \frac{\sin 40°}{4} = \frac{\sin 80°}{c}$$

Solving for b and c, we find that

$$b = \frac{4 \sin 60°}{\sin 40°} \approx 5.39 \qquad c = \frac{4 \sin 80°}{\sin 40°} \approx 6.13 \qquad ■$$

Notice in Example 1 that we found b and c by working with the given side a. This is better than finding b first and working with a rounded value of b to find c.

NOW WORK PROBLEM **1**.

| EXAMPLE 2 | Using the Law of Sines to Solve an ASA Triangle |

Solve the triangle: $\alpha = 35°, \beta = 15°, c = 5$

Solution　Figure 24 illustrates the triangle that we want to solve. Because we know two angles ($\alpha = 35°$ and $\beta = 15°$), we find the third angle using equation (6).

Figure 24

$$\alpha + \beta + \gamma = 180°$$
$$35° + 15° + \gamma = 180°$$
$$\gamma = 130°$$

Now we know the three angles and one side ($c = 5$) of the triangle. To find the remaining two sides a and b, we use the Law of Sines (twice).

$$\frac{\sin \alpha}{a} = \frac{\sin \gamma}{c} \qquad\qquad \frac{\sin \beta}{b} = \frac{\sin \gamma}{c}$$

$$\frac{\sin 35°}{a} = \frac{\sin 130°}{5} \qquad \frac{\sin 15°}{b} = \frac{\sin 130°}{5}$$

$$a = \frac{5 \sin 35°}{\sin 130°} \approx 3.74 \qquad b = \frac{5 \sin 15°}{\sin 130°} \approx 1.69 \qquad ■$$

NOW WORK PROBLEM **15**.

THE AMBIGUOUS CASE

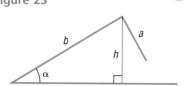

Figure 25

2 Case 2 (SSA), which applies to triangles for which two sides and the angle opposite one of them are known, is referred to as the **ambiguous case,** because the known information may result in one triangle, two triangles, or no triangle at all. Suppose that we are given sides a and b and angle α, as illustrated in Figure 25. The key to determining the possible triangles, if any, that may be formed from the given information lies primarily with the height h and the fact that $h = b \sin \alpha$.

No Triangle   If $a < h = b \sin \alpha$, then side a is not sufficiently long to form a triangle. See Figure 26.

One Right Triangle   If $a = h = b \sin \alpha$, then side a is just long enough to form a right triangle. See Figure 27.

Figure 26
$a < h = b \sin \alpha$

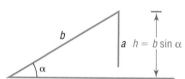

Figure 27
$a = b \sin \alpha$

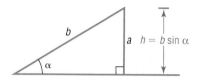

Two Triangles   If $a < b$ and $h = b \sin \alpha < a$, then two distinct triangles can be formed from the given information. See Figure 28.

One Triangle   If $a \geq b$, then only one triangle can be formed. See Figure 29.

Figure 28
$b \sin \alpha < a$ and $a < b$

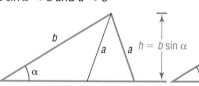

Figure 29
$a \geq b$

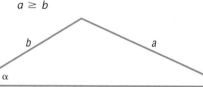

Fortunately, we do not have to rely on an illustration to draw the correct conclusion in the ambiguous case. The Law of Sines will lead us to the correct determination. Let's see how.

EXAMPLE 3

Using the Law of Sines to Solve a SSA Triangle (One Solution)

Solve the triangle: $a = 3, b = 2, \alpha = 40°$

Solution   See Figure 30(a). Because $a = 3$, $b = 2$, and $\alpha = 40°$ are known, we use the Law of Sines to find the angle β.

Figure 30(a)

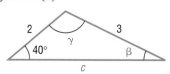

$$\frac{\sin \alpha}{a} = \frac{\sin \beta}{b}$$

Then

$$\frac{\sin 40°}{3} = \frac{\sin \beta}{2}$$

$$\sin \beta = \frac{2 \sin 40°}{3} \approx 0.43$$

There are two angles $\beta, 0° < \beta < 180°$, for which $\sin \beta \approx 0.43$.

$$\beta_1 \approx 25.4° \quad \text{and} \quad \beta_2 \approx 154.6°$$

NOTE: Here we computed β using the stored value of $\sin \beta$. If you use the rounded value, $\sin \beta \approx 0.43$, you will obtain slightly different results.

The second possibility, $\beta_2 \approx 154.6°$, is ruled out, because $\alpha = 40°$, making $\alpha + \beta_2 \approx 194.6° > 180°$. Now, using $\beta_1 \approx 25.4°$, we find that

$$\gamma = 180° - \alpha - \beta_1 \approx 180° - 40° - 25.4° = 114.6°$$

The third side c may now be determined using the Law of Sines.

$$\frac{\sin \alpha}{a} = \frac{\sin \gamma}{c}$$

$$\frac{\sin 40°}{3} = \frac{\sin 114.6°}{c}$$

$$c = \frac{3 \sin 114.6°}{\sin 40°} \approx 4.24$$

Figure 30(b)

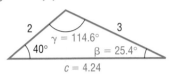

Figure 30(b) illustrates the solved triangle.   ∎

| **EXAMPLE 4** | **Using the Law of Sines to Solve a SSA Triangle (Two Solutions)** |

Solve the triangle: $a = 6, b = 8, \alpha = 35°$

Solution   See Figure 31(a). Because $a = 6, b = 8$, and $\alpha = 35°$ are known, we use the Law of Sines to find the angle β.

Figure 31(a)

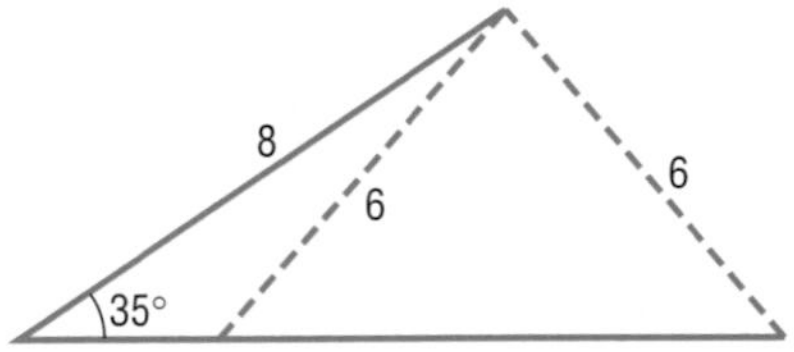

$$\frac{\sin \alpha}{a} = \frac{\sin \beta}{b}$$

Then

$$\frac{\sin 35°}{6} = \frac{\sin \beta}{8}$$

$$\sin \beta = \frac{8 \sin 35°}{6} \approx 0.76$$

$$\beta_1 \approx 49.9° \quad \text{or} \quad \beta_2 \approx 130.1°$$

For both choices of β, we have $\alpha + \beta < 180°$. There are two triangles, one containing the angle $\beta_1 \approx 49.9°$ and the other containing the angle $\beta_2 \approx 130.1°$. The third angle γ is either

$$\gamma_1 = 180° - \alpha - \beta_1 \approx 95.1° \quad \text{or} \quad \gamma_2 = 180° - \alpha - \beta_2 \approx 14.9°$$
$$\qquad\quad\uparrow \qquad\qquad\qquad\qquad\qquad\qquad\quad\uparrow$$
$$\qquad\quad\alpha = 35° \qquad\qquad\qquad\qquad\qquad\quad \alpha = 35°$$
$$\qquad\quad\beta_1 = 49.9° \qquad\qquad\qquad\qquad\qquad \beta_2 = 130.1°$$

The third side c obeys the Law of Sines, so we have

Figure 31(b)

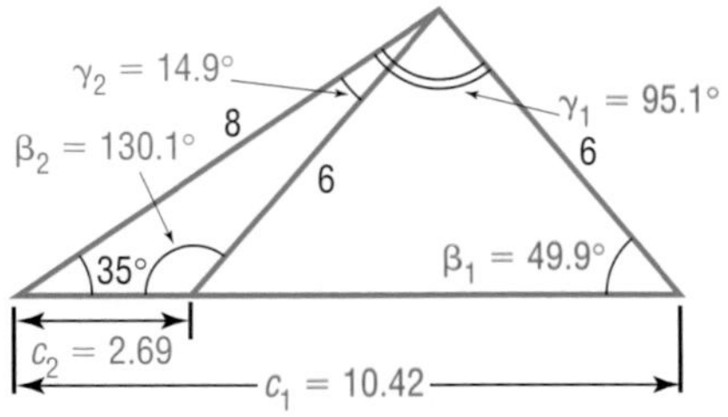

$$\frac{\sin \alpha}{a} = \frac{\sin \gamma_1}{c_1} \qquad\qquad \frac{\sin \alpha}{a} = \frac{\sin \gamma_2}{c_2}$$

$$\frac{\sin 35°}{6} = \frac{\sin 95.1°}{c_1} \qquad\qquad \frac{\sin 35°}{6} = \frac{\sin 14.9°}{c_2}$$

$$c_1 = \frac{6 \sin 95.1°}{\sin 35°} \approx 10.42 \qquad\qquad c_2 = \frac{6 \sin 14.9°}{\sin 35°} \approx 2.69$$

The two solved triangles are illustrated in Figure 31(b).   ∎

EXAMPLE 5	**Using the Law of Sines to Solve a SSA Triangle (No Solution)**

Solve the triangle: $a = 2, c = 1, \gamma = 50°$

Solution   Because $a = 2, c = 1$, and $\gamma = 50°$ are known, we use the Law of Sines to find the angle α.

$$\frac{\sin \alpha}{a} = \frac{\sin \gamma}{c}$$

$$\frac{\sin \alpha}{2} = \frac{\sin 50°}{1}$$

$$\sin \alpha = 2 \sin 50° \approx 1.53$$

Figure 32

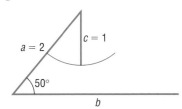

Since there is no angle α for which $\sin \alpha > 1$, there can be no triangle with the given measurements. Figure 32 illustrates the measurements given. Notice that, no matter how we attempt to position side c, it will never touch side b to form a triangle.   ∎

NOW WORK PROBLEMS **17** AND **23**.

APPLICATIONS

③   The Law of Sines is particularly useful for solving certain applied problems.

EXAMPLE 6	**Finding the Height of a Mountain**

To measure the height of a mountain, a surveyor takes two sightings of the peak at a distance 900 meters apart on a direct line to the mountain.* See Figure 33(a). The first observation results in an angle of elevation of 47°, whereas the second results in an angle of elevation of 35°. If the transit is 2 meters high, what is the height h of the mountain?

Figure 33

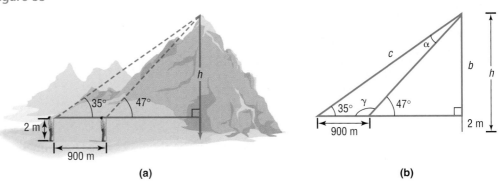

(a)   (b)

Solution   Figure 33(b) shows the triangles that replicate the illustration in Figure 33(a). Since $\gamma + 47° = 180°$, we find that $\gamma = 133°$. Also, since $\alpha + \gamma + 35° = 180°$, we find that $\alpha = 180° - 35° - \gamma = 145° - 133° = 12°$. We use the Law of Sines to find c.

$$\frac{\sin \alpha}{a} = \frac{\sin \gamma}{c}$$

$\alpha = 12°, \gamma = 133°, a = 900$

$$c = \frac{900 \sin 133°}{\sin 12°} = 3165.86$$

*For simplicity, we assume that these sightings are at the same level.

Using the larger right triangle, we have

$$\sin 35° = \frac{b}{c} \qquad c = 3165.86$$

$$b = 3165.86 \sin 35° = 1815.86 \approx 1816 \text{ meters}$$

The height of the peak from ground level is approximately $1816 + 2 = 1818$ meters. ■

 NOW WORK PROBLEM **31.**

| EXAMPLE 7 | **Rescue at Sea** |

Coast Guard Station Zulu is located 120 miles due west of Station X-ray. A ship at sea sends an SOS call that is received by each station. The call to Station Zulu indicates that the bearing of the ship from Zulu is N40°E (40° east of north). The call to Station X-ray indicates that the bearing of the ship from X-ray is N30°W (30° west of north).

(a) How far is each station from the ship?

(b) If a helicopter capable of flying 200 miles per hour is dispatched from the nearest station to the ship, how long will it take to reach the ship?

Solution (a) Figure 34 illustrates the situation. The angle γ is found to be

$$\gamma = 180° - 50° - 60° = 70°$$

Figure 34

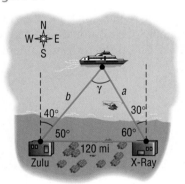

The Law of Sines can now be used to find the two distances a and b that we seek.

$$\frac{\sin 50°}{a} = \frac{\sin 70°}{120}$$

$$a = \frac{120 \sin 50°}{\sin 70°} \approx 97.82 \text{ miles}$$

$$\frac{\sin 60°}{b} = \frac{\sin 70°}{120}$$

$$b = \frac{120 \sin 60°}{\sin 70°} \approx 110.59 \text{ miles}$$

Station Zulu is about 111 miles from the ship, and Station X-ray is about 98 miles from the ship.

(b) The time t needed for the helicopter to reach the ship from Station X-ray is found by using the formula

$$(\text{Velocity, } v)(\text{Time, } t) = \text{Distance, } a$$

Then

$$t = \frac{a}{v} = \frac{97.82}{200} \approx 0.49 \text{ hour} \approx 29 \text{ minutes}$$

It will take about 29 minutes for the helicopter to reach the ship. ■

 NOW WORK PROBLEM **29.**

4.2 EXERCISES

In Problems 1–8, solve each triangle.

1.
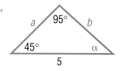

2.
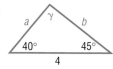

3.
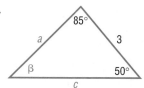

4.
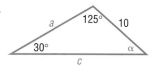

5.
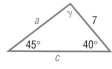

6.
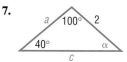

7.

8.

In Problems 9–16, solve each triangle.

9. $\alpha = 40°$, $\beta = 20°$, $a = 2$

10. $\alpha = 50°$, $\gamma = 20°$, $a = 3$

11. $\beta = 70°$, $\gamma = 10°$, $b = 5$

12. $\alpha = 70°$, $\beta = 60°$, $c = 4$

13. $\alpha = 110°$, $\gamma = 30°$, $c = 3$

14. $\beta = 10°$, $\gamma = 100°$, $b = 2$

15. $\alpha = 40°$, $\beta = 40°$, $c = 2$

16. $\beta = 20°$, $\gamma = 70°$, $a = 1$

In Problems 17–28, two sides and an angle are given. Determine whether the given information results in one triangle, two triangles, or no triangle at all. Solve any triangle(s) that results.

17. $a = 3$, $b = 2$, $\alpha = 50°$

18. $b = 4$, $c = 3$, $\beta = 40°$

19. $b = 5$, $c = 3$, $\beta = 100°$

20. $a = 2$, $c = 1$, $\alpha = 120°$

21. $a = 4$, $b = 5$, $\alpha = 60°$

22. $b = 2$, $c = 3$, $\beta = 40°$

23. $b = 4$, $c = 6$, $\beta = 20°$

24. $a = 3$, $b = 7$, $\alpha = 70°$

25. $a = 2$, $c = 1$, $\gamma = 100°$

26. $b = 4$, $c = 5$, $\beta = 95°$

27. $a = 2$, $c = 1$, $\gamma = 25°$

28. $b = 4$, $c = 5$, $\beta = 40°$

---

29. Rescue at Sea Coast Guard Station Able is located 150 miles due south of Station Baker. A ship at sea sends an SOS call that is received by each station. The call to Station Able indicates that the ship is located N55°E; the call to Station Baker indicates that the ship is located S60°E.
(a) How far is each station from the ship?
(b) If a helicopter capable of flying 200 miles per hour is dispatched from the nearest station to the ship, how long will it take to reach the ship?

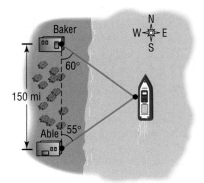

30. Surveying Consult the figure. To find the distance from the house at *A* to the house at *B*, a surveyor measures the angle *BAC* to be 40° and then walks off a distance of 100 feet to *C* and measures the angle *ACB* to be 50°. What is the distance from *A* to *B*?

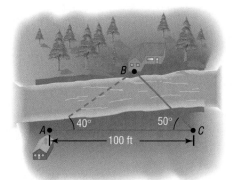

31. Finding the Length of a Ski Lift Consult the figure. To find the length of the span of a proposed ski lift from *A* to *B*, a surveyor measures the angle *DAB* to be 25° and then walks off a distance of 1000 feet to C and measures the angle ACB to be 15°. What is the distance from A to B?

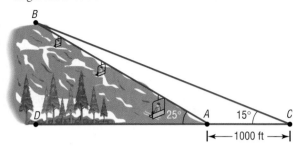

32. Finding the Height of a Mountain Use the illustration in Problem 31 to find the height BD of the mountain at B.

33. Finding the Height of an Airplane An aircraft is spotted by two observers who are 1000 feet apart. As the airplane passes over the line joining them, each observer takes a sighting of the angle of elevation to the plane, as indicated in the figure. How high is the airplane?

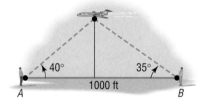

34. Finding the Height of the Bridge over the Royal Gorge The highest bridge in the world is the bridge over the Royal Gorge of the Arkansas River in Colorado.* Sightings to the same point at water level directly under the bridge are taken from each side of the 880-foot-long bridge, as indicated in the figure. How high is the bridge?

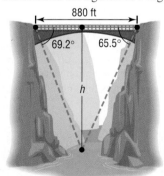

35. Navigation An airplane flies from city A to city B, a distance of 150 miles, and then turns through an angle of 40° and heads toward city C, as shown in the figure.
(a) If the distance between cities A and C is 300 miles, how far is it from city B to city C?

(b) Through what angle should the pilot turn at city C to return to city A?

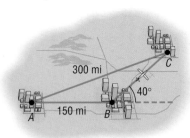

36. Time Lost due to a Navigation Error In attempting to fly from city A to city B, an aircraft followed a course that was 10° in error, as indicated in the figure. After flying a distance of 50 miles, the pilot corrected the course by turning at point C and flying 70 miles farther. If the constant speed of the aircraft was 250 miles per hour, how much time was lost due to the error?

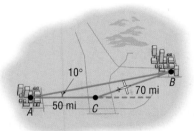

37. Finding the Lean of the Leaning Tower of Pisa The famous Leaning Tower of Pisa was originally 184.5 feet high.[†] At a distance of 123 feet from the base of the tower, the angle of elevation to the top of the tower is found to be 60°. Find the angle CAB indicated in the figure. Also, find the perpendicular distance from C to AB.

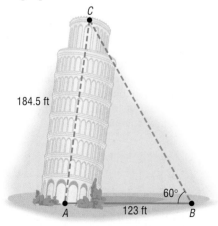

---

* *Source: Guinness Book of World Records.*

[†] In their 1986 report on the fragile seven-century-old bell tower, scientists in Pisa, Italy, said that the Leaning Tower of Pisa had increased its famous lean by 1 millimeter, or 0.04 inch. This is about the annual average, although the tilting had slowed to about half that much in the previous 2 years. (*Source:* United Press International, June 29, 1986.)

PISA, ITALY. September 1995. The Leaning Tower of Pisa has suddenly shifted, jeopardizing years of preservation work to stabilize it, Italian newspapers said Sunday. The tower, built on shifting subsoil between 1174 and 1350 as a belfry for the nearby cathedral, recently moved 0.07 inch in one night. The tower has been closed to tourists since 1990.

38. Crankshafts on Cars   On a certain automobile, the crankshaft is 3 inches long and the connecting rod is 9 inches long (see the figure). At the time when the angle OPA is 15°, how far is the piston (P) from the center (O) of the crankshaft?

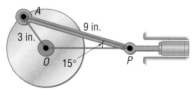

39. Constructing a Highway   U.S. 41, a highway whose primary directions are north–south, is being constructed along the west coast of Florida. Near Naples, a bay obstructs the straight path of the road. Since the cost of a bridge is prohibitive, engineers decide to go around the bay. The illustration shows the path that they decide on and the measurements taken. What is the length of highway needed to go around the bay?

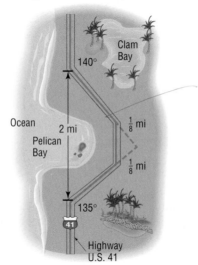

40. Determining Distances at Sea   The navigator of a ship at sea spots two lighthouses that she knows to be 3 miles apart along a straight seashore. She determines that the angles formed between two line-of-sight observations of the lighthouses and the line from the ship directly to shore are 15° and 35°. See the illustration.
(a) How far is the ship from lighthouse A?
(b) How far is the ship from lighthouse B?
(c) How far is the ship from shore?

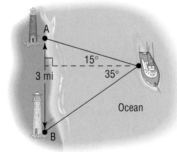

41. Calculating Distance at Sea   The navigator of a ship at sea has the harbor in sight at which the ship is to dock. She spots a lighthouse that she knows is 1 mile up the coast from the mouth of the harbor, and she measures the angle between the line-of-sight observations of the harbor and lighthouse to be 20°. With the ship heading directly toward the harbor, she repeats this measurement after 5 minutes of traveling at 12 miles per hour. If the new angle is 30°, how far is the ship from the harbor?

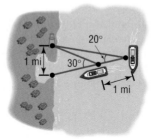

42. Finding Distances   A forest ranger is walking on a path inclined at 5° to the horizontal directly toward a 100-foot-tall fire observation tower. The angle of elevation from the path to the top of the tower is 40°. How far is the ranger from the tower at this time?

43. Great Pyramid of Cheops   One of the original Seven Wonders of the World, the Great Pyramid of Cheops was built about 2580 BC. Its original height was 480 feet 11 inches, but due to the loss of its topmost stones, it is now shorter.* Find the current height of the Great Pyramid, using the information given in the illustration.

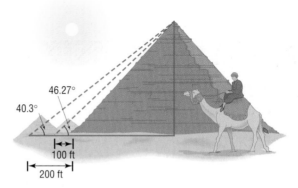

44. Determining the Height of an Aircraft   Two sensors are spaced 700 feet apart along the approach to a small airport. When an aircraft is nearing the airport, the angle of elevation from the first sensor to the aircraft is 20°, and from the second sensor to the aircraft it is 15°. Determine how high the aircraft is at this time.

*Source: Guinness Book of World Records.

45. Landscaping   Pat needs to determine the height of a tree before cutting it down to be sure that it will not fall on a nearby fence. The angle of elevation of the tree from one position on a flat path from the tree is 30°, and from a second position 40 feet farther along this path it is 20°. What is the height of the tree?

46. Construction   A loading ramp 10 feet long that makes an angle of 18° with the horizontal is to be replaced by one that makes an angle of 12° with the horizontal. How long is the new ramp?

47. Finding the Height of a Helicopter   Two observers simultaneously measure the angle of elevation of a helicopter. One angle is measured as 25°, the other as 40° (see the figure). If the observers are 100 feet apart and the helicopter lies over the line joining them, how high is the helicopter?

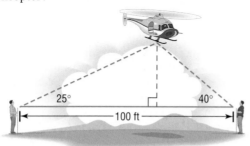

48. Mollweide's Formula   For any triangle, Mollweide's Formula (named after Karl Mollweide, 1774–1825) states that

$$\frac{a + b}{c} = \frac{\cos\left[\frac{1}{2}(\alpha - \beta)\right]}{\sin\left(\frac{1}{2}\gamma\right)}$$

Derive it.
[**Hint:** Use the Law of Sines and then a sum-to-product formula. Notice that this formula involves all six parts of a triangle. As a result, it is sometimes used to check the solution of a triangle.]

49. Mollweide's Formula   Another form of Mollweide's Formula is

$$\frac{a - b}{c} = \frac{\sin\left[\frac{1}{2}(\alpha - \beta)\right]}{\cos\left(\frac{1}{2}\gamma\right)}$$

Derive it.

50. For any triangle, derive the formula

$$a = b\cos\gamma + c\cos\beta$$

[**Hint:** Use the fact that $\sin\alpha = \sin(180° - \beta - \gamma)$.]

51. Law of Tangents   For any triangle, derive the Law of Tangents.

$$\frac{a - b}{a + b} = \frac{\tan\left[\frac{1}{2}(\alpha - \beta)\right]}{\tan\left[\frac{1}{2}(\alpha + \beta)\right]}$$

[**Hint:** Use Mollweide's Formula.]

52. Circumscribing a Triangle   Show that

$$\frac{\sin\alpha}{a} = \frac{\sin\beta}{b} = \frac{\sin\gamma}{c} = \frac{1}{2r}$$

where r is the radius of the circle circumscribing the triangle ABC whose sides are a, b, and c, as shown in the figure.
[**Hint:** Draw the diameter AB'. Then β = angle ABC = angle $AB'C$, and angle $ACB' = 90°$.]

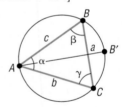

53. Make up three problems involving oblique triangles. One should result in one triangle, the second in two triangles, and the third in no triangle.

PREPARING FOR THIS SECTION

Before getting started, review the following:

✓ Trigonometric Equations (I) (Section 3.7, pp. 204–207)

4.3   THE LAW OF COSINES

OBJECTIVES   **1** Solve SAS Triangles
2 Solve SSS Triangles
3 Solve Applied Problems

In Section 4.2, we used the Law of Sines to solve Case 1 (SAA or ASA) and Case 2 (SSA) of an oblique triangle. In this section, we derive the Law of Cosines and use it to solve the remaining cases, 3 and 4.

CASE 3: Two sides and the included angle are known (SAS).

CASE 4: Three sides are known (SSS).

Theorem   Law of Cosines

For a triangle with sides a, b, c and opposite angles α, β, γ, respectively,

$$c^2 = a^2 + b^2 - 2ab\cos\gamma \qquad \textbf{(1)}$$

$$b^2 = a^2 + c^2 - 2ac\cos\beta \qquad \textbf{(2)}$$

$$a^2 = b^2 + c^2 - 2bc\cos\alpha \qquad \textbf{(3)}$$

Proof   We will prove only formula (1) here. Formulas (2) and (3) may be proved using the same argument.

We begin by strategically placing a triangle on a rectangular coordinate system so that the vertex of angle γ is at the origin and side b lies along the positive x-axis. Regardless of whether γ is acute, as in Figure 35(a), or obtuse, as in Figure 35(b), the vertex B has coordinates $(a\cos\gamma, a\sin\gamma)$. Vertex A has coordinates $(b, 0)$.

We can now use the distance formula to compute c^2.

$$\begin{aligned}
c^2 &= (b - a\cos\gamma)^2 + (0 - a\sin\gamma)^2 \\
&= b^2 - 2ab\cos\gamma + a^2\cos^2\gamma + a^2\sin^2\gamma \\
&= b^2 - 2ab\cos\gamma + a^2(\cos^2\gamma + \sin^2\gamma) \\
&= a^2 + b^2 - 2ab\cos\gamma
\end{aligned}$$

Each of formulas (1), (2), and (3) may be stated in words as follows:

Figure 35

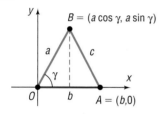

(a) Angle γ is acute

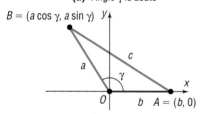

(b) Angle γ is obtuse

Theorem   Law of Cosines

The square of one side of a triangle equals the sum of the squares of the other two sides minus twice their product times the cosine of their included angle.

Observe that if the triangle is a right triangle (so that, say, $\gamma = 90°$) then formula (1) becomes the familiar Pythagorean Theorem: $c^2 = a^2 + b^2$. Thus, the Pythagorean Theorem is a special case of the Law of Cosines.

Let's see how to use the Law of Cosines to solve Case 3 (SAS), which applies to triangles for which two sides and the included angle are known.

EXAMPLE 1   **Using the Law of Cosines to Solve a SAS Triangle**

Solve the triangle: $a = 2, b = 3, \gamma = 60°$.

Solution See Figure 36. The Law of Cosines makes it easy to find the third side, c.

Figure 36

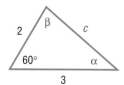

$$c^2 = a^2 + b^2 - 2ab\cos\gamma$$
$$= 4 + 9 - 2 \cdot 2 \cdot 3 \cdot \cos 60°$$
$$= 13 - \left(12 \cdot \frac{1}{2}\right) = 7$$
$$c = \sqrt{7}$$

Side c is of length $\sqrt{7}$. To find the angles α and β, we may use either the Law of Sines or the Law of Cosines. It is preferable to use the Law of Cosines, since it will lead to an equation with one solution. Using the Law of Sines would lead to an equation with two solutions that would need to be checked to determine which solution fits the given data. We choose to use formulas (2) and (3) of the Law of Cosines to find α and β.

For α:

$$a^2 = b^2 + c^2 - 2bc\cos\alpha$$
$$2bc\cos\alpha = b^2 + c^2 - a^2$$
$$\cos\alpha = \frac{b^2 + c^2 - a^2}{2bc} = \frac{9 + 7 - 4}{2 \cdot 3\sqrt{7}} = \frac{12}{6\sqrt{7}} = \frac{2\sqrt{7}}{7}$$
$$\alpha \approx 40.9°$$

For β:

$$b^2 = a^2 + c^2 - 2ac\cos\beta$$
$$\cos\beta = \frac{a^2 + c^2 - b^2}{2ac} = \frac{4 + 7 - 9}{4\sqrt{7}} = \frac{1}{2\sqrt{7}} = \frac{\sqrt{7}}{14}$$
$$\beta \approx 79.1°$$

Notice that $\alpha + \beta + \gamma = 40.9° + 79.1° + 60° = 180°$, as required. ■

NOW WORK PROBLEM **1**.

② The next example illustrates how the Law of Cosines is used when three sides of a triangle are known, Case 4 (SSS).

EXAMPLE 2 **Using the Law of Cosines to Solve a SSS Triangle**

Solve the triangle: $a = 4, b = 3, c = 6$.

Solution See Figure 37. To find the angles α, β, and γ, we proceed as we did in the latter part of the solution to Example 1.

Figure 37

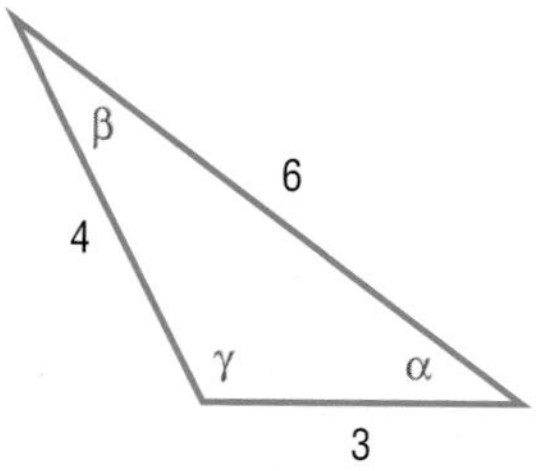

For α:

$$\cos\alpha = \frac{b^2 + c^2 - a^2}{2bc} = \frac{9 + 36 - 16}{2 \cdot 3 \cdot 6} = \frac{29}{36}$$
$$\alpha \approx 36.3°$$

For β:

$$\cos\beta = \frac{a^2 + c^2 - b^2}{2ac} = \frac{16 + 36 - 9}{2 \cdot 4 \cdot 6} = \frac{43}{48}$$
$$\beta \approx 26.4°$$

Since we know α and β,

$$\gamma = 180° - \alpha - \beta \approx 180° - 36.3° - 26.4° = 117.3°$$

NOW WORK PROBLEM 7.

EXAMPLE 3

3

Correcting a Navigational Error

A motorized sailboat leaves Naples, Florida, bound for Key West, 150 miles away. Maintaining a constant speed of 15 miles per hour, but encountering heavy crosswinds and strong currents, the crew finds, after 4 hours, that the sailboat is off course by 20°.

(a) How far is the sailboat from Key West at this time?

(b) Through what angle should the sailboat turn to correct its course?

(c) How much time has been added to the trip because of this? (Assume that the speed remains at 15 miles per hour.)

Figure 38

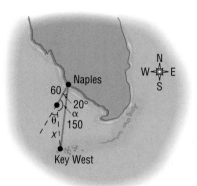

Solution   See Figure 38. With a speed of 15 miles per hour, the sailboat has gone 60 miles after 4 hours. We seek the distance x of the sailboat from Key West. We also seek the angle θ that the sailboat should turn through to correct its course.

(a) To find x, we use the Law of Cosines, since we know two sides and the included angle.

$$x^2 = 150^2 + 60^2 - 2(150)(60)\cos 20° \approx 9186$$

$$x \approx 95.8$$

The sailboat is about 96 miles from Key West.

(b) We now know three sides of the triangle, so we can use the Law of Cosines again to find the angle α opposite the side of length 150 miles.

$$150^2 = 96^2 + 60^2 - 2(96)(60)\cos\alpha$$

$$9684 = -11,520\cos\alpha$$

$$\cos\alpha \approx -0.8406$$

$$\alpha \approx 147.2°$$

The sailboat should turn through an angle of

$$\theta = 180° - \alpha \approx 180° - 147.2° = 32.8°$$

The sailboat should turn through an angle of about 33° to correct its course.

(c) The total length of the trip is now $60 + 96 = 156$ miles. The extra 6 miles will only require about 0.4 hour or 24 minutes more if the speed of 15 miles per hour is maintained.

NOW WORK PROBLEM 27.

HISTORICAL FEATURE

The Law of Sines was known vaguely long before it was explicitly stated by Nasîr Eddîn (about AD 1250). Ptolemy (about AD 150) was aware of it in a form using a chord function instead of the sine function. But it was first clearly stated in Europe by Regiomontanus, writing in 1464.

The Law of Cosines appears first in Euclid's *Elements* (Book II), but in a well-disguised form in which squares built on the sides of triangles are added and a rectangle representing the cosine term is subtracted. It was thus known to all mathematicians be-

cause of their familiarity with Euclid's work. An early modern form of the Law of Cosines, that for finding the angle when the sides are known, was stated by François Viète (in 1593).

The Law of Tangents (see Problem 51 of Exercise 4.2) has become obsolete. In the past it was used in place of the Law of Cosines, because the Law of Cosines was very inconvenient for calculation with logarithms or slide rules. Mixing of addition and multiplication is now very easy on a calculator, however, and the Law of Tangents has been shelved along with the slide rule.

4.3 EXERCISES

In Problems 1–8, solve each triangle.

1.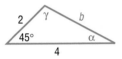

2.

3.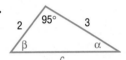

4.

5.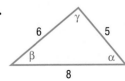

6.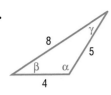

7.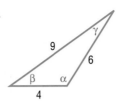

8.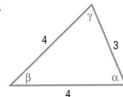

In Problems 9–24, solve each triangle.

9. $a = 3$, $b = 4$, $\gamma = 40°$

10. $a = 2$, $c = 1$, $\beta = 10°$

11. $b = 1$, $c = 3$, $\alpha = 80°$

12. $a = 6$, $b = 4$, $\gamma = 60°$

13. $a = 3$, $c = 2$, $\beta = 110°$

14. $b = 4$, $c = 1$, $\alpha = 120°$

15. $a = 2$, $b = 2$, $\gamma = 50°$

16. $a = 3$, $c = 2$, $\beta = 90°$

17. $a = 12$, $b = 13$, $c = 5$

18. $a = 4$, $b = 5$, $c = 3$

19. $a = 2$, $b = 2$, $c = 2$

20. $a = 3$, $b = 3$, $c = 2$

21. $a = 5$, $b = 8$, $c = 9$

22. $a = 4$, $b = 3$, $c = 6$

23. $a = 10$, $b = 8$, $c = 5$

24. $a = 9$, $b = 7$, $c = 10$

25. Surveying   Consult the figure. To find the distance from the house at A to the house at B, a surveyor measures the angle ACB, which is found to be $70°$, and then walks off the distance to each house, 50 feet and 70 feet, respectively. How far apart are the houses?

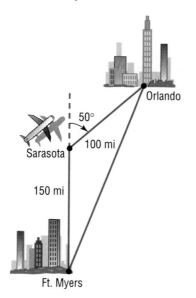

26. Navigation   An airplane flies from Ft. Myers to Sarasota, a distance of 150 miles, and then turns through an angle of $50°$ and flies to Orlando, a distance of 100 miles (see the figure).
(a) How far is it from Ft. Myers to Orlando?
(b) Through what angle should the pilot turn at Orlando to return to Ft. Myers?

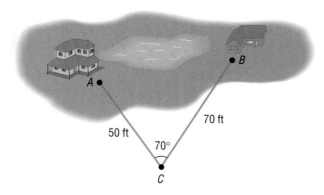

27. Revising a Flight Plan   In attempting to fly from Chicago to Louisville, a distance of 330 miles, a pilot inadvertently took a course that was $10°$ in error, as indicated in the figure.
(a) If the aircraft maintains an average speed of 220 miles per hour and if the error in direction is discovered after 15 minutes, through what angle should the pilot turn to head toward Louisville?
(b) What new average speed should the pilot maintain so that the total time of the trip is 90 minutes?

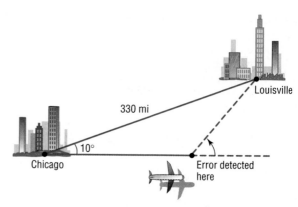

28. Avoiding a Tropical Storm   A cruise ship maintains an average speed of 15 knots in going from San Juan, Puerto Rico, to Barbados, West Indies, a distance of 600 nautical miles. To avoid a tropical storm, the captain heads out of San Juan in a direction of $20°$ off a direct heading to Barbados. The captain maintains the 15-knot speed for 10 hours, after which time the path to Barbados becomes clear of storms.
(a) Through what angle should the captain turn to head directly to Barbados?
(b) Once the turn is made, how long will it be before the ship reaches Barbados if the same 15-knot speed is maintained?

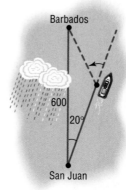

29. Major League Baseball Field   A Major League baseball diamond is actually a square 90 feet on a side. The pitching rubber is located 60.5 feet from home plate on a line joining home plate and second base.
(a) How far is it from the pitching rubber to first base?
(b) How far is it from the pitching rubber to second base?
(c) If a pitcher faces home plate, through what angle does he need to turn to face first base?

30. Little League Baseball Field   According to Little League baseball official regulations, the diamond is a square 60 feet on a side. The pitching rubber is located 46 feet from home plate on a line joining home plate and second base.
(a) How far is it from the pitching rubber to first base?
(b) How far is it from the pitching rubber to second base?
(c) If a pitcher faces home plate, through what angle does he need to turn to face first base?

31. Finding the Length of a Guy Wire The height of a radio tower is 500 feet, and the ground on one side of the tower slopes upward at an angle of 10° (see the figure).
(a) How long should a guy wire be if it is to connect to the top of the tower and be secured at a point on the sloped side 100 feet from the base of the tower?
(b) How long should a second guy wire be if it is to connect to the middle of the tower and be secured at a point 100 feet from the base on the flat side?

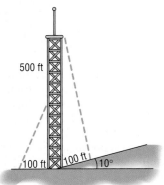

32. Finding the Length of a Guy Wire A radio tower 500 feet high is located on the side of a hill with an inclination to the horizontal of 5° (see the figure). How long should two guy wires be if they are to connect to the top of the tower and be secured at two points 100 feet directly above and directly below the base of the tower?

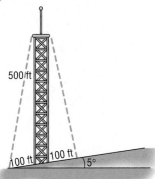

33. Wrigley Field, Home of the Chicago Cubs The distance from home plate to the fence in dead center in Wrigley Field is 400 feet (see the figure). How far is it from the fence in dead center to third base?

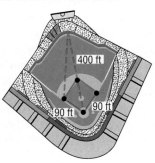

34. Little League Baseball The distance from home plate to the fence in dead center at the Oak Lawn Little League field is 280 feet. How far is it from the fence in dead center to third base?
[**Hint:** The distance between the bases in Little League is 60 feet.]

35. Rods and Pistons Rod OA (see the figure) rotates about the fixed point O so that point A travels on a circle of radius r. Connected to point A is another rod AB of length $L > 2r$, and point B is connected to a piston. Show that the distance x between point O and point B is given by

$$x = r\cos\theta + \sqrt{r^2\cos\theta + L^2 - r^2}$$

where θ is the angle of rotation of rod OA.

36. Geometry Show that the length d of a chord of a circle of radius r is given by the formula

$$d = 2r\sin\frac{\theta}{2}$$

where θ is the central angle formed by the radii to the ends of the chord (see the figure). Use this result to derive the fact that $\sin\theta < \theta$, where $\theta > 0$ is measured in radians.

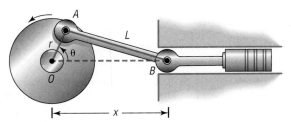

37. For any triangle, show that

$$\cos\frac{\gamma}{2} = \sqrt{\frac{s(s-c)}{ab}}$$

where $s = \frac{1}{2}(a+b+c)$.
[**Hint:** Use a Half-angle Formula and the Law of Cosines.]

38. For any triangle show that

$$\sin\frac{\gamma}{2} = \sqrt{\frac{(s-a)(s-b)}{ab}}$$

where $s = \frac{1}{2}(a+b+c)$.

39. Use the Law of Cosines to prove the identity

$$\frac{\cos\alpha}{a} + \frac{\cos\beta}{b} + \frac{\cos\gamma}{c} = \frac{a^2+b^2+c^2}{2abc}$$

 40. Write down your strategy for solving an oblique triangle.

PREPARING FOR THIS SECTION

Before getting started, review the following:

✓ Geometry Review (Appendix A, Section A.2, pp. 499–500)

4.4   THE AREA OF A TRIANGLE

OBJECTIVES   ① Find the Area of SAS Triangles
② Find the Area of SSS Triangles

In this section, we will derive several formulas for calculating the area A of a triangle. The most familiar of these is the following:

Theorem

The area A of a triangle is

$$A = \tfrac{1}{2}bh \qquad (1)$$

where b is the base and h is an altitude drawn to that base.

Figure 39
$A = \tfrac{1}{2}bh$

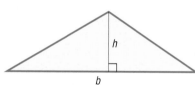

Proof   The derivation of this formula is rather easy once a rectangle of base b and height h is constructed around the triangle. See Figures 39 and 40.

Triangles 1 and 2 in Figure 40 are equal in area, as are triangles 3 and 4. Consequently, the area of the triangle with base b and altitude h is exactly half the area of the rectangle, which is bh.   ■

Figure 40

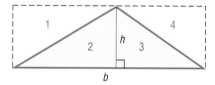

If the base b and altitude h to that base are known, then we can find the area of such a triangle using formula (1). Usually, though, the information required to use formula (1) is not given. Suppose, for example, that we know two sides a and b and the included angle γ (see Figure 41). Then the altitude h can be found by noting that

$$\frac{h}{a} = \sin\gamma$$

so that

$$h = a\sin\gamma$$

Figure 41
$h = a\sin\gamma$

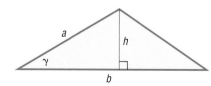

Using this fact in formula (1) produces

$$A = \tfrac{1}{2}bh = \tfrac{1}{2}b(a\sin\gamma) = \tfrac{1}{2}ab\sin\gamma$$

We now have the formula

$$A = \tfrac{1}{2}ab\sin\gamma \qquad (2)$$

By dropping altitudes from the other two vertices of the triangle, we obtain the following corresponding formulas:

$$A = \tfrac{1}{2}bc\sin\alpha \qquad (3)$$

$$A = \tfrac{1}{2}ac\sin\beta \qquad (4)$$

It is easiest to remember these formulas using the following wording:

Theorem   The area A of a triangle equals one-half the product of two of its sides times the sine of their included angle.

EXAMPLE 1   **Finding the Area of a SAS Triangle**

① Find the area A of the triangle for which $a = 8$, $b = 6$, and $\gamma = 30°$.

Figure 42

Solution   See Figure 42. We use formula (2) to get

$$A = \tfrac{1}{2}ab \sin \gamma = \tfrac{1}{2} \cdot 8 \cdot 6 \sin 30° = 12$$

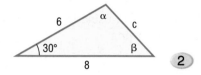

NOW WORK PROBLEM **1**.

② If the three sides of a triangle are known, another formula, called **Heron's Formula** (named after Heron of Alexandria), can be used to find the area of a triangle.

Theorem   **Heron's Formula**

The area A of a triangle with sides a, b, and c is

$$A = \sqrt{s(s - a)(s - b)(s - c)} \tag{5}$$

where $s = \tfrac{1}{2}(a + b + c)$.

Proof   The proof that we shall give uses the Law of Cosines and is quite different from the proof given by Heron.

From the Law of Cosines,

$$c^2 = a^2 + b^2 - 2ab \cos \gamma$$

and the two half-angle formulas,

$$\cos^2 \frac{\gamma}{2} = \frac{1 + \cos \gamma}{2} \qquad \sin^2 \frac{\gamma}{2} = \frac{1 - \cos \gamma}{2}$$

and, using $2s = a + b + c$, we find that

$$\cos^2 \frac{\gamma}{2} = \frac{1 + \cos \gamma}{2} = \frac{1 + \dfrac{a^2 + b^2 - c^2}{2ab}}{2}$$

$$= \frac{a^2 + 2ab + b^2 - c^2}{4ab} = \frac{(a + b)^2 - c^2}{4ab}$$

$$= \frac{(a + b - c)(a + b + c)}{4ab} = \frac{2(s - c) \cdot 2s}{4ab} = \frac{s(s - c)}{ab} \tag{6}$$

$$\underset{\substack{\uparrow \\ a + b - c = a + b + c - 2c \\ = 2s - 2c}}{}$$

Similarly,

$$\sin^2 \frac{\gamma}{2} = \frac{(s - a)(s - b)}{ab} \tag{7}$$

Now we use formula (2) for the area.

$$A = \frac{1}{2} ab \sin \gamma$$

$$= \frac{1}{2} ab \cdot 2 \sin \frac{\gamma}{2} \cos \frac{\gamma}{2} \qquad \sin \gamma = \sin\left[2\left(\frac{\gamma}{2}\right)\right] = 2 \sin \frac{\gamma}{2} \cos \frac{\gamma}{2}$$

$$= ab \sqrt{\frac{(s-a)(s-b)}{ab}} \sqrt{\frac{s(s-c)}{ab}} \qquad \text{Use equations (6) and (7).}$$

$$= \sqrt{s(s-a)(s-b)(s-c)} \qquad \blacksquare$$

EXAMPLE 2 **Finding the Area of a SSS Triangle**

Find the area of a triangle whose sides are 4, 5, and 7.

Solution We let $a = 4$, $b = 5$, and $c = 7$. Then

$$s = \tfrac{1}{2}(a+b+c) = \tfrac{1}{2}(4+5+7) = 8$$

Heron's Formula then gives the area A as

$$A = \sqrt{s(s-a)(s-b)(s-c)} = \sqrt{8 \cdot 4 \cdot 3 \cdot 1} = \sqrt{96} = 4\sqrt{6} \quad \blacksquare$$

NOW WORK PROBLEM 7.

HISTORICAL FEATURE

Heron's formula (also known as *Hero's Formula*) is due to Heron of Alexandria (first century AD), who had, besides his mathematical talents, a good deal of engineering skills. In various temples his mechanical devices produced effects that seemed supernatural, and visitors presumably were thus influenced to generosity. Heron's book *Metrica*, on making such devices, has

survived and was discovered in 1896 in the city of Constantinople.

Heron's Formulas for the area of a triangle caused some mild discomfort in Greek mathematics, because a product with two factors was an area, while one with three factors was a volume, but four factors seemed contradictory in Heron's time.

4.4 EXERCISES

In Problems 1–8, find the area of each triangle. Round answers to two decimal places.

1.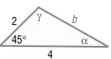
2.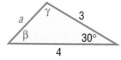
3.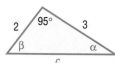
4.
5.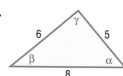
6.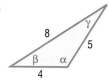
7.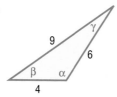
8.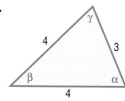

In Problems 9–24, find the area of each triangle. Round answers to two decimal places.

9. $a = 3$, $b = 4$, $\gamma = 40°$

10. $a = 2$, $c = 1$, $\beta = 10°$

11. $b = 1$, $c = 3$, $\alpha = 80°$

12. $a = 6$, $b = 4$, $\gamma = 60°$

13. $a = 3$, $c = 2$, $\beta = 110°$

14. $b = 4$, $c = 1$, $\alpha = 120°$

15. $a = 2$, $b = 2$, $\gamma = 50°$

16. $a = 3$, $c = 2$, $\beta = 90°$

17. $a = 12$, $b = 13$, $c = 5$

18. $a = 4$, $b = 5$, $c = 3$

19. $a = 2$, $b = 2$, $c = 2$

20. $a = 3$, $b = 3$, $c = 2$

21. $a = 5$, $b = 8$, $c = 9$

22. $a = 4$, $b = 3$, $c = 6$

23. $a = 10$, $b = 8$, $c = 5$

24. $a = 9$, $b = 7$, $c = 10$

---

25. **Area of a Segment** Find the area of the segment (shaded in blue in the figure) of a circle whose radius is 8 feet, formed by a central angle of 70°.
[**Hint:** Subtract the area of the triangle from the area of the sector to obtain the area of the segment.]

26. **Area of a Segment** Find the area of the segment of a circle whose radius is 5 inches, formed by a central angle of 40°.

27. **Cost of a Triangular Lot** The dimensions of a triangular lot are 100 feet by 50 feet by 75 feet. If the price of such land is $3 per square foot, how much does the lot cost?

28. **Amount of Materials to Make a Tent** A cone-shaped tent is made from a circular piece of canvas 24 feet in diameter by removing a sector with central angle 100° and connecting the ends. What is the surface area of the tent?

29. **Computing Areas** Find the area of the shaded region enclosed in a semicircle of diameter 8 centimeters. The length of the chord AB is 6 centimeters.
[**Hint:** Triangle ABC is a right triangle.]

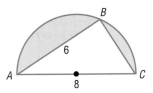

30. **Computing Areas** Find the area of the shaded region enclosed in a semicircle of diameter 10 inches. The length of the chord AB is 8 inches.
[**Hint:** Triangle ABC is a right triangle.]

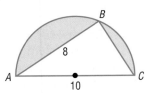

---

31. **Area of a Triangle** Prove that the area A of a triangle is given by the formula

$$A = \frac{a^2 \sin \beta \sin \gamma}{2 \sin \alpha}$$

32. **Area of a Triangle** Prove the two other forms of the formula given in Problem 31.

$$A = \frac{b^2 \sin \alpha \sin \gamma}{2 \sin \beta} \quad \text{and} \quad A = \frac{c^2 \sin \alpha \sin \beta}{2 \sin \gamma}$$

In Problems 33–38, use the results of Problem 31 or 32 to find the area of each triangle. Round answers to two decimal places.

33. $\alpha = 40°$, $\beta = 20°$, $a = 2$

34. $\alpha = 50°$, $\gamma = 20°$, $a = 3$

35. $\beta = 70°$, $\gamma = 10°$, $b = 5$

36. $\alpha = 70°$, $\beta = 60°$, $c = 4$

37. $\alpha = 110°$, $\gamma = 30°$, $c = 3$

38. $\beta = 10°$, $\gamma = 100°$, $b = 2$

39. **Geometry** Consult the figure, which shows a circle of radius r with center at O. Find the area A of the shaded region as a function of the central angle θ.

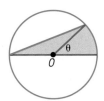

40. **Approximating the Area of a Lake** To approximate the area of a lake, a surveyor walks around the perimeter of the lake, taking the measurements shown in the illustra-

tion. Using this technique, what is the approximate area of the lake?
[**Hint:** Use the Law of Cosines on the three triangles shown and then find the sum of their areas.]

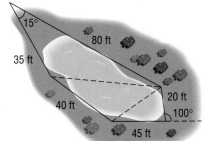

41. The Cow Problem* A cow is tethered to one corner of a square barn, 10 feet by 10 feet, with a rope 100 feet long. What is the maximum grazing area for the cow?
[**Hint:** See the illustration.]

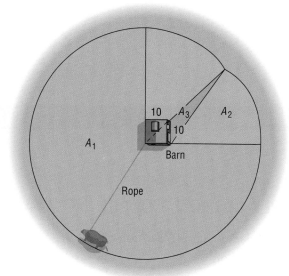

42. Another Cow Problem If the barn in Problem 41 is rectangular, 10 feet by 20 feet, what is the maximum grazing area for the cow?

43. If h_1, h_2, and h_3 are the altitudes dropped from A, B, and C, respectively, in a triangle (see the figure), show that

$$\frac{1}{h_1} + \frac{1}{h_2} + \frac{1}{h_3} = \frac{s}{K}$$

where K is the area of the triangle and $s = \frac{1}{2}(a + b + c)$.
[**Hint:** $h_1 = 2K/a$.]

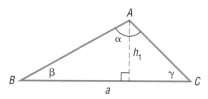

44. Show that a formula for the altitude h from a vertex to the opposite side a of a triangle is

$$h = \frac{a \sin \beta \sin \gamma}{\sin \alpha}$$

Inscribed Circle *For Problems 45–48, the lines that bisect each angle of a triangle meet in a single point O, and the perpendicular distance r from O to each side of the triangle is the*

same. The circle with center at O and radius r is called the **inscribed circle** *of the triangle (see the figure).*

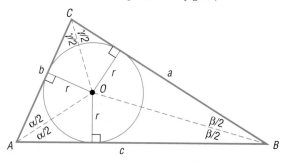

45. Apply Problem 44 to triangle OAB to show that

$$r = \frac{c \sin \dfrac{\alpha}{2} \sin \dfrac{\beta}{2}}{\cos \dfrac{\gamma}{2}}$$

46. Use the results of Problem 45 and Problem 38 in Section 4.3 to show that

$$\cot \frac{\gamma}{2} = \frac{s - c}{r}$$

47. Show that

$$\cot \frac{\alpha}{2} + \cot \frac{\beta}{2} + \cot \frac{\gamma}{2} = \frac{s}{r}$$

48. Show that the area K of triangle ABC is $K = rs$. Then show that

$$r = \sqrt{\frac{(s - a)(s - b)(s - c)}{s}}$$

where $s = \frac{1}{2}(a + b + c)$.

49. Refer to the figure. If $|OA| = 1$, show that:
(a) Area $\triangle OAC = \frac{1}{2}\sin \alpha \cos \alpha$
(b) Area $\triangle OCB = \frac{1}{2}|OB|^2 \sin \beta \cos \beta$
(c) Area $\triangle OAB = \frac{1}{2}|OB| \sin (\alpha + \beta)$
(d) $|OB| = \dfrac{\cos \alpha}{\cos \beta}$
(e) $\sin (\alpha + \beta) = \sin \alpha \cos \beta + \cos \alpha \sin \beta$
[**Hint:** Area $\triangle OAB$ = Area $\triangle OAC$ + Area $\triangle OCB$.]

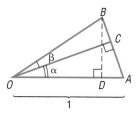

50. Refer to the figure, in which a unit circle is drawn. The line DB is tangent to the circle.
(a) Express the area of $\triangle OBC$ in terms of $\sin \theta$ and $\cos \theta$.

* Suggested by Professor Teddy Koukounas of SUNY at Old Westbury, who learned of it from an old farmer in Virginia. Solution provided by Professor Kathleen Miranda of SUNY at Old Westbury.

(b) Express the area of $\triangle OBD$ in terms of $\sin\theta$ and $\cos\theta$.

(c) The area of the sector $\overset{\frown}{OBC}$ of the circle is $\frac{1}{2}\theta$, where θ is measured in radians. Use the results of parts (a) and (b) and the fact that

$$\text{Area } \triangle OBC < \text{Area } \overset{\frown}{OBC} < \text{Area } \triangle OBD$$

to show that

$$1 < \frac{\theta}{\sin\theta} < \frac{1}{\cos\theta}$$

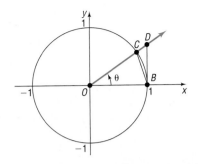

PREPARING FOR THIS SECTION

Before getting started, review the following:

✓ Sinusoidal Graphs (Section 2.4, pp. 127–133)

4.5 SIMPLE HARMONIC MOTION; DAMPED MOTION; COMBINING WAVES

OBJECTIVES
1. Find an Equation for an Object in Simple Harmonic Motion
2. Analyze Simple Harmonic Motion
3. Analyze an Object in Damped Motion
4. Graph the Sum of Two Functions

SIMPLE HARMONIC MOTION

Many physical phenomena can be described as simple harmonic motion. Radio and television waves, light waves, sound waves, and water waves exhibit motion that is simple harmonic.

The swinging of a pendulum, the vibrations of a tuning fork, and the bobbing of a weight attached to a coiled spring are examples of vibrational motion. In this type of motion, an object swings back and forth over the same path. In each illustration in Figure 43, the point B is the **equilibrium (rest) position** of the vibrating object. The **amplitude** of vibration is the distance from the object's rest position to its point of greatest displacement (either point A or point C in Figure 43). The **period** of a vibrating object is the time required to complete one vibration, that is, the time it takes to go from, say, point A through B to C and back to A.

Figure 43

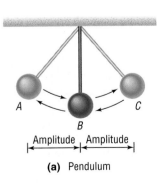

(a) Pendulum

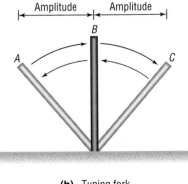

(b) Tuning fork

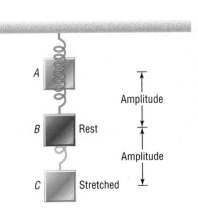

(c) Coiled spring

Simple harmonic motion is a special kind of vibrational motion in which the acceleration a of the object is directly proportional to the negative of its displacement d from its rest position. That is, $a = -kd, k > 0$.

For example, when the mass hanging from the spring in Figure 43(c) is pulled down from its rest position B to the point C, the force of the spring tries to restore the mass to its rest position. Assuming that there is no frictional force* to retard the motion, the amplitude will remain constant. The force increases in direct proportion to the distance that the mass is pulled from its rest position. Since the force increases directly, the acceleration of the mass of the object must do likewise, because (by Newton's Second Law of Motion) force is directly proportional to acceleration. Thus, the acceleration of the object varies directly with its displacement, and the motion is an example of simple harmonic motion.

Simple harmonic motion is related to circular motion. To see this relationship, consider a circle of radius a, with center at $(0, 0)$. See Figure 44. Suppose that an object initially placed at $(a, 0)$ moves counterclockwise around the circle at constant angular speed ω. Suppose further that after time t has elapsed the object is at the point $P = (x, y)$ on the circle. The angle θ, in radians, swept out by the ray $\overrightarrow{OP}$ in this time t is

$$\theta = \omega t$$

The coordinates of the point P at time t are

$$x = a \cos \theta = a \cos (\omega t)$$
$$y = a \sin \theta = a \sin (\omega t)$$

Corresponding to each position $P = (x, y)$ of the object moving about the circle, there is the point $Q = (x, 0)$, called the **projection of P on the x-axis.** As P moves around the circle at a constant rate, the point Q moves back and forth between the points $(a, 0)$ and $(-a, 0)$ along the x-axis with a motion that is simple harmonic. Similarly, for each point P there is a point $Q' = (0, y)$, called the **projection of P on the y-axis.** As P moves around the circle, the point Q' moves back and forth between the points $(0, a)$ and $(0, -a)$ on the y-axis with a motion that is simple harmonic. Simple harmonic motion can be described as the projection of constant circular motion on a coordinate axis.

To put it another way, again consider a mass hanging from a spring where the mass is pulled down from its rest position to the point C and then released. See Figure 45(a). The graph shown in Figure 45(b) describes the

Figure 44

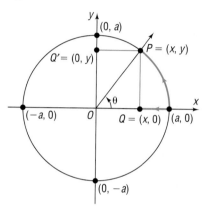

Figure 45

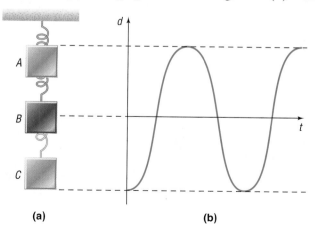

(a)   (b)

---
* If friction is present, the amplitude will decrease with time to 0. This type of motion is an example of **damped motion,** which is discussed later in this section.

displacement d of the object from its rest position as a function of time t, assuming that no frictional force is present.

Theorem | **Simple Harmonic Motion**

An object that moves on a coordinate axis so that its distance d from the origin at time t is given by either

$$d = a\cos(\omega t) \quad \text{or} \quad d = a\sin(\omega t)$$

where a and $\omega > 0$ are constants, moves with simple harmonic motion. The motion has amplitude $|a|$ and period $2\pi/\omega$.

The **frequency** f of an object in simple harmonic motion is the number of oscillations per unit time. Since the period is the time required for one oscillation, it follows that frequency is the reciprocal of the period; that is,

$$f = \frac{\omega}{2\pi}, \quad \omega > 0$$

EXAMPLE 1   Finding an Equation for an Object in Harmonic Motion

① Suppose that an object attached to a coiled spring is pulled down a distance of 5 inches from its rest position and then released. If the time for one oscillation is 3 seconds, write an equation that relates the displacement d of the object from its rest position after time t (in seconds). Assume no friction.

Solution   The motion of the object is simple harmonic. See Figure 46. When the object is released ($t = 0$), the displacement of the object from the rest position is -5 units (since the object was pulled down). Because $d = -5$ when $t = 0$, it is easier to use the cosine function*

$$d = a\cos(\omega t)$$

to describe the motion. Now the amplitude is $|-5| = 5$ and the period is 3, so that

$$a = -5 \quad \text{and} \quad \frac{2\pi}{\omega} = \text{period} = 3, \quad \omega = \frac{2\pi}{3}$$

An equation of the motion of the object is

$$d = -5\cos\left[\frac{2\pi}{3}t\right]$$

Figure 46

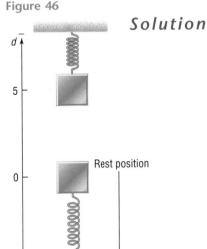

NOTE: In the solution to Example 1, we let $a = -5$, since the initial motion is down. If the initial direction were up, we would let $a = 5$.

✏️━━━► NOW WORK PROBLEM 1.

EXAMPLE 2   Analyzing the Motion of an Object

② Suppose that the displacement d (in meters) of an object at time t (in seconds) satisfies the equation

$$d = 10\sin(5t)$$

---

* No phase shift is required if a cosine function is used.

(a) Describe the motion of the object.
(b) What is the maximum displacement from its resting position?
(c) What is the time required for one oscillation?
(d) What is the frequency?

Solution   We observe that the given equation is of the form

$$d = a \sin(\omega t) \qquad d = 10 \sin(5t)$$

where $a = 10$ and $\omega = 5$.

(a) The motion is simple harmonic.
(b) The maximum displacement of the object from its resting position is the amplitude: $|a| = 10$ meters.
(c) The time required for one oscillation is the period:

$$\text{Period} = \frac{2\pi}{\omega} = \frac{2\pi}{5} \text{ seconds}$$

(d) The frequency is the reciprocal of the period. Thus,

$$\text{Frequency} = f = \frac{5}{2\pi} \text{ oscillations per second}$$

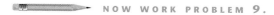 NOW WORK PROBLEM **9**.

DAMPED MOTION

③ Most physical phenomena are affected by friction or other resistive forces. These forces remove energy from a moving system and thereby damp its motion. For example, when a mass hanging from a spring is pulled down a distance a and released, the friction in the spring causes the distance that the mass moves from its at-rest position to decrease over time. Thus, the amplitude of any real oscillating spring or swinging pendulum decreases with time due to air resistance, friction, and so forth. See Figure 47.

Figure 47

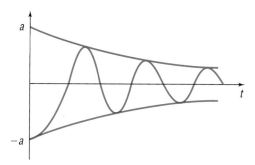

A function that describes this phenomenon maintains a sinusoidal component, but the amplitude of this component will decrease with time in order to account for the damping effect. In addition, the period of the oscillating component will be affected by the damping. The next result, from physics, describes damped motion.

Theorem

Damped Motion

The displacement d of an oscillating object from its at-rest position at time t is given by

$$d(t) = ae^{-bt/2m} \cos\left(\sqrt{\omega^2 - \frac{b^2}{4m^2}}\, t\right)$$

where b is a **damping factor** (most physics texts call this a **damping coefficient**) and m is the mass of the oscillating object.

Notice for $b = 0$ (zero damping) that we have the formula for simple harmonic motion with amplitude $|a|$ and period $\dfrac{2\pi}{\omega}$.

EXAMPLE 3   Analyzing a Damped Vibration Curve

Analyze the damped vibration curve

$$d(t) = e^{-t/\pi} \cos t, \quad t \geq 0$$

Solution   The displacement d is the product of $y = e^{-t/\pi}$ and $y = \cos t$. Using properties of absolute value and the fact that $|\cos t| \leq 1$, we find that

$$|d(t)| = \left|e^{-t/\pi} \cos t\right| = \left|e^{-t/\pi}\right||\cos t| \leq \left|e^{-t/\pi}\right| = e^{-t/\pi}$$
$$\uparrow$$
$$e^{-t/\pi} > 0$$

As a result,

$$-e^{-t/\pi} \leq d(t) \leq e^{-t/\pi}$$

This means that the graph of d will lie between the graphs of $y = e^{-t/\pi}$ and $y = -e^{-t/\pi}$, the **bounding curves** of d.

Also, the graph of d will touch these graphs when $|\cos t| = 1$, that is, when $t = 0, \pi, 2\pi$, and so on. The x-intercepts of the graph of d occur when $\cos t = 0$ at $t = \dfrac{\pi}{2}, \dfrac{3\pi}{2}$, and so on. See Table 1.

		TABLE 1				
t		0	$\dfrac{\pi}{2}$	π	$\dfrac{3\pi}{2}$	2π
$e^{-t/\pi}$		1	$e^{-1/2}$	e^{-1}	$e^{-3/2}$	e^{-2}
$\cos t$		1	0	-1	0	1
$d(t) = e^{-t/\pi}\cos t$		1	0	$-e^{-1}$	0	e^{-2}
Point on Graph of d		$(0, 1)$	$\left(\dfrac{\pi}{2}, 0\right)$	$(\pi, -e^{-1})$	$\left(\dfrac{3\pi}{2}, 0\right)$	$(2\pi, e^{-2})$

We graph $y = \cos t$, $y = e^{-t/\pi}$, $y = -e^{-t/\pi}$, and $d(t) = e^{-t/\pi} \cos t$ in Figure 48.

Figure 48

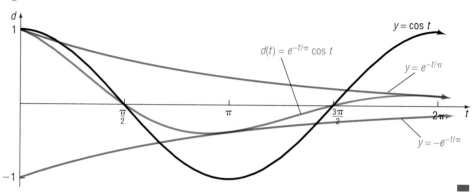

EXPLORATION Graph $Y_1 = e^{-x/\pi} \cos x$, along with $Y_2 = e^{-x/\pi}$, and $Y_3 = -e^{-x/\pi}$, for $0 \le x \le 2\pi$. Determine where Y_1 has its first turning point (local minimum). Compare this to where Y_1 intersects Y_3.

SOLUTION Figure 49 shows the graphs of $Y_1 = e^{-x/\pi} \cos x$, $Y_2 = e^{-x/\pi}$, and $Y_3 = -e^{-x/\pi}$. Using MINIMUM, the first turning point occurs at $x \approx 2.83$; Y_1 INTERSECTS Y_3 at $x = \pi \approx 3.14$.

Figure 49

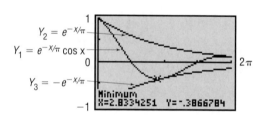

COMBINING WAVES

4 Many physical and biological applications require the graph of the sum of two functions, such as

$$f(x) = x + \sin x \quad \text{or} \quad g(x) = \sin x + \cos 2x$$

For example, if two tones are emitted, the sound produced is the sum of the waves produced by the two tones. See Problem 31 for an explanation of Touch-Tone phones.

To graph the sum of two (or more) functions, we can use the method of adding y-coordinates described next.

EXAMPLE 4

Graphing the Sum of Two Functions

Use the method of adding y-coordinates to graph $f(x) = x + \sin x$.

Solution First, we graph the component functions,

$$y = f_1(x) = x \qquad y = f_2(x) = \sin x$$

in the same coordinate system. See Figure 50(a). Now, select several values of x, say, $x = 0$, $x = \dfrac{\pi}{2}$, $x = \pi$, $x = \dfrac{3\pi}{2}$, and $x = 2\pi$, at which we compute

$f(x) = f_1(x) + f_2(x)$. Table 2 shows the computation. We plot these points and connect them to get the graph, as shown in Figure 50(b).

TABLE 2

x	0	$\dfrac{\pi}{2}$	π	$\dfrac{3\pi}{2}$	2π
$y = f_1(x) = x$	0	$\dfrac{\pi}{2}$	π	$\dfrac{3\pi}{2}$	2π
$y = f_2(x) = \sin x$	0	1	0	-1	0
$f(x) = x + \sin x$	0	$\dfrac{\pi}{2} + 1 \approx 2.57$	π	$\dfrac{3\pi}{2} - 1 \approx 3.71$	2π
Point on Graph of f	$(0, 0)$	$\left(\dfrac{\pi}{2}, 2.57\right)$	(π, π)	$\left(\dfrac{3\pi}{2}, 3.71\right)$	$(2\pi, 2\pi)$

Figure 50

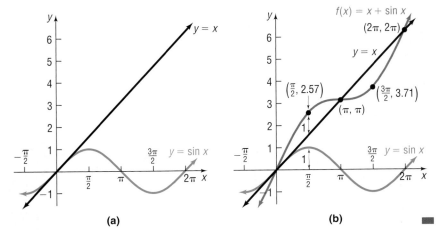

(a)　　　　　　　　(b)

In Example 4, note that the graph of $f(x) = x + \sin x$ intersects the line $y = x$ whenever $\sin x = 0$. Also, notice that the graph of f is not periodic.

 Check: Graph $y = x + \sin x$ and compare the result with Figure 50(b). Use INTERSECT to verify that the graphs intersect when $\sin x = 0$. ■

The next example shows a periodic graph of the sum of two functions.

EXAMPLE 5 ━━━━━━ **Graphing the Sum of two Sinusoidal Functions**

Use the method of adding y-coordinates to graph
$$f(x) = \sin x + \cos(2x)$$

Solution Table 3 shows the steps for computing several points on the graph of f. Figure 51 illustrates the graphs of the component functions, $y = f_1(x) = \sin x$ and $y = f_2(x) = \cos(2x)$, and the graph of $f(x) = \sin x + \cos(2x)$, which is shown in red.

TABLE 3

x	$-\dfrac{\pi}{2}$	0	$\dfrac{\pi}{2}$	π	$\dfrac{3\pi}{2}$	2π
$y = f_1(x) = \sin x$	-1	0	1	0	-1	0
$y = f_2(x) = \cos(2x)$	-1	1	-1	1	-1	1
$f(x) = \sin x + \cos(2x)$	-2	1	0	1	-2	1
Point on Graph of f	$\left(-\dfrac{\pi}{2}, -2\right)$	$(0, 1)$	$\left(\dfrac{\pi}{2}, 0\right)$	$(\pi, 1)$	$\left(\dfrac{3\pi}{2}, -2\right)$	$(2\pi, 1)$

Figure 51
$f(x) = \sin x + \cos(2x)$

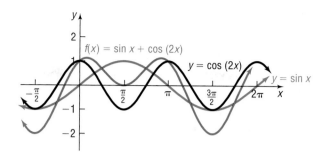

Check: Graph $y = \sin x + \cos(2x)$ and compare the result with Figure 51. ■

4.5 EXERCISES

In Problems 1–4, an object attached to a coiled spring is pulled down a distance a from its rest position and then released. Assuming that the motion is simple harmonic with period T, write an equation that relates the displacement d of the object from its rest position after t seconds. Also assume that the positive direction of the motion is up.

1. $a = 5$; $T = 2$ seconds

2. $a = 10$; $T = 3$ seconds

3. $a = 6$; $T = \pi$ seconds

4. $a = 4$; $T = \dfrac{\pi}{2}$ seconds

5. Rework Problem 1 under the same conditions except that, at time $t = 0$, the object is at its resting position and moving down.

6. Rework Problem 2 under the same conditions except that, at time $t = 0$, the object is at its resting position and moving down.

7. Rework Problem 3 under the same conditions except that, at time $t = 0$, the object is at its resting position and moving down.

8. Rework Problem 4 under the same conditions except that, at time $t = 0$, the object is at its resting position and moving down.

In Problems 9–16, the displacement d (in meters) of an object at time t (in seconds) is given.
   (a) *Describe the motion of the object.*
   (b) *What is the maximum displacement from its resting position?*
   (c) *What is the time required for one oscillation?*
   (d) *What is the frequency?*

9. $d = 5 \sin(3t)$

10. $d = 4 \sin(2t)$

11. $d = 6 \cos(\pi t)$

12. $d = 5 \cos\dfrac{\pi}{2} t$

13. $d = -3 \sin\left(\dfrac{1}{2}t\right)$

14. $d = -2 \cos(2t)$

15. $d = 6 + 2 \cos(2\pi t)$

16. $d = 4 + 3 \sin(\pi t)$

In Problems 17–20, graph each damped vibration curve for $0 \le t \le 2\pi$.

17. $d(t) = e^{-t/\pi} \cos(2t)$

18. $d(t) = e^{-t/2\pi} \cos(2t)$

19. $d(t) = e^{-t/2\pi} \cos t$

20. $d(t) = e^{-t/4\pi} \cos t$

In Problem 21–28, use the method of adding y-coordinates to graph each function.

21. $f(x) = x + \cos x$

22. $f(x) = x + \cos(2x)$

23. $f(x) = x - \sin x$

24. $f(x) = x - \cos x$

25. $f(x) = \sin x + \cos x$

26. $f(x) = \sin(2x) + \cos x$

27. $g(x) = \sin x + \sin(2x)$

28. $g(x) = \cos(2x) + \cos x$

29. **Charging a Capacitor**   If a charged capacitor is connected to a coil by closing a switch (see the figure), energy is transferred to the coil and then back to the capacitor in an oscillatory motion. The voltage V (in volts) across the capacitor will gradually diminish to 0 with time t (in seconds).

(a) Graph the equation relating V and t:
$$V(t) = e^{-t/3} \cos(\pi t), \qquad 0 \le t \le 3$$

(b) At what times t will the graph of V touch the graph of $y = e^{-t/3}$? When does V touch the graph of $y = -e^{-t/3}$?

(c) When will the voltage V be between -0.4 and 0.4 volt?

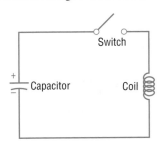

Switch

+ Capacitor   Coil

30. The Sawtooth Curve   An oscilloscope often displays a *sawtooth curve*. This curve can be approximated by sinusoidal curves of varying periods and amplitudes.

(a) Graph the following function, which can be used to approximate the sawtooth curve.

$$f(x) = \tfrac{1}{2}\sin(2\pi x) + \tfrac{1}{4}\sin(4\pi x), \qquad 0 \le x \le 2$$

(b) A better approximation to the sawtooth curve is given by

$$f(x) = \tfrac{1}{2}\sin(2\pi x) + \tfrac{1}{4}\sin(4\pi x) + \tfrac{1}{8}\sin(8\pi x)$$

Graph this function for $0 \le x \le 4$ and compare the result to the graph obtained in part (a).

(c) A third and even better approximation to the sawtooth curve is given by

$$f(x) = \tfrac{1}{2}\sin(2\pi x) + \tfrac{1}{4}\sin(4\pi x) + \tfrac{1}{8}\sin(8\pi x) + \tfrac{1}{16}\sin(16\pi x)$$

Graph this function for $0 \le x \le 4$ and compare the result to the graphs obtained in part (a) and (b).

(d) What do you think the next approximation to the sawtooth curve is?

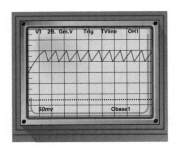

31. Touch-Tone Phones   On a Touch-Tone phone, each button produces a unique sound. The sound produced is the sum of two tones, given by

$$y = \sin(2\pi lt) \quad \text{and} \quad y = \sin(2\pi ht)$$

where l and h are the low and high frequencies (cycles per second) shown on the illustration. For example, if you touch 7, the low frequency is $l = 852$ cycles per second and the high frequency is $h = 1209$ cycles per second. The sound emitted by touching 7 is

$$y = \sin[2\pi(852)t] + \sin[2\pi(1209)t]$$

Graph the sound emitted by touching 7.

Touch-Tone phone

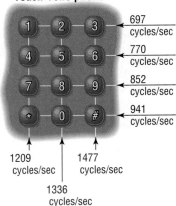

32. Graph the sound emitted by the * key on a Touch-Tone phone. See Problem 31.

33. Graph the function $f(x) = \dfrac{\sin x}{x}$, $x > 0$. Based on the graph, what do you conjecture about the value of $\dfrac{\sin x}{x}$ for x close to 0?

34. Graph $y = x\sin x$, $y = x^2\sin x$, and $y = x^3\sin x$ for $x > 0$. What patterns do you observe?

35. Graph $y = \dfrac{1}{x}\sin x$, $y = \dfrac{1}{x^2}\sin x$, and $y = \dfrac{1}{x^3}\sin x$ for $x > 0$. What patterns do you observe?

36. How would you explain to a friend what simple harmonic motion is? How would you explain damped motion?

CHAPTER REVIEW

Things To Know

Acute angle (p. 222)   An angle θ whose measure is $0° < \theta < 90°$ $\left(\text{or } 0 < \theta < \dfrac{\pi}{2}\right)$

Complementary angles (p. 224)   Two acute angles whose sum is $90°$ $\left(\dfrac{\pi}{2}\right)$

Cofunction (p. 224)   The following pairs of functions are cofunctions of each other: sine and cosine; tangent and cotangent; secant and cosecant

Formulas

Law of Sines (p. 236) $\dfrac{\sin\alpha}{a} = \dfrac{\sin\beta}{b} = \dfrac{\sin\gamma}{c}$

Law of Cosines (p. 247)

$$c^2 = a^2 + b^2 - 2ab \cos \gamma$$
$$b^2 = a^2 + c^2 - 2ac \cos \beta$$
$$a^2 = b^2 + c^2 - 2bc \cos \alpha$$

Area of a triangle (pp. 253–254)

$$A = \tfrac{1}{2} bh$$
$$A = \tfrac{1}{2} ab \sin \gamma$$
$$A = \tfrac{1}{2} bc \sin \alpha$$
$$A = \tfrac{1}{2} ac \sin \beta$$
$$A = \sqrt{s(s-a)(s-b)(s-c)}, \quad \text{where} \quad s = \tfrac{1}{2}(a+b+c)$$

Objectives

You should be able to:

Find the value of trigonometric function of acute angles (p. 222)

Use the complementary angle theorem (p. 224)

Solve right triangles (p. 225)

Solve applied problems using right triangle trigonometry (p. 226)

Solve SAA or ASA triangles (p. 237)

Solve SSA triangles (p. 239)

Solve applied problems using the Law of Sines (p. 241)

Solve SAS triangles (p. 247)

Solve SSS triangles (p. 248)

Solve applied problems using the Law of Cosines (p. 249)

Find the area of SAS triangles (p. 254)

Find the area of SSS triangles (p. 254)

Find an equation for an object in simple harmonic motion (p. 260)

Analyze simple harmonic motion (p. 260)

Analyze an object in damped motion (p. 261)

Graph the sum of two functions (p. 263)

Fill-in-the-Blank Items

1. Two acute angles whose sum is a right angle are called _____ .
2. The sine and _____ functions are confunctions.
3. If two sides and the angle opposite one of them are known, the Law of _____ is used to determine whether the known information results in no triangle, one triangle, or two triangles.
4. If three sides of a triangle are given, the Law of _____ is used to solve the triangle.
5. If three sides of a triangle are given, _____ Formula is used to find the area of the triangle.
6. The motion of an object obeys the equation $d = 4 \cos(6t)$. Such motion is described as _____ _____ .

True/False Items

T   F   **1.** $\tan 62° = \cot 38°$.
T   F   **2.** $\tan 182° = \cot 2°$.
T   F   **3.**   An oblique triangle in which two sides and an angle are given always results in at least one triangle.
T   F   **4.**   Given three sides of a triangle, there is a formula for finding its area.
T   F   **5.**   In a right triangle, if two sides are known, we can solve the triangle.
T   F   **6.**   The ambiguous case refers to the fact that, when two sides and the angle opposite one of them is known, sometimes the Law of Sines cannot be used.

Review Exercises

Blue problem numbers indicate the author's suggestions for use in a Practice Test.

In Problems 1–4, solve each triangle.

1.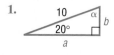

2.

3.

4.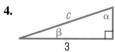

In Problems 5–24, find the remaining angle(s) and side(s) of each triangle, if it (they) exists. If no triangle exists, say "No triangle."

5. $\alpha = 50°$, $\beta = 30°$, $a = 1$

6. $\alpha = 10°$, $\gamma = 40°$, $c = 2$

7. $\alpha = 100°$, $a = 5$, $c = 2$

8. $a = 2$, $c = 5$, $\alpha = 60°$

9. $a = 3$, $c = 1$, $\gamma = 110°$

10. $a = 3$, $c = 1$, $\gamma = 20°$

11. $a = 3$, $c = 1$, $\beta = 100°$

12. $a = 3$, $b = 5$, $\beta = 80°$

13. $a = 2$, $b = 3$, $c = 1$

14. $a = 10$, $b = 7$, $c = 8$

15. $a = 1$, $b = 3$, $\gamma = 40°$

16. $a = 4$, $b = 1$, $\gamma = 100°$

17. $a = 5$, $b = 3$, $\alpha = 80°$

18. $a = 2$, $b = 3$, $\alpha = 20°$

19. $a = 1$, $b = \dfrac{1}{2}$, $c = \dfrac{4}{3}$

20. $a = 3$, $b = 2$, $c = 2$

21. $a = 3$, $\alpha = 10°$, $b = 4$

22. $a = 4$, $\alpha = 20°$, $\beta = 100°$

23. $c = 5$, $b = 4$, $\alpha = 70°$

24. $a = 1$, $b = 2$, $\gamma = 60°$

In Problems 25–34, find the area of each triangle.

25. $a = 2$, $b = 3$, $\gamma = 40°$

26. $b = 5$, $c = 5$, $\alpha = 20°$

27. $b = 4$, $c = 10$, $\alpha = 70°$

28. $a = 2$, $b = 1$, $\gamma = 100°$

29. $a = 4$, $b = 3$, $c = 5$

30. $a = 10$, $b = 7$, $c = 8$

31. $a = 4$, $b = 2$, $c = 5$

32. $a = 3$, $b = 2$, $c = 2$

33. $\alpha = 50°$, $\beta = 30°$, $a = 1$

34. $\alpha = 10°$, $\gamma = 40°$, $c = 3$

---

35. Measuring the Length of a Lake   From a stationary hot-air balloon 500 feet above the ground, two sightings of a lake are made (see the figure). How long is the lake?

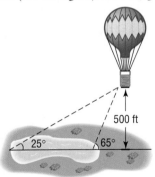

36. Finding the Speed of a Glider   From a glider 200 feet above the ground, two sightings of a stationary object directly in front are taken 1 minute apart (see the figure). What is the speed of the glider?

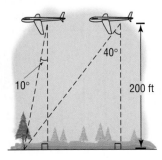

37. Finding the Width of a River   Find the distance from A to C across the river illustrated in the figure.

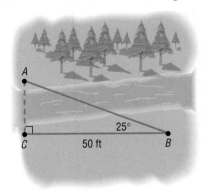

38. Finding the Height of a Building   Find the height of the building shown in the figure.

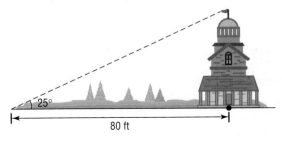

39. Finding the Distance to Shore   The Sears Tower in Chicago is 1454 feet tall and is situated about 1 mile inland from the shore of Lake Michigan, as indicated in the figure. An observer in a pleasure boat on the lake directly in front of the Sears Tower looks at the top of the tower and measures the angle of elevation as 5°. How far offshore is the boat?

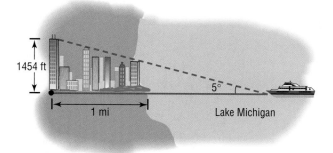

40. **Finding the Grade of a Mountain Trail**   A straight trail with a uniform inclination leads from a hotel, elevation 5000 feet, to a lake in a valley, elevation 4100 feet. The length of the trail is 4100 feet. What is the inclination (grade) of the trail?

41. **Navigation**   An airplane flies from city A to city B, a distance of 100 miles, and then turns through an angle of 20° and heads toward city C, as indicated in the figure. If the distance from A to C is 300 miles, how far is it from city B to city C?

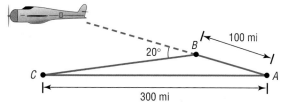

42. **Correcting a Navigation Error**   Two cities A and B are 300 miles apart. In flying from city A to city B, a pilot inadvertently took a course that was 5° in error.
    (a) If the error was discovered after flying 10 minutes at a constant speed of 420 miles per hour, through what angle should the pilot turn to correct the course? (Consult the figure.)
    (b) What new constant speed should be maintained so that no time is lost due to the error? (Assume that the speed would have been a constant 420 miles per hour if no error had occurred.)

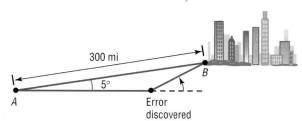

43. **Determining Distances at Sea**   Rebecca, the navigator of a ship at sea, spots two lighthouses that she knows to be 2 miles apart along a straight shoreline. She determines that the angles formed between two line-of-sight observations of the lighthouses and the line from the ship di-

rectly to shore are 12° and 30°. See the illustration.
    (a) How far is the ship from lighthouse A?
    (b) How far is the ship from lighthouse B?
    (c) How far is the ship from shore?

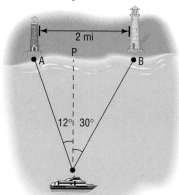

44. **Constructing a Highway**   A highway whose primary directions are north–south is being constructed along the west coast of Florida. Near Naples, a bay obstructs the straight path of the road. Since the cost of a bridge is prohibitive, engineers decide to go around the bay. The illustration shows the path that they decide on and the measurements taken. What is the length of highway needed to go around the bay?

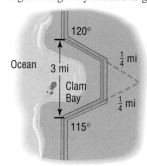

45. **Correcting a Navigational Error**   A sailboat leaves St. Thomas bound for an island in the British West Indies, 200 miles away. Maintaining a constant speed of 18 miles per hour, but encountering heavy crosswinds and strong currents, the crew finds after 4 hours that the sailboat is off course by 15°.
    (a) How far is the sailboat from the island at this time?
    (b) Through what angle should the sailboat turn to correct its course?
    (c) How much time has been added to the trip because of this? (Assume that the speed remains at 18 miles per hour.)

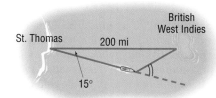

46. Surveying Two homes are located on opposite sides of a small hill. See the illustration. To measure the distance between them, a surveyor walks a distance of 50 feet from house A to point C, uses a transit to measure the angle ACB, which is found to be 80°, and then walks to house B, a distance of 60 feet. How far apart are the houses?

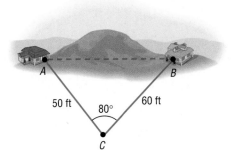

47. Approximating the Area of a Lake To approximate the area of a lake, Cindy walks around the perimeter of the lake, taking the measurements shown in the illustration. Using this technique, what is the approximate area of the lake?

[**Hint:** Use the Law of Cosines on the three triangles shown and then find the sum of their areas.]

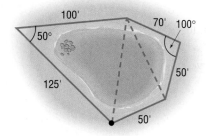

48. Calculating the Cost of Land The irregular parcel of land shown in the figure is being sold for $100 per square foot. What is the cost of this parcel?

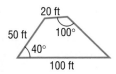

49. Area of a Segment Find the area of the segment of a circle whose radius is 6 inches formed by a central angle of 50°.

50. Finding the Bearing of a Ship The *Majesty* leaves the Port at Boston for Bermuda with a bearing of S80°E at an average speed of 10 knots. After 1 hour, the ship turns 90° toward the southwest. After 2 hours at an average speed of 20 knots, what is the bearing of the ship from Boston?

51. The drive wheel of an engine is 13 inches in diameter, and the pulley on the rotary pump is 5 inches in diameter. If the shafts of the drive wheel and the pulley are 2 feet apart, what length of belt is required to join them as shown in the figure?

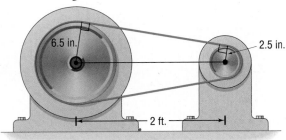

52. Rework Problem 51 if the belt is crossed, as shown in the figure.

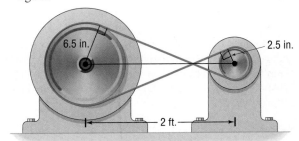

In Problems 53–56, the distance d (in feet) that an object travels in time t (in seconds) is given.

(a) *Describe the motion of the object.*
(b) *What is the maximum displacement from its resting position?*
(c) *What is the time required for one oscillation?*
(d) *What is the frequency?*

53. $d = 6 \sin(2t)$

54. $d = 2 \cos(4t)$

55. $d = -2 \cos(\pi t)$

56. $d = -3 \sin\left(\dfrac{\pi}{2} t\right)$

In Problems 57–62, graph each function.

57. $y = e^{-x/2\pi} \sin(2x), \quad 0 \le x \le 2\pi$

58. $y = e^{-x/3\pi} \cos(4x), \quad 0 \le x \le 2\pi$

59. $y = x \cos x, \quad 0 \le x \le 2\pi$

60. $y = x \sin(2x), \quad 0 \le x \le 2\pi$

61. $y = 2 \sin x + \cos(2x), \quad 0 \le x \le 2\pi$

62. $y = 2 \cos(2x) + \sin\dfrac{x}{2}, \quad 0 \le x \le 2\pi$

How can you build or analyze vibration profile

Automotive electronics have to be built to one of the most stringent mechanical specifications in the industry. One important element of the requirements is the vibration specification. We have to design our products for a harsh vibration environment and validate our designs by testing. The picture below depicts some of our engine controllers on a horizontal vibration slip table. To properly design our products, we have to understand specific vibration profiles, be able to analyze and modify them and precisely determine their interaction with our engine controller (identify modes of vibration, resonances, and transfer functions).

Fundamental tools for vibration analysis and characterization are the Fourier theorem, Fourier series, and Fourier transformations. The Fourier theorem states that *any physical function that varies periodically with time can be expressed by superposition of elementary sine and cosine components of various frequencies* (see Section 4.5 for simple harmonic motion). The basic expression for Fourier series is as follows:

$$f(t) = a_0 + \sum_{n=1}^{\infty} \left[a_n \cos\left(\frac{2\pi n}{T} t\right) + b_n \sin\left(\frac{2\pi n}{T} t\right) \right]$$

where $f(t)$ is the original source function; a_n and b_n are coefficients (precise determination of these coefficients is beyond the scope of this text, but by working in the synthetic direction it is possible to develop a feel for their influences); n is the index; T is the period of the original source function; and t is time. As an example of the power of superposition, a summation of two sine functions $f(x) = \sin x + 1.2 \sin(1.8x)$ is given in the picture at the top of the right column. We see that the resultant (red color) does not resemble precisely any of the components but creates a new quality.

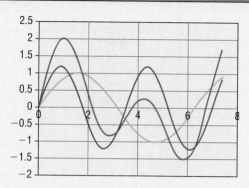

Problem: Build an approximation of a square function.

Given:

$$f_{\text{square}}(t) = \sin(\pi t) + \frac{\sin(3\pi t)}{3}$$
$$+ \frac{\sin(5\pi t)}{5} + \frac{\sin(7\pi t)}{7} + \cdots$$

1. Build each component of a series given as a separate function. For example, $f_1 = \sin(\pi t)$; $f_n = (1/n)\sin(n\pi t)$. Investigate the variability of each function up to $n = 9$.

2. Start adding them together starting from f_1, then $f_1 + f_3$, then $f_1 + f_3 + f_5$, and so on.

3. Observe and analyze what each consecutive component does to the resultant function.

4. Try to squeeze a cosine component, for example, between f_1 and f_3: $f_{13} = (1/2)\cos(2\pi t)$. What does it do to the resultant function?

5. An example of a sum of the first four functions is given in the picture below.

6. Go to Yahoo and search for "Fourier series" and have an adventure with the very interesting interactive animations.

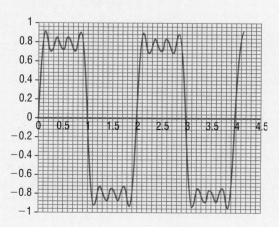

Polar Coordinates; Vectors

*This chapter is in two parts: Polar Coordinates, Sections 5.1–5.3, and Vectors, Sections 5.4–5.7. They are independent of each other and may be covered in any order.

Field Trip to Motorola

Cell phones and pagers receive signals from transmission towers that are often far away. Many of these signals are received directly by the cell phone or pager. Some of the signals received are the result of a reflection of the signal from flat surfaces, in most cases the ground.

Often the cell phone or pager is not stationary; in fact, the user of such devices is typically in motion while receiving calls, in a car or bus or in an elevator that is going up the side of a building. In the latter case, it is important to be able to calculate and analyze the interference caused by the ground on the signal received by the phone as the phone moves along a vertical path. The result of such an analysis is used to establish certain minimum sensitivity requirements for the phone.

PREPARING FOR THIS SECTION

Before getting started, review the following:

✓ Rectangular Coordinates (Section 1.1, pp. 2–5)

✓ Definitions of the Sine and Cosine Functions (Section 2.2, pp. 92–93)

✓ Inverse Tangent Function (Section 3.1, pp. 168–170)

✓ Completing the Square (Appendix A, Section A.3, p. 507)

5.1 POLAR COORDINATES

OBJECTIVES
1. Plot Points Using Polar Coordinates
2. Convert from Polar Coordinates to Rectangular Coordinates
3. Convert from Rectangular Coordinates to Polar Coordinates

So far, we have always used a system of rectangular coordinates to plot points in the plane. Now we are ready to describe another system called *polar coordinates*. As we shall soon see, in many instances polar coordinates offer certain advantages over rectangular coordinates.

In a rectangular coordinate system, you will recall, a point in the plane is represented by an ordered pair of numbers (x, y), where x and y equal the signed distance of the point from the y-axis and x-axis, respectively. In a polar coordinate system, we select a point, called the **pole,** and then a ray with vertex at the pole, called the **polar axis.** Comparing the rectangular and polar coordinate systems, we see (in Figure 1) that the origin in rectangular coordinates coincides with the pole in polar coordinates, and the positive x-axis in rectangular coordinates coincides with the polar axis in polar coordinates.

A point P in a polar coordinate system is represented by an ordered pair of numbers (r, θ). If $r > 0$, then r is the distance of the point from the pole; θ is an angle (in degrees or radians) formed by the polar axis and a ray from the pole through the point. We call the ordered pair (r, θ) the **polar coordinates** of the point. See Figure 2.

As an example, suppose that the polar coordinates of a point P are $\left(2, \dfrac{\pi}{4}\right)$. We locate P by first drawing an angle of $\dfrac{\pi}{4}$ radian, placing its vertex

Figure 1

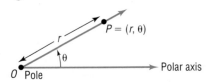

Figure 2

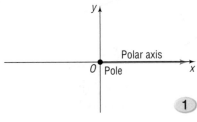

Figure 3

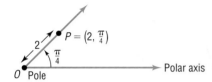

at the pole and its initial side along the polar axis. Then we go out a distance of 2 units along the terminal side of the angle to reach the point P. See Figure 3.

NOW WORK PROBLEM 9.

Recall that an angle measured counterclockwise is positive, whereas an angle measured clockwise is negative. This convention has some interesting consequences relating to polar coordinates. Let's see what these consequences are.

EXAMPLE 1 **Finding Several Polar Coordinates of a Single Point**

Consider again the point P with polar coordinates $\left(2, \dfrac{\pi}{4}\right)$, as shown in Figure 4(a). Because $\dfrac{\pi}{4}, \dfrac{9\pi}{4}$, and $-\dfrac{7\pi}{4}$ all have the same terminal side, we also could have located this point P by using the polar coordinates $\left(2, \dfrac{9\pi}{4}\right)$ or $\left(2, -\dfrac{7\pi}{4}\right)$, as shown in Figures 4(b) and (c).

Figure 4

Figure 5

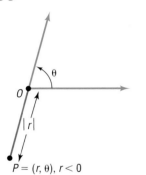

$P = (r, \theta), r < 0$

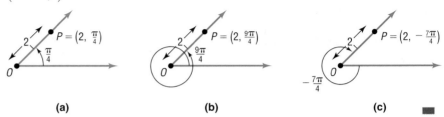

(a)   (b)   (c)

In using polar coordinates (r, θ), it is possible for the first entry r to be negative. When this happens, we follow the convention that the location of the point, instead of being on the terminal side of θ, is on the ray from the pole extending in the direction *opposite* the terminal side of θ at a distance $|r|$ from the pole. See Figure 5 for an illustration.

EXAMPLE 2 **Polar Coordinates $(r, \theta), r < 0$**

Consider again the point P with polar coordinates $\left(2, \dfrac{\pi}{4}\right)$, as shown in Figure 6(a). This same point P can be assigned the polar coordinates $\left(-2, \dfrac{5\pi}{4}\right)$, as indicated in Figure 6(b). To locate the point $\left(-2, \dfrac{5\pi}{4}\right)$, we use the ray in the opposite direction of $\dfrac{5\pi}{4}$ and go out 2 units along that ray to find the point P.

Figure 6

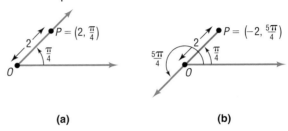

(a)   (b)

These examples show a major difference between rectangular coordinates and polar coordinates. In the former, each point has exactly one pair of rectangular coordinates; in the latter, a point can have infinitely many pairs of polar coordinates.

EXAMPLE 3   Plotting Points Using Polar Coordinates

Plot the points with the following polar coordinates:

(a) $\left(3, \dfrac{5\pi}{3}\right)$ (b) $\left(2, -\dfrac{\pi}{4}\right)$ (c) $(3, 0)$ (d) $\left(-2, \dfrac{\pi}{4}\right)$

Solution   Figure 7 shows the points.

Figure 7

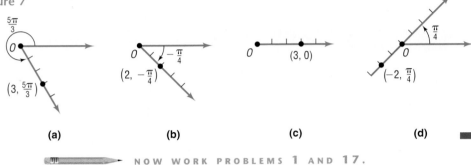

(a)   (b)   (c)   (d)

NOW WORK PROBLEMS **1** AND **17**.

EXAMPLE 4   Finding Other Polar Coordinates of a Given Point

Plot the point P with polar coordinates $\left(3, \dfrac{\pi}{6}\right)$, and find other polar coordinates (r, θ) of this same point for which:

(a) $r > 0, \quad 2\pi \le \theta < 4\pi$ (b) $r < 0, \quad 0 \le \theta < 2\pi$

(c) $r > 0, \quad -2\pi \le \theta < 0$

Figure 8

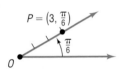

$P = \left(3, \frac{\pi}{6}\right)$

Solution   The point $\left(3, \dfrac{\pi}{6}\right)$ is plotted in Figure 8.

(a) We add 1 revolution (2π radians) to the angle $\dfrac{\pi}{6}$ to get $P = \left(3, \dfrac{\pi}{6} + 2\pi\right) = \left(3, \dfrac{13\pi}{6}\right)$. See Figure 9.

(b) We add $\dfrac{1}{2}$ revolution (π radians) to the angle $\dfrac{\pi}{6}$ and replace 3 by -3 to get $P = \left(-3, \dfrac{\pi}{6} + \pi\right) = \left(-3, \dfrac{7\pi}{6}\right)$. See Figure 10.

(c) We subtract 2π from the angle $\dfrac{\pi}{6}$ to get $P = \left(3, \dfrac{\pi}{6} - 2\pi\right) = \left(3, -\dfrac{11\pi}{6}\right)$. See Figure 11.

Figure 9   Figure 10 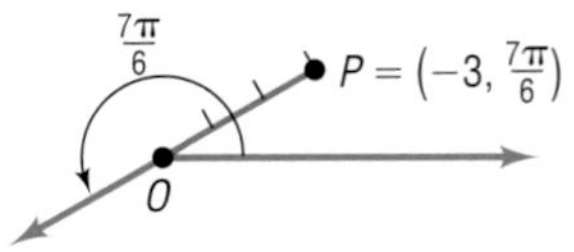   Figure 11

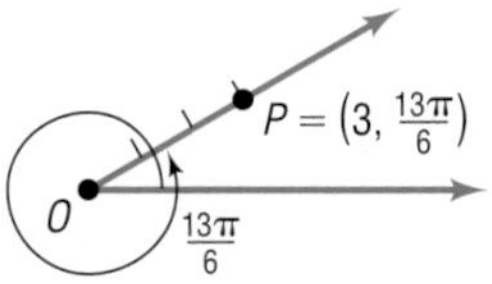

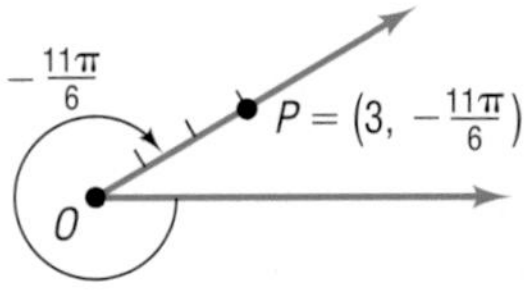

$P = \left(3, \frac{13\pi}{6}\right)$ $P = \left(-3, \frac{7\pi}{6}\right)$ $P = \left(3, -\frac{11\pi}{6}\right)$

NOW WORK PROBLEM **21**.

SUMMARY

A point with polar coordinates (r, θ) also can be represented by either of the following:

$$(r, \theta + 2k\pi) \quad \text{or} \quad (-r, \theta + \pi + 2k\pi), \qquad k \text{ any integer}$$

The polar coordinates of the pole are $(0, \theta)$, where θ can be any angle.

CONVERSION FROM POLAR COORDINATES TO RECTANGULAR COORDINATES, AND VICE VERSA

2 It is sometimes convenient and, indeed, necessary to be able to convert coordinates or equations in rectangular form to polar form, and vice versa. To do this, we recall that the origin in rectangular coordinates is the pole in polar coordinates and that the positive x-axis in rectangular coordinates is the polar axis in polar coordinates.

Theorem

Conversion from Polar Coordinates to Rectangular Coordinates

If P is a point with polar coordinates (r, θ), the rectangular coordinates (x, y) of P are given by

$$x = r \cos \theta \qquad y = r \sin \theta \qquad \text{(1)}$$

Figure 12

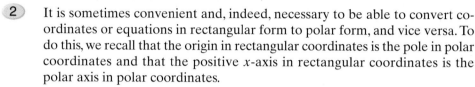

Proof   Suppose that P has the polar coordinates (r, θ). We seek the rectangular coordinates (x, y) of P. Refer to Figure 12.

If $r = 0$, then, regardless of θ, the point P is the pole, for which the rectangular coordinates are $(0, 0)$. Formula (1) is valid for $r = 0$.

If $r > 0$, the point P is on the terminal side of θ, and $r = d(O, P) = \sqrt{x^2 + y^2}$. Since

$$\cos \theta = \frac{x}{r} \qquad \sin \theta = \frac{y}{r}$$

we have

$$x = r \cos \theta \qquad y = r \sin \theta$$

If $r < 0$, then the point $P = (r, \theta)$ can be represented as $(-r, \pi + \theta)$, where $-r > 0$. Since

$$\cos(\pi + \theta) = -\cos \theta = \frac{x}{-r} \qquad \sin(\pi + \theta) = -\sin \theta = \frac{y}{-r}$$

we have

$$x = r \cos \theta \qquad y = r \sin \theta \qquad \blacksquare$$

EXAMPLE 5

Converting from Polar Coordinates to Rectangular Coordinates

Find the rectangular coordinates of the points with the following polar coordinates:

(a) $\left(6, \dfrac{\pi}{6} \right)$ (b) $\left(-4, -\dfrac{\pi}{4} \right)$

Solution We use formula (1): $x = r \cos \theta$ and $y = r \sin \theta$.

Figure 13

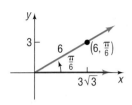

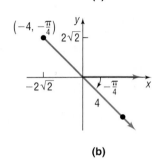

(a) Figure 13(a) shows $\left(6, \dfrac{\pi}{6} \right)$ plotted. With $r = 6$ and $\theta = \dfrac{\pi}{6}$, we have

$$x = r \cos \theta = 6 \cos \frac{\pi}{6} = 6 \cdot \frac{\sqrt{3}}{2} = 3\sqrt{3}$$

$$y = r \sin \theta = 6 \sin \frac{\pi}{6} = 6 \cdot \frac{1}{2} = 3$$

The rectangular coordinates of the point $\left(6, \dfrac{\pi}{6} \right)$ are $(3\sqrt{3}, 3)$.

(b) Figure 13(b) shows $\left(-4, -\dfrac{\pi}{4} \right)$ plotted. With $r = -4$ and $\theta = -\dfrac{\pi}{4}$, we have

$$x = r \cos \theta = -4 \cos \left(-\frac{\pi}{4} \right) = -4 \cdot \frac{\sqrt{2}}{2} = -2\sqrt{2}$$

$$y = r \sin \theta = -4 \sin \left(-\frac{\pi}{4} \right) = -4 \left(-\frac{\sqrt{2}}{2} \right) = 2\sqrt{2}$$

The rectangular coordinates of the point $\left(-4, -\dfrac{\pi}{4} \right)$ are $(-2\sqrt{2}, 2\sqrt{2})$. ∎

NOTE: Most calculators have the capability of converting from polar coordinates to rectangular coordinates. Consult your owner's manual for the proper key strokes. Since in most cases this procedure is tedious, you will find that using formula (1) is faster.

NOW WORK PROBLEMS 29 AND 41.

③   Converting from rectangular coordinates (x, y) to polar coordinates (r, θ) is a little more complicated. Notice that we begin each example by plotting the given rectangular coordinates.

EXAMPLE 6

Converting from Rectangular Coordinates to Polar Coordinates

Find polar coordinates of a point whose rectangular coordinates are $(0, 3)$.

Figure 14

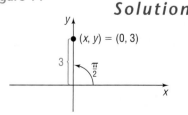

Solution See Figure 14. The point $(0, 3)$ lies on the y-axis a distance of 3 units from the origin (pole), so $r = 3$. A ray with vertex at the pole through $(0, 3)$ forms an angle $\theta = \dfrac{\pi}{2}$ with the polar axis. Polar coordinates for this point can be given by $\left(3, \dfrac{\pi}{2} \right)$. ∎

Figure 15 shows polar coordinates of points that lie on either the x-axis or the y-axis. In each illustration, $a > 0$.

Figure 15

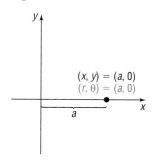

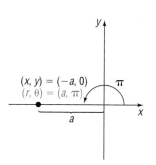

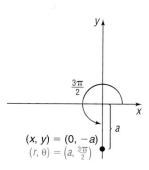

(a) $(x, y) = (a, 0), a > 0$ **(b)** $(x, y) = (0, a), a > 0$ **(c)** $(x, y) = (-a, 0), a > 0$ **(d)** $(x, y) = (0, -a), a > 0$

NOW WORK PROBLEM 45.

EXAMPLE 7   **Converting from Rectangular Coordinates to Polar Coordinates**

Find polar coordinates of a point whose rectangular coordinates are:

(a) $(2, -2)$ (b) $(-1, -\sqrt{3})$

Figure 16

Solution   (a) See Figure 16(a). The distance r from the origin to the point $(2, -2)$ is

$$r = \sqrt{x^2 + y^2} = \sqrt{(2)^2 + (-2)^2} = \sqrt{8} = 2\sqrt{2}$$

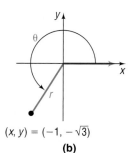

We find θ by recalling that $\tan\theta = \dfrac{y}{x}$, so $\theta = \tan^{-1}\dfrac{y}{x}, -\dfrac{\pi}{2} < \theta < \dfrac{\pi}{2}$. Since $(2, -2)$ lies in quadrant IV, we know that $-\dfrac{\pi}{2} < \theta < 0$. As a result,

$$\theta = \tan^{-1}\frac{y}{x} = \tan^{-1}\left(\frac{-2}{2}\right) = \tan^{-1}(-1) = -\frac{\pi}{4}$$

A set of polar coordinates for this point is $\left(2\sqrt{2}, -\dfrac{\pi}{4}\right)$. Other possible representations include $\left(2\sqrt{2}, \dfrac{7\pi}{4}\right)$ and $\left(-2\sqrt{2}, \dfrac{3\pi}{4}\right)$.

(b) See Figure 16(b). The distance r from the origin to the point $(-1, -\sqrt{3})$ is

$$r = \sqrt{(-1)^2 + (-\sqrt{3})^2} = \sqrt{4} = 2$$

To find θ, we use $\theta = \tan^{-1}\dfrac{y}{x}, -\dfrac{\pi}{2} < \theta < \dfrac{\pi}{2}$. Since the point $(-1, -\sqrt{3})$ lies in quadrant III and the inverse tangent function gives an angle in quadrant I, we add π to the result to obtain an angle in quadrant III.

$$\theta = \pi + \tan^{-1}\left(\frac{-\sqrt{3}}{-1}\right) = \pi + \tan^{-1}\sqrt{3} = \pi + \frac{\pi}{3} = \frac{4\pi}{3}$$

A set of polar coordinates is $\left(2, \dfrac{4\pi}{3}\right)$. Other possible representations include $\left(-2, \dfrac{\pi}{3}\right)$ and $\left(2, -\dfrac{2\pi}{3}\right)$.

Figure 17 shows how to find polar coordinates of a point that lies in a quadrant when its rectangular coordinates (x, y) are given.

Figure 17

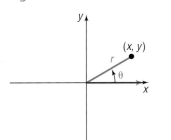

(a) $r = \sqrt{x^2 + y^2}$
$\theta = \tan^{-1}\dfrac{y}{x}$

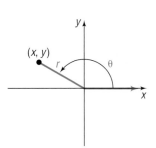

(b) $r = \sqrt{x^2 + y^2}$
$\theta = \pi + \tan^{-1}\dfrac{y}{x}$

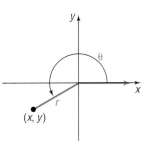

(c) $r = \sqrt{x^2 + y^2}$
$\theta = \pi + \tan^{-1}\dfrac{y}{x}$

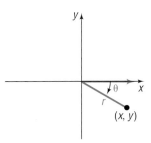

(d) $r = \sqrt{x^2 + y^2}$
$\theta = \tan^{-1}\dfrac{y}{x}$

Based on the preceding discussion, we have the formulas

$$r^2 = x^2 + y^2 \qquad \tan\theta = \frac{y}{x} \qquad \text{if } x \neq 0 \qquad \textbf{(2)}$$

To use formula (2) effectively, follow these steps:

STEPS FOR CONVERTING FROM RECTANGULAR TO POLAR COORDINATES

STEP 1: Always plot the point (x, y) first, as we did in Examples 6 and 7.

STEP 2: To find r, compute the distance from the origin to (x, y).

STEP 3: To find θ, first determine the quadrant that the point lies in.

Quadrant I: $\theta = \tan^{-1}\dfrac{y}{x}$ $\qquad$ Quadrant II: $\theta = \pi + \tan^{-1}\dfrac{y}{x}$

Quadrant III: $\theta = \pi + \tan^{-1}\dfrac{y}{x}$ $\qquad$ Quadrant IV: $\theta = \tan^{-1}\dfrac{y}{x}$

✏️ NOW WORK PROBLEM **49.**

Formulas (1) and (2) may also be used to transform equations.

EXAMPLE 8

Transforming an Equation from Polar to Rectangular Form

Transform the equation $r = 4\sin\theta$ from polar coordinates to rectangular coordinates, and identify the graph.

Solution   If we multiply each side by r, it will be easier to apply formulas (1) and (2).

$$r = 4\sin\theta$$

$$r^2 = 4r\sin\theta \qquad \text{Multiply each side by } r.$$

$$x^2 + y^2 = 4y \qquad r^2 = x^2 + y^2; y = r\sin\theta$$

This is the equation of a circle; we proceed to complete the square to obtain the standard form of the equation.

$$x^2 + (y^2 - 4y) = 0 \quad \text{General form}$$
$$x^2 + (y^2 - 4y + 4) = 4 \quad \text{Complete the square in } y.$$
$$x^2 + (y - 2)^2 = 4 \quad \text{Standard form}$$

The center of the circle is at $(0, 2)$, and its radius is 2. ■

NOW WORK PROBLEM 65.

| EXAMPLE 9 | **Transforming an Equation from Rectangular to Polar Form** |

Transform the equation $4xy = 9$ from rectangular coordinates to polar coordinates.

Solution   We use formula (1).

$$4xy = 9$$
$$4(r\cos\theta)(r\sin\theta) = 9 \quad x = r\cos\theta, \, y = r\sin\theta$$
$$4r^2 \cos\theta \sin\theta = 9$$
$$2r^2 \sin(2\theta) = 9 \quad \text{Double-angle formula.} \quad ■$$

5.1 EXERCISES

In Problems 1–8, match each point in polar coordinates with either A, B, C, or D on the graph.

1. $\left(2, -\dfrac{11\pi}{6}\right)$ **2.** $\left(-2, -\dfrac{\pi}{6}\right)$ **3.** $\left(-2, \dfrac{\pi}{6}\right)$ **4.** $\left(2, \dfrac{7\pi}{6}\right)$

5. $\left(2, \dfrac{5\pi}{6}\right)$ **6.** $\left(-2, \dfrac{5\pi}{6}\right)$ **7.** $\left(-2, \dfrac{7\pi}{6}\right)$ **8.** $\left(2, \dfrac{11\pi}{6}\right)$

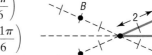

In Problems 9–20, plot each point given in polar coordinates.

9. $(3, 90°)$ **10.** $(4, 270°)$ **11.** $(-2, 0)$ **12.** $(-3, \pi)$

13. $\left(6, \dfrac{\pi}{6}\right)$ **14.** $\left(5, \dfrac{5\pi}{3}\right)$ **15.** $(-2, 135°)$ **16.** $(-3, 120°)$

17. $\left(-1, -\dfrac{\pi}{3}\right)$ **18.** $\left(-3, -\dfrac{3\pi}{4}\right)$ **19.** $(-2, -\pi)$ **20.** $\left(-3, -\dfrac{\pi}{2}\right)$

In Problems 21–28, plot each point given in polar coordinates, and find other polar coordinates (r, θ) of the point for which:
  (a) $r > 0$, $-2\pi \le \theta < 0$ (b) $r < 0$, $0 \le \theta < 2\pi$ (c) $r > 0$, $2\pi \le \theta < 4\pi$

21. $\left(5, \dfrac{2\pi}{3}\right)$ **22.** $\left(4, \dfrac{3\pi}{4}\right)$ **23.** $(-2, 3\pi)$ **24.** $(-3, 4\pi)$

25. $\left(1, \dfrac{\pi}{2}\right)$ **26.** $(2, \pi)$ **27.** $\left(-3, -\dfrac{\pi}{4}\right)$ **28.** $\left(-2, -\dfrac{2\pi}{3}\right)$

In Problems 29–44, polar coordinates of a point are given. Find the rectangular coordinates of each point.

29. $\left(3, \dfrac{\pi}{2}\right)$ **30.** $\left(4, \dfrac{3\pi}{2}\right)$ **31.** $(-2, 0)$ **32.** $(-3, \pi)$

33. $(6, 150°)$ **34.** $(5, 300°)$ **35.** $\left(-2, \dfrac{3\pi}{4}\right)$ **36.** $\left(-2, \dfrac{2\pi}{3}\right)$

37. $\left(-1, -\dfrac{\pi}{3}\right)$ **38.** $\left(-3, -\dfrac{3\pi}{4}\right)$ **39.** $(-2, -180°)$ **40.** $(-3, -90°)$

41. $(7.5, 110°)$ **42.** $(-3.1, 182°)$ **43.** $(6.3, 3.8)$ **44.** $(8.1, 5.2)$

In Problems 45–56, the rectangular coordinates of a point are given. Find polar coordinates for each point.

45. $(3, 0)$ **46.** $(0, 2)$ **47.** $(-1, 0)$ **48.** $(0, -2)$

49. $(1, -1)$ **50.** $(-3, 3)$ **51.** $(\sqrt{3}, 1)$ **52.** $(-2, -2\sqrt{3})$

53. $(1.3, -2.1)$ **54.** $(-0.8, -2.1)$ **55.** $(8.3, 4.2)$ **56.** $(-2.3, 0.2)$

In Problems 57–64, the letters x and y represent rectangular coordinates. Write each equation using polar coordinates (r, θ).

57. $2x^2 + 2y^2 = 3$ **58.** $x^2 + y^2 = x$ **59.** $x^2 = 4y$ **60.** $y^2 = 2x$

61. $2xy = 1$ **62.** $4x^2y = 1$ **63.** $x = 4$ **64.** $y = -3$

In Problems 65–72, the letters r and θ represent polar coordinates. Write each equation using rectangular coordinates (x, y).

65. $r = \cos \theta$ **66.** $r = \sin \theta + 1$ **67.** $r^2 = \cos \theta$ **68.** $r = \sin \theta - \cos \theta$

69. $r = 2$ **70.** $r = 4$ **71.** $r = \dfrac{4}{1 - \cos \theta}$ **72.** $r = \dfrac{3}{3 - \cos \theta}$

73. Show that the formula for the distance d between two points $P_1 = (r_1, \theta_1)$ and $P_2 = (r_2, \theta_2)$ is

$$d = \sqrt{r_1^2 + r_2^2 - 2r_1 r_2 \cos(\theta_2 - \theta_1)}$$

P R E P A R I N G   F O R   T H I S   S E C T I O N

Before getting started, review the following:

✓ Graphs of Equations; Circles (Section 1.2, pp. 9–18)

✓ Even–Odd Properties of Trigonometric Functions (Section 2.3, p. 119)

✓ Difference Formulas (Section 3.4, pp. 182 and 185)

✓ Value of the Sine and Cosine Functions at Certain Angles (Section 2.2, p. 96 and p. 100)

5.2 POLAR EQUATIONS AND GRAPHS

OBJECTIVES   **1** Graph and Identify Polar Equations by Converting to Rectangular Equations

2 Test Polar Equations for Symmetry

3 Graph Polar Equations by Plotting Points

Just as a rectangular grid may be used to plot points given by rectangular coordinates, as in Figure 18(a), we can use a grid consisting of concentric circles (with centers at the pole) and rays (with vertices at the pole) to plot points given by polar coordinates, as shown in Figure 18(b). We shall use such **polar grids** to graph *polar equations*.

Figure 18

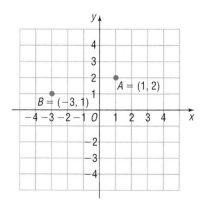

(a) Rectangular grid

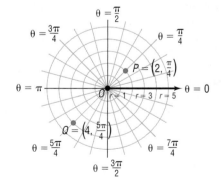

(b) Polar grid

An equation whose variables are polar coordinates is called a **polar equation**. The **graph of a polar equation** consists of all points whose polar coordinates satisfy the equation.

①   One method that we can use to graph a polar equation is to convert the equation to rectangular coordinates. In the discussion that follows, (x, y) represent the rectangular coordinates of a point P, and (r, θ) represent polar coordinates of the point P.

EXAMPLE 1

Identifying and Graphing a Polar Equation (Circle)

Identify and graph the equation: $r = 3$

Solution   We convert the polar equation to a rectangular equation.

$$r = 3$$

$$r^2 = 9 \quad \text{Square both sides.}$$

$$x^2 + y^2 = 9 \quad {\scriptstyle r^2 = x^2 + y^2}$$

The graph of $r = 3$ is a circle, with center at the pole and radius 3. See Figure 19.

Figure 19
$r = 3$ or $x^2 + y^2 = 9$

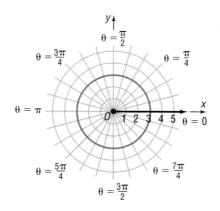

◗─────➤ NOW WORK PROBLEM **1**.

EXAMPLE 2

Identifying and Graphing a Polar Equation (Line)

Identify and graph the equation: $\theta = \dfrac{\pi}{4}$

Solution   We convert the polar equation to a rectangular equation.

$$\theta = \frac{\pi}{4}$$

$$\tan \theta = \tan \frac{\pi}{4} = 1$$

$$\frac{y}{x} = 1 \qquad\qquad {\scriptstyle \tan \theta = \frac{y}{x}}$$

$$y = x$$

The graph of $\theta = \dfrac{\pi}{4}$ is a line passing through the pole making an angle of $\dfrac{\pi}{4}$ with the polar axis. See Figure 20.

Figure 20

$\theta = \dfrac{\pi}{4}$ or $y = x$

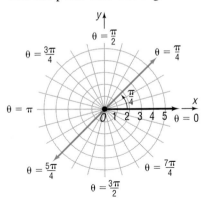

NOW WORK PROBLEM **3.**

EXAMPLE 3   **Identifying and Graphing a Polar Equation (Horizontal Line)**

Identify and graph the equation: $r \sin \theta = 2$

Solution   Since $y = r \sin \theta$, we can write the equation as

$$y = 2$$

We conclude that the graph of $r \sin \theta = 2$ is a horizontal line 2 units above the pole. See Figure 21.

Figure 21
$r \sin \theta = 2$ or $y = 2$

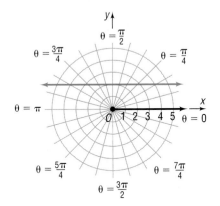

COMMENT: A graphing utility can be used to graph polar equations. Read Using a Graphing Utility to Graph a Polar Equation, Appendix B, Section B.6, page 573.

EXAMPLE 4   **Identifying and Graphing a Polar Equation (Vertical Line)**

Identify and graph the equation: $r \cos \theta = -3$

Solution   Since $x = r \cos \theta$, we can write the equation as

$$x = -3$$

We conclude that the graph of $r \cos \theta = -3$ is a vertical line 3 units to the left of the pole. Figure 22 shows the graph.

Figure 22
$r \cos \theta = -3$ or $x = -3$

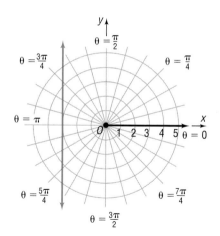

Check: Graph $r = -\dfrac{3}{\cos \theta}$ using θ min $= 0$, θ max $= 2\pi$, and θ step $= \dfrac{\pi}{24}$.
Compare the result to Figure 22.

Based on Examples 3 and 4, we are led to the following results. (The proofs are left as exercises.)

Theorem   Let a be a nonzero real number. Then the graph of the equation

$$r \sin \theta = a$$

is a horizontal line a units above the pole if $a > 0$ and $|a|$ units below the pole if $a < 0$.
   The graph of the equation

$$r \cos \theta = a$$

is a vertical line a units to the right of the pole if $a > 0$ and $|a|$ units to the left of the pole if $a < 0$.

NOW WORK PROBLEM **7**.

EXAMPLE 5   ## Identifying and Graphing a Polar Equation (Circle)

Identify and graph the equation: $r = 4 \sin \theta$

Solution   To transform the equation to rectangular coordinates, we multiply each side by r.

$$r^2 = 4r \sin \theta$$

Now we use the facts that $r^2 = x^2 + y^2$ and $y = r \sin \theta$. Then

$$x^2 + y^2 = 4y$$

$$x^2 + (y^2 - 4y) = 0$$

$$x^2 + (y^2 - 4y + 4) = 4 \qquad \text{Complete the square in } y.$$

$$x^2 + (y - 2)^2 = 4 \qquad \text{Standard equation of a circle}$$

This is the equation of a circle with center at $(0, 2)$ in rectangular coordinates and radius 2. Figure 23 shows the graph.

Figure 23
$r = 4 \sin \theta$ or $x^2 + (y - 2)^2 = 4$

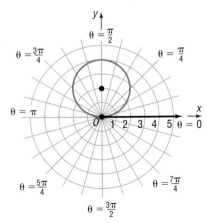

| EXAMPLE 6 | **Identifying and Graphing a Polar Equation (Circle)** |

Identify and graph the equation: $r = -2 \cos \theta$

Solution We proceed as in Example 5.

$$r^2 = -2r \cos \theta \qquad \text{Multiply both sides by } r.$$
$$x^2 + y^2 = -2x \qquad r^2 = x^2 + y^2; \; x = r \cos \theta$$
$$x^2 + 2x + y^2 = 0$$
$$\left(x^2 + 2x + 1\right) + y^2 = 1 \qquad \text{Complete the square in } x.$$
$$(x + 1)^2 + y^2 = 1 \qquad \text{Standard equation of a circle}$$

This is the equation of a circle with center at $(-1, 0)$ in rectangular coordinates and radius 1. Figure 24 shows the graph.

Figure 24
$r = -2\cos \theta$ or $(x + 1)^2 + y^2 = 1$

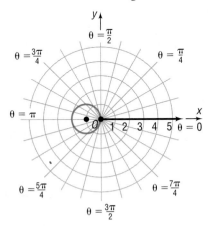

 Check: Graph $r = 4 \sin \theta$ and compare the result with Figure 23. Clear the screen and do the same for $r = -2 \cos \theta$ and compare with Figure 24. Be sure to use a square screen.

EXPLORATION Using a square screen, graph $r_1 = \sin \theta$, $r_2 = 2 \sin \theta$, and $r_3 = 3 \sin \theta$. Do you see the pattern? Clear the screen and graph $r_1 = -\sin \theta$, $r_2 = -2 \sin \theta$, and $r_3 = -3 \sin \theta$. Do you see the pattern? Clear the screen and graph $r_1 = \cos \theta$, $r_2 = 2 \cos \theta$, and $r_3 = 3 \cos \theta$. Do you see the pattern? Clear the screen and graph $r_1 = -\cos \theta$, $r_2 = -2 \cos \theta$, and $r_3 = -3 \cos \theta$. Do you see the pattern?

Based on Examples 5 and 6 and the preceding Exploration, we are led to the following results. (The proofs are left as exercises.)

Theorem   Let a be a positive real number. Then,

Equation	Description
(a) $r = 2a \sin\theta$	Circle: radius a; center at $(0, a)$ in rectangular coordinates
(b) $r = -2a \sin\theta$	Circle: radius a; center at $(0, -a)$ in rectangular coordinates
(c) $r = 2a \cos\theta$	Circle: radius a; center at $(a, 0)$ in rectangular coordinates
(d) $r = -2a \cos\theta$	Circle: radius a; center at $(-a, 0)$ in rectangular coordinates

Each circle passes through the pole.

NOW WORK PROBLEM 9.

The method of converting a polar equation to an identifiable rectangular equation in order to obtain the graph is not always helpful, nor is it always necessary. Usually, we set up a table that lists several points on the graph. By checking for symmetry, it may be possible to reduce the number of points needed to draw the graph.

SYMMETRY

In polar coordinates, the points (r, θ) and $(r, -\theta)$ are symmetric with respect to the polar axis (and to the x-axis). See Figure 25(a). The points (r, θ) and $(r, \pi - \theta)$ are symmetric with respect to the line $\theta = \dfrac{\pi}{2}$ (the y-axis). See Figure 25(b). The points (r, θ) and $(-r, \theta)$ are symmetric with respect to the pole (the origin). See Figure 25(c).

Figure 25

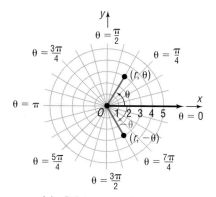

(a) Points symmetric with respect to the polar axis

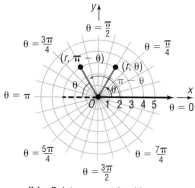

(b) Points symmetric with respect to the line $\theta = \dfrac{\pi}{2}$

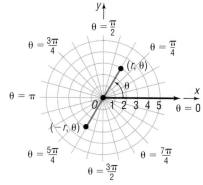

(c) Points symmetric with respect to the pole

The following tests are a consequence of these observations.

Theorem

Tests for Symmetry

Symmetry with Respect to the Polar Axis (x-Axis)

In a polar equation, replace θ by $-\theta$. If an equivalent equation results, the graph is symmetric with respect to the polar axis.

Symmetry with Respect to the Line $\theta = \dfrac{\pi}{2}$ (y-Axis)

In a polar equation, replace θ by $\pi - \theta$. If an equivalent equation results, the graph is symmetric with respect to the line $\theta = \dfrac{\pi}{2}$.

Symmetry with Respect to the Pole (Origin)

In a polar equation, replace r by $-r$. If an equivalent equation results, the graph is symmetric with respect to the pole.

The three tests for symmetry given here are *sufficient* conditions for symmetry, but they are not *necessary* conditions. That is, an equation may fail these tests and still have a graph that is symmetric with respect to the polar axis, the line $\theta = \dfrac{\pi}{2}$, or the pole. For example, the graph of $r = \sin(2\theta)$ turns out to be symmetric with respect to the polar axis, the line $\theta = \dfrac{\pi}{2}$, and the pole, but all three tests given here fail. See also Problems 65, 66, and 67.

EXAMPLE 7

Graphing a Polar Equation (Cardioid)

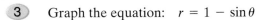

Graph the equation: $r = 1 - \sin\theta$

Solution We check for symmetry first.

Polar Axis: Replace θ by $-\theta$. The result is

$$r = 1 - \sin(-\theta) = 1 + \sin\theta$$

The test fails, so the graph may or may not be symmetric with respect to the polar axis.

The Line $\theta = \dfrac{\pi}{2}$:   Replace θ by $\pi - \theta$. The result is

$$r = 1 - \sin(\pi - \theta) = 1 - (\sin\pi\cos\theta - \cos\pi\sin\theta)$$
$$= 1 - \big[0 \cdot \cos\theta - (-1)\sin\theta\big] = 1 - \sin\theta$$

The test is satisfied, so the graph is symmetric with respect to the line $\theta = \dfrac{\pi}{2}$.

The Pole: Replace r by $-r$. Then the result is $-r = 1 - \sin\theta$, so $r = -1 + \sin\theta$. The test fails, so the graph may or may not be symmetric with respect to the pole.

Next, we identify points on the graph by assigning values to the angle θ and calculating the corresponding values of r. Due to the symmetry with respect to the line $\theta = \dfrac{\pi}{2}$, we only need to assign values to θ from $-\dfrac{\pi}{2}$ to $\dfrac{\pi}{2}$, as given in Table 1.

TABLE 1	
θ	$r = 1 - \sin\theta$
$-\dfrac{\pi}{2}$	$1 - (-1) = 2$
$-\dfrac{\pi}{3}$	$1 - \left(-\dfrac{\sqrt{3}}{2}\right) \approx 1.87$
$-\dfrac{\pi}{6}$	$1 - \left(-\dfrac{1}{2}\right) = \dfrac{3}{2}$
0	$1 - 0 = 1$
$\dfrac{\pi}{6}$	$1 - \dfrac{1}{2} = \dfrac{1}{2}$
$\dfrac{\pi}{3}$	$1 - \dfrac{\sqrt{3}}{2} \approx 0.13$
$\dfrac{\pi}{2}$	$1 - 1 = 0$

Now we plot the points (r, θ) from Table 1 and trace out the graph, beginning at the point $\left(2, -\dfrac{\pi}{2}\right)$ and ending at the point $\left(0, \dfrac{\pi}{2}\right)$. Then we reflect this portion of the graph about the line $\theta = \dfrac{\pi}{2}$ (the y-axis) to obtain the complete graph. Figure 26 shows the graph.

Figure 26
$r = 1 - \sin\theta$

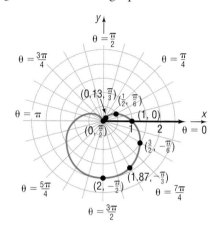

 EXPLORATION Graph $r_1 = 1 + \sin\theta$. Clear the screen and graph $r_1 = 1 - \cos\theta$. Clear the screen and graph $r_1 = 1 + \cos\theta$. Do you see a pattern? ■

The curve in Figure 26 is an example of a *cardioid* (a heart-shaped curve).

> **Cardioids** are characterized by equations of the form
> $$r = a(1 + \cos\theta) \qquad r = a(1 + \sin\theta)$$
> $$r = a(1 - \cos\theta) \qquad r = a(1 - \sin\theta)$$
> where $a > 0$. The graph of a cardioid passes through the pole.

NOW WORK PROBLEM **25.**

EXAMPLE 8   Graphing a Polar Equation (Limaçon without Inner Loop)

Graph the equation: $r = 3 + 2\cos\theta$

Solution   We check for symmetry first.

Polar Axis: Replace θ by $-\theta$. The result is
$$r = 3 + 2\cos(-\theta) = 3 + 2\cos\theta$$
The test is satisfied, so the graph is symmetric with respect to the polar axis.

The Line $\theta = \dfrac{\pi}{2}$:   Replace θ by $\pi - \theta$. The result is
$$r = 3 + 2\cos(\pi - \theta) = 3 + 2(\cos\pi\cos\theta + \sin\pi\sin\theta)$$
$$= 3 - 2\cos\theta$$

The test fails, so the graph may or may not be symmetric with respect to the line $\theta = \dfrac{\pi}{2}$.

The Pole: Replace r by $-r$. The test fails, so the graph may or may not be symmetric with respect to the pole.

Next, we identify points on the graph by assigning values to the angle θ and calculating the corresponding values of r. Due to the symmetry with respect to the polar axis, we only need to assign values to θ from 0 to π, as given in Table 2.

Now we plot the points (r, θ) from Table 2 and trace out the graph, beginning at the point $(5, 0)$ and ending at the point $(1, \pi)$. Then we reflect this portion of the graph about the polar axis (the x-axis) to obtain the complete graph. Figure 27 shows the graph.

TABLE	2
θ	$r = 3 + 2\cos\theta$
0	$3 + 2(1) = 5$
$\dfrac{\pi}{6}$	$3 + 2\left(\dfrac{\sqrt{3}}{2}\right) \approx 4.73$
$\dfrac{\pi}{3}$	$3 + 2\left(\dfrac{1}{2}\right) = 4$
$\dfrac{\pi}{2}$	$3 + 2(0) = 3$
$\dfrac{2\pi}{3}$	$3 + 2\left(-\dfrac{1}{2}\right) = 2$
$\dfrac{5\pi}{6}$	$3 + 2\left(-\dfrac{\sqrt{3}}{2}\right) \approx 1.27$
π	$3 + 2(-1) = 1$

Figure 27
$r = 3 + 2\cos\theta$

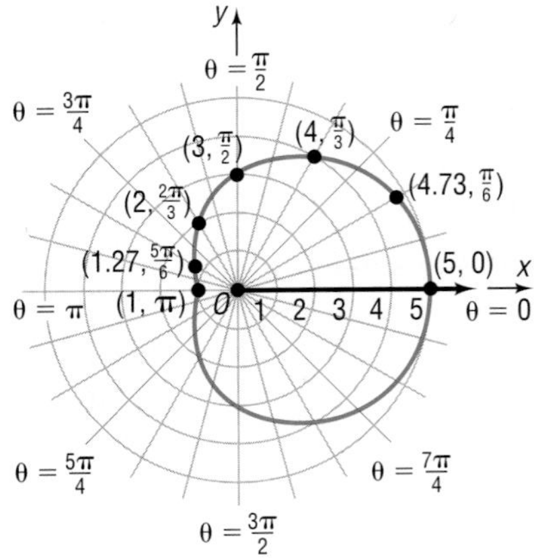

EXPLORATION   Graph $r_1 = 3 - 2\cos\theta$. Clear the screen and graph $r_1 = 3 + 2\sin\theta$. Clear the screen and graph $r_1 = 3 - 2\sin\theta$. Do you see a pattern?

The curve in Figure 27 is an example of a limaçon (the French word for *snail*) without an inner loop.

Limaçons without an inner loop are characterized by equations of the form

$$r = a + b\cos\theta \qquad r = a + b\sin\theta$$

$$r = a - b\cos\theta \qquad r = a - b\sin\theta$$

where $a > 0$, $b > 0$, and $a > b$. The graph of a limaçon without an inner loop does not pass through the pole.

NOW WORK PROBLEM **31.**

EXAMPLE 9   **Graphing a Polar Equation (Limaçon with Inner Loop)**

Graph the equation: $r = 1 + 2\cos\theta$

Solution   First, we check for symmetry.

Polar Axis: Replace θ by $-\theta$. The result is

$$r = 1 + 2\cos(-\theta) = 1 + 2\cos\theta$$

The test is satisfied, so the graph is symmetric with respect to the polar axis.

The Line $\theta = \dfrac{\pi}{2}$: Replace θ by $\pi - \theta$. The result is

$$r = 1 + 2\cos(\pi - \theta) = 1 + 2(\cos\pi\cos\theta + \sin\pi\sin\theta)$$
$$= 1 - 2\cos\theta$$

The test fails, so the graph may or may not be symmetric with respect to the line $\theta = \dfrac{\pi}{2}$.

The Pole: Replace r by $-r$. The test fails, so the graph may or may not be symmetric with respect to the pole.

Next, we identify points on the graph of $r = 1 + 2\cos\theta$ by assigning values to the angle θ and calculating the corresponding values of r. Due to the symmetry with respect to the polar axis, we only need to assign values to θ from 0 to π, as given in Table 3.

Now we plot the points (r, θ) from Table 3, beginning at $(3, 0)$ and ending at $(-1, \pi)$. See Figure 28(a). Finally, we reflect this portion of the graph about the polar axis (the x-axis) to obtain the complete graph. See Figure 28(b).

TABLE 3	
θ	$r = 1 + 2\cos\theta$
0	$1 + 2(1) = 3$
$\dfrac{\pi}{6}$	$1 + 2\left(\dfrac{\sqrt{3}}{2}\right) \approx 2.73$
$\dfrac{\pi}{3}$	$1 + 2\left(\dfrac{1}{2}\right) = 2$
$\dfrac{\pi}{2}$	$1 + 2(0) = 1$
$\dfrac{2\pi}{3}$	$1 + 2\left(-\dfrac{1}{2}\right) = 0$
$\dfrac{5\pi}{6}$	$1 + 2\left(-\dfrac{\sqrt{3}}{2}\right) \approx -0.73$
π	$1 + 2(-1) = -1$

Figure 28

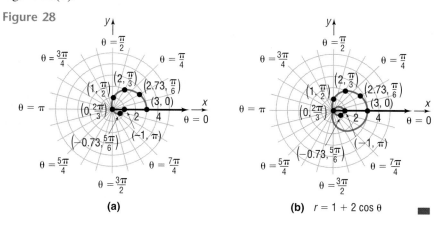

(a)

(b) $r = 1 + 2\cos\theta$

 EXPLORATION   Graph $r_1 = 1 - 2\cos\theta$. Clear the screen and graph $r_1 = 1 + 2\sin\theta$. Clear the screen and graph $r_1 = 1 - 2\sin\theta$. Do you see a pattern?

The curve in Figure 28(b) is an example of a limaçon with an inner loop.

Limaçons with an inner loop are characterized by equations of the form

$$r = a + b\cos\theta \qquad r = a + b\sin\theta$$

$$r = a - b\cos\theta \qquad r = a - b\sin\theta$$

where $a > 0$, $b > 0$, and $a < b$. The graph of a limaçon with an inner loop will pass through the pole twice.

NOW WORK PROBLEM **33.**

| EXAMPLE 10 | Graphing a Polar Equation (Rose) |

Graph the equation: $r = 2\cos(2\theta)$

Solution We check for symmetry.

Polar Axis: If we replace θ by $-\theta$, the result is

$$r = 2\cos[2(-\theta)] = 2\cos(2\theta)$$

The test is satisfied, so the graph is symmetric with respect to the polar axis.

The Line $\theta = \dfrac{\pi}{2}$: If we replace θ by $\pi - \theta$, we obtain

$$r = 2\cos[2(\pi - \theta)] = 2\cos(2\pi - 2\theta) = 2\cos(2\theta)$$

The test is satisfied, so the graph is symmetric with respect to the line $\theta = \dfrac{\pi}{2}$.

The Pole: Since the graph is symmetric with respect to both the polar axis and the line $\theta = \dfrac{\pi}{2}$, it must be symmetric with respect to the pole.

Next, we construct Table 4. Due to the symmetry with respect to the polar axis, the line $\theta = \dfrac{\pi}{2}$, and the pole, we consider only values of θ from 0 to $\dfrac{\pi}{2}$.

We plot and connect these points in Figure 29(a). Finally, because of symmetry, we reflect this portion of the graph first about the polar axis (the x-axis) and then about the line $\theta = \dfrac{\pi}{2}$ (the y-axis) to obtain the complete graph. See Figure 29(b).

T A B L E 4

θ	$r = 2\cos(2\theta)$
0	$2(1) = 2$
$\dfrac{\pi}{6}$	$2\left(\dfrac{1}{2}\right) = 1$
$\dfrac{\pi}{4}$	$2(0) = 0$
$\dfrac{\pi}{3}$	$2\left(-\dfrac{1}{2}\right) = -1$
$\dfrac{\pi}{2}$	$2(-1) = -2$

Figure 29

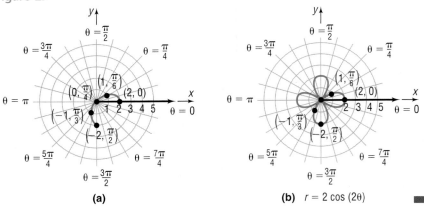

(a)

(b) $r = 2\cos(2\theta)$

EXPLORATION Graph $r = 2\cos(4\theta)$; clear the screen and graph $r = 2\cos(6\theta)$. How many petals did each of these graphs have?

Clear the screen and graph, in order, each on a clear screen, $r = 2\cos(3\theta)$, $r = 2\cos(5\theta)$, and $r = 2\cos(7\theta)$. What do you notice about the number of petals?

The curve in Figure 29(b) is called a *rose* with four petals.

Rose curves are characterized by equations of the form

$$r = a \cos(n\theta), \qquad r = a \sin(n\theta), \qquad a \neq 0$$

and have graphs that are rose shaped. If $n \neq 0$ is even, the rose has $2n$ petals; if $n \neq \pm 1$ is odd, the rose has n petals.

NOW WORK PROBLEM 37.

| EXAMPLE 11 | **Graphing a Polar Equation (Lemniscate)** |

Graph the equation: $r^2 = 4 \sin(2\theta)$

Solution   We leave it to you to verify that the graph is symmetric with respect to the pole. Table 5 lists points on the graph for values of $\theta = 0$ through $\theta = \dfrac{\pi}{2}$. Note that there are no points on the graph for $\dfrac{\pi}{2} < \theta < \pi$ (quadrant II), since $\sin(2\theta) < 0$ for such values. The points from Table 5 where $r \geq 0$ are plotted in Figure 30(a). The remaining points on the graph may be obtained by using symmetry. Figure 30(b) shows the final graph.

TABLE 5		
θ	$r^2 = 4\sin(2\theta)$	r
0	$4(0) = 0$	0
$\dfrac{\pi}{6}$	$4\left(\dfrac{\sqrt{3}}{2}\right) = 2\sqrt{3}$	± 1.9
$\dfrac{\pi}{4}$	$4(1) = 4$	± 2
$\dfrac{\pi}{3}$	$4\left(\dfrac{\sqrt{3}}{2}\right) = 2\sqrt{3}$	± 1.9
$\dfrac{\pi}{2}$	$4(0) = 0$	0

Figure 30

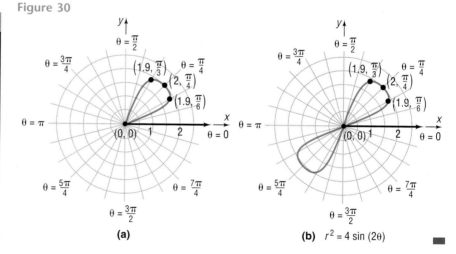

(a)   **(b)** $r^2 = 4\sin(2\theta)$

The curve in Figure 30(b) is an example of a *lemniscate*.

Lemniscates are characterized by equations of the form

$$r^2 = a^2 \sin(2\theta) \qquad r^2 = a^2 \cos(2\theta)$$

where $a \neq 0$, and have graphs that are propeller shaped.

NOW WORK PROBLEM 41.

EXAMPLE 12

Graphing a Polar Equation (Spiral)

Graph the equation: $r = e^{\theta/5}$

Solution The tests for symmetry with respect to the pole, the polar axis, and the line $\theta = \dfrac{\pi}{2}$ fail. Furthermore, there is no number θ for which $r = 0$, so the graph does not pass through the pole. We observe that r is positive for all θ, r increases as θ increases, $r \to 0$ as $\theta \to -\infty$, and $r \to \infty$ as $\theta \to \infty$. With the help of a calculator, we obtain the values in Table 6. See Figure 31 for the graph.

TABLE 6

θ	$r = e^{\theta/5}$
$-\dfrac{3\pi}{2}$	0.39
$-\pi$	0.53
$-\dfrac{\pi}{2}$	0.73
$-\dfrac{\pi}{4}$	0.85
0	1
$\dfrac{\pi}{4}$	1.17
$\dfrac{\pi}{2}$	1.37
π	1.87
$\dfrac{3\pi}{2}$	2.57
2π	3.51

Figure 31
$r = e^{\theta/5}$

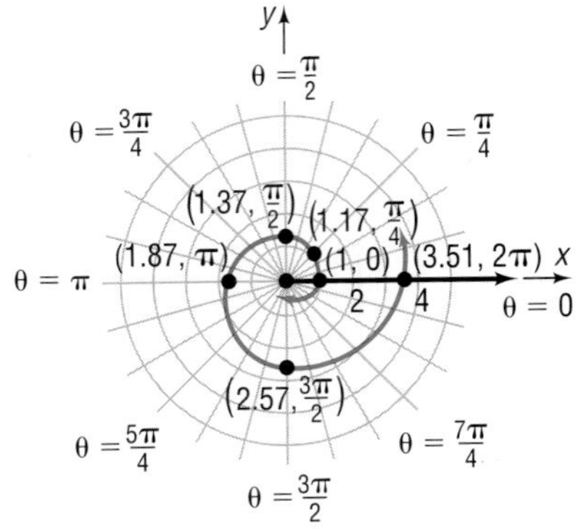

The curve in Figure 31 is called a **logarithmic spiral,** since its equation may be written as $\theta = 5 \ln r$ and it spirals infinitely both toward the pole and away from it. ∎

CLASSIFICATION OF POLAR EQUATIONS

The equations of some lines and circles in polar coordinates and their corresponding equations in rectangular coordinates are given in Table 7 on page 295. Also included are the names and the graphs of a few of the more frequently encountered polar equations.

SKETCHING QUICKLY

If a polar equation only involves a sine (or cosine) function, you can quickly obtain a sketch of its graph by making use of Table 7, periodicity, and a short table.

EXAMPLE 13

Sketching the Graph of a Polar Equation Quickly by Hand

Graph the equation: $r = 2 + 2\sin\theta$

Solution We recognize the polar equation: Its graph is a cardioid. The period of $\sin\theta$ is 2π, so we form a table using $0 \le \theta \le 2\pi$, compute r, plot the points (r, θ), and sketch the graph of a cardioid as θ varies from 0 to 2π. See Table 8 and Figure 32 on page 296.

T A B L E 7

Lines

Description	Line passing through the pole making an angle α with the polar axis	Vertical line	Horizontal line
Rectangular equation	$y = (\tan \alpha)x$	$x = a$	$y = b$
Polar equation	$\theta = \alpha$	$r \cos \theta = a$	$r \sin \theta = b$
Typical graph	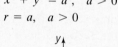		

Circles

Description	Center at the pole, radius a	Passing through the pole, tangent to the line $\theta = \pi/2$, center on the polar axis, radius a	Passing through the pole, tangent to the polar axis, center on the line $\theta = \pi/2$, radius a
Rectangular equation	$x^2 + y^2 = a^2$, $a > 0$	$x^2 + y^2 = \pm 2ax$, $a > 0$	$x^2 + y^2 = \pm 2ay$, $a > 0$
Polar equation	$r = a$, $a > 0$	$r = \pm 2a \cos \theta$, $a > 0$	$r = \pm 2a \sin \theta$, $a > 0$
Typical graph		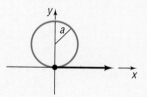	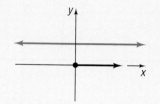

Other Equations

Name	Cardioid	Limaçon without inner loop	Limaçon with inner loop
Polar equations	$r = a \pm a \cos \theta$, $a > 0$ $r = a \pm a \sin \theta$, $a > 0$	$r = a \pm b \cos \theta$, $0 < b < a$ $r = a \pm b \sin \theta$, $0 < b < a$	$r = a \pm b \cos \theta$, $0 < a < b$ $r = a \pm b \sin \theta$, $0 < a < b$
Typical graph	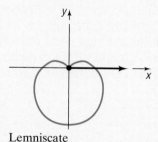	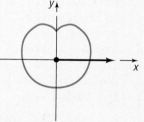	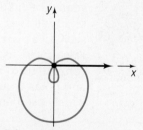

Name	Lemniscate	Rose with three petals	Rose with four petals
Polar equations	$r^2 = a^2 \cos(2\theta)$, $a > 0$ $r^2 = a^2 \sin(2\theta)$, $a > 0$	$r = a \sin(3\theta)$, $a > 0$ $r = a \cos(3\theta)$, $a > 0$	$r = a \sin(2\theta)$, $a > 0$ $r = a \cos(2\theta)$, $a > 0$
Typical graph	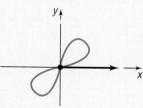	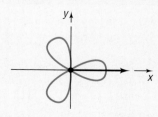	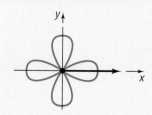

θ	$r = 2 + 2\sin\theta$
0	$2 + 2(0) = 2$
$\dfrac{\pi}{2}$	$2 + 2(1) = 4$
π	$2 + 2(0) = 2$
$\dfrac{3\pi}{2}$	$2 + 2(-1) = 0$
2π	$2 + 2(0) = 2$

TABLE 8

Figure 32

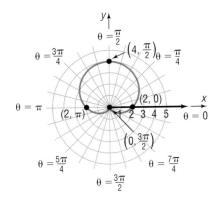

CALCULUS COMMENT

For those of you who are planning to study calculus, a comment about one important role of polar equations is in order.

In rectangular coordinates, the equation $x^2 + y^2 = 1$, whose graph is the unit circle, is not the graph of a function. In fact, it requires two functions to obtain the graph of the unit circle:

$$y_1 = \sqrt{1 - x^2} \quad \text{Upper semicircle} \qquad y_2 = -\sqrt{1 - x^2} \quad \text{Lower semicircle}$$

In polar coordinates, the equation $r = 1$, whose graph is also the unit circle, does define a function. That is, for each choice of θ there is only one corresponding value of r, namely, $r = 1$. Since many uses of calculus require that functions be used, the opportunity to express nonfunctions in rectangular coordinates as functions in polar coordinates becomes extremely useful.

Note also that the vertical-line test for functions is valid only for equations in rectangular coordinates.

HISTORICAL FEATURE

Jakob Bernoulli
1654–1705

Polar coordinates seem to have been invented by Jakob Bernoulli (1654–1705) in about 1691, although, as with most such ideas, earlier traces of the notion exist. Early users of calculus remained committed to rectangular coordinates, and polar coordinates did not become widely used until the early 1800s. Even then, it was mostly geometers who used them for describing odd curves. Finally, about the mid-1800s, applied mathematicians realized the tremendous simplification that polar coordinates make possible in the description of objects with circular or cylindrical symmetry. From then on their use became widespread.

5.2 EXERCISES

In Problems 1–16, transform each polar equation to an equation in rectangular coordinates. Then identify and graph the equation.

1. $r = 4$

2. $r = 2$

3. $\theta = \dfrac{\pi}{3}$

4. $\theta = -\dfrac{\pi}{4}$

5. $r\sin\theta = 4$

6. $r\cos\theta = 4$

7. $r\cos\theta = -2$

8. $r\sin\theta = -2$

9. $r = 2\cos\theta$

10. $r = 2\sin\theta$

11. $r = -4\sin\theta$

12. $r = -4\cos\theta$

13. $r\sec\theta = 4$

14. $r\csc\theta = 8$

15. $r\csc\theta = -2$

16. $r\sec\theta = -4$

In Problems 17–24, match each of the graphs (A) through (H) to one of the following polar equations.

17. $r = 2$

18. $\theta = \dfrac{\pi}{4}$

19. $r = 2\cos\theta$

20. $r\cos\theta = 2$

21. $r = 1 + \cos\theta$

22. $r = 2\sin\theta$

23. $\theta = \dfrac{3\pi}{4}$

24. $r\sin\theta = 2$

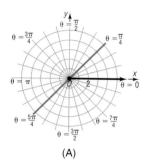

(A)

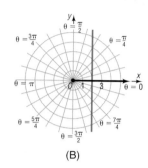

(B)

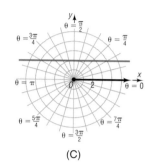

(C)

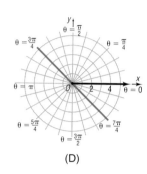

(D)

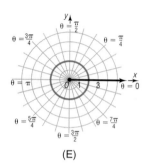

(E)

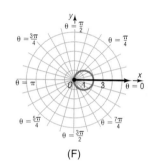

(F)

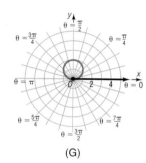

(G)

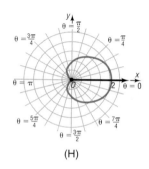

(H)

In Problems 25–48, identify and graph each polar equation.

25. $r = 2 + 2\cos\theta$

26. $r = 1 + \sin\theta$

27. $r = 3 - 3\sin\theta$

28. $r = 2 - 2\cos\theta$

29. $r = 2 + \sin\theta$

30. $r = 2 - \cos\theta$

31. $r = 4 - 2\cos\theta$

32. $r = 4 + 2\sin\theta$

33. $r = 1 + 2\sin\theta$

34. $r = 1 - 2\sin\theta$

35. $r = 2 - 3\cos\theta$

36. $r = 2 + 4\cos\theta$

37. $r = 3\cos(2\theta)$

38. $r = 2\sin(3\theta)$

39. $r = 4\sin(5\theta)$

40. $r = 3\cos(4\theta)$

41. $r^2 = 9\cos(2\theta)$

42. $r^2 = \sin(2\theta)$

43. $r = 2^\theta$

44. $r = 3^\theta$

45. $r = 1 - \cos\theta$

46. $r = 3 + \cos\theta$

47. $r = 1 - 3\cos\theta$

48. $r = 4\cos(3\theta)$

In Problems 49–58, graph each polar equation.

49. $r = \dfrac{2}{1 - \cos\theta}$ (parabola)

50. $r = \dfrac{2}{1 - 2\cos\theta}$ (hyperbola)

51. $r = \dfrac{1}{3 - 2\cos\theta}$ (ellipse)

52. $r = \dfrac{1}{1 - \cos\theta}$ (parabola)

53. $r = \theta,\quad \theta \geq 0$ (spiral of Archimedes)

54. $r = \dfrac{3}{\theta}$ (reciprocal spiral)

55. $r = \csc\theta - 2,\quad 0 < \theta < \pi$ (conchoid)

56. $r = \sin\theta\tan\theta$ (cissoid)

57. $r = \tan\theta,\quad -\dfrac{\pi}{2} < \theta < \dfrac{\pi}{2},$ (kappa curve)

58. $r = \cos\dfrac{\theta}{2}$

59. Show that the graph of the equation $r \sin \theta = a$ is a horizontal line a units above the pole if $a > 0$ and $|a|$ units below the pole if $a < 0$.

60. Show that the graph of the equation $r \cos \theta = a$ is a vertical line a units to the right of the pole if $a > 0$ and $|a|$ units to the left of the pole if $a < 0$.

61. Show that the graph of the equation $r = 2a \sin \theta, a > 0$, is a circle of radius a with center at $(0, a)$ in rectangular coordinates.

62. Show that the graph of the equation $r = -2a \sin \theta, a > 0$, is a circle of radius a with center at $(0, -a)$ in rectangular coordinates.

63. Show that the graph of the equation $r = 2a \cos \theta, a > 0$, is a circle of radius a with center at $(a, 0)$ in rectangular coordinates.

64. Show that the graph of the equation $r = -2a \cos \theta$, $a > 0$, is a circle of radius a with center at $(-a, 0)$ in rectangular coordinates.

65. Explain why the following test for symmetry is valid: Replace r by $-r$ and θ by $-\theta$ in a polar equation. If an equivalent equation results, the graph is symmetric with respect to the line $\theta = \dfrac{\pi}{2}$ (y-axis).

(a) Show that the test on page 288 fails for $r^2 = \cos \theta$, but this new test works.

(b) Show that the test on page 288 works for $r^2 = \sin \theta$, yet this new test fails.

66. Develop a new test for symmetry with respect to the pole.
(a) Find a polar equation for which this new test fails, yet the test on page 288 works.
(b) Find a polar equation for which the test on page 288 fails, yet the new test works.

67. Write down two different tests for symmetry with respect to the polar axis. Find examples in which one test works and the other fails. Which test do you prefer to use? Justify your answer.

PREPARING FOR THIS SECTION

Before getting started, review the following:

✓ Complex Numbers (Appendix A, Section A.4, pp. 513–518)

✓ Definitions of the Sine and Cosine Functions (Section 2.2, pp. 92–93)

✓ Value of the Sine and Cosine Functions at Certain Angles (Section 2.2, p. 96 and 100)

✓ Sum and Difference Formulas for Sine and Cosine (Section 3.4, pp. 182 and 185)

5.3 | THE COMPLEX PLANE; DE MOIVRE'S THEOREM

OBJECTIVES
1. Convert a Complex Number from Rectangular Form to Polar Form
2. Plot Points in the Complex Plane
3. Find Products and Quotients of Complex Numbers in Polar Form
4. Use De Moivre's Theorem
5. Find Complex Roots

Figure 33
Complex plane

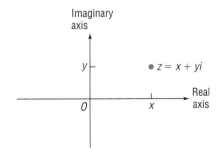

When we first introduced complex numbers, we were not prepared to give a geometric interpretation of a complex number. Now we are ready. Although we could give several interpretations, the one that follows is the easiest to understand.

A complex number $z = x + yi$ can be interpreted geometrically as the point (x, y) in the xy-plane. Each point in the plane corresponds to a complex number and, conversely, each complex number corresponds to a point in the plane. We shall refer to the collection of such points as the **complex plane.** The x-axis will be referred to as the **real axis,** because any point that lies on the real axis is of the form $z = x + 0i = x$, a real number. The y-axis is called the **imaginary axis,** because any point that lies on it is of the form $z = 0 + yi = yi$, a pure imaginary number. See Figure 33.

Let $z = x + yi$ be a complex number. The **magnitude** or **modulus** of z, denoted by $|z|$, is defined as the distance from the origin to the point (x, y). That is,

$$|z| = \sqrt{x^2 + y^2} \tag{1}$$

Figure 34

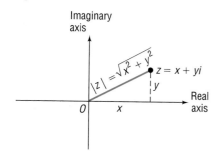

See Figure 34 for an illustration.

This definition for $|z|$ is consistent with the definition for the absolute value of a real number: If $z = x + yi$ is real, then $z = x + 0i$ and

$$|z| = \sqrt{x^2 + 0^2} = \sqrt{x^2} = |x|$$

For this reason, the magnitude of z is sometimes called the absolute value of z.

Recall that if $z = x + yi$ then its **conjugate,** denoted by $\bar{z}$, is $\bar{z} = x - yi$. Because $z\bar{z} = x^2 + y^2$, it follows from equation (1) that the magnitude of z can be written as

$$|z| = \sqrt{z\bar{z}} \tag{2}$$

POLAR FORM OF A COMPLEX NUMBER

① When a complex number is written in the standard form $z = x + yi$, we say that it is in **rectangular,** or **Cartesian, form** because (x, y) are the rectangular coordinates of the corresponding point in the complex plane. Suppose that (r, θ) are the polar coordinates of this point. Then

$$x = r\cos\theta \qquad y = r\sin\theta \tag{3}$$

If $r \geq 0$ and $0 \leq \theta < 2\pi$, the complex number $z = x + yi$ may be written in **polar form** as

$$z = x + yi = (r\cos\theta) + (r\sin\theta)i = r(\cos\theta + i\sin\theta) \tag{4}$$

Figure 35

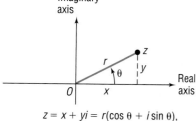

$z = x + yi = r(\cos\theta + i\sin\theta),$
$r \geq 0, 0 \leq \theta < 2\pi$

See Figure 35.

If $z = r(\cos\theta + i\sin\theta)$ is the polar form of a complex number, the angle $\theta, 0 \leq \theta < 2\pi$, is called the **argument of** z.

Also, because $r \geq 0$, we have $r = \sqrt{x^2 + y^2}$. From equation (1) it follows that the magnitude of $z = r(\cos\theta + i\sin\theta)$ is

$$|z| = r$$

EXAMPLE 1

Plotting a Point in the Complex Plane and Writing a Complex Number in Polar Form

Figure 36

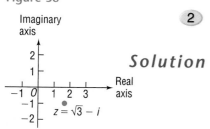

$z = \sqrt{3} - i$

② Plot the point corresponding to $z = \sqrt{3} - i$ in the complex plane, and write an expression for z in polar form.

Solution   The point corresponding to $z = \sqrt{3} - i$ has the rectangular coordinates $(\sqrt{3}, -1)$. The point, located in quadrant IV, is plotted in Figure 36. Because $x = \sqrt{3}$ and $y = -1$, it follows that

$$r = \sqrt{x^2 + y^2} = \sqrt{(\sqrt{3})^2 + (-1)^2} = \sqrt{4} = 2$$

and

$$\sin \theta = \frac{y}{r} = \frac{-1}{2}, \qquad \cos \theta = \frac{x}{r} = \frac{\sqrt{3}}{2}, \qquad 0 \le \theta < 2\pi$$

Then $\theta = \dfrac{11\pi}{6}$ and $r = 2$, so the polar form of $z = \sqrt{3} - i$ is

$$z = r(\cos \theta + i \sin \theta) = 2\left(\cos \frac{11\pi}{6} + i \sin \frac{11\pi}{6} \right)$$ ∎

NOW WORK PROBLEM **1**.

EXAMPLE 2

Plotting a Point in the Complex Plane and Converting from Polar to Rectangular Form

Plot the point corresponding to $z = 2(\cos 30° + i \sin 30°)$ in the complex plane, and write an expression for z in rectangular form.

Figure 37

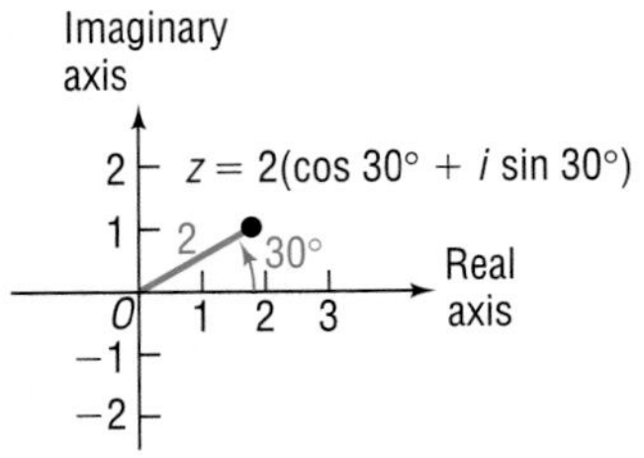

Imaginary axis

$z = 2(\cos 30° + i \sin 30°)$

Solution To plot the complex number $z = 2(\cos 30° + i \sin 30°)$, we plot the point whose polar coordinates are $(r, \theta) = (2, 30°)$, as shown in Figure 37. In rectangular form,

$$z = 2(\cos 30° + i \sin 30°) = 2\left(\frac{\sqrt{3}}{2} + \frac{1}{2} i \right) = \sqrt{3} + i$$

NOW WORK PROBLEM **13**.

3 The polar form of a complex number provides an alternative method for finding products and quotients of complex numbers.

Theorem   Let $z_1 = r_1(\cos \theta_1 + i \sin \theta_1)$ and $z_2 = r_2(\cos \theta_2 + i \sin \theta_2)$ be two complex numbers. Then

$$z_1 z_2 = r_1 r_2 \left[\cos(\theta_1 + \theta_2) + i \sin(\theta_1 + \theta_2) \right] \qquad \textbf{(5)}$$

If $z_2 \ne 0$, then

$$\frac{z_1}{z_2} = \frac{r_1}{r_2} \left[\cos(\theta_1 - \theta_2) + i \sin(\theta_1 - \theta_2) \right] \qquad \textbf{(6)}$$

∎

Proof   We will prove formula (5). The proof of formula (6) is left as an exercise (see Problem 56).

$$z_1 z_2 = \left[r_1(\cos \theta_1 + i \sin \theta_1) \right]\left[r_2(\cos \theta_2 + i \sin \theta_2) \right]$$
$$= r_1 r_2 \left[(\cos \theta_1 + i \sin \theta_1)(\cos \theta_2 + i \sin \theta_2) \right]$$
$$= r_1 r_2 \left[(\cos \theta_1 \cos \theta_2 - \sin \theta_1 \sin \theta_2) + i(\sin \theta_1 \cos \theta_2 + \cos \theta_1 \sin \theta_2) \right]$$
$$= r_1 r_2 \left[\cos(\theta_1 + \theta_2) + i \sin(\theta_1 + \theta_2) \right]$$

∎

Because the magnitude of a complex number z is r and its argument is θ, when $z = r(\cos\theta + i\sin\theta)$, we can restate this theorem as follows:

Theorem

The magnitude of the product (quotient) of two complex numbers equals the product (quotient) of their magnitudes; the argument of the product (quotient) of two complex numbers is determined by the sum (difference) of their arguments.

∎

Let's look at an example of how this theorem can be used.

EXAMPLE 3

Finding Products and Quotients of Complex Numbers in Polar Form

If $z = 3(\cos 20° + i\sin 20°)$ and $w = 5(\cos 100° + i\sin 100°)$, find the following (leave your answers in polar form):

(a) zw

(b) $\dfrac{z}{w}$

Solution

(a) $zw = \left[3(\cos 20° + i\sin 20°)\right]\left[5(\cos 100° + i\sin 100°)\right]$

$= (3 \cdot 5)\left[\cos(20° + 100°) + i\sin(20° + 100°)\right]$

$= 15(\cos 120° + i\sin 120°)$

(b) $\dfrac{z}{w} = \dfrac{3(\cos 20° + i\sin 20°)}{5(\cos 100° + i\sin 100°)}$

$= \frac{3}{5}\left[\cos(20° - 100°) + i\sin(20° - 100°)\right]$

$= \frac{3}{5}\left[\cos(-80°) + i\sin(-80°)\right]$

$= \frac{3}{5}(\cos 280° + i\sin 280°)$ Argument must lie between 0° and 360°.   ∎

$\textbegin{}$ **NOW WORK PROBLEM 23.**

DE MOIVRE'S THEOREM

④ De Moivre's Theorem, stated by Abraham De Moivre (1667–1754) in 1730, but already known to many people by 1710, is important for the following reason: The fundamental processes of algebra are the four operations of addition, subtraction, multiplication, and division, together with powers and the extraction of roots. De Moivre's Theorem allows these latter fundamental algebraic operations to be applied to complex numbers.

De Moivre's Theorem, in its most basic form, is a formula for raising a complex number z to the power n, where $n \geq 1$ is a positive integer. Let's see if we can guess the form of the result.

Let $z = r(\cos\theta + i\sin\theta)$ be a complex number. Then, based on equation (5), we have

$n = 2$: $z^2 = r^2\left[\cos(2\theta) + i\sin(2\theta)\right]$ Equation (5)

$n = 3$: $z^3 = z^2 \cdot z$

$= \left\{r^2\left[\cos(2\theta) + i\sin(2\theta)\right]\right\}\left[r(\cos\theta + i\sin\theta)\right]$

$= r^3\left[\cos(3\theta) + i\sin(3\theta)\right]$ Equation (5)

$n = 4$: $z^4 = z^3 \cdot z$

$= \left\{r^3\left[\cos(3\theta) + i\sin(3\theta)\right]\right\}\left[r(\cos\theta + i\sin\theta)\right]$

$= r^4\left[\cos(4\theta) + i\sin(4\theta)\right]$ Equation (5)

The pattern should now be clear.

Theorem | **De Moivre's Theorem**

If $z = r(\cos\theta + i\sin\theta)$ is a complex number, then

$$z^n = r^n\big[\cos(n\theta) + i\sin(n\theta)\big] \qquad \textbf{(7)}$$

where $n \geq 1$ is a positive integer.

∎

We will not prove De Moivre's Theorem because the proof requires mathematical induction which is not discussed in this book.

Let's look at some examples.

EXAMPLE 4 | **Using De Moivre's Theorem**

Write $\big[2(\cos 20° + i\sin 20°)\big]^3$ in the standard form $a + bi$.

Solution | $\big[2(\cos 20° + i\sin 20°)\big]^3 = 2^3\big[\cos(3 \cdot 20°) + i\sin(3 \cdot 20°)\big]$

$$= 8(\cos 60° + i\sin 60°)$$

$$= 8\left(\frac{1}{2} + \frac{\sqrt{3}}{2}i\right) = 4 + 4\sqrt{3}i$$

∎

NOW WORK PROBLEM **31**.

EXAMPLE 5 | **Using De Moivre's Theorem**

Write $(1 + i)^5$ in the standard form $a + bi$.

Solution | To apply De Moivre's Theorem, we must first write the complex number in polar form. Since the magnitude of $1 + i$ is $\sqrt{1^2 + 1^2} = \sqrt{2}$, we begin by writing

$$1 + i = \sqrt{2}\left(\frac{1}{\sqrt{2}} + \frac{1}{\sqrt{2}}i\right) = \sqrt{2}\left(\cos\frac{\pi}{4} + i\sin\frac{\pi}{4}\right)$$

Now

$$(1 + i)^5 = \left[\sqrt{2}\left(\cos\frac{\pi}{4} + i\sin\frac{\pi}{4}\right)\right]^5$$

$$= (\sqrt{2})^5\left[\cos\left(5 \cdot \frac{\pi}{4}\right) + i\sin\left(5 \cdot \frac{\pi}{4}\right)\right]$$

$$= 4\sqrt{2}\left(\cos\frac{5\pi}{4} + i\sin\frac{5\pi}{4}\right)$$

$$= 4\sqrt{2}\left[-\frac{1}{\sqrt{2}} + \left(-\frac{1}{\sqrt{2}}\right)i\right] = -4 - 4i$$

∎

COMPLEX ROOTS

⑤ Let w be a given complex number, and let $n \geq 2$ denote a positive integer. Any complex number z that satisfies the equation

$$z^n = w$$

is called a **complex nth root** of w. In keeping with previous usage, if $n = 2$, the solutions of the equation $z^2 = w$ are called **complex square roots** of w, and if $n = 3$, the solutions of the equation $z^3 = w$ are called **complex cube roots** of w.

Theorem   **Finding Complex Roots**

Let $w = r(\cos\theta_0 + i\sin\theta_0)$ be a complex number and let $n \geq 2$ be an integer. If $w \neq 0$, there are n distinct complex roots of w, given by the formula

$$z_k = \sqrt[n]{r}\left[\cos\left(\frac{\theta_0}{n} + \frac{2k\pi}{n}\right) + i\sin\left(\frac{\theta_0}{n} + \frac{2k\pi}{n}\right)\right] \qquad \textbf{(8)}$$

where $k = 0, 1, 2, \ldots, n - 1$.

Proof (Outline)   We will not prove this result in its entirety. Instead, we shall show only that each z_k in equation (8) satisfies the equation $z_k^n = w$, proving that each z_k is a complex nth root of w.

$$z_k^n = \left\{\sqrt[n]{r}\left[\cos\left(\frac{\theta_0}{n} + \frac{2k\pi}{n}\right) + i\sin\left(\frac{\theta_0}{n} + \frac{2k\pi}{n}\right)\right]\right\}^n$$

$$= (\sqrt[n]{r})^n\left\{\cos\left[n\left(\frac{\theta_0}{n} + \frac{2k\pi}{n}\right)\right] + i\sin\left[n\left(\frac{\theta_0}{n} + \frac{2k\pi}{n}\right)\right]\right\} \qquad \text{De Moivre's Theorem}$$

$$= r\left[\cos(\theta_0 + 2k\pi) + i\sin(\theta_0 + 2k\pi)\right]$$

$$= r(\cos\theta_0 + i\sin\theta_0) = w \qquad \text{Periodic Property}$$

So, each z_k, $k = 0, 1, \ldots, n - 1$, is a complex nth root of w. To complete the proof, we would need to show that each z_k, $k = 0, 1, \ldots, n - 1$, is, in fact, distinct and that there are no complex nth roots of w other than those given by equation (8). ■

EXAMPLE 6   **Finding Complex Cube Roots**

Find the complex cube roots of $-1 + \sqrt{3}i$. Leave your answers in polar form, with θ in degrees.

Solution   First, we express $-1 + \sqrt{3}i$ in polar form using degrees.

$$-1 + \sqrt{3}i = 2\left(-\frac{1}{2} + \frac{\sqrt{3}}{2}i\right) = 2(\cos 120° + i\sin 120°)$$

So, $r = 2$ and $\theta_0 = 120°$. The three complex cube roots of $-1 + \sqrt{3}i = 2(\cos 120° + i\sin 120°)$ are

$$z_k = \sqrt[3]{2}\left[\cos\left(\frac{120°}{3} + \frac{360°k}{3}\right) + i\sin\left(\frac{120°}{3} + \frac{360°k}{3}\right)\right], \qquad k = 0, 1, 2$$

$$= \sqrt[3]{2}\left[\cos(40° + 120°k) + i\sin(40° + 120°k)\right], \qquad k = 0, 1, 2$$

So,

$$z_0 = \sqrt[3]{2}\left[\cos(40° + 120° \cdot 0) + i\sin(40° + 120° \cdot 0)\right] = \sqrt[3]{2}\left(\cos 40° + i\sin 40°\right)$$
$$z_1 = \sqrt[3]{2}\left[\cos(40° + 120° \cdot 1) + i\sin(40° + 120° \cdot 1)\right] = \sqrt[3]{2}\left(\cos 160° + i\sin 160°\right)$$
$$z_2 = \sqrt[3]{2}\left[\cos(40° + 120° \cdot 2) + i\sin(40° + 120° \cdot 2)\right] = \sqrt[3]{2}\left(\cos 280° + i\sin 280°\right) \quad ∎$$

Notice that each of the three complex roots of $-1 + \sqrt{3}i$ has the same magnitude, $\sqrt[3]{2}$. This means that the points corresponding to each cube root lie the same distance from the origin; that is, the three points lie on a circle with center at the origin and radius $\sqrt[3]{2}$. Furthermore, the arguments of these cube roots are $40°, 160°$, and $280°$, the difference of consecutive pairs being $120° = \dfrac{360°}{3}$.

This means that the three points are equally spaced on the circle, as shown in Figure 38. These results are not coincidental. In fact, you are asked to show that these results hold for complex nth roots in Problems 53 through 55.

Figure 38

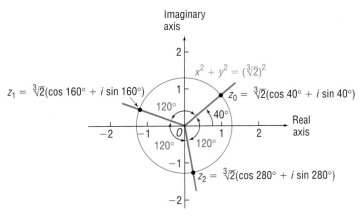

NOW WORK PROBLEM **43**.

HISTORICAL FEATURE

John Wallis

The Babylonians, Greeks, and Arabs considered square roots of negative quantities to be impossible and equations with complex solutions to be unsolvable. The first hint that there was some connection between real solutions of equations and complex numbers came when Girolamo Cardano (1501–1576) and Tartaglia (1499–1557) found *real* roots of cubic equations by taking cube roots of *complex* quantities. For centuries thereafter, mathematicians worked with complex numbers without much belief in their actual existence. In 1673, John Wallis appears to have been the first to suggest the graphical representation of complex numbers, a truly significant idea that was not pursued further until about 1800. Several people, including Karl Friedrich Gauss (1777–1855), then rediscovered the idea, and the graphical representation helped to establish complex numbers as equal members of the number family. In practical applications, complex numbers have found their greatest uses in the area of alternating current, where they are a commonplace tool, and in the area of subatomic physics.

HISTORICAL PROBLEMS

1. The quadratic formula will work perfectly well if the coefficients are complex numbers. Solve the following using De Moivre's Theorem where necessary.

   [**Hint:** The answers are "nice."]

   (a) $z^2 - (2 + 5i)z - 3 + 5i = 0$ (b) $z^2 - (1 + i)z - 2 - i = 0$

5.3 EXERCISES

In Problems 1–12, plot each complex number in the complex plane and write it in polar form. Express the argument in degrees.

1. $1 + i$

2. $-1 + i$

3. $\sqrt{3} - i$

4. $1 - \sqrt{3}i$

5. $-3i$

6. -2

7. $4 - 4i$

8. $9\sqrt{3} + 9i$

9. $3 - 4i$

10. $2 + \sqrt{3}i$

11. $-2 + 3i$

12. $\sqrt{5} - i$

In Problems 13–22, write each complex number in rectangular form.

13. $2(\cos 120° + i \sin 120°)$

14. $3(\cos 210° + i \sin 210°)$

15. $4\left(\cos \dfrac{7\pi}{4} + i \sin \dfrac{7\pi}{4}\right)$

16. $2\left(\cos \dfrac{5\pi}{6} + i \sin \dfrac{5\pi}{6}\right)$

17. $3\left(\cos \dfrac{3\pi}{2} + i \sin \dfrac{3\pi}{2}\right)$

18. $4\left(\cos \dfrac{\pi}{2} + i \sin \dfrac{\pi}{2}\right)$

19. $0.2(\cos 100° + i \sin 100°)$

20. $0.4(\cos 200° + i \sin 200°)$

21. $2\left(\cos \dfrac{\pi}{18} + i \sin \dfrac{\pi}{18}\right)$

22. $3\left(\cos \dfrac{\pi}{10} + i \sin \dfrac{\pi}{10}\right)$

In Problems 23–30, find zw and $\dfrac{z}{w}$. Leave your answers in polar form.

23. $z = 2(\cos 40° + i \sin 40°)$
$w = 4(\cos 20° + i \sin 20°)$

24. $z = \cos 120° + i \sin 120°$
$w = \cos 100° + i \sin 100°$

25. $z = 3(\cos 130° + i \sin 130°)$
$w = 4(\cos 270° + i \sin 270°)$

26. $z = 2(\cos 80° + i \sin 80°)$
$w = 6(\cos 200° + i \sin 200°)$

27. $z = 2\left(\cos \dfrac{\pi}{8} + i \sin \dfrac{\pi}{8}\right)$
$w = 2\left(\cos \dfrac{\pi}{10} + i \sin \dfrac{\pi}{10}\right)$

28. $z = 4\left(\cos \dfrac{3\pi}{8} + i \sin \dfrac{3\pi}{8}\right)$
$w = 2\left(\cos \dfrac{9\pi}{16} + i \sin \dfrac{9\pi}{16}\right)$

29. $z = 2 + 2i$
$w = \sqrt{3} - i$

30. $z = 1 - i$
$w = 1 - \sqrt{3}i$

In Problems 31–42, write each expression in the standard form $a + bi$.

31. $[4(\cos 40° + i \sin 40°)]^3$

32. $[3(\cos 80° + i \sin 80°)]^3$

33. $\left[2\left(\cos \dfrac{\pi}{10} + i \sin \dfrac{\pi}{10}\right)\right]^5$

34. $\left[\sqrt{2}\left(\cos \dfrac{5\pi}{16} + i \sin \dfrac{5\pi}{16}\right)\right]^4$

35. $[\sqrt{3}(\cos 10° + i \sin 10°)]^6$

36. $\left[\tfrac{1}{2}(\cos 72° + i \sin 72°)\right]^5$

37. $\left[\sqrt{5}\left(\cos \dfrac{3\pi}{16} + i \sin \dfrac{3\pi}{16}\right)\right]^4$

38. $\left[\sqrt{3}\left(\cos \dfrac{5\pi}{18} + i \sin \dfrac{5\pi}{18}\right)\right]^6$

39. $(1 - i)^5$

40. $(\sqrt{3} - i)^6$

41. $(\sqrt{2} - i)^6$

42. $(1 - \sqrt{5}i)^8$

In Problems 43–50, find all the complex roots. Leave your answers in polar form with the argument in degrees.

43. The complex cube roots of $1 + i$

44. The complex fourth roots of $\sqrt{3} - i$

45. The complex fourth roots of $4 - 4\sqrt{3}i$

46. The complex cube roots of $-8 - 8i$

47. The complex fourth roots of $-16i$

48. The complex cube roots of -8

49. The complex fifth roots of i

50. The complex fifth roots of $-i$

51. Find the four complex fourth roots of unity (1) and plot each.

52. Find the six complex sixth roots of unity (1) and plot each.

53. Show that each complex nth root of a nonzero complex number w has the same magnitude.

54. Use the result of Problem 53 to draw the conclusion that each complex nth root lies on a circle with center at the origin. What is the radius of this circle?

55. Refer to Problem 54. Show that the complex nth roots of a nonzero complex number w are equally spaced on the circle.

56. Prove formula (6).

PREPARING FOR THIS SECTION

Before getting started, review the following:

✓ Rectangular Coordinates (Section 1.1, pp. 2–5)

✓ Pythagorean Theorem
(Appendix A, Section A.2, pp. 497–499)

5.4 VECTORS

OBJECTIVES
1. Graph Vectors
2. Find a Position Vector
3. Add and Subtract Vectors
4. Find a Scalar Product and the Magnitude of a Vector
5. Find a Unit Vector
6. Find a Vector from Its Direction and Magnitude
7. Work with Objects in Static Equilibrium

In simple terms, a **vector** (derived from the Latin *vehere*, meaning "to carry") is a quantity that has both magnitude and direction. It is customary to represent a vector by using an arrow. The length of the arrow represents the **magnitude** of the vector, and the arrowhead indicates the **direction** of the vector.

Many quantities in physics can be represented by vectors. For example, the velocity of an aircraft can be represented by an arrow that points in the direction of movement; the length of the arrow represents speed. If the aircraft speeds up, we lengthen the arrow; if the aircraft changes direction, we introduce an arrow in the new direction. See Figure 39. Based on this representation, it is not surprising that vectors and directed line segments are somehow related.

Figure 39

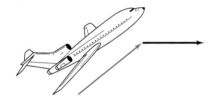

GEOMETRIC VECTORS

If P and Q are two distinct points in the xy-plane, there is exactly one line containing both P and Q [Figure 40(a)]. The points on that part of the line that joins P to Q, including P and Q, form what is called the **line segment** $\overline{PQ}$ [Figure 40(b)]. If we order the points so that they proceed from P to Q, we have a **directed line segment** from P to Q, or a **geometric vector,** which we denote by $\overrightarrow{PQ}$. In a directed line segment $\overrightarrow{PQ}$, we call P the **initial point** and Q the **terminal point,** as indicated in Figure 40(c).

Figure 40

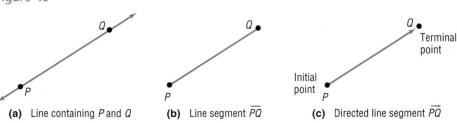

(a) Line containing P and Q **(b)** Line segment $\overline{PQ}$ **(c)** Directed line segment $\overrightarrow{PQ}$

The magnitude of the directed line segment $\overrightarrow{PQ}$ is the distance from the point P to the point Q; that is, it is the length of the line segment.

The direction of $\overrightarrow{PQ}$ is from P to Q. If a vector $\mathbf{v}^*$ has the same magnitude and the same direction as the directed line segment $\overrightarrow{PQ}$, we write

$$\mathbf{v} = \overrightarrow{PQ}$$

The vector $\mathbf{v}$ whose magnitude is 0 is called the **zero vector, 0.** The zero vector is assigned no direction.

Two vectors $\mathbf{v}$ and $\mathbf{w}$ are **equal,** written

$$\mathbf{v} = \mathbf{w}$$

if they have the same magnitude and the same direction.

For example, the vectors shown in Figure 41 have the same magnitude and the same direction, so they are equal, even though they have different initial points and different terminal points. As a result, we find it useful to think of a vector simply as an arrow, keeping in mind that two arrows (vectors) are equal if they have the same direction and the same magnitude (length).

Figure 41

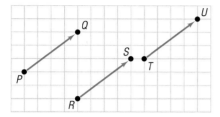

ADDING VECTORS

The **sum $\mathbf{v} + \mathbf{w}$** of two vectors is defined as follows: We position the vectors $\mathbf{v}$ and $\mathbf{w}$ so that the terminal point of $\mathbf{v}$ coincides with the initial point of $\mathbf{w}$, as shown in Figure 42. The vector $\mathbf{v} + \mathbf{w}$ is then the unique vector whose initial point coincides with the initial point of $\mathbf{v}$ and whose terminal point coincides with the terminal point of $\mathbf{w}$.

Vector addition is **commutative.** That is, if $\mathbf{v}$ and $\mathbf{w}$ are any two vectors, then

$$\mathbf{v} + \mathbf{w} = \mathbf{w} + \mathbf{v}$$

Figure 42

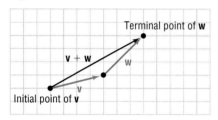

Figure 43 illustrates this fact. (Observe that the commutative property is another way of saying that opposite sides of a parallelogram are equal and parallel.)

Vector addition is also **associative.** That is, if $\mathbf{u}, \mathbf{v},$ and $\mathbf{w}$ are vectors, then

$$\mathbf{u} + (\mathbf{v} + \mathbf{w}) = (\mathbf{u} + \mathbf{v}) + \mathbf{w}$$

Figure 43

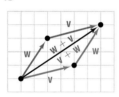

Figure 44 illustrates the associative property for vectors.

The zero vector has the property that

$$\mathbf{v} + \mathbf{0} = \mathbf{0} + \mathbf{v} = \mathbf{v}$$

for any vector $\mathbf{v}$.

If $\mathbf{v}$ is a vector, then $-\mathbf{v}$ is the vector having the same magnitude as $\mathbf{v}$, but whose direction is opposite to $\mathbf{v}$, as shown in Figure 45.

Furthermore,

$$\mathbf{v} + (-\mathbf{v}) = \mathbf{0}$$

Figure 44
$$(\mathbf{u} + \mathbf{v}) + \mathbf{w} = \mathbf{u} + (\mathbf{v} + \mathbf{w})$$

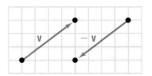

If $\mathbf{v}$ and $\mathbf{w}$ are two vectors, we define the **difference $\mathbf{v} - \mathbf{w}$** as

$$\mathbf{v} - \mathbf{w} = \mathbf{v} + (-\mathbf{w})$$

Figure 45

*Boldface letters will be used to denote vectors, in order to distinguish them from numbers. For handwritten work, an arrow is placed over the letter to signify a vector.

Figure 46

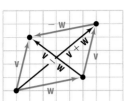

Figure 46 illustrates the relationships among **v, w, v + w,** and **v − w.**

MULTIPLYING VECTORS BY NUMBERS

When dealing with vectors, we refer to real numbers as **scalars.** Scalars are quantities that have only magnitude. Examples from physics of scalar quantities are temperature, speed, and time. We now define how to multiply a vector by a scalar.

If α is a scalar and **v** is a vector, the **scalar product** α**v** is defined as follows:

1. If $\alpha > 0$, the product α**v** is the vector whose magnitude is α times the magnitude of **v** and whose direction is the same as **v.**

2. If $\alpha < 0$, the product α**v** is the vector whose magnitude is $|\alpha|$ times the magnitude of **v** and whose direction is opposite that of **v.**

3. If $\alpha = 0$ or if **v = 0,** then α**v = 0.**

Figure 47

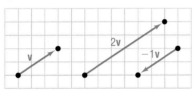

See Figure 47 for some illustrations.

For example, if **a** is the acceleration of an object of mass m due to a force **F** being exerted on it, then, by Newton's second law of motion, **F** = m**a.** Here, m**a** is the product of the scalar m and the vector **a.**

Scalar products have the following properties:

$$0\mathbf{v} = \mathbf{0} \qquad 1\mathbf{v} = \mathbf{v} \qquad -1\mathbf{v} = -\mathbf{v}$$

$$(\alpha + \beta)\mathbf{v} = \alpha\mathbf{v} + \beta\mathbf{v} \qquad \alpha(\mathbf{v} + \mathbf{w}) = \alpha\mathbf{v} + \alpha\mathbf{w}$$

$$\alpha(\beta\mathbf{v}) = (\alpha\beta)\mathbf{v}$$

EXAMPLE 1   ## Graphing Vectors

Figure 48

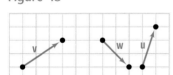

① Use the vectors illustrated in Figure 48 to graph each of the following vectors:

(a) **v − w**   (b) **2v + 3w**   (c) **2v − w + u**

Solution   Figure 49 illustrates each graph.

Figure 49

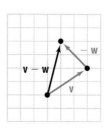

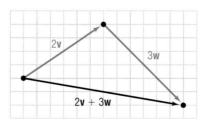

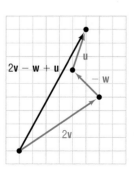

(a) **v − w**   (b) **2v + 3w**   (c) **2v − w + u**

NOW WORK PROBLEMS **1** AND **3.**

MAGNITUDES OF VECTORS

If **v** is a vector, we use the symbol $\|\mathbf{v}\|$ to represent the **magnitude** of **v**. Since $\|\mathbf{v}\|$ equals the length of a directed line segment, it follows that $\|\mathbf{v}\|$ has the following properties:

Theorem

Properties of $\|\mathbf{v}\|$

If **v** is a vector and if α is a scalar, then

(a) $\|\mathbf{v}\| \geq 0$ (b) $\|\mathbf{v}\| = 0$ if and only if $\mathbf{v} = \mathbf{0}$

(c) $\|-\mathbf{v}\| = \|\mathbf{v}\|$ (d) $\|\alpha\mathbf{v}\| = |\alpha|\,\|\mathbf{v}\|$

Property (a) is a consequence of the fact that distance is a nonnegative number. Property (b) follows, because the length of the directed line segment $\overrightarrow{PQ}$ is positive unless P and Q are the same point, in which case the length is 0. Property (c) follows because the length of the line segment $\overline{PQ}$ equals the length of the line segment $\overline{QP}$. Property (d) is a direct consequence of the definition of a scalar product.

A vector **u** for which $\|\mathbf{u}\| = 1$ is called a **unit vector.**

To compute the magnitude and direction of a vector, we need an algebraic way of representing vectors.

ALGEBRAIC VECTORS

② An **algebraic vector v** is represented as

$$\mathbf{v} = \langle a, b \rangle$$

where a and b are real numbers (scalars) called the **components** of the vector **v**.

We use a rectangular coordinate system to represent algebraic vectors in the plane. If $\mathbf{v} = \langle a, b \rangle$ is an algebraic vector whose initial point is at the origin, then **v** is called a **position vector.** See Figure 50. Notice that the terminal point of the position vector $\mathbf{v} = \langle a, b \rangle$ is $P = (a, b)$.

The next result states that any vector whose initial point is not at the origin is equal to a unique position vector.

Figure 50

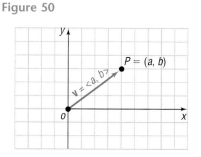

Theorem

Suppose that **v** is a vector with initial point $P_1 = (x_1, y_1)$, not necessarily the origin, and terminal point $P_2 = (x_2, y_2)$. If $\mathbf{v} = \overrightarrow{P_1 P_2}$, then **v** is equal to the position vector

$$\mathbf{v} = \langle x_2 - x_1, y_2 - y_1 \rangle \qquad \textbf{(1)}$$

To see why this is true, look at Figure 51. Triangle OPA and triangle $P_1 P_2 Q$ are congruent. (Do you see why? The line segments have the same magnitude, so $d(O, P) = d(P_1, P_2)$; and they have the same direction, so

Figure 51
$\langle a, b \rangle = \langle x_2 - x_1, y_2 - y_1 \rangle$

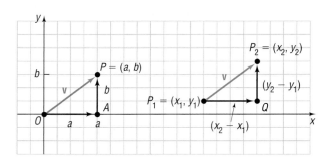

$\angle POA = \angle P_2 P_1 Q$. Since the triangles are right triangles, we have angle–side–angle.) It follows that corresponding sides are equal. As a result, $x_2 - x_1 = a$ and $y_2 - y_1 = b$, so **v** may be written as

$$\mathbf{v} = \langle a, b \rangle = \langle x_2 - x_1, y_2 - y_1 \rangle$$

Because of this result, we can replace any algebraic vector by a unique position vector, and vice versa. This flexibility is one of the main reasons for the wide use of vectors. Unless otherwise specified, from now on the term *vector* will mean the unique position vector equal to it.

EXAMPLE 2

Finding a Position Vector

Find the position vector of the vector $\mathbf{v} = \overrightarrow{P_1 P_2}$ if $P_1 = (-1, 2)$ and $P_2 = (4, 6)$.

Solution
By equation (1), the position vector equal to **v** is

$$\mathbf{v} = \langle 4 - (-1), 6 - 2 \rangle = \langle 5, 4 \rangle$$

See Figure 52.

Figure 52

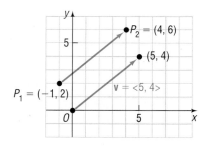

Two position vectors **v** and **w** are equal if and only if the terminal point of **v** is the same as the terminal point of **w**. This leads to the following result:

Theorem

Equality of Vectors

Two vectors **v** and **w** are equal if and only if their corresponding components are equal. That is,

$$\text{If } \mathbf{v} = \langle a_1, b_1 \rangle \quad \text{and} \quad \mathbf{w} = \langle a_2, b_2 \rangle$$

$$\text{then} \quad \mathbf{v} = \mathbf{w} \quad \text{if and only if} \quad a_1 = a_2 \quad \text{and} \quad b_1 = b_2.$$

Figure 53

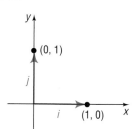

We now present an alternative representation of a vector in the plane that is common in the physical sciences. Let $\mathbf{i}$ denote the unit vector whose direction is along the positive x-axis; let $\mathbf{j}$ denote the unit vector whose direction is along the positive y-axis. Then $\mathbf{i} = \langle 1, 0 \rangle$ and $\mathbf{j} = \langle 0, 1 \rangle$, as shown in Figure 53. Any vector $\mathbf{v} = \langle a, b \rangle$ can be written using the unit vectors $\mathbf{i}$ and $\mathbf{j}$ as follows:

$$\mathbf{v} = \langle a,b \rangle = a\langle 1, 0 \rangle + b\langle 0, 1 \rangle = a\mathbf{i} + b\mathbf{j}$$

We call a and b the **horizontal** and **vertical components** of $\mathbf{v}$, respectively.

NOW WORK PROBLEM **21**.

We define addition, subtraction, scalar product, and magnitude in terms of the components of a vector.

Let $\mathbf{v} = a_1\mathbf{i} + b_1\mathbf{j} = \langle a_1, b_1 \rangle$ and $\mathbf{w} = a_2\mathbf{i} + b_2\mathbf{j} = \langle a_2, b_2 \rangle$ be two vectors, and let α be a scalar. Then

$$\mathbf{v} + \mathbf{w} = (a_1 + a_2)\mathbf{i} + (b_1 + b_2)\mathbf{j} = \langle a_1 + a_2, b_1 + b_2 \rangle \quad \textbf{(2)}$$

$$\mathbf{v} - \mathbf{w} = (a_1 - a_2)\mathbf{i} + (b_1 - b_2)\mathbf{j} = \langle a_1 - a_2, b_1 - b_2 \rangle \quad \textbf{(3)}$$

$$\alpha\mathbf{v} = (\alpha a_1)\mathbf{i} + (\alpha b_1)\mathbf{j} = \langle \alpha a_1, \alpha b_1 \rangle \quad \textbf{(4)}$$

$$\|\mathbf{v}\| = \sqrt{a_1^2 + b_1^2} \quad \textbf{(5)}$$

These definitions are compatible with the geometric definitions given earlier in this section. See Figure 54.

Figure 54

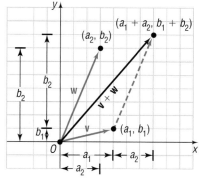

(a) Illustration of property (2)

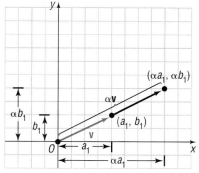

(b) Illustration of property (4), $\alpha > 0$

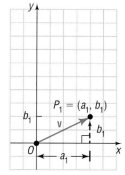

(c) Illustration of property (5):
$\|\mathbf{v}\|$ = Distance from O to P_1
$\|\mathbf{v}\| = \sqrt{a_1^2 + b_1^2}$

To add two vectors, add corresponding components. To subtract two vectors, subtract corresponding components.

EXAMPLE 3 — Adding and Subtracting Vectors

③ If $\mathbf{v} = 2\mathbf{i} + 3\mathbf{j} = \langle 2, 3 \rangle$ and $\mathbf{w} = 3\mathbf{i} - 4\mathbf{j} = \langle 3, -4 \rangle$, find:

(a) $\mathbf{v} + \mathbf{w}$ (b) $\mathbf{v} - \mathbf{w}$

Solution (a) $\mathbf{v} + \mathbf{w} = (2\mathbf{i} + 3\mathbf{j}) + (3\mathbf{i} - 4\mathbf{j}) = (2 + 3)\mathbf{i} + (3 - 4)\mathbf{j} = 5\mathbf{i} - \mathbf{j}$

or

$\mathbf{v} + \mathbf{w} = \langle 2, 3 \rangle + \langle 3, -4 \rangle = \langle 2 + 3, 3 + (-4) \rangle = \langle 5, -1 \rangle$

(b) $\mathbf{v} - \mathbf{w} = (2\mathbf{i} + 3\mathbf{j}) - (3\mathbf{i} - 4\mathbf{j}) = (2 - 3)\mathbf{i} + \left[3 - (-4)\right]\mathbf{j} = -\mathbf{i} + 7\mathbf{j}$

or

$\mathbf{v} - \mathbf{w} = \langle 2, 3 \rangle - \langle 3, -4 \rangle = \langle 2 - 3, 3 - (-4) \rangle = \langle -1, 7 \rangle$ ■

EXAMPLE 4 — Finding Scalar Products and Magnitudes

④ If $\mathbf{v} = 2\mathbf{i} + 3\mathbf{j} = \langle 2, 3 \rangle$ and $\mathbf{w} = 3\mathbf{i} - 4\mathbf{j} = \langle 3, -4 \rangle$, find:

(a) $3\mathbf{v}$ (b) $2\mathbf{v} - 3\mathbf{w}$ (c) $\|\mathbf{v}\|$

Solution (a) $3\mathbf{v} = 3(2\mathbf{i} + 3\mathbf{j}) = 6\mathbf{i} + 9\mathbf{j}$

or

$3\mathbf{v} = 3\langle 2, 3 \rangle = \langle 6, 9 \rangle$

(b) $2\mathbf{v} - 3\mathbf{w} = 2(2\mathbf{i} + 3\mathbf{j}) - 3(3\mathbf{i} - 4\mathbf{j}) = 4\mathbf{i} + 6\mathbf{j} - 9\mathbf{i} + 12\mathbf{j}$

$= -5\mathbf{i} + 18\mathbf{j}$

or

$2\mathbf{v} - 3\mathbf{w} = 2\langle 2, 3 \rangle - 3\langle 3, -4 \rangle = \langle 4, 6 \rangle - \langle 9, -12 \rangle$

$= \langle 4 - 9, 6 - (-12) \rangle = \langle -5, 18 \rangle$

(c) $\|\mathbf{v}\| = \|2\mathbf{i} + 3\mathbf{j}\| = \sqrt{2^2 + 3^2} = \sqrt{13}$ ■

NOW WORK PROBLEMS 27 AND 33.

For the remainder of the section, we will express a vector $\mathbf{v}$ in the form $a\mathbf{i} + b\mathbf{j}$.

⑤   Recall that a unit vector $\mathbf{u}$ is a vector for which $\|\mathbf{u}\| = 1$. In many applications, it is useful to be able to find a unit vector $\mathbf{u}$ that has the same direction as a given vector $\mathbf{v}$.

Theorem   **Unit Vector in the Direction of v**

For any nonzero vector $\mathbf{v}$, the vector

$$\mathbf{u} = \frac{\mathbf{v}}{\|\mathbf{v}\|}$$

is a unit vector that has the same direction as $\mathbf{v}$.

Proof   Let $\mathbf{v} = a\mathbf{i} + b\mathbf{j}$. Then $\|\mathbf{v}\| = \sqrt{a^2 + b^2}$ and

$$\mathbf{u} = \frac{\mathbf{v}}{\|\mathbf{v}\|} = \frac{a\mathbf{i} + b\mathbf{j}}{\sqrt{a^2 + b^2}} = \frac{a}{\sqrt{a^2 + b^2}}\mathbf{i} + \frac{b}{\sqrt{a^2 + b^2}}\mathbf{j}$$

The vector $\mathbf{u}$ is in the same direction as $\mathbf{v}$, since $\|\mathbf{v}\| > 0$. Furthermore,

$$\|\mathbf{u}\| = \sqrt{\frac{a^2}{a^2 + b^2} + \frac{b^2}{a^2 + b^2}} = \sqrt{\frac{a^2 + b^2}{a^2 + b^2}} = 1$$

Thus, $\mathbf{u}$ is a unit vector in the direction of $\mathbf{v}$.   ■

As a consequence of this theorem, if $\mathbf{u}$ is a unit vector in the same direction as a vector $\mathbf{v}$, then $\mathbf{v}$ may be expressed as

$$\mathbf{v} = \|\mathbf{v}\|\,\mathbf{u} \qquad\qquad (6)$$

This way of expressing a vector is useful in many applications.

EXAMPLE 5

Finding a Unit Vector

Find a unit vector in the same direction as $\mathbf{v} = 4\mathbf{i} - 3\mathbf{j}$.

Solution   We find $\|\mathbf{v}\|$ first.

$$\|\mathbf{v}\| = \|4\mathbf{i} - 3\mathbf{j}\| = \sqrt{16 + 9} = 5$$

Now we multiply $\mathbf{v}$ by the scalar $\dfrac{1}{\|\mathbf{v}\|} = \dfrac{1}{5}$. A unit vector in the same direction as $\mathbf{v}$ is

$$\frac{\mathbf{v}}{\|\mathbf{v}\|} = \frac{4\mathbf{i} - 3\mathbf{j}}{5} = \frac{4}{5}\mathbf{i} - \frac{3}{5}\mathbf{j}$$

Check: This vector is, in fact, a unit vector because

$$\left(\tfrac{4}{5}\right)^2 + \left(-\tfrac{3}{5}\right)^2 = \tfrac{16}{25} + \tfrac{9}{25} = \tfrac{25}{25} = 1$$

NOW WORK PROBLEM 43.

WRITING A VECTOR IN TERMS OF ITS MAGNITUDE AND DIRECTION

⑥   If a vector represents the speed and direction of an object, it is called a **velocity vector.** If a vector represents the direction and amount of a force acting on an object, it is called a **force vector.** In many applications, a vector is described in terms of its magnitude and direction, rather than in terms of its components. For example, a ball thrown with an initial speed of 25 miles per hour at an angle $30°$ to the horizontal is a velocity vector.

Suppose that we are given the magnitude $\|\mathbf{v}\|$ of a nonzero vector $\mathbf{v}$ and the angle α, $0° \le \alpha < 360°$, between $\mathbf{v}$ and $\mathbf{i}$. To express $\mathbf{v}$ in terms of $\|\mathbf{v}\|$ and α, we first find the unit vector $\mathbf{u}$ having the same direction as $\mathbf{v}$.

Figure 55

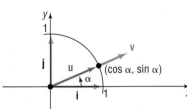

$$\mathbf{u} = \frac{\mathbf{v}}{\|\mathbf{v}\|} \quad \text{or} \quad \mathbf{v} = \|\mathbf{v}\|\mathbf{u} \tag{7}$$

Look at Figure 55. The coordinates of the terminal point of $\mathbf{u}$ are $(\cos\alpha, \sin\alpha)$. Then $\mathbf{u} = \cos\alpha\mathbf{i} + \sin\alpha\mathbf{j}$ and, from (7),

$$\mathbf{v} = \|\mathbf{v}\|(\cos\alpha\mathbf{i} + \sin\alpha\mathbf{j}) \tag{8}$$

where α is the angle between $\mathbf{v}$ and $\mathbf{i}$.

EXAMPLE 6

Writing a Vector When Its Magnitude and Direction Are Given

A ball is thrown with an initial speed of 25 miles per hour in a direction that makes an angle of $30°$ with the positive x-axis. Express the velocity vector $\mathbf{v}$ in terms of $\mathbf{i}$ and $\mathbf{j}$. What is the initial speed in the horizontal direction? What is the initial speed in the vertical direction?

Solution   The magnitude of $\mathbf{v}$ is $\|\mathbf{v}\| = 25$ miles per hour, and the angle between the direction of $\mathbf{v}$ and $\mathbf{i}$, the positive x-axis, is $\alpha = 30°$. By equation (8),

$$\mathbf{v} = \|\mathbf{v}\|(\cos\alpha\mathbf{i} + \sin\alpha\mathbf{j}) = 25(\cos 30°\mathbf{i} + \sin 30°\mathbf{j}) = 25\left(\frac{\sqrt{3}}{2}\mathbf{i} + \frac{1}{2}\mathbf{j}\right) = \frac{25}{2}(\sqrt{3}\mathbf{i} + \mathbf{j})$$

The initial speed of the ball in the horizontal direction is the horizontal component of $\mathbf{v}$, $25\dfrac{\sqrt{3}}{2} \approx 21.65$ miles per hour. The initial speed in the vertical direction is the vertical component of $\mathbf{v}$, $\dfrac{25}{2} = 12.5$ miles per hour. ∎

APPLICATION: STATIC EQUILIBRIUM

⑦ Because forces can be represented by vectors, two forces "combine" the way that vectors "add." If $\mathbf{F}_1$ and $\mathbf{F}_2$ are two forces simultaneously acting on an object, the vector sum $\mathbf{F}_1 + \mathbf{F}_2$ is the **resultant force.** The resultant force produces the same effect on the object as that obtained when the two forces $\mathbf{F}_1$ and $\mathbf{F}_2$ act on the object. See Figure 56. An application of this concept is *static equilibrium.* An object is said to be in **static equilibrium** if (1) the object is at rest and (2) the sum of all forces acting on the object is zero, that is, if the resultant force is 0.

Figure 56

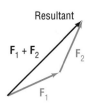

Resultant

$\mathbf{F}_1 + \mathbf{F}_2$ F_2

F_1

EXAMPLE 7

An Object in Static Equilibrium

A box of supplies that weighs 1200 pounds is suspended by two cables attached to the ceiling, as shown in Figure 57. What is the tension in the two cables?

Figure 57

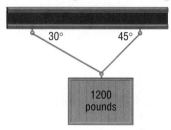

30°   45°

1200 pounds

Solution   We draw a force diagram with the vectors drawn as shown in Figure 58. The tensions in the cables are the magnitudes $\|\mathbf{F}_1\|$ and $\|\mathbf{F}_2\|$ of the force vectors

Figure 58

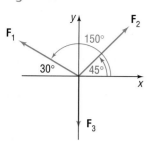

$\mathbf{F}_1$ and $\mathbf{F}_2$. The magnitude of the force vector $\mathbf{F}_3$ equals 1200 pounds, the weight of the box. Now write each force vector in terms of the unit vectors $\mathbf{i}$ and $\mathbf{j}$. For $\mathbf{F}_1$ and $\mathbf{F}_2$, we use equation (8). Remember that α is the angle between the vector and the positive x-axis.

$$\mathbf{F}_1 = \|\mathbf{F}_1\|(\cos 150°\mathbf{i} + \sin 150°\mathbf{j}) = \|\mathbf{F}_1\|\left(-\frac{\sqrt{3}}{2}\mathbf{i} + \frac{1}{2}\mathbf{j}\right) = -\frac{\sqrt{3}}{2}\|\mathbf{F}_1\|\mathbf{i} + \frac{1}{2}\|\mathbf{F}_1\|\mathbf{j}$$

$$\mathbf{F}_2 = \|\mathbf{F}_2\|(\cos 45°\mathbf{i} + \sin 45°\mathbf{j}) = \|\mathbf{F}_2\|\left(\frac{\sqrt{2}}{2}\mathbf{i} + \frac{\sqrt{2}}{2}\mathbf{j}\right) = \frac{\sqrt{2}}{2}\|\mathbf{F}_2\|\mathbf{i} + \frac{\sqrt{2}}{2}\|\mathbf{F}_2\|\mathbf{j}$$

$$\mathbf{F}_3 = -1200\mathbf{j}$$

For static equilibrium, the sum of the force vectors must equal zero.

$$\mathbf{F}_1 + \mathbf{F}_2 + \mathbf{F}_3 = -\frac{\sqrt{3}}{2}\|\mathbf{F}_1\|\mathbf{i} + \frac{1}{2}\|\mathbf{F}_1\|\mathbf{j} + \frac{\sqrt{2}}{2}\|\mathbf{F}_2\|\mathbf{i} + \frac{\sqrt{2}}{2}\|\mathbf{F}_2\|\mathbf{j} - 1200\mathbf{j} = \mathbf{0}$$

The $\mathbf{i}$ component and $\mathbf{j}$ component will each equal zero. This results in the two equations

$$-\frac{\sqrt{3}}{2}\|\mathbf{F}_1\| + \frac{\sqrt{2}}{2}\|\mathbf{F}_2\| = 0 \tag{9}$$

$$\frac{1}{2}\|\mathbf{F}_1\| + \frac{\sqrt{2}}{2}\|\mathbf{F}_2\| - 1200 = 0 \tag{10}$$

We solve equation (9) for $\|\mathbf{F}_2\|$ and obtain

$$\|\mathbf{F}_2\| = \frac{\sqrt{3}}{\sqrt{2}}\|\mathbf{F}_1\| \tag{11}$$

Substituting into equation (10) and solving for $\|\mathbf{F}_1\|$, we obtain

$$\frac{1}{2}\|\mathbf{F}_1\| + \frac{\sqrt{2}}{2}\left(\frac{\sqrt{3}}{\sqrt{2}}\|\mathbf{F}_1\|\right) - 1200 = 0$$

$$\frac{1}{2}\|\mathbf{F}_1\| + \frac{\sqrt{3}}{2}\|\mathbf{F}_1\| - 1200 = 0$$

$$\frac{1 + \sqrt{3}}{2}\|\mathbf{F}_1\| = 1200$$

$$\|\mathbf{F}_1\| = \frac{2400}{1 + \sqrt{3}} \approx 878.5 \text{ pounds}$$

Substituting this value into equation (11) yields $\|\mathbf{F}_2\|$.

$$\|\mathbf{F}_2\| = \frac{\sqrt{3}}{\sqrt{2}}\|\mathbf{F}_1\| = \frac{\sqrt{3}}{\sqrt{2}}\frac{2400}{1 + \sqrt{3}} \approx 1075.9 \text{ pounds}$$

The left cable has tension of approximately 878.5 pounds and the right cable has tension of approximately 1075.9 pounds. ■

HISTORICAL FEATURE

Josiah Gibbs
1839–1903

The history of vectors is surprisingly complicated for such a natural concept. In the *xy*-plane, complex numbers do a good job of imitating vectors. About 1840, mathematicians became interested in finding a system that would do for three dimensions what the complex numbers do for two dimensions. Hermann Grassmann (1809–1877), in Germany, and William Rowan Hamilton (1805–1865), in Ireland, both attempted to find solutions.

Hamilton's system was the *quaternions,* which are best thought of as a real number plus a vector, and do for four dimensions what complex numbers do for two dimensions. In this system the order of multiplication matters; that is, **ab** ≠ **ba.** Also, two products of vectors emerged, the scalar (or dot) product and the vector (or cross) product.

Grassmann's abstract style, although easily read today, was almost impenetrable during the 19th century, and only a few of his ideas were appreciated. Among those few were the same scalar and vector products that Hamilton had found.

About 1880, the American physicist Josiah Willard Gibbs (1839–1903) worked out an algebra involving only the simplest concepts: the vectors and the two products. He then added some calculus, and the resulting system was simple, flexible, and well adapted to expressing a large number of physical laws. This system remains in use essentially unchanged. Hamilton's and Grassmann's more extensive systems each gave birth to much interesting mathematics, but little of this mathematics is seen at elementary levels.

5.4 EXERCISES

In Problems 1–8, use the vectors in the figure at the right to graph each of the following vectors.

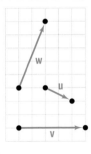

1. **v** + **w**

2. **u** + **v**

3. 3**v**

4. 4**w**

5. **v** − **w**

6. **u** − **v**

7. 3**v** + **u** − 2**w**

8. 2**u** − 3**v** + **w**

In Problems 9–16, use the figure at the right. Determine whether the given statement is true or false.

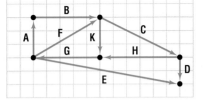

9. **A** + **B** = **F**

10. **K** + **G** = **F**

11. **C** = **D** − **E** + **F**

12. **G** + **H** + **E** = **D**

13. **E** + **D** = **G** + **H**

14. **H** − **C** = **G** − **F**

15. **A** + **B** + **K** + **G** = 0

16. **A** + **B** + **C** + **H** + **G** = 0

17. If ‖**v**‖ = 4, what is ‖3**v**‖?

18. If ‖**v**‖ = 2, what is ‖−4**v**‖?

*In Problems 19–26, the vector **v** has initial point P and terminal point Q. Write **v** in the form a**i** + b**j**, that is, find its position vector.*

19. $P = (0, 0)$; $Q = (3, 4)$

20. $P = (0, 0)$; $Q = (-3, -5)$

21. $P = (3, 2)$; $Q = (5, 6)$

22. $P = (-3, 2)$; $Q = (6, 5)$

23. $P = (-2, -1)$; $Q = (6, -2)$

24. $P = (-1, 4)$; $Q = (6, 2)$

25. $P = (1, 0)$; $Q = (0, 1)$

26. $P = (1, 1)$; $Q = (2, 2)$

*In Problems 27–32, find ‖**v**‖.*

27. **v** = 3**i** − 4**j**

28. **v** = −5**i** + 12**j**

29. **v** = **i** − **j**

30. **v** = −**i** − **j**

31. **v** = −2**i** + 3**j**

32. **v** = 6**i** + 2**j**

In Problems 33–38, find each quantity if $\mathbf{v} = 3\mathbf{i} - 5\mathbf{j}$ *and* $\mathbf{w} = -2\mathbf{i} + 3\mathbf{j}$.

33. $2\mathbf{v} + 3\mathbf{w}$

34. $3\mathbf{v} - 2\mathbf{w}$

35. $\|\mathbf{v} - \mathbf{w}\|$

36. $\|\mathbf{v} + \mathbf{w}\|$

37. $\|\mathbf{v}\| - \|\mathbf{w}\|$

38. $\|\mathbf{v}\| + \|\mathbf{w}\|$

In Problems 39–44, find the unit vector having the same direction as $\mathbf{v}$.

39. $\mathbf{v} = 5\mathbf{i}$

40. $\mathbf{v} = -3\mathbf{j}$

41. $\mathbf{v} = 3\mathbf{i} - 4\mathbf{j}$

42. $\mathbf{v} = -5\mathbf{i} + 12\mathbf{j}$

43. $\mathbf{v} = \mathbf{i} - \mathbf{j}$

44. $\mathbf{v} = 2\mathbf{i} - \mathbf{j}$

45. Find a vector $\mathbf{v}$ whose magnitude is 4 and whose component in the $\mathbf{i}$ direction is twice the component in the $\mathbf{j}$ direction.

46. Find a vector $\mathbf{v}$ whose magnitude is 3 and whose component in the $\mathbf{i}$ direction is equal to the component in the $\mathbf{j}$ direction.

47. If $\mathbf{v} = 2\mathbf{i} - \mathbf{j}$ and $\mathbf{w} = x\mathbf{i} + 3\mathbf{j}$, find all numbers x for which $\|\mathbf{v} + \mathbf{w}\| = 5$.

48. If $P = (-3, 1)$ and $Q = (x, 4)$, find all numbers x such that the vector represented by $\overrightarrow{PQ}$ has length 5.

In Problems 49–54, write the vector $\mathbf{v}$ *in the form* $a\mathbf{i} + b\mathbf{j}$, *given its magnitude* $\|\mathbf{v}\|$ *and the angle* α *it makes with the positive x-axis.*

49. $\|\mathbf{v}\| = 5$, $\alpha = 60°$

50. $\|\mathbf{v}\| = 8$, $\alpha = 45°$

51. $\|\mathbf{v}\| = 14$, $\alpha = 120°$

52. $\|\mathbf{v}\| = 3$, $\alpha = 240°$

53. $\|\mathbf{v}\| = 25$, $\alpha = 330°$

54. $\|\mathbf{v}\| = 15$, $\alpha = 315°$

55. A child pulls a wagon with a force of 40 pounds. The handle of the wagon makes an angle of 30° with the ground. Express the force vector $\mathbf{F}$ in terms of $\mathbf{i}$ and $\mathbf{j}$.

56. A man pushes a wheelbarrow up an incline of 20° with a force of 100 pounds. Express the force vector $\mathbf{F}$ in terms of $\mathbf{i}$ and $\mathbf{j}$.

57. **Resultant Force**   Two forces of magnitude 40 newtons (N) and 60 newtons act on an object at angles of 30° and $-45°$ with the positive x-axis as shown in the figure. Find the direction and magnitude of the resultant force; that is, find $\mathbf{F}_1 + \mathbf{F}_2$.

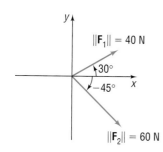

58. **Resultant Force**   Two forces of magnitude 30 newtons (N) and 70 newtons act on an object at angles of 45° and 120° with the positive x-axis as shown in the figure. Find the direction and magnitude of the resultant force; that is, find $\mathbf{F}_1 + \mathbf{F}_2$.

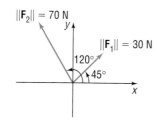

59. **Static Equilibrium**   A weight of 1000 pounds is suspended from two cables as shown in the figure. What is the tension of the two cables?

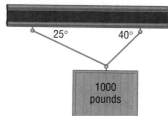

60. **Static Equilibrium**   A weight of 800 pounds is suspended from two cables as shown in the figure. What is the tension of the two cables?

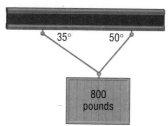

61. **Static Equilibrium**   A tightrope walker located at a certain point deflects the rope as indicated in the figure. If the weight of the tightrope walker is 150 pounds, how much tension is in each part of the slope?

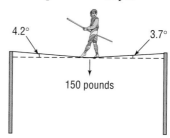

62. Static Equilibrium Repeat Problem 61 if the left angle is 3.8°, the right angle is 2.6°, and the weight of the tightrope walker is 135 pounds.

63. Show on the following graph the force needed for the object at P to be in static equilibrium.

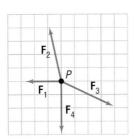

64. Explain in your own words what a vector is. Give an example of a vector.

65. Write a brief paragraph comparing the algebra of complex numbers and the algebra of vectors.

PREPARING FOR THIS SECTION

Before getting started, review the following:

✓ Law of Cosines (Section 4.3, p. 247)

5.5 THE DOT PRODUCT

OBJECTIVES
1. Find the Dot Product of Two Vectors
2. Find the Angle between Two Vectors
3. Determine Whether Two Vectors Are Parallel
4. Determine Whether Two Vectors Are Orthogonal
5. Decompose a Vector into Two Orthogonal Vectors
6. Compute Work

1️⃣ The definition for a product of two vectors is somewhat unexpected. However, such a product has meaning in many geometric and physical applications.

If $\mathbf{v} = a_1\mathbf{i} + b_1\mathbf{j}$ and $\mathbf{w} = a_2\mathbf{i} + b_2\mathbf{j}$ are two vectors, the **dot product** $\mathbf{v} \cdot \mathbf{w}$ is defined as

$$\mathbf{v} \cdot \mathbf{w} = a_1a_2 + b_1b_2 \qquad (1)$$

EXAMPLE 1 **Finding Dot Products**

If $\mathbf{v} = 2\mathbf{i} - 3\mathbf{j}$ and $\mathbf{w} = 5\mathbf{i} + 3\mathbf{j}$, find:

(a) $\mathbf{v} \cdot \mathbf{w}$ (b) $\mathbf{w} \cdot \mathbf{v}$ (c) $\mathbf{v} \cdot \mathbf{v}$

(d) $\mathbf{w} \cdot \mathbf{w}$ (e) $\|\mathbf{v}\|$ (f) $\|\mathbf{w}\|$

Solution (a) $\mathbf{v} \cdot \mathbf{w} = 2(5) + (-3)3 = 1$ (b) $\mathbf{w} \cdot \mathbf{v} = 5(2) + 3(-3) = 1$

(c) $\mathbf{v} \cdot \mathbf{v} = 2(2) + (-3)(-3) = 13$ (d) $\mathbf{w} \cdot \mathbf{w} = 5(5) + 3(3) = 34$

(e) $\|\mathbf{v}\| = \sqrt{2^2 + (-3)^2} = \sqrt{13}$ (f) $\|\mathbf{w}\| = \sqrt{5^2 + 3^2} = \sqrt{34}$ ∎

Since the dot product $\mathbf{v} \cdot \mathbf{w}$ of two vectors $\mathbf{v}$ and $\mathbf{w}$ is a real number (scalar), we sometimes refer to it as the **scalar product.**

PROPERTIES

The results obtained in Example 1 suggest some general properties.

Theorem   **Properties of the Dot Product**

If $\mathbf{u}, \mathbf{v},$ and $\mathbf{w}$ are vectors, then

Commutative Property

$$\mathbf{u} \cdot \mathbf{v} = \mathbf{v} \cdot \mathbf{u} \tag{2}$$

Distributive Property

$$\mathbf{u} \cdot (\mathbf{v} + \mathbf{w}) = \mathbf{u} \cdot \mathbf{v} + \mathbf{u} \cdot \mathbf{w} \tag{3}$$

$$\mathbf{v} \cdot \mathbf{v} = \|\mathbf{v}\|^2 \tag{4}$$

$$\mathbf{0} \cdot \mathbf{v} = 0 \tag{5}$$

■

Proof   We will prove properties (2) and (4) here and leave properties (3) and (5) as exercises (see Problems 33 and 34).

To prove property (2), we let $\mathbf{u} = a_1\mathbf{i} + b_1\mathbf{j}$ and $\mathbf{v} = a_2\mathbf{i} + b_2\mathbf{j}.$ Then

$$\mathbf{u} \cdot \mathbf{v} = a_1a_2 + b_1b_2 = a_2a_1 + b_2b_1 = \mathbf{v} \cdot \mathbf{u}$$

To prove property (4), we let $\mathbf{v} = a\mathbf{i} + b\mathbf{j}.$ Then

$$\mathbf{v} \cdot \mathbf{v} = a^2 + b^2 = \|\mathbf{v}\|^2$$ ■

One use of the dot product is to calculate the angle between two vectors.

ANGLE BETWEEN VECTORS

 Let $\mathbf{u}$ and $\mathbf{v}$ be two vectors with the same initial point $A.$ Then the vectors $\mathbf{u},$ $\mathbf{v},$ and $\mathbf{u} - \mathbf{v}$ form a triangle. The angle θ at vertex A of the triangle is the angle between the vectors $\mathbf{u}$ and $\mathbf{v}.$ See Figure 59. We wish to find a formula for calculating the angle $\theta.$

Figure 59

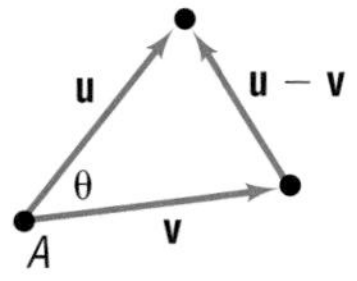

The sides of the triangle have lengths $\|\mathbf{v}\|, \|\mathbf{u}\|,$ and $\|\mathbf{u} - \mathbf{v}\|,$ and θ is the included angle between the sides of length $\|\mathbf{v}\|$ and $\|\mathbf{u}\|.$ The Law of Cosines (Section 4.3) can be used to find the cosine of the included angle.

$$\|\mathbf{u} - \mathbf{v}\|^2 = \|\mathbf{u}\|^2 + \|\mathbf{v}\|^2 - 2\|\mathbf{u}\| \|\mathbf{v}\| \cos\theta$$

Now we use property (4) to rewrite this equation in terms of dot products.

$$(\mathbf{u} - \mathbf{v}) \cdot (\mathbf{u} - \mathbf{v}) = \mathbf{u} \cdot \mathbf{u} + \mathbf{v} \cdot \mathbf{v} - 2\|\mathbf{u}\| \|\mathbf{v}\| \cos\theta \tag{6}$$

Then we apply the distributive property (3) twice on the left side of (6) to obtain

$$(\mathbf{u} - \mathbf{v}) \cdot (\mathbf{u} - \mathbf{v}) = \mathbf{u} \cdot (\mathbf{u} - \mathbf{v}) - \mathbf{v} \cdot (\mathbf{u} - \mathbf{v})$$

$$= \mathbf{u} \cdot \mathbf{u} - \mathbf{u} \cdot \mathbf{v} - \mathbf{v} \cdot \mathbf{u} + \mathbf{v} \cdot \mathbf{v}$$

$$= \mathbf{u} \cdot \mathbf{u} + \mathbf{v} \cdot \mathbf{v} - 2\,\mathbf{u} \cdot \mathbf{v} \qquad \textbf{(7)}$$

↑
Property (2)

Combining equations (6) and (7), we have

$$\mathbf{u} \cdot \mathbf{u} + \mathbf{v} \cdot \mathbf{v} - 2\,\mathbf{u} \cdot \mathbf{v} = \mathbf{u} \cdot \mathbf{u} + \mathbf{v} \cdot \mathbf{v} - 2\|\mathbf{u}\|\,\|\mathbf{v}\|\cos\theta$$

$$\mathbf{u} \cdot \mathbf{v} = \|\mathbf{u}\|\,\|\mathbf{v}\|\cos\theta$$

We have proved the following result:

Theorem **Angle between Vectors**

If **u** and **v** are two nonzero vectors, the angle θ, $0 \le \theta \le \pi$, between **u** and **v** is determined by the formula

$$\cos\theta = \frac{\mathbf{u} \cdot \mathbf{v}}{\|\mathbf{u}\|\,\|\mathbf{v}\|} \qquad \textbf{(8)}$$

EXAMPLE 2 **Finding the Angle θ between Two Vectors**

Find the angle θ between $\mathbf{u} = 4\mathbf{i} - 3\mathbf{j}$ and $\mathbf{v} = 2\mathbf{i} + 5\mathbf{j}$.

Solution We compute the quantities $\mathbf{u} \cdot \mathbf{v}$, $\|\mathbf{u}\|$, and $\|\mathbf{v}\|$.

$$\mathbf{u} \cdot \mathbf{v} = 4(2) + (-3)(5) = -7$$

$$\|\mathbf{u}\| = \sqrt{4^2 + (-3)^2} = 5$$

$$\|\mathbf{v}\| = \sqrt{2^2 + 5^2} = \sqrt{29}$$

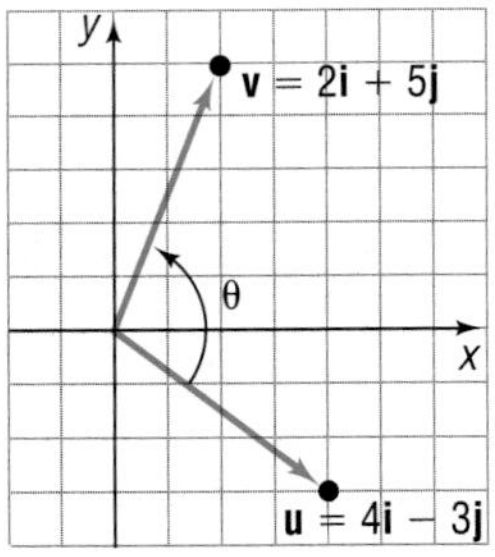

By formula (8), if θ is the angle between **u** and **v**, then

$$\cos\theta = \frac{\mathbf{u} \cdot \mathbf{v}}{\|\mathbf{u}\|\,\|\mathbf{v}\|} = \frac{-7}{5\sqrt{29}} \approx -0.26$$

We find that $\theta \approx 105°$. See Figure 60.

Figure 60

NOW WORK PROBLEM **1(a)** AND **(b)**.

EXAMPLE 3 **Finding the Actual Speed and Direction of an Aircraft**

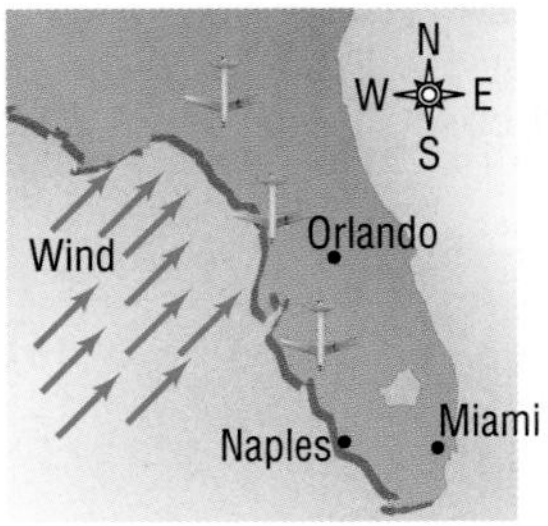

A Boeing 737 aircraft maintains a constant airspeed of 500 miles per hour in the direction due south. The velocity of the jet stream is 80 miles per hour in a northeasterly direction. Find the actual speed and direction of the aircraft relative to the ground.

Figure 61

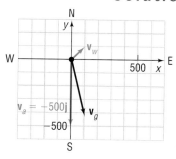

Solution We set up a coordinate system in which north (N) is along the positive *y*-axis. See Figure 61. Let

$$\mathbf{v}_a = \text{velocity of aircraft relative to the air} = -500\mathbf{j}$$
$$\mathbf{v}_g = \text{velocity of aircraft relative to ground}$$
$$\mathbf{v}_w = \text{velocity of jet stream}$$

The velocity of the jet stream $\mathbf{v}_w$ has magnitude 80 and direction NE (northeast), so $\alpha = 45°$. We express $\mathbf{v}_w$ in terms of $\mathbf{i}$ and $\mathbf{j}$ as

$$\mathbf{v}_w = 80(\cos 45°\mathbf{i} + \sin 45°\mathbf{j}) = 80\left(\frac{\sqrt{2}}{2}\mathbf{i} + \frac{\sqrt{2}}{2}\mathbf{j}\right) = 40\sqrt{2}(\mathbf{i} + \mathbf{j})$$

The velocity of the aircraft relative to the ground is

$$\mathbf{v}_g = \mathbf{v}_a + \mathbf{v}_w = -500\mathbf{j} + 40\sqrt{2}(\mathbf{i} + \mathbf{j}) = 40\sqrt{2}\mathbf{i} + (40\sqrt{2} - 500)\mathbf{j}$$

The actual speed of the aircraft is

$$\|\mathbf{v}_g\| = \sqrt{(40\sqrt{2})^2 + (40\sqrt{2} - 500)^2} \approx 447 \text{ miles per hour}$$

The angle θ between $\mathbf{v}_g$ and the vector $\mathbf{v}_a = -500\mathbf{j}$ (the velocity of the aircraft relative to the air) is determined by the equation

$$\cos\theta = \frac{\mathbf{v}_g \cdot \mathbf{v}_a}{\|\mathbf{v}_g\| \|\mathbf{v}_a\|} = \frac{(40\sqrt{2} - 500)(-500)}{(447)(500)} \approx 0.9920$$
$$\theta \approx 7.3°$$

The direction of the aircraft relative to the ground is approximately S7.3°E (about 7.3° east of south). ∎

 NOW WORK PROBLEM 19.

PARALLEL AND ORTHOGONAL VECTORS

③ Two vectors **v** and **w** are said to be **parallel** if there is a nonzero scalar α so that $\mathbf{v} = \alpha\mathbf{w}$. In this case, the angle θ between **v** and **w** is 0 or π.

EXAMPLE 4 **Determining Whether Vectors Are Parallel**

The vectors $\mathbf{v} = 3\mathbf{i} - \mathbf{j}$ and $\mathbf{w} = 6\mathbf{i} - 2\mathbf{j}$ are parallel, since $\mathbf{v} = \frac{1}{2}\mathbf{w}$. Furthermore, since

$$\cos\theta = \frac{\mathbf{v} \cdot \mathbf{w}}{\|\mathbf{v}\| \|\mathbf{w}\|} = \frac{18 + 2}{\sqrt{10}\sqrt{40}} = \frac{20}{\sqrt{400}} = 1$$

the angle θ between **v** and **w** is 0. ∎

④ If the angle θ between two nonzero vectors **v** and **w** is $\dfrac{\pi}{2}$, the vectors **v** and **w** are called **orthogonal.*** See Figure 62.

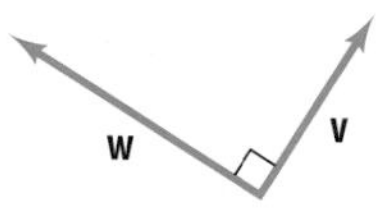

It follows from formula (8) that if **v** and **w** are orthogonal then $\mathbf{v} \cdot \mathbf{w} = 0$, since $\cos\dfrac{\pi}{2} = 0$.

Figure 62
v is orthogonal to **w**

On the other hand, if $\mathbf{v} \cdot \mathbf{w} = 0$, then either $\mathbf{v} = 0$ or $\mathbf{w} = 0$ or $\cos\theta = 0$. In the latter case, $\theta = \dfrac{\pi}{2}$, and **v** and **w** are orthogonal. If **v** or **w** is the zero vector, then, since the zero vector has no specific direction, we adopt the convention that the zero vector is orthogonal to every vector.

---

* *Orthogonal, perpendicular,* and *normal* are all terms that mean "meet at a right angle." It is customary to refer to two vectors as being *orthogonal,* two lines as being *perpendicular,* and a line and a plane or a vector and a plane as being *normal.*

Theorem	Two vectors **v** and **w** are orthogonal if and only if

$$\mathbf{v} \cdot \mathbf{w} = 0$$

EXAMPLE 5   **Determining Whether Two Vectors Are Orthogonal**

Figure 63

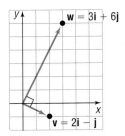

The vectors

$$\mathbf{v} = 2\mathbf{i} - \mathbf{j} \quad \text{and} \quad \mathbf{w} = 3\mathbf{i} + 6\mathbf{j}$$

are orthogonal, since

$$\mathbf{v} \cdot \mathbf{w} = 6 - 6 = 0$$

See Figure 63.

NOW WORK PROBLEM **1(c)**.

PROJECTION OF A VECTOR ONTO ANOTHER VECTOR

Figure 64

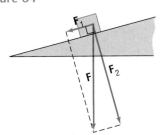

In many physical applications, it is necessary to find "how much" of a vector is applied in a given direction. Look at Figure 64. The force **F** due to gravity is pulling straight down (toward the center of Earth) on the block. To study the effect of gravity on the block, it is necessary to determine how much of **F** is actually pushing the block down the incline $(\mathbf{F}_1)$ and how much is pressing the block against the incline $(\mathbf{F}_2)$, at a right angle to the incline. Knowing the **decomposition** of **F** often will allow us to determine when friction is overcome and the block will slide down the incline.

Suppose that **v** and **w** are two nonzero vectors with the same initial point P. We seek to decompose **v** into two vectors: $\mathbf{v}_1$, which is parallel to **w**, and $\mathbf{v}_2$, which is orthogonal to **w**. See Figure 65(a) and (b). The vector $\mathbf{v}_1$ is called the **vector projection of v onto w.**

Figure 65

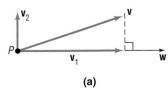

(a)

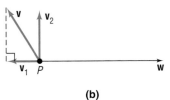

(b)

The vector $\mathbf{v}_1$ is obtained as follows: From the terminal point of **v**, drop a perpendicular to the line containing **w**. The vector $\mathbf{v}_1$ is the vector from P to the foot of this perpendicular. The vector $\mathbf{v}_2$ is given by $\mathbf{v}_2 = \mathbf{v} - \mathbf{v}_1$. Note that $\mathbf{v} = \mathbf{v}_1 + \mathbf{v}_2$, $\mathbf{v}_1$ is parallel to **w**, and $\mathbf{v}_2$ is orthogonal to **w**. This is the decomposition of **v** that we wanted.

Now we seek a formula for $\mathbf{v}_1$ that is based on a knowledge of the vectors **v** and **w**. Since $\mathbf{v} = \mathbf{v}_1 + \mathbf{v}_2$, we have

$$\mathbf{v} \cdot \mathbf{w} = (\mathbf{v}_1 + \mathbf{v}_2) \cdot \mathbf{w} = \mathbf{v}_1 \cdot \mathbf{w} + \mathbf{v}_2 \cdot \mathbf{w} \tag{9}$$

Since $\mathbf{v}_2$ is orthogonal to **w**, we have $\mathbf{v}_2 \cdot \mathbf{w} = 0$. Since $\mathbf{v}_1$ is parallel to **w**, we have $\mathbf{v}_1 = \alpha\mathbf{w}$ for some scalar α. Equation (9) can be written as

$$\mathbf{v} \cdot \mathbf{w} = \alpha\mathbf{w} \cdot \mathbf{w} = \alpha\|\mathbf{w}\|^2$$

$$\alpha = \frac{\mathbf{v} \cdot \mathbf{w}}{\|\mathbf{w}\|^2}$$

Then

$$\mathbf{v}_1 = \alpha\mathbf{w} = \frac{\mathbf{v} \cdot \mathbf{w}}{\|\mathbf{w}\|^2} \mathbf{w}$$

Theorem   If **v** and **w** are two nonzero vectors, the vector projection of **v** onto **w** is

$$\mathbf{v}_1 = \frac{\mathbf{v} \cdot \mathbf{w}}{\|\mathbf{w}\|^2} \mathbf{w} \tag{10}$$

The decomposition of **v** into $\mathbf{v}_1$ and $\mathbf{v}_2$, where $\mathbf{v}_1$ is parallel to **w** and $\mathbf{v}_2$ is perpendicular to **w,** is

$$\mathbf{v}_1 = \frac{\mathbf{v} \cdot \mathbf{w}}{\|\mathbf{w}\|^2} \mathbf{w} \qquad \mathbf{v}_2 = \mathbf{v} - \mathbf{v}_1 \tag{11}$$

EXAMPLE 6   Decomposing a Vector into Two Orthogonal Vectors

Find the vector projection of $\mathbf{v} = \mathbf{i} + 3\mathbf{j}$ onto $\mathbf{w} = \mathbf{i} + \mathbf{j}$. Decompose **v** into two vectors $\mathbf{v}_1$ and $\mathbf{v}_2$, where $\mathbf{v}_1$ is parallel to **w** and $\mathbf{v}_2$ is orthogonal to **w.**

Figure 66

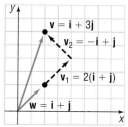

Solution   We use formulas (10) and (11).

$$\mathbf{v}_1 = \frac{\mathbf{v} \cdot \mathbf{w}}{\|\mathbf{w}\|^2} \mathbf{w} = \frac{1 + 3}{\left(\sqrt{2}\right)^2} \mathbf{w} = 2\mathbf{w} = 2(\mathbf{i} + \mathbf{j})$$

$$\mathbf{v}_2 = \mathbf{v} - \mathbf{v}_1 = (\mathbf{i} + 3\mathbf{j}) - 2(\mathbf{i} + \mathbf{j}) = -\mathbf{i} + \mathbf{j}$$

See Figure 66.

NOW WORK PROBLEM **13**.

WORK DONE BY A CONSTANT FORCE

⑥   In elementary physics, the **work** W done by a constant force **F** in moving an object from a point A to a point B is defined as

$$W = (\text{magnitude of force})(\text{distance}) = \|\mathbf{F}\| \, \|\overrightarrow{AB}\|$$

Work is commonly measured in foot-pounds or in newton-meters (joules).

In this definition, it is assumed that the force **F** is applied along the line of motion. If the constant force **F** is not along the line of motion, but, instead, is at an angle θ to the direction of motion, as illustrated in Figure 67, then the **work W done by F** in moving an object from A to B is defined as

$$W = \mathbf{F} \cdot \overrightarrow{AB} \tag{12}$$

Figure 67

This definition is compatible with the force times distance definition given above, since

$$W = (\text{amount of force in the direction of } \overrightarrow{AB})(\text{distance})$$

$$= \|\text{projection of } \mathbf{F} \text{ on } AB\| \, \|\overrightarrow{AB}\| = \frac{\mathbf{F} \cdot \overrightarrow{AB}}{\|\overrightarrow{AB}\|^2} \|\overrightarrow{AB}\| \, \|\overrightarrow{AB}\| = \mathbf{F} \cdot \overrightarrow{AB}$$

EXAMPLE 7   Computing Work

Figure 68(a) shows a girl pulling a wagon with a force of 50 pounds. How much work is done in moving the wagon 100 feet if the handle makes an angle of 30° with the ground?

Figure 68

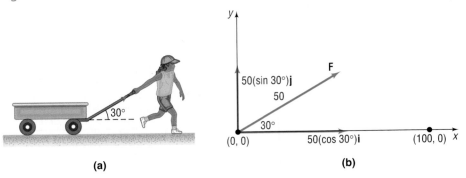

(a)   (b)

Solution   We position the vectors in a coordinate system in such a way that the wagon is moved from $(0, 0)$ to $(100, 0)$. The motion is from $A = (0, 0)$ to $B = (100, 0)$, so $\overrightarrow{AB} = 100\mathbf{i}$. The force vector $\mathbf{F}$, as shown in Figure 68(b), is

$$\mathbf{F} = 50(\cos 30°\mathbf{i} + \sin 30°\mathbf{j}) = 50\left(\frac{\sqrt{3}}{2}\mathbf{i} + \frac{1}{2}\mathbf{j}\right) = 25(\sqrt{3}\mathbf{i} + \mathbf{j})$$

By formula (12), the work done is

$$W = \mathbf{F} \cdot \overrightarrow{AB} = 25(\sqrt{3}\mathbf{i} + \mathbf{j}) \cdot 100\mathbf{i} = 2500\sqrt{3} \text{ foot-pounds} \quad ▬$$

NOW WORK PROBLEM 29.

HISTORICAL FEATURE

1. We stated in an earlier Historical Feature that complex numbers were used as vectors in the plane before the general notion of a vector was clarified. Suppose that we make the correspondence

Vector ↔ Complex number

$$a\mathbf{i} + b\mathbf{j} \leftrightarrow a + bi$$

$$c\mathbf{i} + d\mathbf{j} \leftrightarrow c + di$$

Show that

$$(a\mathbf{i} + b\mathbf{j}) \cdot (c\mathbf{i} + d\mathbf{j}) = \text{real part}\left[\overline{(a + bi)}(c + di)\right]$$

This is how the dot product was found originally. The imaginary part is also interesting. It is a determinant and represents the area of the parallelogram whose edges are the vectors. This is close to some of Hermann Grassmann's ideas and is also connected with the scalar triple product of three-dimensional vectors.

5.5 EXERCISES

In Problems 1–10, (a) find the dot product $\mathbf{v} \cdot \mathbf{w}$; (b) find the angle between $\mathbf{v}$ and $\mathbf{w}$; (c) state whether the vectors are parallel, orthogonal, or neither.

1. $\mathbf{v} = \mathbf{i} - \mathbf{j}, \quad \mathbf{w} = \mathbf{i} + \mathbf{j}$

2. $\mathbf{v} = \mathbf{i} + \mathbf{j}, \quad \mathbf{w} = -\mathbf{i} + \mathbf{j}$

3. $\mathbf{v} = 2\mathbf{i} + \mathbf{j}, \quad \mathbf{w} = \mathbf{i} + 2\mathbf{j}$

4. $\mathbf{v} = 2\mathbf{i} + 2\mathbf{j}, \quad \mathbf{w} = \mathbf{i} + 2\mathbf{j}$

5. $\mathbf{v} = \sqrt{3}\mathbf{i} - \mathbf{j}, \quad \mathbf{w} = \mathbf{i} + \mathbf{j}$

6. $\mathbf{v} = \mathbf{i} + \sqrt{3}\mathbf{j}, \quad \mathbf{w} = \mathbf{i} - \mathbf{j}$

7. $\mathbf{v} = 3\mathbf{i} + 4\mathbf{j}, \quad \mathbf{w} = 4\mathbf{i} + 3\mathbf{j}$

8. $\mathbf{v} = 3\mathbf{i} - 4\mathbf{j}, \quad \mathbf{w} = 4\mathbf{i} - 3\mathbf{j}$

9. $\mathbf{v} = 4\mathbf{i}, \quad \mathbf{w} = \mathbf{j}$

10. $\mathbf{v} = \mathbf{i}, \quad \mathbf{w} = -3\mathbf{j}$

11. Find a so that the vectors $\mathbf{v} = \mathbf{i} - a\mathbf{j}$ and $\mathbf{w} = 2\mathbf{i} + 3\mathbf{j}$ are orthogonal.

12. Find b so that the vectors $\mathbf{v} = \mathbf{i} + \mathbf{j}$ and $\mathbf{w} = \mathbf{i} + b\mathbf{j}$ are orthogonal.

In Problems 13–18, decompose **v** *into two vectors* **v**₁ *and* **v**₂, *where* **v**₁ *is parallel to* **w** *and* **v**₂ *is orthogonal to* **w**.

13. $\mathbf{v} = 2\mathbf{i} - 3\mathbf{j}$, $\mathbf{w} = \mathbf{i} - \mathbf{j}$

14. $\mathbf{v} = -3\mathbf{i} + 2\mathbf{j}$, $\mathbf{w} = 2\mathbf{i} + \mathbf{j}$

15. $\mathbf{v} = \mathbf{i} - \mathbf{j}$, $\mathbf{w} = \mathbf{i} + 2\mathbf{j}$

16. $\mathbf{v} = 2\mathbf{i} - \mathbf{j}$, $\mathbf{w} = \mathbf{i} - 2\mathbf{j}$

17. $\mathbf{v} = 3\mathbf{i} + \mathbf{j}$, $\mathbf{w} = -2\mathbf{i} - \mathbf{j}$

18. $\mathbf{v} = \mathbf{i} - 3\mathbf{j}$, $\mathbf{w} = 4\mathbf{i} - \mathbf{j}$

---

19. Finding the Actual Speed and Direction of an Aircraft
A DC-10 jumbo jet maintains an airspeed of 550 miles per hour in a southwesterly direction. The velocity of the jet stream is a constant 80 miles per hour from the west. Find the actual speed and direction of the aircraft.

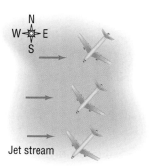

Jet stream

20. Finding the Correct Compass Heading   The pilot of an aircraft wishes to head directly east, but is faced with a wind speed of 40 miles per hour from the northwest. If the pilot maintains an airspeed of 250 miles per hour, what compass heading should be maintained? What is the actual speed of the aircraft?

21. Correct Direction for Crossing a River   A river has a constant current of 3 kilometers per hour. At what angle to a boat dock should a motorboat, capable of maintaining a constant speed of 20 kilometers per hour, be headed in order to reach a point directly opposite the dock? If the river is $\frac{1}{2}$ kilometer wide, how long will it take to cross?

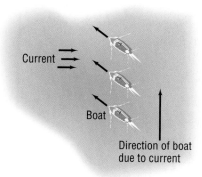

Current

Boat

Direction of boat
due to current

22. Correct Direction for Crossing a River   Repeat Problem 21 if the current is 5 kilometers per hour.

23. Braking Load   A Toyota Sienna with a gross weight of 5300 pounds is parked on a street with a slope of 8°. See

the figure. Find the force required to keep the Sienna from rolling down the hill. What is the force perpendicular to the hill?

Weight = 5300 pounds

24. Braking Load   A Pontiac Bonneville with a gross weight of 4500 pounds is parked on a street with a slope of 10°. Find the force required to keep the Bonneville from rolling down the hill. What is the force perpendicular to the hill?

25. Ground Speed and Direction of an Airplane   An airplane has an airspeed of 500 kilometers per hour bearing N45°E. The wind velocity is 60 kilometers per hour in the direction N30°W. Find the resultant vector representing the path of the plane relative to the ground. What is the ground speed of the plane? What is its direction?

26. Ground Speed and Direction of an Airplane   An airplane has an airspeed of 600 kilometers per hour bearing S30°E. The wind velocity is 40 kilometers per hour in the direction S45°E. Find the resultant vector representing the path of the plane relative to the ground. What is the ground speed of the plane? What is its direction?

27. Crossing a River   A small motorboat in still water maintains a speed of 20 miles per hour. In heading directly across a river (that is, perpendicular to the current) whose current is 3 miles per hour, find a vector representing the speed and direction of the motorboat. What is the true speed of the motorboat? What is its direction?

28. Crossing a River   A small motorboat in still water maintains a speed of 10 miles per hour. In heading directly across a river (that is, perpendicular to the current) whose current is 4 miles per hour, find a vector representing the speed and direction of the motorboat. What is the true speed of the motorboat? What is its direction?

29. Computing Work   Find the work done by a force of 3 pounds acting in the direction 60° to the horizontal in moving an object 2 feet from $(0, 0)$ to $(2, 0)$.

30. Computing Work Find the work done by a force of 1 pound acting in the direction 45° to the horizontal in moving an object 5 feet from $(0, 0)$ to $(5, 0)$.

31. Computing Work A wagon is pulled horizontally by exerting a force of 20 pounds on the handle at an angle of 30° with the horizontal. How much work is done in moving the wagon 100 feet?

32. Find the acute angle that a constant unit force vector makes with the positive x-axis if the work done by the force in moving a particle from $(0, 0)$ to $(4, 0)$ equals 2.

33. Prove the distributive property:

$$\mathbf{u} \cdot (\mathbf{v} + \mathbf{w}) = \mathbf{u} \cdot \mathbf{v} + \mathbf{u} \cdot \mathbf{w}$$

34. Prove property (5), $\mathbf{0} \cdot \mathbf{v} = 0$.

35. If $\mathbf{v}$ is a unit vector and the angle between $\mathbf{v}$ and $\mathbf{i}$ is α, show that $\mathbf{v} = \cos \alpha \mathbf{i} + \sin \alpha \mathbf{j}$.

36. Suppose that $\mathbf{v}$ and $\mathbf{w}$ are unit vectors. If the angle between $\mathbf{v}$ and $\mathbf{i}$ is α and if the angle between $\mathbf{w}$ and $\mathbf{i}$ is β, use the idea of the dot product $\mathbf{v} \cdot \mathbf{w}$ to prove that

$$\cos(\alpha - \beta) = \cos \alpha \cos \beta + \sin \alpha \sin \beta$$

37. Show that the projection of $\mathbf{v}$ onto $\mathbf{i}$ is $(\mathbf{v} \cdot \mathbf{i})\mathbf{i}$. In fact, show that we can always write a vector $\mathbf{v}$ as

$$\mathbf{v} = (\mathbf{v} \cdot \mathbf{i})\mathbf{i} + (\mathbf{v} \cdot \mathbf{j})\mathbf{j}$$

38. (a) If $\mathbf{u}$ and $\mathbf{v}$ have the same magnitude, show that $\mathbf{u} + \mathbf{v}$ and $\mathbf{u} - \mathbf{v}$ are orthogonal.
(b) Use this to prove that an angle inscribed in a semicircle is a right angle (see the figure).

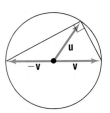

39. Let $\mathbf{v}$ and $\mathbf{w}$ denote two nonzero vectors. Show that the vector $\mathbf{v} - \alpha\mathbf{w}$ is orthogonal to $\mathbf{w}$ if $\alpha = (\mathbf{v} \cdot \mathbf{w})/\|\mathbf{w}\|^2$.

40. Let $\mathbf{v}$ and $\mathbf{w}$ denote two nonzero vectors. Show that the vectors $\|\mathbf{w}\|\mathbf{v} + \|\mathbf{v}\|\mathbf{w}$ and $\|\mathbf{w}\|\mathbf{v} - \|\mathbf{v}\|\mathbf{w}$ are orthogonal.

41. In the definition of work given in this section, what is the work done if $\mathbf{F}$ is orthogonal to $\overrightarrow{AB}$?

42. Prove the **polarization identity,**

$$\|\mathbf{u} + \mathbf{v}\|^2 - \|\mathbf{u} - \mathbf{v}\|^2 = 4(\mathbf{u} \cdot \mathbf{v}).$$

43. Make up an application different from any found in the text that requires the dot product.

P R E P A R I N G   F O R   T H I S   S E C T I O N

Before getting started, review the following:

✓ Distance Formula (Section 1.1, p. 4)

5.6 VECTORS IN SPACE

OBJECTIVES
1. Find the Distance between Two Points
2. Find Position Vectors
3. Perform Operations on Vectors
4. Find the Dot Product
5. Find the Angle between Two Vectors
6. Find the Direction Angles of a Vector

Figure 69

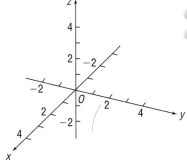

RECTANGULAR COORDINATES IN SPACE

In the plane, each point is associated with an ordered pair of real numbers. In space, each point is associated with an ordered triple of real numbers. Through a fixed point, the **origin**, O, draw three mutually perpendicular lines, the x-axis, the y-axis, and the z-axis. On each of these axes, select an appropriate scale and the positive direction. See Figure 69.

The direction chosen for the positive z-axis in Figure 69 makes the system *right-handed*. This conforms to the *right-hand rule*, which states that, if the index finger of the right hand points in the direction of the positive x-axis and the middle finger points in the direction of the positive y-axis, then the thumb will point in the direction of the positive z-axis. See Figure 70.

Figure 70

Figure 71

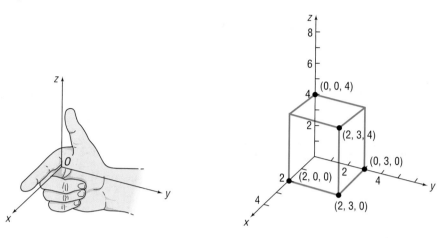

We associate with each point P an ordered triple (x, y, z) of real numbers, the **coordinates of P**. For example, the point $(2, 3, 4)$ is located by starting at the origin and moving 2 units along the positive x-axis, 3 units in the direction of the positive y-axis, and 4 units in the direction of the positive z-axis. See Figure 71.

Figure 71 also shows the location of the points $(2, 0, 0)$, $(0, 3, 0)$, $(0, 0, 4)$, and $(2, 3, 0)$. Points of the form $(x, 0, 0)$ lie on the x-axis, while points of the form $(0, y, 0)$ and $(0, 0, z)$ lie on the y-axis and z-axis, respectively. Points of the form $(x, y, 0)$ lie in a plane, called the **xy-plane.** Its equation is $z = 0$. Similarly, points of the form $(x, 0, z)$ lie in the **xz-plane** (equation $y = 0$) and points of the form $(0, y, z)$ lie in the **yz-plane** (equation $x = 0$). See Figure 72(a). By extension of these ideas, all points obeying the equation $z = 3$ will lie in a plane parallel to and 3 units above the xy-plane. The equation $y = 4$ represents a plane parallel to the xz-plane and 4 units to the right of the plane $y = 0$. See Figure 72(b).

Figure 72

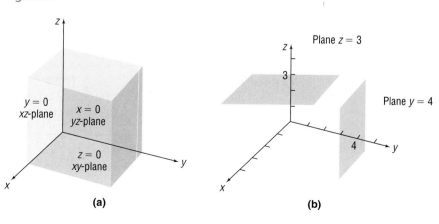

(a)

(b)

NOW WORK PROBLEM 3.

1 The formula for the distance between two points in space is an extension of the Distance Formula for points in the plane given in Chapter 1.

Theorem

Distance Formula in Space

If $P_1 = (x_1, y_1, z_1)$ and $P_2 = (x_2, y_2, z_2)$ are two points in space, the distance d from P_1 to P_2 is

$$d = \sqrt{(x_2 - x_1)^2 + (y_2 - y_1)^2 + (z_2 - z_1)^2} \qquad \textbf{(1)}$$

The proof, which we omit, utilizes a double application of the Pythagorean Theorem.

EXAMPLE 1

Using the Distance Formula

Find the distance from $P_1 = (-1, 3, 2)$ to $P_2 = (4, -2, 5)$.

Solution $d = \sqrt{[4 - (-1)]^2 + [-2 - 3]^2 + [5 - 2]^2} = \sqrt{25 + 25 + 9} = \sqrt{59}$ ∎

NOW WORK PROBLEM **9**.

REPRESENTING VECTORS IN SPACE

2 To represent vectors in space, we introduce the unit vectors **i**, **j**, and **k** whose directions are along the positive x-axis, positive y-axis, and positive z-axis, respectively. See Figure 73. If **v** is a vector with initial point at the origin O and terminal point at $P = (a, b, c)$, then we can represent **v** in terms of the vectors **i**, **j**, and **k** as

Figure 73

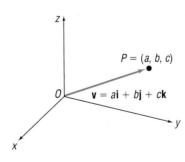

$$\mathbf{v} = a\mathbf{i} + b\mathbf{j} + c\mathbf{k}$$

The scalars a, b, and c are called the **components** of the vector $\mathbf{v} = a\mathbf{i} + b\mathbf{j} + c\mathbf{k}$, with a being the component in the direction **i**, b the component in the direction **j**, and c the component in the direction **k**.

A vector whose initial point is at the origin is called a **position vector.** The next result states that any vector whose initial point is not at the origin is equal to a unique position vector.

Theorem

Suppose that **v** is a vector with initial point $P_1 = (x_1, y_1, z_1)$, not necessarily the origin, and terminal point $P_2 = (x_2, y_2, z_2)$. If $\mathbf{v} = \overrightarrow{P_1 P_2}$, then **v** is equal to the position vector

$$\mathbf{v} = (x_2 - x_1)\mathbf{i} + (y_2 - y_1)\mathbf{j} + (z_2 - z_1)\mathbf{k} \qquad \textbf{(2)}$$

Figure 74 illustrates this result.

Figure 74

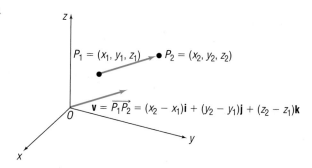

EXAMPLE 2

Finding a Position Vector

Find the position vector of the vector $\mathbf{v} = \overrightarrow{P_1P_2}$ if $P_1 = (-1, 2, 3)$ and $P_2 = (4, 6, 2)$.

Solution   By equation (2), the position vector equal to $\mathbf{v}$ is

$$\mathbf{v} = \left[4 - (-1) \right]\mathbf{i} + (6 - 2)\mathbf{j} + (2 - 3)\mathbf{k} = 5\mathbf{i} + 4\mathbf{j} - \mathbf{k}$$ ■

─── **NOW WORK PROBLEM 23.**

③ Next, we define equality, addition, subtraction, scalar product, and magnitude in terms of the components of a vector.

Let $\mathbf{v} = a_1\mathbf{i} + b_1\mathbf{j} + c_1\mathbf{k}$ and $\mathbf{w} = a_2\mathbf{i} + b_2\mathbf{j} + c_2\mathbf{k}$ be two vectors, and let α be a scalar. Then

$$\mathbf{v} = \mathbf{w} \quad \text{if and only if } a_1 = a_2, b_1 = b_2, \text{ and } c_1 = c_2$$

$$\mathbf{v} + \mathbf{w} = (a_1 + a_2)\mathbf{i} + (b_1 + b_2)\mathbf{j} + (c_1 + c_2)\mathbf{k}$$

$$\mathbf{v} - \mathbf{w} = (a_1 - a_2)\mathbf{i} + (b_1 - b_2)\mathbf{j} + (c_1 - c_2)\mathbf{k}$$

$$\alpha\mathbf{v} = (\alpha a_1)\mathbf{i} + (\alpha b_1)\mathbf{j} + (\alpha c_1)\mathbf{k}$$

$$\|\mathbf{v}\| = \sqrt{a_1^2 + b_1^2 + c_1^2}$$

These definitions are compatible with the geometric ones given earlier in Section 5.4.

EXAMPLE 3

Adding and Subtracting Vectors

If $\mathbf{v} = 2\mathbf{i} + 3\mathbf{j} - 2\mathbf{k}$ and $\mathbf{w} = 3\mathbf{i} - 4\mathbf{j} + 5\mathbf{k}$, find:

(a) $\mathbf{v} + \mathbf{w}$ (b) $\mathbf{v} - \mathbf{w}$

Solution   (a) $\mathbf{v} + \mathbf{w} = (2\mathbf{i} + 3\mathbf{j} - 2\mathbf{k}) + (3\mathbf{i} - 4\mathbf{j} + 5\mathbf{k})$
$$= (2 + 3)\mathbf{i} + (3 - 4)\mathbf{j} + (-2 + 5)\mathbf{k}$$
$$= 5\mathbf{i} - \mathbf{j} + 3\mathbf{k}$$

(b) $\mathbf{v} - \mathbf{w} = (2\mathbf{i} + 3\mathbf{j} - 2\mathbf{k}) - (3\mathbf{i} - 4\mathbf{j} + 5\mathbf{k})$
$$= (2 - 3)\mathbf{i} + \left[3 - (-4) \right]\mathbf{j} + \left[-2 - 5 \right]\mathbf{k}$$
$$= -\mathbf{i} + 7\mathbf{j} - 7\mathbf{k}$$ ■

| EXAMPLE 4 | **Finding Scalar Products and Magnitudes** |

If $\mathbf{v} = 2\mathbf{i} + 3\mathbf{j} - 2\mathbf{k}$ and $\mathbf{w} = 3\mathbf{i} - 4\mathbf{j} + 5\mathbf{k}$, find:

(a) $3\mathbf{v}$ (b) $2\mathbf{v} - 3\mathbf{w}$ (c) $\|\mathbf{v}\|$

Solution (a) $3\mathbf{v} = 3(2\mathbf{i} + 3\mathbf{j} - 2\mathbf{k}) = 6\mathbf{i} + 9\mathbf{j} - 6\mathbf{k}$

(b) $2\mathbf{v} - 3\mathbf{w} = 2(2\mathbf{i} + 3\mathbf{j} - 2\mathbf{k}) - 3(3\mathbf{i} - 4\mathbf{j} + 5\mathbf{k})$
$= 4\mathbf{i} + 6\mathbf{j} - 4\mathbf{k} - 9\mathbf{i} + 12\mathbf{j} - 15\mathbf{k} = -5\mathbf{i} + 18\mathbf{j} - 19\mathbf{k}$

(c) $\|\mathbf{v}\| = \|2\mathbf{i} + 3\mathbf{j} - 2\mathbf{k}\| = \sqrt{2^2 + 3^2 + (-2)^2} = \sqrt{17}$

> ✏️ **NOW WORK PROBLEMS 27 AND 33.**

Recall that a unit vector $\mathbf{u}$ is one for which $\|\mathbf{u}\| = 1$. In many applications, it is useful to be able to find a unit vector $\mathbf{u}$ that has the same direction as a given vector $\mathbf{v}$.

Theorem **Unit Vector in the Direction of v**

For any nonzero vector $\mathbf{v}$, the vector

$$\mathbf{u} = \frac{\mathbf{v}}{\|\mathbf{v}\|}$$

is a unit vector that has the same direction as $\mathbf{v}$.

As a consequence of this theorem, if $\mathbf{u}$ is a unit vector in the same direction as a vector $\mathbf{v}$, then $\mathbf{v}$ may be expressed as

$$\mathbf{v} = \|\mathbf{v}\|\,\mathbf{u}$$

This way of expressing a vector is useful in many applications.

| EXAMPLE 5 | **Finding a Unit Vector** |

Find a unit vector in the same direction as $\mathbf{v} = 2\mathbf{i} - 3\mathbf{j} - 6\mathbf{k}$.

Solution We find $\|\mathbf{v}\|$ first.

$$\|\mathbf{v}\| = \|2\mathbf{i} - 3\mathbf{j} - 6\mathbf{k}\| = \sqrt{4 + 9 + 36} = \sqrt{49} = 7$$

Now we multiply $\mathbf{v}$ by the scalar $\dfrac{1}{\|\mathbf{v}\|} = \dfrac{1}{7}$. The result is the unit vector

$$\mathbf{u} = \frac{\mathbf{v}}{\|\mathbf{v}\|} = \frac{2\mathbf{i} - 3\mathbf{j} - 6\mathbf{k}}{7} = \frac{2}{7}\mathbf{i} - \frac{3}{7}\mathbf{j} - \frac{6}{7}\mathbf{k}$$

> ✏️ **NOW WORK PROBLEM 41.**

DOT PRODUCT

④ The definition of *dot product* is an extension of the definition given for vectors in the plane.

If $\mathbf{v} = a_1\mathbf{i} + b_1\mathbf{j} + c_1\mathbf{k}$ and $\mathbf{w} = a_2\mathbf{i} + b_2\mathbf{j} + c_2\mathbf{k}$ are two vectors, the **dot product $\mathbf{v} \cdot \mathbf{w}$** is defined as

$$\mathbf{v} \cdot \mathbf{w} = a_1 a_2 + b_1 b_2 + c_1 c_2 \tag{3}$$

| EXAMPLE 6 | **Finding Dot Products** |

If $\mathbf{v} = 2\mathbf{i} - 3\mathbf{j} + 6\mathbf{k}$ and $\mathbf{w} = 5\mathbf{i} + 3\mathbf{j} - \mathbf{k}$, find:

(a) $\mathbf{v} \cdot \mathbf{w}$ (b) $\mathbf{w} \cdot \mathbf{v}$ (c) $\mathbf{v} \cdot \mathbf{v}$

(d) $\mathbf{w} \cdot \mathbf{w}$ (e) $\|\mathbf{v}\|$ (f) $\|\mathbf{w}\|$

Solution

(a) $\mathbf{v} \cdot \mathbf{w} = 2(5) + (-3)3 + 6(-1) = -5$

(b) $\mathbf{w} \cdot \mathbf{v} = 5(2) + 3(-3) + (-1)(6) = -5$

(c) $\mathbf{v} \cdot \mathbf{v} = 2(2) + (-3)(-3) + 6(6) = 49$

(d) $\mathbf{w} \cdot \mathbf{w} = 5(5) + 3(3) + (-1)(-1) = 35$

(e) $\|\mathbf{v}\| = \sqrt{2^2 + (-3)^2 + 6^2} = \sqrt{49} = 7$

(f) $\|\mathbf{w}\| = \sqrt{5^2 + 3^2 + (-1)^2} = \sqrt{35}$

The dot product in space has the same properties as the dot product in the plane.

Theorem

Properties of the Dot Product

If $\mathbf{u}, \mathbf{v}$, and $\mathbf{w}$ are vectors, then

Commutative Property

$$\mathbf{u} \cdot \mathbf{v} = \mathbf{v} \cdot \mathbf{u}$$

Distributive Property

$$\mathbf{u} \cdot (\mathbf{v} + \mathbf{w}) = \mathbf{u} \cdot \mathbf{v} + \mathbf{u} \cdot \mathbf{w}$$

$$\mathbf{v} \cdot \mathbf{v} = \|\mathbf{v}\|^2$$

$$\mathbf{0} \cdot \mathbf{v} = 0$$

The angle θ between two vectors in space follows the same formula as for two vectors in the plane.

Theorem

Angle between Vectors

If $\mathbf{u}$ and $\mathbf{v}$ are two nonzero vectors, the angle $\theta, 0 \le \theta \le \pi$, between $\mathbf{u}$ and $\mathbf{v}$ is determined by the formula

$$\cos\theta = \frac{\mathbf{u} \cdot \mathbf{v}}{\|\mathbf{u}\| \|\mathbf{v}\|} \tag{4}$$

EXAMPLE 7 | **Finding the Angle θ between Two Vectors**

Find the angle θ between $\mathbf{u} = 2\mathbf{i} - 3\mathbf{j} + 6\mathbf{k}$ and $\mathbf{v} = 2\mathbf{i} + 5\mathbf{j} - \mathbf{k}.$

Solution We compute the quantities $\mathbf{u} \cdot \mathbf{v}$, $\|\mathbf{u}\|$, and $\|\mathbf{v}\|$.

$$\mathbf{u} \cdot \mathbf{v} = 2(2) + (-3)(5) + 6(-1) = -17$$
$$\|\mathbf{u}\| = \sqrt{2^2 + (-3)^2 + 6^2} = \sqrt{49} = 7$$
$$\|\mathbf{v}\| = \sqrt{2^2 + 5^2 + (-1)^2} = \sqrt{30}$$

By formula (4), if θ is the angle between $\mathbf{u}$ and $\mathbf{v}$, then

$$\cos\theta = \frac{\mathbf{u} \cdot \mathbf{v}}{\|\mathbf{u}\| \, \|\mathbf{v}\|} = \frac{-17}{7\sqrt{30}} \approx -0.443$$

We find that $\theta \approx 116.3°$. ∎

NOW WORK PROBLEM **45.**

DIRECTION ANGLES OF VECTORS IN SPACE

⑥ A nonzero vector $\mathbf{v}$ in space can be described by specifying its magnitude and its three **direction angles** $\alpha, \beta,$ and γ. These direction angles are defined as

$\alpha = $ angle between $\mathbf{v}$ and $\mathbf{i}$, the positive x-axis, $0 \le \alpha \le \pi$

$\beta = $ angle between $\mathbf{v}$ and $\mathbf{j}$, the positive y-axis, $0 \le \beta \le \pi$

$\gamma = $ angle between $\mathbf{v}$ and $\mathbf{k}$, the positive z-axis, $0 \le \gamma \le \pi$

See Figure 75.

Figure 75

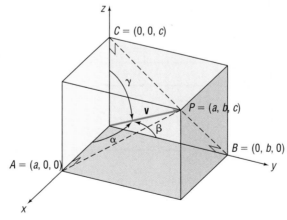

$0 \le \alpha \le \pi, 0 \le \beta \le \pi, 0 \le \gamma \le \pi$

Our first goal is to find an expression for $\alpha, \beta,$ and γ in terms of the components of a vector. Let $\mathbf{v} = a\mathbf{i} + b\mathbf{j} + c\mathbf{k}$ denote a nonzero vector. The angle α between $\mathbf{v}$ and $\mathbf{i}$, the positive x-axis, obeys

$$\cos\alpha = \frac{\mathbf{v} \cdot \mathbf{i}}{\|\mathbf{v}\| \, \|\mathbf{i}\|} = \frac{a}{\|\mathbf{v}\|}$$

Similarly,

$$\cos\beta = \frac{b}{\|\mathbf{v}\|} \qquad \cos\gamma = \frac{c}{\|\mathbf{v}\|}$$

Since $\|\mathbf{v}\| = \sqrt{a^2 + b^2 + c^2}$, we have the following result:

Theorem

Direction Angles

If $\mathbf{v} = a\mathbf{i} + b\mathbf{j} + c\mathbf{k}$ is a nonzero vector in space, the direction angles α, β, and γ obey

$$\cos\alpha = \frac{a}{\sqrt{a^2 + b^2 + c^2}} = \frac{a}{\|\mathbf{v}\|} \qquad \cos\beta = \frac{b}{\sqrt{a^2 + b^2 + c^2}} = \frac{b}{\|\mathbf{v}\|}$$

$$\cos\gamma = \frac{c}{\sqrt{a^2 + b^2 + c^2}} = \frac{c}{\|\mathbf{v}\|} \qquad \textbf{(5)}$$

The numbers $\cos\alpha$, $\cos\beta$, and $\cos\gamma$ are called the **direction cosines** of the vector **v.** They play the same role in space as slope does in the plane.

EXAMPLE 8

Finding the Direction Angles of a Vector

Find the direction angles of $\mathbf{v} = -3\mathbf{i} + 2\mathbf{j} - 6\mathbf{k}.$

Solution

$$\|\mathbf{v}\| = \sqrt{(-3)^2 + 2^2 + (-6)^2} = \sqrt{49} = 7$$

Using the theorem on direction angles, we get

$$\cos\alpha = \frac{-3}{7} \qquad \cos\beta = \frac{2}{7} \qquad \cos\gamma = \frac{-6}{7}$$

$$\alpha \approx 115.4° \qquad \beta \approx 73.4° \qquad \gamma \approx 149.0°$$

Theorem

Property of Direction Cosines

If α, β, and γ are the direction angles of a nonzero vector **v** in space, then

$$\cos^2\alpha + \cos^2\beta + \cos^2\gamma = 1 \qquad \textbf{(6)}$$

The proof is a direct consequence of equations (5).

Based on equation (6), when two direction cosines are known, the third is determined up to its sign. Knowing two direction cosines is not sufficient to uniquely determine the direction of a vector in space.

EXAMPLE 9

Finding the Direction Angle of a Vector

The vector **v** makes an angle of $\alpha = \dfrac{\pi}{3}$ with the positive x-axis, an angle of $\beta = \dfrac{\pi}{3}$ with the positive y-axis, and an acute angle γ with the positive z-axis. Find γ.

Solution   By equation (6), we have

$$\cos^2\left(\frac{\pi}{3}\right) + \cos^2\left(\frac{\pi}{3}\right) + \cos^2\gamma = 1 \qquad 0 \le \gamma \le \pi$$

$$\left(\frac{1}{2}\right)^2 + \left(\frac{1}{2}\right)^2 + \cos^2\gamma = 1$$

$$\cos^2\gamma = \frac{1}{2}$$

$$\cos\gamma = \frac{\sqrt{2}}{2} \quad \text{or} \quad \cos\gamma = -\frac{\sqrt{2}}{2}$$

$$\gamma = \frac{\pi}{4} \quad \text{or} \quad \gamma = \frac{3\pi}{4}$$

Since we are requiring that γ be acute, the answer is $\gamma = \dfrac{\pi}{4}$. ∎

The direction cosines of a vector give information about only the direction of the vector; they provide no information about its magnitude. For example, *any* vector parallel to the xy-plane and making an angle of $\dfrac{\pi}{4}$ radian with the positive x- and y-axes has direction cosines

$$\cos\alpha = \frac{\sqrt{2}}{2} \qquad \cos\beta = \frac{\sqrt{2}}{2} \qquad \cos\gamma = 0$$

However, if the direction angles *and* the magnitude of a vector are known, then the vector is uniquely determined.

EXAMPLE 10   **Writing a Vector in Terms of Its Magnitude and Direction Cosines**

Show that any nonzero vector $\mathbf{v}$ in space can be written in terms of its magnitude and direction cosines as

$$\mathbf{v} = \|\mathbf{v}\|\left[(\cos\alpha)\mathbf{i} + (\cos\beta)\mathbf{j} + (\cos\gamma)\mathbf{k}\right] \tag{7}$$

Solution   Let $\mathbf{v} = a\mathbf{i} + b\mathbf{j} + c\mathbf{k}$. From equation (5), we see that

$$a = \|\mathbf{v}\|\cos\alpha \qquad b = \|\mathbf{v}\|\cos\beta \qquad c = \|\mathbf{v}\|\cos\gamma$$

Substituting, we find that

$$\mathbf{v} = a\mathbf{i} + b\mathbf{j} + c\mathbf{k} = \|\mathbf{v}\|(\cos\alpha)\mathbf{i} + \|\mathbf{v}\|(\cos\beta)\mathbf{j} + \|\mathbf{v}\|(\cos\gamma)\mathbf{k}$$

$$= \|\mathbf{v}\|\left[(\cos\alpha)\mathbf{i} + (\cos\beta)\mathbf{j} + (\cos\gamma)\mathbf{k}\right] \quad ∎$$

Example 10 shows that the direction cosines of a vector $\mathbf{v}$ are also the components of the unit vector in the direction of $\mathbf{v}$.

5.6   EXERCISES

In Problems 1–8, describe the set of points (x, y, z) *defined by the equation.*

1. $y = 0$ **2.** $x = 0$ **3.** $z = 2$ **4.** $y = 3$

5. $x = -4$ **6.** $z = -3$ **7.** $x = 1$ and $y = 2$ **8.** $x = 3$ and $z = 1$

In Problems 9–14, find the distance from P_1 to P_2.

9. $P_1 = (0, 0, 0)$ and $P_2 = (4, 1, 2)$ **10.** $P_1 = (0, 0, 0)$ and $P_2 = (1, -2, 3)$

11. $P_1 = (-1, 2, -3)$ and $P_2 = (0, -2, 1)$ **12.** $P_1 = (-2, 2, 3)$ and $P_2 = (4, 0, -3)$

13. $P_1 = (4, -2, -2)$ and $P_2 = (3, 2, 1)$ **14.** $P_1 = (2, -3, -3)$ and $P_2 = (4, 1, -1)$

In Problems 15–20, opposite vertices of a rectangular box whose edges are parallel to the coordinate axes are given. List the coordinates of the other six vertices of the box.

15. $(0, 0, 0)$; $(2, 1, 3)$ **16.** $(0, 0, 0)$; $(4, 2, 2)$ **17.** $(1, 2, 3)$; $(3, 4, 5)$

18. $(5, 6, 1)$; $(3, 8, 2)$ **19.** $(-1, 0, 2)$; $(4, 2, 5)$ **20.** $(-2, -3, 0)$; $(-6, 7, 1)$

In Problems 21–26, the vector $\mathbf{v}$ has initial point P and terminal point Q. Write $\mathbf{v}$ in the form $a\mathbf{i} + b\mathbf{j} + c\mathbf{k}$; that is, find its position vector.

21. $P = (0, 0, 0)$; $Q = (3, 4, -1)$ **22.** $P = (0, 0, 0)$; $Q = (-3, -5, 4)$

23. $P = (3, 2, -1)$; $Q = (5, 6, 0)$ **24.** $P = (-3, 2, 0)$; $Q = (6, 5, -1)$

25. $P = (-2, -1, 4)$; $Q = (6, -2, 4)$ **26.** $P = (-1, 4, -2)$; $Q = (6, 2, 2)$

In Problems 27–32, find $\|\mathbf{v}\|$.

27. $\mathbf{v} = 3\mathbf{i} - 6\mathbf{j} - 2\mathbf{k}$ **28.** $\mathbf{v} = -6\mathbf{i} + 12\mathbf{j} + 4\mathbf{k}$ **29.** $\mathbf{v} = \mathbf{i} - \mathbf{j} + \mathbf{k}$

30. $\mathbf{v} = -\mathbf{i} - \mathbf{j} + \mathbf{k}$ **31.** $\mathbf{v} = -2\mathbf{i} + 3\mathbf{j} - 3\mathbf{k}$ **32.** $\mathbf{v} = 6\mathbf{i} + 2\mathbf{j} - 2\mathbf{k}$

In Problems 33–38, find each quantity if $\mathbf{v} = 3\mathbf{i} - 5\mathbf{j} + 2\mathbf{k}$ and $\mathbf{w} = -2\mathbf{i} + 3\mathbf{j} - 2\mathbf{k}$.

33. $2\mathbf{v} + 3\mathbf{w}$ **34.** $3\mathbf{v} - 2\mathbf{w}$ **35.** $\|\mathbf{v} - \mathbf{w}\|$

36. $\|\mathbf{v} + \mathbf{w}\|$ **37.** $\|\mathbf{v}\| - \|\mathbf{w}\|$ **38.** $\|\mathbf{v}\| + \|\mathbf{w}\|$

In Problems 39–44, find the unit vector having the same direction as $\mathbf{v}$.

39. $\mathbf{v} = 5\mathbf{i}$ **40.** $\mathbf{v} = -3\mathbf{j}$ **41.** $\mathbf{v} = 3\mathbf{i} - 6\mathbf{j} - 2\mathbf{k}$

42. $\mathbf{v} = -6\mathbf{i} + 12\mathbf{j} + 4\mathbf{k}$ **43.** $\mathbf{v} = \mathbf{i} + \mathbf{j} + \mathbf{k}$ **44.** $\mathbf{v} = 2\mathbf{i} - \mathbf{j} + \mathbf{k}$

In Problems 45–52, find the dot product $\mathbf{v} \cdot \mathbf{w}$ and the angle between $\mathbf{v}$ and $\mathbf{w}$.

45. $\mathbf{v} = \mathbf{i} - \mathbf{j}$, $\mathbf{w} = \mathbf{i} + \mathbf{j} + \mathbf{k}$ **46.** $\mathbf{v} = \mathbf{i} + \mathbf{j}$, $\mathbf{w} = -\mathbf{i} + \mathbf{j} - \mathbf{k}$

47. $\mathbf{v} = 2\mathbf{i} + \mathbf{j} - 3\mathbf{k}$, $\mathbf{w} = \mathbf{i} + 2\mathbf{j} + 2\mathbf{k}$ **48.** $\mathbf{v} = 2\mathbf{i} + 2\mathbf{j} - \mathbf{k}$, $\mathbf{w} = \mathbf{i} + 2\mathbf{j} + 3\mathbf{k}$

49. $\mathbf{v} = 3\mathbf{i} - \mathbf{j} + 2\mathbf{k}$, $\mathbf{w} = \mathbf{i} + \mathbf{j} - \mathbf{k}$ **50.** $\mathbf{v} = \mathbf{i} + 3\mathbf{j} + 2\mathbf{k}$, $\mathbf{w} = \mathbf{i} - \mathbf{j} + \mathbf{k}$

51. $\mathbf{v} = 3\mathbf{i} + 4\mathbf{j} + \mathbf{k}$, $\mathbf{w} = 6\mathbf{i} + 8\mathbf{j} + 2\mathbf{k}$ **52.** $\mathbf{v} = 3\mathbf{i} - 4\mathbf{j} + \mathbf{k}$, $\mathbf{w} = 6\mathbf{i} - 8\mathbf{j} + 2\mathbf{k}$

In Problems 53–60, find the direction angles of each vector. Write each vector in the form of equation (7).

53. $\mathbf{v} = 3\mathbf{i} - 6\mathbf{j} - 2\mathbf{k}$ **54.** $\mathbf{v} = -6\mathbf{i} + 12\mathbf{j} + 4\mathbf{k}$ **55.** $\mathbf{v} = \mathbf{i} + \mathbf{j} + \mathbf{k}$

56. $\mathbf{v} = \mathbf{i} - \mathbf{j} - \mathbf{k}$ **57.** $\mathbf{v} = \mathbf{i} + \mathbf{j}$ **58.** $\mathbf{v} = \mathbf{j} + \mathbf{k}$

59. $\mathbf{v} = 3\mathbf{i} - 5\mathbf{j} + 2\mathbf{k}$ **60.** $\mathbf{v} = 2\mathbf{i} + 3\mathbf{j} - 4\mathbf{k}$

61. The Sphere   In space, the collection of all points that are the same distance from some fixed point is called a **sphere.** See the illustration. The constant distance is called the **radius,** and the fixed point is the **center** of the sphere. Show that the equation of a sphere with center at (x_0, y_0, z_0) and radius r is

$$(x - x_0)^2 + (y - y_0)^2 + (z - z_0)^2 = r^2$$

[**Hint:** Use the Distance Formula (1).]

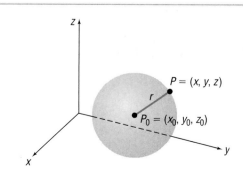

In Problems 62–64, find the equation of a sphere with radius r and center P_0.

62. $r = 1$; $P_0 = (3, 1, 1)$ **63.** $r = 2$; $P_0 = (1, 2, 2)$ **64.** $r = 3$; $P_0 = (-1, 1, 2)$

In Problems 65–70, find the radius and center of each sphere. [**Hint:** *Complete the square in each variable.*]

65. $x^2 + y^2 + z^2 + 2x - 2y = 2$ **66.** $x^2 + y^2 + z^2 + 2x - 2z = -1$

67. $x^2 + y^2 + z^2 - 4x + 4y + 2z = 0$ **68.** $x^2 + y^2 + z^2 - 4x = 0$

69. $2x^2 + 2y^2 + 2z^2 - 8x + 4z = -1$ **70.** $3x^2 + 3y^2 + 3z^2 + 6x - 6y = 3$

The **work** W *done by a constant force* **F** *in moving an object from a point A in space to a point B in space is defined as* $W = \mathbf{F} \cdot \overrightarrow{AB}$. *Use this definition in Problems 71–73.*

71. Work   Find the work done by a force of 3 newtons acting in the direction $2\mathbf{i} + \mathbf{j} + 2\mathbf{k}$ in moving an object 2 meters from $(0, 0, 0)$ to $(0, 2, 0)$.

72. Work   Find the work done by a force of 1 newton acting in the direction $2\mathbf{i} + 2\mathbf{j} + \mathbf{k}$ in moving an object 3 meters from $(0, 0, 0)$ to $(1, 2, 2)$.

73. Work   Find the work done in moving an object along a vector $\mathbf{u} = 3\mathbf{i} + 2\mathbf{j} - 5\mathbf{k}$ if the applied force is $\mathbf{F} = 2\mathbf{i} - \mathbf{j} - \mathbf{k}$.

5.7 THE CROSS PRODUCT

OBJECTIVES   **1**   Find the Cross Product of Two Vectors

           2   Know Algebraic Properties of the Cross Product

           3   Know Geometric Properties of the Cross Product

           4   Find a Vector Orthogonal to Two Given Vectors

           5   Find the Area of a Parallelogram

1   For vectors in space, and only for vectors in space, a second product of two vectors is defined, called the *cross product*. The cross product of two vectors in space is, in fact, also a vector that has applications in both geometry and physics.

If $\mathbf{v} = a_1\mathbf{i} + b_1\mathbf{j} + c_1\mathbf{k}$ and $\mathbf{w} = a_2\mathbf{i} + b_2\mathbf{j} + c_2\mathbf{k}$ are two vectors in space, the **cross product** $\mathbf{v} \times \mathbf{w}$ is defined as the vector

$$\mathbf{v} \times \mathbf{w} = (b_1c_2 - b_2c_1)\mathbf{i} - (a_1c_2 - a_2c_1)\mathbf{j} + (a_1b_2 - a_2b_1)\mathbf{k} \quad \textbf{(1)}$$

Notice that the cross product $\mathbf{v} \times \mathbf{w}$ of two vectors is a vector. Because of this, it is sometimes referred to as the **vector product.**

EXAMPLE 1   **Finding Cross Products Using Equation (1)**

If $\mathbf{v} = 2\mathbf{i} + 3\mathbf{j} + 5\mathbf{k}$ and $\mathbf{w} = \mathbf{i} + 2\mathbf{j} + 3\mathbf{k}$, then an application of equation (1) gives

$$\mathbf{v} \times \mathbf{w} = (3 \cdot 3 - 2 \cdot 5)\mathbf{i} - (2 \cdot 3 - 1 \cdot 5)\mathbf{j} + (2 \cdot 2 - 1 \cdot 3)\mathbf{k}$$

$$= (9 - 10)\mathbf{i} - (6 - 5)\mathbf{j} + (4 - 3)\mathbf{k}$$

$$= -\mathbf{i} - \mathbf{j} + \mathbf{k}$$

∎

Determinants may be used as an aid in computing cross products. A **2 by 2 determinant,** symbolized by

$$\begin{vmatrix} a_1 & b_1 \\ a_2 & b_2 \end{vmatrix}$$

has the value $a_1 b_2 - a_2 b_1$; that is,

$$\begin{vmatrix} a_1 & b_1 \\ a_2 & b_2 \end{vmatrix} = a_1 b_2 - a_2 b_1$$

A **3 by 3 determinant** has the value

$$\begin{vmatrix} A & B & C \\ a_1 & b_1 & c_1 \\ a_2 & b_2 & c_2 \end{vmatrix} = \begin{vmatrix} b_1 & c_1 \\ b_2 & c_2 \end{vmatrix} A - \begin{vmatrix} a_1 & c_1 \\ a_2 & c_2 \end{vmatrix} B + \begin{vmatrix} a_1 & b_1 \\ a_2 & b_2 \end{vmatrix} C$$

EXAMPLE 2

Evaluating Determinants

(a) $\begin{vmatrix} 2 & 3 \\ 1 & 2 \end{vmatrix} = 2 \cdot 2 - 1 \cdot 3 = 4 - 3 = 1$

(b) $\begin{vmatrix} A & B & C \\ 2 & 3 & 5 \\ 1 & 2 & 3 \end{vmatrix} = \begin{vmatrix} 3 & 5 \\ 2 & 3 \end{vmatrix} A - \begin{vmatrix} 2 & 5 \\ 1 & 3 \end{vmatrix} B + \begin{vmatrix} 2 & 3 \\ 1 & 2 \end{vmatrix} C$

$\qquad\qquad\quad = (9 - 10)A - (6 - 5)B + (4 - 3)C$

$\qquad\qquad\quad = -A - B + C$ ∎

> NOW WORK PROBLEM **1.**

The cross product of the vectors $\mathbf{v} = a_1 \mathbf{i} + b_1 \mathbf{j} + c_1 \mathbf{k}$, and $\mathbf{w} = a_2 \mathbf{i} + b_2 \mathbf{j} + c_2 \mathbf{k}$, that is,

$$\mathbf{v} \times \mathbf{w} = (b_1 c_2 - b_2 c_1)\mathbf{i} - (a_1 c_2 - a_2 c_1)\mathbf{j} + (a_1 b_2 - a_2 b_1)\mathbf{k},$$

may be written symbolically using determinants as

$$\mathbf{v} \times \mathbf{w} = \begin{vmatrix} \mathbf{i} & \mathbf{j} & \mathbf{k} \\ a_1 & b_1 & c_1 \\ a_2 & b_2 & c_2 \end{vmatrix} = \begin{vmatrix} b_1 & c_1 \\ b_2 & c_2 \end{vmatrix} \mathbf{i} - \begin{vmatrix} a_1 & c_1 \\ a_2 & c_2 \end{vmatrix} \mathbf{j} + \begin{vmatrix} a_1 & b_1 \\ a_2 & b_2 \end{vmatrix} \mathbf{k}$$

EXAMPLE 3

Using Determinants to Find Cross Products

If $\mathbf{v} = 2\mathbf{i} + 3\mathbf{j} + 5\mathbf{k}$ and $\mathbf{w} = \mathbf{i} + 2\mathbf{j} + 3\mathbf{k}$, find:

(a) $\mathbf{v} \times \mathbf{w}$ (b) $\mathbf{w} \times \mathbf{v}$ (c) $\mathbf{v} \times \mathbf{v}$ (d) $\mathbf{w} \times \mathbf{w}$

Solution (a) $\mathbf{v} \times \mathbf{w} = \begin{vmatrix} \mathbf{i} & \mathbf{j} & \mathbf{k} \\ 2 & 3 & 5 \\ 1 & 2 & 3 \end{vmatrix} = \begin{vmatrix} 3 & 5 \\ 2 & 3 \end{vmatrix} \mathbf{i} - \begin{vmatrix} 2 & 5 \\ 1 & 3 \end{vmatrix} \mathbf{j} + \begin{vmatrix} 2 & 3 \\ 1 & 2 \end{vmatrix} \mathbf{k} = -\mathbf{i} - \mathbf{j} + \mathbf{k}$

(b) $\mathbf{w} \times \mathbf{v} = \begin{vmatrix} \mathbf{i} & \mathbf{j} & \mathbf{k} \\ 1 & 2 & 3 \\ 2 & 3 & 5 \end{vmatrix} = \begin{vmatrix} 2 & 3 \\ 3 & 5 \end{vmatrix} \mathbf{i} - \begin{vmatrix} 1 & 3 \\ 2 & 5 \end{vmatrix} \mathbf{j} + \begin{vmatrix} 1 & 2 \\ 2 & 3 \end{vmatrix} \mathbf{k} = \mathbf{i} + \mathbf{j} - \mathbf{k}$

(c) $\mathbf{v} \times \mathbf{v} = \begin{vmatrix} \mathbf{i} & \mathbf{j} & \mathbf{k} \\ 2 & 3 & 5 \\ 2 & 3 & 5 \end{vmatrix}$

$= \begin{vmatrix} 3 & 5 \\ 3 & 5 \end{vmatrix} \mathbf{i} - \begin{vmatrix} 2 & 5 \\ 2 & 5 \end{vmatrix} \mathbf{j} + \begin{vmatrix} 2 & 3 \\ 2 & 3 \end{vmatrix} \mathbf{k} = 0\mathbf{i} - 0\mathbf{j} + 0\mathbf{k} = \mathbf{0}$

(d) $\mathbf{w} \times \mathbf{w} = \begin{vmatrix} \mathbf{i} & \mathbf{j} & \mathbf{k} \\ 1 & 2 & 3 \\ 1 & 2 & 3 \end{vmatrix}$

$= \begin{vmatrix} 2 & 3 \\ 2 & 3 \end{vmatrix} \mathbf{i} - \begin{vmatrix} 1 & 3 \\ 1 & 3 \end{vmatrix} \mathbf{j} + \begin{vmatrix} 1 & 2 \\ 1 & 2 \end{vmatrix} \mathbf{k} = 0\mathbf{i} - 0\mathbf{j} + 0\mathbf{k} = \mathbf{0}$

NOW WORK PROBLEM **9**.

ALGEBRAIC PROPERTIES OF THE CROSS PRODUCT

② Notice in Example 3(a) and 3(b) that $\mathbf{v} \times \mathbf{w}$ and $\mathbf{w} \times \mathbf{v}$ are negatives of one another. From Examples 3(c) and 3(d), we might conjecture that the cross product of a vector with itself is the zero vector. These and other algebraic properties of cross product are given next.

Theorem

Algebraic Properties of the Cross Product

If $\mathbf{u}, \mathbf{v},$ and $\mathbf{w}$ are vectors in space and if α is a scalar, then

$$\mathbf{u} \times \mathbf{u} = \mathbf{0} \tag{2}$$

$$\mathbf{u} \times \mathbf{v} = -(\mathbf{v} \times \mathbf{u}) \tag{3}$$

$$\alpha(\mathbf{u} \times \mathbf{v}) = (\alpha\mathbf{u}) \times \mathbf{v} = \mathbf{u} \times (\alpha\mathbf{v}) \tag{4}$$

$$\mathbf{u} \times (\mathbf{v} + \mathbf{w}) = (\mathbf{u} \times \mathbf{v}) + (\mathbf{u} \times \mathbf{w}) \tag{5}$$

Proof We will prove properties (2) and (4) here and leave properties (3) and (5) as exercises (see Problems 49 and 50 at the end of this section).
To prove property (2), we let $\mathbf{u} = a_1\mathbf{i} + b_1\mathbf{j} + c_1\mathbf{k}$. Then

$$\mathbf{u} \times \mathbf{u} = \begin{vmatrix} \mathbf{i} & \mathbf{j} & \mathbf{k} \\ a_1 & b_1 & c_1 \\ a_1 & b_1 & c_1 \end{vmatrix} = \begin{vmatrix} b_1 & c_1 \\ b_1 & c_1 \end{vmatrix} \mathbf{i} - \begin{vmatrix} a_1 & c_1 \\ a_1 & c_1 \end{vmatrix} \mathbf{j} + \begin{vmatrix} a_1 & b_1 \\ a_1 & b_1 \end{vmatrix} \mathbf{k}$$

$$= 0\mathbf{i} - 0\mathbf{j} + 0\mathbf{k} = \mathbf{0}$$

To prove property (4), we let $\mathbf{u} = a_1\mathbf{i} + b_1\mathbf{j} + c_1\mathbf{k}$ and $\mathbf{v} = a_2\mathbf{i} + b_2\mathbf{j} + c_2\mathbf{k}$. Then

$$\alpha(\mathbf{u} \times \mathbf{v}) = \alpha\big[(b_1c_2 - b_2c_1)\mathbf{i} - (a_1c_2 - a_2c_1)\mathbf{j} + (a_1b_2 - a_2b_1)\mathbf{k}\big]$$

↑
Apply (1)

$$= \alpha(b_1c_2 - b_2c_1)\mathbf{i} - \alpha(a_1c_2 - a_2c_1)\mathbf{j} + \alpha(a_1b_2 - a_2b_1)\mathbf{k} \tag{6}$$

Since $\alpha\mathbf{u} = \alpha a_1\mathbf{i} + \alpha b_1\mathbf{j} + \alpha c_1\mathbf{k},$ we have

$$(\alpha\mathbf{u}) \times \mathbf{v} = (\alpha b_1 c_2 - b_2\alpha c_1)\mathbf{i} - (\alpha a_1 c_2 - a_2\alpha c_1)\mathbf{j} + (\alpha a_1 b_2 - a_2\alpha b_1)\mathbf{k}$$
$$= \alpha(b_1 c_2 - b_2 c_1)\mathbf{i} - \alpha(a_1 c_2 - a_2 c_1)\mathbf{j} + \alpha(a_1 b_2 - a_2 b_1)\mathbf{k} \qquad \textbf{(7)}$$

Based on equations (6) and (7), the first part of property (4) follows. The second part can be proved in like fashion.   ∎

> **NOW WORK PROBLEM 11.**

3   ## GEOMETRIC PROPERTIES OF THE CROSS PRODUCT

The cross product has several interesting geometric properties.

Theorem   **Geometric Properties of the Cross Product**

Let $\mathbf{u}$ and $\mathbf{v}$ be vectors in space.

$\mathbf{u} \times \mathbf{v}$ is orthogonal to both $\mathbf{u}$ and $\mathbf{v}$. $\qquad$ **(8)**

$\|\mathbf{u} \times \mathbf{v}\| = \|\mathbf{u}\|\,\|\mathbf{v}\|\sin\theta$, where θ is the angle between $\mathbf{u}$ and $\mathbf{v}$. $\qquad$ **(9)**

$\|\mathbf{u} \times \mathbf{v}\|$ is the area of the parallelogram having $\mathbf{u} \neq \mathbf{0}$ and $\mathbf{v} \neq \mathbf{0}$ as adjacent sides. $\qquad$ **(10)**

$\mathbf{u} \times \mathbf{v} = \mathbf{0}$ if and only if $\mathbf{u}$ and $\mathbf{v}$ are parallel. $\qquad$ **(11)**

∎

Proof of Property (8)   Let $\mathbf{u} = a_1\mathbf{i} + b_1\mathbf{j} + c_1\mathbf{k}$ and $\mathbf{v} = a_2\mathbf{i} + b_2\mathbf{j} + c_2\mathbf{k}.$ Then

$$\mathbf{u} \times \mathbf{v} = (b_1 c_2 - b_2 c_1)\mathbf{i} - (a_1 c_2 - a_2 c_1)\mathbf{j} + (a_1 b_2 - a_2 b_1)\mathbf{k}$$

Now we compute the dot product $\mathbf{u} \cdot (\mathbf{u} \times \mathbf{v}).$

$$\mathbf{u} \cdot (\mathbf{u} \times \mathbf{v}) = (a_1\mathbf{i} + b_1\mathbf{j} + c_1\mathbf{k}) \cdot [(b_1 c_2 - b_2 c_1)\mathbf{i} - (a_1 c_2 - a_2 c_1)\mathbf{j} + (a_1 b_2 - a_2 b_1)\mathbf{k}]$$
$$= a_1(b_1 c_2 - b_2 c_1) - b_1(a_1 c_2 - a_2 c_1) + c_1(a_1 b_2 - a_2 b_1) = 0$$

Since two vectors are orthogonal if their dot product is zero, it follows that $\mathbf{u}$ and $\mathbf{u} \times \mathbf{v}$ are orthogonal. Similarly, $\mathbf{v} \cdot (\mathbf{u} \times \mathbf{v}) = 0,$ so $\mathbf{v}$ and $\mathbf{u} \times \mathbf{v}$ are orthogonal.   ∎

4   As long as the vectors $\mathbf{u}$ and $\mathbf{v}$ are not parallel, they will form a plane in space. See Figure 76. Based on property (8), the vector $\mathbf{u} \times \mathbf{v}$ is normal to this plane. As Figure 76 illustrates, there are two vectors normal to the plane containing $\mathbf{u}$ and $\mathbf{v}$. It can be shown that the vector $\mathbf{u} \times \mathbf{v}$ is the one determined by the thumb of the right hand when the other fingers of the right hand are cupped so that they point in a direction from $\mathbf{u}$ to $\mathbf{v}$. See Figure 77.*

Figure 76

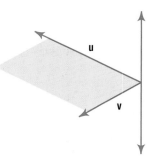

Figure 77

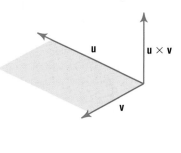

*This is a consequence of using a right-handed coordinate system.

EXAMPLE 4

Finding a Vector Orthogonal to Two Given Vectors

Find a vector that is orthogonal to $\mathbf{u} = 3\mathbf{i} - 2\mathbf{j} + \mathbf{k}$ and $\mathbf{v} = -\mathbf{i} + 3\mathbf{j} - \mathbf{k}$.

Solution   Based on property (8), such a vector is $\mathbf{u} \times \mathbf{v}$.

$$\mathbf{u} \times \mathbf{v} = \begin{vmatrix} \mathbf{i} & \mathbf{j} & \mathbf{k} \\ 3 & -2 & 1 \\ -1 & 3 & -1 \end{vmatrix} = (2 - 3)\mathbf{i} - \left[-3 - (-1)\right]\mathbf{j} + (9 - 2)\mathbf{k} = -\mathbf{i} + 2\mathbf{j} + 7\mathbf{k}$$

The vector $-\mathbf{i} + 2\mathbf{j} + 7\mathbf{k}$ is orthogonal to both $\mathbf{u}$ and $\mathbf{v}$. ∎

Check: Two vectors are orthogonal if their dot product is zero.

$$\mathbf{u} \cdot (-\mathbf{i} + 2\mathbf{j} + 7\mathbf{k}) = (3\mathbf{i} - 2\mathbf{j} + \mathbf{k}) \cdot (-\mathbf{i} + 2\mathbf{j} + 7\mathbf{k}) = -3 - 4 + 7 = 0$$
$$\mathbf{v} \cdot (-\mathbf{i} + 2\mathbf{j} + 7\mathbf{k}) = (-\mathbf{i} + 3\mathbf{j} - \mathbf{k}) \cdot (-\mathbf{i} + 2\mathbf{j} + 7\mathbf{k}) = 1 + 6 - 7 = 0 \quad ∎$$

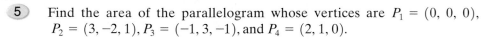 NOW WORK PROBLEM 35.

The proof of property (9) is left as an exercise. See Problem 52.

Figure 78

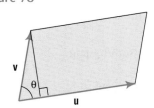

Proof of Property (10)   Suppose that $\mathbf{u}$ and $\mathbf{v}$ are adjacent sides of a parallelogram. See Figure 78. Then the lengths of these sides are $\|\mathbf{u}\|$ and $\|\mathbf{v}\|$. If θ is the angle between $\mathbf{u}$ and $\mathbf{v}$, then the height of the parallelogram is $\|\mathbf{v}\| \sin \theta$ and its area is

$$\text{Area of parallelogram} = \text{Base} \times \text{Height} = \|\mathbf{u}\|\big[\|\mathbf{v}\| \sin \theta\big] \underset{\underset{\text{Property (9)}}{\uparrow}}{=} \|\mathbf{u} \times \mathbf{v}\| \quad ∎$$

EXAMPLE 5

Finding the Area of a Parallelogram

⑤   Find the area of the parallelogram whose vertices are $P_1 = (0, 0, 0)$, $P_2 = (3, -2, 1)$, $P_3 = (-1, 3, -1)$, and $P_4 = (2, 1, 0)$.

Solution   Two adjacent sides* of this parallelogram are

$$\mathbf{u} = \overrightarrow{P_1 P_2} = 3\mathbf{i} - 2\mathbf{j} + \mathbf{k} \quad \text{and} \quad \mathbf{v} = \overrightarrow{P_1 P_3} = -\mathbf{i} + 3\mathbf{j} - \mathbf{k}$$

Since $\mathbf{u} \times \mathbf{v} = -\mathbf{i} + 2\mathbf{j} + 7\mathbf{k}$ (Example 4), the area of the parallelogram is

$$\text{Area of parallelogram} = \|\mathbf{u} \times \mathbf{v}\| = \sqrt{1 + 4 + 49} = \sqrt{54} = 3\sqrt{6} \quad ∎$$

NOW WORK PROBLEM 43.

Proof of Property (11)   The proof requires two parts. If $\mathbf{u}$ and $\mathbf{v}$ are parallel, then there is a scalar α such that $\mathbf{u} = \alpha \mathbf{v}$. Then

$$\mathbf{u} \times \mathbf{v} = (\alpha \mathbf{v}) \times \mathbf{v} = \alpha(\mathbf{v} \times \mathbf{v}) = \mathbf{0}$$
$$\quad\quad\quad\quad\quad \underset{\text{Property (4)}}{\uparrow} \quad\quad \underset{\text{Property (2)}}{\uparrow}$$

If $\mathbf{u} \times \mathbf{v} = \mathbf{0}$, then, by property (9), we have

$$\|\mathbf{u} \times \mathbf{v}\| = \|\mathbf{u}\| \, \|\mathbf{v}\| \sin \theta = 0$$

Since $\mathbf{u} \neq \mathbf{0}$ and $\mathbf{v} \neq \mathbf{0}$, then we must have $\sin \theta = 0$, so $\theta = 0$ or $\theta = \pi$. In either case, since θ is the angle between $\mathbf{u}$ and $\mathbf{v}$, then $\mathbf{u}$ and $\mathbf{v}$ are parallel. ∎

*Be careful! Not all pairs of vertices give rise to a side. For example, $\overrightarrow{P_1 P_4}$ is a diagonal of the parallelogram since $\overrightarrow{P_1 P_3} + \overrightarrow{P_3 P_4} = \overrightarrow{P_1 P_4}$. Also, $\overrightarrow{P_1 P_3}$ and $\overrightarrow{P_2 P_4}$ are not adjacent sides; they are parallel sides.

5.7 EXERCISES

In Problems 1–8, find the value of each determinant.

1. $\begin{vmatrix} 3 & 4 \\ 1 & 2 \end{vmatrix}$

2. $\begin{vmatrix} -2 & 5 \\ 2 & -3 \end{vmatrix}$

3. $\begin{vmatrix} 6 & 5 \\ -2 & -1 \end{vmatrix}$

4. $\begin{vmatrix} -4 & 0 \\ 5 & 3 \end{vmatrix}$

5. $\begin{vmatrix} A & B & C \\ 2 & 1 & 4 \\ 1 & 3 & 1 \end{vmatrix}$

6. $\begin{vmatrix} A & B & C \\ 0 & 2 & 4 \\ 3 & 1 & 3 \end{vmatrix}$

7. $\begin{vmatrix} A & B & C \\ -1 & 3 & 5 \\ 5 & 0 & -2 \end{vmatrix}$

8. $\begin{vmatrix} A & B & C \\ 1 & -2 & -3 \\ 0 & 2 & -2 \end{vmatrix}$

In Problems 9–16, find (a) $\mathbf{v} \times \mathbf{w}$, (b) $\mathbf{w} \times \mathbf{v}$, (c) $\mathbf{w} \times \mathbf{w}$, and (d) $\mathbf{v} \times \mathbf{v}$.

9. $\mathbf{v} = 2\mathbf{i} - 3\mathbf{j} + \mathbf{k}$
   $\mathbf{w} = 3\mathbf{i} - 2\mathbf{j} - \mathbf{k}$

10. $\mathbf{v} = -\mathbf{i} + 3\mathbf{j} + 2\mathbf{k}$
    $\mathbf{w} = 3\mathbf{i} - 2\mathbf{j} - \mathbf{k}$

11. $\mathbf{v} = \mathbf{i} + \mathbf{j}$
    $\mathbf{w} = 2\mathbf{i} + \mathbf{j} + \mathbf{k}$

12. $\mathbf{v} = \mathbf{i} - 4\mathbf{j} + 2\mathbf{k}$
    $\mathbf{w} = 3\mathbf{i} + 2\mathbf{j} + \mathbf{k}$

13. $\mathbf{v} = 2\mathbf{i} - \mathbf{j} + 2\mathbf{k}$
    $\mathbf{w} = \mathbf{j} - \mathbf{k}$

14. $\mathbf{v} = 3\mathbf{i} + \mathbf{j} + 3\mathbf{k}$
    $\mathbf{w} = \mathbf{i} - \mathbf{k}$

15. $\mathbf{v} = \mathbf{i} - \mathbf{j} - \mathbf{k}$
    $\mathbf{w} = 4\mathbf{i} - 3\mathbf{k}$

16. $\mathbf{v} = 2\mathbf{i} - 3\mathbf{j}$
    $\mathbf{w} = 3\mathbf{j} - 2\mathbf{k}$

In Problems 17–38, use the vectors $\mathbf{u}$, $\mathbf{v}$, and $\mathbf{w}$ given next to find each expression.

$$\mathbf{u} = 2\mathbf{i} - 3\mathbf{j} + \mathbf{k} \qquad \mathbf{v} = -3\mathbf{i} + 3\mathbf{j} + 2\mathbf{k} \qquad \mathbf{w} = \mathbf{i} + \mathbf{j} + 3\mathbf{k}$$

17. $\mathbf{u} \times \mathbf{v}$

18. $\mathbf{v} \times \mathbf{w}$

19. $\mathbf{v} \times \mathbf{u}$

20. $\mathbf{w} \times \mathbf{v}$

21. $\mathbf{v} \times \mathbf{v}$

22. $\mathbf{w} \times \mathbf{w}$

23. $(3\mathbf{u}) \times \mathbf{v}$

24. $\mathbf{v} \times (4\mathbf{w})$

25. $\mathbf{u} \times (2\mathbf{v})$

26. $(-3\mathbf{v}) \times \mathbf{w}$

27. $\mathbf{u} \cdot (\mathbf{u} \times \mathbf{v})$

28. $\mathbf{v} \cdot (\mathbf{v} \times \mathbf{w})$

29. $\mathbf{u} \cdot (\mathbf{v} \times \mathbf{w})$

30. $(\mathbf{u} \times \mathbf{v}) \cdot \mathbf{w}$

31. $\mathbf{v} \cdot (\mathbf{u} \times \mathbf{w})$

32. $(\mathbf{v} \times \mathbf{u}) \cdot \mathbf{w}$

33. $\mathbf{u} \times (\mathbf{v} \times \mathbf{v})$

34. $(\mathbf{w} \times \mathbf{w}) \times \mathbf{v}$

35. Find a vector orthogonal to both $\mathbf{u}$ and $\mathbf{v}$.

36. Find a vector orthogonal to both $\mathbf{u}$ and $\mathbf{w}$.

37. Find a vector orthogonal to both $\mathbf{u}$ and $\mathbf{i} + \mathbf{j}$.

38. Find a vector orthogonal to both $\mathbf{u}$ and $\mathbf{j} + \mathbf{k}$.

In Problems 39–42, find the area of the parallelogram with one corner at P_1 and adjacent sides $\overrightarrow{P_1 P_2}$ and $\overrightarrow{P_1 P_3}$.

39. $P_1 = (0,0,0)$, $P_2 = (1,2,3)$, $P_3 = (-2,3,0)$

40. $P_1 = (0,0,0)$, $P_2 = (2,3,1)$, $P_3 = (-2,4,1)$

41. $P_1 = (1,2,0)$, $P_2 = (-2,3,4)$, $P_3 = (0,-2,3)$

42. $P_1 = (-2,0,2)$, $P_2 = (2,1,-1)$, $P_3 = (2,-1,2)$

In Problems 43–46, find the area of the parallelogram with vertices P_1, P_2, P_3, and P_4.

43. $P_1 = (1,1,2)$, $P_2 = (1,2,3)$, $P_3 = (-2,3,0)$, $P_4 = (-2,4,1)$

44. $P_1 = (2,1,1)$, $P_2 = (2,3,1)$, $P_3 = (-2,4,1)$, $P_4 = (-2,6,1)$

45. $P_1 = (1,2,-1)$, $P_2 = (4,2,-3)$, $P_3 = (6,-5,2)$, $P_4 = (9,-5,0)$

46. $P_1 = (-1,1,1)$, $P_2 = (-1,2,2)$, $P_3 = (-3,4,-5)$, $P_4 = (-3,5,-4)$

47. Find a unit vector normal to the plane containing $\mathbf{v} = \mathbf{i} + 3\mathbf{j} - 2\mathbf{k}$ and $\mathbf{w} = -2\mathbf{i} + \mathbf{j} + 3\mathbf{k}$.

48. Find a unit vector normal to the plane containing $\mathbf{v} = 2\mathbf{i} + 3\mathbf{j} - \mathbf{k}$ and $\mathbf{w} = -2\mathbf{i} - 4\mathbf{j} - 3\mathbf{k}$.

49. Prove property (3).

50. Prove property (5).

51. Prove for vectors $\mathbf{u}$ and $\mathbf{v}$ that

$$\|\mathbf{u} \times \mathbf{v}\|^2 = \|\mathbf{u}\|^2 \|\mathbf{v}\|^2 - (\mathbf{u} \cdot \mathbf{v})^2.$$

[**Hint:** Proceed as in the proof of property (4), computing first the left side and then the right side.]

52. Prove property (9).
[**Hint:** Use the result of Problem 51 and the fact that if θ is the angle between $\mathbf{u}$ and $\mathbf{v}$ then $\mathbf{u} \cdot \mathbf{v} = \|\mathbf{u}\| \|\mathbf{v}\| \cos \theta$.]

53. Show that if $\mathbf{u}$ and $\mathbf{v}$ are orthogonal then

$$\|\mathbf{u} \times \mathbf{v}\| = \|\mathbf{u}\| \|\mathbf{v}\|.$$

54. Show that if $\mathbf{u}$ and $\mathbf{v}$ are orthogonal unit vectors then so is $\mathbf{u} \times \mathbf{v}$.

55. If $\mathbf{u} \cdot \mathbf{v} = 0$ and $\mathbf{u} \times \mathbf{v} = \mathbf{0}$, what can you conclude about $\mathbf{u}$ and $\mathbf{v}$?

CHAPTER REVIEW

Things To Know

Relationship between polar coordinates (r, θ) and rectangular coordinates (x, y)	$x = r \cos \theta,\ y = r \sin \theta$ (p. 277)
	$x^2 + y^2 = r^2,\ \tan \theta = \dfrac{y}{x},\ x \neq 0$ (p. 280)
Polar form of a complex number (p. 299)	If $z = x + yi$, then $z = r(\cos \theta + i \sin \theta)$, where $r = |z| = \sqrt{x^2 + y^2},\ \sin \theta = \dfrac{y}{r},\ \cos \theta = \dfrac{x}{r},\ 0 \leq \theta < 2\pi$
DeMoivre's Theorem (p. 302)	If $z = r(\cos \theta + i \sin \theta)$, then $z^n = r^n[\cos(n\theta) + i \sin(n\theta)]$, where $n \geq 1$ is a positive integer
nth root of a complex number $z = r(\cos \theta_0 + i \sin \theta_0)$ (p. 303)	$\sqrt[n]{z} = \sqrt[n]{r}\left[\cos\left(\dfrac{\theta_0}{n} + \dfrac{2k\pi}{n}\right) + i \sin\left(\dfrac{\theta_0}{n} + \dfrac{2k\pi}{n}\right)\right],\ k = 0, \ldots, n - 1$, where $n \geq 2$ is an integer.
Vector (p. 306)	Quantity having magnitude and direction; equivalent to a directed line segment $\overrightarrow{PQ}$
Position vector (p. 309 and 328)	Vector whose initial point is at the origin
Unit vector (p. 312)	Vector whose magnitude is 1
Dot product (p. 318 and 330)	If $\mathbf{v} = a_1\mathbf{i} + b_1\mathbf{j}$ and $\mathbf{w} = a_2\mathbf{i} + b_2\mathbf{j}$, then $\mathbf{v} \cdot \mathbf{w} = a_1 a_2 + b_1 b_2$. If $\mathbf{v} = a_1\mathbf{i} + b_1\mathbf{j} + c_1\mathbf{k}$ and $\mathbf{w} = a_2\mathbf{i} + b_2\mathbf{j} + c_2\mathbf{k}$, then $\mathbf{v} \cdot \mathbf{w} = a_1 a_2 + b_1 b_2 + c_1 c_2$.
Angle θ between two nonzero vectors $\mathbf{u}$ and $\mathbf{v}$ (p. 320 and 331)	$\cos \theta = \dfrac{\mathbf{u} \cdot \mathbf{v}}{\|\mathbf{u}\|\,\|\mathbf{v}\|}$
Vectors in space (p. 333)	If $\mathbf{v} = a\mathbf{i} + b\mathbf{j} + c\mathbf{k}$, then $\mathbf{v} = \|\mathbf{v}\|[(\cos \alpha)\mathbf{i} + (\cos \beta)\mathbf{j} + (\cos \gamma)\mathbf{k}]$, where $\cos \alpha = \dfrac{a}{\|\mathbf{v}\|},\ \cos \beta = \dfrac{b}{\|\mathbf{v}\|},\ \cos \gamma = \dfrac{c}{\|\mathbf{v}\|}$.
Cross Product (p. 336)	If $\mathbf{v} = a_1\mathbf{i} + b_1\mathbf{j} + c_1\mathbf{k}$ and $\mathbf{w} = a_2\mathbf{i} + b_2\mathbf{j} + c_2\mathbf{k}$, then $\mathbf{v} \times \mathbf{w} = [b_1 c_2 - b_2 c_1]\mathbf{i} - [a_1 c_2 - a_2 c_1]\mathbf{j} + [a_1 b_2 - a_2 b_1]\mathbf{k}$
Area of parallelogram (p. 339)	$\|\mathbf{u} \times \mathbf{v}\| = \|\mathbf{u}\|\,\|\mathbf{v}\| \sin \theta$, where θ is the angle between $\mathbf{u}$ and $\mathbf{v}$.

Objectives

You should be able to:

Plot points using polar coordinates (p. 274)

Convert from polar coordinates to rectangular coordinates (p. 277)

Convert from rectangular coordinates to polar coordinates (p. 278)

Graph and identify polar equations by converting to rectangular equations (p. 283)

Test polar equations for symmetry (p. 287)

Graph polar equations by plotting points (p. 288)

Convert a complex number from rectangular form to polar form (p. 299)

Plot points in the complex plane (p. 299)

Find products and quotients of complex numbers in polar form (p. 300)

Use DeMoivre's Theorem (p. 301)

Find complex roots (p. 302)

Graph vectors (p. 308)

Find a position vector (pp. 309 and 328)

Add and subtract vectors (pp. 312 and 329)

Find a scalar product and the magnitude of a vector (pp. 312 and 330)

Find a unit vector (pp. 312 and 330)

Find a vector from its direction and magnitude (p. 313)

Work with objects in static equilibrium (p. 314)

Find the dot product of two vectors (pp. 318 and 331)

Find the angle between two vectors (pp. 319 and 332)

Determine whether two vectors are parallel (p. 321)

Determine whether two vectors are orthogonal (p. 321)

Decompose a vector into two orthogonal vectors (p. 322)

Compute work (p. 323)

Find the distance between two points in space (p. 328)

Find the direction angles of a vector in space (p. 332)

Find the cross product of two vectors in space (p. 336)

Know algebraic properties of the cross product (p. 338)

Know the geometric properties of the cross product (p. 339)

Find a vector orthogonal to two given vectors (p. 339)

Find the area of a parallelogram (p. 340)

Fill-in-the-Blank Items

1. In polar coordinates, the origin is called the _____, and the positive x-axis is referred to as the _____ _____.

2. Another representation in polar coordinates for the point $\left(2, \dfrac{\pi}{3}\right)$ is $\left(\text{_____}, \dfrac{4\pi}{3}\right)$.

3. Using polar coordinates (r, θ), the circle $x^2 + y^2 = 2x$ takes the form _____.

4. In a polar equation, replace θ by $-\theta$. If an equivalent equation results, the graph is symmetric with respect to the _____ _____.

5. When a complex number z is written in the polar form $z = r(\cos\theta + i\sin\theta)$, the nonnegative number r is the _____ or _____ of z, and the angle $\theta, 0 \le \theta < 2\pi$, is the _____ of z.

6. A vector whose magnitude is 1 is called a(n) _____ vector.

7. If the angle between two vectors $\mathbf{v}$ and $\mathbf{w}$ is $\dfrac{\pi}{2}$, then the dot product $\mathbf{v} \cdot \mathbf{w}$ equals _____.

True/False Items

T F   **1.** The polar coordinates of a point are unique.
T F   **2.** The rectangular coordinates of a point are unique.
T F   **3.** The tests for symmetry in polar coordinates are necessary, but not sufficient.
T F   **4.** DeMoivre's Theorem is useful for raising a complex number to a positive integer power.
T F   **5.** Vectors are quantities that have magnitude and direction.
T F   **6.** Force is a physical example of a vector.
T F   **7.** If $\mathbf{u}$ and $\mathbf{v}$ are orthogonal vectors, then $\mathbf{u} \cdot \mathbf{v} = 0$.
T F   **8.** The sum of the squares of the direction cosines of a vector in space equals 1.

Review Exercises

Blue problem numbers indicate the author's suggestions for use in a Practice Test.

In Problems 1–6, plot each point given in polar coordinates, and find its rectangular coordinates.

1. $\left(3, \dfrac{\pi}{6}\right)$
2. $\left(4, \dfrac{2\pi}{3}\right)$
3. $\left(-2, \dfrac{4\pi}{3}\right)$
4. $\left(-1, \dfrac{5\pi}{4}\right)$
5. $\left(-3, -\dfrac{\pi}{2}\right)$
6. $\left(-4, -\dfrac{\pi}{4}\right)$

In Problems 7–12, the rectangular coordinates of a point are given. Find two pairs of polar coordinates (r, θ) for each point, one with $r > 0$ and the other with $r < 0$. Express θ in radians.

7. $(-3, 3)$
8. $(1, -1)$
9. $(0, -2)$
10. $(2, 0)$
11. $(3, 4)$
12. $(-5, 12)$

In Problems 13–18, the letters x and y represent rectangular coordinates. Write each equation using polar coordinates (r, θ).

13. $3x^2 + 3y^2 = 6y$
14. $2x^2 - 2y^2 = 5y$
15. $2x^2 - y^2 = \dfrac{y}{x}$
16. $x^2 + 2y^2 = \dfrac{y}{x}$
17. $x(x^2 + y^2) = 4$
18. $y(x^2 - y^2) = 3$

In Problems 19–24, the letters r and θ represent polar coordinates. Write each polar equation as an equation in rectangular coordinates (x, y).

19. $r = 2\sin\theta$
20. $3r = \sin\theta$
21. $r = 5$
22. $\theta = \dfrac{\pi}{4}$
23. $r\cos\theta + 3r\sin\theta = 6$
24. $r^2\tan\theta = 1$

In Problems 25–30, sketch the graph of each polar equation. Be sure to test for symmetry.

25. $r = 4\cos\theta$
26. $r = 3\sin\theta$
27. $r = 3 - 3\sin\theta$
28. $r = 2 + \cos\theta$
29. $r = 4 - \cos\theta$
30. $r = 1 - 2\sin\theta$

In Problems 31–34, write each complex number in polar form. Express each argument in degrees.

31. $-1 - i$ **32.** $-\sqrt{3} + i$ **33.** $4 - 3i$ **34.** $3 - 2i$

In Problems 35–40, write each complex number in the standard form $a + bi$.

35. $2(\cos 150° + i \sin 150°)$ **36.** $3(\cos 60° + i \sin 60°)$ **37.** $3\left(\cos \dfrac{2\pi}{3} + i \sin \dfrac{2\pi}{3}\right)$

38. $4\left(\cos \dfrac{3\pi}{4} + i \sin \dfrac{3\pi}{4}\right)$ **39.** $0.1(\cos 350° + i \sin 350°)$ **40.** $0.5(\cos 160° + i \sin 160°)$

In Problems 41–46, find zw and $\dfrac{z}{w}$. Leave your answers in polar form.

41. $z = \cos 80° + i \sin 80°$
   $w = \cos 50° + i \sin 50°$

42. $z = \cos 205° + i \sin 205°$
   $w = \cos 85° + i \sin 85°$

43. $z = 3\left(\cos \dfrac{9\pi}{5} + i \sin \dfrac{9\pi}{5}\right)$
   $w = 2\left(\cos \dfrac{\pi}{5} + i \sin \dfrac{\pi}{5}\right)$

44. $z = 2\left(\cos \dfrac{5\pi}{3} + i \sin \dfrac{5\pi}{3}\right)$
   $w = 3\left(\cos \dfrac{\pi}{3} + i \sin \dfrac{\pi}{3}\right)$

45. $z = 5(\cos 10° + i \sin 10°)$
   $w = \cos 355° + i \sin 355°$

46. $z = 4(\cos 50° + i \sin 50°)$
   $w = \cos 340° + i \sin 340°$

In Problems 47–54, write each expression in the standard form $a + bi$.

47. $\left[3(\cos 20° + i \sin 20°)\right]^3$

48. $\left[2(\cos 50° + i \sin 50°)\right]^3$

49. $\left[\sqrt{2}\left(\cos \dfrac{5\pi}{8} + i \sin \dfrac{5\pi}{8}\right)\right]^4$

50. $\left[2\left(\cos \dfrac{5\pi}{16} + i \sin \dfrac{5\pi}{16}\right)\right]^4$

51. $(1 - \sqrt{3}i)^6$ **52.** $(2 - 2i)^8$ **53.** $(3 + 4i)^4$ **54.** $(1 - 2i)^4$

55. Find all the complex cube roots of 27. **56.** Find all the complex fourth roots of -16.

In Problems 57–64, the vector $\mathbf{v}$ is represented by the directed line segment $\overrightarrow{PQ}$. Write $\mathbf{v}$ in the form $a\mathbf{i} + b\mathbf{j}$ or in the form $a\mathbf{i} + b\mathbf{j} + c\mathbf{k}$ and find $\|\mathbf{v}\|$.

57. $P = (1, -2);\quad Q = (3, -6)$ **58.** $P = (-3, 1);\quad Q = (4, -2)$

59. $P = (0, -2);\quad Q = (-1, 1)$ **60.** $P = (3, -4);\quad Q = (-2, 0)$

61. $P = (6, 2, 1);\quad Q = (3, 0, 2)$ **62.** $P = (4, 7, 0);\quad Q = (0, 5, 6)$

63. $P = (-1, 0, 1);\quad Q = (2, 0, 0)$ **64.** $P = (6, 2, 2);\quad Q = (2, 6, 2)$

In Problems 65–72, use the vectors $\mathbf{v} = -2\mathbf{i} + \mathbf{j}$ and $\mathbf{w} = 4\mathbf{i} - 3\mathbf{j}$.

65. Find $4\mathbf{v} - 3\mathbf{w}$. **66.** Find $-\mathbf{v} + 2\mathbf{w}$. **67.** Find $\|\mathbf{v}\|$.

68. Find $\|\mathbf{v} + \mathbf{w}\|$. **69.** Find $\|\mathbf{v}\| + \|\mathbf{w}\|$. **70.** Find $\|2\mathbf{v}\| - 3\|\mathbf{w}\|$.

71. Find a unit vector in the same direction as $\mathbf{v}$. **72.** Find a unit vector in the opposite direction of $\mathbf{w}$.

In Problems 73–82, use the vectors $\mathbf{v} = 3\mathbf{i} + \mathbf{j} - 2\mathbf{k}$ and $\mathbf{w} = -3\mathbf{i} + 2\mathbf{j} - \mathbf{k}$ to find each expression.

73. $4\mathbf{v} - 3\mathbf{w}$ **74.** $-\mathbf{v} + 2\mathbf{w}$ **75.** $\|\mathbf{v} - \mathbf{w}\|$ **76.** $\|\mathbf{v} + \mathbf{w}\|$

77. $\|\mathbf{v}\| - \|\mathbf{w}\|$ **78.** $\|\mathbf{v}\| + \|\mathbf{w}\|$ **79.** $\mathbf{v} \times \mathbf{w}$ **80.** $\mathbf{v} \cdot (\mathbf{v} \times \mathbf{w})$

81. Find a unit vector in the same direction as $\mathbf{v}$ and then in the opposite direction of $\mathbf{v}$.

82. Find a unit vector orthogonal to both $\mathbf{v}$ and $\mathbf{w}$.

In Problems 83–90, find the dot product $\mathbf{v} \cdot \mathbf{w}$ and the angle between $\mathbf{v}$ and $\mathbf{w}$.

83. $\mathbf{v} = -2\mathbf{i} + \mathbf{j},\quad \mathbf{w} = 4\mathbf{i} - 3\mathbf{j}$ **84.** $\mathbf{v} = 3\mathbf{i} - \mathbf{j},\quad \mathbf{w} = \mathbf{i} + \mathbf{j}$

85. $\mathbf{v} = \mathbf{i} - 3\mathbf{j},\quad \mathbf{w} = -\mathbf{i} + \mathbf{j}$ **86.** $\mathbf{v} = \mathbf{i} + 4\mathbf{j},\quad \mathbf{w} = 3\mathbf{i} - 2\mathbf{j}$

87. $\mathbf{v} = \mathbf{i} + \mathbf{j} + \mathbf{k},\quad \mathbf{w} = \mathbf{i} - \mathbf{j} + \mathbf{k}$ **88.** $\mathbf{v} = \mathbf{i} - \mathbf{j} + \mathbf{k},\quad \mathbf{w} = 2\mathbf{i} + \mathbf{j} + \mathbf{k}$

89. $\mathbf{v} = 4\mathbf{i} - \mathbf{j} + 2\mathbf{k},\quad \mathbf{w} = \mathbf{i} - 2\mathbf{j} - 3\mathbf{k}$ **90.** $\mathbf{v} = -\mathbf{i} - 2\mathbf{j} + 3\mathbf{k},\quad \mathbf{w} = 5\mathbf{i} + \mathbf{j} + \mathbf{k}$

91. Find the vector projection of $\mathbf{v} = 2\mathbf{i} + 3\mathbf{j}$ onto $\mathbf{w} = 3\mathbf{i} + \mathbf{j}$.

92. Find the vector projection of $\mathbf{v} = -\mathbf{i} + 2\mathbf{j}$ onto $\mathbf{w} = 3\mathbf{i} - \mathbf{j}$.

93. Find the direction angles of the vector $\mathbf{v} = 3\mathbf{i} - 4\mathbf{j} + 2\mathbf{k}$.

94. Find the direction angles of the vector $\mathbf{v} = \mathbf{i} - \mathbf{j} + 2\mathbf{k}$.

95. Find the area of the parallelogram with vertices $P_1 = (1, 1, 1)$, $P_2 = (2, 3, 4)$, $P_3 = (6, 5, 2)$, and $P_4 = (7, 7, 5)$.

96. Find the area of the parallelogram with vertices $P_1 = (2, -1, 1)$, $P_2 = (5, 1, 4)$, $P_3 = (0, 1, 1)$, and $P_4 = (3, 3, 4)$.

97. Actual Speed and Direction of a Swimmer A swimmer can maintain a constant speed of 5 miles per hour. If the swimmer heads directly across a river that has a current moving at the rate of 2 miles per hour, what is the actual speed of the swimmer? (See the figure.) If the river is 1 mile wide, how far downstream will the swimmer end up from the point directly across the river from the starting point?

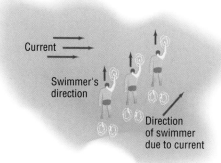

98. Actual Speed and Direction of an Airplane An airplane has an airspeed of 500 kilometers per hour in a northerly direction. The wind velocity is 60 kilometers per hour in a southeasterly direction. Find the actual speed and direction of the plane relative to the ground.

99. Static Equilibrium A weight of 2000 pounds is suspended from two cables as shown in the figure. What are the tensions of each cable?

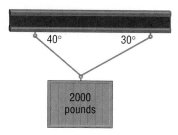

Signal Fades due to interference

Consider an individual wearing a pager in an exterior elevator on a high-rise building. The pager receives direct signals from a transmitting tower far away as well as their reflections from the flat ground (see the figure).

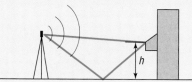

Oversimplifying things, it can be shown that the normalized signal received by the pager is given by

$$E(h)_{\text{norm}} \approx 1 + 0.25\big[\cos(2\pi h) + i\sin(2\pi h)\big]$$

where h is the height of the pager (elevator) from the ground expressed in meters.

1. Show that the magnitude of the normalized signal received by the pager oscillates between a minimum and a maximum value as the elevator moves up the building. This pattern is called a **standing wave pattern.** Calculate the ratio of the maximum to the minimum signal strength that the pager sees. This is called the **Standing Wave Ratio** (SWR).

2. Show that the distance between two consecutive minima is 1 meter.

3. Show that the distance between two consecutive maxima is 1 meter.

4. Show that the distance between a minimum and its nearby maximum is 0.5 meter.

5. What should the sensitivity of the pager be to guarantee reception of the signal regardless of the distance of the elevator from the ground?

6. Draw the graph E_{norm} in the complex plane as h varies. Explain all your answers above using this figure.

Analytic Geometry

Field Trip to Motorola

One of the many activities of the National Aeronautic and Space Administration (NASA) has to do with placing antennas in orbit. These antennas are usually in the shape of a paraboloid of revolution. However, in transporting objects of any kind into orbit, cost is a major factor. One factor that reduces the cost is to keep the weight as low as possible. A second factor is to keep the size of the object as small as possible.

As a result, a parabolic reflector is manufactured by NASA for use on the space shuttle by making the antenna out of material that can be folded. Once in orbit, the folded antenna is deployed and opened to look like a parabolic reflector. One problem associated with this method of deployment is to be sure that when the antenna is opened, it aligns properly with the axis of the parabola.

6.1 CONICS

OBJECTIVE 1 Know the Names of the Conics

1 The word *conic* derives from the word *cone*, which is a geometric figure that can be constructed in the following way: Let *a* and *g* be two distinct lines that intersect at a point *V*. Keep the line *a* fixed. Now rotate the line *g* about *a* while maintaining the same angle between *a* and *g*. The collection of points swept out (generated) by the line *g* is called a (**right circular**) **cone.** See Figure 1. The fixed line *a* is called the **axis** of the cone; the point *V* is called its **vertex;** the lines that pass through *V* and make the same angle with *a* as *g* are called **generators** of the cone. Each generator is a line that lies entirely on the cone. The cone consists of two parts, called **nappes,** that intersect at the vertex.

Figure 1

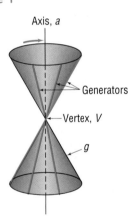

Axis, *a*

Generators

Vertex, *V*

g

Conics, an abbreviation for **conic sections,** are curves that result from the intersection of a (right circular) cone and a plane. The conics we shall study arise when the plane does not contain the vertex, as shown in Figure 2. These conics are **circles** when the plane is perpendicular to the axis of the cone and intersects each generator; **ellipses** when the plane is tilted slightly so that it intersects each generator, but intersects only one nappe of the cone; **parabolas** when the plane is tilted farther so that it is parallel to one (and only one) generator and intersects only one nappe of the cone; and **hyperbolas** when the plane intersects both nappes.

If the plane does contain the vertex, the intersection of the plane and the cone is a point, a line, or a pair of intersecting lines. These are usually called **degenerate conics.**

Figure 2

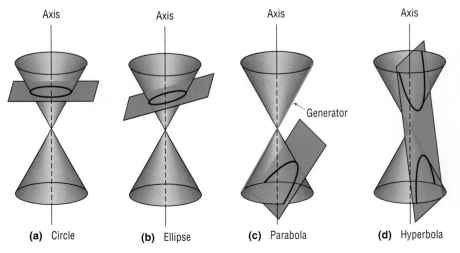

(a) Circle   (b) Ellipse   (c) Parabola   (d) Hyperbola

Before getting started, review the following:

✓ Distance Formula (Section 1.1, p. 4)

✓ Symmetry (Section 1.2, pp. 12–14)

✓ Completing the Square (Appendix A, Section A.3, p. 507)

✓ Graphing Techniques: Transformations (Section 1.5, pp. 47–56)

6.2   THE PARABOLA

OBJECTIVES
1   Find the Equation of a Parabola
2   Graph Parabolas
3   Discuss the Equation of a Parabola
4   Work with Parabolas with Vertex at (h, k)
5   Solve Applied Problems Involving Parabolas

We begin with a geometric definition of a parabola and use it to obtain an equation.

> A **parabola** is the collection of all points P in the plane that are the same distance from a fixed point F as they are from a fixed line D. The point F is called the **focus** of the parabola, and the line D is its **directrix**. As a result, a parabola is the set of points P for which
>
> $$d(F, P) = d(P, D) \tag{1}$$

1   Figure 3 shows a parabola. The line through the focus F and perpendicular to the directrix D is called the **axis of symmetry** of the parabola. The point of intersection of the parabola with its axis of symmetry is called the **vertex V**.

Figure 3

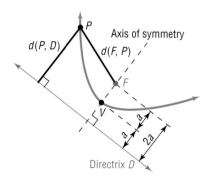

Because the vertex V lies on the parabola, it must satisfy equation (1): $d(F, V) = d(V, D)$. The vertex is midway between the focus and the directrix. We shall let a equal the distance $d(F, V)$ from F to V. Now we are ready to derive an equation for a parabola. To do this, we use a rectangular system of coordinates positioned so that the vertex V, focus F, and directrix D of the parabola are conveniently located. If we choose to locate the vertex V at the origin $(0, 0)$, then we can conveniently position the focus F on either the x-axis or the y-axis.

First, we consider the case where the focus F is on the positive x-axis, as shown in Figure 4. Because the distance from F to V is a, the coordinates of F will be $(a, 0)$ with $a > 0$. Similarly, because the distance from V to the directrix D is also a and because D must be perpendicular to the x-axis (since the x-axis is the axis of symmetry), the equation of the directrix D must be $x = -a$. Now, if $P = (x, y)$ is any point on the parabola, then P must obey equation (1):

$$d(F, P) = d(P, D)$$

Figure 4
$y^2 = 4ax$

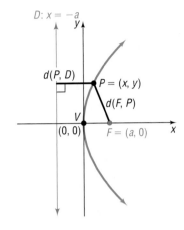

So we have

$$\sqrt{(x - a)^2 + y^2} = |x + a| \qquad \text{Use the distance formula.}$$
$$(x - a)^2 + y^2 = (x + a)^2 \qquad \text{Square both sides.}$$
$$x^2 - 2ax + a^2 + y^2 = x^2 + 2ax + a^2 \qquad \text{Remove parentheses.}$$
$$y^2 = 4ax \qquad \text{Simplify.}$$

Theorem | **Equation of a Parabola; Vertex at (0, 0), Focus at (a, 0), a > 0**

The equation of a parabola with vertex at $(0, 0)$, focus at $(a, 0)$, and directrix $x = -a, a > 0$, is

$$y^2 = 4ax \qquad \qquad \textbf{(2)}$$

EXAMPLE 1 | Finding the Equation of a Parabola and Graphing It

2 Find an equation of the parabola with vertex at $(0, 0)$ and focus at $(3, 0)$. Graph the equation.

Solution The distance from the vertex $(0, 0)$ to the focus $(3, 0)$ is $a = 3$. Based on equation (2), the equation of this parabola is

$$y^2 = 4ax$$
$$y^2 = 12x \qquad a = 3$$

Figure 5
$y^2 = 12x$

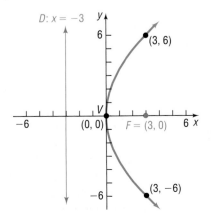

To graph this parabola, it is helpful to plot the two points on the graph above and below the focus. To locate them, we let $x = 3$. Then

$$y^2 = 12x = 12(3) = 36$$
$$y = \pm 6 \qquad \text{Solve for } y.$$

The points on the parabola above and below the focus are $(3, 6)$ and $(3, -6)$. These points help in graphing the parabola because they determine the "opening." See Figure 5.

In general, the points on a parabola $y^2 = 4ax$ that lie above and below the focus $(a, 0)$ are each at a distance $2a$ from the focus. This follows from the fact that if $x = a$ then $y^2 = 4ax = 4a^2$ so $y = \pm 2a$. The line segment joining these two points is called the **latus rectum**; its length is $4a$.

 COMMENT   To graph the parabola $y^2 = 12x$ discussed in Example 1, we need to graph the two functions $Y_1 = \sqrt{12x}$ and $Y_2 = -\sqrt{12x}$. Do this and compare what you see with Figure 5. ■

 NOW WORK PROBLEM **9**.

By reversing the steps we used to obtain equation (2), it follows that the graph of an equation of the form of equation (2), $y^2 = 4ax$, is a parabola; its vertex is at $(0, 0)$, its focus is at $(a, 0)$, its directrix is the line $x = -a$, and its axis of symmetry is the x-axis.

 For the remainder of this section, the direction "Discuss the equation" will mean to find the vertex, focus, and directrix of the parabola and graph it.

EXAMPLE 2   Discussing the Equation of a Parabola

Discuss the equation: $y^2 = 8x$

Solution   The equation $y^2 = 8x$ is of the form $y^2 = 4ax$, where $4a = 8$ so that $a = 2$. Consequently, the graph of the equation is a parabola with vertex at $(0, 0)$ and focus on the positive x-axis at $(2, 0)$. The directrix is the vertical line $x = -2$. The two points defining the latus rectum are obtained by letting $x = 2$. Then $y^2 = 16$ so $y = \pm 4$. See Figure 6. ■

Figure 6
$y^2 = 8x$

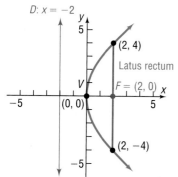

Recall that we arrived at equation (2) after placing the focus on the positive x-axis. If the focus is placed on the negative x-axis, positive y-axis, or negative y-axis, a different form of the equation for the parabola results. The four forms of the equation of a parabola with vertex at $(0, 0)$ and focus on a coordinate axis a distance a from $(0, 0)$ are given in Table 1, and their graphs are given in Figure 7. Notice that each graph is symmetric with respect to its axis of symmetry.

TABLE 1 EQUATIONS OF A PARABOLA: VERTEX AT (0,0); FOCUS ON AN AXIS; $a > 0$				
Vertex	**Focus**	**Directrix**	**Equation**	**Description**
$(0, 0)$	$(a, 0)$	$x = -a$	$y^2 = 4ax$	Parabola, axis of symmetry is the x-axis, opens to right
$(0, 0)$	$(-a, 0)$	$x = a$	$y^2 = -4ax$	Parabola, axis of symmetry is the x-axis, opens to left
$(0, 0)$	$(0, a)$	$y = -a$	$x^2 = 4ay$	Parabola, axis of symmetry is the y-axis, opens up
$(0, 0)$	$(0, -a)$	$y = a$	$x^2 = -4ay$	Parabola, axis of symmetry is the y-axis, opens down

Figure 7

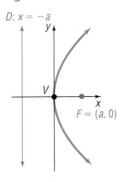

(a) $y^2 = 4ax$

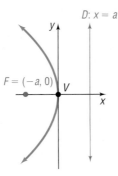

(b) $y^2 = -4ax$

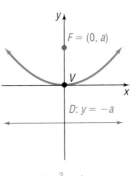

(c) $x^2 = 4ay$

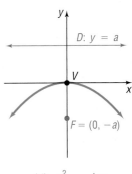

(d) $x^2 = -4ay$

EXAMPLE 3

Discussing the Equation of a Parabola

Discuss the equation: $x^2 = -12y$

Figure 8
$x^2 = -12y$

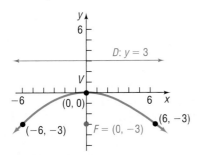

Solution The equation $x^2 = -12y$ is of the form $x^2 = -4ay$, with $a = 3$. Consequently, the graph of the equation is a parabola with vertex at $(0, 0)$, focus at $(0, -3)$ and directrix the line $y = 3$. The parabola opens down, and its axis of symmetry is the y-axis. To obtain the points defining the latus rectum, let $y = -3$. Then $x^2 = 36$ so $x = \pm 6$. See Figure 8. ∎

NOW WORK PROBLEM **27.**

EXAMPLE 4

Finding the Equation of a Parabola

Find the equation of the parabola with focus at $(0, 4)$ and directrix the line $y = -4$. Graph the equation.

Figure 9
$x^2 = 16y$

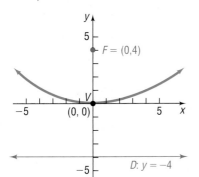

Solution A parabola whose focus is at $(0, 4)$ and whose directrix is the horizontal line $y = -4$ will have its vertex at $(0, 0)$. (Do you see why? The vertex is midway between the focus and the directrix.) Since the focus is on the positive y-axis at $(0, 4)$, the equation of this parabola is of the form $x^2 = 4ay$, with $a = 4$; that is,

$$x^2 = 4ay \underset{\underset{a\,=\,4}{\uparrow}}{=} 4(4)y = 16y$$

Figure 9 shows the graph of $x^2 = 16y$. ∎

| EXAMPLE 5 | Finding the Equation of a Parabola |

Find the equation of a parabola with vertex at $(0, 0)$ if its axis of symmetry is the x-axis and its graph contains the point $\left(-\dfrac{1}{2}, 2\right)$. Find its focus and directrix, and graph the equation.

Solution   The vertex is at the origin, the axis of symmetry is the x-axis, and the graph contains a point in the second quadrant, so the parabola opens to the left. We see from Table 1 that the form of the equation is

$$y^2 = -4ax$$

Because the point $\left(-\dfrac{1}{2}, 2\right)$ is on the parabola, the coordinates $x = -\dfrac{1}{2}$, $y = 2$ must satisfy the equation. Substituting $x = -\dfrac{1}{2}$ and $y = 2$ into the equation, we find that

$$4 = -4a\left(-\dfrac{1}{2}\right)$$
$$a = 2$$

The equation of the parabola is

$$y^2 = -4(2)x = -8x$$

The focus is at $(-2, 0)$ and the directrix is the line $x = 2$. Letting $x = -2$, we find $y^2 = 16$ so $y = \pm 4$. The points $(-2, 4)$ and $(-2, -4)$ define the latus rectum. See Figure 10.   ∎

Figure 10
$y^2 = -8x$

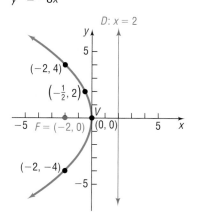

NOW   WORK   PROBLEM   **19.**

VERTEX AT (h, k)

④ If a parabola with vertex at the origin and axis of symmetry along a coordinate axis is shifted horizontally h units and then vertically k units, the result is a parabola with vertex at (h, k) and axis of symmetry parallel to a coordinate axis. The equations of such parabolas have the same forms as those in Table 1, but with x replaced by $x - h$ (the horizontal shift) and y replaced by $y - k$ (the vertical shift). Table 2 gives the forms of the equations of such parabolas. Figure 11(a)–(d) illustrates the graphs for $h > 0, k > 0$.

TABLE 2
PARABOLAS WITH VERTEX AT (h, k); AXIS OF SYMMETRY PARALLEL TO A COORDINATE AXIS, $a > 0$

Vertex	Focus	Directrix	Equation	Description
(h, k)	$(h + a, k)$	$x = h - a$	$(y - k)^2 = 4a(x - h)$	Parabola, axis of symmetry parallel to x-axis, opens to right
(h, k)	$(h - a, k)$	$x = h + a$	$(y - k)^2 = -4a(x - h)$	Parabola, axis of symmetry parallel to x-axis, opens to left
(h, k)	$(h, k + a)$	$y = k - a$	$(x - h)^2 = 4a(y - k)$	Parabola, axis of symmetry parallel to y-axis, opens up
(h, k)	$(h, k - a)$	$y = k + a$	$(x - h)^2 = -4a(y - k)$	Parabola, axis of symmetry parallel to y-axis, opens down

Figure 11

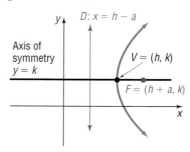

(a) $(y - k)^2 = 4a(x - h)$

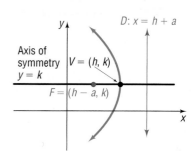

(b) $(y - k)^2 = -4a(x - h)$

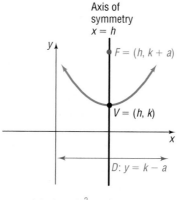

(c) $(x - h)^2 = 4a(y - k)$

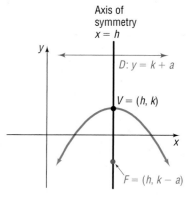

(d) $(x - h)^2 = -4a(y - k)$

EXAMPLE 6

Finding the Equation of a Parabola, Vertex Not at Origin

Find an equation of the parabola with vertex at $(-2, 3)$ and focus at $(0, 3)$. Graph the equation.

Figure 12
$(y - 3)^2 = 8(x + 2)$

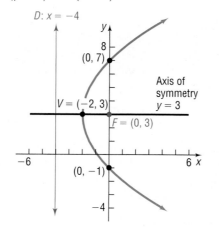

Solution The vertex $(-2, 3)$ and focus $(0, 3)$ both lie on the horizontal line $y = 3$ (the axis of symmetry). The distance a from the vertex $(-2, 3)$ to the focus $(0, 3)$ is $a = 2$. Also, because the focus lies to the right of the vertex, we know that the parabola opens to the right. Consequently, the form of the equation is

$$(y - k)^2 = 4a(x - h)$$

where $(h, k) = (-2, 3)$ and $a = 2$. Therefore, the equation is

$$(y - 3)^2 = 4 \cdot 2[x - (-2)]$$
$$(y - 3)^2 = 8(x + 2)$$

If $x = 0$, then $(y - 3)^2 = 16$. Then, $y - 3 = \pm 4$ so $y = -1$ or $y = 7$. The points $(0, -1)$ and $(0, 7)$ define the latus rectum; the line $x = -4$ is the directrix. See Figure 12. ∎

NOW WORK PROBLEM **17**.

Polynomial equations define parabolas whenever they involve two variables that are quadratic in one variable and linear in the other. To discuss this type of equation, we first complete the square of the variable that is quadratic.

| EXAMPLE 7 | Discussing the Equation of a Parabola |

Discuss the equation: $x^2 + 4x - 4y = 0$

Solution   To discuss the equation $x^2 + 4x - 4y = 0$, we complete the square involving the variable x.

$$x^2 + 4x - 4y = 0$$
$$x^2 + 4x = 4y \qquad \text{Isolate the terms involving } x \text{ on the left side.}$$
$$x^2 + 4x + 4 = 4y + 4 \qquad \text{Complete the square on the left side.}$$
$$(x + 2)^2 = 4(y + 1) \qquad \text{Factor.}$$

This equation is of the form $(x - h)^2 = 4a(y - k)$, with $h = -2$, $k = -1$, and $a = 1$. The graph is a parabola with vertex at $(h, k) = (-2, -1)$ that opens up. The focus is at $(-2, 0)$, and the directrix is the line $y = -2$. See Figure 13.

Figure 13
$x^2 + 4x - 4y = 0$

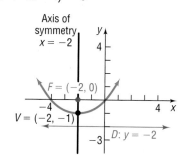

NOW WORK PROBLEM 35.

Parabolas find their way into many applications. For example, suspension bridges have cables in the shape of a parabola. Another property of parabolas that is used in applications is their reflecting property.

REFLECTING PROPERTY

Suppose that a mirror is shaped like a **paraboloid of revolution,** a surface formed by rotating a parabola about its axis of symmetry. If a light (or any other emitting source) is placed at the focus of the parabola, all the rays emanating from the light will reflect off the mirror in lines parallel to the axis of symmetry. This principle is used in the design of searchlights, flashlights, certain automobile headlights, and other such devices. See Figure 14.

Conversely, suppose that rays of light (or other signals) emanate from a distant source so that they are essentially parallel. When these rays strike the surface of a parabolic mirror whose axis of symmetry is parallel to these rays, they are reflected to a single point at the focus. This principle is used in the design of some solar energy devices, satellite dishes, and the mirrors used in some types of telescopes. See Figure 15.

Figure 14
Searchlight

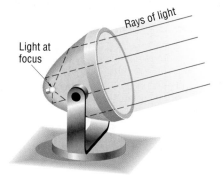

Figure 15
Telescope

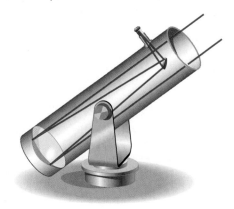

EXAMPLE 8	**Satellite Dish**

A satellite dish is shaped like a paraboloid of revolution. The signals that emanate from a satellite strike the surface of the dish and are reflected to a single point, where the receiver is located. If the dish is 8 feet across at its opening and is 3 feet deep at its center, at what position should the receiver be placed?

Solution Figure 16(a) shows the satellite dish. We draw the parabola used to form the dish on a rectangular coordinate system so that the vertex of the parabola is at the origin and its focus is on the positive *y*-axis. See Figure 16(b).

Figure 16

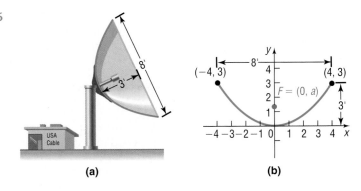

(a)   (b)

The form of the equation of the parabola is

$$x^2 = 4ay$$

and its focus is at $(0, a)$. Since $(4, 3)$ is a point on the graph, we have

$$4^2 = 4a(3)$$

$$a = \frac{4}{3}$$

The receiver should be located $1\frac{1}{3}$ feet from the base of the dish, along its axis of symmetry.

NOW WORK PROBLEM **51**.

6.2 EXERCISES

In Problems 1–8, the graph of a parabola is given. Match each graph to its equation.

A. $y^2 = 4x$
B. $x^2 = 4y$

C. $y^2 = -4x$
D. $x^2 = -4y$

E. $(y - 1)^2 = 4(x - 1)$
F. $(x + 1)^2 = 4(y + 1)$

G. $(y - 1)^2 = -4(x - 1)$
H. $(x + 1)^2 = -4(y + 1)$

1.

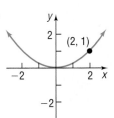

2.

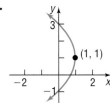

3.

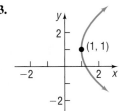

4.

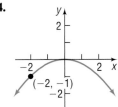

5.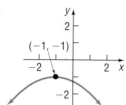

6.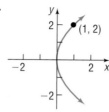

7.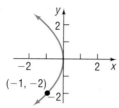

8.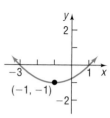

In Problems 9–24, find the equation of the parabola described. Find the two points that define the latus rectum, and graph the equation.

9. Focus at $(4, 0)$;   vertex at $(0, 0)$

10. Focus at $(0, 2)$;   vertex at $(0, 0)$

11. Focus at $(0, -3)$;   vertex at $(0, 0)$

12. Focus at $(-4, 0)$;   vertex at $(0, 0)$

13. Focus at $(-2, 0)$;   directrix the line $x = 2$

14. Focus at $(0, -1)$;   directrix the line $y = 1$

15. Directrix the line $y = -\dfrac{1}{2}$;   vertex at $(0, 0)$

16. Directrix the line $x = -\dfrac{1}{2}$;   vertex at $(0, 0)$

17. Vertex at $(2, -3)$;   focus at $(2, -5)$

18. Vertex at $(4, -2)$;   focus at $(6, -2)$

19. Vertex at $(0, 0)$;   axis of symmetry the y-axis; containing the point $(2, 3)$

20. Vertex at $(0, 0)$;   axis of symmetry the x-axis; containing the point $(2, 3)$

21. Focus at $(-3, 4)$;   directrix the line $y = 2$

22. Focus at $(2, 4)$;   directrix the line $x = -4$

23. Focus at $(-3, -2)$;   directrix the line $x = 1$

24. Focus at $(-4, 4)$;   directrix the line $y = -2$

In Problems 25–42, find the vertex, focus, and directrix of each parabola. Graph the equation.

25. $x^2 = 4y$

26. $y^2 = 8x$

27. $y^2 = -16x$

28. $x^2 = -4y$

29. $(y - 2)^2 = 8(x + 1)$

30. $(x + 4)^2 = 16(y + 2)$

31. $(x - 3)^2 = -(y + 1)$

32. $(y + 1)^2 = -4(x - 2)$

33. $(y + 3)^2 = 8(x - 2)$

34. $(x - 2)^2 = 4(y - 3)$

35. $y^2 - 4y + 4x + 4 = 0$

36. $x^2 + 6x - 4y + 1 = 0$

37. $x^2 + 8x = 4y - 8$

38. $y^2 - 2y = 8x - 1$

39. $y^2 + 2y - x = 0$

40. $x^2 - 4x = 2y$

41. $x^2 - 4x = y + 4$

42. $y^2 + 12y = -x + 1$

In Problems 43–50, write an equation for each parabola.

43.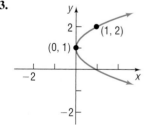

44.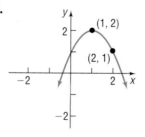

45.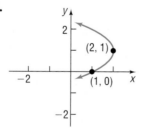

46.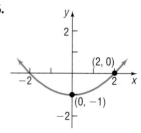

47.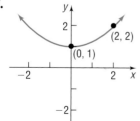

48.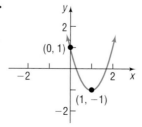

49.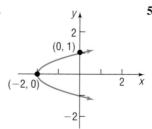

50.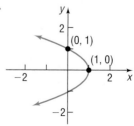

51. **Satellite Dish** A satellite dish is shaped like a paraboloid of revolution. The signals that emanate from a satellite strike the surface of the dish and are reflected to a single point, where the receiver is located. If the dish is 10 feet across at its opening and is 4 feet deep at its center, at what position should the receiver be placed?

52. **Constructing a TV Dish** A cable TV receiving dish is in the shape of a paraboloid of revolution. Find the location of the receiver, which is placed at the focus, if the dish is 6 feet across at its opening and 2 feet deep.

53. **Constructing a Flashlight** The reflector of a flashlight is in the shape of a paraboloid of revolution. Its diameter is 4 inches and its depth is 1 inch. How far from the vertex should the light bulb be placed so that the rays will be reflected parallel to the axis?

54. **Constructing a Headlight** A sealed-beam headlight is in the shape of a paraboloid of revolution. The bulb, which is placed at the focus, is 1 inch from the vertex. If the depth is to be 2 inches, what is the diameter of the headlight at its opening?

55. **Suspension Bridge** The cables of a suspension bridge are in the shape of a parabola, as shown in the figure. The towers supporting the cable are 600 feet apart and 80 feet high. If the cables touch the road surface midway between the towers, what is the height of the cable at a point 150 feet from the center of the bridge?

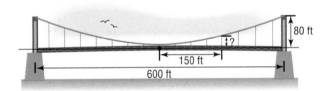

56. **Suspension Bridge** The cables of a suspension bridge are in the shape of a parabola. The towers supporting the cable are 400 feet apart and 100 feet high. If the cables are at a height of 10 feet midway between the towers, what is the height of the cable at a point 50 feet from the center of the bridge?

57. **Searchlight** A searchlight is shaped like a paraboloid of revolution. If the light source is located 2 feet from the base along the axis of symmetry and the opening is 5 feet across, how deep should the searchlight be?

58. **Searchlight** A searchlight is shaped like a paraboloid of revolution. If the light source is located 2 feet from the base along the axis of symmetry and the depth of the searchlight is 4 feet, what should the width of the opening be?

59. **Solar Heat** A mirror is shaped like a paraboloid of revolution and will be used to concentrate the rays of the sun at its focus, creating a heat source. (See the figure.) If the mirror is 20 feet across at its opening and is 6 feet deep, where will the heat source be concentrated?

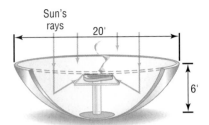

60. **Reflecting Telescope** A reflecting telescope contains a mirror shaped like a paraboloid of revolution. If the mirror is 4 inches across at its opening and is 3 feet deep, where will the collected light be concentrated?

61. **Parabolic Arch Bridge** A bridge is built in the shape of a parabolic arch. The bridge has a span of 120 feet and a maximum height of 25 feet. See the illustration. Choose a suitable rectangular coordinate system and find the height of the arch at distances of 10, 30, and 50 feet from the center.

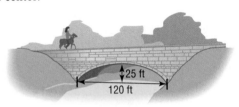

62. **Parabolic Arch Bridge** A bridge is to be built in the shape of a parabolic arch and is to have a span of 100 feet. The height of the arch a distance of 40 feet from the center is to be 10 feet. Find the height of the arch at its center.

63. Show that an equation of the form

$$Ax^2 + Ey = 0, \qquad A \neq 0, E \neq 0$$

is the equation of a parabola with vertex at $(0, 0)$ and axis of symmetry the y-axis. Find its focus and directrix.

64. Show that an equation of the form

$$Cy^2 + Dx = 0, \qquad C \neq 0, D \neq 0$$

is the equation of a parabola with vertex at $(0, 0)$ and axis of symmetry the x-axis. Find its focus and directrix.

65. Show that the graph of an equation of the form

$$Ax^2 + Dx + Ey + F = 0, \qquad A \neq 0$$

(a) Is a parabola if $E \neq 0$.
(b) Is a vertical line if $E = 0$ and $D^2 - 4AF = 0$.
(c) Is two vertical lines if $E = 0$ and $D^2 - 4AF > 0$.
(d) Contains no points if $E = 0$ and $D^2 - 4AF < 0$.

66. Show that the graph of an equation of the form

$$Cy^2 + Dx + Ey + F = 0, \qquad C \neq 0$$

(a) Is a parabola if $D \neq 0$.
(b) Is a horizontal line if $D = 0$ and $E^2 - 4CF = 0$.
(c) Is two horizontal lines if $D = 0$ and $E^2 - 4CF > 0$.
(d) Contains no points if $D = 0$ and $E^2 - 4CF < 0$.

PREPARING FOR THIS SECTION

Before getting started, review the following:

✓ Distance Formula (Section 1.1, p. 4)

✓ Completing the Square
(Appendix A, Section A.3, p. 507)

✓ Intercepts (Section 1.2, pp. 11–12)

✓ Symmetry (Section 1.2, pp. 12–14)

✓ Circles (Section 1.2, pp. 16–18)

✓ Graphing Techniques: Transformations
(Section 1.5, pp. 47–56)

6.3 THE ELLIPSE

OBJECTIVES   **1** Find the Equation of an Ellipse
2 Graph Ellipses
3 Discuss the Equation of an Ellipse
4 Work with Ellipses with Center at (h, k)
5 Solve Applied Problems Involving Ellipses

An **ellipse** is the collection of all points in the plane the sum of whose distances from two fixed points, called the **foci,** is a constant.

Figure 17

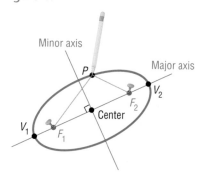

The definition actually contains within it a physical means for drawing an ellipse. Find a piece of string (the length of this string is the constant referred to in the definition). Then take two thumbtacks (the foci) and stick them on a piece of cardboard so that the distance between them is less than the length of the string. Now attach the ends of the string to the thumbtacks and, using the point of a pencil, pull the string taut. See Figure 17. Keeping the string taut, rotate the pencil around the two thumbtacks. The pencil traces out an ellipse, as shown in Figure 17.

In Figure 17, the foci are labeled F_1 and F_2. The line containing the foci is called the **major axis.** The midpoint of the line segment joining the foci is called the **center** of the ellipse. The line through the center and perpendicular to the major axis is called the **minor axis.**

The two points of intersection of the ellipse and the major axis are the **vertices,** V_1 and V_2, of the ellipse. The distance from one vertex to the other is called the **length of the major axis.** The ellipse is symmetric with respect to its major axis and with respect to its minor axis.

With these ideas in mind, we are now ready to find the equation of an ellipse in a rectangular coordinate system. First, we place the center of the ellipse at the origin. Second, we position the ellipse so that its major axis coincides with a coordinate axis. Suppose that the major axis coincides with the x-axis, as shown in Figure 18. If c is the distance from the center to a focus, then one focus will be at $F_1 = (-c, 0)$ and the other at $F_2 = (c, 0)$. As we shall see, it is convenient to let $2a$ denote the constant distance referred to in the definition. Then, if $P = (x, y)$ is any point on the ellipse, we have

Figure 18
$$d(F_1, P) + d(F_2, P) = 2a$$

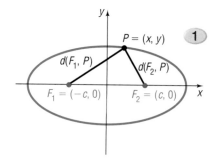

1

$$d(F_1, P) + d(F_2, P) = 2a$$

Sum of the distances from P to
the foci equals a constant, $2a$.

$$\sqrt{(x + c)^2 + y^2} + \sqrt{(x - c)^2 + y^2} = 2a$$

Use the distance formula.

$$\sqrt{(x + c)^2 + y^2} = 2a - \sqrt{(x - c)^2 + y^2}$$

Isolate one radical.

$$(x + c)^2 + y^2 = 4a^2 - 4a\sqrt{(x - c)^2 + y^2} \qquad \text{Square both sides.}$$
$$+ (x - c)^2 + y^2$$
$$x^2 + 2cx + c^2 + y^2 = 4a^2 - 4a\sqrt{(x - c)^2 + y^2} \qquad \text{Remove parentheses.}$$
$$+ x^2 - 2cx + c^2 + y^2$$
$$4cx - 4a^2 = -4a\sqrt{(x - c)^2 + y^2} \qquad \text{Simplify; Isolate the radical.}$$
$$cx - a^2 = -a\sqrt{(x - c)^2 + y^2} \qquad \text{Divide each side by 4.}$$
$$(cx - a^2)^2 = a^2[(x - c)^2 + y^2] \qquad \text{Square both sides again.}$$
$$c^2x^2 - 2a^2cx + a^4 = a^2(x^2 - 2cx + c^2 + y^2) \qquad \text{Remove parentheses.}$$
$$(c^2 - a^2)x^2 - a^2y^2 = a^2c^2 - a^4 \qquad \text{Rearrange the terms.}$$
$$(a^2 - c^2)x^2 + a^2y^2 = a^2(a^2 - c^2) \qquad \begin{array}{l}\text{Multiply each side by }-1;\\ \text{factor } a^2 \text{ on the right side.}\end{array} \qquad \textbf{(1)}$$

To obtain points on the ellipse off the x-axis, it must be that $a > c$. To see why, look again at Figure 18.

$$d(F_1, P) + d(F_2, P) > d(F_1, F_2) \qquad \begin{array}{l}\text{The sum of the lengths of two sides of a triangle}\\ \text{is greater than the length of the third side.}\end{array}$$

$$2a > 2c \qquad d(F_1, P) + d(F_2, P) = 2a; d(F_1, F_2) = 2c.$$

$$a > c$$

Since $a > c$, we also have $a^2 > c^2$, so $a^2 - c^2 > 0$. Let $b^2 = a^2 - c^2, b > 0$. Then $a > b$ and equation (1) can be written as

$$b^2x^2 + a^2y^2 = a^2b^2$$

$$\frac{x^2}{a^2} + \frac{y^2}{b^2} = 1 \qquad \text{Divide each side by } a^2b^2.$$

Theorem   **Equation of an Ellipse; Center at (0, 0); Foci at (± c, 0); Major Axis along the x-Axis**

An equation of the ellipse with center at $(0, 0)$ and foci at $(-c, 0)$ and $(c, 0)$ is

$$\frac{x^2}{a^2} + \frac{y^2}{b^2} = 1, \qquad \text{where } a > b > 0 \text{ and } b^2 = a^2 - c^2 \qquad \textbf{(2)}$$

The major axis is the x-axis. The vertices are at $(-a, 0)$ and $(a, 0)$.

As you can verify, the ellipse defined by equation (2) is symmetric with respect to the x-axis, y-axis, and origin.

To find the vertices of the ellipse defined by equation (2), let $y = 0$. The vertices satisfy the equation $\dfrac{x^2}{a^2} = 1$, the solutions of which are $x = \pm a$. Consequently, the vertices of the ellipse given by equation (2) are $V_1 = (-a, 0)$ and $V_2 = (a, 0)$. The y-intercepts of the ellipse, found by letting $x = 0$, have coordinates $(0, -b)$ and $(0, b)$. These four intercepts, $(a, 0)$, $(-a, 0)$, $(0, b)$, and $(0, -b)$, are used to graph the ellipse. See Figure 19.

Figure 19

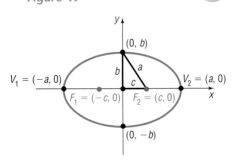

Notice in Figure 19 the right triangle formed with the points $(0, 0)$, $(c, 0)$, and $(0, b)$. Because $b^2 = a^2 - c^2$ (or $b^2 + c^2 = a^2$), the distance from the focus at $(c, 0)$ to the point $(0, b)$ is a.

EXAMPLE 1

Finding an Equation of an Ellipse

Find an equation of the ellipse with center at the origin, one focus at $(3, 0)$, and a vertex at $(-4, 0)$. Graph the equation.

Solution The ellipse has its center at the origin and, since the given focus and vertex lie on the x-axis, the major axis is the x-axis. The distance from the center, $(0, 0)$, to one of the foci, $(3, 0)$, is $c = 3$. The distance from the center, $(0, 0)$, to one of the vertices, $(-4, 0)$, is $a = 4$. From equation (2), it follows that

$$b^2 = a^2 - c^2 = 16 - 9 = 7$$

so an equation of the ellipse is

$$\frac{x^2}{16} + \frac{y^2}{7} = 1$$

Figure 20 shows the graph.   ■

Figure 20

$$\frac{x^2}{16} + \frac{y^2}{7} = 1$$

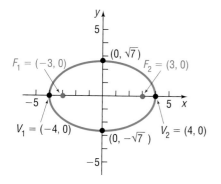

Notice in Figure 20 how we used the intercepts of the equation to graph the ellipse. Following this practice will make it easier for you to obtain an accurate graph of an ellipse.

 COMMENT   The intercepts of the ellipse also provide information about how to set the viewing rectangle. To graph the ellipse

$$\frac{x^2}{16} + \frac{y^2}{7} = 1$$

discussed in Example 1, we would set the viewing rectangle using a square screen that includes the intercepts, perhaps $-4.5 \le x \le 4.5$, $-3 \le y \le 3$. Then we would proceed to solve the equation for y:

$$\frac{x^2}{16} + \frac{y^2}{7} = 1$$

$$\frac{y^2}{7} = 1 - \frac{x^2}{16} \qquad \text{Subtract } \frac{x^2}{16} \text{ from each side.}$$

$$y^2 = 7\left(1 - \frac{x^2}{16}\right) \qquad \text{Multiply both sides by 7.}$$

$$y = \pm\sqrt{7\left(1 - \frac{x^2}{16}\right)} \qquad \text{Take the square root of each side.}$$

Now graph the two functions

$$Y_1 = \sqrt{7\left(1 - \frac{x^2}{16}\right)} \quad \text{and} \quad Y_2 = -\sqrt{7\left(1 - \frac{x^2}{16}\right)}$$

Figure 21

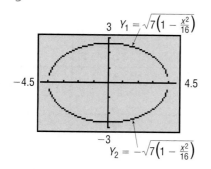

Figure 21 shows the result.   ■

✏ NOW WORK PROBLEM **15.**

An equation of the form of equation (2), with $a > b$, is the equation of an ellipse with center at the origin, foci on the x-axis at $(-c, 0)$ and $(c, 0)$, where $c^2 = a^2 - b^2$, and major axis along the x-axis.

3 For the remainder of this section, the direction "Discuss the equation" will mean to find the center, major axis, foci, and vertices of the ellipse and graph it.

EXAMPLE 2

Discussing the Equation of an Ellipse

Discuss the equation: $\dfrac{x^2}{25} + \dfrac{y^2}{9} = 1$

Solution The given equation is of the form of equation (2), with $a^2 = 25$ and $b^2 = 9$. The equation is that of an ellipse with center $(0, 0)$ and major axis along the x-axis. The vertices are at $(\pm a, 0) = (\pm 5, 0)$. Because $b^2 = a^2 - c^2$, we find that

$$c^2 = a^2 - b^2 = 25 - 9 = 16$$

The foci are at $(\pm c, 0) = (\pm 4, 0)$. Figure 22 shows the graph.

Figure 22

$\dfrac{x^2}{25} + \dfrac{y^2}{9} = 1$

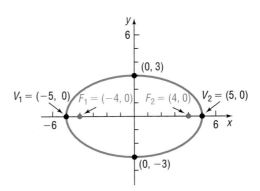

NOW WORK PROBLEM 5.

If the major axis of an ellipse with center at $(0, 0)$ lies on the y-axis, then the foci are at $(0, -c)$ and $(0, c)$. Using the same steps as before, the definition of an ellipse leads to the following result:

Theorem

Equation of an Ellipse; Center at (0, 0); Foci at (0, ± c); Major Axis along the y-Axis

An equation of the ellipse with center at $(0, 0)$ and foci at $(0, -c)$ and $(0, c)$ is

$$\frac{x^2}{b^2} + \frac{y^2}{a^2} = 1, \qquad \text{where } a > b > 0 \text{ and } b^2 = a^2 - c^2 \qquad \textbf{(3)}$$

The major axis is the y-axis; the vertices are at $(0, -a)$ and $(0, a)$.

Figure 23

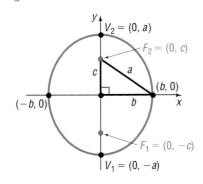

Figure 23 illustrates the graph of such an ellipse. Again, notice the right triangle with the points at $(0, 0)$, $(b, 0)$, and $(0, c)$.

Look closely at equations (2) and (3). Although they may look alike, there is a difference! In equation (2), the larger number, a^2, is in the denominator of the x^2-term, so the major axis of the ellipse is along the x-axis. In equation (3), the larger number, a^2, is in the denominator of the y^2-term, so the major axis is along the y-axis.

| EXAMPLE 3 | Discussing the Equation of an Ellipse |

Discuss the equation: $9x^2 + y^2 = 9$

Solution   To put the equation in proper form, we divide each side by 9.

$$x^2 + \frac{y^2}{9} = 1$$

The larger number, 9, is in the denominator of the y^2-term so, based on equation (3), this is the equation of an ellipse with center at the origin and major axis along the y-axis. Also, we conclude that $a^2 = 9$, $b^2 = 1$, and $c^2 = a^2 - b^2 = 9 - 1 = 8$. The vertices are at $(0, \pm a) = (0, \pm 3)$, and the foci are at $(0, \pm c) = (0, \pm 2\sqrt{2})$. The graph is given in Figure 24.   ▇

Figure 24
$$x^2 + \frac{y^2}{9} = 1$$

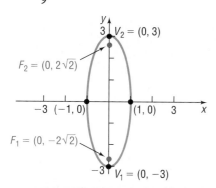

NOW WORK PROBLEM **9**.

| EXAMPLE 4 | Finding an Equation of an Ellipse |

Find an equation of the ellipse having one focus at $(0, 2)$ and vertices at $(0, -3)$ and $(0, 3)$. Graph the equation.

Solution   Because the vertices are at $(0, -3)$ and $(0, 3)$, the center of this ellipse is at their midpoint, the origin. Also, its major axis lies on the y-axis. The distance from the center, $(0, 0)$, to one of the foci, $(0, 2)$, is $c = 2$. The distance from the center, $(0, 0)$, to one of the vertices, $(0, 3)$, is $a = 3$. So $b^2 = a^2 - c^2 = 9 - 4 = 5$. The form of the equation of this ellipse is given by equation (3).

Figure 25
$$\frac{x^2}{5} + \frac{y^2}{9} = 1$$

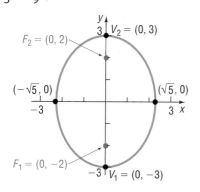

$$\frac{x^2}{b^2} + \frac{y^2}{a^2} = 1$$

$$\frac{x^2}{5} + \frac{y^2}{9} = 1$$

Figure 25 shows the graph.   ▇

NOW WORK PROBLEM **17**.

The circle may be considered a special kind of ellipse. To see why, let $a = b$ in equation (2) or (3). Then

$$\frac{x^2}{a^2} + \frac{y^2}{a^2} = 1$$

$$x^2 + y^2 = a^2$$

This is the equation of a circle with center at the origin and radius a. The value of c is

$$c^2 = a^2 - b^2 = 0$$

We conclude that the closer the two foci of an ellipse are to the center, the more the ellipse will look like a circle.

CENTER AT (h, k)

④ If an ellipse with center at the origin and major axis coinciding with a coordinate axis is shifted horizontally h units and then vertically k units, the result is an ellipse with center at (h, k) and major axis parallel to a coordinate axis. The equations of such ellipses have the same forms as those given in equations (2) and (3), except that x is replaced by $x - h$ (the horizontal shift) and y is replaced by $y - k$ (the vertical shift). Table 3 gives the forms of the equations of such ellipses, and Figure 26 shows their graphs.

TABLE 3
ELLIPSES WITH CENTER AT (b, k) AND MAJOR AXIS PARALLEL TO A COORDINATE AXIS

Center	Major Axis	Foci	Vertices	Equation
(h, k)	Parallel to x-axis	$(h + c, k)$	$(h + a, k)$	$\dfrac{(x - h)^2}{a^2} + \dfrac{(y - k)^2}{b^2} = 1,$
		$(h - c, k)$	$(h - a, k)$	$a > b$ and $b^2 = a^2 - c^2$
(h, k)	Parallel to y-axis	$(h, k + c)$	$(h, k + a)$	$\dfrac{(x - h)^2}{b^2} + \dfrac{(y - k)^2}{a^2} = 1,$
		$(h, k - c)$	$(h, k - a)$	$a > b$ and $b^2 = a^2 - c^2$

Figure 26

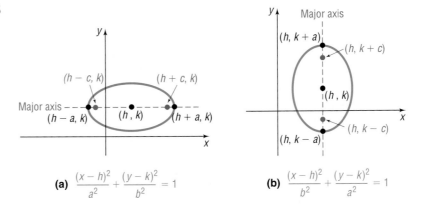

(a) $\dfrac{(x - h)^2}{a^2} + \dfrac{(y - k)^2}{b^2} = 1$ **(b)** $\dfrac{(x - h)^2}{b^2} + \dfrac{(y - k)^2}{a^2} = 1$

EXAMPLE 5

Finding an Equation of an Ellipse, Center Not at the Origin

Find an equation for the ellipse with center at $(2, -3)$, one focus at $(3, -3)$, and one vertex at $(5, -3)$. Graph the equation.

Solution   The center is at $(h, k) = (2, -3)$, so $h = 2$ and $k = -3$. Since the center, focus, and vertex all lie on the line $y = -3$, the major axis is parallel to the x-axis. The distance from the center $(2, -3)$ to a focus $(3, -3)$ is $c = 1$; the distance from the center $(2, -3)$ to a vertex $(5, -3)$ is $a = 3$. Then, $b^2 = a^2 - c^2 = 9 - 1 = 8$. The form of the equation is

$$\frac{(x-h)^2}{a^2} + \frac{(y-k)^2}{b^2} = 1, \quad \text{where } h = 2, k = -3, a = 3, b = 2\sqrt{2}$$

$$\frac{(x-2)^2}{9} + \frac{(y+3)^2}{8} = 1$$

To graph the equation, we use the center $(h, k) = (2, -3)$ to locate the vertices. The major axis is parallel to the x-axis, so the vertices are $a = 3$ units left and right of the center $(2, -3)$. Therefore, the vertices are

$$V_1 = (2 - 3, -3) = (-1, -3) \quad \text{and} \quad V_2 = (2 + 3, -3) = (5, -3)$$

Since $c = 1$ and the major axis is parallel to the x-axis, the foci are 1 unit left and right of the center. Therefore, the foci are

$$F_1 = (2 - 1, -3) = (1, -3) \quad \text{and} \quad F_2 = (2 + 1, -3) = (3, -3)$$

Finally, we use the value of $b = 2\sqrt{2}$ to find the two points above and below the center.

$$(2, -3 - 2\sqrt{2}) \quad \text{and} \quad (2, -3 + 2\sqrt{2})$$

Figure 27 shows the graph.

Figure 27

$$\frac{(x-2)^2}{9} + \frac{(y+3)^2}{8} = 1$$

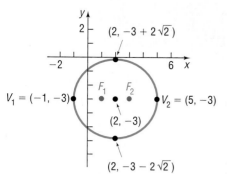

NOW WORK PROBLEM **41**.

EXAMPLE 6   **Discussing the Equation of an Ellipse**

Discuss the equation: $4x^2 + y^2 - 8x + 4y + 4 = 0$

Solution   We proceed to complete the squares in x and in y.

$$4x^2 + y^2 - 8x + 4y + 4 = 0$$

$$4x^2 - 8x + y^2 + 4y = -4 \qquad \text{Group like variables; place the constant on the right side.}$$

$$4(x^2 - 2x) + (y^2 + 4y) = -4 \qquad \text{Factor out 4 from the first two terms.}$$

$$4(x^2 - 2x + 1) + (y^2 + 4y + 4) = -4 + 4 + 4 \qquad \text{Complete each square.}$$

$$4(x - 1)^2 + (y + 2)^2 = 4 \qquad \text{Factor.}$$

$$(x - 1)^2 + \frac{(y + 2)^2}{4} = 1 \qquad \text{Divide each side by 4.}$$

Figure 28

$$(x - 1)^2 + \frac{(y + 2)^2}{4} = 1$$

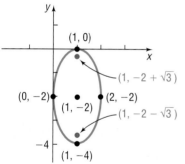

This is the equation of an ellipse with center at $(1, -2)$ and major axis parallel to the y-axis. Since $a^2 = 4$ and $b^2 = 1$, we have $c^2 = a^2 - b^2 = 4 - 1 = 3$. The vertices are at $(h, k \pm a) = (1, -2 \pm 2)$ or $(1, 0)$ and $(1, -4)$. The foci are at $(h, k \pm c) = (1, -2 \pm \sqrt{3})$ or $(1, -2 - \sqrt{3})$ and $(1, -2 + \sqrt{3})$. Figure 28 shows the graph.

NOW WORK PROBLEM **33**.

APPLICATIONS

⑤ Ellipses are found in many applications in science and engineering. For example, the orbits of the planets around the Sun are elliptical, with the Sun's position at a focus. See Figure 29.

Figure 29

Stone and concrete bridges are often shaped as semielliptical arches. Elliptical gears are used in machinery when a variable rate of motion is required.

Ellipses also have an interesting reflection property. If a source of light (or sound) is placed at one focus, the waves transmitted by the source will reflect off the ellipse and concentrate at the other focus. This is the principle behind *whispering galleries*, which are rooms designed with elliptical ceilings. A person standing at one focus of the ellipse can whisper and be heard by a person standing at the other focus, because all the sound waves that reach the ceiling are reflected to the other person.

EXAMPLE 7 | **Whispering Galleries**

Figure 30 shows the specifications for an elliptical ceiling in a hall designed to be a whispering gallery. In a whispering gallery, a person standing at one focus of the ellipse can whisper and be heard by another person standing at the other focus, because all the sound waves that reach the ceiling from one focus are reflected to the other focus. Where are the foci located in the hall?

Solution We set up a rectangular coordinate system so that the center of the ellipse is at the origin and the major axis is along the *x*-axis. See Figure 31. The equation of the ellipse is

$$\frac{x^2}{a^2} + \frac{y^2}{b^2} = 1$$

where $a = 25$ and $b = 20$.

Figure 30

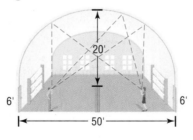

Figure 31

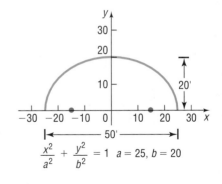

$$\frac{x^2}{a^2} + \frac{y^2}{b^2} = 1 \quad a = 25, b = 20$$

Then, since

$$c^2 = a^2 - b^2 = 25^2 - 20^2 = 625 - 400 = 225$$

we have $c = 15$. The foci are located 15 feet from the center of the ellipse along the major axis. ∎

6.3   EXERCISES

In Problems 1–4, the graph of an ellipse is given. Match each graph to its equation.

A. $\dfrac{x^2}{4} + y^2 = 1$ 　　　B. $x^2 + \dfrac{y^2}{4} = 1$ 　　　C. $\dfrac{x^2}{16} + \dfrac{y^2}{4} = 1$ 　　　D. $\dfrac{x^2}{4} + \dfrac{y^2}{16} = 1$

1.
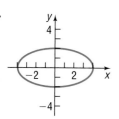

2.
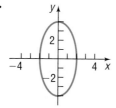

3.
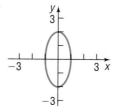

4.
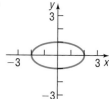

In Problems 5–14, find the vertices and foci of each ellipse. Graph each equation.

5. $\dfrac{x^2}{25} + \dfrac{y^2}{4} = 1$ 　　**6.** $\dfrac{x^2}{9} + \dfrac{y^2}{4} = 1$ 　　**7.** $\dfrac{x^2}{9} + \dfrac{y^2}{25} = 1$ 　　**8.** $x^2 + \dfrac{y^2}{16} = 1$

9. $4x^2 + y^2 = 16$ 　　**10.** $x^2 + 9y^2 = 18$ 　　**11.** $4y^2 + x^2 = 8$ 　　**12.** $4y^2 + 9x^2 = 36$

13. $x^2 + y^2 = 16$ 　　**14.** $x^2 + y^2 = 4$

In Problems 15–24, find an equation for each ellipse. Graph the equation.

15. Center at $(0, 0)$;　focus at $(3, 0)$;　vertex at $(5, 0)$ 　　**16.** Center at $(0, 0)$;　focus at $(-1, 0)$;　vertex at $(3, 0)$

17. Center at $(0, 0)$;　focus at $(0, -4)$;　vertex at $(0, 5)$ 　　**18.** Center at $(0, 0)$;　focus at $(0, 1)$;　vertex at $(0, -2)$

19. Foci at $(\pm 2, 0)$;　length of the major axis is 6 　　**20.** Focus at $(0, -4)$;　vertices at $(0, \pm 8)$

21. Foci at $(0, \pm 3)$;　x-intercepts are ± 2 　　**22.** Foci at $(0, \pm 2)$;　length of the major axis is 8

23. Center at $(0, 0)$;　vertex at $(0, 4)$;　$b = 1$ 　　**24.** Vertices at $(\pm 5, 0)$;　$c = 2$

In Problems 25–28, write an equation for each ellipse.

25.
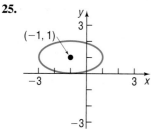

26.
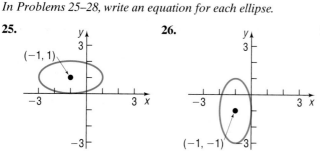

27.
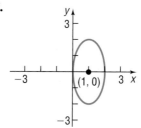

28.
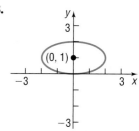

In Problems 29–40, discuss each equation, that is, find the center, foci, and vertices of each ellipse. Graph each equation.

29. $\dfrac{(x - 3)^2}{4} + \dfrac{(y + 1)^2}{9} = 1$ 　　　**30.** $\dfrac{(x + 4)^2}{9} + \dfrac{(y + 2)^2}{4} = 1$

31. $(x + 5)^2 + 4(y - 4)^2 = 16$ 　　　**32.** $9(x - 3)^2 + (y + 2)^2 = 18$

33. $x^2 + 4x + 4y^2 - 8y + 4 = 0$ 　　　**34.** $x^2 + 3y^2 - 12y + 9 = 0$

35. $2x^2 + 3y^2 - 8x + 6y + 5 = 0$ 　　　**36.** $4x^2 + 3y^2 + 8x - 6y = 5$

37. $9x^2 + 4y^2 - 18x + 16y - 11 = 0$ 　　　**38.** $x^2 + 9y^2 + 6x - 18y + 9 = 0$

39. $4x^2 + y^2 + 4y = 0$ 　　　**40.** $9x^2 + y^2 - 18x = 0$

In Problems 41–50, find an equation for each ellipse. Graph the equation.

41. Center at $(2, -2)$; vertex at $(7, -2)$; focus at $(4, -2)$

42. Center at $(-3, 1)$; vertex at $(-3, 3)$; focus at $(-3, 0)$

43. Vertices at $(4, 3)$ and $(4, 9)$; focus at $(4, 8)$

44. Foci at $(1, 2)$ and $(-3, 2)$; vertex at $(-4, 2)$

45. Foci at $(5, 1)$ and $(-1, 1)$; length of the major axis is 8

46. Vertices at $(2, 5)$ and $(2, -1)$; $c = 2$

47. Center at $(1, 2)$; focus at $(4, 2)$; contains the point $(1, 3)$

48. Center at $(1, 2)$; focus at $(1, 4)$; contains the point $(2, 2)$

49. Center at $(1, 2)$; vertex at $(4, 2)$; contains the point $(1, 3)$

50. Center at $(1, 2)$; vertex at $(1, 4)$; contains the point $(2, 2)$

In Problems 51–54, graph each function.
[**Hint:** Notice that each function is half an ellipse.]

51. $f(x) = \sqrt{16 - 4x^2}$
52. $f(x) = \sqrt{9 - 9x^2}$
53. $f(x) = -\sqrt{64 - 16x^2}$
54. $f(x) = -\sqrt{4 - 4x^2}$

---

55. Semielliptical Arch Bridge An arch in the shape of the upper half of an ellipse is used to support a bridge that is to span a river 20 meters wide. The center of the arch is 6 meters above the center of the river (see the figure). Write an equation for the ellipse in which the x-axis coincides with the water level and the y-axis passes through the center of the arch.

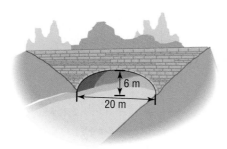

56. Semielliptical Arch Bridge The arch of a bridge is a semiellipse with a horizontal major axis. The span is 30 feet, and the top of the arch is 10 feet above the major axis. The roadway is horizontal and is 2 feet above the top of the arch. Find the vertical distance from the roadway to the arch at 5-foot intervals along the roadway.

57. Whispering Gallery A hall 100 feet in length is to be designed as a whispering gallery. If the foci are located 25 feet from the center, how high will the ceiling be at the center?

58. Whispering Gallery Jim, standing at one focus of a whispering gallery, is 6 feet from the nearest wall. His friend is standing at the other focus, 100 feet away. What is the length of this whispering gallery? How high is its elliptical ceiling at the center?

59. Semielliptical Arch Bridge A bridge is built in the shape of a semielliptical arch. The bridge has a span of 120 feet and a maximum height of 25 feet. Choose a suitable rectangular coordinate system and find the height of the arch at distances of 10, 30, and 50 feet from the center.

60. Semielliptical Arch Bridge A bridge is to be built in the shape of a semielliptical arch and is to have a span of 100 feet. The height of the arch, at a distance of 40 feet from the center, is to be 10 feet. Find the height of the arch at its center.

61. Semielliptical Arch An arch in the form of half an ellipse is 40 feet wide and 15 feet high at the center. Find the height of the arch at intervals of 10 feet along its width.

62. Semielliptical Arch Bridge An arch for a bridge over a highway is in the form of half an ellipse. The top of the arch is 20 feet above the ground level (the major axis). The highway has four lanes, each 12 feet wide; a center safety strip 8 feet wide; and two side strips, each 4 feet wide. What should the span of the bridge be (the length of its major axis) if the height 28 feet from the center is to be 13 feet?

---

*In Problems 63–66, use the fact that the orbit of a planet about the Sun is an ellipse, with the Sun at one focus. The **aphelion** of a planet is its greatest distance from the Sun, and the **perihelion** is its shortest distance. The **mean distance** of a planet from the Sun is the length of the semimajor axis of the elliptical orbit. See the illustration.*

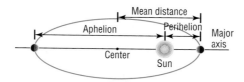

63. Earth The mean distance of Earth from the Sun is 93 million miles. If the aphelion of Earth is 94.5 million miles, what is the perihelion? Write an equation for the orbit of Earth around the Sun.

64. Mars The mean distance of Mars from the Sun is 142 million miles. If the perihelion of Mars is 128.5 million miles, what is the aphelion? Write an equation for the orbit of Mars about the Sun.

65. Jupiter The aphelion of Jupiter is 507 million miles. If the distance from the Sun to the center of its elliptical orbit is 23.2 million miles, what is the perihelion? What is the mean distance? Write an equation for the orbit of Jupiter around the Sun.

66. Pluto The perihelion of Pluto is 4551 million miles, and the distance of the Sun from the center of its elliptical orbit is 897.5 million miles. Find the aphelion of Pluto. What is the mean distance of Pluto from the Sun? Write an equation for the orbit of Pluto about the Sun.

67. Racetrack Design Consult the figure. A racetrack is in the shape of an ellipse, 100 feet long and 50 feet wide. What is the width 10 feet from a vertex?

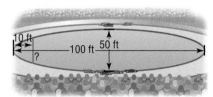

68. Racetrack Design A racetrack is in the shape of an ellipse 80 feet long and 40 feet wide. What is the width 10 feet from a vertex?

69. Show that an equation of the form

$$Ax^2 + Cy^2 + F = 0, \qquad A \neq 0, C \neq 0, F \neq 0$$

where A and C are of the same sign and F is of opposite sign,
(a) Is the equation of an ellipse with center at $(0, 0)$ if $A \neq C$.
(b) Is the equation of a circle with center $(0, 0)$ if $A = C$.

70. Show that the graph of an equation of the form

$$Ax^2 + Cy^2 + Dx + Ey + F = 0, \qquad A \neq 0, C \neq 0$$

where A and C are of the same sign,
(a) Is an ellipse if $(D^2/4A) + (E^2/4C) - F$ is the same sign as A.
(b) Is a point if $(D^2/4A) + (E^2/4C) - F = 0$.
(c) Contains no points if $(D^2/4A) + (E^2/4C) - F$ is of opposite sign to A.

71. The **eccentricity** e of an ellipse is defined as the number c/a, where a and c are the numbers given in equation (2). Because $a > c$, it follows that $e < 1$. Write a brief paragraph about the general shape of each of the following ellipses. Be sure to justify your conclusions.
(a) Eccentricity close to 0
(b) Eccentricity = 0.5
(c) Eccentricity close to 1

PREPARING FOR THIS SECTION

Before getting started, review the following:

✓ Distance Formula (Section 1.1, p. 4)

✓ Completing the Square (Appendix A, Section A.3, p. 507)

✓ Symmetry (Section 1.2, pp. 12–14)

✓ Graphing Techniques: Transformations (Section 1.5, pp. 47–56)

6.4 THE HYPERBOLA

OBJECTIVES
1. Find the Equation of a Hyperbola
2. Graph Hyperbolas
3. Discuss the Equation of a Hyperbola
4. Find the Asymptotes of a Hyperbola
5. Work with Hyperbolas with Center at (h, k)
6. Solve Applied Problems Involving Hyperbolas

A **hyperbola** is the collection of all points in the plane the difference of whose distances from two fixed points, called the **foci,** is a constant.

Figure 32

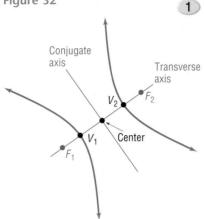

① Figure 32 illustrates a hyperbola with foci F_1 and F_2. The line containing the foci is called the **transverse axis.** The midpoint of the line segment joining the foci is called the **center** of the hyperbola. The line through the center and perpendicular to the transverse axis is called the **conjugate axis.** The hyperbola consists of two separate curves, called **branches,** that are symmetric with respect to the transverse axis, conjugate axis, and center. The two points of intersection of the hyperbola and the transverse axis are the **vertices,** V_1 and V_2, of the hyperbola.

With these ideas in mind, we are now ready to find the equation of a hyperbola in a rectangular coordinate system. First, we place the center at the origin. Next, we position the hyperbola so that its transverse axis coincides with a coordinate axis. Suppose that the transverse axis coincides with the x-axis, as shown in Figure 33.

If c is the distance from the center to a focus, then one focus will be at $F_1 = (-c, 0)$ and the other at $F_2 = (c, 0)$. Now we let the constant difference of the distances from any point $P = (x, y)$ on the hyperbola to the foci F_1 and F_2 be denoted by $\pm 2a$. (If P is on the right branch, the $+$ sign is used; if P is on the left branch, the $-$ sign is used.) The coordinates of P must satisfy the equation

Figure 33
$$d(F_1, P) - d(F_2, P) = \pm 2a$$

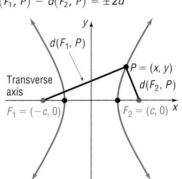

$$d(F_1, P) - d(F_2, P) = \pm 2a \qquad \text{Difference of the distances from } P \text{ to the foci equals } \pm 2a.$$

$$\sqrt{(x + c)^2 + y^2} - \sqrt{(x - c)^2 + y^2} = \pm 2a \qquad \text{Use the distance formula.}$$

$$\sqrt{(x + c)^2 + y^2} = \pm 2a + \sqrt{(x - c)^2 + y^2} \qquad \text{Isolate one radical.}$$

$$(x + c)^2 + y^2 = 4a^2 \pm 4a\sqrt{(x - c)^2 + y^2} \qquad \text{Square both sides.}$$
$$+ (x - c)^2 + y^2$$

Next, we remove the parentheses.

$$x^2 + 2cx + c^2 + y^2 = 4a^2 \pm 4a\sqrt{(x - c)^2 + y^2} + x^2 - 2cx + c^2 + y^2$$

$$4cx - 4a^2 = \pm 4a\sqrt{(x - c)^2 + y^2} \qquad \text{Simplify; Isolate the radical.}$$

$$cx - a^2 = \pm a\sqrt{(x - c)^2 + y^2} \qquad \text{Divide each side by 4.}$$

$$(cx - a^2)^2 = a^2[(x - c)^2 + y^2] \qquad \text{Square both sides.}$$

$$c^2x^2 - 2ca^2x + a^4 = a^2(x^2 - 2cx + c^2 + y^2) \qquad \text{Simplify.}$$

$$c^2x^2 + a^4 = a^2x^2 + a^2c^2 + a^2y^2 \qquad \text{Remove parentheses and simplify.}$$

$$(c^2 - a^2)x^2 - a^2y^2 = a^2c^2 - a^4 \qquad \text{Rearrange terms.}$$

$$(c^2 - a^2)x^2 - a^2y^2 = a^2(c^2 - a^2) \qquad \text{Factor } a^2 \text{ on the right side.} \quad \textbf{(1)}$$

To obtain points on the hyperbola off the x-axis, it must be that $a < c$. To see why, look again at Figure 33.

$$d(F_1, P) < d(F_2, P) + d(F_1, F_2) \quad \text{Use triangle } F_1 P F_2.$$

$$d(F_1, P) - d(F_2, P) < d(F_1, F_2) \qquad\qquad \text{P is on the right branch, so}$$
$$\qquad\qquad\qquad\qquad\qquad\qquad\qquad\qquad d(F_1, P) - d(F_2, P) = 2a.$$

$$2a < 2c$$

$$a < c$$

Since $a < c$, we also have $a^2 < c^2$, so $c^2 - a^2 > 0$. Let $b^2 = c^2 - a^2, b > 0$. Then equation (1) can be written as

$$b^2 x^2 - a^2 y^2 = a^2 b^2$$

$$\frac{x^2}{a^2} - \frac{y^2}{b^2} = 1 \qquad \text{Divide each side by } a^2 b^2.$$

To find the vertices of the hyperbola defined by this equation, let $y = 0$. The vertices satisfy the equation $\dfrac{x^2}{a^2} = 1$, the solutions of which are $x = \pm a$. Consequently, the vertices of the hyperbola are $V_1 = (-a, 0)$ and $V_2 = (a, 0)$. Notice that the distance from the center $(0, 0)$ to either vertex is a.

Theorem

Equation of a Hyperbola; Center at (0, 0); Foci at ($\pm$ c, 0); Vertices at ($\pm$ a, 0); Transverse Axis along the x-Axis

An equation of the hyperbola with center at $(0, 0)$, foci at $(-c, 0)$ and $(c, 0)$, and vertices at $(-a, 0)$ and $(a, 0)$ is

$$\frac{x^2}{a^2} - \frac{y^2}{b^2} = 1, \qquad \text{where } b^2 = c^2 - a^2 \qquad\qquad \textbf{(2)}$$

The transverse axis is the x-axis. The vertices are at $(-a, 0)$ and $(a, 0)$.

■

② As you can verify, the hyperbola defined by equation (2) is symmetric with respect to the x-axis, y-axis, and origin. To find the y-intercepts, if any, let $x = 0$ in equation (2). This results in the equation $\dfrac{y^2}{b^2} = -1$, which has no real solution. We conclude that the hyperbola defined by equation (2) has no y-intercepts. In fact, since $\dfrac{x^2}{a^2} - 1 = \dfrac{y^2}{b^2} \geq 0$, it follows that $x^2/a^2 \geq 1$. There are no points on the graph for $-a < x < a$. See Figure 34.

Figure 34
$\dfrac{x^2}{a^2} - \dfrac{y^2}{b^2} = 1, \quad b^2 = c^2 - a^2$

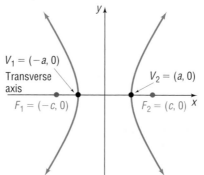

| EXAMPLE 1 | Finding and Graphing an Equation of a Hyperbola |

Find an equation of the hyperbola with center at the origin, one focus at $(3, 0)$, and one vertex at $(-2, 0)$. Graph the equation.

Solution   The hyperbola has its center at the origin, and the transverse axis coincides with the x-axis. One focus is at $(c, 0) = (3, 0)$, so $c = 3$. One vertex is at $(-a, 0) = (-2, 0)$, so $a = 2$. From equation (2), it follows that $b^2 = c^2 - a^2 = 9 - 4 = 5$, so an equation of the hyperbola is

$$\frac{x^2}{4} - \frac{y^2}{5} = 1$$

To graph a hyperbola, it is helpful to locate and plot other points on the graph. For example, to find the points above and below the foci, we let $x = \pm 3$. Then

$$\frac{x^2}{4} - \frac{y^2}{5} = 1$$

$$\frac{(\pm 3)^2}{4} - \frac{y^2}{5} = 1 \qquad x = \pm 3$$

$$\frac{9}{4} - \frac{y^2}{5} = 1$$

$$\frac{y^2}{5} = \frac{5}{4}$$

$$y^2 = \frac{25}{4}$$

$$y = \pm\frac{5}{2}$$

Figure 35
$$\frac{x^2}{4} - \frac{y^2}{5} = 1$$

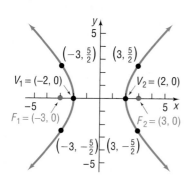

The points above and below the foci are $\left(\pm 3, \dfrac{5}{2}\right)$ and $\left(\pm 3, -\dfrac{5}{2}\right)$. These points help because they determine the "opening" of the hyperbola. See Figure 35. ■

COMMENT   To graph the hyperbola $\left(\dfrac{x^2}{4}\right) - \left(\dfrac{y^2}{5}\right) = 1$ discussed in Example 1, we need to graph the two functions $Y_1 = \sqrt{5}\sqrt{(x^2/4) - 1}$ and $Y_2 = -\sqrt{5}\sqrt{(x^2/4) - 1}$. Do this and compare what you see with Figure 35. ■

✏ NOW WORK PROBLEM 5.

An equation of the form of equation (2) is the equation of a hyperbola with center at the origin; foci on the x-axis at $(-c, 0)$ and $(c, 0)$, where $c^2 = a^2 + b^2$; and transverse axis along the x-axis.

③   For the remainder of this section, the direction "Discuss the equation" will mean to find the center, transverse axis, vertices, and foci of the hyperbola and graph it.

| EXAMPLE 2 | Discussing the Equation of a Hyperbola |

Discuss the equation: $\dfrac{x^2}{16} - \dfrac{y^2}{4} = 1$

Solution   The given equation is of the form of equation (2), with $a^2 = 16$ and $b^2 = 4$. The graph of the equation is a hyperbola with center at $(0, 0)$ and transverse axis along the x-axis. Also, we know that $c^2 = a^2 + b^2 = 16 + 4 = 20$. The vertices are at $(\pm a, 0) = (\pm 4, 0)$, and the foci are at $(\pm c, 0) = (\pm 2\sqrt{5}, 0)$.

To locate the points on the graph above and below the foci, we let $x = \pm 2\sqrt{5}$. Then

$$\frac{x^2}{16} - \frac{y^2}{4} = 1$$

$$\frac{(\pm 2\sqrt{5})^2}{16} - \frac{y^2}{4} = 1 \qquad x = \pm 2\sqrt{5}$$

$$\frac{20}{16} - \frac{y^2}{4} = 1$$

$$\frac{5}{4} - \frac{y^2}{4} = 1$$

$$\frac{y^2}{4} = \frac{1}{4}$$

$$y = \pm 1$$

The points above and below the foci are $(\pm 2\sqrt{5}, 1)$ and $(\pm 2\sqrt{5}, -1)$. See Figure 36.

Figure 36
$$\frac{x^2}{16} - \frac{y^2}{4} = 1$$

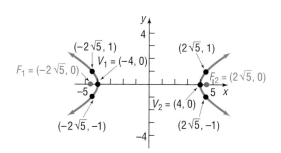

NOW WORK PROBLEM **15.**

The next result gives the form of the equation of a hyperbola with center at the origin and transverse axis along the y-axis.

Theorem   **Equation of a Hyperbola; Center at (0, 0); Foci at (0, ± c); Vertices at (0, ± a); Transverse Axis along the y-Axis**

An equation of the hyperbola with center at $(0, 0)$, foci at $(0, -c)$ and $(0, c)$, and vertices at $(0, -a)$ and $(0, a)$ is

$$\frac{y^2}{a^2} - \frac{x^2}{b^2} = 1, \qquad \text{where } b^2 = c^2 - a^2 \qquad \textbf{(3)}$$

The transverse axis is the y-axis. The vertices are at $(0, -a)$ and $(0, a)$.

Figure 37

$$\frac{y^2}{a^2} - \frac{x^2}{b^2} = 1, \quad b^2 = c^2 - a^2$$

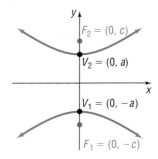

Figure 37 shows the graph of a typical hyperbola defined by equation (3).

An equation of the form of equation (2), $\dfrac{x^2}{a^2} - \dfrac{y^2}{b^2} = 1$, is the equation of a hyperbola with center at the origin, foci on the x-axis at $(-c, 0)$ and $(c, 0)$, where $c^2 = a^2 + b^2$, and transverse axis along the x-axis.

An equation of the form of equation (3), $\dfrac{y^2}{a^2} - \dfrac{x^2}{b^2} = 1$, is the equation of a hyperbola with center at the origin, foci on the y-axis at $(0, -c)$ and $(0, c)$, where $c^2 = a^2 + b^2$, and transverse axis along the y-axis.

Notice the difference in the forms of equations (2) and (3). When the y^2-term is subtracted from the x^2-term, the transverse axis is the x-axis. When the x^2-term is subtracted from the y^2-term, the transverse axis is the y-axis.

EXAMPLE 3

Discussing the Equation of a Hyperbola

Discuss the equation: $y^2 - 4x^2 = 4$

Solution To put the equation in proper form, we divide each side by 4:

$$\frac{y^2}{4} - x^2 = 1$$

Since the x^2-term is subtracted from the y^2-term, the equation is that of a hyperbola with center at the origin and transverse axis along the y-axis. Also, comparing the above equation to equation (3), we find $a^2 = 4$, $b^2 = 1$, and $c^2 = a^2 + b^2 = 5$. The vertices are at $(0, \pm a) = (0, \pm 2)$, and the foci are at $(0, \pm c) = (0, \pm\sqrt{5})$.

To locate other points on the graph, we let $x = \pm 2$. Then

Figure 38

$$\frac{y^2}{4} - x^2 = 1$$

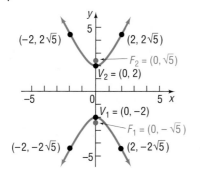

$$y^2 - 4x^2 = 4$$
$$y^2 - 4(\pm 2)^2 = 4 \qquad x = \pm 2$$
$$y^2 - 16 = 4$$
$$y^2 = 20$$
$$y = \pm 2\sqrt{5}$$

Four other points on the graph are $(\pm 2, 2\sqrt{5})$ and $(\pm 2, -2\sqrt{5})$. See Figure 38. ∎

EXAMPLE 4

Finding an Equation of a Hyperbola

Find an equation of the hyperbola having one vertex at $(0, 2)$ and foci at $(0, -3)$ and $(0, 3)$.

Solution Since the foci are at $(0, -3)$ and $(0, 3)$, the center of the hyperbola is at their midpoint, the origin. Also, the transverse axis is along the y-axis. The given information also reveals that $c = 3$, $a = 2$, and $b^2 = c^2 - a^2 = 9 - 4 = 5$. The form of the equation of the hyperbola is given by equation (3):

Figure 39

$$\frac{y^2}{4} - \frac{x^2}{5} = 1$$

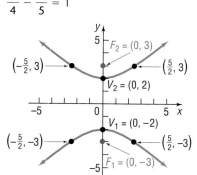

$$\frac{y^2}{a^2} - \frac{x^2}{b^2} = 1$$

$$\frac{y^2}{4} - \frac{x^2}{5} = 1$$

Let $y = \pm 3$ to obtain points on the graph across from the foci. See Figure 39. ■

NOW WORK PROBLEM **9.**

Look at the equations of the hyperbolas in Examples 2 and 4. For the hyperbola in Example 2, $a^2 = 16$ and $b^2 = 4$, so $a > b$; for the hyperbola in Example 4, $a^2 = 4$ and $b^2 = 5$, so $a < b$. We conclude that, for hyperbolas, there are no requirements involving the relative sizes of a and b. Contrast this situation to the case of an ellipse, in which the relative sizes of a and b dictate which axis is the major axis. Hyperbolas have another feature to distinguish them from ellipses and parabolas: Hyperbolas have asymptotes.

ASYMPTOTES

④ An **asymptote** of a graph is a line with the property that the distance from the line to points on the graph approaches 0 as $x \rightarrow -\infty$ or as $x \rightarrow \infty$. The asymptotes provide information about the end behavior of the graph of a hyperbola.

Theorem | **Asymptotes of a Hyperbola**

The hyperbola $\dfrac{x^2}{a^2} - \dfrac{y^2}{b^2} = 1$ has the two asymptotes

$$y = \frac{b}{a}x \quad \text{and} \quad y = -\frac{b}{a}x \qquad (4)$$

■

Proof We begin by solving for y in the equation of the hyperbola.

$$\frac{x^2}{a^2} - \frac{y^2}{b^2} = 1$$

$$\frac{y^2}{b^2} = \frac{x^2}{a^2} - 1$$

$$y^2 = b^2\left(\frac{x^2}{a^2} - 1\right)$$

Since $x \neq 0$, we can rearrange the right side in the form

$$y^2 = \frac{b^2 x^2}{a^2}\left(1 - \frac{a^2}{x^2}\right)$$

$$y = \pm \frac{bx}{a}\sqrt{1 - \frac{a^2}{x^2}}$$

Now, as $x \rightarrow -\infty$ or as $x \rightarrow \infty$, the term $\dfrac{a^2}{x^2}$ approaches 0, so the expression under the radical approaches 1. Thus, as $x \rightarrow -\infty$ or as $x \rightarrow \infty$, the value of y approaches $\pm \dfrac{bx}{a}$; that is, the graph of the hyperbola approaches the lines

$$y = -\frac{b}{a}x \quad \text{and} \quad y = \frac{b}{a}x$$

These lines are asymptotes of the hyperbola.

The asymptotes of a hyperbola are not part of the hyperbola, but they do serve as a guide for graphing a hyperbola. For example, suppose that we want to graph the equation

$$\frac{x^2}{a^2} - \frac{y^2}{b^2} = 1$$

We begin by plotting the vertices $(-a, 0)$ and $(a, 0)$. Then we plot the points $(0, -b)$ and $(0, b)$ and use these four points to construct a rectangle, as shown in Figure 40. The diagonals of this rectangle have slopes $\dfrac{b}{a}$ and $-\dfrac{b}{a}$, and their extensions are the asymptotes $y = \left(\dfrac{b}{a}\right)x$ and $y = -\left(\dfrac{b}{a}\right)x$ of the hyperbola. If we graph the asymptotes, we can use them to establish the "opening" of the hyperbola and avoid plotting other points.

Figure 40

$\dfrac{x^2}{a^2} - \dfrac{y^2}{b^2} = 1$

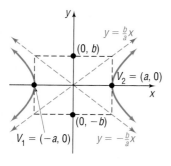

Theorem   **Asymptotes of a Hyperbola**

The hyperbola $\dfrac{y^2}{a^2} - \dfrac{x^2}{b^2} = 1$ has the two asymptotes

$$y = \frac{a}{b}x \quad \text{and} \quad y = -\frac{a}{b}x \qquad \textbf{(5)}$$

You are asked to prove this result in Problem 60.

For the remainder of this section, the direction "Discuss the equation" will mean to find the center, transverse axis, vertices, foci, and asymptotes of the hyperbola and graph it.

EXAMPLE 5   **Discussing the Equation of a Hyperbola**

Figure 41

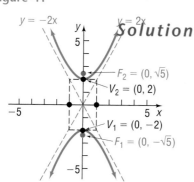

Solution   Discuss the equation: $\dfrac{y^2}{4} - x^2 = 1$

Since the x^2-term is subtracted from the y^2-term, the equation is of the form of equation (3) and is a hyperbola with center at the origin and transverse axis along the y-axis. Also, comparing this equation to equation (3), we find that $a^2 = 4$, $b^2 = 1$, and $c^2 = a^2 + b^2 = 5$. The vertices are at $(0, \pm a) = (0, \pm 2)$, and the foci are at $(0, \pm c) = (0, \pm\sqrt{5})$. Using equation (5), the asymptotes are the lines $y = \dfrac{a}{b}x = 2x$ and $y = -\dfrac{a}{b}x = -2x$. Form the rectangle containing the points $(0, \pm a) = (0, \pm 2)$ and $(\pm b, 0) = (\pm 1, 0)$. The extensions of the diagonals of this rectangle are the asymptotes. Now graph the rectangle, the asymptotes, and the hyperbola. See Figure 41.

EXAMPLE 6	**Discussing the Equation of a Hyperbola**

Discuss the equation: $9x^2 - 4y^2 = 36$

Solution   Divide each side of the equation by 36 to put the equation in proper form.

$$\frac{x^2}{4} - \frac{y^2}{9} = 1$$

Figure 42
$\dfrac{x^2}{4} - \dfrac{y^2}{9} = 1$

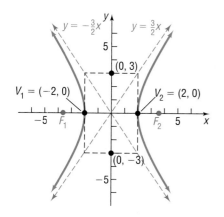

We now proceed to analyze the equation. The center of the hyperbola is the origin. Since the x^2-term is first in the equation, we know that the transverse axis is along the x-axis and the vertices and foci will lie on the x-axis. Using equation (2), we find $a^2 = 4$, $b^2 = 9$, and $c^2 = a^2 + b^2 = 13$. The vertices are $a = 2$ units left and right of the center at $(\pm a, 0) = (\pm 2, 0)$; the foci are $c = \sqrt{13}$ units left and right of the center at $(\pm c, 0) = (\pm\sqrt{13}, 0)$; and the asymptotes have the equations

$$y = \frac{b}{a}x = \frac{3}{2}x \quad \text{and} \quad y = -\frac{b}{a}x = -\frac{3}{2}x$$

To graph the hyperbola, form the rectangle containing the points $(\pm a, 0)$ and $(0, \pm b)$, that is, $(-2, 0)$, $(2, 0)$, $(0, -3)$, and $(0, 3)$. The extensions of the diagonals of this rectangle are the asymptotes. See Figure 42 for the graph.   ■

NOW WORK PROBLEM **17.**

EXPLORATION   Graph the upper portion of the hyperbola $9x^2 - 4y^2 = 36$ discussed in Example 6 and its asymptotes $y = \frac{3}{2}x$ and $y = -\frac{3}{2}x$. Now use ZOOM and TRACE to see what happens as x becomes unbounded in the positive direction. What happens as x becomes unbounded in the negative direction?   ■

Center at (h, k)

⑤   If a hyperbola with center at the origin and transverse axis coinciding with a coordinate axis is shifted horizontally h units and then vertically k units, the result is a hyperbola with center at (h, k) and transverse axis parallel to a coordinate axis. The equations of such hyperbolas have the same forms as those given in equations (2) and (3), except that x is replaced by $x - h$ (the horizontal shift) and y is replaced by $y - k$ (the vertical shift). Table 4 gives the forms of the equations of such hyperbolas. See Figure 43 for the graphs.

	TABLE 4 — HYPERBOLAS WITH CENTER AT (h, k) AND TRANSVERSE AXIS PARALLEL TO A COORDINATE AXIS				
Center	**Transverse Axis**	**Foci**	**Vertices**	**Equation**	**Asymptotes**
(h, k)	Parallel to x-axis	$(h \pm c, k)$	$(h \pm a, k)$	$\dfrac{(x - h)^2}{a^2} - \dfrac{(y - k)^2}{b^2} = 1$, $\quad b^2 = c^2 - a^2$	$y - k = \pm \dfrac{b}{a}(x - h)$
(h, k)	Parallel to y-axis	$(h, k \pm c)$	$(h, k \pm a)$	$\dfrac{(y - k)^2}{a^2} - \dfrac{(x - h)^2}{b^2} = 1$, $\quad b^2 = c^2 - a^2$	$y - k = \pm \dfrac{a}{b}(x - h)$

Figure 43

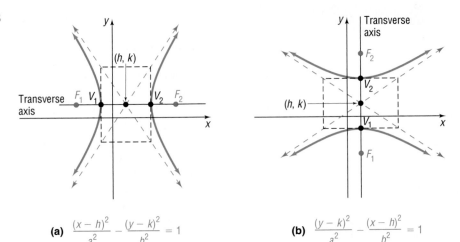

(a) $\dfrac{(x-h)^2}{a^2} - \dfrac{(y-k)^2}{b^2} = 1$

(b) $\dfrac{(y-k)^2}{a^2} - \dfrac{(x-h)^2}{b^2} = 1$

EXAMPLE 7

Finding an Equation of a Hyperbola, Center Not at the Origin

Find an equation for the hyperbola with center at $(1, -2)$, one focus at $(4, -2)$, and one vertex at $(3, -2)$. Graph the equation.

Solution The center is at $(h, k) = (1, -2)$, so $h = 1$ and $k = -2$. Since the center, focus, and vertex all lie on the line $y = -2$, the transverse axis is parallel to the x-axis. The distance from the center $(1, -2)$ to the focus $(4, -2)$ is $c = 3$; the distance from the center $(1, -2)$ to the vertex $(3, -2)$ is $a = 2$. Thus, $b^2 = c^2 - a^2 = 9 - 4 = 5$. The equation is

$$\frac{(x-h)^2}{a^2} - \frac{(y-k)^2}{b^2} = 1$$

$$\frac{(x-1)^2}{4} - \frac{(y+2)^2}{5} = 1$$

See Figure 44.

Figure 44
$$\frac{(x-1)^2}{4} - \frac{(y+2)^2}{5} = 1$$

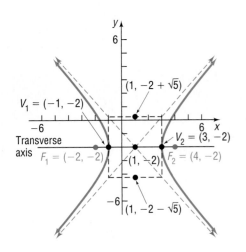

NOW WORK PROBLEM **27.**

| EXAMPLE 8 | **Discussing the Equation of a Hyperbola** |

Discuss the equation: $-x^2 + 4y^2 - 2x - 16y + 11 = 0$

Solution   We complete the squares in x and in y.

$$-x^2 + 4y^2 - 2x - 16y + 11 = 0$$

$$-(x^2 + 2x) + 4(y^2 - 4y) = -11 \qquad \text{Group terms.}$$

$$-(x^2 + 2x + 1) + 4(y^2 - 4y + 4) = -11 - 1 + 16 \qquad \text{Complete each square.}$$

$$-(x + 1)^2 + 4(y - 2)^2 = 4$$

$$(y - 2)^2 - \frac{(x + 1)^2}{4} = 1 \qquad \text{Divide each side by 4.}$$

Figure 45

$$(y - 2)^2 - \frac{(x + 1)^2}{4} = 1$$

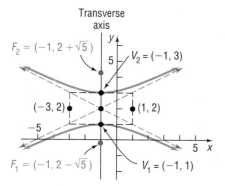

This is the equation of a hyperbola with center at $(-1, 2)$ and transverse axis parallel to the y-axis. Also, $a^2 = 1$ and $b^2 = 4$, so $c^2 = a^2 + b^2 = 5$. Since the transverse axis is parallel to the y-axis, the vertices and foci are located a and c units above and below the center, respectively. The vertices are at $(h, k \pm a) = (-1, 2 \pm 1)$, or $(-1, 1)$ and $(-1, 3)$. The foci are at $(h, k \pm c) = (-1, 2 \pm \sqrt{5})$. The asymptotes are $y - 2 = \frac{1}{2}(x + 1)$ and $y - 2 = -\frac{1}{2}(x + 1)$. Figure 45 shows the graph.   ∎

NOW WORK PROBLEM **41**.

APPLICATIONS

⑥   See Figure 46. Suppose that a gun is fired from an unknown source S. An observer at O_1 hears the report (sound of gun shot) 1 second after another observer at O_2. Because sound travels at about 1100 feet per second, it follows that the point S must be 1100 feet closer to O_2 than to O_1. S lies on one branch of a hyperbola with foci at O_1 and O_2. (Do you see why? The difference of the distances from S to O_1 and from S to O_2 is the constant 1100.) If a third observer at O_3 hears the same report 2 seconds after O_1 hears it, then S will lie on a branch of a second hyperbola with foci at O_1 and O_3. The intersection of the two hyperbolas will pinpoint the location of S.

Figure 46

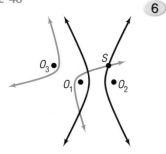

LORAN

In the LOng RAnge Navigation system (LORAN), a master radio sending station and a secondary sending station emit signals that can be received by a ship at sea. See Figure 47. Because a ship monitoring the two signals will usually be nearer to one of the two stations, there will be a difference in the distance that the two signals travel, which will register as a slight time difference between the signals. As long as the time difference remains constant, the difference of the two distances will also be constant. If the ship follows a path corresponding to the fixed time difference, it will follow the path of a hyperbola whose foci are located at the positions of the two sending stations. So for each time difference a different hyperbolic path results, each bringing the ship to a different shore location. Navigation charts show the various hyperbolic paths corresponding to different time differences.

Figure 47

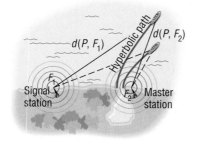

$d(P, F_1) - d(P, F_2) = \text{constant}$

EXAMPLE 9	LORAN

Two LORAN stations are positioned 250 miles apart along a straight shore.

(a) A ship records a time difference of 0.00054 second between the LORAN signals. Set up an appropriate rectangular coordinate system to determine where the ship would reach shore if it were to follow the hyperbola corresponding to this time difference.

(b) If the ship wants to enter a harbor located between the two stations 25 miles from the master station, what time difference should it be looking for?

(c) If the ship is 80 miles offshore when the desired time difference is obtained, what is the approximate location of the ship?

[**NOTE:**   The speed of each radio signal is 186,000 miles per second.]

Solution

Figure 48

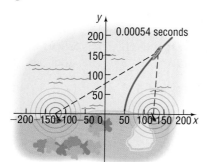

(a) We set up a rectangular coordinate system so that the two stations lie on the x-axis and the origin is midway between them. See Figure 48. The ship lies on a hyperbola whose foci are the locations of the two stations. The reason for this is that the constant time difference of the signals from each station results in a constant difference in the distance of the ship from each station. Since the time difference is 0.00054 second and the speed of the signal is 186,000 miles per second, the difference of the distances from the ship to each station (foci) is

$$\text{Distance} = \text{Speed} \times \text{Time} = 186{,}000 \times 0.00054 \approx 100 \text{ miles}$$

The difference of the distances from the ship to each station, 100, equals $2a$, so $a = 50$ and the vertex of the corresponding hyperbola is at $(50, 0)$. Since the focus is at $(125, 0)$, following this hyperbola the ship would reach shore 75 miles from the master station.

(b) To reach shore 25 miles from the master station, the ship would follow a hyperbola with vertex at $(100, 0)$. For this hyperbola, $a = 100$, so the constant difference of the distances from the ship to each station is $2a = 200$. The time difference that the ship should look for is

$$\text{Time} = \frac{\text{Distance}}{\text{Speed}} = \frac{200}{186{,}000} \approx 0.001075 \text{ second}$$

(c) To find the approximate location of the ship, we need to find the equation of the hyperbola with vertex at $(100, 0)$ and a focus at $(125, 0)$. The form of the equation of this hyperbola is

$$\frac{x^2}{a^2} - \frac{y^2}{b^2} = 1$$

where $a = 100$. Since $c = 125$, we have

$$b^2 = c^2 - a^2 = 125^2 - 100^2 = 5625$$

The equation of the hyperbola is

$$\frac{x^2}{100^2} - \frac{y^2}{5625} = 1$$

Figure 49

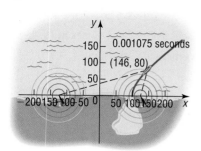

Since the ship is 80 miles from shore, we use $y = 80$ in the equation and solve for x.

$$\frac{x^2}{100^2} - \frac{80^2}{5625} = 1$$

$$\frac{x^2}{100^2} = 1 + \frac{80^2}{5625} \approx 2.14$$

$$x^2 \approx 100^2(2.14)$$

$$x \approx 146$$

The ship is at the position $(146, 80)$. See Figure 49.

NOW WORK PROBLEM **53**.

6.4   EXERCISES

In Problems 1–4, the graph of a hyperbola is given. Match each graph to its equation.

A. $\dfrac{x^2}{4} - y^2 = 1$ B. $x^2 - \dfrac{y^2}{4} = 1$ C. $\dfrac{y^2}{4} - x^2 = 1$ D. $y^2 - \dfrac{x^2}{4} = 1$

1. 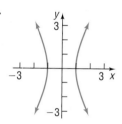   **2.** 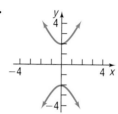   **3.** 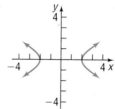   **4.**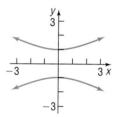

In Problems 5–14, find an equation for the hyperbola described. Graph the equation.

5. Center at $(0, 0)$;   focus at $(3, 0)$;   vertex at $(1, 0)$
6. Center at $(0, 0)$;   focus at $(0, 5)$;   vertex at $(0, 3)$
7. Center at $(0, 0)$;   focus at $(0, -6)$;   vertex at $(0, 4)$
8. Center at $(0, 0)$;   focus at $(-3, 0)$;   vertex at $(2, 0)$
9. Foci at $(-5, 0)$ and $(5, 0)$;   vertex at $(3, 0)$
10. Focus at $(0, 6)$;   vertices at $(0, -2)$ and $(0, 2)$
11. Vertices at $(0, -6)$ and $(0, 6)$;   asymptote the line $y = 2x$
12. Vertices at $(-4, 0)$ and $(4, 0)$;   asymptote the line $y = 2x$
13. Foci at $(-4, 0)$ and $(4, 0)$;   asymptote the line $y = -x$
14. Foci at $(0, -2)$ and $(0, 2)$;   asymptote the line $y = -x$

In Problems 15–22, find the center, transverse axis, vertices, foci, and asymptotes. Graph each equation.

15. $\dfrac{x^2}{25} - \dfrac{y^2}{9} = 1$ **16.** $\dfrac{y^2}{16} - \dfrac{x^2}{4} = 1$ **17.** $4x^2 - y^2 = 16$ **18.** $y^2 - 4x^2 = 16$

19. $y^2 - 9x^2 = 9$ **20.** $x^2 - y^2 = 4$ **21.** $y^2 - x^2 = 25$ **22.** $2x^2 - y^2 = 4$

In Problems 23–26, write an equation for each hyperbola.

23. 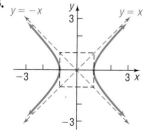   **24.** 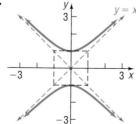   **25.** 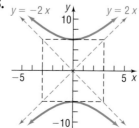   **26.**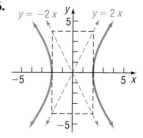

In Problems 27–34, find an equation for the hyperbola described. Graph the equation.

27. Center at $(4, -1)$; focus at $(7, -1)$; vertex at $(6, -1)$

28. Center at $(-3, 1)$; focus at $(-3, 6)$; vertex at $(-3, 4)$

29. Center at $(-3, -4)$; focus at $(-3, -8)$; vertex at $(-3, -2)$

30. Center at $(1, 4)$; focus at $(-2, 4)$; vertex at $(0, 4)$

31. Foci at $(3, 7)$ and $(7, 7)$; vertex at $(6, 7)$

32. Focus at $(-4, 0)$; vertices at $(-4, 4)$ and $(-4, 2)$

33. Vertices at $(-1, -1)$ and $(3, -1)$; asymptote the line
$$y + 1 = \frac{3}{2}(x - 1)$$

34. Vertices at $(1, -3)$ and $(1, 1)$; asymptote the line
$$y + 1 = \frac{3}{2}(x - 1)$$

In Problems 35–48, find the center, transverse axis, vertices, foci, and asymptotes. Graph each equation.

35. $\dfrac{(x - 2)^2}{4} - \dfrac{(y + 3)^2}{9} = 1$

36. $\dfrac{(y + 3)^2}{4} - \dfrac{(x - 2)^2}{9} = 1$

37. $(y - 2)^2 - 4(x + 2)^2 = 4$

38. $(x + 4)^2 - 9(y - 3)^2 = 9$

39. $(x + 1)^2 - (y + 2)^2 = 4$

40. $(y - 3)^2 - (x + 2)^2 = 4$

41. $x^2 - y^2 - 2x - 2y - 1 = 0$

42. $y^2 - x^2 - 4y + 4x - 1 = 0$

43. $y^2 - 4x^2 - 4y - 8x - 4 = 0$

44. $2x^2 - y^2 + 4x + 4y - 4 = 0$

45. $4x^2 - y^2 - 24x - 4y + 16 = 0$

46. $2y^2 - x^2 + 2x + 8y + 3 = 0$

47. $y^2 - 4x^2 - 16x - 2y - 19 = 0$

48. $x^2 - 3y^2 + 8x - 6y + 4 = 0$

In Problems 49–52, graph each function.
[**Hint:** Notice that each function is half a hyperbola.]

49. $f(x) = \sqrt{16 + 4x^2}$

50. $f(x) = -\sqrt{9 + 9x^2}$

51. $f(x) = -\sqrt{-25 + x^2}$

52. $f(x) = \sqrt{-1 + x^2}$

---

53. LORAN Two LORAN stations are positioned 200 miles apart along a straight shore.
  (a) A ship records a time difference of 0.00038 second between the LORAN signals. Set up an appropriate rectangular coordinate system to determine where the ship would reach shore if it were to follow the hyperbola corresponding to this time difference.
  (b) If the ship wants to enter a harbor located between the two stations 20 miles from the master station, what time difference should it be looking for?
  (c) If the ship is 50 miles offshore when the desired time difference is obtained, what is the approximate location of the ship?
  [**Note:** The speed of each radio signal is 186,000 miles per second.]

54. LORAN Two LORAN stations are positioned 100 miles apart along a straight shore.
  (a) A ship records a time difference of 0.00032 second between the LORAN signals. Set up an appropriate rectangular coordinate system to determine where the ship would reach shore if it were to follow the hyperbola corresponding to this time difference.
  (b) If the ship wants to enter a harbor located between the two stations 10 miles from the master station, what time difference should it be looking for?
  (c) If the ship is 20 miles offshore when the desired time difference is obtained, what is the approximate location of the ship?
  [**Note:** The speed of each radio signal is 186,000 miles per second.]

55. Calibrating Instruments In a test of their recording devices, a team of seismologists positioned two of the devices 2000 feet apart, with the device at point A to the west of the device at point B. At a point between the devices and 200 feet from point B, a small amount of explosive was detonated and a note made of the time at which the sound reached each device. A second explosion is to be carried out at a point directly north of point B.
  (a) How far north should the site of the second explosion be chosen so that the measured time difference recorded by the devices for the second detonation is the same as that recorded for the first detonation?

  (b) Explain why this experiment can be used to calibrate the instruments.

56. Explain in your own words the LORAN system of navigation.

57. The **eccentricity** e of a hyperbola is defined as the number c/a, where a and c are the numbers given in equation (2). Because $c > a$, it follows that $e > 1$. Describe the general shape of a hyperbola whose eccentricity is close to 1. What is the shape if e is very large?

58. A hyperbola for which $a = b$ is called an **equilateral hyperbola**. Find the eccentricity e of an equilateral hyperbola.
  [**Note:** The eccentricity of a hyperbola is defined in Problem 57.]

59. Two hyperbolas that have the same set of asymptotes are called **conjugate.** Show that the hyperbolas
$$\frac{x^2}{4} - y^2 = 1 \quad \text{and} \quad y^2 - \frac{x^2}{4} = 1$$
are conjugate. Graph each hyperbola on the same set of coordinate axes.

60. Prove that the hyperbola
$$\frac{y^2}{a^2} - \frac{x^2}{b^2} = 1$$
has the two oblique asymptotes
$$y = \frac{a}{b}x \quad \text{and} \quad y = -\frac{a}{b}x$$

61. Show that the graph of an equation of the form

$$Ax^2 + Cy^2 + F = 0, \qquad A \neq 0, C \neq 0, F \neq 0$$

where A and C are of opposite sign, is a hyperbola with center at $(0, 0)$.

62. Show that the graph of an equation of the form

$$Ax^2 + Cy^2 + Dx + Ey + F = 0, \qquad A \neq 0, C \neq 0$$

where A and C are of opposite sign,
(a) Is a hyperbola if $(D^2/4A) + (E^2/4C) - F \neq 0$.
(b) Is two intersecting lines if

$$(D^2/4A) + (E^2/4C) - F = 0$$

PREPARING FOR THIS SECTION

Before getting started, review the following:

✓ Sum Formulas for Sine and Cosine
(Section 3.4, pp. 182–185)

✓ Half-angle Formulas for Sine and Cosine (Section 3.5, p. 196)

✓ Double-angle Formulas for Sine and Cosine
(Section 3.5, p. 193)

6.5   ROTATION OF AXES; GENERAL FORM OF A CONIC

OBJECTIVES   1   Identify a Conic

   2   Use a Rotation of Axes to Transform Equations

   3   Discuss an Equation Using a Rotation of Axes

   4   Identify Conics without a Rotation of Axes

In this section, we show that the graph of a general second-degree polynomial containing two variables x and y, that is, an equation of the form

$$Ax^2 + Bxy + Cy^2 + Dx + Ey + F = 0 \qquad \textbf{(1)}$$

where A, B, and C are not simultaneously 0, is a conic. We shall not concern ourselves here with the degenerate cases of equation (1), such as $x^2 + y^2 = 0$, whose graph is a single point $(0, 0)$; or $x^2 + 3y^2 + 3 = 0$, whose graph contains no points; or $x^2 - 4y^2 = 0$, whose graph is two lines, $x - 2y = 0$ and $x + 2y = 0$.

We begin with the case where $B = 0$. In this case, the term containing xy is not present, so equation (1) has the form

$$Ax^2 + Cy^2 + Dx + Ey + F = 0$$

where either $A \neq 0$ or $C \neq 0$.

   1   We have already discussed the procedure for identifying the graph of this kind of equation; we complete the squares of the quadratic expressions in x or y, or both. Once this has been done, the conic can be identified by comparing it to one of the forms studied in Sections 6.2 through 6.4.

In fact, though, we can identify the conic directly from the equation without completing the squares.

Theorem

Identifying Conics without Completing the Squares

Excluding degenerate cases, the equation

$$Ax^2 + Cy^2 + Dx + Ey + F = 0 \qquad (2)$$

where A and C cannot both equal zero:

(a) Defines a parabola if $AC = 0$.

(b) Defines an ellipse (or a circle) if $AC > 0$.

(c) Defines a hyperbola if $AC < 0$.

∎

Proof

(a) If $AC = 0$, then either $A = 0$ or $C = 0$, but not both, so the form of equation (2) is either

$$Ax^2 + Dx + Ey + F = 0, \qquad A \neq 0$$

or

$$Cy^2 + Dx + Ey + F = 0, \qquad C \neq 0$$

Using the results of Problems 65 and 66 in Exercise 6.2, it follows that, except for the degenerate cases, the equation is a parabola.

(b) If $AC > 0$, then A and C are of the same sign. Using the results of Problems 69 and 70 in Exercise 6.3, except for the degenerate cases, the equation is an ellipse if $A \neq C$ or a circle if $A = C$.

(c) If $AC < 0$, then A and C are of opposite sign. Using the results of Problems 61 and 62 in Exercise 6.4, except for the degenerate cases, the equation is a hyperbola.

∎

We will not be concerned with the degenerate cases of equation (2). However, in practice, you should be alert to the possibility of degeneracy.

EXAMPLE 1

Identifying a Conic without Completing the Squares

Identify each equation without completing the squares.

(a) $3x^2 + 6y^2 + 6x - 12y = 0$ (b) $2x^2 - 3y^2 + 6y + 4 = 0$

(c) $y^2 - 2x + 4 = 0$

Solution

(a) We compare the given equation to equation (2) and conclude that $A = 3$ and $C = 6$. Since $AC = 18 > 0$, the equation is an ellipse.

(b) Here, $A = 2$ and $C = -3$, so $AC = -6 < 0$. The equation is a hyperbola.

(c) Here, $A = 0$ and $C = 1$, so $AC = 0$. The equation is a parabola. ∎

NOW WORK PROBLEM 1.

Although we can now identify the type of conic represented by any equation of the form of equation (2) without completing the squares, we will still need to complete the squares if we desire additional information about a conic.

Now we turn our attention to equations of the form of equation (1), where $B \neq 0$. To discuss this case, we first need to investigate a new procedure: *rotation of axes*.

ROTATION OF AXES

Figure 50

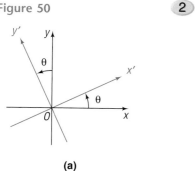

(a)

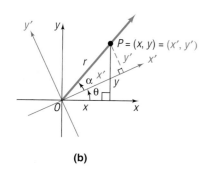

(b)

2   In a **rotation of axes,** the origin remains fixed while the x-axis and y-axis are rotated through an angle θ to a new position; the new positions of the x- and y-axes are denoted by x' and y', respectively, as shown in Figure 50(a).

Now look at Figure 50(b). There the point P has the coordinates (x, y) relative to the xy-plane, while the same point P has coordinates (x', y') relative to the $x'y'$-plane. We seek relationships that will enable us to express x and y in terms of x', y', and θ.

As Figure 50(b) shows, r denotes the distance from the origin O to the point P, and α denotes the angle between the positive x'-axis and the ray from O through P. Then, using the definitions of sine and cosine, we have

$$x' = r\cos\alpha \qquad\qquad y' = r\sin\alpha \qquad\qquad \textbf{(3)}$$

$$x = r\cos(\theta + \alpha) \qquad y = r\sin(\theta + \alpha) \qquad \textbf{(4)}$$

Now

$$x = r\cos(\theta + \alpha)$$
$$= r(\cos\theta\cos\alpha - \sin\theta\sin\alpha) \qquad \text{Sum formula for cosine}$$
$$= (r\cos\alpha)(\cos\theta) - (r\sin\alpha)(\sin\theta)$$
$$= x'\cos\theta - y'\sin\theta \qquad\qquad \text{By equation (3)}$$

Similarly,

$$y = r\sin(\theta + \alpha)$$
$$= r(\sin\theta\cos\alpha + \cos\theta\sin\alpha)$$
$$= x'\sin\theta + y'\cos\theta$$

Theorem

Rotation Formulas

If the x- and y-axes are rotated through an angle θ, the coordinates (x, y) of a point P relative to the xy-plane and the coordinates (x', y') of the same point relative to the new x'- and y'-axes are related by the formulas

$$x = x'\cos\theta - y'\sin\theta \qquad y = x'\sin\theta + y'\cos\theta \qquad \textbf{(5)}$$

EXAMPLE 2

Rotating Axes

Express the equation $xy = 1$ in terms of new $x'y'$-coordinates by rotating the axes through a $45°$ angle. Discuss the new equation.

Solution   Let $\theta = 45°$ in equation (5). Then

$$x = x'\cos 45° - y'\sin 45° = x'\frac{\sqrt{2}}{2} - y'\frac{\sqrt{2}}{2} = \frac{\sqrt{2}}{2}(x' - y')$$

$$y = x'\sin 45° + y'\cos 45° = x'\frac{\sqrt{2}}{2} + y'\frac{\sqrt{2}}{2} = \frac{\sqrt{2}}{2}(x' + y')$$

Figure 51

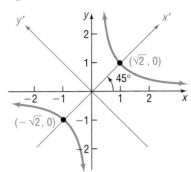

Substituting these expressions for x and y in $xy = 1$ gives

$$\left[\frac{\sqrt{2}}{2}(x' - y')\right]\left[\frac{\sqrt{2}}{2}(x' + y')\right] = 1$$

$$\frac{1}{2}(x'^2 - y'^2) = 1$$

$$\frac{x'^2}{2} - \frac{y'^2}{2} = 1$$

This is the equation of a hyperbola with center at $(0, 0)$ and transverse axis along the x'-axis. The vertices are at $(\pm\sqrt{2}, 0)$ on the x'-axis; the asymptotes are $y' = x'$ and $y' = -x'$ (which correspond to the original x- and y-axes). See Figure 51 for the graph. ∎

As Example 2 illustrates, a rotation of axes through an appropriate angle can transform a second-degree equation in x and y containing an xy-term into one in x' and y' in which no $x'y'$-term appears. In fact, we will show that a rotation of axes through an appropriate angle will transform any equation of the form of equation (1) into an equation in x' and y' without an $x'y'$-term.

To find the formula for choosing an appropriate angle θ through which to rotate the axes, we begin with equation (1),

$$Ax^2 + Bxy + Cy^2 + Dx + Ey + F = 0, \qquad B \neq 0$$

Next we rotate through an angle θ using rotation formulas (5).

$$A(x' \cos\theta - y' \sin\theta)^2 + B(x' \cos\theta - y' \sin\theta)(x' \sin\theta + y' \cos\theta)$$
$$+ C(x' \sin\theta + y' \cos\theta)^2 + D(x' \cos\theta - y' \sin\theta)$$
$$+ E(x' \sin\theta + y' \cos\theta) + F = 0$$

By expanding and collecting like terms, we obtain

$$(A \cos^2\theta + B \sin\theta \cos\theta + C \sin^2\theta)x'^2 + \left[B(\cos^2\theta - \sin^2\theta) + 2(C - A)(\sin\theta \cos\theta)\right]x'y'$$
$$+ (A \sin^2\theta - B \sin\theta \cos\theta + C \cos^2\theta)y'^2$$
$$+ (D \cos\theta + E \sin\theta)x'$$
$$+ (-D \sin\theta + E \cos\theta)y' + F = 0 \qquad \textbf{(6)}$$

In equation (6), the coefficient of $x'y'$ is

$$2(C - A)(\sin\theta \cos\theta) + B(\cos^2\theta - \sin^2\theta)$$

Since we want to eliminate the $x'y'$-term, we select an angle θ so that

$$2(C - A)(\sin\theta \cos\theta) + B(\cos^2\theta - \sin^2\theta) = 0$$

$$(C - A)(\sin(2\theta)) + B \cos(2\theta) = 0 \qquad \text{Double-angle formulas}$$

$$B \cos(2\theta) = (A - C)(\sin(2\theta))$$

$$\cot(2\theta) = \frac{A - C}{B}, \qquad B \neq 0$$

Theorem   To transform the equation

$$Ax^2 + Bxy + Cy^2 + Dx + Ey + F = 0, \qquad B \neq 0$$

into an equation in x' and y' without an $x'y'$-term, rotate the axes through an angle θ that satisfies the equation

$$\cot(2\theta) = \frac{A - C}{B} \qquad \textbf{(7)}$$

Equation (7) has an infinite number of solutions for θ. We shall adopt the convention of choosing the acute angle θ that satisfies (7). Then we have the following two possibilities:

> If $\cot(2\theta) \geq 0$, then $0° < 2\theta \leq 90°$ so that $0° < \theta \leq 45°$.
> If $\cot(2\theta) < 0$, then $90° < 2\theta < 180°$ so that $45° < \theta < 90°$.

Each of these results in a counterclockwise rotation of the axes through an acute angle θ.*

WARNING: Be careful if you use a calculator to solve equation (7).

1. If $\cot(2\theta) = 0$, then $2\theta = 90°$ and $\theta = 45°$.
2. If $\cot(2\theta) \neq 0$, first find $\cos(2\theta)$. Then use the inverse cosine function key(s) to obtain 2θ, $0° < 2\theta < 180°$. Finally, divide by 2 to obtain the correct acute angle θ.

EXAMPLE 3   Discussing an Equation Using a Rotation of Axes

   Discuss the equation: $x^2 + \sqrt{3}xy + 2y^2 - 10 = 0$

Solution   Since an xy-term is present, we must rotate the axes. Using $A = 1$, $B = \sqrt{3}$, and $C = 2$ in equation (7), the appropriate acute angle θ through which to rotate the axes satisfies the equation

$$\cot(2\theta) = \frac{A - C}{B} = \frac{-1}{\sqrt{3}} = -\frac{\sqrt{3}}{3}, \qquad 0° < 2\theta < 180°$$

Since $\cot(2\theta) = -\dfrac{\sqrt{3}}{3}$, we find $2\theta = 120°$, so $\theta = 60°$. Using $\theta = 60°$ in rotation formulas (5), we find

$$x = \frac{1}{2}x' - \frac{\sqrt{3}}{2}y' = \frac{1}{2}\left(x' - \sqrt{3}y'\right)$$

$$y = \frac{\sqrt{3}}{2}x' + \frac{1}{2}y' = \frac{1}{2}\left(\sqrt{3}x' + y'\right)$$

Substituting these values into the original equation and simplifying, we have

$$x^2 + \sqrt{3}xy + 2y^2 - 10 = 0$$

$$\frac{1}{4}\left(x' - \sqrt{3}y'\right)^2 + \sqrt{3}\left[\frac{1}{2}\left(x' - \sqrt{3}y'\right)\right]\left[\frac{1}{2}\left(\sqrt{3}x' + y'\right)\right] + 2\left[\frac{1}{4}\left(\sqrt{3}x' + y'\right)^2\right] = 10$$

---

*Any rotation (clockwise or counterclockwise) through an angle θ that satisfies $\cot(2\theta) = (A - C)/B$ will eliminate the $x'y'$-term. However, the final form of the transformed equation may be different (but equivalent), depending on the angle chosen.

Multiply both sides by 4 and expand to obtain

$$x'^2 - 2\sqrt{3}x'y' + 3y'^2 + \sqrt{3}\left(\sqrt{3}x'^2 - 2x'y' - \sqrt{3}y'^2\right) + 2\left(3x'^2 + 2\sqrt{3}x'y' + y'^2\right) = 40$$

$$10x'^2 + 2y'^2 = 40$$

$$\frac{x'^2}{4} + \frac{y'^2}{20} = 1$$

Figure 52

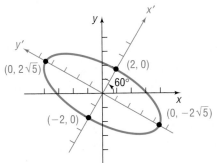

This is the equation of an ellipse with center at $(0, 0)$ and major axis along the y'-axis. The vertices are at $(0, \pm 2\sqrt{5})$ on the y'-axis. See Figure 52 for the graph.   ∎

NOW WORK PROBLEM **21**.

In Example 3, the acute angle θ through which to rotate the axes was easy to find because of the numbers that we used in the given equation. In general, the equation $\cot(2\theta) = \dfrac{A - C}{B}$ will not have such a "nice" solution. As the next example shows, we can still find the appropriate rotation formulas without using a calculator approximation by applying half-angle formulas.

EXAMPLE 4

Discussing an Equation Using a Rotation of Axes

Discuss the equation: $4x^2 - 4xy + y^2 + 5\sqrt{5}x + 5 = 0$

Solution   Letting $A = 4$, $B = -4$, and $C = 1$ in equation (7), the appropriate angle θ through which to rotate the axes satisfies

$$\cot(2\theta) = \frac{A - C}{B} = \frac{3}{-4} = -\frac{3}{4}$$

In order to use rotation formulas (5), we need to know the values of $\sin\theta$ and $\cos\theta$. Since we seek an acute angle θ, we know that $\sin\theta > 0$ and $\cos\theta > 0$. We use the half-angle formulas in the form

$$\sin\theta = \sqrt{\frac{1 - \cos(2\theta)}{2}} \qquad \cos\theta = \sqrt{\frac{1 + \cos(2\theta)}{2}}$$

Now we need to find the value of $\cos(2\theta)$. Since $\cot(2\theta) = -\dfrac{3}{4}$ and $90° < 2\theta < 180°$ (Do you know why?), it follows that $\cos(2\theta) = -\dfrac{3}{5}$. Then

$$\sin\theta = \sqrt{\frac{1 - \cos(2\theta)}{2}} = \sqrt{\frac{1 - \left(-\frac{3}{5}\right)}{2}} = \sqrt{\frac{4}{5}} = \frac{2}{\sqrt{5}} = \frac{2\sqrt{5}}{5}$$

$$\cos\theta = \sqrt{\frac{1 + \cos(2\theta)}{2}} = \sqrt{\frac{1 + \left(-\frac{3}{5}\right)}{2}} = \sqrt{\frac{1}{5}} = \frac{1}{\sqrt{5}} = \frac{\sqrt{5}}{5}$$

With these values, the rotation formulas (5) are

$$x = \frac{\sqrt{5}}{5}x' - \frac{2\sqrt{5}}{5}y' = \frac{\sqrt{5}}{5}(x' - 2y')$$

$$y = \frac{2\sqrt{5}}{5}x' + \frac{\sqrt{5}}{5}y' = \frac{\sqrt{5}}{5}(2x' + y')$$

Substituting these values in the original equation and simplifying, we obtain

$$4x^2 - 4xy + y^2 + 5\sqrt{5}x + 5 = 0$$

$$4\left[\frac{\sqrt{5}}{5}(x' - 2y')\right]^2 - 4\left[\frac{\sqrt{5}}{5}(x' - 2y')\right]\left[\frac{\sqrt{5}}{5}(2x' + y')\right]$$

$$+ \left[\frac{\sqrt{5}}{5}(2x' + y')\right]^2 + 5\sqrt{5}\left[\frac{\sqrt{5}}{5}(x' - 2y')\right] = -5$$

Multiply both sides by 5 and expand to obtain

$$4(x'^2 - 4x'y' + 4y'^2) - 4(2x'^2 - 3x'y' - 2y'^2)$$

$$+ 4x'^2 + 4x'y' + y'^2 + 25(x' - 2y') = -25$$

$$25y'^2 - 50y' + 25x' = -25 \quad \text{Combine like terms}$$

$$y'^2 - 2y' + x' = -1 \quad \text{Divide by 25}$$

$$y'^2 - 2y' + 1 = -x' \quad \text{Complete the square in } y'.$$

$$(y' - 1)^2 = -x'$$

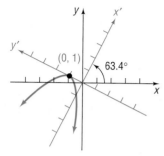

Figure 53

This is the equation of a parabola with vertex at $(0, 1)$ in the $x'y'$-plane. The axis of symmetry is parallel to the x'-axis. Using a calculator to solve $\sin\theta = \dfrac{2\sqrt{5}}{5}$, we find that $\theta \approx 63.4°$. See Figure 53 for the graph.   ■

NOW WORK PROBLEM 27.

IDENTIFYING CONICS WITHOUT A ROTATION OF AXES

④ Suppose that we are required only to identify (rather than discuss) an equation of the form

$$Ax^2 + Bxy + Cy^2 + Dx + Ey + F = 0, \qquad B \neq 0 \qquad \textbf{(8)}$$

If we apply rotation formulas (5) to this equation, we obtain an equation of the form

$$A'x'^2 + B'x'y' + C'y'^2 + D'x' + E'y' + F' = 0 \qquad \textbf{(9)}$$

where $A', B', C', D', E',$ and F' can be expressed in terms of $A, B, C, D, E,$ $F,$ and the angle θ of rotation (see Problem 43). It can be shown that the value of $B^2 - 4AC$ in equation (8) and the value of $B'^2 - 4A'C'$ in equation (9) are equal no matter what angle θ of rotation is chosen (see Problem 45). In particular, if the angle θ of rotation satisfies equation (7), then $B' = 0$ in equation (9), and $B^2 - 4AC = -4A'C'$. Since equation (9) then has the form of equation (2),

$$A'x'^2 + C'y'^2 + D'x' + E'y' + F' = 0$$

we can identify it without completing the squares, as we did in the beginning of this section. In fact, now we can identify the conic described by any equation of the form of equation (8) without a rotation of axes.

Theorem   **Identifying Conics without a Rotation of Axes**

Except for degenerate cases, the equation

$$Ax^2 + Bxy + Cy^2 + Dx + Ey + F = 0$$

(a) Defines a parabola if $B^2 - 4AC = 0$.

(b) Defines an ellipse (or a circle) if $B^2 - 4AC < 0$.

(c) Defines a hyperbola if $B^2 - 4AC > 0$.

You are asked to prove this theorem in Problem 46.

EXAMPLE 5   **Identifying a Conic without a Rotation of Axes**

Identify the equation: $8x^2 - 12xy + 17y^2 - 4\sqrt{5}x - 2\sqrt{5}y - 15 = 0$

Solution   Here $A = 8$, $B = -12$, and $C = 17$, so $B^2 - 4AC = -400$. Since $B^2 - 4AC < 0$, the equation defines an ellipse.

NOW WORK PROBLEM 33.

6.5 EXERCISES

In Problems 1–10, identify each equation without completing the squares.

1. $x^2 + 4x + y + 3 = 0$
2. $2y^2 - 3y + 3x = 0$
3. $6x^2 + 3y^2 - 12x + 6y = 0$
4. $2x^2 + y^2 - 8x + 4y + 2 = 0$
5. $3x^2 - 2y^2 + 6x + 4 = 0$
6. $4x^2 - 3y^2 - 8x + 6y + 1 = 0$
7. $2y^2 - x^2 - y + x = 0$
8. $y^2 - 8x^2 - 2x - y = 0$
9. $x^2 + y^2 - 8x + 4y = 0$
10. $2x^2 + 2y^2 - 8x + 8y = 0$

In Problems 11–20, determine the appropriate rotation formulas to use so that the new equation contains no xy-term.

11. $x^2 + 4xy + y^2 - 3 = 0$
12. $x^2 - 4xy + y^2 - 3 = 0$
13. $5x^2 + 6xy + 5y^2 - 8 = 0$
14. $3x^2 - 10xy + 3y^2 - 32 = 0$
15. $13x^2 - 6\sqrt{3}xy + 7y^2 - 16 = 0$
16. $11x^2 + 10\sqrt{3}xy + y^2 - 4 = 0$
17. $4x^2 - 4xy + y^2 - 8\sqrt{5}x - 16\sqrt{5}y = 0$
18. $x^2 + 4xy + 4y^2 + 5\sqrt{5}y + 5 = 0$
19. $25x^2 - 36xy + 40y^2 - 12\sqrt{13}x - 8\sqrt{13}y = 0$
20. $34x^2 - 24xy + 41y^2 - 25 = 0$

In Problems 21–32, rotate the axes so that the new equation contains no xy-term. Discuss and graph the new equation. Refer to Problems 11–20 for Problems 21–30.

21. $x^2 + 4xy + y^2 - 3 = 0$
22. $x^2 - 4xy + y^2 - 3 = 0$
23. $5x^2 + 6xy + 5y^2 - 8 = 0$
24. $3x^2 - 10xy + 3y^2 - 32 = 0$
25. $13x^2 - 6\sqrt{3}xy + 7y^2 - 16 = 0$
26. $11x^2 + 10\sqrt{3}xy + y^2 - 4 = 0$
27. $4x^2 - 4xy + y^2 - 8\sqrt{5}x - 16\sqrt{5}y = 0$
28. $x^2 + 4xy + 4y^2 + 5\sqrt{5}y + 5 = 0$
29. $25x^2 - 36xy + 40y^2 - 12\sqrt{13}x - 8\sqrt{13}y = 0$
30. $34x^2 - 24xy + 41y^2 - 25 = 0$
31. $16x^2 + 24xy + 9y^2 - 130x + 90y = 0$
32. $16x^2 + 24xy + 9y^2 - 60x + 80y = 0$

In Problems 33–42, identify each equation without applying a rotation of axes.

33. $x^2 + 3xy - 2y^2 + 3x + 2y + 5 = 0$
34. $2x^2 - 3xy + 4y^2 + 2x + 3y - 5 = 0$
35. $x^2 - 7xy + 3y^2 - y - 10 = 0$
36. $2x^2 - 3xy + 2y^2 - 4x - 2 = 0$
37. $9x^2 + 12xy + 4y^2 - x - y = 0$
38. $10x^2 + 12xy + 4y^2 - x - y + 10 = 0$
39. $10x^2 - 12xy + 4y^2 - x - y - 10 = 0$
40. $4x^2 + 12xy + 9y^2 - x - y = 0$
41. $3x^2 - 2xy + y^2 + 4x + 2y - 1 = 0$
42. $3x^2 + 2xy + y^2 + 4x - 2y + 10 = 0$

In Problems 43–45, apply rotation formulas (5) to

$$Ax^2 + Bxy + Cy^2 + Dx + Ey + F = 0$$

to obtain the equation

$$A'x'^2 + B'x'y' + C'y'^2 + D'x' + E'y' + F' = 0$$

43. Express A', B', C', D', E', and F' in terms of $A, B, C, D,$ E, F, and the angle θ of rotation. [**Hint:** Refer to Equation (6)].

44. Show that $A + C = A' + C'$, and thus show that $A + C$ is **invariant;** that is, its value does not change under a rotation of axes.

45. Refer to Problem 44. Show that $B^2 - 4AC$ is invariant.

46. Prove that, except for degenerate cases, the equation

$$Ax^2 + Bxy + Cy^2 + Dx + Ey + F = 0$$

  (a) Defines a parabola if $B^2 - 4AC = 0$.
  (b) Defines an ellipse (or a circle) if $B^2 - 4AC < 0$.
  (c) Defines a hyperbola if $B^2 - 4AC > 0$.

47. Use rotation formulas (5) to show that distance is invariant under a rotation of axes. That is, show that the distance from $P_1 = (x_1, y_1)$ to $P_2 = (x_2, y_2)$ in the xy-plane equals the distance from $P_1 = (x_1', y_1')$ to $P_2 = (x_2', y_2')$ in the $x'y'$-plane.

48. Show that the graph of the equation $x^{1/2} + y^{1/2} = a^{1/2}$ is part of the graph of a parabola.

49. Formulate a strategy for discussing and graphing an equation of the form

$$Ax^2 + Cy^2 + Dx + Ey + F = 0$$

How does your strategy change if the equation is of the form

$$Ax^2 + Bxy + Cy^2 + Dx + Ey + F = 0$$

PREPARING FOR THIS SECTION

Before getting started, review the following:

✓ Polar Coordinates (Section 5.1, pp. 274–281)

6.6   POLAR EQUATIONS OF CONICS

OBJECTIVES   **1**   Discuss and Graph Polar Equations of Conics
          2   Convert a Polar Equation of a Conic to a Rectangular Equation

1   In Sections 6.2 through 6.4, we gave separate definitions for the parabola, ellipse, and hyperbola based on geometric properties and the distance formula. In this section, we present an alternative definition that simultaneously defines all these conics. As we shall see, this approach is well suited to polar coordinate representation. (Refer to Section 5.1.)

Let D denote a fixed line called the **directrix;** let F denote a fixed point called the **focus,** which is not on D; and let e be a fixed positive number called the **eccentricity.** A **conic** is the set of points P in the plane such that the ratio of the distance from F to P to the distance from D to P equals e. That is, a conic is the collection of points P for which

$$\frac{d(F, P)}{d(D, P)} = e \qquad (1)$$

If $e = 1$, the conic is a **parabola.**
If $e < 1$, the conic is an **ellipse.**
If $e > 1$, the conic is a **hyperbola.**

Observe that if $e = 1$ the definition of a parabola in equation (1) is exactly the same as the definition used earlier in Section 6.2.

In the case of an ellipse, the **major axis** is a line through the focus perpendicular to the directrix. In the case of a hyperbola, the **transverse axis** is a line through the focus perpendicular to the directrix. For both an ellipse and a hyperbola, the eccentricity e satisfies

$$e = \frac{c}{a} \qquad (2)$$

where c is the distance from the center to the focus and a is the distance from the center to a vertex.

Just as we did earlier using rectangular coordinates, we derive equations for the conics in polar coordinates by choosing a convenient position for the focus F and the directrix D. The focus F is positioned at the pole, and the directrix D is either parallel to the polar axis or perpendicular to it.

Suppose that we start with the directrix D perpendicular to the polar axis at a distance p units to the left of the pole (the focus F). See Figure 54.

If $P = (r, \theta)$ is any point on the conic, then, by equation (1),

$$\frac{d(F, P)}{d(D, P)} = e \quad \text{or} \quad d(F, P) = e \cdot d(D, P) \qquad (3)$$

Figure 54

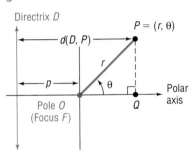

Directrix D

Now we use the point Q obtained by dropping the perpendicular from P to the polar axis to calculate $d(D, P)$.

$$d(D, P) = p + d(O, Q) = p + r \cos \theta$$

Using this expression and the fact that $d(F, P) = d(O, P) = r$ in equation (3), we get

$$d(F, P) = e \cdot d(D, P)$$
$$r = e(p + r \cos \theta)$$
$$r = ep + er \cos \theta$$
$$r - er \cos \theta = ep$$
$$r(1 - e \cos \theta) = ep$$
$$r = \frac{ep}{1 - e \cos \theta}$$

Theorem

Polar Equation of a Conic; Focus at Pole; Directrix Perpendicular to Polar Axis a Distance p to the Left of the Pole

The polar equation of a conic with focus at the pole and directrix perpendicular to the polar axis at a distance p to the left of the pole is

$$r = \frac{ep}{1 - e \cos \theta} \qquad (4)$$

where e is the eccentricity of the conic.

| EXAMPLE 1 | Discussing and Graphing the Polar Equation of a Conic |

Discuss and graph the equation: $r = \dfrac{4}{2 - \cos\theta}$

Solution   The given equation is not quite in the form of equation (4), since the first term in the denominator is 2 instead of 1. We divide the numerator and denominator by 2 to obtain

$$r = \frac{2}{1 - \frac{1}{2}\cos\theta} \qquad r = \frac{ep}{1 - e\cos\theta}$$

This equation is in the form of equation (4), with

$$e = \tfrac{1}{2} \quad \text{and} \quad ep = \tfrac{1}{2}p = 2, \quad p = 4$$

We conclude that the conic is an ellipse, since $e = \frac{1}{2} < 1$. One focus is at the pole, and the directrix is perpendicular to the polar axis, a distance of $p = 4$ units to the left of the pole. It follows that the major axis is along the polar axis. To find the vertices, we let $\theta = 0$ and $\theta = \pi$. The vertices of the ellipse are $(4, 0)$ and $\left(\tfrac{4}{3}, \pi\right)$. The midpoint of the vertices, $\left(\tfrac{4}{3}, 0\right)$ in polar coordinates, is the center of the ellipse. [Do you see why? The vertices $(4, 0)$ and $\left(\tfrac{4}{3}, \pi\right)$ in polar coordinates are $(4, 0)$ and $\left(-\tfrac{4}{3}, 0\right)$ in rectangular coordinates. The midpoint in rectangular coordinates is $\left(\tfrac{4}{3}, 0\right)$, which is also $\left(\tfrac{4}{3}, 0\right)$ in polar coordinates.] Then $a = $ distance from the center to a vertex $= \tfrac{8}{3}$. Using $a = \tfrac{8}{3}$ and $e = \tfrac{1}{2}$ in equation (2), $e = \dfrac{c}{a}$, we find $c = \tfrac{4}{3}$. Finally, using $a = \tfrac{8}{3}$ and $c = \tfrac{4}{3}$ in $b^2 = a^2 - c^2$, we have

$$b^2 = a^2 - c^2 = \frac{64}{9} - \frac{16}{9} = \frac{48}{9}$$

$$b = \frac{4\sqrt{3}}{3}$$

Figure 55

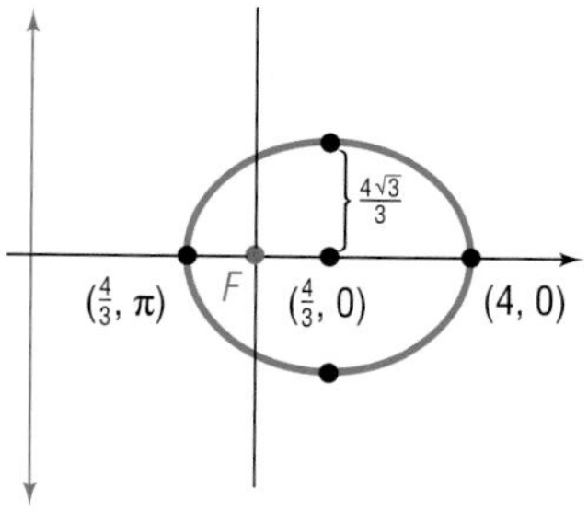

Figure 55 shows the graph.   ■

⌨ Check: In POLar mode with $\theta\,\min = 0$, $\theta\,\max = 2\pi$, and $\theta\,\text{step} = \pi/24$, graph $r_1 = \dfrac{4}{2 - \cos\theta}$ and compare the result with Figure 55.   ■

⌨ **EXPLORATION**   Graph $r_1 = \dfrac{4}{2 + \cos\theta}$ and compare the result with Figure 55. What do you conclude? Clear the screen and graph $r_1 = \dfrac{4}{2 - \sin\theta}$ and then $r_1 = \dfrac{4}{2 + \sin\theta}$. Compare each of these graphs with Figure 55. What do you conclude?   ■

✏ NOW WORK PROBLEM **5**.

Equation (4) was obtained under the assumption that the directrix was perpendicular to the polar axis at a distance p units to the left of the pole. A similar derivation (see Problem 37), in which the directrix is perpendicular to the polar axis at a distance p units to the right of the pole, results in the equation

$$r = \frac{ep}{1 + e\cos\theta}$$

In Problems 38 and 39 you are asked to derive the polar equations of conics with focus at the pole and directrix parallel to the polar axis. Table 5 summarizes the polar equations of conics.

TABLE 5	
POLAR EQUATIONS OF CONICS (FOCUS AT THE POLE, ECCENTRICITY e)	
Equation	**Description**
(a) $r = \dfrac{ep}{1 - e\cos\theta}$	Directrix is perpendicular to the polar axis at a distance p units to the left of the pole.
(b) $r = \dfrac{ep}{1 + e\cos\theta}$	Directrix is perpendicular to the polar axis at a distance p units to the right of the pole.
(c) $r = \dfrac{ep}{1 + e\sin\theta}$	Directrix is parallel to the polar axis at a distance p units above the pole.
(d) $r = \dfrac{ep}{1 - e\sin\theta}$	Directrix is parallel to the polar axis at a distance p units below the pole.
Eccentricity	
If $e = 1$, the conic is a parabola; the axis of symmetry is perpendicular to the directrix.	
If $e < 1$, the conic is an ellipse; the major axis is perpendicular to the directrix.	
If $e > 1$, the conic is a hyperbola; the transverse axis is perpendicular to the directrix.	

EXAMPLE 2   **Discussing and Graphing the Polar Equation of a Conic**

Discuss and graph the equation: $r = \dfrac{6}{3 + 3\sin\theta}$

Solution   To place the equation in proper form, we divide the numerator and denominator by 3 to get

$$r = \frac{2}{1 + \sin\theta}$$

Referring to Table 5, we conclude that this equation is in the form of equation (c) with

$$e = 1 \quad \text{and} \quad ep = 2$$
$$p = 2$$

The conic is a parabola with focus at the pole. The directrix is parallel to the polar axis at a distance 2 units above the pole; the axis of symmetry is perpendicular to the polar axis. The vertex of the parabola is at $\left(1, \dfrac{\pi}{2}\right)$. (Do you see why?) See Figure 56 for the graph. Notice that we plotted two additional points, $(2, 0)$ and $(2, \pi)$, to assist in graphing.   ■

Figure 56

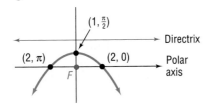

$(1, \frac{\pi}{2})$

$(2, \pi)$ $(2, 0)$ Directrix

F Polar axis

NOW WORK PROBLEM 7.

EXAMPLE 3   **Discussing and Graphing the Polar Equation of a Conic**

Discuss and graph the equation: $r = \dfrac{3}{1 + 3\cos\theta}$

Solution This equation is in the form of equation (b) in Table 5. We conclude that

$$e = 3 \quad \text{and} \quad ep = 3p = 3$$
$$p = 1$$

This is the equation of a hyperbola with a focus at the pole. The directrix is perpendicular to the polar axis, 1 unit to the right of the pole. The transverse axis is along the polar axis. To find the vertices, we let $\theta = 0$ and $\theta = \pi$. The vertices are $\left(\frac{3}{4}, 0\right)$ and $\left(-\frac{3}{2}, \pi\right)$. The center, which is at the midpoint of $\left(\frac{3}{4}, 0\right)$ and $\left(-\frac{3}{2}, \pi\right)$, is $\left(\frac{9}{8}, 0\right)$. Then $c = $ distance from the center to a focus $= \frac{9}{8}$. Since $e = 3$, it follows from equation (2), $e = \dfrac{c}{a}$, that $a = \frac{3}{8}$. Finally, using $a = \frac{3}{8}$ and $c = \frac{9}{8}$ in $b^2 = c^2 - a^2$, we find

Figure 57

$$b^2 = c^2 - a^2 = \frac{81}{64} - \frac{9}{64} = \frac{72}{64} = \frac{9}{8}$$

$$b = \frac{3}{2\sqrt{2}} = \frac{3\sqrt{2}}{4}$$

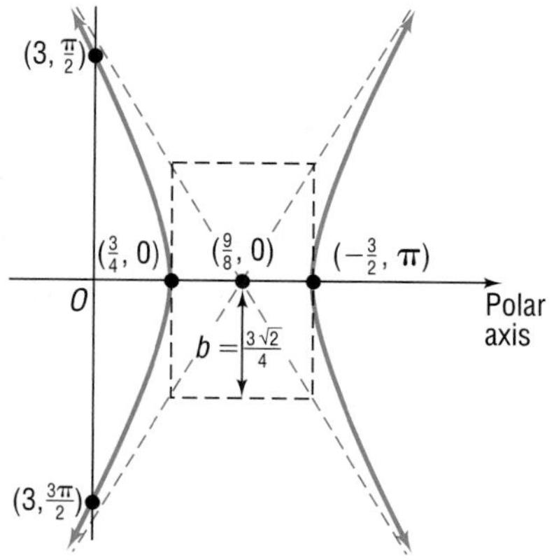

Figure 57 shows the graph. Notice that we plotted two additional points, $\left(3, \dfrac{\pi}{2}\right)$ and $\left(3, \dfrac{3\pi}{2}\right)$, on the left branch and used symmetry to obtain the right branch. The asymptotes of this hyperbola were found in the usual way by constructing the rectangle shown.   ∎

Check: Graph $r_1 = \dfrac{3}{1 + 3\cos\theta}$ and compare the result with Figure 57.   ∎

NOW WORK PROBLEM 11.

EXAMPLE 4   **Converting a Polar Equation to a Rectangular Equation**

② Convert the polar equation

$$r = \frac{1}{3 - 3\cos\theta}$$

to a rectangular equation.

Solution The strategy here is first to rearrange the equation and square each side before using the transformation equations.

$$r = \frac{1}{3 - 3\cos\theta}$$

$$3r - 3r\cos\theta = 1$$

$$3r = 1 + 3r\cos\theta \qquad \text{Rearrange the equation.}$$

$$9r^2 = (1 + 3r\cos\theta)^2 \qquad \text{Square each side.}$$

$$9(x^2 + y^2) = (1 + 3x)^2 \qquad x^2 + y^2 = r^2;\ x = r\cos\theta$$

$$9x^2 + 9y^2 = 9x^2 + 6x + 1$$

$$9y^2 = 6x + 1$$

This is the equation of a parabola in rectangular coordinates.   ∎

NOW WORK PROBLEM 19.

6.6 EXERCISES

In Problems 1–6, identify the conic that each polar equation represents. Also, give the position of the directrix.

1. $r = \dfrac{1}{1 + \cos\theta}$

2. $r = \dfrac{3}{1 - \sin\theta}$

3. $r = \dfrac{4}{2 - 3\sin\theta}$

4. $r = \dfrac{2}{1 + 2\cos\theta}$

5. $r = \dfrac{3}{4 - 2\cos\theta}$

6. $r = \dfrac{6}{8 + 2\sin\theta}$

In Problems 7–18, discuss each equation and graph it.

7. $r = \dfrac{1}{1 + \cos\theta}$

8. $r = \dfrac{3}{1 - \sin\theta}$

9. $r = \dfrac{8}{4 + 3\sin\theta}$

10. $r = \dfrac{10}{5 + 4\cos\theta}$

11. $r = \dfrac{9}{3 - 6\cos\theta}$

12. $r = \dfrac{12}{4 + 8\sin\theta}$

13. $r = \dfrac{8}{2 - \sin\theta}$

14. $r = \dfrac{8}{2 + 4\cos\theta}$

15. $r(3 - 2\sin\theta) = 6$

16. $r(2 - \cos\theta) = 2$

17. $r = \dfrac{6\sec\theta}{2\sec\theta - 1}$

18. $r = \dfrac{3\csc\theta}{\csc\theta - 1}$

In Problems 19–30, convert each polar equation to a rectangular equation.

19. $r = \dfrac{1}{1 + \cos\theta}$

20. $r = \dfrac{3}{1 - \sin\theta}$

21. $r = \dfrac{8}{4 + 3\sin\theta}$

22. $r = \dfrac{10}{5 + 4\cos\theta}$

23. $r = \dfrac{9}{3 - 6\cos\theta}$

24. $r = \dfrac{12}{4 + 8\sin\theta}$

25. $r = \dfrac{8}{2 - \sin\theta}$

26. $r = \dfrac{8}{2 + 4\cos\theta}$

27. $r(3 - 2\sin\theta) = 6$

28. $r(2 - \cos\theta) = 2$

29. $r = \dfrac{6\sec\theta}{2\sec\theta - 1}$

30. $r = \dfrac{3\csc\theta}{\csc\theta - 1}$

In Problems 31–36, find a polar equation for each conic. For each, a focus is at the pole.

31. $e = 1$; directrix is parallel to the polar axis 1 unit above the pole

32. $e = 1$; directrix is parallel to the polar axis 2 units below the pole

33. $e = \dfrac{4}{5}$; directrix is perpendicular to the polar axis 3 units to the left of the pole

34. $e = \dfrac{2}{3}$; directrix is parallel to the polar axis 3 units above the pole

35. $e = 6$; directrix is parallel to the polar axis 2 units below the pole

36. $e = 5$; directrix is perpendicular to the polar axis 5 units to the right of the pole

37. Derive equation (b) in Table 5:
$$r = \dfrac{ep}{1 + e\cos\theta}$$

38. Derive equation (c) in Table 5:
$$r = \dfrac{ep}{1 + e\sin\theta}$$

39. Derive equation (d) in Table 5:
$$r = \dfrac{ep}{1 - e\sin\theta}$$

40. Orbit of Mercury The planet Mercury travels around the Sun in an elliptical orbit given approximately by
$$r = \dfrac{(3.442)10^7}{1 - 0.206\cos\theta}$$
where r is measured in miles and the Sun is at the pole. Find the distance from Mercury to the Sun at *aphelion* (greatest distance from the Sun) and at *perihelion* (shortest distance from the Sun). See the figure. Use the aphelion and perihelion to graph the orbit of Mercury using a graphing utility.

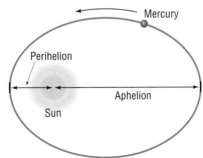

PREPARING FOR THIS SECTION

Before getting started, review the following:

✓ Amplitude and Period of Sinusoidal Graphs (Section 2.4, pp. 127–129)

6.7   PLANE CURVES AND PARAMETRIC EQUATIONS

OBJECTIVES   **1** Graph Parametric Equations
2 Find a Rectangular Equation for a Curve Defined Parametrically
3 Use Time as a Parameter in Parametric Equations
4 Find Parametric Equations for Curves Defined by Rectangular Equations

Equations of the form $y = f(x)$, where f is a function, have graphs that are intersected no more than once by any vertical line. The graphs of many of the conics and certain other, more complicated graphs do not have this characteristic. Yet each graph, like the graph of a function, is a collection of points (x, y) in the xy-plane; that is, each is a *plane curve*. In this section, we discuss another way of representing such graphs.

Let $x = f(t)$ and $y = g(t)$, where f and g are two functions whose common domain is some interval I. The collection of points defined by

$$(x, y) = (f(t), g(t))$$

is called a **plane curve.** The equations

$$x = f(t) \qquad y = g(t)$$

where t is in I, are called **parametric equations** of the curve. The variable t is called a **parameter.**

1   Parametric equations are particularly useful in describing movement along a curve. Suppose that a curve is defined by the parametric equations

$$x = f(t), \qquad y = g(t), \qquad a \leq t \leq b$$

where f and g are each defined over the interval $a \leq t \leq b$. For a given value of t, we can find the value of $x = f(t)$ and $y = g(t)$, obtaining a point (x, y) on the curve. In fact, as t varies over the interval from $t = a$ to $t = b$, successive values of t give rise to a directed movement along the curve; that is, the curve is traced out in a certain direction by the corresponding succession of points (x, y). See Figure 58. The arrows show the direction, or **orientation,** along the curve as t varies from a to b.

Figure 58

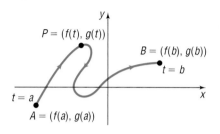

EXAMPLE 1

Discussing a Curve Defined by Parametric Equations

Discuss the curve defined by the parametric equations

$$x = 3t^2, \qquad y = 2t, \qquad -2 \leq t \leq 2 \qquad \textbf{(1)}$$

Solution   For each number $t, -2 \leq t \leq 2$, there corresponds a number x and a number y. For example, when $t = -2$, then $x = 12$ and $y = -4$. When $t = 0$, then

$x = 0$ and $y = 0$. Indeed, we can set up a table listing various choices of the parameter t and the corresponding values for x and y, as shown in Table 6. Plotting these points and connecting them with a smooth curve leads to Figure 59. The arrows in Figure 59 are used to indicate the orientation.

TABLE 6			
t	x	y	(x, y)
-2	12	-4	$(12, -4)$
-1	3	-2	$(3, -2)$
0	0	0	$(0, 0)$
1	3	2	$(3, 2)$
2	12	4	$(12, 4)$

Figure 59

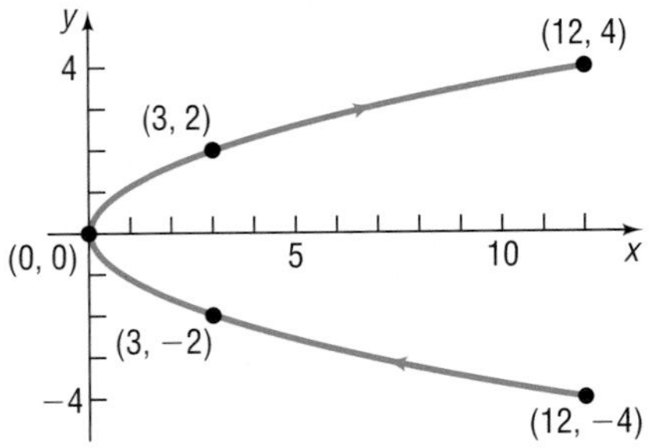

COMMENT: Most graphing utilities have the capability of graphing parametric equations. See Section B.7 in Appendix B.

The curve given in Example 1 should be familiar. To identify it accurately, we find the corresponding rectangular equation by eliminating the parameter t from the parametric equations (1) given in Example 1:

$$x = 3t^2, \qquad y = 2t, \qquad -2 \le t \le 2$$

Noting that we can readily solve for t in $y = 2t$, obtaining $t = \dfrac{y}{2}$, we substitute this expression in the other equation.

$$x = 3t^2 \underset{\underset{t = \frac{y}{2}}{\uparrow}}{=} 3\left(\frac{y}{2}\right)^2 = \frac{3y^2}{4}, \qquad -4 \le y \le 4$$

This equation, $x = \dfrac{3y^2}{4}$, is the equation of a parabola with vertex at $(0, 0)$ and axis of symmetry along the x-axis.

Note that the parameterized curve defined by equation (1) and shown in Figure 59 is only a part of the parabola $x = \dfrac{3y^2}{4}$. The graph of the rectangular equation obtained by eliminating the parameter will, in general, contain more points than the original parameterized curve. Care must therefore be taken when a parameterized curve is sketched by hand after eliminating the parameter. Even so, the process of eliminating the parameter t of a parameterized curve in order to identify it accurately is sometimes a better approach than merely plotting points. However, the elimination process sometimes requires a little ingenuity.

EXAMPLE 2 Finding the Rectangular Equation of a Curve Defined Parametrically

Find the rectangular equation of the curve whose parametric equations are

$$x = a \cos t \qquad y = a \sin t$$

where $a > 0$ is a constant. Graph this curve, indicating its orientation.

Solution The presence of sines and cosines in the parametric equations suggests that we use a Pythagorean identity. In fact, since

$$\cos t = \frac{x}{a} \qquad \sin t = \frac{y}{a}$$

we find that

$$\cos^2 t + \sin^2 t = 1$$

$$\left(\frac{x}{a}\right)^2 + \left(\frac{y}{a}\right)^2 = 1$$

$$x^2 + y^2 = a^2$$

Figure 60

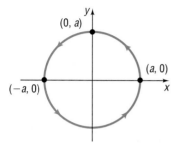

The curve is a circle with center at $(0, 0)$ and radius a. As the parameter t increases, say from $t = 0$ [the point $(a, 0)$] to $t = \dfrac{\pi}{2}$ [the point $(0, a)$] to $t = \pi$ [the point $(-a, 0)$], we see that the corresponding points are traced in a counterclockwise direction around the circle. The orientation is as indicated in Figure 60. ■

NOW WORK PROBLEMS **1** AND **13**.

Let's discuss the curve in Example 2 further. The domain of each parametric equation is $-\infty < t < \infty$. Thus, the graph in Figure 60 is actually being repeated each time that t increases by 2π.

If we wanted the curve to consist of exactly 1 revolution in the counterclockwise direction, we could write

$$x = a \cos t, \qquad y = a \sin t, \qquad 0 \le t \le 2\pi$$

This curve starts at $t = 0$ [the point $(a, 0)$] and, proceeding counterclockwise around the circle, ends at $t = 2\pi$ [also the point $(a, 0)$].

If we wanted the curve to consist of exactly three revolutions in the counterclockwise direction, we could write

$$x = a \cos t, \qquad y = a \sin t, \qquad -2\pi \le t \le 4\pi$$

or

$$x = a \cos t, \qquad y = a \sin t, \qquad 0 \le t \le 6\pi$$

or

$$x = a \cos t, \qquad y = a \sin t, \qquad 2\pi \le t \le 8\pi$$

EXAMPLE 3 **Describing Parametric Equations**

Find rectangular equations for and graph the following curves defined by parametric equations.

(a) $x = a \cos t, \quad y = a \sin t, \quad 0 \le t \le \pi, \quad a > 0$
(b) $x = -a \sin t, \quad y = -a \cos t, \quad 0 \le t \le \pi, \quad a > 0$

Solution (a) We eliminate the parameter t using a Pythagorean identity.

$$\left(\frac{x}{a}\right)^2 + \left(\frac{y}{a}\right)^2 = \cos^2 t + \sin^2 t = 1$$

$$x^2 + y^2 = a^2$$

The curve defined by these parametric equations is a circle, with radius a and center at $(0, 0)$. The circle begins at the point $(a, 0)$, $t = 0$;

Figure 61

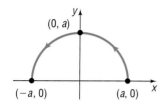

Figure 62

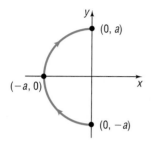

passes through the point $(0, a)$, $t = \dfrac{\pi}{2}$; and ends at the point $(-a, 0)$, $t = \pi$. The parametric equations define an upper semicircle of radius a with a counterclockwise orientation. See Figure 61. The rectangular equation is

$$y = a\sqrt{1 - \left(\frac{x}{a}\right)^2}, \qquad -a \le x \le a$$

(b) We eliminate the parameter t using a Pythagorean identity.

$$\left(\frac{x}{-a}\right)^2 + \left(\frac{y}{-a}\right)^2 = \sin^2 t + \cos^2 t = 1$$

$$x^2 + y^2 = a^2$$

The curve defined by these parametric equations is a circle, with radius a and center at $(0, 0)$. The circle begins at the point $(0, -a)$, $t = 0$; passes through the point $(-a, 0)$, $t = \dfrac{\pi}{2}$; and ends at the point $(0, a)$, $t = \pi$. The parametric equations define a left semicircle of radius a with a clockwise orientation. See Figure 62. The rectangular equation is

$$x = -a\sqrt{1 - \left(\frac{y}{a}\right)^2}, \qquad -a \le y \le a$$

Example 3 illustrates the versatility of parametric equations for replacing complicated rectangular equations, while providing additional information about orientation. These characteristics make parametric equations very useful in applications, such as projectile motion.

 SEEING THE CONCEPT Graph $x = \cos t$, $y = \sin t$ for $0 \le t \le 2\pi$. Compare to Figure 60. Graph $x = \cos t$, $y = \sin t$ for $0 \le t \le \pi$. Compare to Figure 61. Graph $x = -\sin t$, $y = -\cos t$ for $0 \le t \le \pi$. Compare to Figure 62.

TIME AS A PARAMETER: PROJECTILE MOTION; SIMULATED MOTION

③ If we think of the parameter t as time, then the parametric equations $x = f(t)$ and $y = g(t)$ of a curve C specify how the x- and y-coordinates of a moving point vary with time.

For example, we can use parametric equations to describe the motion of an object, sometimes referred to as **curvilinear motion.** Using parametric equations, we can specify not only where the object travels, that is, its location (x, y), but also when it gets there, that is, the time t.

When an object is propelled upward at an inclination θ to the horizontal with initial speed v_0, the resulting motion is called **projectile motion.** See Figure 63(a).

In calculus it is shown that the parametric equations of the path of a projectile fired at an inclination θ to the horizontal, with an initial speed v_0, from a height h above the horizontal are

$$x = (v_0 \cos \theta)t \qquad y = -\frac{1}{2}gt^2 + (v_0 \sin \theta)t + h \qquad \textbf{(2)}$$

where t is the time and g is the constant acceleration due to gravity (approximately 32 ft/sec/sec or 9.8 m/sec/sec). See Figure 63(b).

Figure 63

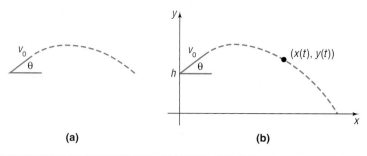

(a)   (b)

| EXAMPLE 4 | **Projectile Motion** |

Figure 64

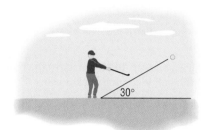

Suppose that Jim hit a golf ball with an initial velocity of 150 feet per second at an angle of 30° to the horizontal. See Figure 64.

(a) Find parametric equations that describe the position of the ball as a function of time.

(b) How long is the golf ball in the air?

(c) When is the ball at its maximum height? Determine the maximum height of the ball.

(d) Determine the distance that the ball traveled.

(e) Using a graphing utility, simulate the motion of the golf ball by simultaneously graphing the equations found in part (a).

Solution (a) We have $v_0 = 150$, $\theta = 30°$, $h = 0$ (the ball is on the ground), and $g = 32$ (since units are in feet and seconds). Substituting these values into equations (2), we find that

$$x = (v_0 \cos \theta)t = (150 \cos 30°)t = 75\sqrt{3}t$$

$$y = -\frac{1}{2}gt^2 + (v_0 \sin \theta)t + h \quad = -\frac{1}{2}(32)t^2 + (150 \sin 30°)t + 0$$

$$= -16t^2 + 75t$$

(b) To determine the length of time that the ball is in the air, we solve the equation $y = 0$.

$$-16t^2 + 75t = 0$$

$$t(-16t + 75) = 0$$

$$t = 0 \text{ sec} \quad \text{or} \quad t = \frac{75}{16} = 4.6875 \text{ sec}$$

The ball will strike the ground after 4.6875 seconds.

(c) Notice that the height y of the ball is a quadratic function of t, so the maximum height of the ball can be found by determining the vertex of $y = -16t^2 + 75t$. The value of t at the vertex is

$$t = \frac{-b}{2a} = \frac{-75}{-32} = 2.34375 \text{ sec}$$

The ball is at its maximum height after 2.34375 seconds. The maximum height of the ball is found by evaluating the function y at $t = 2.34375$ seconds.

Maximum height $= -16(2.34375)^2 + (75)2.34375 \approx 87.89$ feet

Figure 65

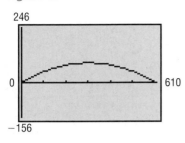

(d) Since the ball is in the air for 4.6875 seconds, the horizontal distance that the ball travels is

$$x = (75\sqrt{3})4.6875 \approx 608.92 \text{ feet}$$

 (e) We enter the equations from part (a) into a graphing utility with $T\text{min} = 0$, $T\text{max} = 4.7$, and T step $= 0.1$. We use ZOOM-SQUARE to avoid any distortion to the angle of elevation. See Figure 65. ■

EXPLORATION Simulate the motion of a ball thrown straight up with an initial speed of 100 feet per second from a height of 5 feet above the ground. Use PARametric mode with $T\text{min} = 0$, $T\text{max} = 6.5$, $T\text{step} = 0.1$, $X\text{min} = 0$, $X\text{max} = 5$, $Y\text{min} = 0$, and $Y\text{max} = 180$. What happens to the speed with which the graph is drawn as the ball goes up and then comes back down? How do you interpret this physically? Repeat the experiment using other values for Tstep. How does this affect the experiment?

[**Hint:** In the projectile motion equations, let $\theta = 90°$, $v_0 = 100$, $h = 5$, and $g = 32$. We use $x = 3$ instead of $x = 0$ to see the vertical motion better.]

RESULT See Figure 66. In Figure 66(a) the ball is going up. In Figure 66(b) the ball is near its highest point. Finally, in Figure 66(c) the ball is coming back down.

Notice that, as the ball goes up, its speed decreases, until at the highest point it is zero. Then the speed increases as the ball comes back down.

Figure 66

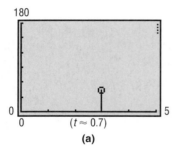 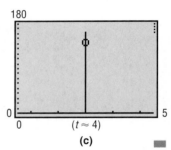

(a) (t ≈ 0.7)   (b) (t ≈ 3)   (c) (t ≈ 4)

■

 NOW WORK PROBLEM **27.**

A graphing utility can be used to simulate other kinds of motion as well.

EXAMPLE 5   **Simulating Motion**

Tanya, who is a long distance runner, runs at an average velocity of 8 miles per hour. Two hours after Tanya leaves your house, you leave in your Honda and follow the same route. If your average velocity is 40 miles per hour, how long will it be before you catch up to Tanya? See Figure 67. Use a simulation of the two motions to verify the answer.

Figure 67

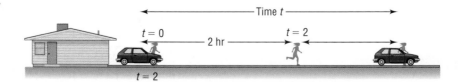

Solution   We begin with two sets of parametric equations: one to describe Tanya's motion, the other to describe the motion of the Honda. We choose time $t = 0$ to be when Tanya leaves the house. If we choose $y_1 = 2$ as Tanya's path, then we can use $y_2 = 4$ as the parallel path of the Honda. The horizontal distances traversed in time t (Distance = Velocity × Time) are

$$\text{Tanya:} \quad x_1 = 8t \qquad \text{Honda:} \quad x_2 = 40(t - 2)$$

The Honda catches up to Tanya when $x_1 = x_2$.

$$8t = 40(t - 2)$$
$$8t = 40t - 80$$
$$-32t = -80$$
$$t = \frac{-80}{-32} = 2.5$$

The Honda catches up to Tanya 2.5 hours after Tanya leaves the house.
In PARametric mode with Tstep = 0.01, we simultaneously graph

$$\text{Tanya:} \quad x_1 = 8t \qquad\qquad \text{Honda:} \quad x_2 = 40(t - 2)$$
$$y_1 = 2 \qquad\qquad\qquad\qquad y_2 = 4$$

for $0 \le t \le 3$.

Figure 68 shows the relative position of Tanya and the Honda for $t = 0$, $t = 2$, $t = 2.25$, $t = 2.5$, and $t = 2.75$.

Figure 68

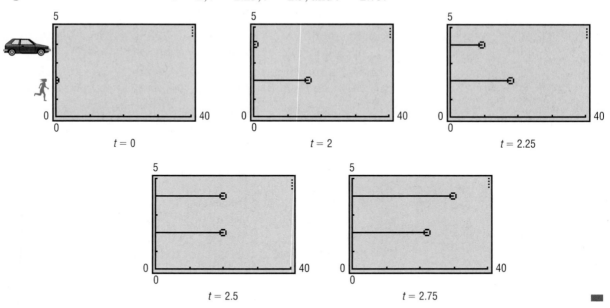

$t = 0$ $t = 2$ $t = 2.25$

$t = 2.5$ $t = 2.75$

FINDING PARAMETRIC EQUATIONS

We now take up the question of how to find parametric equations of a given curve.

④   If a curve is defined by the equation $y = f(x)$, where f is a function, one way of finding parametric equations is to let $x = t$. Then $y = f(t)$ and

$$x = t, \quad y = f(t), \qquad t \text{ in the domain of } f$$

are parametric equations of the curve.

EXAMPLE 6	**Finding Parametric Equations for a Curve Defined by a Rectangular Equation**

Find parametric equations for the equation $y = x^2 - 4$.

Solution Let $x = t$. Then the parametric equations are

$$x = t, \quad y = t^2 - 4, \quad -\infty < t < \infty$$ ∎

Another less obvious approach to Example 6 is to let $x = t^3$. Then the parametric equations become

$$x = t^3, \quad y = t^6 - 4, \quad -\infty < t < \infty$$

Care must be taken when using this approach, since the substitution for x must be a function that allows x to take on all the values stipulated by the domain of f. For example, letting $x = t^2$ so that $y = t^4 - 4$ does not result in equivalent parametric equations for $y = x^2 - 4$, since only points for which $x \geq 0$ are obtained.

EXAMPLE 7	**Finding Parametric Equations for an Object in Motion**

Find parametric equations for the ellipse

$$x^2 + \frac{y^2}{9} = 1$$

where the parameter t is time (in seconds) and

(a) The motion around the ellipse is clockwise, begins at the point $(0, 3)$, and requires 1 second for a complete revolution.

(b) The motion around the ellipse is counterclockwise, begins at the point $(1, 0)$, and requires 2 seconds for a complete revolution.

Solution (a) See Figure 69. Since the motion begins at the point $(0, 3)$, we want $x = 0$ and $y = 3$ when $t = 0$. Furthermore, since the given equation is an ellipse, we begin by letting

$$x = \sin(\omega t) \qquad \frac{y}{3} = \cos(\omega t)$$

Figure 69

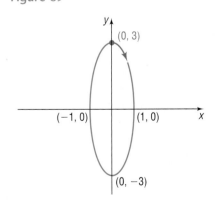

for some constant ω. These parametric equations satisfy the equation of the ellipse. Furthermore, with this choice, when $t = 0$, we have $x = 0$ and $y = 3$.

For the motion to be clockwise, the motion will have to begin with the value of x increasing and y decreasing as t increases. This requires that $\omega > 0$. [Do you know why? If $\omega > 0$, then $x = \sin(\omega t)$ is increasing when $t > 0$ is near zero and $y = 3\cos(\omega t)$ is decreasing when $t > 0$ is near zero]. See the red part of the graph in Figure 69.

Finally, since 1 revolution requires 1 second, the period $\dfrac{2\pi}{\omega} = 1$, so $\omega = 2\pi$. Parametric equations that satisfy the conditions stipulated are

$$x = \sin(2\pi t), \quad y = 3\cos(2\pi t), \quad 0 \leq t \leq 1 \qquad \textbf{(3)}$$

(b) See Figure 70. Since the motion begins at the point $(1, 0)$, we want $x = 1$ and $y = 0$ when $t = 0$. Furthermore, since the given equation is an ellipse, we begin by letting

$$x = \cos(\omega t) \qquad \frac{y}{3} = \sin(\omega t)$$

Figure 70

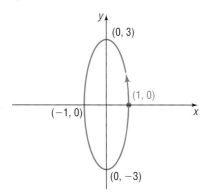

for some constant ω. These parametric equations satisfy the equation of the ellipse. Furthermore, with this choice, when $t = 0$, we have $x = 1$ and $y = 0$.

For the motion to be counterclockwise, the motion will have to begin with the value of x decreasing and y increasing as t increases. This requires that $\omega > 0$. [Do you know why?] Finally, since 1 revolution requires 2 seconds, the period is $\dfrac{2\pi}{\omega} = 2$, so $\omega = \pi$. The parametric equations that satisfy the conditions stipulated are

$$x = \cos(\pi t), \quad y = 3\sin(\pi t), \qquad 0 \le t \le 2 \qquad \textbf{(4)} \quad \blacksquare$$

Either of equations (3) or (4) can serve as parametric equations for the ellipse $\dfrac{x^2 + y^2}{9} = 1$ given in Example 7. The direction of the motion, the beginning point, and the time for 1 revolution merely serve to help us arrive at a particular parametric representation.

✏ **NOW WORK PROBLEM 43.**

THE CYCLOID

Suppose that a circle of radius a rolls along a horizontal line without slipping. As the circle rolls along the line, a point P on the circle will trace out a curve called a **cycloid** (see Figure 71). We now seek parametric equations* for a cycloid.

Figure 71

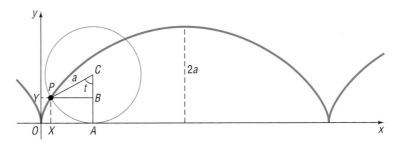

We begin with a circle of radius a and take the fixed line on which the circle rolls as the x-axis. Let the origin be one of the points at which the point P comes in contact with the x-axis. Figure 71 illustrates the position of this point P after the circle has rolled somewhat. The angle t (in radians) measures the angle through which the circle has rolled.

Since we require no slippage, it follows that

$$\text{Arc } AP = d(O, A)$$

The length of the arc AP is given by $s = r\theta$, where $r = a$ and $\theta = t$ radians. Then,

$$at = d(O, A) \quad \text{{\small $s = r\theta$, where $r = a$ and $\theta = t$}}$$

The x-coordinate of the point P is

$$d(O, X) = d(O, A) - d(X, A) = at - a\sin t = a(t - \sin t)$$

---

* Any attempt to derive the rectangular equation of a cycloid would soon demonstrate how complicated the task is.

The y-coordinate of the point P is equal to

$$d(O, Y) = d(A, C) - d(B, C) = a - a \cos t = a(1 - \cos t)$$

The parametric equations of the cycloid are

$$x = a(t - \sin t) \qquad y = a(1 - \cos t) \qquad \textbf{(5)}$$

 EXPLORATION Graph $x = t - \sin t$, $y = 1 - \cos t$, $0 \le t \le 3\pi$, using your graphing utility with Tstep $= \pi/36$ and a square screen. Compare your results with Figure 71. ∎

APPLICATIONS TO MECHANICS

If a is negative in equation (5), we obtain an inverted cycloid, as shown in Figure 72(a). The inverted cycloid occurs as a result of some remarkable applications in the field of mechanics. We shall mention two of them: the *brachistochrone* and the *tautochrone*.*

Figure 72

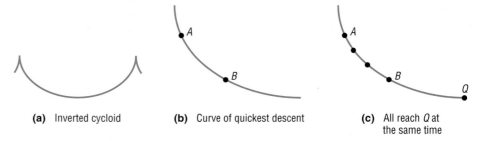

(a) Inverted cycloid   (b) Curve of quickest descent   (c) All reach Q at the same time

 The **brachistochrone** is the curve of quickest descent. If a particle is constrained to follow some path from one point A to a lower point B (not on the same vertical line) and is acted on only by gravity, the time needed to make the descent is least if the path is an inverted cycloid. See Figure 72(b). This remarkable discovery, which is attributed to many famous mathematicians (including Johann Bernoulli and Blaise Pascal), was a significant step in creating the branch of mathematics known as the *calculus of variations*.

Figure 73

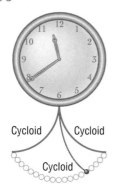

Cycloid   Cycloid

Cycloid

To define the **tautochrone,** let Q be the lowest point on an inverted cycloid. If several particles placed at various positions on an inverted cycloid simultaneously begin to slide down the cycloid, they will reach the point Q at the same time, as indicated in Figure 72(c). The tautochrone property of the cycloid was used by Christiaan Huygens (1629–1695), the Dutch mathematician, physicist, and astronomer, to construct a pendulum clock with a bob that swings along a cycloid (see Figure 73). In Huygen's clock, the bob was made to swing along a cycloid by suspending the bob on a thin wire constrained by two plates shaped like cycloids. In a clock of this design, the period of the pendulum is independent of its amplitude.

* In Greek, *brachistochrone* means "the shortest time" and *tautochrone* means "equal time."

6.7   EXERCISES

In Problems 1–20, graph the curve whose parametric equations are given and show its orientation. Find the rectangular equation of each curve.

1. $x = 3t + 2$, $y = t + 1$; $0 \le t \le 4$

2. $x = t - 3$, $y = 2t + 4$; $0 \le t \le 2$

3. $x = t + 2$, $y = \sqrt{t}$; $t \ge 0$

4. $x = \sqrt{2t}$, $y = 4t$; $t \ge 0$

5. $x = t^2 + 4$, $y = t^2 - 4$; $-\infty < t < \infty$

6. $x = \sqrt{t} + 4$, $y = \sqrt{t} - 4$; $t \ge 0$

7. $x = 3t^2$, $y = t + 1$; $-\infty < t < \infty$

8. $x = 2t - 4$, $y = 4t^2$; $-\infty < t < \infty$

9. $x = 2e^t$, $y = 1 + e^t$; $t \ge 0$

10. $x = e^t$, $y = e^{-t}$; $t \ge 0$

11. $x = \sqrt{t}$, $y = t^{3/2}$; $t \ge 0$

12. $x = t^{3/2} + 1$, $y = \sqrt{t}$; $t \ge 0$

13. $x = 2\cos t$, $y = 3\sin t$; $0 \le t \le 2\pi$

14. $x = 2\cos t$, $y = 3\sin t$; $0 \le t \le \pi$

15. $x = 2\cos t$, $y = 3\sin t$; $-\pi \le t \le 0$

16. $x = 2\cos t$, $y = \sin t$; $0 \le t \le \dfrac{\pi}{2}$

17. $x = \sec t$, $y = \tan t$; $0 \le t \le \dfrac{\pi}{4}$

18. $x = \csc t$, $y = \cot t$; $\dfrac{\pi}{4} \le t \le \dfrac{\pi}{2}$

19. $x = \sin^2 t$, $y = \cos^2 t$; $0 \le t \le 2\pi$

20. $x = t^2$, $y = \ln t$; $t > 0$

---

21. **Projectile Motion**   Bob throws a ball straight up with an initial speed of 50 feet per second from a height of 6 feet.
    (a) Find parametric equations that describe the motion of the ball as a function of time.
    (b) How long is the ball in the air?
    (c) When is the ball at its maximum height? Determine the maximum height of the ball.
    (d) Simulate the motion of the ball by graphing the equations found in part (a).

22. **Projectile Motion**   Alice throws a ball straight up with an initial speed of 40 feet per second from a height of 5 feet.
    (a) Find parametric equations that describe the motion of the ball as a function of time.
    (b) How long is the ball in the air?
    (c) When is the ball at its maximum height? Determine the maximum height of the ball.
    (d) Simulate the motion of the ball by graphing the equations found in part (a).

23. **Catching a Train**   Bill's train leaves at 8:06 AM and accelerates at the rate of 2 meters per second per second. Bill, who can run 5 meters per second, arrives at the train station 5 seconds after the train has left.
    (a) Find parametric equations that describe the motion of the train and Bill as a function of time.
    (b) Determine algebraically whether Bill will catch the train. If so, when?
    (c) Simulate the motion of the train and Bill by simultaneously graphing the equations found in part (a).

24. **Catching a Bus**   Jodi's bus leaves at 5:30 PM and accelerates at the rate of 3 meters per second per second. Jodi, who can run 5 meters per second, arrives at the bus station 2 seconds after the bus has left.
    (a) Find parametric equations that describe the motion of the bus and Jodi as a function of time.
    (b) Determine algebraically whether Jodi will catch the bus. If so, when?

    (c) Simulate the motion of the bus and Jodi by simultaneously graphing the equations found in part (a).

25. **Projectile Motion**   Nolan Ryan throws a baseball with an initial speed of 145 feet per second at an angle of 20° to the horizontal. The ball leaves Nolan Ryan's hand at a height of 5 feet.
    (a) Find parametric equations that describe the position of the ball as a function of time.
    (b) How long is the ball in the air?
    (c) When is the ball at its maximum height? Determine the maximum height of the ball.
    (d) Determine the distance that the ball traveled.
    (e) Using a graphing utility, simultaneously graph the equations found in part (a).

26. **Projectile Motion**   Mark McGwire hit a baseball with an initial speed of 180 feet per second at an angle of 40° to the horizontal. The ball was hit at a height of 3 feet off the ground.
    (a) Find parametric equations that describe the position of the ball as a function of time.
    (b) How long is the ball in the air?
    (c) When is the ball at its maximum height? Determine the maximum height of the ball.
    (d) Determine the distance that the ball traveled.
    (e) Using a graphing utility, simultaneously graph the equations found in part (a).

27. **Projectile Motion**   Suppose that Adam throws a tennis ball off a cliff 300 meters high with an initial speed of 40 meters per second at an angle of 45° to the horizontal.
    (a) Find parametric equations that describe the position of the ball as a function of time.
    (b) How long is the ball in the air?
    (c) When is the ball at its maximum height? Determine the maximum height of the ball.
    (d) Determine the distance that the ball traveled.
    (e) Using a graphing utility, simultaneously graph the equations found in part (a).

28. Projectile Motion Suppose that Adam throws a tennis ball off a cliff 300 meters high with an initial speed of 40 meters per second at an angle of 45° to the horizontal on the Moon (gravity on the Moon is one-sixth of that on Earth).

(a) Find parametric equations that describe the position of the ball as a function of time.

(b) How long is the ball in the air?

(c) When is the ball at its maximum height? Determine the maximum height of the ball.

(d) Determine the distance that the ball traveled.

(e) Using a graphing utility, simultaneously graph the equations found in part (a).

29. Uniform Motion A Toyota Paseo (traveling east at 40 mph) and Pontiac Bonneville (traveling north at 30 mph) are heading toward the same intersection. The Paseo is 5 miles from the intersection when the Bonneville is 4 miles from the intersection. See the figure.

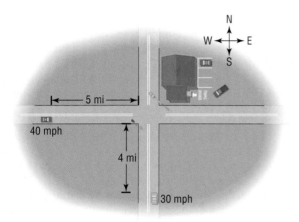
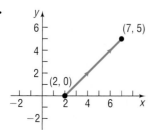

(a) Find parametric equations that describe the motion of the Paseo and Bonneville.

(b) Find a formula for the distance between the cars as a function of time.

(c) Graph the function in part (b) using a graphing utility.

(d) What is the minimum distance between the cars? When are the cars closest?

(e) Simulate the motion of the cars by simultaneously graphing the equations found in part (a).

30. Uniform Motion A Cessna (heading south at 120 mph) and a Boeing 747 (heading west at 600 mph) are flying toward the same point at the same altitude. The Cessna is 100 miles from the point where the flight patterns intersect and the 747 is 550 miles from this intersection point. See the figure.

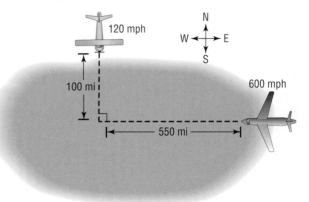
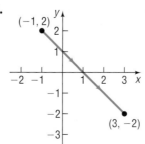

(a) Find parametric equations that describe the motion of the Cessna and 747.

(b) Find a formula for the distance between the planes as a function of time.

(c) Graph the function in part (b) using a graphing utility.

(d) What is the minimum distance between the planes? When are the planes closest?

(e) Simulate the motion of the planes by simultaneously graphing the equations found in part (a).

In Problems 31–38, find two different parametric equations for each rectangular equation.

31. $y = 4x - 1$

32. $y = -8x + 3$

33. $y = x^2 + 1$

34. $y = -2x^2 + 1$

35. $y = x^3$

36. $y = x^4 + 1$

37. $x = y^{3/2}$

38. $x = \sqrt{y}$

In Problems 39–42, find parametric equations that define the curve shown.

39.

40.

41.

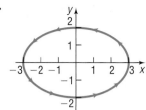

42.

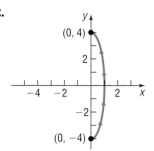

In Problems 43–46, find parametric equations for an object that moves along the ellipse $\dfrac{x^2}{4} + \dfrac{y^2}{9} = 1$ with the motion described.

43. The motion begins at $(2, 0)$, is clockwise, and requires 2 seconds for a complete revolution.

44. The motion begins at $(0, 3)$, is counterclockwise, and requires 1 second for a complete revolution.

45. The motion begins at $(0, 3)$, is clockwise, and requires 1 second for a complete revolution.

46. The motion begins at $(2, 0)$, is counterclockwise, and requires 3 seconds for a complete revolution.

In Problems 47 and 48, the parametric equations of four curves are given. Graph each of them, indicating the orientation.

47. C_1: $x = t$, $y = t^2$; $-4 \le t \le 4$
C_2: $x = \cos t$, $y = 1 - \sin^2 t$; $0 \le t \le \pi$
C_3: $x = e^t$, $y = e^{2t}$; $0 \le t \le \ln 4$
C_4: $x = \sqrt{t}$, $y = t$; $0 \le t \le 16$

48. C_1: $x = t$, $y = \sqrt{1 - t^2}$; $-1 \le t \le 1$
C_2: $x = \sin t$, $y = \cos t$; $0 \le t \le 2\pi$
C_3: $x = \cos t$, $y = \sin t$; $0 \le t \le 2\pi$
C_4: $x = \sqrt{1 - t^2}$, $y = t$; $-1 \le t \le 1$

49. Show that the parametric equations for a line passing through the points (x_1, y_1) and (x_2, y_2) are

$$x = (x_2 - x_1)t + x_1$$

$$y = (y_2 - y_1)t + y_1, \quad -\infty < t < \infty$$

What is the orientation of this line?

50. Projectile Motion The position of a projectile fired with an initial velocity v_0 feet per second and at an angle θ to the horizontal at the end of t seconds is given by the parametric equations

$$x = (v_0 \cos \theta)t \qquad y = (v_0 \sin \theta)t - 16t^2$$

See the following illustration.

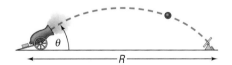

(a) Obtain the rectangular equation of the trajectory and identify the curve.
(b) Show that the projectile hits the ground $(y = 0)$ when $t = \frac{1}{16} v_0 \sin \theta$.
(c) How far has the projectile traveled (horizontally) when it strikes the ground? In other words, find the range R.
(d) Find the time t when $x = y$. Then find the horizontal distance x and the vertical distance y traveled by the projectile in this time. Then compute $\sqrt{x^2 + y^2}$. This is the distance R, the range, that the projectile travels up a plane inclined at $45°$ to the horizontal $(x = y)$. See the following illustration.

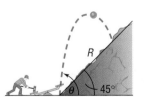

 In Problems 51–54, use a graphing utility to graph the curve defined by the given parametric equations.

51. $x = t \sin t$, $y = t \cos t$

52. $x = \sin t + \cos t$, $y = \sin t - \cos t$

53. $x = 4 \sin t - 2 \sin(2t)$
$y = 4 \cos t - 2 \cos(2t)$

54. $x = 4 \sin t + 2 \sin(2t)$
$y = 4 \cos t + 2 \cos(2t)$

55. Hypocycloid The hypocycloid is a curve defined by the parametric equations

$$x(t) = \cos^3 t, \quad y(t) = \sin^3 t, \quad 0 \le t \le 2\pi$$

(a) Graph the hypocycloid using a graphing utility.
(b) Find rectangular equations of the hypocycloid.

56. In Problem 55, we graphed the hypocycloid. Now graph the rectangular equations of the hypocycloid. Did you obtain a complete graph? If not, experiment until you do.

57. Look up the curves called *hypocycloid* and *epicycloid*. Write a report on what you find. Be sure to draw comparisons with the cycloid.

CHAPTER REVIEW

Things To Know

Equations

Parabola	See Tables 1 and 2 (pp. 351 and 353).
Ellipse	See Table 3 (p. 364).
Hyperbola	See Table 4 (p. 377).
General equation of a conic (p. 390)	$Ax^2 + Bxy + Cy^2 + Dx + Ey + F = 0$ Parabola if $B^2 - 4AC = 0$ Ellipse (or circle) if $B^2 - 4AC < 0$ Hyperbola if $B^2 - 4AC > 0$
Polar equations of a conic with focus at the pole	See Table 5 (p. 394).
Parametric equations of a curve (p. 397)	$x = f(t), y = g(t), t$ is the parameter

Definitions

Parabola (p. 349)	Set of points P in the plane for which $d(F, P) = d(P, D)$, where F is the focus and D is the directrix
Ellipse (p. 359)	Set of points P in the plane, the sum of whose distances from two fixed points (the foci) is a constant
Hyperbola (p. 369)	Set of points P in the plane, the difference of whose distances from two fixed points (the foci) is a constant
Conic in polar coordinates (p. 391)	$\dfrac{d(F, P)}{d(P, D)} = e$ Parabola if $e = 1$ Ellipse if $e < 1$ Hyperbola if $e > 1$

Formulas

Rotation formulas (p. 385)	$x = x' \cos\theta - y' \sin\theta$ $y = x' \sin\theta + y' \cos\theta$
Angle θ of rotation that eliminates the $x'y'$-term (p. 387)	$\cot(2\theta) = \dfrac{A - C}{B}, \quad 0° < \theta < 90°$

Objectives

You should be able to:

Know the names of the conics (p. 348)

Find the equation of a parabola (p. 349)

Graph parabolas (p. 350)

Discuss the equation of a parabola (p. 351)

Work with parabolas with vertex at (h, k) (p. 353)

Solve applied problems involving parabolas (p. 355)

Find the equation of an ellipse (p. 359)

Graph ellipses (p. 360)

Discuss the equation of an ellipse (p. 362)

Work with ellipses with center at (h, k) (p. 364)

Solve applied problems involving ellipses (p. 366)

Find the equation of a hyperbola (p. 370)

Graph hyperbolas (p. 371)

Discuss the equation of a hyperbola (p. 372)

Find the asymptotes of a hyperbola (p. 375)

Work with hyperbolas with center at (h, k) (p. 377)

Solve applied problems involving hyperbolas (p. 379)

Identify a conic (p. 383)

Use a rotation of axes to transform equations (p. 385)

Discuss an equation using a rotation of axes (p. 387)

Identify conics without a rotation of axes (p. 388)

Discuss and graph polar equations of conics (p. 391)

Convert a polar equation of a conic to a rectangular equation (p. 395)

Graph parametric equations (p. 397)

Find a rectangular equation for a curve defined parametrically (p. 398)

Use time as a parameter in parametric equations (p. 400)

Find parametric equations for curves defined by rectangular equations (p. 403)

Fill-in-the-Blank Items

1. A(n) _____ is the collection of all points in the plane such that the distance from each point to a fixed point equals its distance to a fixed line.

2. A(n) _____ is the collection of all points in the plane the sum of whose distances from two fixed points is a constant.

3. A(n) _____ is the collection of all points in the plane the difference of whose distances from two fixed points is a constant.

4. For an ellipse, the foci lie on the _____ axis; for a hyperbola, the foci lie on the _____ axis.

5. For the ellipse $\dfrac{x^2}{9} + \dfrac{y^2}{16} = 1$, the major axis is along the _____.

6. The equations of the asymptotes of the hyperbola $\dfrac{y^2}{9} - \dfrac{x^2}{4} = 1$ are _____ and _____.

7. To transform the equation

$$Ax^2 + Bxy + Cy^2 + Dx + Ey + F = 0, \qquad B \neq 0$$

into one in x' and y' without an $x'y'$-term, rotate the axes through an acute angle θ that satisfies the equation _____.

8. The polar equation $r = \dfrac{8}{4 - 2\sin\theta}$ is a conic whose eccentricity is _____. It is a(n) _____ whose directrix is _____ to the polar axis at a distance _____ units _____ the pole.

9. The parametric equations $x = 2\sin t$ and $y = 3\cos t$ represent a(n) _____.

True/False Items

T F   **1.** On a parabola, the distance from any point to the focus equals the distance from that point to the directrix.

T F   **2.** The foci of an ellipse lie on its minor axis.

T F   **3.** The foci of a hyperbola lie on its transverse axis.

T F   **4.** Hyperbolas always have asymptotes, and ellipses never have asymptotes.

T F   **5.** A hyperbola never intersects its conjugate axis.

T F   **6.** A hyperbola always intersects its transverse axis.

T F   **7.** The equation $ax^2 + 6y^2 - 12y = 0$ defines an ellipse if $a > 0$.

T F   **8.** The equation $3x^2 + bxy + 12y^2 = 10$ defines a parabola if $b = -12$.

T F   **9.** If (r, θ) are polar coordinates, the equation $r = \dfrac{2}{2 + 3\sin\theta}$ defines a hyperbola.

T F   **10.** Parametric equations defining a curve are unique.

Review Exercises

Blue problem numbers indicate the author's suggestions for use in a Practice Test.

In Problems 1–20, identify each equation. If it is a parabola, gives its vertex, focus, and directrix; if it is an ellipse, give its center, vertices, and foci; if it is a hyperbola, give its center, vertices, foci, and asymptotes.

1. $y^2 = -16x$

2. $16x^2 = y$

3. $\dfrac{x^2}{25} - y^2 = 1$

4. $\dfrac{y^2}{25} - x^2 = 1$

5. $\dfrac{y^2}{25} + \dfrac{x^2}{16} = 1$

6. $\dfrac{x^2}{9} + \dfrac{y^2}{16} = 1$

7. $x^2 + 4y = 4$

8. $3y^2 - x^2 = 9$

9. $4x^2 - y^2 = 8$

10. $9x^2 + 4y^2 = 36$

11. $x^2 - 4x = 2y$

12. $2y^2 - 4y = x - 2$

13. $y^2 - 4y - 4x^2 + 8x = 4$

14. $4x^2 + y^2 + 8x - 4y + 4 = 0$

15. $4x^2 + 9y^2 - 16x - 18y = 11$

16. $4x^2 + 9y^2 - 16x + 18y = 11$

17. $4x^2 - 16x + 16y + 32 = 0$

18. $4y^2 + 3x - 16y + 19 = 0$

19. $9x^2 + 4y^2 - 18x + 8y = 23$

20. $x^2 - y^2 - 2x - 2y = 1$

In Problems 21–36, obtain an equation of the conic described. Graph the equation.

21. Parabola; focus at $(-2, 0)$; directrix the line $x = 2$

22. Ellipse; center at $(0, 0)$; focus at $(0, 3)$; vertex at $(0, 5)$

23. Hyperbola; center at $(0, 0)$; focus at $(0, 4)$; vertex at $(0, -2)$

24. Parabola; vertex at $(0, 0)$; directrix the line $y = -3$

25. Ellipse; foci at $(-3, 0)$ and $(3, 0)$; vertex at $(4, 0)$

26. Hyperbola; vertices at $(-2, 0)$ and $(2, 0)$; focus at $(4, 0)$

27. Parabola; vertex at $(2, -3)$; focus at $(2, -4)$

28. Ellipse; center at $(-1, 2)$; focus at $(0, 2)$; vertex at $(2, 2)$

29. Hyperbola; center at $(-2, -3)$; focus at $(-4, -3)$; vertex at $(-3, -3)$

30. Parabola; focus at $(3, 6)$; directrix the line $y = 8$

31. Ellipse; foci at $(-4, 2)$ and $(-4, 8)$; vertex at $(-4, 10)$

32. Hyperbola; vertices at $(-3, 3)$ and $(5, 3)$; focus at $(7, 3)$

33. Center at $(-1, 2)$; $a = 3$; $c = 4$; transverse axis parallel to the x-axis

34. Center at $(4, -2)$; $a = 1$; $c = 4$; transverse axis parallel to the y-axis

35. Vertices at $(0, 1)$ and $(6, 1)$; asymptote the line $3y + 2x = 9$

36. Vertices at $(4, 0)$ and $(4, 4)$; asymptote the line $y + 2x = 10$

In Problems 37–46, identify each conic without completing the squares and without applying a rotation of axes.

37. $y^2 + 4x + 3y - 8 = 0$

38. $2x^2 - y + 8x = 0$

39. $x^2 + 2y^2 + 4x - 8y + 2 = 0$

40. $x^2 - 8y^2 - x - 2y = 0$

41. $9x^2 - 12xy + 4y^2 + 8x + 12y = 0$

42. $4x^2 + 4xy + y^2 - 8\sqrt{5}x + 16\sqrt{5}y = 0$

43. $4x^2 + 10xy + 4y^2 - 9 = 0$

44. $4x^2 - 10xy + 4y^2 - 9 = 0$

45. $x^2 - 2xy + 3y^2 + 2x + 4y - 1 = 0$

46. $4x^2 + 12xy - 10y^2 + x + y - 10 = 0$

In Problems 47–52, rotate the axes so that the new equation contains no xy-term. Discuss and graph the new equation.

47. $2x^2 + 5xy + 2y^2 - \frac{9}{2} = 0$

48. $2x^2 - 5xy + 2y^2 - \frac{9}{2} = 0$

49. $6x^2 + 4xy + 9y^2 - 20 = 0$

50. $x^2 + 4xy + 4y^2 + 16\sqrt{5}x - 8\sqrt{5}y = 0$

51. $4x^2 - 12xy + 9y^2 + 12x + 8y = 0$

52. $9x^2 - 24xy + 16y^2 + 80x + 60y = 0$

In Problems 53–58, identify the conic that each polar equation represents and graph it.

53. $r = \dfrac{4}{1 - \cos\theta}$

54. $r = \dfrac{6}{1 + \sin\theta}$

55. $r = \dfrac{6}{2 - \sin\theta}$

56. $r = \dfrac{2}{3 + 2\cos\theta}$

57. $r = \dfrac{8}{4 + 8\cos\theta}$

58. $r = \dfrac{10}{5 + 20\sin\theta}$

In Problems 59–62, convert each polar equation to a rectangular equation.

59. $r = \dfrac{4}{1 - \cos\theta}$

60. $r = \dfrac{6}{2 - \sin\theta}$

61. $r = \dfrac{8}{4 + 8\cos\theta}$

62. $r = \dfrac{2}{3 + 2\cos\theta}$

In Problems 63–68, graph the curve whose parametric equations are given and show its orientation. Find the rectangular equation of each curve.

63. $x = 4t - 2$, $\quad y = 1 - t$; $\quad -\infty < t < \infty$

64. $x = 2t^2 + 6$, $\quad y = 5 - t$; $\quad -\infty < t < \infty$

65. $x = 3\sin t$, $\quad y = 4\cos t + 2$; $\quad 0 \le t \le 2\pi$

66. $x = \ln t$, $\quad y = t^3$; $\quad t > 0$

67. $x = \sec^2 t$, $\quad y = \tan^2 t$; $\quad 0 \le t \le \dfrac{\pi}{4}$

68. $x = t^{3/2}$, $\quad y = 2t + 4$; $\quad t \ge 0$

---

69. Find an equation of the hyperbola whose foci are the vertices of the ellipse $4x^2 + 9y^2 = 36$ and whose vertices are the foci of this ellipse.

70. Find an equation of the ellipse whose foci are the vertices of the hyperbola $x^2 - 4y^2 = 16$ and whose vertices are the foci of this hyperbola.

71. Describe the collection of points in a plane so that the distance from each point to the point $(3, 0)$ is three-fourths of its distance from the line $x = \frac{16}{3}$.

72. Describe the collection of points in a plane so that the distance from each point to the point $(5, 0)$ is five-fourths of its distance from the line $x = \frac{16}{5}$.

73. Mirrors A mirror is shaped like a paraboloid of revolution. If a light source is located 1 foot from the base along the axis of symmetry and the opening is 2 feet across, how deep should the mirror be?

74. Parabolic Arch Bridge A bridge is built in the shape of a parabolic arch. The bridge has a span of 60 feet and a maximum height of 20 feet. Find the height of the arch at distances of 5, 10, and 20 feet from the center.

75. Semi-elliptical Arch Bridge A bridge is built in the shape of a semi-elliptical arch. The bridge has a span of 60 feet and a maximum height of 20 feet. Find the height of the arch at distances of 5, 10, and 20 feet from the center.

76. Whispering Galleries The figure shows the specifications for an elliptical ceiling in a hall designed to be a whispering gallery. Where are the foci located in the hall?

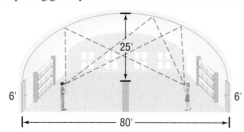

77. LORAN Two LORAN stations are positioned 150 miles apart along a straight shore.

(a) A ship records a time difference of 0.00032 second between the LORAN signals. Set up an appropriate rectangular coordinate system to determine where the ship would reach shore if it were to follow the hyperbola corresponding to this time difference.

(b) If the ship wants to enter a harbor located between the two stations 15 miles from the master station, what time difference should it be looking for?

(c) If the ship is 20 miles offshore when the desired time difference is obtained, what is the approximate location of the ship?

[**NOTE:** The speed of each radio signal is 186,000 miles per second.]

78. Uniform Motion Mary's train leaves at 7:15 AM and accelerates at the rate of 3 meters per second per second. Mary, who can run 6 meters per second, arrives at the train station 2 seconds after the train has left.

(a) Find parametric equations that describe the motion of the train and Mary as a function of time.

(b) Determine algebraically whether Mary will catch the train. If so, when?

(c) Simulate the motion of the train and Mary by simultaneously graphing the equations found in part (a).

79. Projectile Motion Drew Bledsoe throws a football with an initial speed of 100 feet per second at an angle of 35° to the horizontal. The ball leaves Drew Bledsoe's hand at a height of 6 feet.

(a) Find parametric equations that describe the position of the ball as a function of time.

(b) How long is the ball in the air?

(c) When is the ball at its maximum height? Determine the maximum height of the ball.

(d) Determine the distance that the ball travels.

(e) Using a graphing utility, simultaneously graph the equations found in part (a).

80. Formulate a strategy for discussing and graphing an equation of the form

$$Ax^2 + Bxy + Cy^2 + Dx + Ey + F = 0$$

Project at Motorola

Distorted Deployable Space Reflector Antennas

A certain parabolic reflector antenna is to be used in a space application. To avoid excessive costs transporting the antenna into orbit, NASA makes these antennas from a very thin metallic cloth material attached on a frame that is deployed while in space to open up and take the form of a parabolic umbrella. In the correct configuration, the axis of the antenna aligns with the z-axis. To ensure that the deployed antenna is of the right shape, the following scheme is devised. Eight small, light-reflecting targets are placed on the cloth with the coordinates below (when the antenna is correctly deployed). Four of them are on the xz plane and four are on the yz plane. Tension wires are used to adjust the position of these targets when distorted, that is, when the actual values of x, y, and z do not equal the correct values. When in the cargo bay of the space shuttle, a laser placed on the z-axis and at a distance $L = 10$ m from the vertex performs measurements and records its distance from the targets and their angle θ (see the figure). Table 1 contains the correct values of x, y, and z for the parabolic reflector.

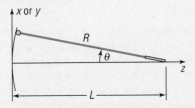

TABLE 1			
Target	x (m)	y (m)	z (m)
T_1	0	−2	0.5
T_2	0	−1	0.125
T_3	0	1	0.125
T_4	0	2	0.5
T_5	−2	0	0.5
T_6	−1	0	0.125
T_7	1	0	0.125
T_8	2	0	0.5

The laser measurements of the targets on the antenna immediately after deployment are tabulated in Table 2.

TABLE 2			
Target	x (m)	R (m)	θ (deg)
T_1	0	9.551	−11.78
T_2	0	9.948	−5.65
T_3	0	9.928	5.90
T_4	0	9.708	11.89
Target	y (m)	R (m)	θ (deg)
T_5	0	9.722	−11.99
T_6	0	9.917	−5.85
T_7	0	9.925	5.78
T_8	0	9.551	11.78

Using any of the target points from Table 1 and the fact that the equation of the parabolic reflector (assuming vertex at the origin) is $z = 4ax^2 + 4ay^2$:

1. Find the distance from the focus to the vertex of the undistorted parabolic reflector.

2. Use the tabulated measurements of the target points (Table 2) to convert the R and θ coordinates of the targets into Cartesian coordinates (z and x or y, as appropriate). That is, find the actual values of x, y, and z.

4. Find how much the targets T_1 through T_4 have to be moved in the y and z directions and T_5 through T_8 in the x and z directions in order to correct the distortions of the antenna.

NOTE: In actuality, instead of adjusting the tension wires, the laser measurements are used to determine how much the feed antenna at the focus of the parabola needs to change its electronic distribution so as to correct for the distortions (phase conjugation).

Exponential and Logarithmic Functions

Field Trip to Motorola

Many physical and chemical processes exhibit exponential or logarithmic behavior. These functions are the cornerstones of science and engineering. For example, chemical reactions often can be modeled by an exponential function. Metal fatigue, which is very complex when we introduce temperature and the rate of cycling, can be described with logarithmic functions.

PREPARING FOR THIS SECTION

Before getting started, review the following:

✓ Integer Exponents (Appendix A, Section A.1, pp. 492–493)

✓ Graphing Techniques: Transformations (Section 1.5, pp. 47–56)

✓ Rational Exponents (Appendix A, Section A.6, pp. 534–536)

✓ Solving Equations (Appendix A, Section A.3, pp. 502–511)

7.1 EXPONENTIAL FUNCTIONS

OBJECTIVES
1. Evaluate Exponential Functions
2. Graph Exponential Functions
3. Define the Number e
4. Solve Exponential Equations

1. In Appendix A, Section A.6, we give a definition for raising a real number a to a rational power. Based on that discussion, we gave meaning to expressions of the form

$$a^r$$

where the base a is a positive real number and the exponent r is a rational number.

But what is the meaning of a^x, where the base a is a positive real number and the exponent x is an irrational number? Although a rigorous definition requires methods discussed in calculus, the basis for the definition is easy to follow: Select a rational number r that is formed by truncating (removing) all but a finite number of digits from the irrational number x. Then it is reasonable to expect that

$$a^x \approx a^r$$

For example, take the irrational number $\pi = 3.14159\ldots$. Then, an approximation to a^π is

$$a^\pi \approx a^{3.14}$$

where the digits after the hundredths position have been removed from the value for π. A better approximation would be

$$a^\pi \approx a^{3.14159}$$

where the digits after the hundred-thousandths position have been removed. Continuing in this way, we can obtain approximations to a^π to any desired degree of accuracy.

Most calculators have an $\boxed{x^y}$ key or a caret key $\boxed{\wedge}$ for working with exponents. To evaluate expressions of the form a^x, enter the base a, then press the $\boxed{x^y}$ key (or the $\boxed{\wedge}$ key), enter the exponent x, and press $\boxed{=}$ (or $\boxed{\text{enter}}$).

EXAMPLE 1 **Using a Calculator to Evaluate Powers of 2**

Using a calculator, evaluate:

(a) $2^{1.4}$ (b) $2^{1.41}$ (c) $2^{1.414}$ (d) $2^{1.4142}$ (e) $2^{\sqrt{2}}$

Solution (a) $2^{1.4} \approx 2.639015822$ (b) $2^{1.41} \approx 2.657371628$
          (c) $2^{1.414} \approx 2.66474965$ (d) $2^{1.4142} \approx 2.665119089$
          (e) $2^{\sqrt{2}} \approx 2.665144143$ ∎

NOW WORK PROBLEM **1**.

It can be shown that the familiar laws for rational exponents hold for real exponents.

Theorem **Laws of Exponents**

If s, t, a, and b are real numbers, with $a > 0$ and $b > 0$, then

$$a^s \cdot a^t = a^{s+t} \qquad \left(a^s\right)^t = a^{st} \qquad (ab)^s = a^s \cdot b^s$$

$$1^s = 1 \qquad a^{-s} = \frac{1}{a^s} = \left(\frac{1}{a}\right)^s \qquad a^0 = 1 \qquad \textbf{(1)}$$

∎

We are now ready for the following definition:

An **exponential function** is a function of the form

$$f(x) = a^x$$

where a is a positive real number $(a > 0)$ and $a \neq 1$. The domain of f is the set of all real numbers.

We exclude the base $a = 1$, because this function is simply the constant function $f(x) = 1^x = 1$. We also need to exclude the bases that are negative, because, otherwise, we would have to exclude many values of x from the domain, such as $x = \dfrac{1}{2}$ and $x = \dfrac{3}{4}$. [Recall that $(-2)^{1/2} = \sqrt{-2}, (-3)^{3/4} = \sqrt[4]{(-3)^3} = \sqrt[4]{-27}$, and so on, are not defined in the system of real numbers.]

GRAPHS OF EXPONENTIAL FUNCTIONS

② First, we graph the exponential function $f(x) = 2^x$.

EXAMPLE 2 **Graphing an Exponential Function**

Graph the exponential function: $f(x) = 2^x$

Solution The domain of $f(x) = 2^x$ consists of all real numbers. We begin by locating some points on the graph of $f(x) = 2^x$, as listed in Table 1 (page 418).

Since $2^x > 0$ for all x, the range of f is the interval $(0, \infty)$. From this, we conclude that the graph has no x-intercepts, and, in fact, the graph will lie above the x-axis. As Table 1 indicates, the y-intercept is 1. Table 1 also indicates that as $x \rightarrow -\infty$ the value of $f(x) = 2^x$ gets closer and closer to 0. We conclude that the x-axis is a horizontal asymptote to the graph as $x \rightarrow -\infty$. This gives us the end behavior of the graph for x large and negative.

To determine the end behavior for x large and positive, look again at Table 1. As $x \rightarrow \infty$, $f(x) = 2^x$ grows very quickly, causing the graph of $f(x) = 2^x$ to rise very rapidly. It is apparent that f is an increasing function and hence is one-to-one.

Using all this information, we plot some of the points from Table 1 and connect them with a smooth, continuous curve, as shown in Figure 1.

TABLE 1	
x	$f(x) = 2^x$
-10	$2^{-10} \approx 0.00098$
-3	$2^{-3} = \dfrac{1}{8}$
-2	$2^{-2} = \dfrac{1}{4}$
-1	$2^{-1} = \dfrac{1}{2}$
0	$2^0 = 1$
1	$2^1 = 2$
2	$2^2 = 4$
3	$2^3 = 8$
10	$2^{10} = 1024$

Figure 1
$y = 2^x$

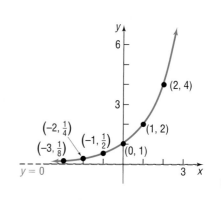

As we shall see, graphs that look like the one in Figure 1 occur very frequently in a variety of situations. For example, look at the graph in Figure 2, which illustrates the closing price of a share of Dell Computer stock.

Figure 2

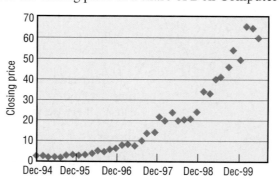

Investors might conclude from this graph that the price of Dell Computer is *behaving exponentially*; that is, the graph exhibits rapid, or exponential, growth. We shall have more to say about situations that lead to exponential growth later in this chapter. For now, we continue to seek properties of the exponential functions.

The graph of $f(x) = 2^x$ in Figure 1 is typical of all exponential functions that have a base larger than 1. Such functions are increasing functions and hence are one-to-one. Their graphs lie above the x-axis, pass through the point $(0, 1)$, and thereafter rise rapidly as $x \rightarrow \infty$. As $x \rightarrow -\infty$, the x-axis

Figure 3

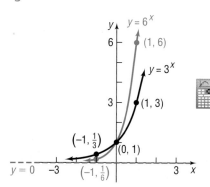

$(y = 0)$ is a horizontal asymptote. There are no vertical asymptotes. Finally, the graphs are smooth and continuous, with no corners or gaps.

Figure 3 illustrates the graphs of two more exponential functions whose bases are larger than 1. Notice that for the larger base the graph is steeper when $x > 0$ and is closer to the x-axis when $x < 0$.

SEEING THE CONCEPT Graph $y = 2^x$ and compare what you see to Figure 1. Clear the screen and graph $y = 3^x$ and $y = 6^x$ and compare what you see to Figure 3. Clear the screen and graph $y = 10^x$ and $y = 100^x$. What viewing rectangle seems to work best? ■

The following display summarizes the information that we have about $f(x) = a^x, a > 1$.

Figure 4
$f(x) = a^x, a > 1$

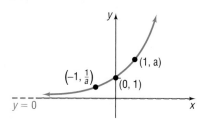

Properties of the Graph of an Exponential Function $f(x) = a^x, a > 1$

1. The domain is all real numbers; the range is the set of positive real numbers.
2. There are no x-intercepts; the y-intercept is 1.
3. The x-axis $(y = 0)$ is a horizontal asymptote as $x \to -\infty$.
4. $f(x) = a^x, a > 1$, is an increasing function and is one-to-one.
5. The graph of f contains the points $(0, 1), (1, a)$, and $\left(-1, \dfrac{1}{a}\right)$.
6. The graph of f is smooth and continuous, with no corners or gaps. See Figure 4.

Now we consider $f(x) = a^x$ when $0 < a < 1$.

EXAMPLE 3 | **Graphing an Exponential Function**

Graph the exponential function: $f(x) = \left(\dfrac{1}{2}\right)^x$

Solution The domain of $f(x) = \left(\dfrac{1}{2}\right)^x$ consists of all real numbers. As before, we locate some points on the graph by creating Table 2. Since $\left(\dfrac{1}{2}\right)^x > 0$ for all x, the range of f is the interval $(0, \infty)$. The graph lies above the x-axis and so has no x-intercepts. The y-intercept is 1. As $x \to -\infty$, $f(x) = \left(\dfrac{1}{2}\right)^x$ grows very quickly. As $x \to \infty$, the values of $f(x)$ approach 0. The x-axis $(y = 0)$ is a horizontal asymptote as $x \to \infty$. It is apparent that f is a decreasing function and hence is one-to-one. Figure 5 illustrates the graph.

TABLE 2

x	$f(x) = \left(\tfrac{1}{2}\right)^x$
-10	$\left(\dfrac{1}{2}\right)^{-10} = 1024$
-3	$\left(\dfrac{1}{2}\right)^{-3} = 8$
-2	$\left(\dfrac{1}{2}\right)^{-2} = 4$
-1	$\left(\dfrac{1}{2}\right)^{-1} = 2$
0	$\left(\dfrac{1}{2}\right)^{0} = 1$
1	$\left(\dfrac{1}{2}\right)^{1} = \dfrac{1}{2}$
2	$\left(\dfrac{1}{2}\right)^{2} = \dfrac{1}{4}$
3	$\left(\dfrac{1}{2}\right)^{3} = \dfrac{1}{8}$
10	$\left(\dfrac{1}{2}\right)^{10} \approx 0.00098$

Figure 5
$y = \left(\dfrac{1}{2}\right)^x$

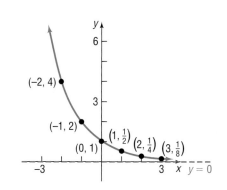

We could have obtained the graph of $y = \left(\dfrac{1}{2}\right)^x$ from the graph of $y = 2^x$ using transformations. If $f(x) = 2^x$, then $f(-x) = 2^{-x} = \dfrac{1}{2^x} = \left(\dfrac{1}{2}\right)^x$. The graph of $y = \left(\dfrac{1}{2}\right)^x = 2^{-x}$ is a reflection about the y-axis of the graph of $y = 2^x$. See Figures 6(a) and (b).

Figure 6

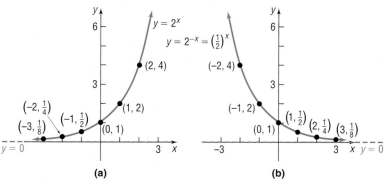

(a)   (b)

SEEING THE CONCEPT   Using a graphing utility, simultaneously graph

(a) $Y_1 = 3^x$, $Y_2 = \left(\dfrac{1}{3}\right)^x$

(b) $Y_1 = 6^x$, $Y_2 = \left(\dfrac{1}{6}\right)^x$

Conclude that the graph of $Y_2 = \left(\dfrac{1}{a}\right)^x$, for $a > 0$, is the reflection about the y-axis of the graph of $Y_1 = a^x$.   ∎

Figure 7

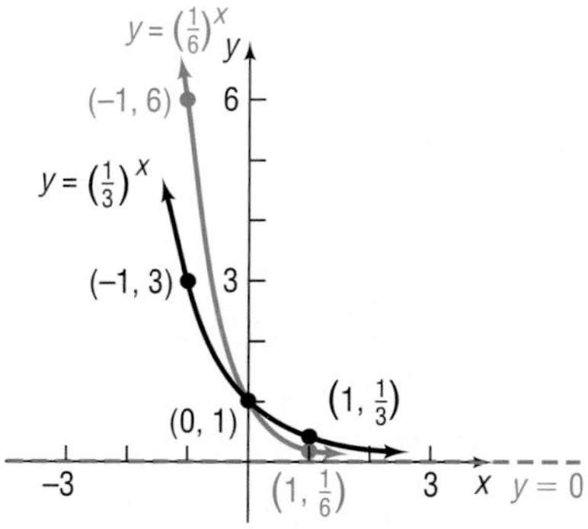

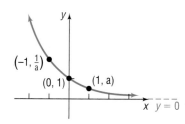

The graph of $f(x) = \left(\dfrac{1}{2}\right)^x$ in Figure 5 is typical of all exponential functions that have a base between 0 and 1. Such functions are decreasing and one-to-one. Their graphs lie above the x-axis and pass through the point $(0, 1)$. The graphs rise rapidly as $x \to -\infty$. As $x \to \infty$, the x-axis is a horizontal asymptote. There are no vertical asymptotes. Finally, the graphs are smooth and continuous, with no corners or gaps.

Figure 7 illustrates the graphs of two more exponential functions whose bases are between 0 and 1. Notice that the choice of a base closer to 0 results in a graph that is steeper when $x < 0$ and closer to the x-axis when $x > 0$.

SEEING THE CONCEPT   Graph $y = \left(\tfrac{1}{2}\right)^x$ and compare what you see to Figure 5. Clear the screen and graph $y = \left(\tfrac{1}{3}\right)^x$ and $y = \left(\tfrac{1}{6}\right)^x$ and compare what you see to Figure 7. Clear the screen and graph $y = \left(\tfrac{1}{10}\right)^x$ and $y = \left(\tfrac{1}{100}\right)^x$. What viewing rectangle seems to work best?   ∎

The following display summarizes the information that we have about the function $f(x) = a^x, 0 < a < 1$.

> **Properties of the Graph of an Exponential Function**
> $f(x) = a^x, 0 < a < 1$

Figure 8
$f(x) = a^x, 0 < a < 1$

1. The domain is all real numbers; the range is the set of positive real numbers.
2. There are no x-intercepts; the y-intercept is 1.
3. The x-axis $(y = 0)$ is a horizontal asymptote as $x \to \infty$.
4. $f(x) = a^x, 0 < a < 1$, is a decreasing function and is one-to-one.
5. The graph of f contains the points $(0, 1)$, $(1, a)$, and $\left(-1, \dfrac{1}{a}\right)$.
6. The graph of f is smooth and continuous, with no corners or gaps. See Figure 8.

EXAMPLE 4   ## Graphing Exponential Functions Using Transformations

Graph $f(x) = 2^{-x} - 3$ and determine the domain, range, and horizontal asymptote of f.

Solution   We begin with the graph of $y = 2^x$. Figure 9 shows the various steps.

Figure 9

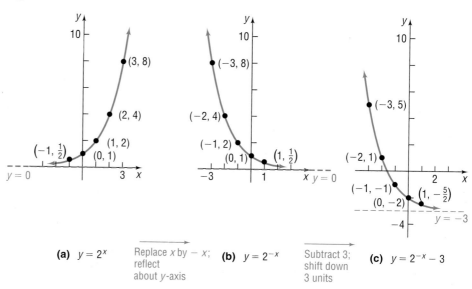

(a) $y = 2^x$ Replace x by $-x$;   **(b)** $y = 2^{-x}$ Subtract 3;   **(c)** $y = 2^{-x} - 3$
reflect   shift down
about y-axis   3 units

As Figure 9(c) illustrates, the domain of $f(x) = 2^{-x} - 3$ is the interval $(-\infty, \infty)$ and the range is the interval $(-3, \infty)$. The horizontal asymptote of f is the line $y = -3$.   ■

 NOW WORK PROBLEM 19.

THE BASE e

③   As we shall see shortly, many problems that occur in nature require the use of an exponential function whose base is a certain irrational number, symbolized by the letter e.

Let's look now at one way of arriving at this important number e.

The **number** e is defined as the number that the expression

$$\left(1 + \frac{1}{n}\right)^n \tag{2}$$

approaches as $n \to \infty$. In calculus, this is expressed using limit notation as

$$e = \lim_{n \to \infty}\left(1 + \frac{1}{n}\right)^n$$

Table 3 illustrates what happens to the defining expression (2) as n takes on increasingly large values. The last number in the last column in the table is correct to nine decimal places and is the same as the entry given for e on your calculator (if expressed correctly to nine decimal places).

TABLE 3

n	$\dfrac{1}{n}$	$1 + \dfrac{1}{n}$	$\left(1 + \dfrac{1}{n}\right)^n$
1	1	2	2
2	0.5	1.5	2.25
5	0.2	1.2	2.48832
10	0.1	1.1	2.59374246
100	0.01	1.01	2.704813829
1,000	0.001	1.001	2.716923932
10,000	0.0001	1.0001	2.718145927
100,000	0.00001	1.00001	2.718268237
1,000,000	0.000001	1.000001	2.718280469
1,000,000,000	10^{-9}	$1 + 10^{-9}$	2.718281827

The exponential function $f(x) = e^x$, whose base is the number e, occurs with such frequency in applications that it is usually referred to as *the* exponential function. Indeed, most calculators have the key* $\boxed{e^x}$ or $\boxed{\exp(x)}$, which may be used to evaluate the exponential function for a given value of x.

Now use your calculator to approximate e^x for $x = -2$, $x = -1$, $x = 0$, $x = 1$, and $x = 2$, as we have done to create Table 4. The graph of the exponential function $f(x) = e^x$ is given in Figure 10. Since $2 < e < 3$, the graph of $y = e^x$ lies between the graphs of $y = 2^x$ and $y = 3^x$. Do you see why? (Refer to Figures 1 and 3.)

TABLE 4

x	e^x
-2	0.14
-1	0.37
0	1
1	2.72
2	7.39

Figure 10
$y = e^x$

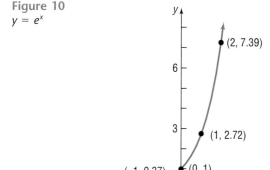

*If your calculator does not have this key, e^x is the second function of the $\boxed{\ln}$ key. You can display the number e as follows:

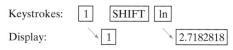

Keystrokes: $\boxed{1}$ $\boxed{\text{SHIFT}}$ $\boxed{\ln}$

Display: $\boxed{1}$ $\boxed{2.7182818}$

The reason this works will become clear in Section 7.2.

SEEING THE CONCEPT   Graph $Y_1 = e^x$ and compare what you see to Figure 10. Use eVALUEate or TABLE to verify the points on the graph shown in Figure 10. Now graph $Y_2 = 2^x$ and $Y_3 = 3^x$ on the same screen as $Y_1 = e^x$. Notice that the graph of $Y_1 = e^x$ lies between these two graphs. ▬

| EXAMPLE 5 | Graphing Exponential Functions Using Transformations |

Graph $f(x) = -e^{x-3}$ and determine the domain, range, and horizontal asymptote of f.

Solution   We begin with the graph of $y = e^x$. Figure 11 shows the various steps.

Figure 11

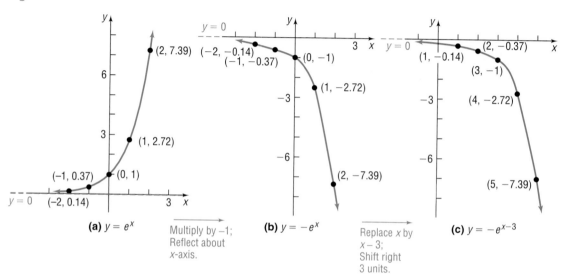

(a) $y = e^x$ Multiply by –1; Reflect about x-axis.   **(b)** $y = -e^x$ Replace x by $x - 3$; Shift right 3 units.   **(c)** $y = -e^{x-3}$

As Figure 11(c) illustrates, the domain of $f(x) = -e^{x-3}$ is the interval $(-\infty, \infty)$ and the range is the interval $(-\infty, 0)$. The horizontal asymptote is the line $y = 0$. ▬

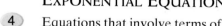   NOW WORK PROBLEM **27**.

EXPONENTIAL EQUATIONS

④ Equations that involve terms of the form $a^x, a > 0, a \neq 1$, are often referred to as **exponential equations.** Such equations can sometimes be solved by appropriately applying the Laws of Exponents and property (3).

$$\text{If } a^u = a^v, \quad \text{then } u = v \qquad (3)$$

Property (3) is a consequence of the fact that exponential functions are one-to-one. To use property (3), each side of the equality must be written with the same base.

EXAMPLE 6

Solving an Exponential Equation

Solve: $3^{x+1} = 81$

Solution

Since $81 = 3^4$, we can write the equation as

$$3^{x+1} = 81 = 3^4$$

Now we have the same base, 3, on each side, so we can apply property (3) to obtain

$$x + 1 = 4$$
$$x = 3$$ ∎

NOW WORK PROBLEM 35.

EXAMPLE 7

Solving an Exponential Equation

Solve: $e^{-x^2} = \left(e^x\right)^2 \cdot \dfrac{1}{e^3}$

Solution

We use Laws of Exponents first to get the base e on the right side.

$$\left(e^x\right)^2 \cdot \frac{1}{e^3} = e^{2x} \cdot e^{-3} = e^{2x-3}$$

As a result,

$$e^{-x^2} = e^{2x-3}$$
$$-x^2 = 2x - 3 \qquad \text{Apply property (3).}$$
$$x^2 + 2x - 3 = 0 \qquad \text{Place the quadratic equation in standard form.}$$
$$(x + 3)(x - 1) = 0 \qquad \text{Factor.}$$
$$x = -3 \quad \text{or} \quad x = 1 \qquad \text{Use the Zero-Product Property.}$$

The solution set is $\{-3, 1\}$. ∎

APPLICATION

Many applications involve the exponential functions. Let's look at one.

EXAMPLE 8

Exponential Probability

Between 9:00 PM and 10:00 PM cars arrive at Burger King's drive-thru at the rate of 12 cars per hour (0.2 car per minute). The following formula from probability can be used to determine the probability that a car will arrive within t minutes of 9:00 PM.

$$F(t) = 1 - e^{-0.2t}$$

(a) Determine the probability that a car will arrive within 5 minutes of 9 PM (that is, before 9:05 PM).

(b) Determine the probability that a car will arrive within 30 minutes of 9 PM (before 9:30 PM).

(c) What value does F approach as t becomes unbounded in the positive direction?

  (d) Graph $F(t) = 1 - e^{-0.2t}, t > 0$. Use eVALUEate or TABLE to compare the values of F at $t = 5$ [part (a)] and at $t = 30$ [part (b)].

(e) Within how many minutes of 9 PM will the probability of a car arriving equal 50%? [**Hint:** Use TRACE or TABLE].

Solution   (a) The probability that a car will arrive within 5 minutes is found by evaluating $F(t)$ at $t = 5$.

$$F(5) = 1 - e^{-0.2(5)} \approx 0.63212$$

↑
Use a calculator

We conclude that there is a 63% probability that a car will arrive within 5 minutes.

(b) The probability that a car will arrive within 30 minutes is found by evaluating $F(t)$ at $t = 30$.

$$F(30) = 1 - e^{-0.2(30)} \approx 0.9975$$

↑
Use a calculator

There is a 99.75% probability that a car will arrive within 30 minutes.

(c) As time passes, the probability that a car will arrive increases. The value that F approaches can be found by letting $t \to \infty$. Since $e^{-0.2t} = \dfrac{1}{e^{0.2t}}$, it follows that $e^{-0.2t} \to 0$ as $t \to \infty$. Thus, F approaches 1 as t gets large.

(d) See Figure 12 for the graph of F.

(e) Within 3.5 minutes of 9 PM, the probability of a car arriving equals 50%.

Figure 12

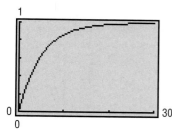

SUMMARY

Properties of the Exponential Function

$f(x) = a^x, \quad a > 1$	Domain: the interval $(-\infty, \infty)$; Range: the interval $(0, \infty)$; x-intercepts: none; y-intercept: 1; horizontal asymptote: x-axis as $x \to -\infty$; increasing; one-to-one; smooth; continuous See Figure 4 for a typical graph.
$f(x) = a^x, \quad 0 < a < 1$	Domain: the interval $(-\infty, \infty)$; Range: the interval $(0, \infty)$; x-intercepts: none; y-intercept: 1; horizontal asymptote: x-axis as $x \to \infty$; decreasing; one-to-one; smooth; continuous See Figure 8 for a typical graph.

If $a^u = a^v$, then $u = v$.

7.1 EXERCISES

In Problems 1–10, approximate each number using a calculator. Express your answer rounded to three decimal places.

1. (a) $3^{2.2}$ (b) $3^{2.23}$ (c) $3^{2.236}$ (d) $3^{\sqrt{5}}$

2. (a) $5^{1.7}$ (b) $5^{1.73}$ (c) $5^{1.732}$ (d) $5^{\sqrt{3}}$

3. (a) $2^{3.14}$ (b) $2^{3.141}$ (c) $2^{3.1415}$ (d) 2^{π}

4. (a) $2^{2.7}$ (b) $2^{2.71}$ (c) $2^{2.718}$ (d) 2^{e}

5. (a) $3.1^{2.7}$ (b) $3.14^{2.71}$ (c) $3.141^{2.718}$ (d) π^{e}

6. (a) $2.7^{3.1}$ (b) $2.71^{3.14}$ (c) $2.718^{3.141}$ (d) e^{π}

7. $e^{1.2}$ 8. $e^{-1.3}$ 9. $e^{-0.85}$ 10. $e^{2.1}$

In Problems 11–18, the graph of an exponential function is given. Match each graph to one of the following functions.

A. $y = 3^x$ B. $y = 3^{-x}$ C. $y = -3^x$ D. $y = -3^{-x}$

E. $y = 3^x - 1$ F. $y = 3^{x-1}$ G. $y = 3^{1-x}$ H. $y = 1 - 3^x$

11.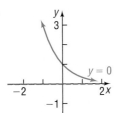

12.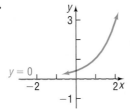

13.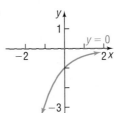

14.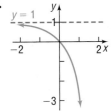

15.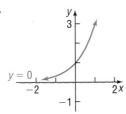

16.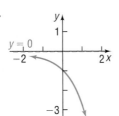

17.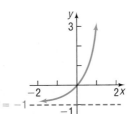

18.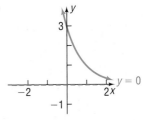

In Problems 19–26, use transformations to graph each function. Determine the domain, range, and horizontal asymptote of each function.

19. $f(x) = 2^x + 1$ **20.** $f(x) = 2^{x+2}$ **21.** $f(x) = 3^{-x} - 2$ **22.** $f(x) = -3^x + 1$

23. $f(x) = 2 + 3(4^x)$ **24.** $f(x) = 1 - 3(2^x)$ **25.** $f(x) = 2 + 3^{x/2}$ **26.** $f(x) = 1 - 2^{-x/3}$

In Problems 27–34, begin with the graph of $y = e^x$ (Figure 20) and use transformations to graph each function. Determine the domain, range, and horizontal asymptote of each function.

27. $f(x) = e^{-x}$ **28.** $f(x) = -e^x$ **29.** $f(x) = e^{x+2}$ **30.** $f(x) = e^x - 1$

31. $f(x) = 5 - e^{-x}$ **32.** $f(x) = 9 - 3e^{-x}$ **33.** $f(x) = 2 - e^{-x/2}$ **34.** $f(x) = 7 - 3e^{2x}$

In Problems 35–48, solve each equation.

35. $2^{2x+1} = 4$ **36.** $5^{1-2x} = \dfrac{1}{5}$ **37.** $3^{x^3} = 9^x$ **38.** $4^{x^2} = 2^x$

39. $8^{x^2-2x} = \dfrac{1}{2}$ **40.** $9^{-x} = \dfrac{1}{3}$ **41.** $2^x \cdot 8^{-x} = 4^x$ **42.** $\left(\dfrac{1}{2}\right)^{1-x} = 4$

43. $\left(\dfrac{1}{5}\right)^{2-x} = 25$ **44.** $4^x - 2^x = 0$ **45.** $4^x = 8$ **46.** $9^{2x} = 27$

47. $e^{x^2} = \left(e^{3x}\right) \cdot \dfrac{1}{e^2}$ **48.** $\left(e^4\right)^x \cdot e^{x^2} = e^{12}$

49. If $4^x = 7$, what does 4^{-2x} equal?   **50.** If $2^x = 3$, what does 4^{-x} equal?

51. If $3^{-x} = 2$, what does 3^{2x} equal?   **52.** If $5^{-x} = 3$, what does 5^{3x} equal?

In Problems 53–56, graph each function f. Based on the graph, state the domain, range, and intercepts, if any, of f.

53. $f(x) = \begin{cases} e^{-x} & \text{if } x < 0 \\ e^x & \text{if } x \geq 0 \end{cases}$

54. $f(x) = \begin{cases} e^x & \text{if } x < 0 \\ e^{-x} & \text{if } x \geq 0 \end{cases}$

55. $f(x) = \begin{cases} -e^x & \text{if } x < 0 \\ -e^{-x} & \text{if } x \geq 0 \end{cases}$

56. $f(x) = \begin{cases} -e^{-x} & \text{if } x < 0 \\ -e^x & \text{if } x \geq 0 \end{cases}$

---

57. Optics If a single pane of glass obliterates 3% of the light passing through it, then the percent p of light that passes through n successive panes is given approximately by the function

$$p(n) = 100e^{-0.03n}$$

(a) What percent of light will pass through 10 panes?
(b) What percent of light will pass through 25 panes?

58. Atmospheric Pressure The atmospheric pressure p on a balloon or plane decreases with increasing height. This pressure, measured in millimeters of mercury, is related to the number of kilometers h above sea level by the function

$$p(h) = 760e^{-0.145h}$$

(a) Find the atmospheric pressure at a height of 2 kilometers (over 1 mile).
(b) What is it at a height of 10 kilometers (over 30,000 feet)?

59. Space Satellites The number of watts w provided by a space satellite's power supply after a period of d days is given by the function

$$w(d) = 50e^{-0.004d}$$

(a) How much power will be available after 30 days?
(b) How much power will be available after 1 year (365 days)?

60. Healing of Wounds The normal healing of wounds can be modeled by an exponential function. If A_0 represents the original area of the wound and if A equals the area of the wound after n days, then the function

$$A(n) = A_0e^{-0.35n}$$

describes the area of a wound on the nth day following an injury when no infection is present to retard the healing. Suppose that a wound initially had an area of 100 square millimeters.

(a) If healing is taking place, how large will the area of the wound be after 3 days?
(b) How large will it be after 10 days?

61. Drug Medication The function

$$D(h) = 5e^{-0.4h}$$

can be used to find the number of milligrams D of a certain drug that is in a patient's bloodstream h hours after the drug has been administered. How many milligrams will be present after 1 hour? After 6 hours?

62. Spreading of Rumors A model for the number of people N in a college community who have heard a certain rumor is

$$N = P(1 - e^{-0.15d})$$

where P is the total population of the community and d is the number of days that have elapsed since the rumor began. In a community of 1000 students, how many students will have heard the rumor after 3 days?

63. Exponential Probability Between 12:00 PM and 1:00 PM, cars arrive at Citibank's drive-thru at the rate of 6 cars per hour (0.1 car per minute). The following formula from probability can be used to determine the probability that a car will arrive within t minutes of 12:00 PM:

$$F(t) = 1 - e^{-0.1t}$$

(a) Determine the probability that a car will arrive within 10 minutes of 12:00 PM (that is, before 12:10 PM).
(b) Determine the probability that a car will arrive within 40 minutes of 12:00 PM (before 12:40 PM).
(c) What value does F approach as t becomes unbounded in the positive direction?
(d) Graph F using your graphing utility.
(e) Using TRACE, determine how many minutes are needed for the probability to reach 50%?

64. Exponential Probability Between 5:00 PM and 6:00 PM, cars arrive at Jiffy Lube at the rate of 9 cars per hour (0.15 car per minute). The following formula from probability can be used to determine the probability that a car will arrive within t minutes of 5:00 PM:

$$F(t) = 1 - e^{-0.15t}$$

(a) Determine the probability that a car will arrive within 15 minutes of 5:00 PM (that is, before 5:15 PM).
(b) Determine the probability that a car will arrive within 30 minutes of 5:00 PM (before 5:30 PM).
(c) What value does F approach as t becomes unbounded in the positive direction?
(d) Graph F using your graphing utility.
(e) Using TRACE, determine how many minutes are needed for the probability to reach 60%?

65. Poisson Probability Between 5:00 PM and 6:00 PM, cars arrive at McDonald's drive-thru at the rate of 20 cars per hour. The following formula from probability can be used to determine the probability that x cars will arrive between 5:00 PM and 6:00 PM.

$$P(x) = \frac{20^x e^{-20}}{x!}$$

where

$$x! = x \cdot (x-1) \cdot (x-2) \cdot \cdots \cdot 3 \cdot 2 \cdot 1$$

(a) Determine the probability that $x = 15$ cars will arrive between 5:00 PM and 6:00 PM.

(b) Determine the probability that $x = 20$ cars will arrive between 5:00 PM and 6:00 PM.

66. Poisson Probability People enter a line for the *Demon Roller Coaster* at the rate of 4 per minute. The following formula from probability can be used to determine the probability that x people will arrive within the next minute.

$$P(x) = \frac{4^x e^{-4}}{x!}$$

where

$$x! = x \cdot (x-1) \cdot (x-2) \cdot \cdots \cdot 3 \cdot 2 \cdot 1$$

(a) Determine the probability that $x = 5$ people will arrive within the next minute.

(b) Determine the probability that $x = 8$ people will arrive within the next minute.

67. Relative Humidity The relative humidity is the ratio (expressed as a percent) of the amount of water vapor in the air to the maximum amount that the air can hold at a specific temperature. The relative humidity, R, is found using the following formula:

$$R = 10^{\left(\frac{2345}{T} - \frac{2345}{D} + 2\right)}$$

where T is the air temperature (in Kelvins) and D is the dew point temperature (in Kelvins).

[**NOTE:** Kelvins are found by adding 273 to Celsius degrees.]

(a) Determine the relative humidity if the air temperature is 10° Celsius and the dew point temperature is 5° Celsius.

(b) Determine the relative humidity if the air temperature is 20° Celsius and the dew point temperature is 15° Celsius.

(c) What is the relative humidity if the air temperature and the dew point temperature are the same?

68. Learning Curve Suppose that a student has 500 vocabulary words to learn. If the student learns 15 words after 5 minutes, the function

$$L(t) = 500\left(1 - e^{-0.0061t}\right)$$

approximates the number of words L that the student will learn after t minutes.

(a) How many words will the student learn after 30 minutes?

(b) How many words will the student learn after 60 minutes?

69. Alternating Current in a RL Circuit The equation governing the amount of current I (in amperes) after time t (in seconds) in a single RL circuit consisting of a resistance R (in ohms), an inductance L (in henrys), and an electromotive force E (in volts) is

$$I = \frac{E}{R}\left[1 - e^{-(R/L)t}\right]$$

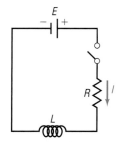

(a) If $E = 120$ volts, $R = 10$ ohms, and $L = 5$ henrys, how much current I_1 is flowing after 0.3 second? After 0.5 second? After 1 second?

(b) What is the maximum current?

(c) Graph this function $I = I_1(t)$, measuring I along the y-axis and t along the x-axis.

(d) If $E = 120$ volts, $R = 5$ ohms, and $L = 10$ henrys, how much current I_2 is flowing after 0.3 second? After 0.5 second? After 1 second?

(e) What is the maximum current?

(f) Graph this function $I = I_2(t)$ on the same coordinate axes as $I_1(t)$.

70. Alternating Current in a RC Circuit The equation governing the amount of current I (in amperes) after time t (in microseconds) in a single RC circuit consisting of a resistance R (in ohms), a capacitance C (in microfarads), and an electromotive force E (in volts) is

$$I = \frac{E}{R} e^{-t/(RC)}$$

(a) If $E = 120$ volts, $R = 2000$ ohms, and $C = 1.0$ microfarad, how much current I_1 is flowing initially ($t = 0$)? After 1000 microseconds? After 3000 microseconds?

(b) What is the maximum current?
(c) Graph this function $I = I_1(t)$, measuring I along the y-axis and t along the x-axis.
(d) If $E = 120$ volts, $R = 1000$ ohms, and $C = 2.0$ microfarads, how much current I_2 is flowing initially? After 1000 microseconds? After 3000 microseconds?
(e) What is the maximum current?
(f) Graph this function $I = I_2(t)$ on the same coordinate axes as $I_1(t)$.

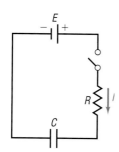

71. Another Formula for e Use a calculator to compute the values of

$$2 + \frac{1}{2!} + \frac{1}{3!} + \cdots + \frac{1}{n!}$$

for $n = 4, 6, 8,$ and 10. Compare each result with e.
[**Hint:** $1! = 1, 2! = 2 \cdot 1, 3! = 3 \cdot 2 \cdot 1,$
$n! = n(n - 1) \cdot \cdots \cdot (3)(2)(1)$]

72. Another Formula for e Use a calculator to compute the various values of the expression. Compare the values to e.

$$2 + \cfrac{1}{1 + \cfrac{1}{2 + \cfrac{2}{3 + \cfrac{3}{4 + \cfrac{4}{\text{etc.}}}}}}$$

73. Difference Quotient If $f(x) = a^x$, show that

$$\frac{f(x + h) - f(x)}{h} = a^x \left(\frac{a^h - 1}{h} \right)$$

74. If $f(x) = a^x$, show that $f(A + B) = f(A) \cdot f(B)$.

75. If $f(x) = a^x$, show that $f(-x) = \dfrac{1}{f(x)}$.

76. If $f(x) = a^x$, show that $f(\alpha x) = [f(x)]^\alpha$.

77. The *Challenger* Disaster* After the *Challenger* disaster in 1986, a study of the 23 launches that preceded the fatal flight was made. A mathematical model was developed involving the relationship between the Fahrenheit temperature x around the O-rings and the number y of eroded or leaky primary O-rings. The model stated that

$$y = \frac{6}{1 + e^{-(5.085 - 0.1156x)}}$$

where the number 6 indicates the 6 primary O-rings on the spacecraft.

(a) What is the predicted number of eroded or leaky primary O-rings at a temperature of 100°F?
(b) What is the predicted number of eroded or leaky primary O-rings at a temperature of 60°F?
(c) What is the predicted number of eroded or leaky primary O-rings at a temperature of 30°F?
(d) Graph the equation and TRACE. At what temperature is the predicted number of eroded or leaky O-rings 1? 3? 5?

78. Historical Problem Pierre de Fermat (1601–1665) conjectured that the function

$$f(x) = 2^{(2^x)} + 1$$

for $x = 1, 2, 3, \ldots$, would always have a value equal to a prime number. But Leonhard Euler (1707–1783) showed that this formula fails for $x = 5$. Use a calculator to determine the prime numbers produced by f for $x = 1, 2, 3, 4$. Then show that $f(5) = 641 \times 6,700,417$, which is not prime.

79. The bacteria in a 4-liter container double every minute. After 60 minutes the container is full. How long did it take to fill half the container?

80. Explain in your own words what the number e is. Provide at least two applications that require the use of this number.

81. Do you think that there is a power function that increases more rapidly than an exponential function whose base is greater than 1? Explain.

*Linda Tappin, "Analyzing Data Relating to the *Challenger* Disaster," *Mathematics Teacher*, Vol. 87, No. 6, September 1994, pp. 423–426.

PREPARING FOR THIS SECTION

Before getting started, review the following:

✓ Solving Inequalities
(Appendix A, Section A.5, pp. 524–525)

✓ One-to-One Functions; Inverse Functions
(Section 1.6, pp. 60–69)

7.2 LOGARITHMIC FUNCTIONS

OBJECTIVES
1. Change Exponential Expressions to Logarithmic Expressions
2. Change Logarithmic Expressions to Exponential Expressions
3. Evaluate Logarithmic Functions
4. Determine the Domain of a Logarithmic Function
5. Graph Logarithmic Functions
6. Solve Logarithmic Equations

Recall that a one-to-one function $y = f(x)$ has an inverse function that is defined implicitly by the equation $x = f(y)$. In particular, the exponential function $y = f(x) = a^x$, $a > 0$, $a \neq 1$, is one-to-one and hence has an inverse function that is defined implicitly by the equation

$$x = a^y, \qquad a > 0, \quad a \neq 1$$

This inverse function is so important that it is given a name, the *logarithmic function*.

> The **logarithmic function to the base a,** where $a > 0$ and $a \neq 1$, is denoted by $y = \log_a x$ (read as "y is the logarithm to the base a of x") and is defined by
>
> $$y = \log_a x \quad \text{if and only if} \quad x = a^y$$
>
> The domain of the logarithmic function $y = \log_a x$ is $x > 0$.

A *logarithm* is merely a name for a certain exponent.

EXAMPLE 1 | Relating Logarithms to Exponents

(a) If $y = \log_3 x$, then $x = 3^y$. For example, $2 = \log_3 9$ is equivalent to $9 = 3^2$.

(b) If $y = \log_5 x$, then $x = 5^y$. For example, $-1 = \log_5\left(\dfrac{1}{5}\right)$ is equivalent to $\dfrac{1}{5} = 5^{-1}$. ∎

EXAMPLE 2 | Changing Exponential Expressions to Logarithmic Expressions

1. Change each exponential expression to an equivalent expression involving a logarithm.

(a) $1.2^3 = m$ (b) $e^b = 9$ (c) $a^4 = 24$

Solution   We use the fact that $y = \log_a x$ and $x = a^y, a > 0, a \neq 1$, are equivalent.

(a) If $1.2^3 = m$, then $3 = \log_{1.2} m$.   (b) If $e^b = 9$, then $b = \log_e 9$.

(c) If $a^4 = 24$, then $4 = \log_a 24$.   ■

▬▬▬▬▬► NOW WORK PROBLEM **1**.

EXAMPLE 3   **Changing Logarithmic Expressions to Exponential Expressions**

② Change each logarithmic expression to an equivalent expression involving an exponent.

(a) $\log_a 4 = 5$ (b) $\log_e b = -3$ (c) $\log_3 5 = c$

Solution   (a) If $\log_a 4 = 5$, then $a^5 = 4$.   (b) If $\log_e b = -3$, then $e^{-3} = b$.

(c) If $\log_3 5 = c$, then $3^c = 5$.   ■

▬▬▬▬▬► NOW WORK PROBLEM **13**.

③ To find the exact value of a logarithm, we write the logarithm in exponential notation and use the fact that if $a^u = a^v$ then $u = v$.

EXAMPLE 4   **Finding the Exact Value of a Logarithmic Expression**

Find the exact value of

(a) $\log_2 16$ (b) $\log_3 \dfrac{1}{27}$

Solution   (a)

$$y = \log_2 16$$
$$2^y = 16 \qquad \text{Change to exponential form.}$$
$$2^y = 2^4 \qquad 16 = 2^4$$
$$y = 4 \qquad \text{Equate exponents.}$$

Therefore, $\log_2 16 = 4$.

(b)

$$y = \log_3 \frac{1}{27}$$

$$3^y = \frac{1}{27} \qquad \text{Change to exponential form.}$$

$$3^y = 3^{-3} \qquad \frac{1}{27} = \frac{1}{3^3} = 3^{-3}$$

$$y = -3 \qquad \text{Equate exponents.}$$

Therefore, $\log_3 \dfrac{1}{27} = -3$.   ■

▬▬▬▬▬► NOW WORK PROBLEM **25**.

DOMAIN OF A LOGARITHMIC FUNCTION

④ The logarithmic function $y = \log_a x$ has been defined as the inverse of the exponential function $y = a^x$. That is, if $f(x) = a^x$, then $f^{-1}(x) = \log_a x$. Based on the discussion given in Section 1.6 on inverse functions, we know that, for a function f and its inverse f^{-1},

Domain of f^{-1} = Range of f and   Range of f^{-1} = Domain of f

Consequently, it follows that

> Domain of logarithmic function = Range of exponential function = $(0, \infty)$
>
> Range of logarithmic function = Domain of exponential function = $(-\infty, \infty)$

In the next box, we summarize some properties of the logarithmic function:

> $y = \log_a x$ (defining equation: $x = a^y$)
>
> Domain: $0 < x < \infty$ Range: $-\infty < y < \infty$

The domain of a logarithmic function consists of the *positive* real numbers, so the argument of a logarithmic function must be greater than zero.

EXAMPLE 5   **Finding the Domain of a Logarithmic Function**

Find the domain of each logarithmic function.

(a) $F(x) = \log_2(1 - x)$ (b) $g(x) = \log_5\left(\dfrac{1 + x}{1 - x}\right)$

(c) $h(x) = \log_{1/2}|x|$

Solution (a) The domain of F consists of all x for which $1 - x > 0$, that is, all $x < 1$, or using interval notation, $(-\infty, 1)$.

(b) The domain of g is restricted to

$$\frac{1 + x}{1 - x} > 0$$

Solving this inequality, we find that the domain of g consists of all x between -1 and 1, that is, $-1 < x < 1$, or using interval notation, $(-1, 1)$.

(c) Since $|x| > 0$, provided that $x \neq 0$, the domain of h consists of all nonzero real numbers, or using interval notation, $(-\infty, 0)$ or $(0, \infty)$. ∎

NOW WORK PROBLEMS **39** AND **45**.

GRAPHS OF LOGARITHMIC FUNCTIONS

⑤ Since exponential functions and logarithmic functions are inverses of each other, the graph of a logarithmic function $y = \log_a x$ is the reflection about the line $y = x$ of the graph of the exponential function $y = a^x$, as shown in Figure 13.

Figure 13

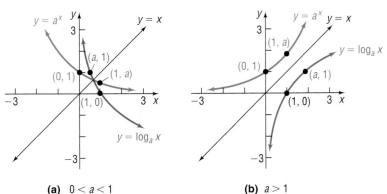

(a) $0 < a < 1$ (b) $a > 1$

Properties of the Graph of a Logarithmic Function $f(x) = \log_a x$

1. The domain is the set of positive real numbers; the range is all real numbers.
2. The x-intercept of the graph is 1. There is no y-intercept.
3. The y-axis ($x = 0$) is a vertical asymptote of the graph.
4. A logarithmic function is decreasing if $0 < a < 1$ and increasing if $a > 1$.
5. The graph of f contains the points $(1, 0)$, $(a, 1)$, and $\left(\dfrac{1}{a}, -1\right)$.
6. The graph is smooth and continuous, with no corners or gaps.

If the base of a logarithmic function is the number e, then we have the **natural logarithm function.** This function occurs so frequently in applications that it is given a special symbol, **ln** (from the Latin, *logarithmus naturalis*). Thus,

$$y = \log_e x = \ln x \quad \text{if and only if} \quad x = e^y \qquad \textbf{(1)}$$

Since $y = \ln x$ and the exponential function $y = e^x$ are inverse functions, we can obtain the graph of $y = \ln x$ by reflecting the graph of $y = e^x$ about the line $y = x$. See Figure 14.

Figure 14

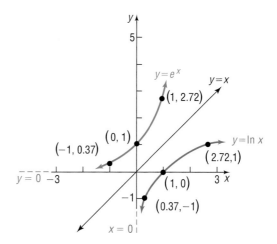

Using a calculator with an $\boxed{\text{In}}$ key, we can obtain other points on the graph of $f(x) = \ln x$. See Table 5.

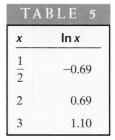

TABLE 5

x	$\ln x$
$\dfrac{1}{2}$	-0.69
2	0.69
3	1.10

 SEEING THE CONCEPT   Graph $Y_1 = e^x$ and $Y_2 = \ln x$ on the same square screen. Use eVALUEate to verify the points on the graph given in Figure 14. Do you see the symmetry of the two graphs with respect to the line $y = x$?

EXAMPLE 6   **Graphing Logarithmic Functions Using Transformations**

Graph $f(x) = -\ln(x + 2)$ by starting with the graph of $y = \ln x$. Determine the domain, range, and vertical asymptote of f.

Solution   The domain of f consists of all x for which

$$x + 2 > 0 \quad \text{or} \quad x > -2$$

To obtain the graph of $y = -\ln(x + 2)$, we use the steps illustrated in Figure 15.

Figure 15

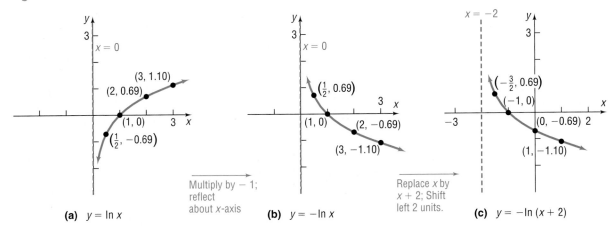

(a) $y = \ln x$ **(b)** $y = -\ln x$ **(c)** $y = -\ln(x + 2)$

The range of $f(x) = -\ln(x + 2)$ is the interval $(-\infty, \infty)$, and the vertical asymptote is $x = -2$. [Do you see why? The original asymptote $(x = 0)$ is shifted to the left 2 units.] ∎

EXAMPLE 7   **Graphing Logarithmic Functions Using Transformations**

Graph $f(x) = \ln(1 - x)$. Determine the domain, range, and vertical asymptote of f.

Solution   The domain of f consists of all x for which

$$1 - x > 0 \quad \text{or} \quad x < 1$$

To obtain the graph of $y = \ln(1 - x)$, we use the steps illustrated in Figure 16.

Figure 16

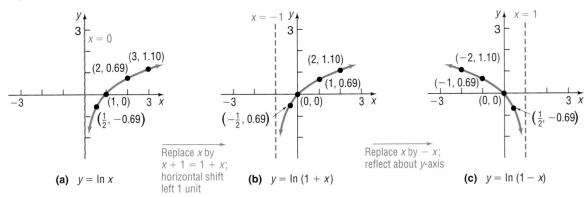

(a) $y = \ln x$ **(b)** $y = \ln(1 + x)$ **(c)** $y = \ln(1 - x)$

The range of $f(x) = \ln(1 - x)$ is the interval $(-\infty, \infty)$, and the vertical asymptote is $x = 1$. ∎

NOW WORK PROBLEM **61.**

LOGARITHMIC EQUATIONS

⑥ Equations that contain logarithms are called **logarithmic equations.** Care must be taken when solving logarithmic equations algebraically. Be sure to check each apparent solution in the original equation and discard any that are extraneous. In the expression $\log_a M$, remember that a and M are positive and $a \neq 1$.

    Some logarithmic equations can be solved by changing from a logarithmic expression to an exponential expression.

EXAMPLE 8

Solving a Logarithmic Equation

Solve: (a) $\log_3(4x - 7) = 2$ (b) $\log_x 64 = 2$

Solution   (a) We can obtain an exact solution by changing the logarithm to exponential form.

$$\log_3(4x - 7) = 2$$
$$4x - 7 = 3^2 \qquad \text{Change to exponential form.}$$
$$4x - 7 = 9$$
$$4x = 16$$
$$x = 4$$

(b) We can obtain an exact solution by changing the logarithm to exponential form.

$$\log_x 64 = 2$$
$$x^2 = 64 \qquad \text{Change to exponential form.}$$
$$x = \pm\sqrt{64} = \pm 8$$

The base of a logarithm is always positive. As a result, we discard -8; the only solution is 8. ∎

EXAMPLE 9

Using Logarithms to Solve Exponential Equations

Solve: $e^{2x} = 5$

Solution We can obtain an exact solution by changing the exponential equation to logarithmic form.

$$e^{2x} = 5$$
$$\ln 5 = 2x \qquad \text{Change to a logarithmic expression using (1).}$$
$$x = \frac{\ln 5}{2} \approx 0.805$$

∎

       ✏️   NOW WORK PROBLEMS **73** AND **85.**

EXAMPLE 10

Alcohol and Driving

The concentration of alcohol in a person's blood is measurable. Recent medical research suggests that the risk R (given as a percent) of having an accident while driving a car can be modeled by the equation

$$R = 6e^{kx}$$

where x is the variable concentration of alcohol in the blood and k is a constant.

(a) Suppose that a concentration of alcohol in the blood of 0.04 results in a 10% risk $(R = 10)$ of an accident. Find the constant k in the equation.

(b) Using this value of k, what is the risk if the concentration is 0.17?

(c) Using the same value of k, what concentration of alcohol corresponds to a risk of 100%?

(d) If the law asserts that anyone with a risk of having an accident of 20% or more should not have driving privileges, at what concentration of alcohol in the blood should a driver be arrested and charged with a DUI (Driving Under the Influence)?

Solution (a) For a concentration of alcohol in the blood of 0.04 and a risk of 10%, we let $x = 0.04$ and $R = 10$ in the equation and solve for k.

$$R = 6e^{kx}$$
$$10 = 6e^{k(0.04)}$$
$$\frac{10}{6} = e^{0.04k} \qquad \text{Divide both sides by 6.}$$
$$0.04k = \ln\frac{10}{6} \qquad \text{Change to a logarithmic expression.}$$
$$k = \frac{\ln(10/6)}{0.04} \qquad \text{Solve for } k.$$
$$k \approx 12.77$$

(b) Using $k = 12.77$ and $x = 0.17$ in the equation, we find the risk R to be

$$R = 6e^{kx} = 6e^{(12.77)(0.17)} = 52.6$$

For a concentration of alcohol in the blood of 0.17, the risk of an accident is about 52.6%.

(c) Using $k = 12.77$ and $R = 100$ in the equation, we find the concentration x of alcohol in the blood to be

$$R = 6e^{kx}$$
$$100 = 6e^{12.77x}$$
$$\frac{100}{6} = e^{12.77x} \qquad \text{Divide both sides by 6.}$$
$$12.77x = \ln\frac{100}{6} \qquad \text{Change to a logarithmic expression.}$$
$$x = \frac{\ln(100/6)}{12.77} \qquad \text{Solve for } x.$$
$$x \approx 0.22$$

For a concentration of alcohol in the blood of 0.22, the risk of an accident is 100%.

(d) Using $k = 12.77$ and $R = 20$ in the equation, we find the concentration x of alcohol in the blood to be

$$R = 6e^{kx}$$
$$20 = 6e^{12.77x}$$
$$\frac{20}{6} = e^{12.77x}$$
$$12.77x = \ln\frac{20}{6}$$
$$x = \frac{\ln(20/6)}{12.77}$$
$$x \approx 0.094$$

A driver with a concentration of alcohol in the blood of 0.094 or more should be arrested and charged with DUI.   ∎

[**NOTE:**   Most states use 0.08 or 0.10 as the blood alcohol content at which a DUI citation is given.]

SUMMARY

Properties of the Logarithmic Function

$f(x) = \log_a x, \quad a > 1$
$(y = \log_a x$ means $x = a^y)$

Domain: the interval $(0, \infty)$; Range: the interval $(-\infty, \infty)$;
x-intercept: 1; y-intercept: none; vertical asymptote: $x = 0$ (y-axis);
increasing; one-to-one

See Figure 17(a) for a typical graph.

$f(x) = \log_a x, \quad 0 < a < 1$
$(y = \log_a x$ means $x = a^y)$

Domain: the interval $(0, \infty)$; Range: the interval $(-\infty, \infty)$;
x-intercept: 1; y-intercept: none; vertical asymptote: $x = 0$ (y-axis);
decreasing; one-to-one

See Figure 17(b) for a typical graph.

Figure 17

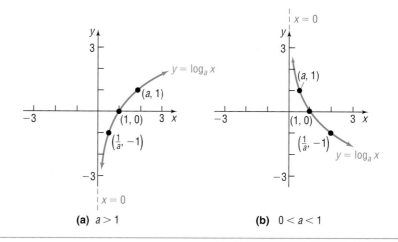

(a) $a > 1$ **(b)** $0 < a < 1$

7.2 EXERCISES

In Problems 1–12, change each exponential expression to an equivalent expression involving a logarithm.

1. $9 = 3^2$ **2.** $16 = 4^2$ **3.** $a^2 = 1.6$ **4.** $a^3 = 2.1$

5. $1.1^2 = M$ **6.** $2.2^3 = N$ **7.** $2^x = 7.2$ **8.** $3^x = 4.6$

9. $x^{\sqrt{2}} = \pi$ **10.** $x^\pi = e$ **11.** $e^x = 8$ **12.** $e^{2.2} = M$

In Problems 13–24, change each logarithmic expression to an equivalent expression involving an exponent.

13. $\log_2 8 = 3$

14. $\log_3 \left(\dfrac{1}{9} \right) = -2$

15. $\log_a 3 = 6$

16. $\log_b 4 = 2$

17. $\log_3 2 = x$

18. $\log_2 6 = x$

19. $\log_2 M = 1.3$

20. $\log_3 N = 2.1$

21. $\log_{\sqrt{2}} \pi = x$

22. $\log_\pi x = \frac{1}{2}$

23. $\ln 4 = x$

24. $\ln x = 4$

In Problems 25–36, find the exact value of each logarithm without using a calculator.

25. $\log_2 1$

26. $\log_8 8$

27. $\log_5 25$

28. $\log_3 \left(\dfrac{1}{9} \right)$

29. $\log_{1/2} 16$

30. $\log_{1/3} 9$

31. $\log_{10} \sqrt{10}$

32. $\log_5 \sqrt[3]{25}$

33. $\log_{\sqrt{2}} 4$

34. $\log_{\sqrt{3}} 9$

35. $\ln \sqrt{e}$

36. $\ln e^3$

In Problems 37–46, find the domain of each function.

37. $f(x) = \ln(x - 3)$

38. $g(x) = \ln(x - 1)$

39. $F(x) = \log_2 x^2$

40. $H(x) = \log_5 x^3$

41. $h(x) = \log_{1/2}(x^2 - 2x + 1)$

42. $G(x) = \log_{1/2}(x^2 - 1)$

43. $f(x) = \ln \left(\dfrac{1}{x + 1} \right)$

44. $g(x) = \ln \left(\dfrac{1}{x - 5} \right)$

45. $g(x) = \log_5 \left(\dfrac{x + 1}{x} \right)$

46. $h(x) = \log_3 \left(\dfrac{x}{x - 1} \right)$

In Problems 47–50, use a calculator to evaluate each expression. Round your answer to three decimal places.

47. $\ln \dfrac{5}{3}$

48. $\dfrac{\ln 5}{3}$

49. $\dfrac{\ln \left(\dfrac{10}{3} \right)}{0.04}$

50. $\dfrac{\ln \left(\dfrac{2}{3} \right)}{-0.1}$

51. Find a so that the graph of $f(x) = \log_a x$ contains the point $(2, 2)$.

52. Find a so that the graph of $f(x) = \log_a x$ contains the point $\left(\frac{1}{2}, -4 \right)$.

In Problems 53–60, the graph of a logarithmic function is given. Match each graph to one of the following functions:

A. $y = \log_3 x$

B. $y = \log_3(-x)$

C. $y = -\log_3 x$

D. $y = -\log_3(-x)$

E. $y = \log_3 x - 1$

F. $y = \log_3(x - 1)$

G. $y = \log_3(1 - x)$

H. $y = 1 - \log_3 x$

53.

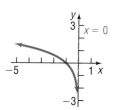

54.

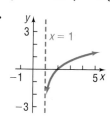

55.

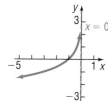

56.

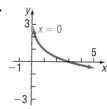

57.

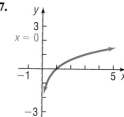

58.

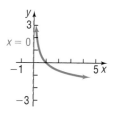

59.

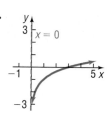

60.

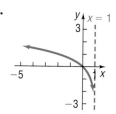

In Problems 61–72, use transformations to graph each function. Determine the domain, range, and vertical asymptote of each function.

61. $f(x) = \ln(x + 4)$ **62.** $f(x) = \ln(x - 3)$ **63.** $f(x) = \ln(-x)$ **64.** $f(x) = -\ln(-x)$

65. $g(x) = \ln(2x)$ **66.** $h(x) = \ln\left(\dfrac{1}{2}x\right)$ **67.** $f(x) = 3\ln x$ **68.** $f(x) = -2\ln x$

69. $g(x) = \ln(3 - x)$ **70.** $h(x) = \ln(4 - x)$ **71.** $f(x) = -\ln(x - 1)$ **72.** $f(x) = 2 - \ln x$

In Problems 73–92, solve each equation.

73. $\log_3 x = 2$ **74.** $\log_5 x = 3$ **75.** $\log_2(2x + 1) = 3$ **76.** $\log_3(3x - 2) = 2$

77. $\log_x 4 = 2$ **78.** $\log_x\left(\dfrac{1}{8}\right) = 3$ **79.** $\ln e^x = 5$ **80.** $\ln e^{-2x} = 8$

81. $\log_4 64 = x$ **82.** $\log_5 625 = x$ **83.** $\log_3 243 = 2x + 1$ **84.** $\log_6 36 = 5x + 3$

85. $e^{3x} = 10$ **86.** $e^{-2x} = \dfrac{1}{3}$ **87.** $e^{2x+5} = 8$ **88.** $e^{-2x+1} = 13$

89. $\log_3(x^2 + 1) = 2$ **90.** $\log_5(x^2 + x + 4) = 2$ **91.** $\log_2 8^x = -3$ **92.** $\log_3 3^x = -1$

In Problems 93–96, graph each function f. Based on the graph, state the domain, range, and intercepts, if any, of f.

93. $f(x) = \begin{cases} \ln(-x) & \text{if } x < 0 \\ \ln x & \text{if } x > 0 \end{cases}$

94. $f(x) = \begin{cases} \ln(-x) & \text{if } x \le -1 \\ -\ln(-x) & \text{if } -1 < x < 0 \end{cases}$

95. $f(x) = \begin{cases} -\ln x & \text{if } 0 < x < 1 \\ \ln x & \text{if } x \ge 1 \end{cases}$

96. $f(x) = \begin{cases} \ln x & \text{if } 0 < x < 1 \\ -\ln x & \text{if } x \ge 1 \end{cases}$

---

97. Optics   If a single pane of glass obliterates 10% of the light passing through it, then the percent P of light that passes through n successive panes is given approximately by the equation

$$P = 100e^{-0.1n}$$

(a) How many panes are necessary to block at least 50% of the light?

(b) How many panes are necessary to block at least 75% of the light?

98. Chemistry   The pH of a chemical solution is given by the formula

$$\text{pH} = -\log_{10}[\text{H}^+]$$

where $[\text{H}^+]$ is the concentration of hydrogen ions in moles per liter. Values of pH range from 0 (acidic) to 14 (alkaline).

(a) Find the pH of a 1-liter container of water with 0.0000001 mole of hydrogen ion.

(b) Find the hydrogen ion concentration of a mildly acidic solution with a pH of 4.2.

99. Space Satellites   The number of watts w provided by a space satellite's power supply after d days is given by the formula

$$w = 50e^{-0.004d}$$

(a) How long will it take for the available power to drop to 30 watts?

(b) How long will it take for the available power to drop to only 5 watts?

100. Healing of Wounds   The normal healing of wounds can be modeled by an exponential function. If A_0 represents the original area of the wound and if A equals the area of the wound after n days, then the formula

$$A = A_0 e^{-0.35n}$$

describes the area of a wound on the nth day following an injury when no infection is present to retard the healing. Suppose that a wound initially had an area of 100 square millimeters.

(a) If healing is taking place, how many days will pass before the wound is one-half its original size?

(b) How long before the wound is 10% of its original size?

101. Exponential Probability Between 12:00 PM and 1:00 PM, cars arrive at Citibank's drive-thru at the rate of 6 cars per hour (0.1 car per minute). The following formula from probability can be used to determine the probability that a car will arrive within t minutes of 12:00 PM.

$$F(t) = 1 - e^{-0.1t}$$

(a) Determine how many minutes are needed for the probability to reach 50%.
(b) Determine how many minutes are needed for the probability to reach 80%.
(c) Is it possible for the probability to equal 100%? Explain.

102. Exponential Probability Between 5:00 PM and 6:00 PM, cars arrive at Jiffy Lube at the rate of 9 cars per hour (0.15 car per minute). The following formula from probability can be used to determine the probability that a car will arrive within t minutes of 5:00 PM.

$$F(t) = 1 - e^{-0.15t}$$

(a) Determine how many minutes are needed for the probability to reach 50%.
(b) Determine how many minutes are needed for the probability to reach 80%.

103. Drug Medication The formula

$$D = 5e^{-0.4h}$$

can be used to find the number of milligrams D of a certain drug that is in a patient's bloodstream h hours after the drug has been administered. When the number of milligrams reaches 2, the drug is to be administered again. What is the time between injections?

104. Spreading of Rumors A model for the number of people N in a college community who have heard a certain rumor is

$$N = P(1 - e^{-0.15d})$$

where P is the total population of the community and d is the number of days that have elapsed since the rumor began. In a community of 1000 students, how many days will elapse before 450 students have heard the rumor?

105. Current in a *RL* Circuit The equation governing the amount of current I (in amperes) after time t (in seconds) in a simple *RL* circuit consisting of a resistance R (in ohms), an inductance L (in henrys), and an electromotive force E (in volts) is

$$I = \frac{E}{R}\left[1 - e^{-(R/L)t}\right]$$

If $E = 12$ volts, $R = 10$ ohms, and $L = 5$ henrys, how long does it take to obtain a current of 0.5 ampere? Of 1.0 ampere? Graph the equation.

106. Learning Curve Psychologists sometimes use the function

$$L(t) = A(1 - e^{-kt})$$

to measure the amount L learned at time t. The number A represents the amount to be learned, and the number k measures the rate of learning. Suppose that a student has an amount A of 200 vocabulary words to learn. A psychologist determines that the student learned 20 vocabulary words after 5 minutes.

(a) Determine the rate of learning k.
(b) Approximately how many words will the student have learned after 10 minutes?
(c) After 15 minutes?
(d) How long does it take for the student to learn 180 words?

107. Alcohol and Driving The concentration of alcohol in a person's blood is measurable. Suppose that the risk R (given as a percent) of having an accident while driving a car can be modeled by the equation

$$R = 3e^{kx}$$

where x is the variable concentration of alcohol in the blood and k is a constant.

(a) Suppose that a concentration of alcohol in the blood of 0.06 results in a 10% risk ($R = 10$) of an accident. Find the constant k in the equation.
(b) Using this value of k, what is the risk if the concentration is 0.17?
(c) Using the same value of k, what concentration of alcohol corresponds to a risk of 100%?
(d) If the law asserts that anyone with a risk of having an accident of 15% or more should not have driving privileges, at what concentration of alcohol in the blood should a driver be arrested and charged with a DUI?
 (e) Compare this situation with that of Example 10. If you were a lawmaker, which situation would you support? Give your reasons.

108. Is there any function of the form $y = x^\alpha, 0 < \alpha < 1$, that increases more slowly than a logarithmic function whose base is greater than 1? Explain.

109. Critical Thinking In buying a new car, one consideration might be how well the price of the car holds up over time. Different makes of cars have different depreciation rates. One way to compute a depreciation rate for a car is given here. Suppose that the current prices of a certain Mercedes automobile are as follows:

	Age in Years				
New	1	2	3	4	5
$38,000	$36,600	$32,400	$28,750	$25,400	$21,200

Use the formula New $= \text{Old}(e^{Rt})$ to find R, the annual depreciation rate, for a specific time t. When might be the best time to trade in the car? Consult the NADA ("blue") book and compare two like models that you are interested in. Which has the better depreciation rate?

PREPARING FOR THIS SECTION

Before getting started, review the following:

✓ One-to-One Functions; Inverse Functions (Section 1.6, pp. 60–69)

7.3   PROPERTIES OF LOGARITHMS; EXPONENTIAL AND LOGARITHMIC MODELS

OBJECTIVES   1   Work with the Properties of Logarithms
2   Write a Logarithmic Expression as a Sum or Difference of Logarithms
3   Write a Logarithmic Expression as a Single Logarithm
4   Evaluate Logarithms Whose Base Is Neither 10 nor e
5   Work with Exponential and Logarithmic Models

1   Logarithms have some very useful properties that can be derived directly from the definition and the laws of exponents.

EXAMPLE 1   **Establishing Properties of Logarithms**

(a) Show that $\log_a 1 = 0$.   (b) Show that $\log_a a = 1$.

Solution   (a) This fact was established when we graphed $y = \log_a x$ (see Figure 23). To show the result algebraically, let $y = \log_a 1$. Then

$$y = \log_a 1$$

$$a^y = 1 \qquad \text{Change to an exponent.}$$

$$a^y = a^0 \qquad a^0 = 1$$

$$y = 0 \qquad \text{Solve for } y.$$

$$\log_a 1 = 0 \qquad y = \log_a 1$$

(b) Let $y = \log_a a$. Then

$$y = \log_a a$$

$$a^y = a \qquad \text{Change to an exponent.}$$

$$a^y = a^1 \qquad a^1 = a$$

$$y = 1 \qquad \text{Solve for } y.$$

$$\log_a a = 1 \qquad y = \log_a a \qquad ∎$$

To summarize:

$$\log_a 1 = 0 \qquad \log_a a = 1$$

Theorem   **Properties of Logarithms**

In the properties given next, M and a are positive real numbers, with $a \neq 1$, and r is any real number.

The number $\log_a M$ is the exponent to which a must be raised to obtain M. That is,

$$a^{\log_a M} = M \tag{1}$$

The logarithm to the base a of a raised to a power equals that power. That is,

$$\log_a a^r = r \tag{2}$$

The proof uses the fact that $y = a^x$ and $y = \log_a x$ are inverses.

Proof of Property (1)   For inverse functions,

$$f\bigl(f^{-1}(x)\bigr) = x$$

Using $f(x) = a^x$ and $f^{-1}(x) = \log_a x$, we find

$$f\bigl(f^{-1}(x)\bigr) = a^{\log_a x} = x$$

Now let $x = M$ to obtain $a^{\log_a M} = M$.   ∎

Proof of Property (2)   For inverse functions,

$$f^{-1}\bigl(f(x)\bigr) = x$$

Using $f(x) = a^x$ and $f^{-1}(x) = \log_a x$, we find

$$f^{-1}\bigl(f(x)\bigr) = \log_a a^x = x$$

Now let $x = r$ to obtain $\log_a a^r = r$.   ∎

EXAMPLE 2   **Using Properties (1) and (2)**

(a) $2^{\log_2 \pi} = \pi$ (b) $\log_{0.2} 0.2^{-\sqrt{2}} = -\sqrt{2}$ (c) $\ln e^{kt} = kt$ ∎

NOW WORK PROBLEM 3.

Other useful properties of logarithms are given next.

Theorem   **Properties of Logarithms**

In the following properties, M, N, and a are positive real numbers, with $a \neq 1$, and r is any real number.

The Log of a Product Equals the Sum of the Logs

$$\log_a(MN) = \log_a M + \log_a N \tag{3}$$

The Log of a Quotient Equals the Difference of the Logs

$$\log_a\left(\frac{M}{N}\right) = \log_a M - \log_a N \qquad \textbf{(4)}$$

The Log of a Power Equals the Product of the Power and the Log

$$\log_a M^r = r \log_a M \qquad \textbf{(5)}$$

We shall derive properties (3) and (5) and leave the derivation of property (4) as an exercise (see Problem 87).

Proof of Property (3)   Let $A = \log_a M$ and let $B = \log_a N$. These expressions are equivalent to the exponential expressions

$$a^A = M \quad \text{and} \quad a^B = N$$

Now

$$\log_a(MN) = \log_a(a^A a^B) = \log_a a^{A+B} \qquad \text{Law of Exponents}$$

$$= A + B \qquad \text{Property (2) of logarithms}$$

$$= \log_a M + \log_a N$$

Proof of Property (5)   Let $A = \log_a M$. This expression is equivalent to

$$a^A = M$$

Now

$$\log_a M^r = \log_a(a^A)^r = \log_a a^{rA} \qquad \text{Law of Exponents}$$

$$= rA \qquad \text{Property (2) of logarithms}$$

$$= r \log_a M$$

NOW WORK PROBLEM 7.

2   Logarithms can be used to transform products into sums, quotients into differences, and powers into factors. Such transformations prove useful in certain types of calculus problems.

EXAMPLE 3   **Writing a Logarithmic Expression as a Sum of Logarithms**

Write $\log_a(x\sqrt{x^2 + 1})$ as a sum of logarithms. Express all powers as factors.

Solution $\log_a(x\sqrt{x^2 + 1}) = \log_a x + \log_a \sqrt{x^2 + 1} \qquad \text{Property (3)}$

$$= \log_a x + \log_a(x^2 + 1)^{1/2}$$

$$= \log_a x + \tfrac{1}{2}\log_a(x^2 + 1) \qquad \text{Property (5)}$$

EXAMPLE 4 **Writing a Logarithmic Expression as a Difference of Logarithms**

Write

$$\ln \frac{x^2}{(x-1)^3}$$

as a difference of logarithms. Express all powers as factors.

Solution

$$\ln \frac{x^2}{(x-1)^3} = \underset{\uparrow}{\ln x^2 - \ln(x-1)^3} = \underset{\uparrow}{2\ln x - 3\ln(x-1)}$$

Property (4) Property (5) ■

EXAMPLE 5 **Writing a Logarithmic Expression as a Sum and Difference of Logarithms**

Write

$$\log_a \frac{x^3\sqrt{x^2+1}}{(x+1)^4}$$

as a sum and difference of logarithms. Express all powers as factors.

Solution
$$\log_a \frac{x^3\sqrt{x^2+1}}{(x+1)^4} = \log_a(x^3\sqrt{x^2+1}) - \log_a(x+1)^4 \qquad \text{Property (4)}$$

$$= \log_a x^3 + \log_a \sqrt{x^2+1} - \log_a(x+1)^4 \qquad \text{Property (3)}$$

$$= \log_a x^3 + \log_a(x^2+1)^{1/2} - \log_a(x+1)^4$$

$$= 3\log_a x + \tfrac{1}{2}\log_a(x^2+1) - 4\log_a(x+1) \qquad \text{Property (5)}$$

■

 NOW WORK PROBLEM 35.

③ Another use of properties (3) through (5) is to write sums and/or differences of logarithms with the same base as a single logarithm. This skill will be needed to solve certain logarithmic equations discussed in the next section.

EXAMPLE 6 **Writing Expressions as a Single Logarithm**

Write each of the following as a single logarithm.

(a) $\log_a 7 + 4\log_a 3$ 　　　　　　　　　(b) $\dfrac{2}{3}\ln 8 - \ln(3^4 - 8)$

(c) $\log_a x + \log_a 9 + \log_a(x^2+1) - \log_a 5$

Solution　(a)　$\log_a 7 + 4\log_a 3 = \log_a 7 + \log_a 3^4$ 　　Property (5)

$$= \log_a 7 + \log_a 81$$

$$= \log_a(7 \cdot 81) \qquad \text{Property (3)}$$

$$= \log_a 567$$

(b)　$\tfrac{2}{3}\ln 8 - \ln(3^4 - 8) = \ln 8^{2/3} - \ln(81 - 8)$ 　　Property (5)

$$= \ln 4 - \ln 73$$

$$= \ln\left(\frac{4}{73}\right) \qquad \text{Property (4)}$$

(c) $\log_a x + \log_a 9 + \log_a(x^2 + 1) - \log_a 5 = \log_a(9x) + \log_a(x^2 + 1) - \log_a 5$

$$= \log_a[9x(x^2 + 1)] - \log_a 5$$

$$= \log_a\left[\frac{9x(x^2 + 1)}{5}\right]$$

WARNING: A common error made by some students is to express the logarithm of a sum as the sum of logarithms.

$$\log_a(M + N) \quad \text{is not equal to} \quad \log_a M + \log_a N$$

Correct statement $\log_a(MN) = \log_a M + \log_a N$ Property (3)

Another common error is to express the difference of logarithms as the quotient of logarithms.

$$\log_a M - \log_a N \quad \text{is not equal to} \quad \frac{\log_a M}{\log_a N}$$

Correct statement $\log_a M - \log_a N = \log_a\left(\dfrac{M}{N}\right)$ Property (4)

A third common error is to express a logarithm raised to a power as the product of the power times the logarithm.

$$(\log_a M)^r \quad \text{is not equal to} \quad r\log_a M$$

Correct statement $\log_a M^r = r\log_a M$ Property (5)

✎ **NOW WORK PROBLEM 41.**

Two other properties of logarithms that we need to know are consequences of the fact that the logarithmic function $y = \log_a x$ is one-to-one.

Theorem

In the following properties, M, N, and a are positive real numbers, with $a \neq 1$.

> If $M = N$, then $\log_a M = \log_a N$.   **(6)**
>
> If $\log_a M = \log_a N$, then $M = N$.   **(7)**

When property (6) is used, we start with the equation $M = N$ and say "take the logarithm of both sides" to obtain $\log_a M = \log_a N$.

Properties (6) and (7) are useful for solving *exponential and logarithmic equations*, a topic discussed in the next section.

USING A CALCULATOR TO EVALUATE LOGARITHMS WITH BASES OTHER THAN 10 OR e

④ Logarithms to the base 10, called **common logarithms,** were used to facilitate arithmetic computations before the widespread use of calculators. (See the Historical Feature at the end of this section.) Natural logarithms, that is, logarithms whose base is the number e, remain very important because they arise frequently in the study of natural phenomena.

Common logarithms are usually abbreviated by writing **log,** with the base understood to be 10, just as natural logarithms are abbreviated by **ln,** with the base understood to be e.

Most calculators have both ▢log and ▢ln keys to calculate the common logarithm and natural logarithm of a number. Let's look at an example to see how to approximate logarithms having a base other than 10 or e.

EXAMPLE 7

Approximating Logarithms Whose Base Is Neither 10 nor e

Approximate $\log_2 7$.
Round the answer to four decimal places.

Solution Let $y = \log_2 7$. Then $2^y = 7$, so

$$2^y = 7$$
$$\ln 2^y = \ln 7 \qquad \text{Property (6)}$$
$$y \ln 2 = \ln 7 \qquad \text{Property (5)}$$
$$y = \frac{\ln 7}{\ln 2} \qquad \text{Solve for } y.$$
$$y \approx 2.8074 \qquad \text{Use calculator (}\boxed{\ln}\text{ key) and round to four decimal places.}$$

∎

Example 7 shows how to approximate a logarithm whose base is 2 by changing to logarithms involving the base e. In general, we use the **Change-of-Base Formula**.

Theorem

Change-of-Base Formula

If $a \neq 1$, $b \neq 1$, and M are positive real numbers, then

$$\log_a M = \frac{\log_b M}{\log_b a} \qquad (8)$$

∎

Proof We derive this formula as follows: Let $y = \log_a M$. Then

$$a^y = M$$
$$\log_b a^y = \log_b M \qquad \text{Property (6)}$$
$$y \log_b a = \log_b M \qquad \text{Property (5)}$$
$$y = \frac{\log_b M}{\log_b a} \qquad \text{Solve for } y.$$
$$\log_a M = \frac{\log_b M}{\log_b a} \qquad y = \log_a M$$

∎

Since calculators have only keys for ▢log and ▢ln, in practice, the Change-of-Base Formula uses either $b = 10$ or $b = e$. Thus,

$$\log_a M = \frac{\log M}{\log a} \quad \text{and} \quad \log_a M = \frac{\ln M}{\ln a} \qquad (9)$$

EXAMPLE 8

Using the Change-of-Base Formula

Approximate: (a) $\log_5 89$ (b) $\log_{\sqrt{2}} \sqrt{5}$
Round answers to four decimal places.

Solution (a) $\log_5 89 = \dfrac{\log 89}{\log 5} \approx \dfrac{1.949390007}{0.6989700043} \approx 2.7889$

or

$\log_5 89 = \dfrac{\ln 89}{\ln 5} \approx \dfrac{4.48863637}{1.609437912} \approx 2.7889$

(b) $\log_{\sqrt{2}} \sqrt{5} = \dfrac{\log \sqrt{5}}{\log \sqrt{2}} = \dfrac{\frac{1}{2}\log 5}{\frac{1}{2}\log 2} \approx 2.3219$

or

$\log_{\sqrt{2}} \sqrt{5} = \dfrac{\ln \sqrt{5}}{\ln \sqrt{2}} = \dfrac{\frac{1}{2}\ln 5}{\frac{1}{2}\ln 2} \approx 2.3219$ ∎

COMMENT: To graph logarithmic functions when the base is different from e or 10 requires the Change-of-Base Formula. For example, to graph $y = \log_2 x$, we would instead graph $y = (\ln x)/(\ln 2)$. Try it.   ∎

 NOW WORK PROBLEMS 11 AND 51.

EXPONENTIAL AND LOGARITHMIC MODELS

⑤   When placed in a scatter diagram, data sometimes indicate a relation between the variables that is exponential or logarithmic. Figure 18 shows typical scatter diagrams for data that are exponential or logarithmic.

Figure 18

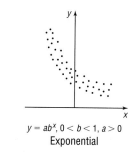

$y = ab^x, a > 0, b > 1$
Exponential

$y = ab^x, 0 < b < 1, a > 0$
Exponential

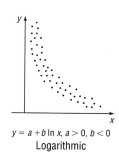

$y = a + b \ln x, a > 0, b < 0$
Logarithmic

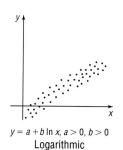

$y = a + b \ln x, a > 0, b > 0$
Logarithmic

EXAMPLE 9	**Fitting a Curve to an Exponential Model**

Beth is interested in finding a function that explains the closing price (adjusted for stock splits) of Dell Computer stock at the end of each year. She obtains the data in the table:

Year	Closing Price
1989	0.0573
1990	0.1927
1991	0.2669
1992	0.75
1993	0.3535
1994	0.6406
1995	1.082
1996	3.3203
1997	10.5
1998	36.5938
1999	51

Source: NASDAQ

(a) Draw a scatter diagram using year as the independent variable. Use $x = 1$ for 1989, $x = 2$ for 1990, and so on.

(b) The exponential function of best fit to the data is found to be

$$y = 0.03028(1.8905)^x$$

Express the function of best fit in the form $A = A_0 e^{kt}$, where $x = t$.

(c) Use the solution to part (b) to predict the closing price of Dell Computer stock at the end of the year 2000.

(d) Use a graphing utility to verify the exponential function of best fit.

(e) Use a graphing utility to draw a scatter diagram of the data and then graph the exponential function of best fit on it.

Solution (a) Letting 1 represent 1989, 2 represent 1990, and so on, we obtain the scatter diagram shown in Figure 19.

Figure 19

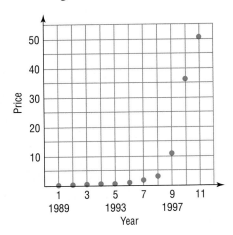

(b) To express $y = ab^x$ in the form $A = A_0 e^{kt}$, where $x = t$, we proceed as follows:

$$ab^x = A_0 e^{kt}, \qquad x = t$$

We set the coefficients equal to each other and the exponential expressions equal to each other.

$$a = A_0, \qquad b^x = e^{kt}, \qquad x = t$$

$$b^x = \left(e^k\right)^t, \qquad x = t$$

$$b = e^k$$

Since $y = ab^x = 0.03028(1.8905)^x$, we find that $a = 0.03028$ and $b = 1.8905$. Thus,

$$a = A_0 = 0.03028 \quad \text{and} \quad b = 1.8905 = e^k$$

$$k = \ln(1.8905) = 0.6368$$

As a result, $A = A_0 e^{kt} = 0.03028 e^{0.6368t}$.

Figure 20

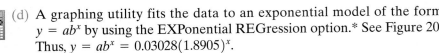

(c) Let $t = 12$ (the year 2000) in the function found in part (b). The predicted closing price of Dell Computer stock at the end of the year 2000 is

$$A = 0.03028 e^{0.6368(12)} = \$63.08$$

(d) A graphing utility fits the data to an exponential model of the form $y = ab^x$ by using the EXPonential REGression option.* See Figure 20. Thus, $y = ab^x = 0.03028(1.8905)^x$.

---

*If your utility does not have such an option but does have a LINear REGression option, you can transform the exponential model to a linear model by the following technique:

$y = ab^x$	Exponential form
$\ln y = \ln(ab^x)$	Take natural logs of each side.
$\ln y = \ln a + \ln b^x$	Property of logarithms
$\ln y = \ln a + (\ln b)x$	Property of logarithms
$Y = \ln a + (\ln b)x$	

Now apply the LINear REGression techniques discussed in Appendix A, Section A.8. Be careful though. The dependent variable is now $Y = \ln y$, while the independent variable remains x. Once the line of best fit is obtained, you can find a and b and obtain the exponential curve $y = ab^x$.

Figure 21

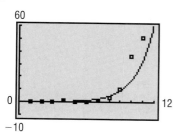

(e) Figure 21 shows the graph of $y = 0.03028(1.8905)^x$ on a scatter diagram of the data. ∎

✏ N O W   W O R K   P R O B L E M   **89.**

There are relations between variables that do not follow an exponential model, but, instead, follow a logarithmic model in which the independent variable is related to the dependent variable by a logarithmic function.

| EXAMPLE 10 | **Fitting a Curve to a Logarithmic Model** |

Jodi, a meteorologist, is interested in finding a function that explains the relation between the height of a weather balloon (in kilometers) and the atmospheric pressure (measured in millimeters of mercury) on the balloon. She collects the data in the table to the left.

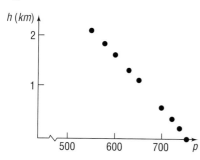

Atmospheric Pressure, p	Height, h
760	0
740	0.184
725	0.328
700	0.565
650	1.079
630	1.291
600	1.634
580	1.862
550	2.235

(a) Draw a scatter diagram of the data with atmospheric pressure as the independent variable.

(b) The logarithmic function of best fit to the data is found to be

$$h = 45.8 - 6.9 \ln p$$

where h is the height of the weather balloon and p is the atmospheric pressure. Use the logarithmic function of best fit to predict the height of the balloon if the atmospheric pressure is 560 millimeters of mercury.

(c) Use a graphing utility to verify the logarithmic function of best fit.

(d) Use a graphing utility to draw a scatter diagram of the data, and then graph the logarithmic function of best fit on it.

Solution (a) See Figure 22.

Figure 22

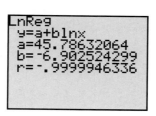

(b) Using the given function, Jodi predicts the height of the weather balloon when the atmospheric pressure is 560 to be

$$h = 45.8 - 6.9 \ln 560 \approx 2.14 \text{ kilometers}$$

(c) A graphing utility fits the data to a logarithmic model of the form $y = a + b \ln x$ by using the Logarithm REGression option. See Figure 23. Notice that $|r|$ is close to 1, indicating a good fit.

(d) Figure 24 shows the graph of $h = 45.8 - 6.9 \ln p$ on the scatter diagram.

Figure 23

Figure 24

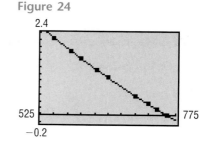

SUMMARY

Properties of Logarithms

In the list that follows, $a > 0$, $a \neq 1$, and $b > 0$, $b \neq 1$; also, $M > 0$ and $N > 0$.

Definition	$y = \log_a x$ means $x = a^y$
Properties of logarithms	$\log_a 1 = 0; \quad \log_a a = 1$
	$a^{\log_a M} = M; \quad \log_a a^r = r$
	$\log_a(MN) = \log_a M + \log_a N$
	$\log_a\left(\dfrac{M}{N}\right) = \log_a M - \log_a N$
	$\log_a M^r = r \log_a M$
	If $M = N$, then $\log_a M = \log_a N$.
	If $\log_a M = \log_a N$, then $M = N$.
Change-of-Base Formula	$\log_a M = \dfrac{\log_b M}{\log_b a}$

HISTORICAL FEATURE

John Napier
(1550–1617)

Logarithms were invented about 1590 by John Napier (1550–1617) and Jobst Bürgi (1552–1632), working independently. Napier, whose work had the greater influence, was a Scottish lord, a secretive man whose neighbors were inclined to believe him to be in league with the devil. His approach to logarithms was very different from ours; it was based on the relationship between arithmetic and geometric sequences, and not on the inverse function relationship of logarithms to exponential functions (described in Section 7.2). Napier's tables, published in 1614, listed what would now be called *natural*

logarithms of sines and were rather difficult to use. A London professor, Henry Briggs, became interested in the tables and visited Napier. In their conversations, they developed the idea of common logarithms, which were published in 1617. Their importance for calculation was immediately recognized, and by 1650 they were being printed as far away as China. They remained an important calculation tool until the advent of the inexpensive handheld calculator about 1972, which has decreased their calculational, but not their theoretical, importance.

A side effect of the invention of logarithms was the popularization of the decimal system of notation for real numbers.

7.3 EXERCISES

In Problems 1–16, use properties of logarithms to find the exact value of each expression. Do not use a calculator.

1. $\log_3 3^{71}$

2. $\log_2 2^{-13}$

3. $\ln e^{-4}$

4. $\ln e^{\sqrt{2}}$

5. $2^{\log_2 7}$

6. $e^{\ln 8}$

7. $\log_8 2 + \log_8 4$

8. $\log_6 9 + \log_6 4$

9. $\log_6 18 - \log_6 3$

10. $\log_8 16 - \log_8 2$

11. $\log_2 6 \cdot \log_6 4$

12. $\log_3 8 \cdot \log_8 9$

13. $3^{\log_3 5 - \log_3 4}$

14. $5^{\log_5 6 + \log_5 7}$

15. $e^{\log_{e^2} 16}$

16. $e^{\log_{e^2} 9}$

In Problems 17–24, suppose that $\ln 2 = a$ *and* $\ln 3 = b$. *Use properties of logarithms to write each logarithm in terms of a and b.*

17. $\ln 6$

18. $\ln \dfrac{2}{3}$

19. $\ln 1.5$

20. $\ln 0.5$

21. $\ln 8$

22. $\ln 27$

23. $\ln \sqrt[5]{6}$

24. $\ln \sqrt[4]{\dfrac{2}{3}}$

In Problems 25–40, write each expression as a sum and/or difference of logarithms. Express powers as factors.

25. $\log_a(u^2 v^3)$

26. $\log_2\left(\dfrac{a}{b^2}\right)$

27. $\log \dfrac{1}{M^3}$

28. $\log(10u^2)$

29. $\log_5 \sqrt{\dfrac{a^3}{b}}$

30. $\log_6\left(\dfrac{ab^4}{\sqrt[3]{c^2}}\right)$

31. $\ln(x^2\sqrt{1-x})$

32. $\ln(x\sqrt{1+x^2})$

33. $\log_2\left(\dfrac{x^3}{x-3}\right)$

34. $\log_5\left(\dfrac{\sqrt[3]{x^2+1}}{x^2-1}\right)$

35. $\log\left[\dfrac{x(x+2)}{(x+3)^2}\right]$

36. $\log\left[\dfrac{x^3\sqrt{x+1}}{(x-2)^2}\right]$

37. $\ln\left[\dfrac{x^2-x-2}{(x+4)^2}\right]^{1/3}$

38. $\ln\left[\dfrac{(x-4)^2}{x^2-1}\right]^{2/3}$

39. $\ln \dfrac{5x\sqrt{1-3x}}{(x-4)^3}$

40. $\ln\left[\dfrac{5x^2\sqrt[3]{1-x}}{4(x+1)^2}\right]$

In Problems 41–50, write each expression as a single logarithm.

41. $3\log_5 u + 4\log_5 v$

42. $\log_3 u^2 - \log_3 v$

43. $\log_{1/2}\sqrt{x} - \log_{1/2}x^3$

44. $\log_2\left(\dfrac{1}{x}\right) + \log_2\left(\dfrac{1}{x^2}\right)$

45. $\ln\left(\dfrac{x}{x-1}\right) + \ln\left(\dfrac{x+1}{x}\right) - \ln(x^2-1)$

46. $\log\left(\dfrac{x^2+2x-3}{x^2-4}\right) - \log\left(\dfrac{x^2+7x+6}{x+2}\right)$

47. $8\log_2\sqrt{3x-2} - \log_2\left(\dfrac{4}{x}\right) + \log_2 4$

48. $21\log_3\sqrt[3]{x} + \log_3(9x^2) - \log_3 25$

49. $2\log_a(5x^3) - \dfrac{1}{2}\log_a(2x+3)$

50. $\dfrac{1}{3}\log(x^3+1) + \dfrac{1}{2}\log(x^2+1)$

In Problems 51–58, use the Change-of-Base Formula and a calculator to approximate each logarithm. Round your answer to three decimal places.

51. $\log_3 21$

52. $\log_5 18$

53. $\log_{1/3} 71$

54. $\log_{1/2} 15$

55. $\log_{\sqrt{2}} 7$

56. $\log_{\sqrt{5}} 8$

57. $\log_\pi e$

58. $\log_\pi \sqrt{2}$

In Problems 59–62, find the exact value of each expression. Do not use a calculator.

59. $\log_2 3 \cdot \log_3 4 \cdot \log_4 5 \cdot \log_5 6 \cdot \log_6 7 \cdot \log_7 8$

60. $\log_2 4 \cdot \log_4 6 \cdot \log_6 8$

61. $\log_2 3 \cdot \log_3 4 \cdot \ldots \cdot \log_n(n+1) \cdot \log_{n+1} 2$

62. $\log_2 2 \cdot \log_2 4 \cdot \log_2 8 \cdot \ldots \cdot \log_2 2^n$

In Problems 63–68, graph each function using a graphing utility and the Change-of-Base Formula.

63. $y = \log_4 x$

64. $y = \log_5 x$

65. $y = \log_2(x+2)$

66. $y = \log_4(x-3)$

67. $y = \log_{x-1}(x+1)$

68. $y = \log_{x+2}(x-2)$

In Problems 69–78, express y as a function of x. The constant C is a positive number.

69. $\ln y = \ln x + \ln C$

70. $\ln y = \ln(x+C)$

71. $\ln y = \ln x + \ln(x+1) + \ln C$

72. $\ln y = 2\ln x - \ln(x+1) + \ln C$

73. $\ln y = 3x + \ln C$

74. $\ln y = -2x + \ln C$

75. $\ln(y - 3) = -4x + \ln C$

76. $\ln(y + 4) = 5x + \ln C$

77. $3 \ln y = \frac{1}{2}\ln(2x + 1) - \frac{1}{3}\ln(x + 4) + \ln C$

78. $2 \ln y = -\frac{1}{2}\ln x + \frac{1}{3}\ln(x^2 + 1) + \ln C$

79. Show that $\log_a(x + \sqrt{x^2 - 1}) + \log_a(x - \sqrt{x^2 - 1}) = 0$.

80. Show that $\log_a(\sqrt{x} + \sqrt{x - 1}) + \log_a(\sqrt{x} - \sqrt{x - 1}) = 0$.

81. Show that $\ln(1 + e^{2x}) = 2x + \ln(1 + e^{-2x})$.

82. If $f(x) = \log_a x$, show that $\dfrac{f(x + h) - f(x)}{h} = \log_a\left(1 + \dfrac{h}{x}\right)^{1/h}$, $h \neq 0$.

83. If $f(x) = \log_a x$, show that $-f(x) = \log_{1/a} x$.

84. If $f(x) = \log_a x$, show that $f(AB) = f(A) + f(B)$.

85. If $f(x) = \log_a x$, show that $f\left(\dfrac{1}{x}\right) = -f(x)$.

86. If $f(x) = \log_a x$, show that $f(x^\alpha) = \alpha f(x)$.

87. Show that $\log_a\left(\dfrac{M}{N}\right) = \log_a M - \log_a N$, where a, M, and N are positive real numbers, with $a \neq 1$.

88. Show that $\log_a\left(\dfrac{1}{N}\right) = -\log_a N$, where a and N are positive real numbers, with $a \neq 1$.

---

89. Chemistry A chemist has a 100-gram sample of a radioactive material. He records the amount of radioactive material every week for 6 weeks and obtains the following data:

Week	Weight (in Grams)
0	100.0
1	88.3
2	75.9
3	69.4
4	59.1
5	51.8
6	45.5

(a) Draw a scatter diagram with week as the independent variable.

(b) The exponential function of best fit to the data is found to be

$$y = 100(0.88)^x$$

Express the function of best fit in the form $A = A_0 e^{kt}$, where $x = t$.

(c) Use the function in part (b) to estimate the time it takes until 50 grams of material is left. (Since 50 grams is one-half of the original amount, this time is referred to as the **half-life** of the radioactive material).

(d) Use the function in part (b) to predict how much radioactive material will be left after 50 weeks.

(e) Use a graphing utility to verify the exponential function of best fit.

(f) Use a graphing utility to draw a scatter diagram of the data and then graph the exponential function of best fit on it.

90. Chemistry A chemist has a 1000-gram sample of a radioactive material. She records the amount of radioactive material remaining in the sample every day for a week and obtains the following data:

Day	Weight (in Grams)
0	1000.0
1	897.1
2	802.5
3	719.8
4	651.1
5	583.4
6	521.7
7	468.3

(a) Draw a scatter diagram with day as the independent variable.

(b) The exponential function of best fit to the data is found to be

$$y = 999(0.898)^x$$

Express the function of best fit in the form $A = A_0 e^{kt}$, where $x = t$.

(c) Use the function in part (b) to estimate the time it takes until 500 grams of material is left. (Since 500 grams is one-half of the original amount, this time is referred to as the **half-life** of the radioactive material.)

(d) Use the function in part (b) to predict how much radioactive material is left after 20 days.

(e) Use a graphing utility to verify the exponential function of best fit.

(f) Use a graphing utility to draw a scatter diagram of the data and then graph the exponential function of best fit on it.

91. Economics and Marketing   A store manager collected the following data regarding price and quantity demanded of shoes:

Price ($/Unit)	Quantity Demanded
79	10
67	20
54	30
46	40
38	50
31	60

(a) Draw a scatter diagram with quantity demanded on the x-axis and price on the y-axis.

(b) The exponential function of best fit to the data is found to be

$$y = 96(0.98)^x$$

Express the function of best fit in the form $y = A_0 e^{kt}$, where $x = t$.

(c) Use the function in part (b) to predict the quantity demanded if the price is $60.

(d) Use a graphing utility to verify the exponential function of best fit.

(e) Use a graphing utility to draw a scatter diagram of the data and then graph the exponential function of best fit on it.

92. Economics and Marketing   A store manager collected the following data regarding price and quantity supplied of dresses:

Price ($/Unit)	Quantity Supplied
25	10
32	20
40	30
46	40
60	50
74	60

(a) Draw a scatter diagram with quantity supplied on the x-axis and price on the y-axis.

(b) The exponential function of best fit to the data is found to be

$$y = 20.524(1.022)^x$$

Express the function of best fit in the form $A = A_0 e^{kx}$, where $x = t$.

(c) Use the function in part (b) to predict the quantity supplied if the price is $45.

(d) Use a graphing utility to verify the exponential function of best fit.

(e) Use a graphing utility to draw a scatter diagram of the data and then graph the exponential function of best fit on it.

93. Economics and Marketing   The following data represent the price and quantity demanded in 1997 for IBM personal computers at Best Buy.

Price ($/Computer)	Quantity Demanded
2300	152
2000	159
1700	164
1500	171
1300	176
1200	180
1000	189

(a) Draw a scatter diagram of the data with price as the dependent variable.

(b) The logarithmic function of best fit to the data is found to be

$$y = 32{,}741 - 6071 \ln x$$

Use it to predict the number of IBM personal computers that would be demanded if the price were $1650.

(c) Use a graphing utility to verify the logarithmic function of best fit.

(d) Use a graphing utility to draw a scatter diagram of the data and then graph the logarithmic function of best fit on it.

94. Find the domain of $f(x) = \log_a x^2$ and the domain of $g(x) = 2 \log_a x$. Since $\log_a x^2 = 2 \log_a x$, how do you reconcile the fact that the domains are not equal? Write a brief explanation.

7.4 LOGARITHMIC AND EXPONENTIAL EQUATIONS

OBJECTIVES
1. Solve Logarithmic Equations Using the Properties of Logarithms
2. Solve Exponential Equations
3. Solve Logarithmic and Exponential Equations Using a Graphing Utility

LOGARITHMIC EQUATIONS

1. In Section 7.2 we solved logarithmic equations by changing a logarithm to exponential form. Often, however, some manipulation of the equation (usually using the properties of logarithms) is required before we can change to exponential form.

Our practice will be to solve equations, whenever possible, by finding exact solutions using algebraic methods. When algebraic methods cannot be used, approximate solutions will be obtained using a graphing utility. The reader is encouraged to pay particular attention to the form of equations for which exact solutions are possible.

EXAMPLE 1 **Solving a Logarithmic Equation**

Solve: $2 \log_5 x = \log_5 9$

Solution Because each logarithm has the same base, 5, we can obtain an exact solution as follows:

$$2 \log_5 x = \log_5 9$$
$$\log_5 x^2 = \log_5 9 \qquad \log_a M^r = r \log_a M$$
$$x^2 = 9 \qquad \text{If } \log_a M = \log_a N, \text{ then } M = N$$
$$x = 3 \quad \text{or} \quad \cancel{x = -3} \qquad \text{Recall that logarithms of negative numbers are not defined, so, in the expression } 2 \log_5 x, \ x \text{ must be positive.}$$
$$\text{Therefore, } -3 \text{ is extraneous and we discard it.}$$

The equation has only one solution, 3. ■

➤ NOW WORK PROBLEM 5.

EXAMPLE 2 **Solving a Logarithmic Equation**

Solve: $\log_4(x + 3) + \log_4(2 - x) = 1$

Solution To obtain an exact solution, we need to express the left side as a single logarithm. Then we will change the expression to exponential form.

$$\log_4(x + 3) + \log_4(2 - x) = 1$$
$$\log_4[(x + 3)(2 - x)] = 1 \qquad \log_a M + \log_a N = \log_a(MN)$$
$$(x + 3)(2 - x) = 4^1 = 4 \qquad \text{Change to an exponential expression.}$$
$$-x^2 - x + 6 = 4 \qquad \text{Simplify.}$$
$$x^2 + x - 2 = 0 \qquad \text{Place the quadratic equation in standard form.}$$
$$(x + 2)(x - 1) = 0 \qquad \text{Factor.}$$
$$x = -2 \quad \text{or} \quad x = 1 \qquad \text{Zero-Product Property}$$

Since the arguments of each logarithmic expression in the equation are positive for both $x = -2$ and $x = 1$, neither is extraneous. The solution set is $\{-2, 1\}$. ∎

NOW WORK PROBLEM **9**.

EXPONENTIAL EQUATIONS

② In Sections 7.1 and 7.2, we solved certain exponential equations by expressing each side of the equation with the same base. However, many exponential equations cannot be rewritten so that each side has the same base. In such cases, properties of logarithms along with algebraic techniques can sometimes be used to obtain a solution.

EXAMPLE 3   Solving an Exponential Equation

Solve: $4^x - 2^x - 12 = 0$

Solution   We note that $4^x = (2^2)^x = 2^{2x} = (2^x)^2$, so the equation is actually quadratic in form, and we can rewrite it as

$$(2^x)^2 - 2^x - 12 = 0 \qquad \text{Let } u = 2^x; \text{ then } u^2 - u - 12 = 0.$$

Now we can factor as usual.

$$(2^x - 4)(2^x + 3) = 0 \qquad (u - 4)(u + 3) = 0$$

$$2^x - 4 = 0 \quad \text{or} \quad 2^x + 3 = 0 \qquad u - 4 = 0 \quad \text{or} \quad u + 3 = 0$$

$$2^x = 4 \qquad\qquad 2^x = -3 \qquad u = 2^x = 4 \qquad u = 2^x = -3$$

The equation on the left has the solution $x = 2$, since $2^x = 4 = 2^2$; the equation on the right has no solution, since $2^x > 0$ for all x. The only solution is 2. ∎

In Example 3, we were able to write the exponential expression using the same base after utilizing some algebra, obtaining an exact solution to the equation. When this is not possible, logarithms can sometimes be used to obtain the solution.

EXAMPLE 4   Solving an Exponential Equation

Solve: $2^x = 5$

Solution   We write the exponential equation as the equivalent logarithmic equation.

$$2^x = 5$$

$$x = \log_2 5 = \frac{\ln 5}{\ln 2}$$

$\uparrow$
Change-of-Base Formula (9), Section 7.3

Alternatively, we can solve the equation $2^x = 5$ by taking the logarithm of each side (refer to Property (6), Section 7.3). Taking the natural logarithm,

$$2^x = 5$$
$$\ln 2^x = \ln 5 \qquad \text{If } M = N, \log_a M = \log_a N$$
$$x \ln 2 = \ln 5 \qquad \log_a M^r = r \log_a M$$
$$x = \frac{\ln 5}{\ln 2}$$

Using a calculator, the solution, rounded to three decimal places, is

$$x = \frac{\ln 5}{\ln 2} \approx 2.322$$

NOW WORK PROBLEM **17.**

| EXAMPLE 5 | **Solving an Exponential Equation** |

Solve: $8 \cdot 3^x = 5$

Solution
$$8 \cdot 3^x = 5$$
$$3^x = \frac{5}{8} \qquad \text{Solve for } 3^x.$$
$$x = \log_3\left(\frac{5}{8}\right) = \frac{\ln \frac{5}{8}}{\ln 3} \qquad \text{Solve for } x.$$

The solution, rounded to three decimal places, is

$$x = \frac{\ln \dfrac{5}{8}}{\ln 3} \approx -0.428$$

| EXAMPLE 6 | **Solving an Exponential Equation** |

Solve: $5^{x-2} = 3^{3x+2}$

Solution   Because the bases are different, we first apply Property (6), Section 7.3 (taking the natural logarithm of each side) and then use appropriate properties of logarithms. The result is an equation in x that we can solve.

$$5^{x-2} = 3^{3x+2}$$
$$\ln 5^{x-2} = \ln 3^{3x+2} \qquad \text{If } M = N, \log_a M = \log_a N$$
$$(x - 2)\ln 5 = (3x + 2)\ln 3 \qquad \log_a M^r = r \log_a M$$
$$x \ln 5 - 2 \ln 5 = 3x \ln 3 + 2 \ln 3 \qquad \text{Distribute.}$$
$$x \ln 5 - 3x \ln 3 = 2 \ln 3 + 2 \ln 5 \qquad \text{Place terms involving } x \text{ on the left.}$$
$$(\ln 5 - 3 \ln 3)x = 2 (\ln 3 + \ln 5) \qquad \text{Factor.}$$
$$x = \frac{2(\ln 3 + \ln 5)}{\ln 5 - 3 \ln 3} \approx -3.212$$

NOW WORK PROBLEM **25.**

GRAPHING UTILITY SOLUTIONS

 ③ The techniques introduced in this section apply only to certain types of logarithmic and exponential equations. Solutions for other types are usually studied in calculus, using numerical methods. However, we can use a graphing utility to approximate the solution.

| EXAMPLE 7 | **Solving Equations Using a Graphing Utility** |

Solve: $\log_3 x + \log_4 x = 4$
Express the solution(s) rounded to two decimal places.

Figure 25

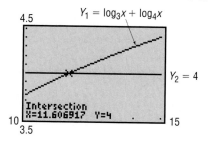

Solution   The solution is found by graphing

$$Y_1 = \log_3 x + \log_4 x = \frac{\log x}{\log 3} + \frac{\log x}{\log 4} \quad \text{and} \quad Y_2 = 4$$

(Remember that you must use the Change-of-Base Formula to graph Y_1.) Y_1 is an increasing function (do you know why?), and so there is only one point of intersection for Y_1 and Y_2. Figure 25 shows the graphs of Y_1 and Y_2. Using the INTERSECT command, the solution is 11.61, rounded to two decimal places.

Can you discover an algebraic solution to Example 7?
[**Hint:** Factor $\log x$ from Y_1.]

| EXAMPLE 8 | **Solving Equations Using a Graphing Utility** |

Solve: $x + e^x = 2$
Express the solution(s) rounded to two decimal places.

Figure 26

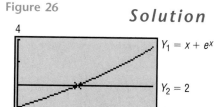

Solution   The solution is found by graphing $Y_1 = x + e^x$ and $Y_2 = 2$. Y_1 is an increasing function (do you know why?), and so there is only one point of intersection for Y_1 and Y_2. Figure 26 shows the graphs of Y_1 and Y_2. Using the INTERSECT command, the solution is 0.44, rounded to two decimal places.

7.4   EXERCISES

In Problems 1–44, solve each equation.

1. $\log_4(x + 2) = \log_4 8$

2. $\log_5(2x + 3) = \log_5 3$

3. $\frac{1}{2}\log_3 x = 2\log_3 2$

4. $-2\log_4 x = \log_4 9$

5. $2\log_5 x = 3\log_5 4$

6. $3\log_2 x = -\log_2 27$

7. $3\log_2(x - 1) + \log_2 4 = 5$

8. $2\log_3(x + 4) - \log_3 9 = 2$

9. $\log x + \log(x + 15) = 2$

10. $\log_4 x + \log_4(x - 3) = 1$

11. $\ln x + \ln(x + 2) = 4$

12. $\ln(x + 1) - \ln x = 2$

13. $2^{2x} + 2^x - 12 = 0$

14. $3^{2x} + 3^x - 2 = 0$

15. $3^{2x} + 3^{x+1} - 4 = 0$

16. $2^{2x} + 2^{x+2} - 12 = 0$

17. $2^x = 10$

18. $3^x = 14$

19. $8^{-x} = 1.2$

20. $2^{-x} = 1.5$

21. $3^{1-2x} = 4^x$

22. $2^{x+1} = 5^{1-2x}$

23. $\left(\frac{3}{5}\right)^x = 7^{1-x}$

24. $\left(\frac{4}{3}\right)^{1-x} = 5^x$

25. $1.2^x = (0.5)^{-x}$

26. $(0.3)^{1+x} = 1.7^{2x-1}$

27. $\pi^{1-x} = e^x$

28. $e^{x+3} = \pi^x$

29. $5(2^{3x}) = 8$

30. $0.3(4^{0.2x}) = 0.2$

31. $\log_a(x - 1) - \log_a(x + 6) = \log_a(x - 2) - \log_a(x + 3)$

32. $\log_a x + \log_a(x - 2) = \log_a(x + 4)$

33. $\log_{1/3}(x^2 + x) - \log_{1/3}(x^2 - x) = -1$

34. $\log_4(x^2 - 9) - \log_4(x + 3) = 3$

35. $\log_2(x + 1) - \log_4 x = 1$
[**Hint:** Change $\log_4 x^2$ to base 2.]

36. $\log_2(3x + 2) - \log_4 x = 3$

37. $\log_{16} x + \log_4 x + \log_2 x = 7$

38. $\log_9 x + 3\log_3 x = 14$

39. $\left(\sqrt[3]{2}\right)^{2-x} = 2^{x^2}$

40. $\log_2 x^{\log_2 x} = 4$

41. $\dfrac{e^x + e^{-x}}{2} = 1$

42. $\dfrac{e^x + e^{-x}}{2} = 3$

43. $\dfrac{e^x - e^{-x}}{2} = 2$

44. $\dfrac{e^x - e^{-x}}{2} = -2$

[**Hint:** Multiply each side by e^x.]

In Problems 45–60, use a graphing utility to solve each equation. Express your answer rounded to two decimal places.

45. $\log_5 x + \log_3 x = 1$

46. $\log_2 x + \log_6 x = 3$

47. $\log_5(x + 1) - \log_4(x - 2) = 1$

48. $\log_2(x - 1) - \log_6(x + 2) = 2$

49. $e^x = -x$

50. $e^{2x} = x + 2$

51. $e^x = x^2$

52. $e^x = x^3$

53. $\ln x = -x$

54. $\ln(2x) = -x + 2$

55. $\ln x = x^3 - 1$

56. $\ln x = -x^2$

57. $e^x + \ln x = 4$

58. $e^x - \ln x = 4$

59. $e^{-x} = \ln x$

60. $e^{-x} = -\ln x$

7.5 COMPOUND INTEREST

OBJECTIVES

1. Determine the Future Value of a Lump Sum of Money
2. Calculate Effective Rates of Return
3. Determine the Present Value of a Lump Sum of Money
4. Determine the Time Required to Double or Triple a Lump Sum of Money

1 Interest is money paid for the use of money. The total amount borrowed (whether by an individual from a bank in the form of a loan or by a bank from an individual in the form of a savings account) is called the **principal.** The **rate of interest,** expressed as a percent, is the amount charged for the use of the principal for a given period of time, usually on a yearly (that is, per annum) basis.

Simple Interest Formula

If a principal of P dollars is borrowed for a period of t years at a per annum interest rate r, expressed as a decimal, the interest I charged is

$$I = Prt \qquad \qquad (1)$$

Interest charged according to formula (1) is called **simple interest.**

In working with problems involving interest, we define the term **payment period** as follows:

Annually	Once per year	Monthly	12 times per year
Semiannually	Twice per year	Daily	365 times per year*
Quarterly	Four times per year		

When the interest due at the end of a payment period is added to the principal so that the interest computed at the end of the next payment period is based on this new principal amount (old principal + interest), the interest is said to have been **compounded. Compound interest** is interest paid on principal and previously earned interest.

EXAMPLE 1	**Computing Compound Interest**

A credit union pays interest of 8% per annum compounded quarterly on a certain savings plan. If $1000 is deposited in such a plan and the interest is left to accumulate, how much is in the account after 1 year?

Solution   We use the simple interest formula, $I = Prt$. The principal P is $1000 and the rate of interest is 8% = 0.08. After the first quarter of a year, the time t is $\frac{1}{4}$ year, so the interest earned is

$$I = Prt = (\$1000)(0.08)\left(\frac{1}{4}\right) = \$20$$

The new principal is $P + I = \$1000 + \$20 = \$1020$. At the end of the second quarter, the interest on this principal is

$$I = (\$1020)(0.08)\left(\frac{1}{4}\right) = \$20.40$$

At the end of the third quarter, the interest on the new principal of $1020 + $20.40 = $1040.40 is

$$I = (\$1040.40)(0.08)\left(\frac{1}{4}\right) = \$20.81$$

After the fourth quarter, the interest is

$$I = (\$1061.21)(0.08)\left(\frac{1}{4}\right) = \$21.22$$

After 1 year the account contains $1061.21 + $21.22 = $1082.43.   ■

The pattern of the calculations performed in Example 1 leads to a general formula for compound interest. To fix our ideas, let P represent the principal to be invested at a per annum interest rate r that is compounded n times per year, so the time of each compounding period is $\frac{1}{n}$ years. (For computing purposes, r is expressed as a decimal.) The interest earned after each compounding period is the principal $P \times$ rate $r \times$ time $\frac{1}{n} = P \cdot \frac{r}{n}$. The amount A after one compounding period is

$$A = P + P\left(\frac{r}{n}\right) = P\left(1 + \frac{r}{n}\right)$$

*Most banks use a 360-day "year." Why do you think they do?

After two compounding periods, the amount A, based on the new principal $P\left(1 + \dfrac{r}{n}\right)$, is

$$A = \underbrace{P\left(1 + \frac{r}{n}\right)}_{\substack{\text{New}\\ \text{principal}}} + \underbrace{P\left(1 + \frac{r}{n}\right)\left(\frac{r}{n}\right)}_{\substack{\text{Interest on}\\ \text{new principal}}} = P\left(1 + \frac{r}{n}\right)\left(1 + \frac{r}{n}\right) = P\left(1 + \frac{r}{n}\right)^2$$

After three compounding periods,

$$A = P\left(1 + \frac{r}{n}\right)^2 + P\left(1 + \frac{r}{n}\right)^2\left(\frac{r}{n}\right) = P\left(1 + \frac{r}{n}\right)^2\left(1 + \frac{r}{n}\right) = P\left(1 + \frac{r}{n}\right)^3$$

Continuing this way, after n compounding periods (1 year),

$$A = P\left(1 + \frac{r}{n}\right)^n$$

Because t years will contain $n \cdot t$ compounding periods, after t years we have

$$A = P\left(1 + \frac{r}{n}\right)^{nt}$$

Theorem

Compound Interest Formula

The amount A after t years due to a principal P invested at an annual interest rate r compounded n times per year is

$$A = P\left(1 + \frac{r}{n}\right)^{nt} \qquad\qquad \textbf{(2)}$$

For example, to rework Example 1, we would use $P = \$1000, r = 0.08,$ $n = 4$, (quarterly compounding) and $t = 1$ year to obtain

$$A = 1000\left(1 + \frac{0.08}{4}\right)^4 = \$1082.43$$

In Equation (2), the amount A is typically referred to as the **accumulated value or future value** of the account, while P is called the **present value.**

 EXPLORATION To see the effects of compounding interest monthly on an initial deposit of \$1, graph $Y_1 = \left(1 + \frac{r}{12}\right)^{12x}$ with $r = 0.06$ and $r = 0.12$ for $0 \le x \le 30$. What is the future value of \$1 in 30 years when the interest rate per annum is $r = 0.06$ (6%)? What is the future value of \$1 in 30 years when the interest rate per annum is $r = 0.12$ (12%)? Does doubling the interest rate double the future value? ∎

NOTE: In using your calculator, be sure to use stored values, rather than approximations, in order to avoid round-off errors. At the final step, round money to the nearest cent.

 NOW WORK PROBLEM 1.

EXAMPLE 2

Comparing Investments Using Different Compounding Periods

Investing \$1000 at an annual rate of 10% compounded annually, quarterly, monthly, and daily will yield the following amounts after 1 year:

Annual compounding:
$$A = P(1 + r)$$
$$= (\$1000)(1 + 0.10) = \$1100.00$$

Quarterly compounding:
$$A = P\left(1 + \frac{r}{4}\right)^4$$
$$= (\$1000)(1 + 0.025)^4 = \$1103.81$$

Monthly compounding:
$$A = P\left(1 + \frac{r}{12}\right)^{12}$$
$$= (\$1000)(1 + 0.00833)^{12} = \$1104.71$$

Daily compounding:
$$A = P\left(1 + \frac{r}{365}\right)^{365}$$
$$= (\$1000)(1 + 0.000274)^{365} = \$1105.16$$

∎

From Example 2 we can see that the effect of compounding more frequently is that the amount after 1 year is higher: $1000 compounded 4 times a year at 10% results in $1103.81; $1000 compounded 12 times a year at 10% results in $1104.71; and $1000 compounded 365 times a year at 10% results in $1105.16. This leads to the following question: What would happen to the amount after 1 year if the number of times that the interest is compounded were increased without bound?

Let's find the answer. Suppose that P is the principal, r is the per annum interest rate, and n is the number of times that the interest is compounded each year. The amount after 1 year is

$$A = P\left(1 + \frac{r}{n}\right)^n$$

Rewrite this expression as follows:

$$A = P\left(1 + \frac{r}{n}\right)^n = P\left(1 + \frac{1}{n/r}\right)^n = P\left[\left(1 + \frac{1}{n/r}\right)^{n/r}\right]^r = P\left[\left(1 + \frac{1}{h}\right)^h\right]^r \quad (3)$$

$$h = \frac{n}{r}$$

Now suppose that the number n of times that the interest is compounded per year gets larger and larger; that is, suppose that $n \to \infty$. Then $h = \dfrac{n}{r} \to \infty$, and the expression in brackets equals e. [Refer to equation (2), p. 421.] That is, $A \to Pe^r$.

Table 6 compares $\left(1 + \dfrac{r}{n}\right)^n$, for large values of n, to e^r for $r = 0.05$, $r = 0.10$, $r = 0.15$, and $r = 1$. The larger that n gets, the closer $\left(1 + \dfrac{r}{n}\right)^n$ gets to e^r. No matter how frequent the compounding, the amount after 1 year has the definite ceiling Pe^r.

TABLE 6

$$\left(1 + \frac{r}{n}\right)^n$$

	$n = 100$	$n = 1000$	$n = 10,000$	e^r
$r = 0.05$	1.0512580	1.0512698	1.051271	1.0512711
$r = 0.10$	1.1051157	1.1051654	1.1051704	1.1051709
$r = 0.15$	1.1617037	1.1618212	1.1618329	1.1618342
$r = 1$	2.7048138	2.7169239	2.7181459	2.7182818

When interest is compounded so that the amount after 1 year is Pe^r, we say the interest is **compounded continuously.**

Theorem

Continuous Compounding

The amount A after t years due to a principal P invested at an annual interest rate r compounded continuously is

$$A = Pe^{rt} \qquad \text{(4)}$$

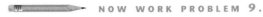

EXAMPLE 3

Using Continuous Compounding

The amount A that results from investing a principal P of $1000 at an annual rate r of 10% compounded continuously for a time t of 1 year is

$$A = \$1000e^{0.10} = (\$1000)(1.10517) = \$1105.17$$

 NOW WORK PROBLEM **9.**

2 The **effective rate of interest** is the equivalent annual simple rate of interest that would yield the same amount as compounding after 1 year. For example, based on Example 3, a principal of $1000 will result in $1105.17 at a rate of 10% compounded continuously. To get this same amount using a simple rate of interest would require that interest of $1105.17 − $1000.00 = $105.17 be earned on the principal. Since $105.17 is 10.517% of $1000, a simple rate of interest of 10.517% is needed to equal 10% compounded continuously. The effective rate of interest of 10% compounded continuously is 10.517%.

Based on the results of Examples 2 and 3, we find the following comparisons:

	Annual Rate	Effective Rate
Annual compounding	10%	10%
Quarterly compounding	10%	10.381%
Monthly compounding	10%	10.471%
Daily compounding	10%	10.516%
Continuous compounding	10%	10.517%

NOW WORK PROBLEM **21.**

EXAMPLE 4

Computing the Value of an IRA

On January 2, 2000, $2000 is placed in an Individual Retirement Account (IRA) that will pay interest of 10% per annum compounded continuously. What will the IRA be worth on January 1, 2020?

Solution

The amount A after 20 years is

$$A = Pe^{rt} = \$2000e^{(0.10)(20)} = \$14,778.11$$

 EXPLORATION How long will it be until A = $4000? $6000?
[**Hint:** Graph $Y_1 = 2000e^{0.1x}$ and $Y_2 = 4000$. Use INTERSECT to find x.]

③ When people engaged in finance speak of the "time value of money," they are usually referring to the *present value* of money. The **present value** of A dollars to be received at a future date is the principal that you would need to invest now so that it would grow to A dollars in the specified time period. The present value of money to be received at a future date is always less than the amount to be received, since the amount to be received will equal the present value (money invested now) *plus* the interest accrued over the time period.

We use the compound interest formula (2) to get a formula for present value. If P is the present value of A dollars to be received after t years at a per annum interest rate r compounded n times per year, then, by formula (2),

$$A = P\left(1 + \frac{r}{n}\right)^{nt}$$

To solve for P, we divide both sides by $\left(1 + \dfrac{r}{n}\right)^{nt}$. The result is

$$\frac{A}{\left(1 + \dfrac{r}{n}\right)^{nt}} = P \quad \text{or} \quad P = A\left(1 + \frac{r}{n}\right)^{-nt}$$

Time is money

Theorem   **Present Value Formulas**

The present value P of A dollars to be received after t years, assuming a per annum interest rate r compounded n times per year, is

$$P = A\left(1 + \frac{r}{n}\right)^{-nt} \tag{5}$$

If the interest is compounded continuously, then

$$P = Ae^{-rt} \tag{6}$$

To prove (6), solve formula (4) for P.

EXAMPLE 5   Computing the Value of a Zero-Coupon Bond

A zero-coupon (noninterest-bearing) bond can be redeemed in 10 years for $1000. How much should you be willing to pay for it now if you want a return of

(a) 8% compounded monthly?   (b) 7% compounded continuously?

Solution   (a) We are seeking the present value of $1000. We use formula (5) with A = $1000, n = 12, r = 0.08, and t = 10.

$$P = A\left(1 + \frac{r}{n}\right)^{-nt}$$

$$= \$1000\left(1 + \frac{0.08}{12}\right)^{-12(10)}$$

$$= \$450.52$$

For a return of 8% compounded monthly, you should pay $450.52 for the bond.

(b) Here we use formula (6) with $A = \$1000$, $r = 0.07$, and $t = 10$.

$$P = Ae^{-rt}$$
$$= \$1000e^{-(0.07)(10)}$$
$$= \$496.59$$

For a return of 7% compounded continuously, you should pay $496.59 for the bond. ∎

NOW WORK PROBLEM 11.

EXAMPLE 6

Rate of Interest Required to Double an Investment

④ What annual rate of interest compounded annually should you seek if you want to double your investment in 5 years?

Solution   If P is the principal and we want P to double, the amount A will be $2P$. We use the compound interest formula with $n = 1$ and $t = 5$ to find r.

$$2P = P(1 + r)^5$$
$$2 = (1 + r)^5$$
$$1 + r = \sqrt[5]{2}$$
$$r = \sqrt[5]{2} - 1 = 1.148698 - 1 = 0.148698$$

The annual rate of interest needed to double the principal in 5 years is 14.87%. ∎

NOW WORK PROBLEM 23.

EXAMPLE 7

Doubling and Tripling Time for an Investment

(a) How long will it take for an investment to double in value if it earns 5% compounded continuously?

(b) How long will it take to triple at this rate?

Solution   (a) If P is the initial investment and we want P to double, the amount A will be $2P$. We use formula (4) for continuously compounded interest with $r = 0.05$. Then

$$A = Pe^{rt}$$
$$2P = Pe^{0.05t}$$
$$2 = e^{0.05t}$$
$$0.05t = \ln 2$$
$$t = \frac{\ln 2}{0.05} = 13.86$$

It will take about 14 years to double the investment.

(b) To triple the investment, we set $A = 3P$ in formula (4).

$$A = Pe^{rt}$$

$$3P = Pe^{0.05t}$$

$$3 = e^{0.05t}$$

$$0.05t = \ln 3$$

$$t = \frac{\ln 3}{0.05} = 21.97$$

It will take about 22 years to triple the investment.

NOW WORK PROBLEM **29.**

7.5   EXERCISES

In Problems 1–10, find the amount that results from each investment.

1. $100 invested at 4% compounded quarterly after a period of 2 years

2. $50 invested at 6% compounded monthly after a period of 3 years

3. $500 invested at 8% compounded quarterly after a period of $2\frac{1}{2}$ years

4. $300 invested at 12% compounded monthly after a period of $1\frac{1}{2}$ years

5. $600 invested at 5% compounded daily after a period of 3 years

6. $700 invested at 6% compounded daily after a period of 2 years

7. $10 invested at 11% compounded continuously after a period of 2 years

8. $40 invested at 7% compounded continuously after a period of 3 years

9. $100 invested at 10% compounded continuously after a period of $2\frac{1}{4}$ years

10. $100 invested at 12% compounded continuously after a period of $3\frac{3}{4}$ years

In Problems 11–20, find the principal needed now to get each amount; that is, find the present value.

11. To get $100 after 2 years at 6% compounded monthly

12. To get $75 after 3 years at 8% compounded quarterly

13. To get $1000 after $2\frac{1}{2}$ years at 6% compounded daily

14. To get $800 after $3\frac{1}{2}$ years at 7% compounded monthly

15. To get $600 after 2 years at 4% compounded quarterly

16. To get $300 after 4 years at 3% compounded daily

17. To get $80 after $3\frac{1}{4}$ years at 9% compounded continuously

18. To get $800 after $2\frac{1}{2}$ years at 8% compounded continuously

19. To get $400 after 1 year at 10% compounded continuously

20. To get $1000 after 1 year at 12% compounded continuously

21. Find the effective rate of interest for $5\frac{1}{4}$% compounded quarterly.

22. What interest rate compounded quarterly will give an effective interest rate of 7%?

23. What rate of interest compounded annually is required to double an investment in 3 years?

24. What rate of interest compounded annually is required to double an investment in 10 years?

In Problems 25–28, which of the two rates would yield the larger amount in 1 year?
[**Hint:** Start with a principal of $10,000 in each instance.]

25. 6% compounded quarterly or $6\frac{1}{4}$% compounded annually

26. 9% compounded quarterly or $9\frac{1}{4}$% compounded annually

27. 9% compounded monthly or 8.8% compounded daily

28. 8% compounded semiannually or 7.9% compounded daily

29. How long does it take for an investment to double in value if it is invested at 8% per annum compounded monthly? Compounded continuously?

30. How long does it take for an investment to double in value if it is invested at 10% per annum compounded monthly? Compounded continuously?

31. If Tanisha has $100 to invest at 8% per annum compounded monthly, how long will it be before she has $150? If the compounding is continuous, how long will it be?

32. If Angela has $100 to invest at 10% per annum compounded monthly, how long will it be before she has $175? If the compounding is continuous, how long will it be?

33. How many years will it take for an initial investment of $10,000 to grow to $25,000? Assume a rate of interest of 6% compounded continuously.

34. How many years will it take for an initial investment of $25,000 to grow to $80,000? Assume a rate of interest of 7% compounded continuously.

35. What will a $90,000 house cost 5 years from now if the inflation rate over that period averages 3% compounded annually?

36. Sears charges 1.25% per month on the unpaid balance for customers with charge accounts (interest is compounded monthly). A customer charges $200 and does not pay her bill for 6 months. What is the bill at that time?

37. Jerome will be buying a used car for $15,000 in 3 years. How much money should he ask his parents for now so that, if he invests it at 5% compounded continuously, he will have enough to buy the car?

38. John will require $3000 in 6 months to pay off a loan that has no prepayment privileges. If he has the $3000 now, how much of it should he save in an account paying 3% compounded monthly so that in 6 months he will have exactly $3000?

39. George is contemplating the purchase of 100 shares of a stock selling for $15 per share. The stock pays no dividends. The history of the stock indicates that it should grow at an annual rate of 15% per year. How much will the 100 shares of stock be worth in 5 years?

40. Tracy is contemplating the purchase of 100 shares of a stock selling for $15 per share. The stock pays no dividends. Her broker says that the stock will be worth $20 per share in 2 years. What is the annual rate of return on this investment?

41. A business purchased for $650,000 in 1994 is sold in 1997 for $850,000. What is the annual rate of return for this investment?

42. Tanya has just inherited a diamond ring appraised at $5000. If diamonds have appreciated in value at an an-

nual rate of 8%, what was the value of the ring 10 years ago when the ring was purchased?

43. Jim places $1000 in a bank account that pays 5.6% compounded continuously. After 1 year, will he have enough money to buy a computer system that costs $1060? If another bank will pay Jim 5.9% compounded monthly, is this a better deal?

44. On January 1, Kim places $1000 in a certificate of deposit that pays 6.8% compounded continuously and matures in 3 months. Then Kim places the $1000 and the interest in a passbook account that pays 5.25% compounded monthly. How much does Kim have in the passbook account on May 1?

45. Will invests $2000 in a bond trust that pays 9% interest compounded semiannually. His friend Henry invests $2000 in a certificate of deposit (CD) that pays $8\frac{1}{2}$% compounded continuously. Who has more money after 20 years, Will or Henry?

46. Suppose that April has access to an investment that will pay 10% interest compounded continuously. Which is better: To be given $1000 now so that she can take advantage of this investment opportunity or to be given $1325 after 3 years?

47. Colleen and Bill have just purchased a house for $150,000, with the seller holding a second mortgage of $50,000. They promise to pay the seller $50,000 plus all accrued interest 5 years from now. The seller offers them three interest options on the second mortgage:
(a) Simple interest at 12% per annum
(b) $11\frac{1}{2}$% interest compounded monthly
(c) $11\frac{1}{4}$% interest compounded continuously
Which option is best; that is, which results in the least interest on the loan?

48. The First National Bank advertises that it pays interest on savings accounts at the rate of 4.25% compounded daily. Find the effective rate if the bank uses (a) 360 days or (b) 365 days in determining the daily rate.

Problems 49–52 involve zero-coupon bonds. A zero-coupon bond is a bond that is sold now at a discount and will pay its face value at the time when it matures; no interest payments are made.

49. A zero-coupon bond can be redeemed in 20 years for $10,000. How much should you be willing to pay for it now if you want a return of:
(a) 10% compounded monthly?
(b) 10% compounded continuously?

50. A child's grandparents are considering buying a $40,000 face value zero-coupon bond at birth so that she will have enough money for her college education 17 years later. If they want a rate of return of 8% compounded annually, what should they pay for the bond?

51. How much should a $10,000 face value zero-coupon bond, maturing in 10 years, be sold for now if its rate of return is to be 8% compounded annually?

52. If Pat pays $12,485.52 for a $25,000 face value zero-coupon bond that matures in 8 years, what is his annual rate of return?

53. Time to Double or Triple an Investment The formula

$$t = \frac{\ln m}{n \ln\left(1 + \dfrac{r}{n}\right)}$$

can be used to find the number of years t required to multiply an investment m times when r is the per annum interest rate compounded n times a year.
(a) How many years will it take to double the value of an IRA that compounds annually at the rate of 12%?
(b) How many years will it take to triple the value of a savings account that compounds quarterly at an annual rate of 6%?
(c) Give a derivation of this formula.

54. Time to Reach an Investment Goal   The formula

$$t = \frac{\ln A - \ln P}{r}$$

can be used to find the number of years t required for an investment P to grow to a value A when compounded continuously at an annual rate r.

(a) How long will it take to increase an initial investment of $1000 to $8000 at an annual rate of 10%?

(b) What annual rate is required to increase the value of a $2000 IRA to $30,000 in 35 years?

(c) Give a derivation of this formula.

55. Explain in your own words what the term *compound interest* means. What does *continuous compounding* mean?

56. Explain in your own words the meaning of *present value*.

57. Critical Thinking   You have just contracted to buy a house and will seek financing in the amount of $100,000. You go to several banks. Bank 1 will lend you $100,000 at the rate of 8.75% amortized over 30 years with a loan origination fee of 1.75%. Bank 2 will lend you $100,000 at the rate of 8.375% amortized over 15 years with a loan origination fee of 1.5%. Bank 3 will lend you $100,000 at the rate of 9.125% amortized over 30 years with no loan origination fee. Bank 4 will lend you $100,000 at the rate of 8.625% amortized over 15 years with no loan origination fee. Which loan would you take? Why? Be sure to have sound reasons for your choice. Use the information in the table to assist you. If the amount of the monthly payment does not matter to you, which loan would you take? Again, have sound reasons for your choice. Compare your final decision with others in the class. Discuss.

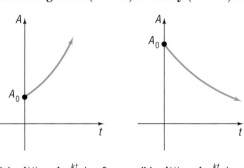

	Monthly Payment	Loan Origination Fee
Bank 1	$786.70	$1,750.00
Bank 2	$977.42	$1,500.00
Bank 3	$813.63	$0.00
Bank 4	$990.68	$0.00

7.6   GROWTH AND DECAY; NEWTON'S LAW; LOGISTIC MODELS

OBJECTIVES
1 Find Equations of Populations That Obey the Law of Uninhibited Growth
2 Find Equations of Populations That Obey the Law of Uninhibited Decay
3 Use Newton's Law of Cooling
4 Use Logistic Growth Models

1   Many natural phenomena have been found to follow the law that an amount A varies with time t according to

$$A(t) = A_0 e^{kt} \tag{1}$$

Here $A_0 = A(0)$ is the original amount $(t = 0)$ and $k \neq 0$ is a constant.

If $k > 0$, then equation (1) states that the amount A is increasing over time; if $k < 0$, the amount A is decreasing over time. In either case, when an amount A varies over time according to equation (1), it is said to follow the **exponential law** or the **law of uninhibited growth** $(k > 0)$ **or decay** $(k < 0)$. See Figure 27.

Figure 27

(a) $A(t) = A_0 e^{kt}, k > 0$ (b) $A(t) = A_0 e^{kt}, k < 0$

For example, we saw in Section 7.5 that continuously compounded interest follows the law of uninhibited growth. In this section we shall look at three additional phenomena that follow the exponential law.

UNINHIBITED GROWTH

Cell division is a process in the growth of many living organisms, such as amoebas, plants, and human skin cells. Based on an ideal situation in which no cells die and no by-products are produced, the number of cells present at a given time follows the law of uninhibited growth. Actually, however, after enough time has passed, growth at an exponential rate will cease due to the influence of factors such as lack of living space and dwindling food supply. The law of uninhibited growth accurately reflects only the early stages of the cell division process.

The cell division process begins with a culture containing N_0 cells. Each cell in the culture grows for a certain period of time and then divides into two identical cells. We assume that the time needed for each cell to divide in two is constant and does not change as the number of cells increases. These new cells then grow, and eventually each divides in two, and so on.

A model that gives the number N of cells in the culture after a time t has passed (in the early stages of growth) is given next.

Uninhibited Growth of Cells

$$N(t) = N_0 e^{kt}, \qquad k > 0 \qquad\qquad (2)$$

Here $N_0 = N(0)$ is the initial number of cells and k is a positive constant that represents the growth rate of the cells.

In using formula (2) to model the growth of cells, we are using a function that yields positive real numbers, even though we are counting the number of cells, which must be an integer. This is a common practice in many applications.

EXAMPLE 1   **Uninhibited Growth**

A certain culture of cells increases according to the equation

$$N(t) = 75e^{0.32t}$$

where t is time in hours.

(a) How many cells are in the culture initially?
(b) What is the growth rate of the culture?
(c) How many cells are in the culture after 4 hours?

Solution   (a) The initial number of cells is obtained when $t = 0$.

$$N_0 = N(0) = 75e^{0.32(0)} = 75$$

There are 75 cells present initially.

(b) Compare $N(t) = 75e^{0.32t}$ to $N(t) = 75e^{kt}$. The value of k, 0.32, indicates a growth rate of 32% per hour.

(c) After 4 hours, $t = 4$, we have

$$N(4) = 75e^{0.32(4)} \approx 269.75$$

There are about 270 cells after 4 hours. ■

EXAMPLE 2   **Bacterial Growth**

A colony of bacteria increases according to the law of uninhibited growth.

(a) If the number of bacteria doubles in 3 hours, find k in formula (2) and express N as a function of t.

(b) How long will it take for the size of the colony to triple?

(c) How long does it take for the population to double a second time (that is, increase four times).

Solution   (a) Using formula (2), the number N of cells at a time t is

$$N(t) = N_0 e^{kt}$$

where N_0 is the initial number of bacteria present and k is a positive number. We first seek the number k. The number of cells doubles in 3 hours, so we have

$$N(3) = 2N_0$$

But $N(3) = N_0 e^{k(3)}$, so

$$N_0 e^{k(3)} = 2N_0$$

$$e^{3k} = 2$$

$$3k = \ln 2 \qquad \text{Write the exponential equation as a logarithm.}$$

$$k = \tfrac{1}{3} \ln 2 \approx 0.2310$$

Formula (2) for this growth process is therefore

$$N(t) = N_0 e^{0.2310t}$$

(b) The time t needed for the size of the colony to triple requires that $N(t) = 3N_0$. We substitute $3N_0$ for $N(t)$ to get

$$3N_0 = N_0 e^{0.2310t}$$

$$3 = e^{0.2310t}$$

$$0.2310t = \ln 3$$

$$t = \frac{\ln 3}{0.2310} \approx 4.756 \text{ hours}$$

It will take about 4.756 hours for the size of the colony to triple.

(c) If a population doubles in 3 hours, it will double a second time in 3 more hours, for a total time of 6 hours. ■

> ─── **NOW WORK PROBLEM 1.**

RADIOACTIVE DECAY

2   Radioactive materials follow the law of uninhibited decay. The amount A of a radioactive material present at time t is given by the following model:

Uninhibited Radioactive Decay

$$A(t) = A_0 e^{kt}, \qquad k < 0 \qquad \text{(3)}$$

Here $A_0 = A(0)$ is the original amount of radioactive material and k is a negative number that represents the rate of decay.

All radioactive substances have a specific **half-life,** which is the time required for half of the radioactive substance to decay. In **carbon dating,** we use the fact that all living organisms contain two kinds of carbon, carbon 12 (a stable carbon) and carbon 14 (a radioactive carbon, with a half-life of 5600 years). While an organism is living, the ratio of carbon 12 to carbon 14 is constant. But when an organism dies, the original amount of carbon 12 present remains unchanged, whereas the amount of carbon 14 begins to decrease. This change in the amount of carbon 14 present relative to the amount of carbon 12 present makes it possible to calculate when an organism died.

EXAMPLE 3   ## Estimating the Age of Ancient Tools

Traces of burned wood along with ancient stone tools in an archeological dig in Chile were found to contain approximately 1.67% of the original amount of carbon 14. If the half-life of carbon 14 is 5600 years, approximately when was the tree cut and burned?

Solution   Using formula (3), the amount A of carbon 14 present at time t is

$$A(t) = A_0 e^{kt}$$

where A_0 is the original amount of carbon 14 present and k is a negative number. We first seek the number k. To find it, we use the fact that after 5600 years, half of the original amount of carbon 14 remains, so $A(5600) = \dfrac{1}{2} A_0$. Thus,

$$\frac{1}{2} A_0 = A_0 e^{k(5600)}$$

$$\frac{1}{2} = e^{5600k}$$

$$5600k = \ln \frac{1}{2}$$

$$k = \frac{\ln \frac{1}{2}}{5600} \approx -0.000124$$

Formula (3) therefore becomes

$$A(t) = A_0 e^{-0.000124t}$$

If the amount A of carbon 14 now present is 1.67% of the original amount, it follows that

$$0.0167 A_0 = A_0 e^{-0.000124t}$$

$$0.0167 = e^{-0.000124t}$$

$$-0.000124t = \ln 0.0167$$

$$t = \frac{\ln 0.0167}{-0.000124} \approx 33,000 \text{ years}$$

The tree was cut and burned about 33,000 years ago. Some archeologists use this conclusion to argue that humans lived in the Americas 33,000 years ago, much earlier than is generally accepted.   ■

NOW WORK PROBLEM 3.

NEWTON'S LAW OF COOLING

③   **Newton's Law of Cooling*** states that the temperature of a heated object decreases exponentially over time toward the temperature of the surrounding medium.

Newton's Law of Cooling

The temperature u of a heated object at a given time t can be modeled by the following function:

$$u(t) = T + (u_0 - T)e^{kt}, \quad k < 0 \tag{4}$$

where T is the constant temperature of the surrounding medium, u_0 is the initial temperature of the heated object, and k is a negative constant.

EXAMPLE 4   **Using Newton's Law of Cooling**

An object is heated to 100°C (degrees Celsius) and is then allowed to cool in a room whose air temperature is 30°C.

(a) If the temperature of the object is 80°C after 5 minutes, when will its temperature be 50°C?

(b) Determine the elapsed time before the temperature of the object is 35°C.

✍ (c) What do you notice about $u(t)$, the temperature, as t, time, passes?

Solution   (a) Using formula (4) with $T = 30$ and $u_0 = 100$, the temperature (in degrees Celsius) of the object at time t (in minutes) is

$$u(t) = 30 + (100 - 30)e^{kt} = 30 + 70e^{kt} \tag{5}$$

where k is a negative constant. To find k, we use the fact that $u = 80$ when $t = 5$. Then

$$u(t) = 30 + 70e^{kt}$$
$$80 = 30 + 70e^{k(5)} \qquad t = 5; u(5) = 80$$
$$50 = 70e^{5k}$$
$$e^{5k} = \frac{50}{70}$$
$$5k = \ln\frac{5}{7}$$
$$k = \frac{\ln\frac{5}{7}}{5} \approx -0.0673$$

Formula (5) therefore becomes

$$u(t) = 30 + 70e^{-0.0673t} \tag{6}$$

△ *Named after Sir Isaac Newton (1642–1727), one of the cofounders of calculus.

We want to find t when $u = 50°C$, so

$$50 = 30 + 70e^{-0.0673t}$$

$$20 = 70e^{-0.0673t}$$

$$e^{-0.0673t} = \frac{20}{70}$$

$$-0.0673t = \ln\frac{2}{7}$$

$$t = \frac{\ln\frac{2}{7}}{-0.0673} \approx 18.6 \text{ minutes}$$

The temperature of the object will be 50°C after about 18.6 minutes.

(b) If $u = 35°C$, then, based on equation (6), we have

$$35 = 30 + 70e^{-0.0673t}$$

$$5 = 70e^{-0.0673t}$$

$$e^{-0.0673t} = \frac{5}{70}$$

$$-0.0673t = \ln\frac{5}{70}$$

$$t = \frac{\ln\frac{5}{70}}{-0.0673} \approx 39.2 \text{ minutes}$$

The object will reach a temperature of 35°C after about 39.2 minutes.

(c) Refer to equation (6). As time passes, the value of t increases, the value of $e^{-0.0673t}$ approaches zero, and the value of $u(t)$ approaches 30°C. ■

NOW WORK PROBLEM **13**.

Logistic Models

④ The exponential growth model $A(t) = A_0e^{kt}$, $k > 0$, assumes uninhibited growth, meaning that the value of the function grows without limit. Earlier we stated that cell division could be modeled using this function, assuming that no cells die and no by-products are produced. However, cell division would eventually be limited by factors such as living space and food supply. The **logistic growth model** is an exponential function that can model situations where the growth of the dependent variable is limited.

Other situations that lead to a logistic growth model include population growth and the sales of a product due to advertising. See Problems 21 through 24. The logistic growth model is given next.

Logistic Growth Model

$$P(t) = \frac{c}{1 + ae^{-bt}}$$

Here a, b, and c are constants with $c > 0$ and $b > 0$.

The number c is called the **carrying capacity** because the value $P(t)$ approaches c as t approaches infinity; that is, $\lim_{t \to \infty} P(t) = c$.

| EXAMPLE 5 | **Fruit Fly Population** |

Fruit flies are placed in a half-pint milk bottle with a banana (for food) and yeast plants (for food and to provide a stimulus to lay eggs). Suppose that the fruit fly population after t days is given by

$$P(t) = \frac{230}{1 + 56.5e^{-0.37t}}$$

(a) What is the carrying capacity of the half-pint bottle? That is, what is $P(t)$ as $t \to \infty$?

(b) How many fruit flies were initially placed in the half-pint bottle?

(c) When will the population of fruit flies be 180?

 (d) Using a graphing utility, graph $P(t)$.

Solution   (a) As $t \to \infty$, $e^{-0.37t} \to 0$ and $P(t) \to 230/1$. The carrying capacity of the half-pint bottle is 230 fruit flies.

(b) To find the initial number of fruit flies in the half-pint bottle, we evaluate $P(0)$.

$$P(0) = \frac{230}{1 + 56.5e^{-0.37(0)}}$$

$$= \frac{230}{1 + 56.5}$$

$$= 4$$

So initially there were four fruit flies in the half-pint bottle.

(c) To determine when the population of fruit flies will be 180, we solve the equation

$$\frac{230}{1 + 56.5e^{-0.37t}} = 180$$

$$230 = 180\left(1 + 56.5e^{-0.37t}\right)$$

$$1.2778 = 1 + 56.5e^{-0.37t} \qquad \text{Divide both sides by 180.}$$

$$0.2778 = 56.5e^{-0.37t} \qquad \text{Subtract 1 from both sides.}$$

$$0.0049 = e^{-0.37t} \qquad \text{Divide both sides by 56.5.}$$

$$\ln(0.0049) = -0.37t \qquad \text{Rewrite as a logarithmic expression.}$$

$$t \approx 14.4 \text{ days} \qquad \text{Divide both sides by } -0.37.$$

Figure 28

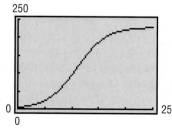

It will take approximately 14.4 days for the population to reach 180 fruit flies.

(d) See Figure 28 for the graph of $P(t)$.   ∎

 EXPLORATION   On the same viewing rectangle, graph $Y_1 = \dfrac{500}{1 + 24e^{-0.03t}}$ and $Y_2 = \dfrac{500}{1 + 24e^{-0.08t}}$. What effect does b have on the logistic growth function?   ∎

7.6 EXERCISES

1. **Growth of an Insect Population** The size P of a certain insect population at time t (in days) obeys the function $P(t) = 500e^{0.02t}$. After how many days will the population reach 1000? 2000?

2. **Growth of Bacteria** The number N of bacteria present in a culture at time t (in hours) obeys the function $N(t) = 1000e^{0.01t}$. After how many hours will the population equal 1500? 2000?

3. **Radioactive Decay** Strontium 90 is a radioactive material that decays according to the function $A(t) = A_0 e^{-0.0244t}$, where A_0 is the initial amount present and A is the amount present at time t (in years).
   (a) What is the half-life of strontium 90?
   (b) Determine how long it takes for 100 grams of strontium 90 to decay to 10 grams.

4. **Radioactive Decay** Iodine 131 is a radioactive material that decays according to the function $A(t) = A_0 e^{-0.087t}$, where A_0 is the initial amount present and A is the amount present at time t (in days).
   (a) What is the half-life of iodine 131?
   (b) Determine how long it takes for 100 grams of iodine 131 to decay to 10 grams.

5. **Growth of a Colony of Mosquitoes** The population of a colony of mosquitoes obeys the law of uninhibited growth. If there are 1000 mosquitoes initially and there are 1800 after 1 day, what is the size of the colony after 3 days? How long is it until there are 10,000 mosquitoes?

6. **Bacterial Growth** A culture of bacteria obeys the law of uninhibited growth. If 500 bacteria are present initially and there are 800 after 1 hour, how many will be present in the culture after 5 hours? How long is it until there are 20,000 bacteria?

7. **Population Growth** The population of a southern city follows the exponential law. If the population doubled in size over an 18-month period and the current population is 10,000, what will the population be 2 years from now?

8. **Population Growth** The population of a midwestern city follows the exponential law. If the population decreased from 900,000 to 800,000 from 1993 to 1995, what will the population be in 1997?

9. **Radioactive Decay** The half-life of radium is 1690 years. If 10 grams is present now, how much will be present in 50 years?

10. **Radioactive Decay** The half-life of radioactive potassium is 1.3 billion years. If 10 grams is present now, how much will be present in 100 years? In 1000 years?

11. **Estimating the Age of a Tree** A piece of charcoal is found to contain 30% of the carbon 14 that it originally had. When did the tree from which the charcoal came die? Use 5600 years as the half-life of carbon 14.

12. **Estimating the Age of a Fossil** A fossilized leaf contains 70% of its normal amount of carbon 14. How old is the fossil?

13. **Cooling Time of a Pizza** A pizza baked at 450°F is removed from the oven at 5:00 PM into a room that is a constant 70°F. After 5 minutes, the pizza is at 300°F.
   (a) At what time can you begin eating the pizza if you want its temperature to be 135°F?
   (b) Determine the time that needs to elapse before the pizza is 160°F.
   (c) What do you notice about the temperature as time passes?

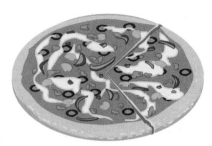

14. **Newton's Law of Cooling** A thermometer reading 72°F is placed in a refrigerator where the temperature is a constant 38°F.
   (a) If the thermometer reads 60°F after 2 minutes, what will it read after 7 minutes?
   (b) How long will it take before the thermometer reads 39°F?
   (c) Determine the time needed to elapse before the thermometer reads 45°F.
   (d) What do you notice about the temperature as time passes?

15. **Newton's Law** A thermometer reading 8°C is brought into a room with a constant temperature of 35°C. If the thermometer reads 15°C after 3 minutes, what will it read after being in the room for 5 minutes? For 10 minutes?
   [**Hint:** You need to construct a formula similar to equation (4).]

16. **Thawing Time of a Steak** A frozen steak has a temperature of 28°F. It is placed in a room with a constant temperature of 70°F. After 10 minutes, the temperature of the steak has risen to 35°F. What will the temperature of the steak be after 30 minutes? How long will it take the steak to thaw to a temperature of 45°F? [See the hint given for Problem 15.]

17. **Decomposition of Salt in Water** Salt (NaCl) decomposes in water into sodium (NA^+) and chloride (Cl^-) ions according to the law of uninhibited decay. If the initial amount of salt is 25 kilograms and, after 10 hours, 15 kilograms of salt is left, how much salt is left after 1 day? How long does it take until $\frac{1}{2}$ kilogram of salt is left?

18. **Voltage of a Conductor** The voltage of a certain conductor decreases over time according to the law of uninhibited decay. If the initial voltage is 40 volts, and 2 seconds later it is 10 volts, what is the voltage after 5 seconds?

19. **Radioactivity from Chernobyl**   After the release of radioactive material into the atmosphere from a nuclear power plant at Chernobyl (Ukraine) in 1986, the hay in Austria was contaminated by iodine 131 (half-life 8 days). If it is all right to feed the hay to cows when 10% of the iodine 131 remains, how long do the farmers need to wait to use this hay?

20. **Pig Roasts**   The hotel Bora-Bora is having a pig roast. At noon, the chef put the pig in a large earthen oven. The pig's original temperature was 75°F. At 2:00 PM the chef checked the pig's temperature and was upset because it had reached only 100°F. If the oven's temperature remains a constant 325°F, at what time may the hotel serve its guests, assuming that pork is done when it reaches 175°F?

21. **Proportion of the Population That Owns a VCR**   The logistic growth model

$$P(t) = \frac{0.9}{1 + 6e^{-0.32t}}$$

relates the proportion of U.S. households that own a VCR to the year. Let $t = 0$ represent 1984, $t = 1$ represent 1985, and so on.
(a) What proportion of the U.S. households owned a VCR in 1984?
(b) Determine the maximum proportion of households that will own a VCR.
(c) When will 0.8 (80%) of U.S. households own a VCR?

22. **Market Penetration of Intel's Coprocessor**   The logistic growth model

$$P(t) = \frac{0.90}{1 + 3.5e^{-0.339t}}$$

relates the proportion of new personal computers sold at Best Buy that have Intel's latest coprocessor t months after it has been introduced.
(a) What proportion of new personal computers sold at Best Buy will have Intel's latest coprocessor when it is first introduced (that is, at $t = 0$)?
(b) Determine the maximum proportion of new personal computers sold at Best Buy that will have Intel's latest coprocessor.
(c) When will 0.75 (75%) of new personal computers sold at Best Buy have Intel's latest coprocessor?

23. **Population of a Bacteria Culture**   The logistic growth model

$$P(t) = \frac{1000}{1 + 32.33e^{-0.439t}}$$

represents the population of a bacterium after t hours.
(a) What is the carrying capacity of the environment?
(b) What was the initial amount of bacteria in the population?
(c) When will the amount of bacteria be 800?

24. **Population of a Endangered Species**   Often environmentalists will capture an endangered species and transport the species to a controlled environment where the species can produce offspring and regenerate its population. Suppose that six American bald eagles are captured, transported to Montana, and set free. Based on experience, the environmentalists expect the population to grow according to the model

$$P(t) = \frac{500}{1 + 83.33e^{-0.162t}}$$

where $P(t)$ is the population after t years.
(a) What is the carrying capacity of the environment?
(b) What is the predicted population of the American bald eagle in 20 years?
(c) When will the population be 300?

7.7   LOGARITHMIC SCALES

OBJECTIVES   ① Work with Decibels (Loudness of Sound)
② Work with the Richter Scale (Earthquake Magnitude)

Common logarithms often appear in the measurement of quantities, because they provide a way to scale down positive numbers that vary from very small to very large. For example, if a certain quantity can take on values from $0.0000000001 = 10^{-10}$ to $10,000,000,000 = 10^{10}$, the common logarithms of such numbers would be between -10 and 10, respectively.

LOUDNESS OF SOUND

 Our first application utilizes a logarithmic scale to measure the loudness of a sound. Physicists define the **intensity of a sound wave** as the amount of energy that the wave transmits through a given area. For example, the least intense sound that a human ear can detect at a frequency of 100 hertz is about 10^{-12} watt per square meter. The **loudness** $L(x)$, measured in **decibels** (named in honor of Alexander Graham Bell), of a sound of intensity x (measured in watts per square meter) is defined as

$$L(x) = 10 \log \frac{x}{I_0} \qquad \textbf{(1)}$$

where $I_0 = 10^{-12}$ watt per square meter is the least intense sound that a human ear can detect. If we let $x = I_0$ in equation (1), we get

$$L(I_0) = 10 \log \frac{I_0}{I_0} = 10 \log 1 = 0$$

At the threshold of human hearing, the loudness is 0 decibels. Table 7 gives the loudness of some common sounds.

TABLE 7
LOUDNESS OF COMMON SOUNDS (IN DECIBELS)

Decibels	Common Sound	Result
140	Shotgun blast, jet 100 feet away at takeoff	Pain
130	Motor test chamber	Human ear pain threshold
120	Firecrackers, severe thunder, pneumatic jackhammer, hockey crowd	Uncomfortably loud
110	Amplified rock music	
100	Textile loom, subway train, elevated train, farm tractor, power lawn mower, newspaper press	Loud
90	Heavy city traffic, noisy factory	
80	Diesel truck going 40 mi/hr 50 feet away, crowded restaurant, garbage disposal, average factory, vacuum cleaner	Moderately loud
70	Passenger car going 50 mi/hr 50 feet away	
60	Quiet typewriter, singing birds, window air conditioner, quiet automobile	Quiet
50	Normal conversation, average office	
40	Household refrigerator, quiet office	Very quiet
30	Average home, dripping faucet, whisper 5 feet away	
20	Light rainfall, rustle of leaves	Average person's threshold of hearing
10	Whisper across room	Just audible
0		Threshold for acute hearing

Note that a decibel is not a linear unit like the meter. For example, a noise level of 10 decibels is 10 times as great as a noise level of 0 decibels. [If $L(x) = 10$, then $x = 10I_0$.] A noise level of 20 decibels is 100 times as great as a noise level of 0 decibels. [If $L(x) = 20$, then $x = 100I_0$.] A noise level of 30 decibels is 1000 times as great as a noise level of 0 decibels, and so on.

EXAMPLE 1   Finding the Intensity of a Sound

Use Table 7 to find the intensity of the sound of a dripping faucet.

Solution From Table 7 we see that the loudness of the sound of a dripping faucet is 30 decibels. By equation (1), its intensity x may be found as follows:

$$L(x) = 10 \log\left(\frac{x}{I_0}\right) \qquad \text{Equation (1)}$$

$$30 = 10 \log\left(\frac{x}{I_0}\right) \qquad L(x) = 30$$

$$3 = \log\left(\frac{x}{I_0}\right) \qquad \text{Divide by 10.}$$

$$\frac{x}{I_0} = 10^3 \qquad \text{Write in exponential form.}$$

$$x = 1000 I_0$$

where $I_0 = 10^{-12}$ watt per square meter. The intensity of the sound of a dripping faucet is 1000 times as great as a noise level of 0 decibels; that is, such a sound has an intensity of $1000 \cdot 10^{-12} = 10^{-9}$ watt per square meter. ■

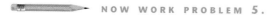 **NOW WORK PROBLEM 5.**

EXAMPLE 2   Finding the Loudness of a Sound

Use Table 7 to determine the loudness of a subway train if it is known that this sound is 10 times as intense as the sound due to heavy city traffic.

Solution The sound due to heavy city traffic has a loudness of 90 decibels. Its intensity, therefore, is the value of x in the equation

$$90 = 10 \log\left(\frac{x}{I_0}\right)$$

A sound 10 times as intense as x has loudness $L(10x)$. The loudness of the subway train is

$$L(10x) = 10 \log\left(\frac{10x}{I_0}\right) \qquad \text{Substitute } 10x \text{ for } x.$$

$$= 10 \log\left(10 \cdot \frac{x}{I_0}\right)$$

$$= 10\left[\log 10 + \log\left(\frac{x}{I_0}\right)\right] \qquad \text{Log of product} = \text{sum of logs}$$

$$= 10 \log 10 + 10 \log\left(\frac{x}{I_0}\right)$$

$$= 10 + 90 = 100 \text{ decibels} \qquad \log 10 = 1; 90 = 10 \log(x/I_0) \qquad ■$$

Magnitude of an Earthquake

(2) Our second application uses a logarithmic scale to measure the magnitude of an earthquake.

The **Richter scale**[*] is one way of converting seismographic readings into numbers that provide an easy reference for measuring the magnitude M of an earthquake. All earthquakes are compared to a **zero-level earthquake** whose seismographic reading measures 0.001 millimeter at a distance of 100 kilometers from the epicenter. An earthquake whose seismographic reading measures x millimeters has **magnitude** $M(x)$, given by

$$M(x) = \log \frac{x}{x_0} = \log \frac{x}{10^{-3}} \qquad (2)$$

where $x_0 = 10^{-3}$ is the reading of a zero-level earthquake the same distance from its epicenter.

EXAMPLE 3	Finding the Magnitude of an Earthquake

What is the magnitude of an earthquake whose seismographic reading is 0.1 millimeter at a distance of 100 kilometers from its epicenter?

Solution   If $x = 0.1$, the magnitude $M(x)$ of this earthquake is

$$M(0.1) = \log \frac{x}{x_0} = \log \left(\frac{0.1}{10^{-3}} \right) = \log \frac{10^{-1}}{10^{-3}} = \log 10^2 = 2$$

This earthquake measures 2.0 on the Richter scale. ∎

NOW WORK PROBLEM 7.

Based on formula (2), we define the **intensity of an earthquake** as the ratio of x to x_0. For example, the intensity of the earthquake described in Example 3 is $\frac{0.1}{0.001} = 10^2 = 100$. That is, it is 100 times as intense as a zero-level earthquake.

EXAMPLE 4	Comparing the Intensity of Two Earthquakes

The devastating San Francisco earthquake of 1906 measured 6.9 on the Richter scale. How did the intensity of that earthquake compare to the Papua, New Guinea, earthquake of 1988, which measured 6.7 on the Richter scale?

Figure 29

---

[*]Named after the American scientist, C. F. Richter, who devised it in 1935.

Solution   Let x_1 and x_2 denote the seismographic readings, respectively, of the 1906 San Francisco earthquake and the 1988 Papua, New Guinea, earthquake. Then, based on formula (2),

$$6.9 = \log \frac{x_1}{x_0} \qquad 6.7 = \log \frac{x_2}{x_0}$$

Consequently,

$$\frac{x_1}{x_0} = 10^{6.9} \qquad \frac{x_2}{x_0} = 10^{6.7}$$

The 1906 San Francisco earthquake was $10^{6.9}$ times as intense as a zero-level earthquake. The Papua, New Guinea, earthquake was $10^{6.7}$ times as intense as a zero-level earthquake. Form the ratio $\dfrac{x_1}{x_2}$ to compare the two earthquakes,

$$\frac{x_1}{x_2} = \frac{10^{6.9} x_0}{10^{6.7} x_0} = 10^{0.2} \approx 1.58$$

$$x_1 \approx 1.58 x_2$$

The San Francisco earthquake was 1.58 times as intense as the Papua, New Guinea, earthquake.   ■

Example 4 demonstrates that the relative intensity of two earthquakes can be found by raising 10 to a power equal to the difference of their readings on the Richter scale.

7.7 EXERCISES

1. **Loudness of a Dishwasher**   Find the loudness of a dishwasher that operates at an intensity of 10^{-5} watt per square meter. Express your answer in decibels.

2. **Loudness of a Diesel Engine**   Find the loudness of a diesel engine that operates at an intensity of 10^{-3} watt per square meter. Express your answer in decibels.

3. **Loudness of a Jet Engine**   With engines at full throttle, a Boeing 727 jetliner produces noise at an intensity of 0.15 watt per square meter. Find the loudness of the engines in decibels.

4. **Loudness of a Whisper**   A whisper produces noise at an intensity of $10^{-9.8}$ watt per square meter. What is the loudness of a whisper in decibels?

5. **Intensity of a Sound at the Threshold of Pain**   For humans, the threshold of pain due to sound averages 130 decibels. What is the intensity of such a sound in watts per square meter?

6. **Comparing Sounds**   If one sound is 50 times as intense as another, what is the difference in the loudness of the two sounds? Express your answer in decibels.

7. **Magnitude of an Earthquake**   Find the magnitude of an earthquake whose seismographic reading is 10.0 mil-

limeters at a distance of 100 kilometers from its epicenter.

8. **Magnitude of an Earthquake**   Find the magnitude of an earthquake whose seismographic reading is 1210 millimeters at a distance of 100 kilometers from its epicenter.

9. **Comparing Earthquakes**   The Mexico City earthquake of 1985 registered 8.1 on the Richter scale. What would a seismograph 100 kilometers from the epicenter have measured for this earthquake? How does this earthquake compare in intensity to the 1906 San Francisco earthquake, which registered 6.9 on the Richter scale?

10. **Comparing Earthquakes**   Two earthquakes differ by 1.0 when measured on the Richter scale. How would the seismographic readings differ at a distance of 100 kilometers from the epicenter? How do their intensities compare?

11. **NBA Finals 1997**   In game 5 of the NBA Finals between the Chicago Bulls and the Utah Jazz at the Delta Center, the crowd noise measured 110 decibels. NBA guidelines say sound levels are not to exceed 95 decibels. Compute the ratio of the intensities of these two sounds to determine by how much the crowd noise exceeded guidelines.

CHAPTER REVIEW

Things To Know

Properties of the exponential function (pp. 419 and 420)

$f(x) = a^x, \quad a > 1$

Domain: the interval $(-\infty, \infty)$; Range: the interval $(0, \infty)$;
x-intercepts: none; y-intercept: 1;
horizontal asymptote: x-axis $(y = 0)$ as $x \to -\infty$;
increasing; one-to-one; smooth; continuous

See Figure 4 for a typical graph.

$f(x) = a^x, \quad 0 < a < 1$

Domain: the interval $(-\infty, \infty)$; Range: the interval $(0, \infty)$;
x-intercepts: none; y-intercept: 1;
horizontal asymptote: x-axis $(y = 0)$ as $x \to \infty$;
decreasing; one-to-one; smooth; continuous

See Figure 8 for a typical graph.

Number e (p. 421)

Value approached by the expression $\left(1 + \dfrac{1}{n}\right)^n$ as $n \to \infty$; that is, $\displaystyle\lim_{n \to \infty}\left(1 + \dfrac{1}{n}\right)^n = e$.

Property of exponents (p. 423)

If $a^u = a^v$, then $u = v$.

Properties of the logarithmic function (pp. 433 and 437)

$f(x) = \log_a x, \quad a > 1$
$(y = \log_a x \text{ means } x = a^y)$

Domain: the interval $(0, \infty)$; Range: the interval $(-\infty, \infty)$;
x-intercept: 1; y-intercept: none;
vertical asymptote: $x = 0$ (y-axis);
increasing; one-to-one; smooth; continuous

See Figure 17(a) for a typical graph.

$f(x) = \log_a x, \quad 0 < a < 1$
$(y = \log_a x \text{ means } x = a^y)$

Domain: the interval $(0, \infty)$; Range: the interval $(-\infty, \infty)$;
x-intercept: 1; y-intercept: none;
vertical asymptote: $x = 0$ (y-axis);
decreasing; one-to-one; smooth; continuous

See Figure 17(b) for a typical graph.

Natural logarithm (p. 433)

$y = \ln x$ means $x = e^y$.

Properties of logarithms (pp. 441–443)

$\log_a 1 = 0 \qquad \log_a a = 1 \qquad a^{\log_a M} = M \qquad \log_a a^r = r$

$\log_a(MN) = \log_a M + \log_a N \qquad \log_a\left(\dfrac{M}{N}\right) = \log_a M - \log_a N$
$\log_a M^r = r \log_a M$

(p. 445)

If $M = N$, then $\log_a M = \log_a N$.
If $\log_a M = \log_a N$, then $M = N$.

Formulas

Change-of-Base Formula (p. 446)

$\log_a M = \dfrac{\log_b M}{\log_b a}$

Compound Interest Formula (p. 460)

$A = P\left(1 + \dfrac{r}{n}\right)^{nt}$

Continuous compounding (p. 462)

$A = Pe^{rt}$

Present Value Formulas (p. 463)

$P = A\left(1 + \dfrac{r}{n}\right)^{-nt} \quad \text{and} \quad P = Ae^{-rt}$

Growth and Decay (p. 457) $\qquad A(t) = A_0 e^{kt}$

Logistic Growth Model (p. 472) $\qquad P(t) = \dfrac{c}{1 + ae^{-bt}}$

Objectives

You should be able to:

Evaluate exponential functions (p. 416)

Graph exponential functions (p. 417)

Define the number e (p. 421)

Solve exponential equations (p. 423)

Change exponential expressions to logarithmic expressions (p. 430)

Change logarithmic expressions to exponential expressions (p. 431)

Evaluate logarithmic functions (p. 431)

Determine the domain of a logarithmic function (p. 431)

Graph logarithmic functions (p. 432)

Solve logarithmic equations (p. 435)

Work with the properties of logarithms (p. 441)

Write a logarithmic expression as a sum or difference of logarithms (p. 443)

Write a logarithmic expression as a single logarithm (p. 444)

Evaluate logarithms whose base is neither 10 nor e (p. 445)

Work with exponential and logarithmic models (p. 447)

Solve logarithmic equations using the properties of logarithms (p. 454)

Solve exponential equations (p. 455)

Solve logarithmic and exponential equations using a graphing utility (p. 457)

Determine the future value of a lump sum of money (p. 458)

Calculate effective rates of return (p. 462)

Determine the present value of a lump sum of money (p. 463)

Determine the time required to double or triple a lump sum of money (p. 464)

Find equations of populations that obey the law of uninhibited growth (p. 467)

Find equations of populations that obey the law of uninhibited decay (p. 469)

Use Newton's Law of Cooling (p. 471)

Use logistic growth models (p. 472)

Work with decibels (p. 476)

Work with the Richter scale (p. 478)

Fill-in-the-Blank Items

1. The graph of every exponential function $f(x) = a^x, a > 0, a \neq 1$, passes through the three points _____.

2. If the graph of an exponential function $f(x) = a^x, a > 0, a \neq 1$, is decreasing, then its base must be less than _____.

3. If $3^x = 3^4$, then $x = $ _____.

4. The logarithm of a product equals the _____ of the logarithms.

5. For every base, the logarithm of _____ equals 0.

6. If $\log_8 M = \dfrac{\log_5 7}{\log_5 8}$, then $M = $ _____.

7. The domain of the logarithmic function $f(x) = \log_a x$ consists of _____.

8. The graph of every logarithmic function $f(x) = \log_a x, a > 0, a \neq 1$, passes through the three points _____.

9. If the graph of a logarithmic function $f(x) = \log_a x, a > 0, a \neq 1$, is increasing, then its base must be larger than _____.

10. If $\log_3 x = \log_3 7$, then $x = $ _____.

True/False Items

T F **1.** The graph of every exponential function $f(x) = a^x, a > 0, a \neq 1$, will contain the points $(0, 1)$ and $(1, a)$.

T F **2.** The graphs of $y = 3^{-x}$ and $y = \left(\dfrac{1}{3}\right)^x$ are identical.

T F **3.** The present value of $1000 to be received after 2 years at 10% per annum compounded continuously is approximately $1205.

T F **4.** If $y = \log_a x$, then $y = a^x$.

T F **5.** The graph of every logarithmic function $f(x) = \log_a x, a > 0, a \neq 1$, will contain the points $(1, 0)$ and $(a, 1)$.

T F **6.** $a^{\log_M a} = M$, where $a > 0, a \neq 1, M > 0$.

T F **7.** $\log_a (M + N) = \log_a M + \log_a N$, where $a > 0, a \neq 1, M > 0, N > 0$.

T F **8.** $\log_a M - \log_a N = \log_a \left(\dfrac{M}{N}\right)$, where $a > 0, a \neq 1, M > 0, N > 0$.

Review Exercises

Blue problem numbers indicate the author's suggestions for use in a Practice Test.

In Problems 1–6, find the exact value of each expression. Do not use a calculator.

1. $\log_2 \left(\dfrac{1}{8}\right)$

2. $\log_3 81$

3. $\ln e^{\sqrt{2}}$

4. $e^{\ln 0.1}$

5. $2^{\log_2 0.4}$

6. $\log_2 2^{\sqrt{3}}$

In Problems 7–12, write each expression as the sum and/or difference of logarithms. Express powers as factors.

7. $\log_3 \left(\dfrac{uv^2}{w}\right)$

8. $\log_2 (a^2 \sqrt{b})^4$

9. $\log(x^2 \sqrt{x^3 + 1})$

10. $\log_5 \left(\dfrac{x^2 + 2x + 1}{x^2}\right)$

11. $\ln \left(\dfrac{x\sqrt[3]{x^2 + 1}}{x - 3}\right)$

12. $\ln \left(\dfrac{2x + 3}{x^2 - 3x + 2}\right)^2$

In Problems 13–18, write each expression as a single logarithm.

13. $3 \log_4 x^2 + \dfrac{1}{2} \log_4 \sqrt{x}$

14. $-2 \log_3 \left(\dfrac{1}{x}\right) + \dfrac{1}{3} \log_3 \sqrt{x}$

15. $\ln \left(\dfrac{x - 1}{x}\right) + \ln \left(\dfrac{x}{x + 1}\right) - \ln(x^2 - 1)$

16. $\log(x^2 - 9) - \log(x^2 + 7x + 12)$

17. $2 \log 2 + 3 \log x - \dfrac{1}{2} [\log(x + 3) + \log(x - 2)]$

18. $\dfrac{1}{2} \ln(x^2 + 1) - 4 \ln \dfrac{1}{2} - \dfrac{1}{2} [\ln(x - 4) + \ln x]$

In Problems 19 and 20, use the Change-of-Base Formula and a calculator to evaluate each logarithm. Round your answer to three decimal places.

19. $\log_4 19$

20. $\log_2 21$

In Problems 21–26, find y as a function of x. The constant C is a positive number.

21. $\ln y = 2x^2 + \ln C$

22. $\ln(y - 3) = \ln 2x^2 + \ln C$

23. $\ln(y - 3) + \ln(y + 3) = x + C$

24. $\ln(y - 1) + \ln(y + 1) = -x + C$

25. $e^{y+C} = x^2 + 4$

26. $e^{3y-C} = (x + 4)^2$

In Problems 27–36, use transformations to graph each function. Determine the domain, range, and any asymptotes.

27. $f(x) = 2^{x-3}$

28. $f(x) = -2^x + 3$

29. $f(x) = \frac{1}{2}(3^{-x})$

30. $f(x) = 1 + 3^{2x}$

31. $f(x) = 1 - e^x$

32. $f(x) = 3e^x$

33. $f(x) = 3 + \ln x$

34. $f(x) = \frac{1}{2} \ln x$

35. $f(x) = 3 - e^{-x}$

36. $f(x) = 4 - \ln(-x)$

In Problems 37–56, solve each equation.

37. $4^{1-2x} = 2$

38. $8^{6+3x} = 4$

39. $3^{x^2+x} = \sqrt{3}$

40. $4^{x-x^2} = \dfrac{1}{2}$

41. $\log_x 64 = -3$

42. $\log_{\sqrt{2}} x = -6$

43. $5^x = 3^{x+2}$

44. $5^{x+2} = 7^{x-2}$

45 $9^{2x} = 27^{3x-4}$

46. $25^{2x} = 5^{x^2-12}$

47. $\log_3 \sqrt{x - 2} = 2$

48. $2^{x+1} \cdot 8^{-x} = 4$

49. $8 = 4^{x^2} \cdot 2^{5x}$

50. $2^x \cdot 5 = 10^x$

51. $\log_6(x + 3) + \log_6(x + 4) = 1$

52. $\log_{10}(7x - 12) = 2 \log_{10} x$

53. $e^{1-x} = 5$

54. $e^{1-2x} = 4$

55. $2^{3x} = 3^{2x+1}$

56. $2^{x^3} = 3^{x^2}$

In Problems 57–60, use the following result: If x is the atmospheric pressure (measured in millimeters of mercury), then the formula for the altitude $h(x)$ (measured in meters above sea level) is

$$h(x) = (30T + 8000) \log\left(\frac{P_0}{x}\right)$$

where T is the temperature (in degrees Celsius) and P_0 is the atmospheric pressure at sea level, which is approximately 760 millimeters of mercury.

57. Finding the Altitude of an Airplane At what height is a Piper Cub whose instruments record an outside temperature of 0°C and a barometric pressure of 300 millimeters of mercury?

58. Finding the Height of a Mountain How high is a mountain if instruments placed on its peak record a temperature of 5°C and a barometric pressure of 500 millimeters of mercury?

59. Atmospheric Pressure Outside an Airplane What is the atmospheric pressure outside a Boeing 737 flying at an altitude of 10,000 meters if the outside air temperature is −100°C?

60. Atmospheric Pressure at High Altitudes What is the atmospheric pressure (in millimeters of mercury) on

Mt. Everest, which has an altitude of approximately 8900 meters, if the air temperature is 5°C?

61. Amplifying Sound An amplifier's power output P (in watts) is related to its decibel voltage gain d by the formula $P = 25e^{0.1d}$.

(a) Find the power output for a decibel voltage gain of 4 decibels.

(b) For a power output of 50 watts, what is the decibel voltage gain?

62. Limiting Magnitude of a Telescope A telescope is limited in its usefulness by the brightness of the star it is aimed at and by the diameter of its lens. One measure of a star's brightness is its *magnitude*: the dimmer the star, the larger its magnitude. A formula for the limiting magnitude L of a telescope, that is, the magnitude of the dimmest star that it can be used to view, is given by

$$L = 9 + 5.1 \log d$$

where d is the diameter (in inches) of the lens.
(a) What is the limiting magnitude of a 3.5-inch telescope?
(b) What diameter is required to view a star of magnitude 14?

63. Product Demand The demand for a new product increases rapidly at first and then levels off. The percent P of actual purchases of this product after it has been on the market t months is

$$P = 90 - 80\left(\frac{3}{4}\right)^t$$

(a) What is the percent of purchases of the product after 5 months?
(b) What is the percent of purchases of the product after 10 months?
(c) What is the maximum percent of purchases of the product?
(d) How many months does it take before 40% of purchases occur?
(e) How many months before 70% of purchases occur?

64. Disseminating Information A survey of a certain community of 10,000 residents shows that the number of residents N who have heard a piece of information after m months is given by the formula

$$m = 55.3 - 6\ln(10,000 - N)$$

How many months will it take for half of the citizens to learn about a community program of free blood pressure readings?

65. Salvage Value The number of years n for a piece of machinery to depreciate to a known salvage value can be found using the formula

$$n = \frac{\log s - \log i}{\log(1 - d)}$$

where s is the salvage value of the machinery, i is its initial value, and d is the annual rate of depreciation.
(a) How many years will it take for a piece of machinery to decline in value from $90,000 to $10,000 if the annual rate of depreciation is 0.20 (20%)?
(b) How many years will it take for a piece of machinery to lose half of its value if the annual rate of depreciation is 15%?

66. Funding a College Education A child's grandparents purchase a $10,000 bond fund that matures in 18 years to be used for her college education. The bond fund pays 4% interest compounded semiannually. How much will the bond fund be worth at maturity?

67. Funding a College Education A child's grandparents wish to purchase a bond fund that matures in 18 years to be used for her college education. The bond fund pays 4% interest compounded semiannually. How much should they purchase so that the bond fund will be worth $85,000 at maturity?

68. Funding an IRA First Colonial Bankshares Corporation advertised the following IRA investment plans.

Target IRA Plans

For each $5000 Maturity Value Desired	
Deposit:	**At a Term of:**
$620.17	20 Years
$1045.02	15 Years
$1760.92	10 Years
$2967.26	5 Years

(a) Assuming continuous compounding, what was the annual rate of interest they offered?
(b) First Colonial Bankshares claims that $4000 invested today will have a value of over $32,000 in 20 years. Use the answer found in part (a) to find the actual value of $4000 in 20 years. Assume continuous compounding.

69. Loudness of a Garbage Disposal Find the loudness of a garbage disposal unit that operates at an intensity of 10^{-4} watt per square meter. Express your answer in decibels.

70. Comparing Earthquakes On September 9, 1985, the western suburbs of Chicago experienced a mild earthquake that registered 3.0 on the Richter scale. How did this earthquake compare in intensity to the great San Francisco earthquake of 1906, which registered 6.9 on the Richter scale?

71. Estimating the Date on Which a Prehistoric Man Died The bones of a prehistoric man found in the desert of New Mexico contain approximately 5% of the original amount of carbon 14. If the half-life of carbon 14 is 5600 years, approximately how long ago did the man die?

72. Temperature of a Skillet A skillet is removed from an oven whose temperature is 450°F and placed in a room whose temperature is 70°F. After 5 minutes, the temperature of the skillet is 400°F. How long will it be until its temperature is 150°F?

73. Biology A certain bacteria initially increases according to the law of uninhibited growth. A biologist collects the following data for this bacteria:

Time (Hours)	Population
0	1000
1	1415
2	2000
3	2828
4	4000
5	5656
6	8000

(a) Draw a scatter diagram.
(b) The exponential function of best fit to the data is found to be

$$y = 1000(\sqrt{2})^x$$

Express the function of best fit in the form $N = N_0 e^{kt}$.
(c) Use the solution to part (b) to predict the population at $t = 7$ hours.
(d) Use a graphing utility to verify the exponential function of best fit.
(e) Use a graphing utility to draw a scatter diagram of the data and then graph the exponential function of best fit on it.

74. Finance The following data represent the amount of money an investor has in an investment account each year for 10 years. She wishes to determine the effective rate of return on her investment.

Year	Value of Account
1985	$10,000
1986	$10,573
1987	$11,260
1988	$11,733
1989	$12,424
1990	$13,269
1991	$13,968
1992	$14,823
1993	$15,297
1994	$16,539

(a) Draw a scatter diagram with the number of years after the initial investment as the independent variable and the value of the account as the dependent variable.
(b) The exponential function of best fit to the data is found to be

$$y = 10,014(1.057)^x$$

Express the function of best fit in the form $A = A_0 e^{kt}$, where $x = t$.
(c) Use the solution to part (b) to estimate the value of the account in the year 2020.
(d) Use a graphing utility to verify the exponential function of best fit.
(e) Use a graphing utility to draw a scatter diagram of the data and then graph the exponential function of best fit on it.

75. World Population According to the U.S. Census Bureau, the growth rate of the world's population in 1997 was $k = 1.33\% = 0.0133$. The population of the world in 1997 was 5,840,445,216. Letting $t = 0$ represent 1997, use the uninhibited growth model to predict the world's population in the year 2002.

76. Radioactive Decay The half-life of radioactive cobalt is 5.27 years. If 100 grams of radioactive cobalt is present now, how much will be present in 20 years? In 40 years?

77. Logistic Growth The logistic growth model

$$P(t) = \frac{0.8}{1 + 1.67e^{-0.16t}}$$

represents the proportion of new computers sold that utilize the Microsoft Windows 98 operating system. Let $t = 0$ represent 1998, $t = 1$ represent 1999, and so on.
(a) What proportion of new computers sold in 1998 utilized Windows 98?
(b) Determine the maximum proportion of new computers sold that will utilize Windows 98.
(c) Using a graphing utility, graph $P(t)$.
(d) When will 75% of new computers sold utilize Windows 98?

Thermal Fatigue of Solder Interconnects

What happens to an electronic package when it is subjected to repeated temperature changes? This question is important if you want your electronic product to be reliable. Every time you use your cell phone, pager, computer, or start your car, the electronics inside begin to warm up. When materials warm up, they expand. However, different materials expand at different rates. For example, the glass–epoxy laminate, called a printed circuit board, has a coefficient of thermal expansion (CTE) of 12 to $15 \times 10^{-6}/°C$, while the silicon IC located inside an electronic package has a CTE equal to $2.5 \times 10^{-6}/°C$. These differences will induce inelastic deformation to the solder interconnect. The solder interconnect makes the electrical connection from the PCB to the electronic package and is usually made of a low-melting-point (183°C) alloy comprised of tin and lead. After using your portable product or computer or turning the car engine off, the electronics will cool off. These temperature cycles result in repeated expansion and contraction of the material used to make the electronic assemblies. The greater the temperature change or the greater the difference in CTE between materials, the greater will be the inelastic strain imparted to the solder joint. A decrease in strain will increase fatigue life. Let's look at a typical situation.

Example of an electronic package (top view) soldered to a PCB used to evaluate fatigue life.

EXPERIMENTAL FATIGUE DATA

Solder Joint Strain, εp	Fatigue Cycles, Nf
0.01	10,000
0.035	1000
0.1	100
0.4	10
1.5	1

Using the experimental fatigue life data given in the table, answer the following:

1. Draw a scatter diagram of the data with solder joint strain as the independent variable.

2. Let $X = \ln(\varepsilon p)$. Draw a scatter diagram of the transformed data with X as the independent variable and fatigue cycles as the dependent variable. What happens to the shape of the scatter diagram?

3. Let $Y = \ln(Nf)$. Draw a scatter diagram of the transformed data with X as the independent variable and Y as the dependent variable. What type of relation would best describe the data?

4. Find the line of best fit to the transformed data using a graphing utility. Graph the line of best fit on the scatter diagram drawn in problem 3.

5. Write the equation from problem 4 in the form $Nf = e^{b}(\varepsilon p)^{m}$.
   [**Hint:** The line of best fit is $\ln(Nf) = m\ln(\varepsilon p) + b$. Think of a property of logarithms that will eliminate the natural logarithm.]

6. If the solder joint strain is 0.02, what is the expected fatigue life? What is the inelastic solder joint strain if the fatigue life is 3000 cycles?

7. Rewrite the function from problem 5 so that the inelastic solder joint strain is a function of the fatigue life. Compute the solder joint strain when the fatigue life equals 3000 cycles

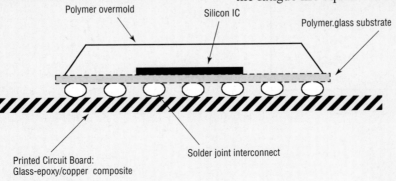

Polymer overmold

Silicon IC

Polymer.glass substrate

Printed Circuit Board:
Glass-epoxy/copper composite

Solder joint interconnect

Review

PREPARING FOR THIS BOOK

Before getting started

✓ Read the Preface to the Student, page xv

A.1 ALGEBRA REVIEW

OBJECTIVES   **1** Evaluate Algebraic Expressions
2 Determine the Domain of a Variable
3 Graph Inequalities
4 Find Distance on the Real Number Line
5 Use the Laws of Exponents
6 Evaluate Square Roots

SETS

When we want to treat a collection of similar but distinct objects as a whole, we use the idea of a **set.** For example, the set of *digits* consists of the collection of numbers 0, 1, 2, 3, 4, 5, 6, 7, 8, and 9. If we use the symbol D to denote the set of digits, then we can write

$$D = \{0, 1, 2, 3, 4, 5, 6, 7, 8, 9\}$$

In this notation, the braces { } are used to enclose the objects, or **elements,** in the set. This method of denoting a set is called the **roster method.** A second way to denote a set is to use **set-builder notation,** where the set D of digits is written as

$$D = \{\quad x \quad | \quad x \text{ is a digit}\}$$

Read as "D is the set of all x such that x is a digit."

487

EXAMPLE 1

Using Set-builder Notation and the Roster Method

(a) $E = \{x \mid x \text{ is an even digit}\} = \{0, 2, 4, 6, 8\}$

(b) $O = \{x \mid x \text{ is an odd digit}\} = \{1, 3, 5, 7, 9\}$ ∎

In listing the elements of a set, we do not list an element more than once because the elements of a set are distinct. Also, the order in which the elements are listed is not relevant. For example, $\{2, 3\}$ and $\{3, 2\}$ both represent the same set.

If every element of a set A is also an element of a set B, then we say that A is a **subset** of B. If two sets A and B have the same elements, then we say that A **equals** B. For example, $\{1, 2, 3\}$ is a subset of $\{1, 2, 3, 4, 5\}$; and $\{1, 2, 3\}$ equals $\{2, 3, 1\}$.

Finally, if a set has no elements, it is called the **empty set,** or the **null set,** and it is denoted by the symbol $\varnothing$.

REAL NUMBERS

Real numbers are represented by symbols such as

$$25, \quad 0, \quad -3, \quad \frac{1}{2}, \quad -\frac{5}{4}, \quad 0.125, \quad \sqrt{2}, \quad \pi, \quad \sqrt[3]{-2}, \quad 0.666 \ldots$$

The set of **counting numbers,** or **natural numbers,** contains the numbers in the set $\{1, 2, 3, 4, \ldots\}$. (The three dots, called an **ellipsis,** indicate that the pattern continues indefinitely.) The set of **integers** contains the numbers in the set $\{\ldots, -3, -2, -1, 0, 1, 2, 3, \ldots\}$. A **rational number** is a number that can be expressed as a *quotient* $\dfrac{a}{b}$ of two integers, where the integer b cannot be 0. Examples of rational numbers are $\dfrac{3}{4}, \dfrac{5}{2}, \dfrac{0}{4}$, and $-\dfrac{2}{3}$. Since $\dfrac{a}{1} = a$ for any integer a, every integer is also a rational number. Real numbers that are not rational are called **irrational.** Examples of irrational numbers are $\sqrt{2}$ and π (the Greek letter pi), which equals the constant ratio of the circumference to the diameter of a circle. See Figure 1.

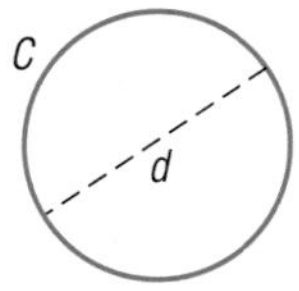

Real numbers can be represented as **decimals.** Rational real numbers have decimal representations that either **terminate** or are nonterminating with **repeating** blocks of digits. For example, $\dfrac{3}{4} = 0.75$ which terminates; and $\dfrac{2}{3} = 0.666 \ldots$, in which the digit 6 repeats indefinitely. Irrational real numbers have decimal representations that neither repeat nor terminate. For example, $\sqrt{2} = 1.414213 \ldots$ and $\pi = 3.14159 \ldots$. In practice, irrational numbers are generally represented by approximations. We use the symbol $\approx$ (read as "approximately equal to") to write $\sqrt{2} \approx 1.4142$ and $\pi \approx 3.1416$.

Two properties of real numbers that we shall use often are given next. Suppose that a, b, and c are real numbers.

Figure 1

$$\pi = \frac{C}{d}$$

Distributive Property

$$a \cdot (b + c) = ab + ac$$

Zero-Product Property

If $ab = 0$, then either $a = 0$ or $b = 0$ or both equal 0.

The Distributive Property can be used to remove parentheses: $2(x + 3) = 2x + 2 \cdot 3 = 2x + 6$.

The Zero-Product Property will be used to solve equations (Section A.3). If $2x = 0$, then $2 = 0$ or $x = 0$. Since $2 \neq 0$, it follows that $x = 0$.

CONSTANTS AND VARIABLES

In algebra we use letters to represent numbers. If the letter used is to represent *any* number from a given set of numbers. it is called a **variable.** A **constant** is either a fixed number, such as 5 or $\sqrt{3}$, or a letter that represents a fixed (possibly unspecified) number.

Constants and variables are combined using the operations of addition, subtraction, multiplication, and division to form *algebraic expressions*. Examples of algebraic expressions include

$$x + 3 \qquad \frac{3}{1 - t} \qquad 7x - 2y$$

1 To evaluate an algebraic expression, substitute for each variable its numerical value.

EXAMPLE 2 ### Evaluating an Algebraic Expression

Evaluate each expression if $x = 3$ and $y = -1$.

(a) $x + 3y$ (b) $5xy$ (c) $\dfrac{3y}{2 - 2x}$

Solution (a) Substitute 3 for x and -1 for y in the expression $x + 3y$.

$$x + 3y = 3 + 3(-1) = 3 + (-3) = 0$$
$$\uparrow$$
$$x = 3, y = -1$$

(b) If $x = 3$ and $y = -1$, then

$$5xy = 5(3)(-1) = -15$$

(c) If $x = 3$ and $y = -1$, then

$$\frac{3y}{2 - 2x} = \frac{3(-1)}{2 - 2(3)} = \frac{-3}{2 - 6} = \frac{-3}{-4} = \frac{3}{4}$$

◾

NOW WORK PROBLEM **1.**

2 In working with expressions or formulas involving variables, the variables may be allowed to take on values from only a certain set of numbers. For example, in the formula for the area A of a circle of radius r, $A = \pi r^2$, the variable r is necessarily restricted to the positive real numbers. In the expression $\dfrac{1}{x}$, the variable x cannot take on the value 0, since division by 0 is not defined.

The set of values that a variable in an expression may assume is called the **domain of the variable.**

EXAMPLE 3 ### Finding the Domain of a Variable

The domain of the variable x in the rational expression

$$\frac{5}{x - 2}$$

is $\{x \mid x \neq 2\}$, since, if $x = 2$, the denominator becomes 0, which is not defined.

◾

| EXAMPLE 4 | **Circumference of a Circle** |

In the formula for the circumference C of a circle of radius r,

$$C = 2\pi r$$

the domain of the variable r, representing the radius of the circle, is the set of positive real numbers. The domain of the variable C, representing the circumference of the circle, is also the set of positive real numbers.   ■

In describing the domain of a variable, we may use either set notation or words, whichever is more convenient.

➤ NOW WORK PROBLEM **9**.

THE REAL NUMBER LINE

The real numbers can be represented by points on a line called the **real number line.** There is a one-to-one correspondence between real numbers and points on a line. That is, every real number corresponds to a point on the line, and each point on the line has a unique real number associated with it.

Pick a point on the line somewhere in the center, and label it O. This point, called the **origin,** corresponds to the real number 0. See Figure 2. The point 1 unit to the right of O corresponds to the number 1. The distance between 0 and 1 determines the scale of the number line. For example, the point associated with the number 2 is twice as far from O as 1 is. Notice that an arrowhead on the right end of the line indicates the direction in which the numbers increase. Figure 2 also shows the points associated with the irrational numbers $\sqrt{2}$ and π. Points to the left of the origin correspond to the real numbers $-1, -2$, and so on.

Figure 2
Real number line.

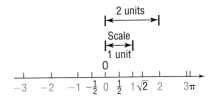

The real number associated with a point P is called the **coordinate** of P, and the line whose points have been assigned coordinates is called the **real number line.**

➤ NOW WORK PROBLEM **21**.

The real number line consists of three classes of real numbers, as shown in Figure 3.

Figure 3

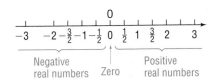

1. The **negative real numbers** are the coordinates of points to the left of the origin O.
2. The real number **zero** is the coordinate of the origin O.
3. The **positive real numbers** are the coordinates of points to the right of the origin O.

INEQUALITIES

An important property of the real number line follows from the fact that, given two numbers (points) a and b, either a is to the left of b, a is at the same location as b, or a is to the right of b. See Figure 4.

If a is to the left of b, we say that "a is less than b" and write $a < b$. If a is to the right of b, we say that "a is greater than b" and write $a > b$. If a is

Figure 4

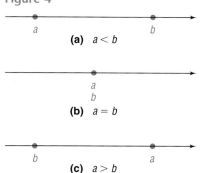

(a) $a < b$

(b) $a = b$

(c) $a > b$

at the same location as b, then $a = b$. If a is either less than or equal to b, we write $a \leq b$. Similarly, $a \geq b$ means that a is either greater than or equal to b. Collectively, the symbols $<, >,\ \leq,$ and $\geq$ are called **inequality symbols.**

Note that $a < b$ and $b > a$ mean the same thing. It does not matter whether we write $2 < 3$ or $3 > 2$.

Furthermore, if $a < b$ or if $b > a$, then the difference $b - a$ is positive. Do you see why?

An **inequality** is a statement in which two expressions are related by an inequality symbol. The expressions are referred to as the **sides** of the inequality. Statements of the form $a < b$ or $b > a$ are called **strict inequalities,** while statements of the form $a \leq b$ or $b \geq a$ are called **nonstrict inequalities.**

Based on the discussion thus far, we conclude that

$a > 0$	is equivalent to	a is positive
$a < 0$	is equivalent to	a is negative

We sometimes read $a > 0$ by saying that "a is positive." If $a \geq 0$, then either $a > 0$ or $a = 0$, and we may read this as "a is nonnegative."

✏ NOW WORK PROBLEMS **25** AND **35.**

③ We shall find it useful in later work to graph inequalities on the real number line.

EXAMPLE 5

Graphing Inequalities

(a) On the real number line, graph all numbers x for which $x > 4$.

(b) On the real number line, graph all numbers x for which $x \leq 5$.

Figure 5

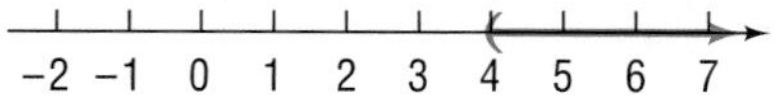

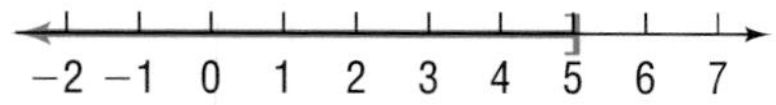

Solution   (a) See Figure 5. Notice that we use a left parenthesis to indicate that the number 4 is not part of the graph.

Figure 6

(b) See Figure 6. Notice that we use a right bracket to indicate that the number 5 is part of the graph.   ∎

✏ NOW WORK PROBLEM **41.**

ABSOLUTE VALUE

Figure 7

The *absolute value* of a number a is the distance from 0 to a on the number line. For example, -4 is 4 units from 0; and 3 is 3 units from 0. See Figure 7. Thus, the absolute value of -4 is 4, and the absolute value of 3 is 3.

A more formal definition of absolute value is given next.

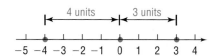

The **absolute value** of a real number a, denoted by the symbol $|a|$, is defined by the rules

$$|a| = a \quad \text{if } a \geq 0 \qquad \text{and} \qquad |a| = -a \quad \text{if } a < 0$$

For example, since $-4 < 0$, the second rule must be used to get $|-4| = -(-4) = 4$.

| **EXAMPLE 6** | **Computing Absolute Value** |

(a) $|8| = 8$ (b) $|0| = 0$ (c) $|-15| = -(-15) = 15$ ∎

═══╺ NOW WORK PROBLEM **43**.

④   Look again at Figure 7. The distance from -4 to 3 is 7 units. This distance is the difference $3 - (-4)$, obtained by subtracting the smaller coordinate from the larger. However, since $|3 - (-4)| = |7| = 7$ and $|-4 - 3| = |-7| = 7$, we can use absolute value to calculate the distance between two points without being concerned about which is smaller.

> If P and Q are two points on a real number line with coordinates a and b, respectively, the **distance between P and Q,** denoted by $d(P, Q)$, is
>
> $$d(P, Q) = |b - a|$$
>
> Since $|b - a| = |a - b|$, it follows that $d(P, Q) = d(Q, P)$.

| **EXAMPLE 7** | **Finding Distance on a Number Line** |

Let $P, Q,$ and R be points on a real number line with coordinates $-5, 7,$ and -3, respectively. Find the distance

(a) between P and Q (b) between Q and R

Solution   See Figure 8.

Figure 8

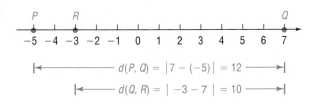

(a) $d(P, Q) = |7 - (-5)| = |12| = 12$
(b) $d(Q, R) = |-3 - 7| = |-10| = 10$ ∎

═══╺ NOW WORK PROBLEM **57**.

EXPONENTS

⑤   Integer exponents provide a shorthand device for representing repeated multiplications of a real number.
    If a is a real number and n is a positive integer, then the symbol a^n represents the product of n factors of a. That is,

$$a^n = \underbrace{a \cdot a \cdot \ldots \cdot a}_{n \text{ factors}}$$

where it is understood that $a^1 = a$. Then, $a^2 = a \cdot a, a^3 = a \cdot a \cdot a$, and so on. In the expression a^n, a is called the **base** and n is called the **exponent,** or

power. We read a^n as "a raised to the power n" or as "a to the nth power." We usually read a^2 as "a squared" and a^3 as "a cubed."

Care must be taken when parentheses are used in conjunction with exponents. For example, $-2^4 = -(2 \cdot 2 \cdot 2 \cdot 2) = -16$, whereas $(-2)^4 = (-2) \cdot (-2) \cdot (-2) \cdot (-2) = 16$. Notice the difference: The exponent applies only to the number or parenthetical expression immediately preceding it.

If $a \neq 0$, we define

$$a^0 = 1 \qquad \text{if } a \neq 0$$

If $a \neq 0$ and if n is a positive integer, then we define

$$a^{-n} = \frac{1}{a^n} \qquad \text{if } a \neq 0$$

With these definitions, the symbol a^n is defined for any integer n.

The following properties, called the **laws of exponents,** can be proved using the preceding definitions. In the list, a and b are real numbers, and m and n are integers.

Laws of Exponents

$$a^m a^n = a^{m+n} \qquad \left(a^m\right)^n = a^{mn} \qquad (ab)^n = a^n b^n$$

$$\frac{a^m}{a^n} = a^{m-n} = \frac{1}{a^{n-m}}, \text{if } a \neq 0 \qquad \left(\frac{a}{b}\right)^n = \frac{a^n}{b^n}, \text{if } b \neq 0$$

EXAMPLE 8   **Using the Laws of Exponents**

Write each expression so that all exponents are positive.

(a) $\dfrac{x^5 y^{-2}}{x^3 y}$, $x \neq 0, y \neq 0$ 　　　　　(b) $\left(\dfrac{x^{-3}}{3y^{-1}}\right)^{-2}$, $x \neq 0, y \neq 0$

Solution (a) $\dfrac{x^5 y^{-2}}{x^3 y} = \dfrac{x^5}{x^3} \cdot \dfrac{y^{-2}}{y} = x^{5-3} \cdot y^{-2-1} = x^2 y^{-3} = x^2 \cdot \dfrac{1}{y^3} = \dfrac{x^2}{y^3}$

(b) $\left(\dfrac{x^{-3}}{3y^{-1}}\right)^{-2} = \dfrac{\left(x^{-3}\right)^{-2}}{\left(3y^{-1}\right)^{-2}} = \dfrac{x^6}{3^{-2}\left(y^{-1}\right)^{-2}} = \dfrac{x^6}{\frac{1}{9}y^2} = \dfrac{9x^6}{y^2}$ ■

NOW WORK PROBLEMS **63** AND **73**.

SQUARE ROOTS

⑥ A real number is squared when it is raised to the power 2. The inverse of squaring is finding a **square root.** For example, since $6^2 = 36$ and $(-6)^2 = 36$, the numbers 6 and -6 are square roots of 36.

The symbol $\sqrt{}$, called a **radical sign,** is used to denote the **principal,** or nonnegative, square root. Thus, $\sqrt{36} = 6$.

In general, if a is a nonnegative real number, the nonnegative number b such that $b^2 = a$ is **the principal square root** of a and is denoted by $b = \sqrt{a}$.

The following comments are noteworthy:

1. Negative numbers do not have square roots (in the real number system), because the square of any real number is *nonnegative*. For example, $\sqrt{-4}$ is not a real number, because there is no real number whose square is -4.
2. The principal square root of 0 is 0, since $0^2 = 0$. That is, $\sqrt{0} = 0$.
3. The principal square root of a positive number is positive.
4. If $c \geq 0$, then $\left(\sqrt{c}\right)^2 = c$. For example, $\left(\sqrt{2}\right)^2 = 2$ and $\left(\sqrt{3}\right)^2 = 3$.

EXAMPLE 9

Evaluating Square Roots

(a) $\sqrt{64} = 8$

(b) $\sqrt{\dfrac{1}{16}} = \dfrac{1}{4}$

(c) $\left(\sqrt{1.4}\right)^2 = 1.4$

(d) $\sqrt{(-3)^2} = |-3| = 3$

Examples 9(a) and (b) are examples of square roots of perfect squares, since $64 = 8^2$ and $\dfrac{1}{16} = \left(\dfrac{1}{4}\right)^2$.

Notice the need for the absolute value in Example 9(d). Since $a^2 \geq 0$, the principal square root of a^2 is defined whether $a > 0$ or $a < 0$. However, since the principal square root is nonnegative, we need the absolute value to ensure the nonnegative result.

In general, we have

$$\sqrt{a^2} = |a| \qquad \qquad \textbf{(1)}$$

EXAMPLE 10

Using Equation (1)

(a) $\sqrt{(2.3)^2} = |2.3| = 2.3$

(b) $\sqrt{(-2.3)^2} = |-2.3| = 2.3$

(c) $\sqrt{x^2} = |x|$

NOW WORK PROBLEM 69.

CALCULATORS

Calculators are finite machines. As a result, they are incapable of displaying decimals that contain a large number of digits. For example, some calculators are capable of displaying only eight digits. When a number requires more than eight digits, the calculator either truncates or rounds. To see how your calculator handles decimals, divide 2 by 3. How many digits do you see? Is the last digit a 6 or a 7? If it is a 6, your calculator truncates; if it is a 7, your calculator rounds.

There are different kinds of calculators. An **arithmetic** calculator can only add, subtract, multiply, and divide numbers; therefore, this type is not adequate for this course. **Scientific** calculators have all the capabilities of arithmetic calculators and also contain **function keys** labeled ln, log, sin, cos, tan,

x^y, inv, and so on. As you proceed through this text, you will discover how to use many of the function keys. **Graphing** calculators have all the capabilities of scientific calculators and contain a screen on which graphs can be displayed.

For those who have access to a graphing calculator, we have included comments, examples, and exercises marked with a 📟, indicating that a graphing calculator is required. We have also included an appendix that explains some of the capabilities of a graphing calculator. The 📟 comments, examples, and exercises may be omitted without loss of continuity, if so desired.

A.1 EXERCISES

In Problems 1–8, find the value of each expression if $x = -2$ and $y = 3$.

1. $x + 2y$ **2.** $3x + y$ **3.** $5xy + 2$ **4.** $-2x + xy$

5. $\dfrac{2x}{x - y}$ **6.** $\dfrac{x + y}{x - y}$ **7.** $\dfrac{3x + 2y}{2 + y}$ **8.** $\dfrac{2x - 3}{y}$

In Problems 9–16, determine which of the value(s) given below, if any, must be excluded from the domain of the variable in each expression.

    (a) $x = 3$ (b) $x = 1$ (c) $x = 0$ (d) $x = -1$

9. $\dfrac{x^2 - 1}{x}$ **10.** $\dfrac{x^2 + 1}{x}$ **11.** $\dfrac{x}{x^2 - 9}$ **12.** $\dfrac{x}{x^2 + 9}$

13. $\dfrac{x^2}{x^2 + 1}$ **14.** $\dfrac{x^3}{x^2 - 1}$ **15.** $\dfrac{x^2 + 5x - 10}{x^3 - x}$ **16.** $\dfrac{-9x^2 - x + 1}{x^3 + x}$

In Problems 17–20, determine the domain of the variable x in each expression.

17. $\dfrac{4}{x - 5}$ **18.** $\dfrac{-6}{x + 4}$ **19.** $\dfrac{x}{x + 4}$ **20.** $\dfrac{x - 2}{x - 6}$

21. On the real number line, label the points with coordinates $0, 1, -1, \dfrac{5}{2}, -2.5, \dfrac{3}{4}$, and 0.25.

22. Repeat Problem 21 for the coordinates $0, -2, 2, -1.5, \dfrac{3}{2}, \dfrac{1}{3}$, and $\dfrac{2}{3}$.

In Problems 23–32, replace the question mark by $<, >,$ or $=$, whichever is correct.

23. $\dfrac{1}{2}$? 0 **24.** 5 ? 6 **25.** -1 ? -2 **26.** -3 ? $-\dfrac{5}{2}$ **27.** π ? 3.14

28. $\sqrt{2}$? 1.41 **29.** $\dfrac{1}{2}$? 0.5 **30.** $\dfrac{1}{3}$? 0.33 **31.** $\dfrac{2}{3}$? 0.67 **32.** $\dfrac{1}{4}$? 0.25

In Problems 33–38, write each statement as an inequality.

33. x is positive   **34.** z is negative   **35.** x is less than 2   **36.** y is greater than -5

37. x is less than or equal to 1   **38.** x is greater than or equal to 2

In Problems 39–42, graph the numbers x on the real number line.

39. $x \geq -2$ **40.** $x < 4$ **41.** $x > -1$ **42.** $x \leq 7$

In Problems 43–52, find the value of each expression if $x = 3$ and $y = -2$.

43. $|x + y|$ **44.** $|x - y|$ **45.** $|x| + |y|$ **46.** $|x| - |y|$ **47.** $\dfrac{|x|}{x}$

48. $\dfrac{|y|}{y}$ **49.** $|4x - 5y|$ **50.** $|3x + 2y|$ **51.** $||4x| - |5y||$ **52.** $3|x| + 2|y|$

In Problems 53–58, use the real number line below to compute each distance.

53. $d(C, D)$ **54.** $d(C, A)$ **55.** $d(D, E)$ **56.** $d(C, E)$ **57.** $d(A, E)$ **58.** $d(D, B)$

In Problems 59–70, simplify each expression.

59. $(-4)^2$ **60.** -4^2 **61.** 4^{-2} **62.** -4^{-2} **63.** $3^{-6} \cdot 3^4$ **64.** $4^{-2} \cdot 4^3$

65. $(3^{-2})^{-1}$ **66.** $(2^{-1})^{-3}$ **67.** $\sqrt{25}$ **68.** $\sqrt{36}$ **69.** $\sqrt{(-4)^2}$ **70.** $\sqrt{(-3)^2}$

In Problems 71–80, simplify each expression. Express the answer so that all exponents are positive. Whenever an exponent is 0 or negative, we assume that the base is not 0.

71. $(8x^3)^{-2}$ **72.** $(-4x^2)^{-1}$ **73.** $(x^2 y^{-1})^2$ **74.** $(x^{-1}y)^3$ **75.** $\dfrac{x^{-2}y^3}{xy^4}$

76. $\dfrac{x^{-2}y}{xy^2}$ **77.** $\dfrac{(-2)^3 x^4 (yz)^2}{3^2 xy^3 z}$ **78.** $\dfrac{4x^{-2}(yz)^{-1}}{2^3 x^4 y}$ **79.** $\left(\dfrac{3x^{-1}}{4y^{-1}}\right)^{-2}$ **80.** $\left(\dfrac{5x^{-2}}{6y^{-2}}\right)^{-3}$

In Problems 81–92, express each statement as an equation involving the indicated variables.

81. Area of a Rectangle The area A of a rectangle is the product of its length l times its width w.

82. Perimeter of a Rectangle The perimeter P of a rectangle is twice the sum of its length l and its width w.

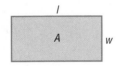

83. Circumference of a Circle The circumference C of a circle is the product of π times its diameter d.

84. Area of a Triangle The area A of a triangle is one-half the product of its base b and its height h.

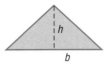

85. Area of an Equilateral Triangle The area A of an equilateral triangle is $\dfrac{\sqrt{3}}{4}$ times the square of the length x of one side.

86. Perimeter of an Equilateral Triangle The perimeter P of an equilateral triangle is 3 times the length x of one side.

87. Volume of a Sphere The volume V of a sphere is $\dfrac{4}{3}$ times π times the cube of the radius r.

88. Surface Area of a Sphere The surface area S of a sphere is 4 times π times the square of the radius r.

89. Volume of a Cube The volume V of a cube is the cube of the length x of a side.

90. Surface Area of a Cube The surface area S of a cube is 6 times the square of the length x of a side.

91. U.S. Voltage In the United States, normal household voltage is 115 volts. It is acceptable for the actual voltage x to differ from normal by at most 5 volts. A formula that describes this is

$$|x - 115| \le 5$$

(a) Show that a voltage of 113 volts is acceptable.
(b) Show that a voltage of 109 volts is not acceptable.

92. Foreign Voltage In other countries, normal household voltage is 220 volts. It is acceptable for the actual voltage x to differ from normal by at most 8 volts. A formula that describes this is

$$|x - 220| \le 8$$

(a) Show that a voltage of 214 volts is acceptable.
(b) Show that a voltage of 209 volts is not acceptable.

93. Making Precision Ball Bearings The FireBall Company manufactures ball bearings for precision equipment. One of their products is a ball bearing with a stated radius of 3 centimeters (cm). Only ball bearings with a radius within 0.01 cm of this stated radius are acceptable. If x is the radius of a ball bearing, a formula describing this situation is

$$|x - 3| \leq 0.01$$

(a) Is a ball bearing of radius $x = 2.999$ acceptable?
(b) Is a ball bearing of radius $x = 2.89$ acceptable?

94. Body Temperature Normal human body temperature is 98.6°F. A temperature x that differs from normal by at least 1.5°F is considered unhealthy. A formula that describes this is

$$|x - 98.6| \geq 1.5$$

(a) Show that a temperature of 97°F is unhealthy.
(b) Show that a temperature of 100°F is not unhealthy.

95. Does $\dfrac{1}{3}$ equal 0.333? If not, which is larger? By how much?

96. Does $\dfrac{2}{3}$ equal 0.666? If not, which is larger? By how much?

97. Is there a positive real number "closest" to 0?

98. I'm thinking of a number! It lies between 1 and 10; its square is rational and lies between 1 and 10. The number is larger than π. Correct to two decimal places, name the number. Now think of your own number, describe it, and challenge a fellow student to name it.

99. Write a brief paragraph that illustrates the similarities and differences between "less than" ($<$) and "less than or equal" ($\leq$).

$\boxed{\text{A.2}}$ GEOMETRY REVIEW

OBJECTIVES   **1** Use the Pythagorean Theorem and Its Converse
           2 Know Geometry Formulas

In this section we review some topics studied in geometry that we shall need for our study of algebra.

PYTHAGOREAN THEOREM

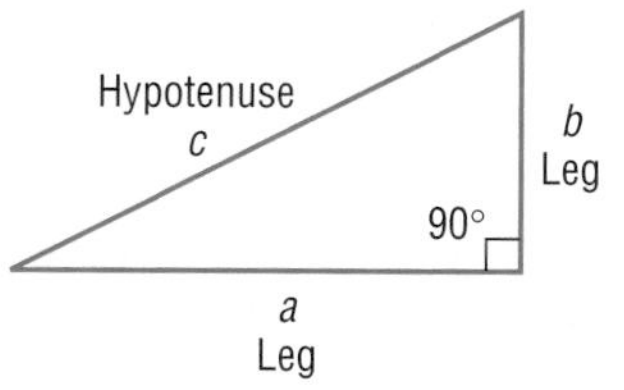

1 The *Pythagorean Theorem* is a statement about *right triangles*. A **right triangle** is one that contains a **right angle,** that is, an angle of 90°. The side of the triangle opposite the 90° angle is called the **hypotenuse;** the remaining two sides are called **legs.** In Figure 9 we have used c to represent the length of the hypotenuse and a and b to represent the lengths of the legs. Notice the use of the symbol ⌐ to show the 90° angle. We now state the Pythagorean Theorem.

Figure 9

Pythagorean Theorem

In a right triangle, the square of the length of the hypotenuse is equal to the sum of the squares of the lengths of the legs. That is, in the right triangle shown in Figure 9,

$$c^2 = a^2 + b^2 \qquad \textbf{(1)}$$

EXAMPLE 1   **Finding the Hypotenuse of a Right Triangle**

In a right triangle, one leg is of length 4 and the other is of length 3. What is the length of the hypotenuse?

Solution   Since the triangle is a right triangle, we use the Pythagorean Theorem with $a = 4$ and $b = 3$ to find the length c of the hypotenuse. From equation (1), we have

$$c^2 = a^2 + b^2$$
$$c^2 = 4^2 + 3^2 = 16 + 9 = 25$$
$$c = \sqrt{25} = 5$$

NOW WORK PROBLEM 3.

The converse of the Pythagorean Theorem is also true.

Converse of the Pythagorean Theorem

In a triangle, if the square of the length of one side equals the sum of the squares of the lengths of the other two sides, then the triangle is a right triangle. The 90° angle is opposite the longest side.

EXAMPLE 2	Verifying That a Triangle Is a Right Triangle

Show that a triangle whose sides are of lengths 5, 12, and 13 is a right triangle. Identify the hypotenuse.

Figure 10

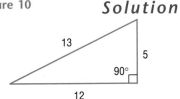

Solution   We square the lengths of the sides.

$$5^2 = 25, \qquad 12^2 = 144, \qquad 13^2 = 169$$

Notice that the sum of the first two squares (25 and 144) equals the third square (169). Hence, the triangle is a right triangle. The longest side, 13, is the hypotenuse. See Figure 10.

NOW WORK PROBLEM 11.

EXAMPLE 3	Applying the Pythagorean Theorem

The tallest inhabited building in the world is the Sears Tower in Chicago.* If the observation tower is 1450 feet above ground level, how far can a person standing in the observation tower see (with the aid of a telescope)? Use 3960 miles for the radius of Earth. See Figure 11.

Figure 11

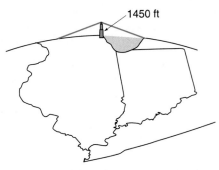

[**NOTE:**   1 mile = 5280 feet]

Source: Council on Tall Buildings and Urban Habitat (1997): Sears Tower No. 1 for tallest roof (1450 ft) and tallest occupied floor (1431 ft).

Figure 12

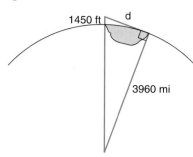

Solution   From the center of Earth, draw two radii: one through the Sears Tower and the other to the farthest point a person can see from the tower. See Figure 12. Apply the Pythagorean Theorem to the right triangle.

Since 1450 feet $= \dfrac{1450}{5280}$ miles, we have

$$d^2 + (3960)^2 = \left(3960 + \frac{1450}{5280}\right)^2$$

$$d^2 = \left(3960 + \frac{1450}{5280}\right)^2 - (3960)^2 \approx 2175.08$$

$$d \approx 46.64$$

A person can see about 47 miles from the observation tower.   ■

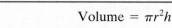 NOW WORK PROBLEM **37.**

GEOMETRY FORMULAS

2   Certain formulas from geometry are useful in solving algebra problems. We list some of these formulas next.

For a rectangle of length l and width w,

Area $= lw$ Perimeter $= 2l + 2w$

For a triangle with base b and altitude h,

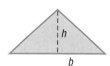

Area $= \dfrac{1}{2}bh$

For a circle of radius r (diameter $d = 2r$),

Area $= \pi r^2$ Circumference $= 2\pi r = \pi d$

For a rectangular box of length l, width w, and height h,

Volume $= lwh$

For a sphere of radius r,

Volume $= \dfrac{4}{3}\pi r^3$ Surface area $= 4\pi r^2$

For a right circular cylinder of height h and radius r,

Volume $= \pi r^2 h$

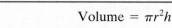 NOW WORK PROBLEM **19.**

| EXAMPLE 4 | **Using Geometry Formulas** |

A Christmas tree ornament is in the shape of a semicircle on top of a triangle. How many square centimeters (cm) of copper is required to make the ornament if the height of the triangle is 6 cm and the base is 4 cm?

Figure 13

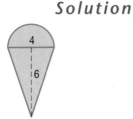

Solution   See Figure 13. The amount of copper required equals the shaded area. This area is the sum of the area of the triangle and the semicircle. The triangle has height $h = 6$ and base $b = 4$. The semicircle has diameter $d = 4$, so its radius is $r = 2$.

$$\text{Area} = \text{Area of triangle} + \text{Area of semicircle}$$

$$= \frac{1}{2}bh + \frac{1}{2}\pi r^2 = \frac{1}{2}(4)(6) + \frac{1}{2}\pi 2^2 \qquad b = 4; h = 6; r = 2.$$

$$= 12 + 2\pi \approx 18.28 \text{ cm}^2$$

About 18.28 cm^2 of copper is required. ∎

NOW WORK PROBLEM 33.

A.2 EXERCISES

In Problems 1–6, the lengths of the legs of a right triangle are given. Find the hypotenuse.

1. $a = 5$, $b = 12$ **2.** $a = 6$, $b = 8$ **3.** $a = 10$, $b = 24$

4. $a = 4$, $b = 3$ **5.** $a = 7$, $b = 24$ **6.** $a = 14$, $b = 48$

In Problems 7–14, the lengths of the sides of a triangle are given. Determine which are right triangles. For those that are, identify the hypotenuse.

7. $3, 4, 5$ **8.** $6, 8, 10$ **9.** $4, 5, 6$ **10.** $2, 2, 3$

11. $7, 24, 25$ **12.** $10, 24, 26$ **13.** $6, 4, 3$ **14.** $5, 4, 7$

15. Find the area A of a rectangle with length 4 inches and width 2 inches.

16. Find the area A of a rectangle with length 9 centimeters and width 4 centimeters.

17. Find the area A of a triangle with height 4 inches and base 2 inches.

18. Find the area A of a triangle with height 9 centimeters and base 4 centimeters.

19. Find the area A and circumference C of a circle of radius 5 meters.

20. Find the area A and circumference C of a circle of radius 2 feet.

21. Find the volume V of a rectangular box with length 8 feet, width 4 feet, and height 7 feet.

22. Find the volume V of a rectangular box with length 9 inches, width 4 inches, and height 8 inches.

23. Find the volume V and surface area S of a sphere of radius 4 centimeters.

24. Find the volume V and surface area S of a sphere of radius 3 feet.

25. Find the volume V of a right circular cylinder with radius 9 inches and height 8 inches.

26. Find the volume V of a right circular cylinder with radius 8 inches and height 9 inches.

In Problems 27–30, find the area of the shaded region.

27.

28.

29.

30.

31. How many feet does a wheel with a diameter of 16 inches travel after four revolutions?

32. How many revolutions will a circular disk with a diameter of 4 feet have completed after it has rolled 20 feet?

33. In the figure shown, *ABCD* is a square, with each side of length 6 feet. The width of the border (shaded portion) between the outer square *EFGH* and *ABCD* is 2 feet. Find the area of the border.

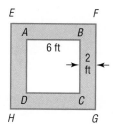

34. Refer to the figure. Square *ABCD* has an area of 100 square feet; square *BEFG* has an area of 16 square feet. What is the area of the triangle *CGF*?

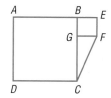

35. Architecture A Norman window consists of a rectangle surmounted by a semicircle. Find the area of the Norman window shown in the illustration. How much wood frame is needed to enclose the window?

36. Construction A circular swimming pool, 20 feet in diameter, is enclosed by a wooden deck that is 3 feet wide. What is the area of the deck? How much fence is required to enclose the deck?

In Problems 37–39, use the facts that the radius of Earth is 3960 miles and 1 mile = 5280 feet.

37. How Far Can You See? The conning tower of the U.S.S. *Silversides*, a World War II submarine now permanently stationed in Muskegon, Michigan, is approximately 20 feet above sea level. How far can you see from the conning tower?

38. How Far Can You See? A person who is 6 feet tall is standing on the beach in Fort Lauderdale, Florida, and looks out onto the Atlantic Ocean. Suddenly, a ship appears on the horizon. How far is the ship from shore?

39. How Far Can You See? The deck of a destroyer is 100 feet above sea level. How far can a person see from the deck? How far can a person see from the bridge, which is 150 feet above sea level?

40. Suppose that *m* and *n* are positive integers with $m > n$. If $a = m^2 - n^2$, $b = 2mn$, and $c = m^2 + n^2$, show that a, b, and c are the lengths of the sides of a right triangle. (This formula can be used to find the sides of a right triangle that are integers, such as 3, 4, 5; 5, 12, 13; and so on. Such triplets of integers are called **Pythagorean triples.**)

41. You have 1000 feet of flexible pool siding and wish to construct a swimming pool. Experiment with rectangular-shaped pools with perimeters of 1000 feet. How do their areas vary? What is the shape of the rectangle with the largest area? Now compute the area enclosed by a circular pool with a perimeter (circumference) of 1000 feet. What would be your choice of shape for the pool? If rectangular, what is your preference for dimensions? Justify your choice. If your only consideration is to have a pool that encloses the most area, what shape should you use?

42. The Gibb's Hill Lighthouse, Southampton, Bermuda, in operation since 1846, stands 117 feet high on a hill 245 feet high, so its beam of light is 362 feet above sea level. A brochure states that the light itself can be seen on the horizon about 26 miles distant. Verify the correctness of this information. The brochure further states that ships 40 miles away can see the light and planes flying at 10,000 feet can see it 120 miles away. Verify the accuracy of these statements. What assumption did the brochure make about the height of the ship?

PREPARING FOR THIS SECTION

Before getting started, review the following:

✓ Zero-Product Property (Appendix A, Section A.1, p. 488)   ✓ Square Roots (Appendix A, Section A.1, pp. 493–494)

$\boxed{\text{A.3}}$ SOLVING EQUATIONS

OBJECTIVES **1** Solve Equations by Factoring
  2 Solve Quadratic Equations by Factoring
  3 Know How to Complete the Square
  4 Solve a Quadratic Equation by Completing the Square
  5 Solve a Quadratic Equation Using the Quadratic Formula

An **equation in one variable** is a statement in which two expressions, at least one containing the variable, are equal. The expressions are called the **sides** of the equation. Since an equation is a statement, it may be true or false, depending on the value of the variable. Unless otherwise restricted, the admissible values of the variable are those in the domain of the variable. Those admissible values of the variable, if any, that result in a true statement are called **solutions,** or **roots,** of the equation. To **solve an equation** means to find all the solutions of the equation.

For example, the following are all equations in one variable, x:

$$x + 5 = 9 \qquad x^2 + 5x = 2x - 2 \qquad \frac{x^2 - 4}{x + 1} = 0 \qquad x^2 + 9 = 5$$

The first of these statements, $x + 5 = 9$, is true when $x = 4$ and false for any other choice of x. Thus, 4 is a solution of the equation $x + 5 = 9$. We also say that 4 **satisfies** the equation $x + 5 = 9$, because, when we substitute 4 for x, a true statement results.

Sometimes an equation will have more than one solution. For example, the equation

$$\frac{x^2 - 4}{x + 1} = 0$$

has $x = -2$ and $x = 2$ as solutions.

Usually, we will write the solution of an equation in set notation. This set is called the **solution set** of the equation. For example, the solution set of the equation $x^2 - 9 = 0$ is $\{-3, 3\}$.

Unless indicated otherwise, we will limit ourselves to real solutions, that is, solutions that are real numbers. Some equations have no real solution. For example, $x^2 + 9 = 5$ has no real solution, because there is no real number whose square when added to 9 equals 5.

An equation that is satisfied for every choice of the variable for which both sides are defined is called an **identity.** For example, the equation

$$3x + 5 = x + 3 + 2x + 2$$

is an identity, because this statement is true for any real number x.

Two or more equations that have precisely the same solutions are called **equivalent equations.**

For example, all the following equations are equivalent, because each has only the solution $x = 5$:

$$2x + 3 = 13$$
$$2x = 10$$
$$x = 5$$

These three equations illustrate one method for solving many types of equations: Replace the original equation by an equivalent equation, and continue until an equation with an obvious solution, such as $x = 5$, is reached. The question, though, is "How do I obtain an equivalent equation?" In general, there are five ways to do so.

PROCEDURES THAT RESULT IN EQUIVALENT EQUATIONS

1. Interchange the two sides of the equation:
   Replace $3 = x$ by $x = 3$
2. Simplify the sides of the equation by combining like terms, eliminating parentheses, and so on:
   Replace $(x + 2) + 6 = 2x + (x + 1)$
   by $x + 8 = 3x + 1$
3. Add or subtract the same expression on both sides of the equation:
   Replace $3x - 5 = 4$
   by $(3x - 5) + 5 = 4 + 5$
4. Multiply or divide both sides of the equation by the same nonzero expression:

   Replace $\dfrac{3x}{x - 1} = \dfrac{6}{x - 1},$ $x \neq 1$

   by $\dfrac{3x}{x - 1} \cdot (x - 1) = \dfrac{6}{x - 1} \cdot (x - 1)$

5. If one side of the equation is 0 and the other side can be factored, then we may use the Zero-Product Property* and set each factor equal to 0:
   Replace $x(x - 3) = 0$
   by $x = 0$ or $x - 3 = 0$

WARNING: Squaring both sides of an equation does not necessarily lead to an equivalent equation. ∎

*The Zero-Product Property says that if $ab = 0$ then $a = 0$ or $b = 0$ or both equal 0.

Whenever it is possible to solve an equation in your head, do so. For example:

The solution of $2x = 8$ is $x = 4$.

The solution of $3x - 15 = 0$ is $x = 5$.

Often, though, some rearrangement is necessary.

EXAMPLE 1 **Solving an Equation**

Solve the equation: $3x - 5 = 4$

Solution We replace the original equation by a succession of equivalent equations.

$$3x - 5 = 4$$
$$(3x - 5) + 5 = 4 + 5 \qquad \text{Add 5 to both sides.}$$
$$3x = 9 \qquad \text{Simplify.}$$
$$\frac{3x}{3} = \frac{9}{3} \qquad \text{Divide both sides by 3.}$$
$$x = 3 \qquad \text{Simplify.}$$

The last equation, $x = 3$, has the single solution 3. All these equations are equivalent, so 3 is the only solution of the original equation, $3x - 5 = 4$. ∎

Check: It is a good practice to check the solution by substituting 3 for x in the original equation.

$$3x - 5 = 4$$
$$3(3) - 5 \overset{?}{=} 4$$
$$9 - 5 \overset{?}{=} 4$$
$$4 = 4$$

The solution checks. ∎

✎ NOW WORK PROBLEMS **15** AND **21**.

① In the next examples, we use the Zero-Product Property.

EXAMPLE 2 **Solving Equations by Factoring**

Solve the equations: (a) $x^2 = 4x$ (b) $x^3 - x^2 - 4x + 4 = 0$

Solution (a) We begin by collecting all terms on one side. This results in 0 on one side and an expression to be factored on the other.

$$x^2 = 4x$$
$$x^2 - 4x = 0$$
$$x(x - 4) = 0 \qquad \text{Factor.}$$
$$x = 0 \quad \text{or} \quad x - 4 = 0 \qquad \text{Apply the Zero-Product Property.}$$
$$x = 4$$

The solution set is $\{0, 4\}$. ∎

Check: $x = 0$: $0^2 = 4 \cdot 0$ So 0 is a solution.

$x = 4$: $4^2 = 4 \cdot 4$ So 4 is a solution. ▪

(b) We group the terms of $x^3 - x^2 - 4x + 4 = 0$ as follows:

$$\left(x^3 - x^2\right) - (4x - 4) = 0$$

Factor out x^2 from the first grouping and 4 from the second.

$$x^2(x - 1) - 4(x - 1) = 0$$

This reveals the common factor $(x - 1)$, so we have

$$\left(x^2 - 4\right)(x - 1) = 0$$

$(x - 2)(x + 2)(x - 1) = 0$ Factor again.

$x - 2 = 0$ or $x + 2 = 0$ or $x - 1 = 0$ Set each factor equal to 0.

$x = 2$ $x = -2$ $x = 1$ Solve.

The solution set is $\{-2, 1, 2\}$.

Check: $x = -2$: $(-2)^3 - (-2)^2 - 4(-2) + 4 = -8 - 4 + 8 + 4 = 0$ −2 is a solution.

$x = 1$: $1^3 - 1^2 - 4(1) + 4 = 1 - 1 - 4 + 4 = 0$ 1 is a solution.

$x = 2$: $2^3 - 2^2 - 4(2) + 4 = 8 - 4 - 8 + 4 = 0$ 2 is a solution. ▪

✏ NOW WORK PROBLEM 25.

There are two points whose distance from the origin is 5 units, −5 and 5, so the equation $|x| = 5$ will have the solution set $\{-5, 5\}$.

EXAMPLE 3   **Solving an Equation Involving Absolute Value**

Solve the equation: $|x + 4| = 13$

Solution   There are two possibilities:

$$x + 4 = 13 \quad \text{or} \quad x + 4 = -13$$

$$x = 9 \qquad\qquad x = -17$$

The solution set is $\{-17, 9\}$. ▪

✏ NOW WORK PROBLEM 37.

QUADRATIC EQUATIONS

A **quadratic equation** is an equation equivalent to one written in the **standard form** $ax^2 + bx + c = 0$, where a, b, and c are real numbers and $a \neq 0$.

② When a quadratic equation is written in the standard form, $ax^2 + bx + c = 0$, it may be possible to factor the expression on the left side as the product of two first-degree polynomials.

| EXAMPLE 4 | Solving a Quadratic Equation by Factoring |

Solve the equation: $2x^2 = x + 3$

Solution   We put the equation in standard form by adding $-x - 3$ to both sides.

$$2x^2 = x + 3 \quad \text{Add } -x - 3 \text{ to both sides.}$$
$$2x^2 - x - 3 = 0$$

The left side may now be factored as

$$(2x - 3)(x + 1) = 0$$

so that

$$2x - 3 = 0 \quad \text{or} \quad x + 1 = 0$$
$$x = \frac{3}{2} \qquad\qquad x = -1$$

The solution set is $\left\{-1, \dfrac{3}{2}\right\}$.   ■

When the left side factors into two linear equations with the same solution, the quadratic equation is said to have a **repeated solution.** We also call this solution a **root of multiplicity 2,** or a **double root.**

| EXAMPLE 5 | Solving a Quadratic Equation by Factoring |

Solve the equation: $9x^2 - 6x + 1 = 0$

Solution   This equation is already in standard form, and the left side can be factored.

$$9x^2 - 6x + 1 = 0$$
$$(3x - 1)(3x - 1) = 0$$

so

$$x = \frac{1}{3} \quad \text{or} \quad x = \frac{1}{3}$$

This equation has only the repeated solution $\dfrac{1}{3}$.   ■

✏ NOW WORK PROBLEM 55.

THE SQUARE ROOT METHOD

Suppose that we wish to solve the quadratic equation

$$x^2 = p \qquad\qquad\qquad \textbf{(1)}$$

where $p \geq 0$ is a nonnegative number. We proceed as in the earlier examples.

$$x^2 - p = 0 \qquad \text{Put in standard form.}$$
$$\left(x - \sqrt{p}\right)\left(x + \sqrt{p}\right) = 0 \qquad \text{Factor (over the real numbers).}$$
$$x = \sqrt{p} \quad \text{or} \quad x = -\sqrt{p} \qquad \text{Solve.}$$

We have the following result:

$$\text{If } x^2 = p \text{ and } p \geq 0, \text{ then } x = \sqrt{p} \text{ or } x = -\sqrt{p}. \qquad \textbf{(2)}$$

When statement (2) is used, it is called the **Square Root Method.** In statement (2), note that if $p > 0$ the equation $x^2 = p$ has two solutions, $x = \sqrt{p}$ and $x = -\sqrt{p}$. We usually abbreviate these solutions as $x = \pm\sqrt{p}$, read as "x equals plus or minus the square root of p."

For example, the two solutions of the equation

$$x^2 = 4$$

are

$$x = \pm\sqrt{4} \qquad \text{Use the Square Root Method.}$$

and, since $\sqrt{4} = 2$, we have

$$x = \pm 2$$

The solution set is $\{-2, 2\}$.

NOW WORK PROBLEM 69.

COMPLETING THE SQUARE

③ We now introduce the method of **completing the square.** The idea behind this method is to *adjust* the left side of a quadratic equation, $ax^2 + bx + c = 0$, so that it becomes a perfect square, that is, the square of a first-degree polynomial. For example, $x^2 + 6x + 9$ and $x^2 - 4x + 4$ are perfect squares because

$$x^2 + 6x + 9 = (x + 3)^2 \quad \text{and} \quad x^2 - 4x + 4 = (x - 2)^2$$

How do we adjust the left side? We do it by adding the appropriate number to the left side to create a perfect square. For example, to make $x^2 + 6x$ a perfect square, we add 9.

Let's look at several examples of completing the square when the coefficient of x^2 is 1:

Start	Add	Result
$x^2 + 4x$	4	$x^2 + 4x + 4 = (x + 2)^2$
$x^2 + 12x$	36	$x^2 + 12x + 36 = (x + 6)^2$
$x^2 - 6x$	9	$x^2 - 6x + 9 = (x - 3)^2$
$x^2 + x$	$\dfrac{1}{4}$	$x^2 + x + \dfrac{1}{4} = \left(x + \dfrac{1}{2}\right)^2$

Do you see the pattern? Provided that the coefficient of x^2 is 1, we complete the square by adding the square of $\dfrac{1}{2}$ of the coefficient of x.

Start	Add	Result
$x^2 + mx$	$\left(\dfrac{m}{2}\right)^2$	$x^2 + mx + \left(\dfrac{m}{2}\right)^2 = \left(x + \dfrac{m}{2}\right)^2$

NOW WORK PROBLEM 73.

④ The next example illustrates how the procedure of completing the square can be used to solve a quadratic equation.

| EXAMPLE 6 | **Solving a Quadratic Equation by Completing the Square** |

Solve by completing the square: $2x^2 - 8x - 5 = 0$

Solution   First, we rewrite the equation.

$$2x^2 - 8x - 5 = 0$$
$$2x^2 - 8x = 5$$

Next, we divide both sides by 2 so that the coefficient of x^2 is 1. (This enables us to complete the square at the next step.)

$$x^2 - 4x = \frac{5}{2}$$

Finally, we complete the square by adding 4 to both sides.

$$x^2 - 4x + 4 = \frac{5}{2} + 4$$

$$(x - 2)^2 = \frac{13}{2}$$

$$x - 2 = \pm\sqrt{\frac{13}{2}} \qquad \text{Use the Square Root Method.}$$

$$x - 2 = \pm\frac{\sqrt{26}}{2} \qquad \sqrt{\frac{13}{2}} = \frac{\sqrt{13}}{\sqrt{2}} \cdot \frac{\sqrt{2}}{\sqrt{2}} = \frac{\sqrt{26}}{2}$$

$$x = 2 \pm \frac{\sqrt{26}}{2}$$

The solution set is $\left\{2 - \dfrac{\sqrt{26}}{2}, 2 + \dfrac{\sqrt{26}}{2}\right\}$. ∎

NOTE: If we wanted an approximation, say rounded to two decimal places, of these solutions, we would use a calculator to get $\{-0.55, 4.55\}$.

NOW WORK PROBLEM **79.**

THE QUADRATIC FORMULA

⑤ We can use the method of completing the square to obtain a general formula for solving the quadratic equation.

$$ax^2 + bx + c = 0, \qquad a \neq 0$$

NOTE: There is no loss in generality to assume that $a > 0$, since if $a < 0$ we can multiply by -1 to obtain an equivalent equation with a positive leading coefficient.

As in Example 6, we rearrange the terms as

$$ax^2 + bx = -c \quad a > 0$$

Since $a > 0$, we can divide both sides by a to get

$$x^2 + \frac{b}{a}x = -\frac{c}{a}$$

Now the coefficient of x^2 is 1. To complete the square on the left side, add the square of $\dfrac{1}{2}$ of the coefficient of x; that is, add

$$\left(\frac{1}{2} \cdot \frac{b}{a}\right)^2 = \frac{b^2}{4a^2}$$

to both sides. Then

$$x^2 + \frac{b}{a}x + \frac{b^2}{4a^2} = \frac{b^2}{4a^2} - \frac{c}{a}$$

$$\left(x + \frac{b}{2a}\right)^2 = \frac{b^2 - 4ac}{4a^2} \qquad \frac{b^2}{4a^2} - \frac{c}{a} = \frac{b^2}{4a^2} - \frac{4ac}{4a^2} = \frac{b^2 - 4ac}{4a^2} \qquad \textbf{(3)}$$

Provided that $b^2 - 4ac \geq 0$, we now can use the Square Root Method to get

$$x + \frac{b}{2a} = \pm\sqrt{\frac{b^2 - 4ac}{4a^2}}$$

$$x + \frac{b}{2a} = \frac{\pm\sqrt{b^2 - 4ac}}{2a} \qquad \text{The square root of a quotient equals the quotient of the square roots. Also, } \sqrt{4a^2} = 2a \text{ since } a > 0.$$

$$x = -\frac{b}{2a} \pm \frac{\sqrt{b^2 - 4ac}}{2a} \qquad \text{Add } -\frac{b}{2a} \text{ to both sides.}$$

$$x = \frac{-b \pm \sqrt{b^2 - 4ac}}{2a} \qquad \text{Combine the quotients on the right.}$$

What if $b^2 - 4ac$ is negative? Then equation (3) states that the left expression (a real number squared) equals the right expression (a negative number). Since this occurrence is impossible for real numbers, we conclude that if $b^2 - 4ac < 0$ the quadratic equation has no *real* solution. (We discuss quadratic equations for which the quantity $b^2 - 4ac < 0$ in detail in the next section."

We now state the *quadratic formula*.

Theorem   Consider the quadratic equation

$$ax^2 + bx + c = 0, \qquad a \neq 0$$

If $b^2 - 4ac < 0$, this equation has no real solution.
If $b^2 - 4ac \geq 0$, the real solution(s) of this equation is (are) given by the **quadratic formula.**

Quadratic Formula

$$x = \frac{-b \pm \sqrt{b^2 - 4ac}}{2a} \qquad \textbf{(4)}$$

The quantity $b^2 - 4ac$ is called the **discriminant** of the quadratic equation, because its value tells us whether the equation has real solutions. In fact, it also tells us how many solutions to expect.

> **DISCRIMINANT OF A QUADRATIC EQUATION**
>
> For a quadratic equation $ax^2 + bx + c = 0$:
>
> 1. If $b^2 - 4ac > 0$, there are two unequal real solutions.
> 2. If $b^2 - 4ac = 0$, there is a repeated solution, a root of multiplicity 2.
> 3. If $b^2 - 4ac < 0$, there is no real solution.

When asked to find the real solutions, if any, of a quadratic equation, always evaluate the discriminant first to see how many real solutions there are.

EXAMPLE 7

Solving a Quadratic Equation Using the Quadratic Formula

Use the quadratic formula to find the real solutions, if any, of the equation

$$3x^2 - 5x + 1 = 0$$

Solution The equation is in standard form, so we compare it to $ax^2 + bx + c = 0$ to find a, b, and c.

$$3x^2 - 5x + 1 = 0$$
$$ax^2 + bx + c = 0 \quad a = 3, b = -5, c = 1$$

With $a = 3$, $b = -5$, and $c = 1$, we evaluate the discriminant $b^2 - 4ac$.

$$b^2 - 4ac = (-5)^2 - 4(3)(1) = 25 - 12 = 13$$

Since $b^2 - 4ac > 0$, there are two real solutions, which can be found using the quadratic formula.

$$x = \frac{-b \pm \sqrt{b^2 - 4ac}}{2a} = \frac{-(-5) \pm \sqrt{13}}{2(3)} = \frac{5 \pm \sqrt{13}}{6}$$

The solution set is $\left\{ \dfrac{5 - \sqrt{13}}{6}, \dfrac{5 + \sqrt{13}}{6} \right\}$. ∎

EXAMPLE 8

Solving a Quadratic Equation Using the Quadratic Formula

Use the quadratic formula to find the real solutions, if any, of the equation

$$3x^2 + 2 = 4x$$

Solution The equation, as given, is not in standard form.

$$3x^2 + 2 = 4x$$
$$3x^2 - 4x + 2 = 0 \qquad \text{Put in standard form.}$$
$$ax^2 + bx + c = 0 \qquad \text{Compare to standard form.}$$

With $a = 3$, $b = -4$, and $c = 2$, we find

$$b^2 - 4ac = (-4)^2 - 4(3)(2) = 16 - 24 = -8$$

Since $b^2 - 4ac < 0$, the equation has no real solution. ∎

NOW WORK PROBLEMS **85** AND **91**.

SUMMARY

Procedure for Solving a Quadratic Equation

To solve a quadratic equation, first put it in standard form:

$$ax^2 + bx + c = 0$$

Then:

STEP 1: Identify a, b, and c.

STEP 2: Evaluate the discriminant, $b^2 - 4ac$.

STEP 3: (a) If the discriminant is negative, the equation has no real solution.
(b) If the discriminant is zero, the equation has one real solution, a repeated root.
(c) If the discriminant is positive, the equation has two distinct real solutions. If you can easily spot factors, use the factoring method to solve the equation. Otherwise, use the quadratic formula or the method of completing the square.

A.3 EXERCISES

In Problems 1–66, solve each equation.

1. $3x = 21$

2. $3x = -24$

3. $5x + 15 = 0$

4. $3x + 18 = 0$

5. $2x - 3 = 5$

6. $3x + 4 = -8$

7. $\dfrac{1}{3}x = \dfrac{5}{12}$

8. $\dfrac{2}{3}x = \dfrac{9}{2}$

9. $6 - x = 2x + 9$

10. $3 - 2x = 2 - x$

11. $2(3 + 2x) = 3(x - 4)$

12. $3(2 - x) = 2x - 1$

13. $8x - (2x + 1) = 3x - 10$

14. $5 - (2x - 1) = 10$

15. $\dfrac{1}{2}x - 4 = \dfrac{3}{4}x$

16. $1 - \dfrac{1}{2}x = 5$

17. $0.9t = 0.4 + 0.1t$

18. $0.9t = 1 + t$

19. $\dfrac{2}{y} + \dfrac{4}{y} = 3$

20. $\dfrac{4}{y} - 5 = \dfrac{5}{2y}$

21. $(x + 7)(x - 1) = (x + 1)^2$

22. $(x + 2)(x - 3) = (x - 3)^2$

23. $z(z^2 + 1) = 3 + z^3$

24. $w(4 - w^2) = 8 - w^3$

25. $x^2 = 9x$

26. $x^3 = x^2$

27. $t^3 - 9t^2 = 0$

28. $4z^3 - 8z^2 = 0$

29. $\dfrac{3}{2x - 3} = \dfrac{2}{x + 5}$

30. $\dfrac{-2}{x + 4} = \dfrac{-3}{x + 1}$

31. $(x + 2)(3x) = (x + 2)(6)$

32. $(x - 5)(2x) = (x - 5)(4)$

33. $\dfrac{2}{x - 2} = \dfrac{3}{x + 5} + \dfrac{10}{(x + 5)(x - 2)}$

34. $\dfrac{1}{2x + 3} + \dfrac{1}{x - 1} = \dfrac{1}{(2x + 3)(x - 1)}$

35. $|2x| = 6$

36. $|3x| = 12$

37. $|2x + 3| = 5$

38. $|3x - 1| = 2$

39. $|1 - 4t| = 5$

40. $|1 - 2z| = 3$

41. $|-2x| = 8$

42. $|-x| = 1$

43. $|-2|x = 4$

44. $|3|x = 9$

45. $|x - 2| = -\dfrac{1}{2}$

46. $|2 - x| = -1$

47. $|x^2 - 4| = 0$

48. $|x^2 - 9| = 0$

49. $|x^2 - 2x| = 3$

50. $|x^2 + x| = 12$

51. $|x^2 + x - 1| = 1$

52. $|x^2 + 3x - 2| = 2$

53. $x^2 = 4x$

54. $x^2 = -8x$

55. $z^2 + 4z - 12 = 0$

56. $v^2 + 7v + 12 = 0$

57. $2x^2 - 5x - 3 = 0$

58. $3x^2 + 5x + 2 = 0$

59. $x(x - 7) + 12 = 0$

60. $x(x + 1) = 12$

61. $4x^2 + 9 = 12x$

62. $25x^2 + 16 = 40x$

63. $6x - 5 = \dfrac{6}{x}$

64. $x + \dfrac{12}{x} = 7$

65. $\dfrac{4(x - 2)}{x - 3} + \dfrac{3}{x} = \dfrac{-3}{x(x - 3)}$

66. $\dfrac{5}{x + 4} = 4 + \dfrac{3}{x - 2}$

In Problems 67–72, solve each equation by the Square Root Method.

67. $x^2 = 25$

68. $x^2 = 36$

69. $(x - 1)^2 = 4$

70. $(x + 2)^2 = 1$

71. $(2x + 3)^2 = 9$

72. $(3x - 2)^2 = 4$

In Problems 73–78, what number should be added to complete the square of each expression?

73. $x^2 + 8x$

74. $x^2 - 4x$

75. $x^2 + \dfrac{1}{2}x$

76. $x^2 - \dfrac{1}{3}x$

77. $x^2 - \dfrac{2}{3}x$

78. $x^2 - \dfrac{2}{5}x$

In Problems 79–84, solve each equation by completing the square.

79. $x^2 + 4x = 21$

80. $x^2 - 6x = 13$

81. $x^2 - \dfrac{1}{2}x - \dfrac{3}{16} = 0$

82. $x^2 + \dfrac{2}{3}x - \dfrac{1}{3} = 0$

83. $3x^2 + x - \dfrac{1}{2} = 0$

84. $2x^2 - 3x - 1 = 0$

In Problems 85–96, find the real solutions, if any, of each equation. Use the quadratic formula.

85. $x^2 - 4x + 2 = 0$

86. $x^2 + 4x + 2 = 0$

87. $x^2 - 5x - 1 = 0$

88. $x^2 + 5x + 3 = 0$

89. $2x^2 - 5x + 3 = 0$

90. $2x^2 + 5x + 3 = 0$

91. $4y^2 - y + 2 = 0$

92. $4t^2 + t + 1 = 0$

93. $4x^2 = 1 - 2x$

94. $2x^2 = 1 - 2x$

95. $x^2 + \sqrt{3}x - 3 = 0$

96. $x^2 + \sqrt{2}x - 2 = 0$

In Problems 97–102, use the discriminant to determine whether each quadratic equation has two unequal real solutions, a repeated real solution, or no real solution, without solving the equation.

97. $x^2 - 5x + 7 = 0$

98. $x^2 + 5x + 7 = 0$

99. $9x^2 - 30x + 25 = 0$

100. $25x^2 - 20x + 4 = 0$

101. $3x^2 + 5x - 8 = 0$

102. $2x^2 - 3x - 4 = 0$

In Problems 103–108, solve each equation. The letters a, b, and c are constants.

103. $ax - b = c, \quad a \neq 0$

104. $1 - ax = b, \quad a \neq 0$

105. $\dfrac{x}{a} + \dfrac{x}{b} = c, \quad a \neq 0, b \neq 0, a \neq -b$

106. $\dfrac{a}{x} + \dfrac{b}{x} = c, \quad c \neq 0$

107. $\dfrac{1}{x - a} + \dfrac{1}{x + a} = \dfrac{2}{x - 1}$

108. $\dfrac{b + c}{x + a} = \dfrac{b - c}{x - a}, \quad c \neq 0, a \neq 0$

Problems 109–114 list some formulas that occur in applications. Solve each formula for the indicated variable.

109. Electricity $\dfrac{1}{R} = \dfrac{1}{R_1} + \dfrac{1}{R_2}$ for R

110. Finance $A = P(1 + rt)$ for r

111. Mechanics $F = \dfrac{mv^2}{R}$ for R

112. Chemistry $PV = nRT$ for T

113. Mathematics $S = \dfrac{a}{1 - r}$ for r

114. Mechanics $v = -gt + v_0$ for t

---

115. Show that the sum of the roots of a quadratic equation is $-\dfrac{b}{a}$.

116. Show that the product of the roots of a quadratic equation is $\dfrac{c}{a}$.

117. Find k such that the equation $kx^2 + x + k = 0$ has a repeated real solution.

118. Find k such that the equation $x^2 - kx + 4 = 0$ has a repeated real solution.

119. Show that the real solutions of the equation $ax^2 + bx + c = 0$ are the negatives of the real solutions of the equation $ax^2 - bx + c = 0$. Assume that $b^2 - 4ac \geq 0$.

120. Show that the real solutions of the equation $ax^2 + bx + c = 0$ are the reciprocals of the real solutions of the equation $cx^2 + bx + a = 0$. Assume that $b^2 - 4ac \geq 0$.

121. Which of the following pairs of equations are equivalent? Explain.
(a) $x^2 = 9; \quad x = 3$
(b) $x = \sqrt{9}; \quad x = 3$
(c) $(x - 1)(x - 2) = (x - 1)^2; \quad x - 2 = x - 1$

122. The equation
$$\frac{5}{x + 3} + 3 = \frac{8 + x}{x + 3}$$
has no solution, yet when we go through the process of solving it we obtain $x = -3$. Write a brief paragraph to explain what causes this to happen.

123. Make up an equation that has no solution and give it to a fellow student to solve. Ask the fellow student to write a critique of your equation.

124. Describe three ways you might solve a quadratic equation. State your preferred method; explain why you chose it.

125. Explain the benefits of evaluating the discriminant of a quadratic equation before attempting to solve it.

126. Make up three quadratic equations: one having two distinct solutions, one having no real solution, and one having exactly one real solution.

127. The word *quadratic* seems to imply four (*quad*), yet a quadratic equation is an equation that involves a polynomial of degree 2. Investigate the origin of the term *quadratic* as it is used in the expression *quadratic equation*. Write a brief essay on your findings.

A.4 COMPLEX NUMBERS; QUADRATIC EQUATIONS WITH A NEGATIVE DISCRIMINANT

OBJECTIVES 1 Add, Subtract, Multiply, and Divide Complex Numbers
2 Solve Quadratic Equations with a Negative Discriminant

One property of a real number is that its square is nonnegative. For example, there is no real number x for which

$$x^2 = -1$$

To remedy this situation, we introduce a number i, called the **imaginary unit,** whose square is -1; that is,

$$i^2 = -1$$

This should not surprise you. If our universe were to consist only of integers, there would be no number x for which $2x = 1$. This unfortunate circumstance was remedied by introducing numbers such as $\frac{1}{2}$ and $\frac{2}{3}$, the *rational numbers*. If our universe were to consist only of rational numbers, there would be no x whose square equals 2. That is, there would be no number x for which $x^2 = 2$. To remedy this, we introduced numbers such as $\sqrt{2}$ and $\sqrt[3]{5}$, the *irrational numbers*. The *real numbers,* you will recall, consist of the rational numbers and the irrational numbers. Now, if our universe were to consist only of real numbers, then there would be no number x whose square is -1. To remedy this, we introduce a number i, whose square is -1.

In the progression outlined, each time that we encountered a situation that was unsuitable, we introduced new numbers to remedy this situation. And each resulting new number system contained the earlier number system as a subset. The number system that results from introducing the number i is called the **complex number system.**

Complex numbers are numbers of the form $a + bi$, where a and b are real numbers. The real number a is called the **real part** of the number $a + bi$; the real number b is called the **imaginary part** of $a + bi$.

For example, the complex number $-5 + 6i$ has the real part -5 and the imaginary part 6.

When a complex number is written in the form $a + bi$, where a and b are real numbers, we say it is in **standard form.** However, if the imaginary part of a complex number is negative, such as in the complex number $3 + (-2)i$, we agree to write it instead in the form $3 - 2i$.

Also, the complex number $a + 0i$ is usually written merely as a. This serves to remind us that the real numbers are a subset of the complex numbers. The complex number $0 + bi$ is usually written as bi. Sometimes the complex number bi is called a **pure imaginary number.**

1   Equality, addition, subtraction, and multiplication of complex numbers are defined so as to preserve the familiar rules of algebra for real numbers.

Two complex numbers are equal if and only if their real parts are equal and their imaginary parts are equal. That is,

Equality of Complex Numbers

$$a + bi = c + di \quad \text{if and only if } a = c \text{ and } b = d \quad \textbf{(1)}$$

Two complex numbers are added by forming the complex number whose real part is the sum of the real parts and whose imaginary part is the sum of the imaginary parts. That is,

Sum of Complex Numbers

$$(a + bi) + (c + di) = (a + c) + (b + d)i \quad \textbf{(2)}$$

To subtract two complex numbers, we use this rule:

Difference of Complex Numbers

$$(a + bi) - (c + di) = (a - c) + (b - d)i \quad \textbf{(3)}$$

EXAMPLE 1

Adding and Subtracting Complex Numbers

(a) $(3 + 5i) + (-2 + 3i) = [3 + (-2)] + (5 + 3)i = 1 + 8i$

(b) $(6 + 4i) - (3 + 6i) = (6 - 3) + (4 - 6)i = 3 + (-2)i = 3 - 2i$ ∎

NOW WORK PROBLEM **5**.

Products of complex numbers are calculated as illustrated in Example 2.

EXAMPLE 2

Multiplying Complex Numbers

$$(5 + 3i) \cdot (2 + 7i) = 5 \cdot (2 + 7i) + 3i(2 + 7i) = 10 + 35i + 6i + 21i^2$$

$\uparrow$ Distributive property $\qquad\qquad$ $\uparrow$ Distributive property

$$= 10 + 41i + 21(-1)$$

$\uparrow$ $i^2 = -1$

$$= -11 + 41i \quad ∎$$

Based on the procedure of Example 2, we define the **product** of two complex numbers by the following formula:

Product of Complex Numbers

$$(a + bi) \cdot (c + di) = (ac - bd) + (ad + bc)i \quad \textbf{(4)}$$

Do not bother to memorize formula (4). Instead, whenever it is necessary to multiply two complex numbers, follow the usual rules for multiplying two binomials, as in Example 2, remembering that $i^2 = -1$. For example,

$$(2i)(2i) = 4i^2 = -4$$

$$(2 + i)(1 - i) = 2 - 2i + i - i^2 = 3 - i$$

NOW WORK PROBLEM 11.

Algebraic properties for addition and multiplication, such as the commutative, associative, and distributive properties, hold for complex numbers. The property that every nonzero complex number has a multiplicative inverse, or reciprocal, requires a closer look.

CONJUGATES

If $z = a + bi$ is a complex number, then its **conjugate,** denoted by $\bar{z}$, is defined as

$$\bar{z} = \overline{a + bi} = a - bi$$

For example, $\overline{2 + 3i} = 2 - 3i$ and $\overline{-6 - 2i} = -6 + 2i$.

| EXAMPLE 3 | Multiplying a Complex Number by Its Conjugate |

Find the product of the complex number $z = 3 + 4i$ and its conjugate $\bar{z}$.

Solution Since $\bar{z} = 3 - 4i$, we have

$$z\bar{z} = (3 + 4i)(3 - 4i) = 9 - 12i + 12i - 16i^2 = 9 + 16 = 25 \quad \blacksquare$$

The result obtained in Example 3 has an important generalization.

Theorem   The product of a complex number and its conjugate is a nonnegative real number. That is, if $z = a + bi$, then

$$z\bar{z} = a^2 + b^2 \tag{5}$$

Proof   If $z = a + bi$, then

$$z\bar{z} = (a + bi)(a - bi) = a^2 - (bi)^2 = a^2 - b^2i^2 = a^2 + b^2 \quad \blacksquare$$

To express the reciprocal of a nonzero complex number z in standard form, multiply the numerator and denominator of $\dfrac{1}{z}$ by $\bar{z}$. That is, if $z = a + bi$ is a nonzero complex number, then

$$\frac{1}{a + bi} = \frac{1}{z} = \frac{1}{z} \cdot \frac{\bar{z}}{\bar{z}} = \frac{\bar{z}}{z\bar{z}} \underset{\substack{\uparrow \\ \text{Use (5).}}}{=} \frac{a - bi}{a^2 + b^2}$$

$$= \frac{a}{a^2 + b^2} - \frac{b}{a^2 + b^2}i$$

EXAMPLE 4   **Writing the Reciprocal of a Complex Number in Standard Form**

Write $\dfrac{1}{3 + 4i}$ in standard form $a + bi$; that is, find the reciprocal of $3 + 4i$.

Solution The idea is to multiply the numerator and denominator by the conjugate of $3 + 4i$, that is, the complex number $3 - 4i$. The result is

$$\frac{1}{3 + 4i} = \frac{1}{3 + 4i} \cdot \frac{3 - 4i}{3 - 4i} = \frac{3 - 4i}{9 + 16} = \frac{3}{25} - \frac{4}{25}i \quad \blacksquare$$

To express the quotient of two complex numbers in standard form, we multiply the numerator and denominator of the quotient by the conjugate of the denominator.

EXAMPLE 5   **Writing the Quotient of Complex Numbers in Standard Form**

Write each of the following in standard form.

(a) $\dfrac{1 + 4i}{5 - 12i}$ (b) $\dfrac{2 - 3i}{4 - 3i}$

Solution (a) $\dfrac{1 + 4i}{5 - 12i} = \dfrac{1 + 4i}{5 - 12i} \cdot \dfrac{5 + 12i}{5 + 12i} = \dfrac{5 + 12i + 20i + 48i^2}{25 + 144}$

$$= \frac{-43 + 32i}{169} = -\frac{43}{169} + \frac{32}{169}i$$

(b) $\dfrac{2 - 3i}{4 - 3i} = \dfrac{2 - 3i}{4 - 3i} \cdot \dfrac{4 + 3i}{4 + 3i} = \dfrac{8 + 6i - 12i - 9i^2}{16 + 9}$

$$= \frac{17 - 6i}{25} = \frac{17}{25} - \frac{6}{25}i \quad \blacksquare$$

NOW WORK PROBLEM 19.

EXAMPLE 6   **Writing Other Expressions in Standard Form**

If $z = 2 - 3i$ and $w = 5 + 2i$, write each of the following expressions in standard form.

(a) $\dfrac{z}{w}$ (b) $\overline{z + w}$ (c) $z + \overline{z}$

Solution (a) $\dfrac{z}{w} = \dfrac{z \cdot \overline{w}}{w \cdot \overline{w}} = \dfrac{(2 - 3i)(5 - 2i)}{(5 + 2i)(5 - 2i)} = \dfrac{10 - 4i - 15i + 6i^2}{25 + 4}$

$$= \frac{4 - 19i}{29} = \frac{4}{29} - \frac{19}{29}i$$

(b) $\overline{z + w} = \overline{(2 - 3i) + (5 + 2i)} = \overline{7 - i} = 7 + i$

(c) $z + \overline{z} = (2 - 3i) + (2 + 3i) = 4 \quad \blacksquare$

The conjugate of a complex number has certain general properties that we shall find useful.

For a real number $a = a + 0i$, the conjugate is $\bar{a} = \overline{a + 0i} = a - 0i = a$. That is,

Theorem   The conjugate of a real number is the real number itself.

∎

Other properties of the conjugate that are direct consequences of the definition are given next. In each statement, z and w represent complex numbers.

Theorem   The conjugate of the conjugate of a complex number is the complex number itself.

$$\bar{\bar{z}} = z \tag{6}$$

The conjugate of the sum of two complex numbers equals the sum of their conjugates.

$$\overline{z + w} = \bar{z} + \bar{w} \tag{7}$$

The conjugate of the product of two complex numbers equals the product of their conjugates.

$$\overline{z \cdot w} = \bar{z} \cdot \bar{w} \tag{8}$$

∎

We leave the proofs of equations (6), (7), and (8) as exercises.

POWERS OF i

The **powers of i** follow a pattern that is useful to know.

$$i^1 = i \qquad\qquad\qquad i^5 = i^4 \cdot i = 1 \cdot i = i$$
$$i^2 = -1 \qquad\qquad\quad i^6 = i^4 \cdot i^2 = -1$$
$$i^3 = i^2 \cdot i = -i \qquad\quad i^7 = i^4 \cdot i^3 = -i$$
$$i^4 = i^2 \cdot i^2 = (-1)(-1) = 1 \qquad i^8 = i^4 \cdot i^4 = 1$$

And so on. The powers of i repeat with every fourth power.

EXAMPLE 7   Evaluating Powers of i

(a) $i^{27} = i^{24} \cdot i^3 = \left(i^4\right)^6 \cdot i^3 = 1^6 \cdot i^3 = -i$

(b) $i^{101} = i^{100} \cdot i^1 = \left(i^4\right)^{25} \cdot i = 1^{25} \cdot i = i$

∎

| EXAMPLE 8 | Writing the Power of a Complex Number in Standard Form |

Write $(2 + i)^3$ in standard form.

Solution   We use the special product formula for $(x + a)^3$.

$$(x + a)^3 = x^3 + 3ax^2 + 3a^2x + a^3$$

Using this special product formula,

$$(2 + i)^3 = 2^3 + 3 \cdot i \cdot 2^2 + 3 \cdot i^2 \cdot 2 + i^3$$
$$= 8 + 12i + 6(-1) + (-i)$$
$$= 2 + 11i$$ ∎

NOW WORK PROBLEM **33.**

QUADRATIC EQUATIONS WITH A NEGATIVE DISCRIMINANT

②   Quadratic equations with a negative discriminant have no real number solution. However, if we extend our number system to allow complex numbers, quadratic equations will always have a solution. Since the solution to a quadratic equation involves the square root of the discriminant, we begin with a discussion of square roots of negative numbers.

If N is a positive real number, we define the **principal square root of** $-N$, denoted by $\sqrt{-N}$, as

$$\sqrt{-N} = \sqrt{N}i$$

where i is the imaginary unit and $i^2 = -1$.

WARNING: In writing $\sqrt{-N} = \sqrt{N}i$, be sure to place i outside the $\sqrt{}$ symbol. ∎

| EXAMPLE 9 | Evaluating the Square Root of a Negative Number |

(a) $\sqrt{-1} = \sqrt{1}\,i = i$ (b) $\sqrt{-4} = \sqrt{4}\,i = 2i$

(c) $\sqrt{-8} = \sqrt{8}\,i = 2\sqrt{2}\,i$ ∎

| EXAMPLE 10 | Solving Equations |

Solve each equation in the complex number system.

(a) $x^2 = 4$ (b) $x^2 = -9$

Solution   (a) $x^2 = 4$

$x = \pm\sqrt{4} = \pm 2$

The equation has two solutions, -2 and 2.

(b) $x^2 = -9$

$$x = \pm\sqrt{-9} = \pm\sqrt{9}\,i = \pm 3i$$

The equation has two solutions, $-3i$ and $3i$. ∎

NOW WORK PROBLEMS **41** AND **45**.

WARNING: When working with square roots of negative numbers, do not set the square root of a product equal to the product of the square roots (which can be done with positive numbers). To see why, look at this calculation: We know that $\sqrt{100} = 10$. However, it is also true that $100 = (-25)(-4)$, so

$$10 = \sqrt{100} = \sqrt{(-25)(-4)} \neq \sqrt{-25}\,\sqrt{-4} = (\sqrt{25}\,i)(\sqrt{4}\,i) = (5i)(2i) = 10i^2 = -10$$

↑
Here is the error. ∎

Because we have defined the square root of a negative number, we can now restate the quadratic formula without restriction.

Theorem

> In the complex number system, the solutions of the quadratic equation $ax^2 + bx + c = 0$, where a, b, and c are real numbers and $a \neq 0$, are given by the formula
>
> $$x = \frac{-b \pm \sqrt{b^2 - 4ac}}{2a} \tag{9}$$

∎

EXAMPLE 11

Solving Quadratic Equations in the Complex Number System

Solve the equation $x^2 - 4x + 8 = 0$ in the complex number system.

Solution Here $a = 1$, $b = -4$, $c = 8$, and $b^2 - 4ac = 16 - 4(1)(8) = -16$. Using equation (9), we find that

$$x = \frac{-(-4) \pm \sqrt{-16}}{2(1)} = \frac{4 \pm \sqrt{16}\,i}{2} = \frac{4 \pm 4i}{2} = 2 \pm 2i$$

The equation has the solution set $\{2 - 2i, 2 + 2i\}$. ∎

Check:

$2 + 2i$: $(2 + 2i)^2 - 4(2 + 2i) + 8 = 4 + 8i + 4i^2 - 8 - 8i + 8$

$$= 4 - 4 = 0$$

$2 - 2i$: $(2 - 2i)^2 - 4(2 - 2i) + 8 = 4 - 8i + 4i^2 - 8 + 8i + 8$

$$= 4 - 4 = 0$$ ∎

NOW WORK PROBLEM **51**.

The discriminant $b^2 - 4ac$ of a quadratic equation still serves as a way to determine the character of the solutions.

CHARACTER OF THE SOLUTIONS OF A QUADRATIC EQUATION

In the complex number system, consider a quadratic equation $ax^2 + bx + c = 0$ with real coefficients.

1. If $b^2 - 4ac > 0$, the equation has two unequal real solutions.
2. If $b^2 - 4ac = 0$, the equation has a repeated real solution, a double root.
3. If $b^2 - 4ac < 0$, the equation has two complex solutions that are not real. The solutions are conjugates of each other.

The third conclusion in the display is a consequence of the fact that, if $b^2 - 4ac = -N < 0$, then, by the quadratic formula, the solutions are

$$x = \frac{-b + \sqrt{b^2 - 4ac}}{2a} = \frac{-b + \sqrt{-N}}{2a} = \frac{-b + \sqrt{N}i}{2a} = \frac{-b}{2a} + \frac{\sqrt{N}}{2a}i$$

and

$$x = \frac{-b - \sqrt{b^2 - 4ac}}{2a} = \frac{-b - \sqrt{-N}}{2a} = \frac{-b - \sqrt{N}i}{2a} = \frac{-b}{2a} - \frac{\sqrt{N}}{2a}i$$

which are conjugates of each other.

EXAMPLE 12

Determining the Character of the Solution of a Quadratic Equation

Without solving, determine the character of the solution of each equation.

(a) $3x^2 + 4x + 5 = 0$
(b) $2x^2 + 4x + 1 = 0$
(c) $9x^2 - 6x + 1 = 0$

Solution

(a) Here $a = 3$, $b = 4$, and $c = 5$, so $b^2 - 4ac = 16 - 4(3)(5) = -44$. The solutions are two complex numbers that are not real and are conjugates of each other.

(b) Here $a = 2$, $b = 4$, and $c = 1$, so $b^2 - 4ac = 16 - 8 = 8$. The solutions are two unequal real numbers.

(c) Here $a = 9$, $b = -6$, and $c = 1$, so $b^2 - 4ac = 36 - 4(9)(1) = 0$. The solution is a repeated real number, that is, a double root. ∎

NOW WORK PROBLEM 65.

A.4 EXERCISES

In Problems 1–38, write each expression in the standard form $a + bi$.

1. $(2 - 3i) + (6 + 8i)$ **2.** $(4 + 5i) + (-8 + 2i)$ **3.** $(-3 + 2i) - (4 - 4i)$ **4.** $(3 - 4i) - (-3 - 4i)$

5. $(2 - 5i) - (8 + 6i)$ **6.** $(-8 + 4i) - (2 - 2i)$ **7.** $3(2 - 6i)$ **8.** $-4(2 + 8i)$

9. $2i(2 - 3i)$ **10.** $3i(-3 + 4i)$ **11.** $(3 - 4i)(2 + i)$ **12.** $(5 + 3i)(2 - i)$

13. $(-6 + i)(-6 - i)$ **14.** $(-3 + i)(3 + i)$ **15.** $\dfrac{10}{3 - 4i}$ **16.** $\dfrac{13}{5 - 12i}$

17. $\dfrac{2 + i}{i}$ **18.** $\dfrac{2 - i}{-2i}$ **19.** $\dfrac{6 - i}{1 + i}$ **20.** $\dfrac{2 + 3i}{1 - i}$

21. $\left(\dfrac{1}{2} + \dfrac{\sqrt{3}}{2}i\right)^2$ **22.** $\left(\dfrac{\sqrt{3}}{2} - \dfrac{1}{2}i\right)^2$ **23.** $(1 + i)^2$ **24.** $(1 - i)^2$

25. i^{23} **26.** i^{14} **27.** i^{-15} **28.** i^{-23}

29. $i^6 - 5$ **30.** $4 + i^3$ **31.** $6i^3 - 4i^5$ **32.** $4i^3 - 2i^2 + 1$

33. $(1 + i)^3$ **34.** $(3i)^4 + 1$ **35.** $i^7(1 + i^2)$ **36.** $2i^4(1 + i^2)$

37. $i^6 + i^4 + i^2 + 1$ **38.** $i^7 + i^5 + i^3 + i$

In Problems 39–44, perform the indicated operations and express your answer in the form $a + bi$.

39. $\sqrt{-4}$ **40.** $\sqrt{-9}$ **41.** $\sqrt{-25}$

42. $\sqrt{-64}$ **43.** $\sqrt{(3 + 4i)(4i - 3)}$ **44.** $\sqrt{(4 + 3i)(3i - 4)}$

In Problems 45–64, solve each equation in the complex number system.

45. $x^2 + 4 = 0$ **46.** $x^2 - 4 = 0$ **47.** $x^2 - 16 = 0$ **48.** $x^2 + 25 = 0$

49. $x^2 - 6x + 13 = 0$ **50.** $x^2 + 4x + 8 = 0$ **51.** $x^2 - 6x + 10 = 0$ **52.** $x^2 - 2x + 5 = 0$

53. $8x^2 - 4x + 1 = 0$ **54.** $10x^2 + 6x + 1 = 0$ **55.** $5x^2 + 1 = 2x$ **56.** $13x^2 + 1 = 6x$

57. $x^2 + x + 1 = 0$ **58.** $x^2 - x + 1 = 0$ **59.** $x^3 - 8 = 0$ **60.** $x^3 + 27 = 0$

61. $x^4 = 16$ **62.** $x^4 = 1$ **63.** $x^4 + 13x^2 + 36 = 0$ **64.** $x^4 + 3x^2 - 4 = 0$

In Problems 65–70, without solving, determine the character of the solutions of each equation in the complex number system.

65. $3x^2 - 3x + 4 = 0$ **66.** $2x^2 - 4x + 1 = 0$ **67.** $2x^2 + 3x = 4$

68. $x^2 + 6 = 2x$ **69.** $9x^2 - 12x + 4 = 0$ **70.** $4x^2 + 12x + 9 = 0$

71. $2 + 3i$ is a solution of a quadratic equation with real coefficients. Find the other solution.

72. $4 - i$ is a solution of a quadratic equation with real coefficients. Find the other solution.

In Problems 73–76, $z = 3 - 4i$ and $w = 8 + 3i$. Write each expression in the standard form $a + bi$.

73. $z + \bar{z}$ **74.** $w - \bar{w}$ **75.** $z\bar{z}$ **76.** $\overline{z - w}$

77. Use $z = a + bi$ to show that $z + \bar{z} = 2a$ and $z - \bar{z} = 2bi$.

78. Use $z = a + bi$ to show that $\bar{\bar{z}} = z$.

79. Use $z = a + bi$ and $w = c + di$ to show that $\overline{z + w} = \bar{z} + \bar{w}$.

80. Use $z = a + bi$ and $w = c + di$ to show that $\overline{z \cdot w} = \bar{z} \cdot \bar{w}$.

81. Explain to a friend how you would add two complex numbers and how you would multiply two complex numbers. Explain any differences in the two explanations.

82. Write a brief paragraph that compares the method used to rationalize the denominator of a rational expression and the method used to write the quotient of two complex numbers in standard form.

PREPARING FOR THIS SECTION

Before getting started, review the following:

✓ Real Number Line, Inequalities, Absolute Value (Appendix A, Section A.1, pp. 490–492)

A.5   INEQUALITIES

OBJECTIVES

  1 Use Interval Notation

  2 Use Properties of Inequalities

  3 Solve Inequalities

  4 Solve Combined Inequalities

  5 Solve Inequalities Involving Absolute Value

Suppose that a and b are two real numbers and $a < b$. We shall use the notation $a < x < b$ to mean that x is a number *between* a and b. The expression $a < x < b$ is equivalent to the two inequalities $a < x$ and $x < b$. Similarly, the expression $a \le x \le b$ is equivalent to the two inequalities $a \le x$ and $x \le b$. The remaining two possibilities, $a \le x < b$ and $a < x \le b$, are defined similarly.

Although it is acceptable to write $3 \ge x \ge 2$, it is preferable to reverse the inequality symbols and write instead $2 \le x \le 3$ so that, as you read from left to right, the values go from smaller to larger.

A statement such as $2 \leq x \leq 1$ is false because there is no number x for which $2 \leq x$ and $x \leq 1$. Finally, we never mix inequality symbols, as in $2 \leq x \geq 3$.

INTERVALS

① Let a and b represent two real numbers with $a < b$.

> A **closed interval,** denoted by **[a, b],** consists of all real numbers x for which $a \leq x \leq b$.
>
> An **open interval,** denoted by **(a, b),** consists of all real numbers x for which $a < x < b$.
>
> The **half-open,** or **half-closed, intervals** are **(a, b],** consisting of all real numbers x for which $a < x \leq b$, and **[a, b),** consisting of all real numbers x for which $a \leq x < b$.

In each of these definitions, a is called the **left endpoint** and b the **right endpoint** of the interval.

The symbol ∞ (read as "infinity") is not a real number, but a notational device used to indicate unboundedness in the positive direction. The symbol $-\infty$ (read as "negative infinity") also is not a real number, but a notational device used to indicate unboundedness in the negative direction. Using the symbols ∞ and $-\infty$, we can define five other kinds of intervals:

$[a, \infty)$	Consists of all real numbers x for which $x \geq a$ $(a \leq x < \infty)$
(a, ∞)	Consists of all real numbers x for which $x > a$ $(a < x < \infty)$
$(-\infty, a]$	Consists of all real numbers x for which $x \leq a$ $(-\infty < x \leq a)$
$(-\infty, a)$	Consists of all real numbers x for which $x < a$ $(-\infty < x < a)$
$(-\infty, \infty)$	Consists of all real numbers x $(-\infty < x < \infty)$

Note that ∞ and $-\infty$ are never included as endpoints, since neither is a real number.

Table 1 summarizes interval notation, corresponding inequality notation, and their graphs.

TABLE 1

Interval	Inequality	Graph
The open interval (a, b)	$a < x < b$	
The closed interval $[a, b]$	$a \leq x \leq b$	
The half-open interval $[a, b)$	$a \leq x < b$	
The half-open interval $(a, b]$	$a < x \leq b$	
The interval $[a, \infty)$	$x \geq a$	
The interval (a, ∞)	$x > a$	
The interval $(-\infty, a]$	$x \leq a$	
The interval $(-\infty, a)$	$x < a$	
The interval $(-\infty, \infty)$	All real numbers	

EXAMPLE 1	**Writing Inequalities Using Interval Notation**

Write each inequality using interval notation.

(a) $1 \leq x \leq 3$ (b) $-4 < x < 0$ (c) $x > 5$ (d) $x \leq 1$

Solution (a) $1 \leq x \leq 3$ describes all numbers x between 1 and 3, inclusive. In interval notation, we write $[1, 3]$.

(b) In interval notation, $-4 < x < 0$ is written $(-4, 0)$.

(c) $x > 5$ consists of all numbers x greater than 5. In interval notation, we write $(5, \infty)$.

(d) In interval notation, $x \leq 1$ is written $(-\infty, 1]$. ∎

EXAMPLE 2	**Writing Intervals Using Inequality Notation**

Write each interval as an inequality involving x.

(a) $[1, 4)$ (b) $(2, \infty)$ (c) $[2, 3]$ (d) $(-\infty, -3]$

Solution (a) $[1, 4)$ consists of all numbers x for which $1 \leq x < 4$.

(b) $(2, \infty)$ consists of all numbers x for which $x > 2$ $(2 < x < \infty)$.

(c) $[2, 3]$ consists of all numbers x for which $2 \leq x \leq 3$.

(d) $(-\infty, -3]$ consists of all numbers x for which $x \leq -3$ $(-\infty < x \leq -3)$. ∎

NOW WORK PROBLEMS **1, 7,** AND **15.**

PROPERTIES OF INEQUALITIES

② The product of two positive real numbers is positive, the product of two negative real numbers is positive, and the product of 0 and 0 is 0. For any real number a, the value of a^2 is 0 or positive; that is, a^2 is nonnegative. This is called the **nonnegative property.**

For any real number a, we have the following:

Nonnegative Property

$$a^2 \geq 0 \tag{1}$$

If we add the same number to both sides of an inequality, we obtain an equivalent inequality. For example, since $3 < 5$, then $3 + 4 < 5 + 4$ or $7 < 9$. This is called the **addition property** of inequalities.

Addition Property of Inequalities

$$\text{If } a < b, \text{ then } a + c < b + c. \tag{2a}$$

$$\text{If } a > b, \text{ then } a + c > b + c. \tag{2b}$$

The addition property states that the sense, or direction, of an inequality remains unchanged if the same number is added to each side.

Let's see what happens if we multiply both sides of an inequality by a positive number and by a negative number.

Begin with $3 < 7$ and multiply each side by 2. The numbers 6 and 14 that result yield the inequality $6 < 14$.

Begin with $9 > 2$ and multiply each side by -4. The numbers -36 and -8 that result yield the inequality $-36 < -8$.

Note that the effect of multiplying both sides of $9 > 2$ by the negative number -4 is that the direction of the inequality symbol is reversed. We are led to the following general **multiplication properties** for inequalities:

Multiplication Properties for Inequalities

If $a < b$ and if $c > 0$, then $ac < bc$.	
If $a < b$ and if $c < 0$, then $ac > bc$.	**(3a)**
If $a > b$ and if $c > 0$, then $ac > bc$.	
If $a > b$ and if $c < 0$, then $ac < bc$.	**(3b)**

The multiplication properties state that the sense, or direction, of an inequality *remains the same* if each side is multiplied by a *positive* real number, whereas the direction is *reversed* if each side is multiplied by a *negative* real number.

✏ NOW WORK PROBLEMS 29 AND 35.

The **reciprocal property** states that the reciprocal of a positive real number is positive and that the reciprocal of a negative real number is negative.

Reciprocal Property for Inequalities

If $a > 0$, then $\dfrac{1}{a} > 0$.	**(4a)**
If $a < 0$, then $\dfrac{1}{a} < 0$.	**(4b)**

SOLVING INEQUALITIES

③ An **inequality in one variable** is a statement involving two expressions, at least one containing the variable, separated by one of the inequality symbols $<, \leq, >,$ or $\geq$. To **solve an inequality** means to find all values of the variable for which the statement is true. These values are called **solutions** of the inequality.

For example, the following are all inequalities involving one variable, x:

$$x + 5 < 8 \qquad 2x - 3 \geq 4 \qquad x^2 - 1 \leq 3 \qquad \frac{x + 1}{x - 2} > 0$$

Two inequalities having exactly the same solution set are called **equivalent inequalities.** As with equations, one method for solving an inequality is to replace it by a series of equivalent inequalities until an inequality with an obvious solution, such as $x < 3$, is obtained. We obtain equivalent inequalities by applying some of the same properties as those used to find equivalent equations. The addition property and the multiplication properties form the basis for the following procedures.

PROCEDURES THAT LEAVE
THE INEQUALITY SYMBOL UNCHANGED

1. Simplify both sides of the inequality by combining like terms and eliminating parentheses:

   Replace $(x + 2) + 6 > 2x + 5(x + 1)$

   by $x + 8 > 7x + 5$

2. Add or subtract the same expression on both sides of the inequality:

   Replace $3x - 5 < 4$

   by $(3x - 5) + 5 < 4 + 5$

3. Multiply or divide both sides of the inequality by the same positive expression:

   Replace $4x > 16$ by $\dfrac{4x}{4} > \dfrac{16}{4}$

PROCEDURES THAT REVERSE THE SENSE
OR DIRECTION OF THE INEQUALITY SYMBOL

1. Interchange the two sides of the inequality:

   Replace $3 < x$ by $x > 3$

2. Multiply or divide both sides of the inequality by the same *negative* expression.

   Replace $-2x > 6$ by $\dfrac{-2x}{-2} < \dfrac{6}{-2}$

As the examples that follow illustrate, we solve inequalities using many of the same steps that we would use to solve equations. In writing the solution of an inequality, we may use either set notation or interval notation, whichever is more convenient.

| EXAMPLE 3 | **Solving an Inequality** |

Solve the inequality: $4x + 7 \geq 2x - 3$
Graph the solution set.

Solution

$$4x + 7 \geq 2x - 3$$
$$4x + 7 - 7 \geq 2x - 3 - 7 \qquad \text{Subtract 7 from both sides.}$$
$$4x \geq 2x - 10 \qquad \text{Simplify.}$$
$$4x - 2x \geq 2x - 10 - 2x \qquad \text{Subtract } 2x \text{ from both sides.}$$
$$2x \geq -10 \qquad \text{Simplify.}$$
$$\frac{2x}{2} \geq \frac{-10}{2} \qquad \text{Divide both sides by 2. (The sense of the inequality symbol is unchanged).}$$
$$x \geq -5 \qquad \text{Simplify.}$$

Figure 14
$x \geq -5$ or $[-5, \infty)$

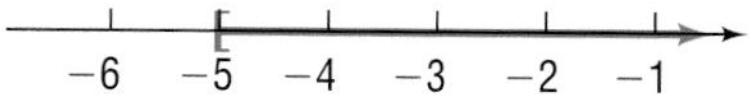

The solution set is $\{x \mid x \geq -5\}$ or, using interval notation, all numbers in the interval $[-5, \infty)$. See Figure 14 for the graph.   ■

NOW WORK PROBLEM 43.

| EXAMPLE 4 | Solving Combined Inequalities |

4 Solve the inequality: $-5 < 3x - 2 < 1$
Graph the solution set.

Solution Recall that the inequality

$$-5 < 3x - 2 < 1$$

is equivalent to the two inequalities

$$-5 < 3x - 2 \quad \text{and} \quad 3x - 2 < 1$$

We will solve each of these inequalities separately.

$-5 < 3x - 2$		$3x - 2 < 1$
$-5 + 2 < 3x - 2 + 2$	Add 2 to both sides.	$3x - 2 + 2 < 1 + 2$
$-3 < 3x$	Simplify.	$3x < 3$
$\dfrac{-3}{3} < \dfrac{3x}{3}$	Divide both sides by 3.	$\dfrac{3x}{3} < \dfrac{3}{3}$
$-1 < x$	Simplify.	$x < 1$

The solution set of the original pair of inequalities consists of all x for which

$$-1 < x \quad \text{and} \quad x < 1$$

Figure 15
$-1 < x < 1$ or $(-1, 1)$

This may be written more compactly as $\{x \mid -1 < x < 1\}$. In interval notation, the solution is $(-1, 1)$. See Figure 15 for the graph. ■

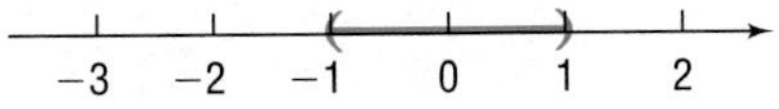

✏ NOW WORK PROBLEM 63.

| EXAMPLE 5 | Using the Reciprocal Property to Solve an Inequality |

Solve the inequality: $(4x - 1)^{-1} > 0$
Graph the solution set.

Solution Since $(4x - 1)^{-1} = \dfrac{1}{4x - 1}$ and since the Reciprocal Property states that when $\dfrac{1}{a} > 0$ then $a > 0$, we have

$$(4x - 1)^{-1} > 0$$

$$\frac{1}{4x - 1} > 0$$

$$4x - 1 > 0 \qquad \text{Reciprocal Property}$$

$$4x > 1$$

$$x > \frac{1}{4}$$

Figure 16
$x > \frac{1}{4}$ or $\left(\frac{1}{4}, \infty\right)$

The solution set is $\left\{x \mid x > \dfrac{1}{4}\right\}$; that is, all x in the interval $\left(\dfrac{1}{4}, \infty\right)$. Figure 16 illustrates the graph. ■

✏ NOW WORK PROBLEM 67.

EXAMPLE 6   Solving an Inequality Involving Absolute Value

Solve the inequality $|x| < 4$, and graph the solution set.

Solution   We are looking for all points whose coordinate x is a distance less than 4 units from the origin. See Figure 17 for an illustration. Because any x between -4 and 4 satisfies the condition $|x| < 4$, the solution set consists of all numbers x for which $-4 < x < 4$, that is, all x in $(-4, 4)$. ■

Figure 17
$-4 < x < 4$ or $(-4, 4)$

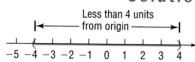
Less than 4 units
from origin

-5 -4 -3 -2 -1 0 1 2 3 4

EXAMPLE 7   Solving an Inequality Involving Absolute Value

Solve the inequality $|x| > 3$, and graph the solution set.

Solution   We are looking for all points whose coordinate x is a distance greater than 3 units from the origin. Figure 18 illustrates the situation. We conclude that any x less than -3 or greater than 3 satisfies the condition $|x| > 3$. Consequently, the solution set consists of all numbers x for which $x < -3$ or $x > 3$, that is, all x in the interval $(-\infty, -3)$ or in the interval $(3, \infty)$. ■

Figure 18
$x < -3$ or $x > 3$;
$(-\infty, -3)$ or $(3, \infty)$

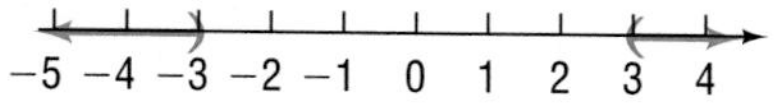

-5 -4 -3 -2 -1 0 1 2 3 4

We are led to the following results:

Theorem

If a is any positive number, then

$\|u\| < a$	is equivalent to	$-a < u < a$	(5)
$\|u\| \le a$	is equivalent to	$-a \le u \le a$	(6)
$\|u\| > a$	is equivalent to	$u < -a$ or $u > a$	(7)
$\|u\| \ge a$	is equivalent to	$u \le -a$ or $u \ge a$	(8)

■

EXAMPLE 8   Solving an Inequality Involving Absolute Value

Solve the inequality $|2x + 4| \le 3$, and graph the solution set.

Solution

$$|2x + 4| \le 3$$ This follows the form of equation (6); the expression $u = 2x + 4$ is inside the absolute value bars.

$$-3 \le 2x + 4 \le 3$$ Apply equation (6).

$$-3 - 4 \le 2x + 4 - 4 \le 3 - 4$$ Subtract 4 from each part.

$$-7 \le 2x \le -1$$ Simplify.

$$\frac{-7}{2} \le \frac{2x}{2} \le \frac{-1}{2}$$ Divide each part by 2.

$$-\frac{7}{2} \le x \le -\frac{1}{2}$$ Simplify.

The solution set is $\left\{ x \mid -\dfrac{7}{2} \le x \le -\dfrac{1}{2} \right\}$, that is, all x in the interval $\left[-\dfrac{7}{2}, -\dfrac{1}{2} \right]$. See Figure 19. ■

Figure 19
$-\frac{7}{2} \le x \le -\frac{1}{2}$ or $\left[-\frac{7}{2}, -\frac{1}{2} \right]$

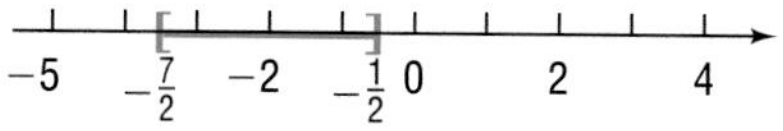

-5 $-\frac{7}{2}$ -2 $-\frac{1}{2}$ 0   2   4

NOW WORK PROBLEM **73.**

EXAMPLE 9	Solving an Inequality Involving Absolute Value

Solve the inequality $|2x - 5| > 3$, and graph the solution set.

Solution

$$|2x - 5| > 3$$ This follows the form of equation (7); the expression
$u = 2x - 5$ is inside the absolute value bars.

$2x - 5 < -3$	or	$2x - 5 > 3$		Apply equation (7).
$2x - 5 + 5 < -3 + 5$	or	$2x - 5 + 5 > 3 + 5$		Add 5 to each part.
$2x < 2$	or	$2x > 8$		Simplify.
$\dfrac{2x}{2} < \dfrac{2}{2}$	or	$\dfrac{2x}{2} > \dfrac{8}{2}$		Divide each part by 2.
$x < 1$	or	$x > 4$		Simplify.

Figure 20
$x < 1$ or $x > 4$;
$(-\infty, 1)$ or $(4, \infty)$

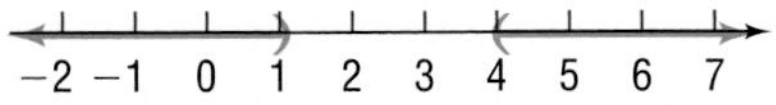

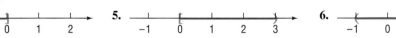

The solution set is $\{x \,|\, x < 1 \text{ or } x > 4\}$, that is, all x in the interval $(-\infty, 1)$ or in the interval $(4, \infty)$. See Figure 20. ∎

WARNING   A common error to be avoided is to attempt to write the solution $x < 1$ or $x > 4$ as $1 > x > 4$, which is incorrect, since there are no numbers x for which $x < 1$ *and* $x > 4$. Another common error is to "mix" the symbols and write $1 < x > 4$, which, of course, makes no sense. ∎

NOW WORK PROBLEM **79**.

A.5 EXERCISES

In Problems 1–6, express the graph shown in color using interval notation. Also express each as an inequality involving x.

1. [number line from −1 to 3]

2. [number line from −1 to 3]

3. [number line from −2 to 2]

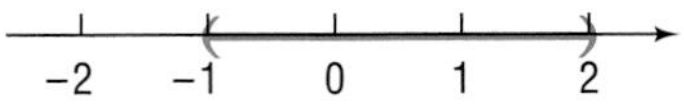

4. 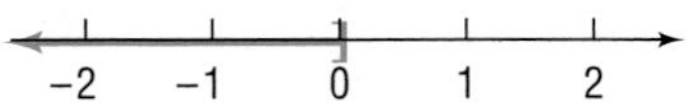 [number line from −2 to 2]

5. [number line from −1 to 3]

6.  [number line from −1 to 3]

In Problems 7–14, write each inequality using interval notation, and illustrate each inequality using the real number line.

7. $0 \le x \le 4$ **8.** $-1 < x < 5$ **9.** $4 \le x < 6$ **10.** $-2 < x < 0$

11. $x \ge 4$ **12.** $x \le 5$ **13.** $x < -4$ **14.** $x > 1$

In Problems 15–22, write each interval as an inequality involving x, and illustrate each inequality using the real number line.

15. $[2, 5]$ **16.** $(1, 2)$ **17.** $(-3, -2)$ **18.** $[0, 1)$

19. $[4, \infty)$ **20.** $(-\infty, 2]$ **21.** $(-\infty, -3)$ **22.** $(-8, \infty)$

In Problems 23–28, an inequality is given. Write the inequality obtained by:
  (a) *Adding 3 to each side of the given inequality.*
  (b) *Subtracting 5 from each side of the given inequality.*
  (c) *Multiplying each side of the given inequality by 3.*
  (d) *Multiplying each side of the given inequality by −2.*

23. $3 < 5$ **24.** $2 > 1$ **25.** $4 > -3$

26. $-3 > -5$ **27.** $2x + 1 < 2$ **28.** $1 - 2x > 5$

In Problems 29–42, fill in the blank with the correct inequality symbol.

29. If $x < 5$, then $x - 5$ _____ 0.   **30.** If $x < -4$, then $x + 4$ _____ 0.

31. If $x > -4$, then $x + 4$ _____ 0.   **32.** If $x > 6$, then $x - 6$ _____ 0.

33. If $x \ge -4$, then $3x$ _____ -12.   **34.** If $x \le 3$, then $2x$ _____ 6.

35. If $x > 6$, then $-2x$ _____ -12.   **36.** If $x > -2$, then $-4x$ _____ 8.

37. If $x \ge 5$, then $-4x$ _____ -20.   **38.** If $x \le -4$, then $-3x$ _____ 12.

39. If $2x > 6$, then x _____ 3.

40. If $3x \le 12$, then x _____ 4.

41. If $-\dfrac{1}{2}x \le 3$, then x _____ -6.

42. If $-\dfrac{1}{4}x > 1$, then x _____ -4.

In Problems 43–84, solve each inequality. Express your answer using set notation or interval notation. Graph the solution set.

43. $x + 1 < 5$

44. $x - 6 < 1$

45. $1 - 2x \le 3$

46. $2 - 3x \le 5$

47. $3x - 7 > 2$

48. $2x + 5 > 1$

49. $3x - 1 \ge 3 + x$

50. $2x - 2 \ge 3 + x$

51. $-2(x + 3) < 8$

52. $-3(1 - x) < 12$

53. $4 - 3(1 - x) \le 3$

54. $8 - 4(2 - x) \le -2x$

55. $\dfrac{1}{2}(x - 4) > x + 8$

56. $3x + 4 > \dfrac{1}{3}(x - 2)$

57. $\dfrac{x}{2} \ge 1 - \dfrac{x}{4}$

58. $\dfrac{x}{3} \ge 2 + \dfrac{x}{6}$

59. $0 \le 2x - 6 \le 4$

60. $4 \le 2x + 2 \le 10$

61. $-5 \le 4 - 3x \le 2$

62. $-3 \le 3 - 2x \le 9$

63. $-3 < \dfrac{2x - 1}{4} < 0$

64. $0 < \dfrac{3x + 2}{2} < 4$

65. $1 < 1 - \dfrac{1}{2}x < 4$

66. $0 < 1 - \dfrac{1}{3}x < 1$

67. $(4x + 2)^{-1} < 0$

68. $(2x - 1)^{-1} > 0$

69. $0 < \dfrac{2}{x} < \dfrac{3}{5}$

70. $0 < \dfrac{4}{x} < \dfrac{2}{3}$

71. $0 < (2x - 4)^{-1} < \dfrac{1}{2}$

72. $0 < (3x + 6)^{-1} < \dfrac{1}{3}$

73. $|2x| < 8$

74. $|3x| < 12$

75. $|3x| > 12$

76. $|2x| > 6$

77. $|2x - 1| \le 1$

78. $|2x + 5| \le 7$

79. $|1 - 2x| > 3$

80. $|2 - 3x| > 1$

81. $|-4x| + |-5| \le 9$

82. $|-x| - |4| \le 2$

83. $|-2x| \ge |-4|$

84. $|-x - 2| \ge 1$

85. Express the fact that x differs from 2 by less than $\dfrac{1}{2}$ as an inequality involving an absolute value. Solve for x.

86. Express the fact that x differs from -1 by less than 1 as an inequality involving an absolute value. Solve for x.

87. Express the fact that x differs from -3 by more than 2 as an inequality involving an absolute value. Solve for x.

88. Express the fact that x differs from 2 by more than 3 as an inequality involving an absolute value. Solve for x.

89. A young adult may be defined as someone older than 21, but less than 30 years of age. Express this statement using inequalities.

90. Middle-aged may be defined as being 40 or more and less than 60. Express the statement using inequalities.

91. Body Temperature "Normal" human body temperature is 98.6°F. If a temperature x that differs from normal by at least 1.5° is considered unhealthy, write the condition for an unhealthy temperature x as an inequality involving an absolute value, and solve for x.

92. Household Voltage In the United States, normal household voltage is 115 volts. However, it is not uncommon for actual voltage to differ from normal voltage by at most 5 volts. Express this situation as an inequality involving an absolute value. Use x as the actual voltage and solve for x.

93. Life Expectancy Metropolitan Life Insurance Co. reported that an average 25-year-old male in 1996 could expect to live at least 48.4 more years and an average 25-year-old female in 1996 could expect to live at least 54.7 more years.

(a) To what age can an average 25-year-old male expect to live? Express your answer as an inequality.

(b) To what age can an average 25-year-old female expect to live? Express your answer as an inequality.

(c) Who can expect to live longer, a male or a female? By how many years?

94. General Chemistry   For a certain ideal gas, the volume V (in cubic centimeters) equals 20 times the temperature T (in degrees Celsius). If the temperature varies from 80° to 120°C inclusive, what is the corresponding range of the volume of the gas?

95. Real Estate   A real estate agent agrees to sell a large apartment complex according to the following commission schedule: $45,000 plus 25% of the selling price in excess of $900,000. Assuming that the complex will sell at some price between $900,000 and $1,100,000 inclusive, over what range does the agent's commission vary? How does the commission vary as a percent of selling price?

96. Sales Commission   A used car salesperson is paid a commission of $25 plus 40% of the selling price in excess of owner's cost. The owner claims that used cars typically sell for at least owner's cost plus $70 and at most owner's cost plus $300. For each sale made, over what range can the salesperson expect the commission to vary?

97. Federal Tax Withholding   The percentage method of withholding for federal income tax (1998)* states that a single person whose weekly wages, after subtracting withholding allowances, are over $517, but not over $1105, shall have $69.90 plus 28% of the excess over $517 withheld. Over what range does the amount withheld vary if the weekly wages vary from $525 to $600 inclusive?

98. Federal Tax Withholding   Rework Problem 97 if the weekly wages vary from $600 to $700 inclusive.

99. Electricity Rates   Commonwealth Edison Company's summer charge for electricity is 10.494¢ per kilowatt-hour.[†] In addition, each monthly bill contains a customer charge of $9.36. If last summer's bills ranged from a low of $80.24 to a high of $271.80, over what range did usage vary (in kilowatt-hours)?

100. Water Bills   The Village of Oak Lawn charges homeowners $21.60 per quarter-year plus $1.70 per 1000 gallons for water usage in excess of 12,000 gallons.[‡] In 2000, one homeowner's quarterly bill ranged from a high of $65.75 to a low of $28.40. Over what range did water usage vary?

101. Markup of a New Car   The markup over dealer's cost of a new car ranges from 12% to 18%. If the sticker price is $8800, over what range will the dealer's cost vary?

102. IQ Tests   A standard intelligence test has an average score of 100. According to statistical theory, of the people who take the test, the 2.5% with the highest scores will have scores of more than 1.96σ above the average, where σ (sigma, a number called the **standard deviation**) depends on the nature of the test. If $\sigma = 12$ for this test and there is (in principle) no upper limit to the score possible on the test, write the interval of possible test scores of the people in the top 2.5%.

103. Computing Grades   In your Economics 101 class, you have scores of 68, 82, 87, and 89 on the first four of five tests. To get a grade of B, the average of the first five test scores must be greater than or equal to 80 and less than 90. Solve an inequality to find the range of the score that you need on the last test to get a B.

What do I need to get a B?

104. Computing Grades   Repeat Problem 103 if the fifth test counts double.

105. Arithmetic Mean   If $a < b$, show that $a < \dfrac{a + b}{2} < b$. The number $\dfrac{a + b}{2}$ is called the **arithmetic mean** of a and b.

106. Refer to Problem 105. Show that the arithmetic mean of a and b is equidistant from a and b.

107. Geometric Mean   If $0 < a < b$, show that $a < \sqrt{ab} < b$. The number $\sqrt{ab}$ is called the **geometric mean** of a and b.

108. Refer to Problems 105 and 107. Show that the geometric mean of a and b is less than the arithmetic mean of a and b.

109. Harmonic Mean   For $0 < a < b$, let h be defined by

$$\frac{1}{h} = \frac{1}{2}\left(\frac{1}{a} + \frac{1}{b}\right)$$

Show that $a < h < b$. The number h is called the **harmonic mean** of a and b.

110. Refer to Problems 105, 107, and 109. Show that the harmonic mean of a and b equals the geometric mean squared divided by the arithmetic mean.

111. Make up an inequality that has no solution. Make up one that has exactly one solution.

---

*Source: *Employer's Tax Guide.* Department of the Treasury, Internal Revenue Service, 1998.
[†] *Source:* Commonwealth Edison Co., Chicago, Illinois, 2000.
[‡] *Source:* Village of Oak Lawn, Illinois, 2000.

112. The inequality $x^2 + 1 < -5$ has no solution. Explain why.

113. Do you prefer to use inequality notation or interval notation to express the solution to an inequality? Give your reasons. Are there particular circumstances when you prefer one to the other? Cite examples.

114. How would you explain to a fellow student the underlying reason for the multiplication properties for inequalities (page 524); that is, the sense or direction of an inequality remains the same if each side is multiplied by a positive real number, whereas the direction is reversed if each side is multiplied by a negative real number.

PREPARING FOR THIS SECTION

Before getting started, review the following:

✓ Exponents, Square Roots (Appendix A, Section A.1, pp. 492–494)

A.6 *n*th ROOTS; RATIONAL EXPONENTS

OBJECTIVES
1. Work with *n*th Roots
2. Simplify Radicals
3. Rationalize Denominators
4. Solve Radical Equations
5. Simplify Expressions with Rational Exponents

*n*TH ROOTS

The **principal *n*th root of a number *a*,** symbolized by $\sqrt[n]{a}$, where $n \geq 2$ is an integer, is defined as follows:

$$\sqrt[n]{a} = b \quad \text{means} \quad a = b^n$$

where $a \geq 0$ and $b \geq 0$ if $n \geq 2$ is even, and a, b are any real numbers if $n \geq 3$ is odd.

Notice that if a is negative and n is even then $\sqrt[n]{a}$ is not defined. When it is defined, the principal *n*th root of a number is unique.

 The symbol $\sqrt[n]{a}$ for the principal *n*th root of a is sometimes called a **radical;** the integer n is called the **index,** and a is called the **radicand.** If the index of a radical is 2, we call $\sqrt[n]{a}$ the **square root** of a and omit the index 2 by simply writing $\sqrt{a}$. If the index is 3, we call $\sqrt[3]{a}$ the **cube root** of a.

EXAMPLE 1   **Evaluating Principal *n*th Roots**

(a) $\sqrt[3]{8} = \sqrt[3]{2^3} = 2$

(b) $\sqrt[3]{-64} = \sqrt[3]{(-4)^3} = -4$

(c) $\sqrt[4]{\dfrac{1}{16}} = \sqrt[4]{\left(\dfrac{1}{2}\right)^4} = \dfrac{1}{2}$

(d) $\sqrt{(-2)^6} = |-2| = 2$ ∎

These are examples of **perfect roots,** since each simplifies to a rational number. Notice the absolute value in Example 1(d). If n is even, the principal *n*th root must be nonnegative.

In general, if $n \geq 2$ is a positive integer and a is a real number, we have

$$\sqrt[n]{a^n} = a, \qquad \text{if } n \geq 3 \text{ is odd} \qquad \textbf{(1a)}$$

$$\sqrt[n]{a^n} = |a|, \qquad \text{if } n \geq 2 \text{ is even} \qquad \textbf{(1b)}$$

∎

NOW WORK PROBLEM **1**.

PROPERTIES OF RADICALS

Let $n \geq 2$ and $m \geq 2$ denote positive integers, and let a and b represent real numbers. Assuming that all radicals are defined, we have the following properties:

$$\sqrt[n]{ab} = \sqrt[n]{a}\,\sqrt[n]{b} \qquad \textbf{(2a)}$$

$$\sqrt[n]{\dfrac{a}{b}} = \dfrac{\sqrt[n]{a}}{\sqrt[n]{b}} \qquad \textbf{(2b)}$$

$$\sqrt[n]{a^m} = \left(\sqrt[n]{a}\right)^m \qquad \textbf{(2c)}$$

②   When used in reference to radicals, the direction to "simplify" will mean to remove from the radicals any perfect roots that occur as factors. Let's look at some examples of how the preceding rules are applied to simplify radicals.

EXAMPLE 2   **Simplifying Radicals**

(a) $\sqrt{32} = \sqrt{16 \cdot 2} = \sqrt{16} \cdot \sqrt{2} = 4\sqrt{2}$

   ↑
   16 is a perfect square.

(b) $\sqrt[3]{16} = \sqrt[3]{8 \cdot 2} = \sqrt[3]{8} \cdot \sqrt[3]{2} = 2\sqrt[3]{2}$

   ↑   ↑
   8 is a perfect cube.   (2a)

(c) $\sqrt[3]{-16x^4} = \sqrt[3]{-8 \cdot 2 \cdot x^3 \cdot x} = \sqrt[3]{(-8x^3)(2x)}$

   ↑   ↑
   Factor perfect   Combine perfect
   cubes inside radical.   cubes.

   $= \sqrt[3]{(-2x)^3 \cdot 2x} = \sqrt[3]{(-2x)^3} \cdot \sqrt[3]{2x}$

   ↑
   $= -2x\sqrt[3]{2x}$ (2a)

∎

NOW WORK PROBLEM **7**.

EXAMPLE 3   **Combining Like Radicals**

(a) $-8\sqrt{12} + \sqrt{3} = -8\sqrt{4 \cdot 3} + \sqrt{3} = -8 \cdot \sqrt{4}\,\sqrt{3} + \sqrt{3}$

$$= -16\sqrt{3} + \sqrt{3} = -15\sqrt{3}$$

(b) $\sqrt[3]{8x^4} + \sqrt[3]{-x} + 4\sqrt[3]{27x} = \sqrt[3]{2^3 x^3 x} + \sqrt[3]{-1 \cdot x} + 4\sqrt[3]{3^3 x}$

$$= \sqrt[3]{(2x)^3} \cdot \sqrt[3]{x} + \sqrt[3]{-1} \cdot \sqrt[3]{x} + 4\sqrt[3]{3^3} \cdot \sqrt[3]{x}$$

$$= 2x\sqrt[3]{x} - 1 \cdot \sqrt[3]{x} + 12\sqrt[3]{x}$$

$$= (2x + 11)\sqrt[3]{x}$$

∎

NOW WORK PROBLEM **25**.

RATIONALIZING

3   When radicals occur in quotients, it is customary to rewrite the quotient so that the denominator contains no square roots. This process is referred to as **rationalizing the denominator.**

The idea is to multiply by an appropriate expression so that the new denominator contains no radicals. For example:

If Denominator Contains the Factor	Multiply By	To Obtain Denominator Free of Radicals
$\sqrt{3}$	$\sqrt{3}$	$\left(\sqrt{3}\right)^2 = 3$
$\sqrt{3} + 1$	$\sqrt{3} - 1$	$\left(\sqrt{3}\right)^2 - 1^2 = 3 - 1 = 2$
$\sqrt{2} - 3$	$\sqrt{2} + 3$	$\left(\sqrt{2}\right)^2 - 3^2 = 2 - 9 = -7$
$\sqrt{5} - \sqrt{3}$	$\sqrt{5} + \sqrt{3}$	$\left(\sqrt{5}\right)^2 - \left(\sqrt{3}\right)^2 = 5 - 3 = 2$
$\sqrt[3]{4}$	$\sqrt[3]{2}$	$\sqrt[3]{4} \cdot \sqrt[3]{2} = \sqrt[3]{8} = 2$

In rationalizing the denominator of a quotient, be sure to multiply both the numerator and the denominator by the expression.

EXAMPLE 4   **Rationalizing Denominators**

Rationalize the denominator of each expression.

(a) $\dfrac{4}{\sqrt{2}}$ (b) $\dfrac{\sqrt{3}}{\sqrt[3]{2}}$ (c) $\dfrac{\sqrt{x} - 2}{\sqrt{x} + 2}, \ x \geq 0$

Solution   (a) $\dfrac{4}{\sqrt{2}} = \dfrac{4}{\sqrt{2}} \cdot \dfrac{\sqrt{2}}{\sqrt{2}} = \dfrac{4\sqrt{2}}{\left(\sqrt{2}\right)^2} = \dfrac{4\sqrt{2}}{2} = 2\sqrt{2}$

Multiply by $\dfrac{\sqrt{2}}{\sqrt{2}}$.

(b) $\dfrac{\sqrt{3}}{\sqrt[3]{2}} = \dfrac{\sqrt{3}}{\sqrt[3]{2}} \cdot \dfrac{\sqrt[3]{4}}{\sqrt[3]{4}} = \dfrac{\sqrt{3} \, \sqrt[3]{4}}{\sqrt[3]{8}} = \dfrac{\sqrt{3} \, \sqrt[3]{4}}{2}$

Multiply by $\dfrac{\sqrt[3]{4}}{\sqrt[3]{4}}$.

(c) $\dfrac{\sqrt{x} - 2}{\sqrt{x} + 2} = \dfrac{\sqrt{x} - 2}{\sqrt{x} + 2} \cdot \dfrac{\sqrt{x} - 2}{\sqrt{x} - 2} = \dfrac{\left(\sqrt{x} - 2\right)^2}{\left(\sqrt{x}\right)^2 - 2^2}$

$= \dfrac{\left(\sqrt{x}\right)^2 - 4\sqrt{x} + 4}{x - 4} = \dfrac{x - 4\sqrt{x} + 4}{x - 4}$ ■

NOW WORK PROBLEM 33.

EQUATIONS CONTAINING RADICALS

4   When the variable in an equation occurs in a square root, cube root, and so on, that is, when it occurs under a radical, the equation is called a **radical equation.** Sometimes a suitable operation will change a radical equation to one that is linear or quadratic. The most commonly used procedure is to isolate the most complicated radical on one side of the equation and then eliminate it by raising each side to a power equal to the index of the radical. Care must be taken, because extraneous solutions may result. Thus, when working with radical equations, we always check apparent solutions. Let's look at an example.

EXAMPLE 5	**Solving Radical Equations**

Solve the equation: $\sqrt[3]{2x - 4} - 2 = 0$

Solution The equation contains a radical whose index is 3. We isolate it on the left side.

$$\sqrt[3]{2x - 4} - 2 = 0$$
$$\sqrt[3]{2x - 4} = 2$$

Now raise each side to the third power (since the index of the radical is 3) and solve.

$$\left(\sqrt[3]{2x - 4}\right)^3 = 2^3 \qquad \text{Raise each side to the 3rd power.}$$
$$2x - 4 = 8 \qquad \text{Simplify.}$$
$$2x = 12 \qquad \text{Solve for } x.$$
$$x = 6$$ ■

Check: $\sqrt[3]{2(6) - 4} - 2 = \sqrt[3]{12 - 4} - 2 = \sqrt[3]{8} - 2 = 2 - 2 = 0.$

The solution is $x = 6$. ■

➤ NOW WORK PROBLEM **41**.

RATIONAL EXPONENTS

Radicals are used to define rational exponents.

If a is a real number and $n \geq 2$ is an integer, then

$$a^{1/n} = \sqrt[n]{a} \qquad \qquad \textbf{(3)}$$

provided that $\sqrt[n]{a}$ exists.

EXAMPLE 6	**Using Equation (3)**

(a) $4^{1/2} = \sqrt{4} = 2$ (b) $(-27)^{1/3} = \sqrt[3]{-27} = -3$

(c) $8^{1/2} = \sqrt{8} = 2\sqrt{2}$ (d) $16^{1/3} = \sqrt[3]{16} = 2\sqrt[3]{2}$ ■

If a is a real number and m and n are integers containing no common factors with $n \geq 2$, then

$$a^{m/n} = \sqrt[n]{a^m} = \left(\sqrt[n]{a}\right)^m \qquad \qquad \textbf{(4)}$$

provided that $\sqrt[n]{a}$ exists.

We have two comments about equation (4):

1. The exponent m/n must be in lowest terms and n must be positive.
2. In simplifying $a^{m/n}$, either $\sqrt[n]{a^m}$ or $\left(\sqrt[n]{a}\right)^m$ may be used. Generally, taking the root first, as in $\left(\sqrt[n]{a}\right)^m$, is easier.

EXAMPLE 7	Using Equation (4)

(a) $4^{3/2} = \left(\sqrt{4}\right)^3 = 2^3 = 8$ (b) $(-8)^{4/3} = \left(\sqrt[3]{-8}\right)^4 = (-2)^4 = 16$

(c) $(32)^{-2/5} = \left(\sqrt[5]{32}\right)^{-2} = 2^{-2} = \dfrac{1}{4}$ ∎

NOW WORK PROBLEM 45.

It can be shown that the laws of exponents hold for rational exponents.

EXAMPLE 8	Simplifying Expressions with Rational Exponents

⑤ Simplify each expression. Express your answer so that only positive exponents occur. Assume that the variables are positive.

(a) $\left(\dfrac{2x^{1/3}}{y^{2/3}}\right)^{-3}$ (b) $\left(x^{2/3}y\right)\left(x^{-2}y\right)^{1/2}$

Solution (a) $\left(\dfrac{2x^{1/3}}{y^{2/3}}\right)^{-3} = \left(\dfrac{y^{2/3}}{2x^{1/3}}\right)^{3} = \dfrac{\left(y^{2/3}\right)^3}{\left(2x^{1/3}\right)^3} = \dfrac{y^2}{2^3\left(x^{1/3}\right)^3} = \dfrac{y^2}{8x}$

(b) $\left(x^{2/3}y\right)\left(x^{-2}y\right)^{1/2} = \left(x^{2/3}y\right)\left[\left(x^{-2}\right)^{1/2}y^{1/2}\right]$

$\qquad\qquad = x^{2/3}yx^{-1}y^{1/2} = \left(x^{2/3}x^{-1}\right)\left(y \cdot y^{1/2}\right)$

$\qquad\qquad = x^{-1/3}y^{3/2} = \dfrac{y^{3/2}}{x^{1/3}}$ ∎

NOW WORK PROBLEM 61.

 The next two examples illustrate some algebra that you will need to know for certain calculus problems.

EXAMPLE 9	Writing an Expression as a Single Quotient

Write the following expression as a single quotient in which only positive exponents appear.

$$\left(x^2 + 1\right)^{1/2} + x \cdot \frac{1}{2}\left(x^2 + 1\right)^{-1/2} \cdot 2x$$

Solution $\left(x^2 + 1\right)^{1/2} + x \cdot \dfrac{1}{2}\left(x^2 + 1\right)^{-1/2} \cdot 2x = \left(x^2 + 1\right)^{1/2} + \dfrac{x^2}{\left(x^2 + 1\right)^{1/2}}$

$$= \frac{\left(x^2 + 1\right)^{1/2}\left(x^2 + 1\right)^{1/2} + x^2}{\left(x^2 + 1\right)^{1/2}}$$

$$= \frac{\left(x^2 + 1\right) + x^2}{\left(x^2 + 1\right)^{1/2}}$$

$$= \frac{2x^2 + 1}{\left(x^2 + 1\right)^{1/2}}$$ ∎

NOW WORK PROBLEM 65.

| EXAMPLE 10 | Factoring an Expression Containing Rational Exponents |

Factor: $4x^{1/3}(2x + 1) + 2x^{4/3}$

Solution We begin by looking for factors that are common to the two terms. Notice that 2 and $x^{1/3}$ are common factors. Then,

$$4x^{1/3}(2x + 1) + 2x^{4/3} = 2x^{1/3}\big[2(2x + 1) + x\big]$$
$$= 2x^{1/3}(5x + 2)$$ ■

A.6 EXERCISES

In Problems 1–28, simplify each expression. Assume that all variables are positive when they appear.

1. $\sqrt[3]{27}$　　　　　　2. $\sqrt[4]{16}$　　　　　　3. $\sqrt[3]{-8}$　　　　　　4. $\sqrt[3]{-1}$

5. $\sqrt{8}$　　　　　　6. $\sqrt[3]{54}$　　　　　　7. $\sqrt[3]{-8x^4}$　　　　　　8. $\sqrt[4]{48x^5}$

9. $\sqrt[4]{x^{12}y^8}$　　　10. $\sqrt[5]{x^{10}y^5}$　　　11. $\sqrt[4]{\dfrac{x^9 y^7}{xy^3}}$　　　12. $\sqrt[3]{\dfrac{3xy^2}{81x^4y^2}}$

13. $\sqrt{36x}$　　　14. $\sqrt{9x^5}$　　　15. $\sqrt{3x^2}\sqrt{12x}$　　　16. $\sqrt{5x}\sqrt{20x^3}$

17. $\left(\sqrt{5}\sqrt[3]{9}\right)^2$　　18. $\left(\sqrt[3]{3}\sqrt{10}\right)^4$　　19. $(3\sqrt{6})(2\sqrt{2})$　　20. $(5\sqrt{8})(-3\sqrt{3})$

21. $(\sqrt{3} + 3)(\sqrt{3} - 1)$　　22. $(\sqrt{5} - 2)(\sqrt{5} + 3)$　　23. $(\sqrt{x} - 1)^2$　　24. $(\sqrt{x} + \sqrt{5})^2$

25. $3\sqrt{2} - 4\sqrt{8}$　　26. $\sqrt[3]{-x^4} + \sqrt[3]{8x}$　　27. $\sqrt[3]{16x^4} - \sqrt[3]{2x}$　　28. $\sqrt[4]{32x} + \sqrt[4]{2x^5}$

In Problems 29–40, rationalize the denominator of each expression. Assume that all variables are positive when they appear.

29. $\dfrac{1}{\sqrt{2}}$　　30. $\dfrac{6}{\sqrt[3]{4}}$　　31. $\dfrac{-\sqrt{3}}{\sqrt{5}}$　　32. $\dfrac{-\sqrt[3]{3}}{\sqrt{8}}$

33. $\dfrac{\sqrt{3}}{5 - \sqrt{2}}$　　34. $\dfrac{\sqrt{2}}{\sqrt{7} + 2}$　　35. $\dfrac{2 - \sqrt{5}}{2 + 3\sqrt{5}}$　　36. $\dfrac{\sqrt{3} - 1}{2\sqrt{3} + 3}$

37. $\dfrac{5}{\sqrt[3]{2}}$　　38. $\dfrac{-2}{\sqrt[3]{9}}$　　39. $\dfrac{\sqrt{x + h} - \sqrt{x}}{\sqrt{x + h} + \sqrt{x}}$　　40. $\dfrac{\sqrt{x + h} + \sqrt{x - h}}{\sqrt{x + h} - \sqrt{x - h}}$

In Problems 41–44, solve each equation.

41. $\sqrt[3]{2t - 1} = 2$　　42. $\sqrt[3]{3t + 1} = -2$　　43. $\sqrt{15 - 2x} = x$　　44. $\sqrt{12 - x} = x$

In Problems 45–56, simplify each expression.

45. $8^{2/3}$　　46. $4^{3/2}$　　47. $(-27)^{1/3}$　　48. $16^{3/4}$　　49. $16^{3/2}$　　50. $64^{3/2}$

51. $9^{-3/2}$　　52. $25^{-5/2}$　　53. $\left(\dfrac{9}{8}\right)^{3/2}$　　54. $\left(\dfrac{27}{8}\right)^{2/3}$　　55. $\left(\dfrac{8}{9}\right)^{-3/2}$　　56. $\left(\dfrac{8}{27}\right)^{-2/3}$

In Problems 57–64 simplify each expression. Express your answer so that only positive exponents occur. Assume that the variables are positive.

57. $x^{3/4}x^{1/3}x^{-1/2}$　　58. $x^{2/3}x^{1/2}x^{-1/4}$　　59. $(x^3y^6)^{1/3}$　　60. $(x^4y^8)^{3/4}$

61. $(x^2y)^{1/3}(xy^2)^{2/3}$　　62. $(xy)^{1/4}(x^2y^2)^{1/2}$　　63. $(16x^2y^{-1/3})^{3/4}$　　64. $(4x^{-1}y^{1/3})^{3/2}$

In Problems 65–20, write each expression as a single quotient in which only positive exponents and/or radicals appear.

65. $\dfrac{x}{(1 + x)^{1/2}} + 2(1 + x)^{1/2}$　　66. $\dfrac{1 + x}{2x^{1/2}} + x^{1/2}$　　67. $\dfrac{\sqrt{1 + x} - x \cdot \dfrac{1}{2\sqrt{1 + x}}}{1 + x}$

68. $\dfrac{\sqrt{x^2 + 1} - x \cdot \dfrac{2x}{2\sqrt{x^2 + 1}}}{x^2 + 1}$　　69. $\dfrac{(x + 4)^{1/2} - 2x(x + 4)^{-1/2}}{x + 4}$　　70. $\dfrac{(9 - x^2)^{1/2} + x^2(9 - x^2)^{-1/2}}{9 - x^2}$

In Problems 71–76, factor each expression.

71. $(x + 1)^{3/2} + x \cdot \frac{3}{2}(x + 1)^{1/2}$　　72. $(x^2 + 4)^{4/3} + x \cdot \frac{4}{3}(x^2 + 4)^{1/3} \cdot 2x$　　73. $6x^{1/2}(x^2 + x) - 8x^{3/2} - 8x^{1/2}$

74. $6x^{1/2}(2x + 3) + x^{3/2} \cdot 8$　　75. $x(\frac{1}{2})(8 - x^2)^{-1/2}(-2x) + (8 - x^2)^{1/2}$　　76. $2x(1 - x^2)^{3/2} + x^2(\frac{3}{2})(1 - x^2)^{1/2}(-2x)$

A.7 LINES

OBJECTIVES   1 Calculate and Interpret the Slope of a Line
2 Graph Lines Given a Point and the Slope
3 Find the Equation of Vertical Lines
4 Use the Point-Slope Form of a Line; Identify Horizontal Lines
5 Find the Equation of a Line Given Two Points
6 Write the Equation of a Line in Slope-Intercept Form
7 Identify the Slope and *y*-Intercept of a Line from Its Equation
8 Write the Equation of a Line in General Form
9 Find Equations of Parallel Lines
10 Find Equations of Perpendicular Lines

In this section we study a certain type of equation that contains two variables, called a *linear equation,* and its graph, a *line.*

SLOPE OF A LINE

1   Consider the staircase illustrated in Figure 21. Each step contains exactly the same horizontal **run** and the same vertical **rise.** The ratio of the rise to the run, called the *slope,* is a numerical measure of the steepness of the staircase. For example, if the run is increased and the rise remains the same, the staircase becomes less steep. If the run is kept the same, but the rise is increased, the staircase becomes more steep. The slope of a line is best defined using rectangular coordinates.

Figure 21

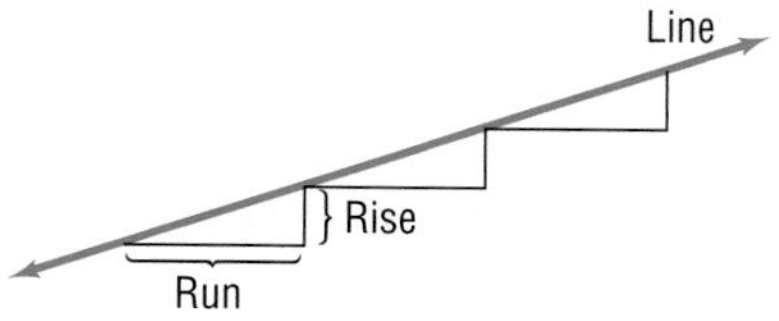

Let $P = (x_1, y_1)$ and $Q = (x_2, y_2)$ be two distinct points. If $x_1 \neq x_2$, the **slope *m*** of the nonvertical line L containing P and Q is defined by the formula

$$m = \frac{y_2 - y_1}{x_2 - x_1} \qquad x_1 \neq x_2 \qquad \textbf{(1)}$$

If $x_1 = x_2$, L is a **vertical line** and the slope m of L is **undefined** (since this results in division by 0).

Figure 22(a) provides an illustration of the slope of a nonvertical line; Figure 22(b) illustrates a vertical line.

Figure 22

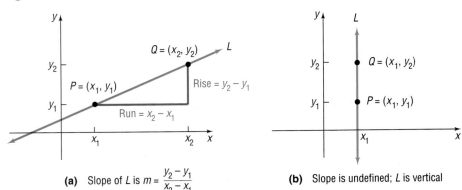

(a) Slope of L is $m = \dfrac{y_2 - y_1}{x_2 - x_1}$

(b) Slope is undefined; L is vertical

As Figure 22(a) illustrates, the slope m of a nonvertical line may be viewed as

$$m = \frac{y_2 - y_1}{x_2 - x_1} = \frac{\text{Rise}}{\text{Run}}$$

We can also express the slope m of a nonvertical line as

$$m = \frac{y_2 - y_1}{x_2 - x_1} = \frac{\text{Change in } y}{\text{Change in } x} = \frac{\Delta y}{\Delta x}$$

That is, the slope m of a nonvertical line L measures how y changes as x changes from x_1 to x_2. This is called the **average rate of change** of y with respect to x.

Two comments about computing the slope of a nonvertical line may prove helpful.

1. Any two distinct points on the line can be used to compute the slope of the line. (See Figure 23 for justification.)

Figure 23
Triangles ABC and PQR are similar (equal angles). Hence, ratios of corresponding sides are proportional so that

Slope using P and $Q = \dfrac{y_2 - y_1}{x_2 - x_1}$

$= $ Slope using A and $B = \dfrac{d(B, C)}{d(A, C)}$

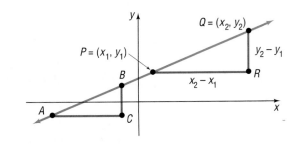

2. The slope of a line may be computed from $P = (x_1, y_1)$ to $Q = (x_2, y_2)$ or from Q to P, because

$$\frac{y_2 - y_1}{x_2 - x_1} = \frac{y_1 - y_2}{x_1 - x_2}$$

| EXAMPLE 1 | **Finding and Interpreting the Slope of a Line Containing Two Points** |

The slope m of the line containing the points $(1, 2)$ and $(5, -3)$ may be computed as

$$m = \frac{-3 - 2}{5 - 1} = \frac{-5}{4} = -\frac{5}{4} \quad \text{or as} \quad m = \frac{2 - (-3)}{1 - 5} = \frac{5}{-4} = -\frac{5}{4}$$

For every 4-unit change in x, y will change by -5 units. That is, if x increases by 4 units, then y decreases by 5 units. The average rate of change of y with respect to x is $-\frac{5}{4}$.   ∎

 **NOW WORK PROBLEMS 1 AND 7.**

To get a better idea of the meaning of the slope m of a line L, consider the following example.

| EXAMPLE 2 | **Finding the Slopes of Various Lines Containing the Same Point (2, 3)** |

Compute the slopes of the lines L_1, L_2, L_3, and L_4 containing the following pairs of points. Graph all four lines on the same set of coordinate axes.

$$L_1: \quad P = (2, 3) \qquad Q_1 = (-1, -2)$$

$$L_2: \quad P = (2, 3) \qquad Q_2 = (3, -1)$$

$$L_3: \quad P = (2, 3) \qquad Q_3 = (5, 3)$$

$$L_4: \quad P = (2, 3) \qquad Q_4 = (2, 5)$$

Figure 24

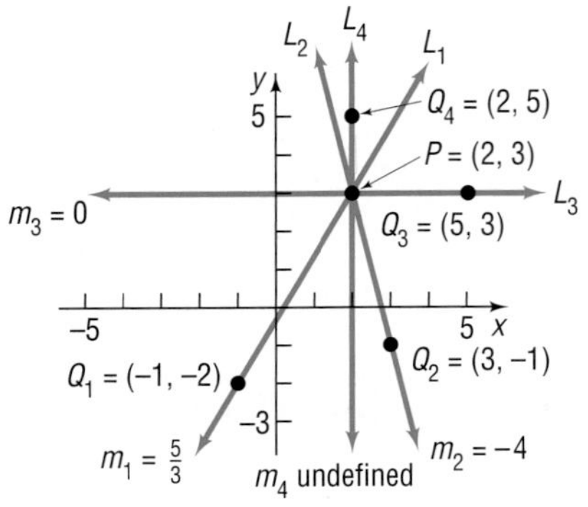

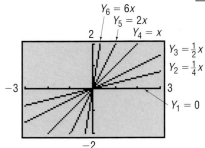

$m_3 = 0$

$Q_1 = (-1, -2)$

$m_1 = \frac{5}{3}$ m_4 undefined $m_2 = -4$

Solution   Let m_1, m_2, m_3, and m_4 denote the slopes of the lines L_1, L_2, L_3, and L_4, respectively. Then

$$m_1 = \frac{-2 - 3}{-1 - 2} = \frac{-5}{-3} = \frac{5}{3}$$ A rise of 5 divided by a run of 3

$$m_2 = \frac{-1 - 3}{3 - 2} = \frac{-4}{1} = -4$$

$$m_3 = \frac{3 - 3}{5 - 2} = \frac{0}{3} = 0$$

m_4 is undefined

The graphs of these lines are given in Figure 24.   ■

Figure 24 illustrates the following facts:

1. When the slope of a line is positive, the line slants upward from left to right (L_1).
2. When the slope of a line is negative, the line slants downward from left to right (L_2).
3. When the slope is 0, the line is horizontal (L_3).
4. When the slope is undefined, the line is vertical (L_4).

Figure 25

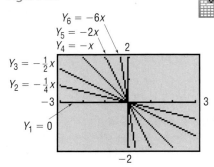

SEEING THE CONCEPT: On the same square screen, graph the following equations:

$Y_1 = 0$ Slope of line is 0.

$Y_2 = \frac{1}{4}x$ Slope of line is $\frac{1}{4}$.

$Y_3 = \frac{1}{2}x$ Slope of line is $\frac{1}{2}$.

$Y_4 = x$ Slope of line is 1.

$Y_5 = 2x$ Slope of line is 2.

$Y_6 = 6x$ Slope of line is 6.

See Figure 25.   ■

Figure 26

SEEING THE CONCEPT: On the same square screen, graph the following equations:

$Y_1 = 0$ Slope of line is 0.

$Y_2 = -\frac{1}{4}x$ Slope of line is $-\frac{1}{4}$.

$Y_3 = -\frac{1}{2}x$ Slope of line is $-\frac{1}{2}$.

$Y_4 = -x$ Slope of line is -1.

$Y_5 = -2x$ Slope of line is -2.

$Y_6 = -6x$ Slope of line is -6.

See Figure 26.   ■

Figures 25 and 26 illustrate that the closer the line is to the vertical position, the greater the magnitude of the slope.

2 The next example illustrates how the slope of a line can be used to graph the line.

EXAMPLE 3   Graphing a Line Given a Point and a Slope

Draw a graph of the line that contains the point $(3, 2)$ and has a slope of

(a) $\dfrac{3}{4}$ (b) $-\dfrac{4}{5}$

Figure 27 Slope $= \frac{3}{4}$

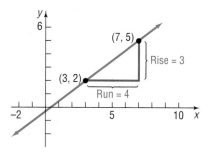

Solution (a) Slope = Rise/Run. The fact that the slope is $\dfrac{3}{4}$ means that for every horizontal movement (run) of 4 units to the right there will be a vertical movement (rise) of 3 units. If we start at the given point $(3, 2)$ and move 4 units to the right and 3 units up, we reach the point $(7, 5)$. By drawing the line through this point and the point $(3, 2)$, we have the graph. See Figure 27.

(b) The fact that the slope is

$$-\frac{4}{5} = \frac{-4}{5} = \frac{\text{Rise}}{\text{Run}}$$

Figure 28 Slope $= -\frac{4}{5}$

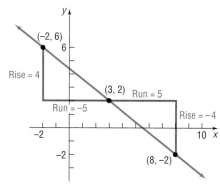

means that for every horizontal movement of 5 units (run = 5) to the right there will be a corresponding vertical movement of -4 units (rise = -4, a downward movement of 4 units). If we start at the given point $(3, 2)$ and move 5 units to the right and then 4 units down, we arrive at the point $(8, -2)$. By drawing the line through these points, we have the graph. See Figure 28.

Alternatively, we can set

$$-\frac{4}{5} = \frac{4}{-5} = \frac{\text{Rise}}{\text{Run}}$$

so that for every horizontal movement of -5 units (a movement to the left) there will be a corresponding vertical movement of 4 units (upward). This approach brings us to the point $(-2, 6)$, which is also on the graph shown in Figure 28. ∎

NOW WORK PROBLEM **15**.

EQUATIONS OF LINES

3 Now that we have discussed the slope of a line, we are ready to derive equations of lines. As we shall see, there are several forms of the equation of a line. Let's start with an example.

EXAMPLE 4   Graphing a Line

Figure 29

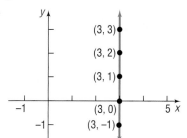

Graph the equation: $x = 3$

Solution We are looking for all points (x, y) in the plane for which $x = 3$. No matter what y-coordinate is used, the corresponding x-coordinate always equals 3. Consequently, the graph of the equation $x = 3$ is a vertical line with x-intercept 3 and undefined slope. See Figure 29. ∎

As suggested by Example 4, we have the following result:

Theorem

Equation of a Vertical Line

A vertical line is given by an equation of the form

$$x = a$$

where a is the x-intercept.

COMMENT: To graph an equation using a graphing utility, we need to express the equation in the form $y =$ expression in x. But $x = 3$ cannot be put in this form. To overcome this, most graphing utilities have special ways for drawing vertical lines. LINE, PLOT, and VERT are among the more common ones. Consult your manual to determine the correct methodology for your graphing utility.

Figure 30

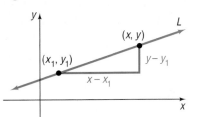

④ Now let L be a nonvertical line with slope m containing the point (x_1, y_1). See Figure 30. For any other point (x, y) on L, we have

$$m = \frac{y - y_1}{x - x_1} \quad \text{or} \quad y - y_1 = m(x - x_1)$$

Theorem

Point-Slope Form of an Equation of a Line

An equation of a nonvertical line of slope m that contains the point (x_1, y_1) is

$$y - y_1 = m(x - x_1) \tag{2}$$

EXAMPLE 5

Using the Point-Slope Form of a Line

Figure 31

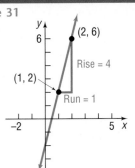

An equation of the line with slope 4 and containing the point $(1, 2)$ can be found by using the point–slope form with $m = 4$, $x_1 = 1$, and $y_1 = 2$.

$$y - y_1 = m(x - x_1)$$
$$y - 2 = 4(x - 1) \qquad \text{\small $m = 4, x_1 = 1, y_1 = 2$}$$

See Figure 31.

EXAMPLE 6

Finding the Equation of a Horizontal Line

Find an equation of the horizontal line containing the point $(3, 2)$.

Figure 32 $y = 2$

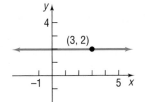

Solution The slope of a horizontal line is 0. To get an equation, we use the point-slope form with $m = 0$, $x_1 = 3$, and $y_1 = 2$.

$$y - y_1 = m(x - x_1)$$
$$y - 2 = 0 \cdot (x - 3) \qquad \text{\small $m = 0, x_1 = 3, y_1 = 2$}$$
$$y - 2 = 0$$
$$y = 2$$

See Figure 32 for the graph.

As suggested by Example 6, we have the following result:

Theorem | **Equation of a Horizontal Line**

A horizontal line is given by an equation of the form

$$y = b$$

where b is the y-intercept.

EXAMPLE 7 | **Finding an Equation of a Line Given Two Points**

⑤ Find an equation of the line L containing the points $(2, 3)$ and $(-4, 5)$. Graph the line L.

Solution Since two points are given, we first compute the slope of the line.

$$m = \frac{5 - 3}{-4 - 2} = \frac{2}{-6} = -\frac{1}{3}$$

We use the point $(2, 3)$ and the fact that the slope $m = -\frac{1}{3}$ to get the point-slope form of the equation of the line.

$$y - 3 = -\frac{1}{3}(x - 2)$$

See Figure 33 for the graph.

Figure 33

$$y - 3 = -\frac{1}{3}(x - 2)$$

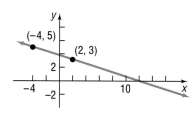

In the solution to Example 7, we could have used the other point, $(-4, 5)$, instead of the point $(2, 3)$. The equation that results, although it looks different, is equivalent to the equation that we obtained in the example. (Try it for yourself.)

⑥ Another useful equation of a line is obtained when the slope m and y-intercept b are known. In this event, we know both the slope m of the line and a point $(0, b)$ on the line; we use the point–slope form, equation (2), to obtain the following equation:

$$y - b = m(x - 0) \quad \text{or} \quad y = mx + b$$

Theorem | **Slope-Intercept Form of an Equation of a Line**

An equation of a line L with slope m and y-intercept b is

$$y = mx + b \qquad \text{(3)}$$

NOW WORK PROBLEM **21.**

Figure 34 $y = mx + 2$

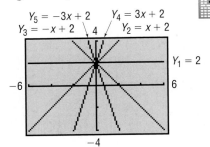

$Y_5 = -3x + 2$ $Y_4 = 3x + 2$
$Y_3 = -x + 2$ $Y_2 = x + 2$

$Y_1 = 2$

SEEING THE CONCEPT: To see the role that the slope m plays, graph the following lines on the same square screen.

$$Y_1 = 2$$
$$Y_2 = x + 2$$
$$Y_3 = -x + 2$$
$$Y_4 = 3x + 2$$
$$Y_5 = -3x + 2$$

See Figure 34. What do you conclude about the lines $y = mx + 2$?

Figure 35 $y = 2x + b$

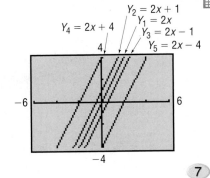

$Y_4 = 2x + 4$
$Y_2 = 2x + 1$
$Y_1 = 2x$
$Y_3 = 2x - 1$
$Y_5 = 2x - 4$

SEEING THE CONCEPT: To see the role of the y-intercept b, graph the following lines on the same square screen.

$$Y_1 = 2x$$
$$Y_2 = 2x + 1$$
$$Y_3 = 2x - 1$$
$$Y_4 = 2x + 4$$
$$Y_5 = 2x - 4$$

See Figure 35. What do you conclude about the lines $y = 2x + b$?

(7) When the equation of a line is written in slope–intercept form, it is easy to find the slope m and y-intercept b of the line. For example, suppose that the equation of a line is

$$y = -2x + 3$$

Compare it to $y = mx + b$:

$$y = -2x + 3$$
$$\uparrow \qquad \uparrow$$
$$y = mx + b$$

The slope of this line is -2 and its y-intercept is 3.

EXAMPLE 8   **Finding the Slope and y-Intercept**

Find the slope m and y-intercept b of the line with equation $2x + 4y = 8$. Graph the equation.

Solution   To obtain the slope and y-intercept, we transform the equation into its slope-intercept form by solving the equation for y.

$$2x + 4y = 8$$
$$4y = -2x + 8$$
$$y = -\frac{1}{2}x + 2 \qquad y = mx + b$$

The coefficient of x, $-\dfrac{1}{2}$, is the slope, and the y-intercept is 2.

We can graph the line in two ways:

Figure 36
$2x + 4y = 8$

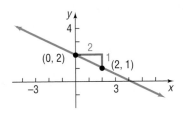

1. Use the fact that the y-intercept is 2 and the slope is $-\dfrac{1}{2}$. Then, starting at the point $(0, 2)$, go to the right 2 units and then down 1 unit to the point $(2, 1)$. See Figure 36.

2. Locate the intercepts. Because the y-intercept is 2, we know that one intercept is $(0, 2)$. To obtain the x-intercept, let $y = 0$ and solve for x. When $y = 0$, we have

$$2x + 4y = 8$$
$$2x + 4 \cdot 0 = 8 \qquad y = 0$$
$$2x = 8$$
$$x = 4$$

The intercepts are $(4, 0)$ and $(0, 2)$. See Figure 37.   ■

Figure 37 $2x + 4y = 8$

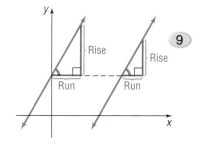

NOW WORK PROBLEM **59.**

8   The form of the equation of the line in Example 8, $2x + 4y = 8$, is called the *general form.*

> The equation of a line L is in **general form** when it is written as
>
> $$Ax + By = C \qquad\qquad \textbf{(4)}$$
>
> where A, B, and C are real numbers and A and B are not both 0.

Every line has an equation that is equivalent to an equation written in general form. For example, a vertical line whose equation is

$$x = a$$

can be written in the general form

$$1 \cdot x + 0 \cdot y = a \qquad A = 1, B = 0, C = a$$

A horizontal line whose equation is

$$y = b$$

can be written in the general form

$$0 \cdot x + 1 \cdot y = b \qquad A = 0, B = 1, C = b$$

Lines that are neither vertical nor horizontal have general equations of the form

$$Ax + By = C \qquad A \neq 0 \text{ and } B \neq 0$$

Because the equation of every line can be written in general form, any equation equivalent to (4) is called a **linear equation.**

NOW WORK PROBLEM **33.**

Figure 38
Two lines are parallel if and only if their slopes are equal.

PARALLEL AND PERPENDICULAR LINES

9   When two lines (in the plane) do not intersect (that is, they have no points in common), they are said to be **parallel.** Look at Figure 38. There we have drawn two lines and have constructed two right triangles by drawing sides parallel to the coordinate axes. These lines are parallel if and only if the right triangles are similar. (Do you see why? Two angles are equal.) But the triangles are similar if and only if the ratios of corresponding sides are equal.

This suggests the following result:

Theorem

Criterion for Parallel Lines

Two nonvertical lines are parallel if and only if their slopes are equal and they have different y-intercepts.

■

The use of the words "if and only if" in the preceding theorem means that actually two statements are being made, one the converse of the other.

If two nonvertical lines are parallel, then their slopes are equal and they have different y-intercepts.

If two nonvertical lines have equal slopes and different y-intercepts, then they are parallel.

Refer to **SEEING THE CONCEPT** and Figure 35, $y = 2x + b$, on page 544.

EXAMPLE 9

Showing That Two Lines Are Parallel

Show that the lines given by the following equations are parallel:

$$L_1: \quad 2x + 3y = 6 \qquad L_2: \quad 4x + 6y = 0$$

Figure 39
Parallel lines

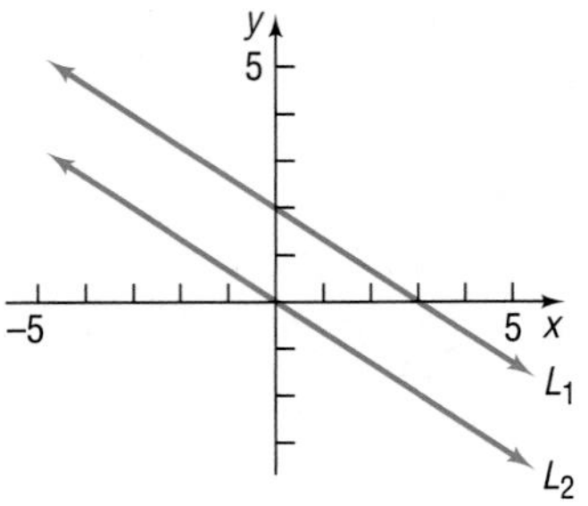

Solution

To determine whether these lines have equal slopes, we write each equation in slope–intercept form:

$$L_1: \quad 2x + 3y = 6 \qquad\qquad L_2: \quad 4x + 6y = 0$$
$$3y = -2x + 6 \qquad\qquad\qquad 6y = -4x$$
$$y = -\frac{2}{3}x + 2 \qquad\qquad\qquad y = -\frac{2}{3}x$$
$$\text{Slope} = -\frac{2}{3} \qquad\qquad\qquad \text{Slope} = -\frac{2}{3}$$
$$y\text{-intercept} = 2 \qquad\qquad\qquad y\text{-intercept} = 0$$

Because these lines have the same slope, $-\dfrac{2}{3}$, but different y-intercepts, the lines are parallel. See Figure 39. ■

EXAMPLE 10

Finding a Line That Is Parallel to a Given Line

Find an equation for the line that contains the point $(2, -3)$ and is parallel to the line $2x + y = 6$.

Figure 40

Solution

The slope of the line that we seek equals the slope of the line $2x + y = 6$, since the two lines are to be parallel. We begin by writing the equation of the line $2x + y = 6$ in slope–intercept form.

$$2x + y = 6$$
$$y = -2x + 6$$

The slope is -2. Since the line that we seek contains the point $(2, -3)$, we use the point–slope form to obtain

$$y + 3 = -2(x - 2) \qquad \text{Point–slope form}$$
$$y + 3 = -2x + 4$$
$$y = -2x + 1 \qquad \text{Slope–intercept form}$$
$$2x + y = 1 \qquad \text{General form}$$

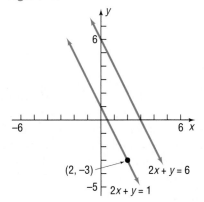

This line is parallel to the line $2x + y = 6$ and contains the point $(2, -3)$. See Figure 40. ■

NOW WORK PROBLEM **41.**

Figure 41
Perpendicular lines

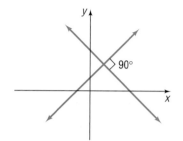

⑩   When two lines intersect at a right angle (90°), they are said to be **perpendicular.** See Figure 41.

The following result gives a condition, in terms of their slopes, for two lines to be perpendicular.

Theorem   Criterion for Perpendicular Lines

Two nonvertical lines are perpendicular if and only if the product of their slopes is −1.

∎

Here, we shall prove the "only if" part of the statement:

If two nonvertical lines are perpendicular, then the product of their slopes is −1.

In Problem 84, you are asked to prove the "if" part of the theorem; that is:

If two nonvertical lines have slopes whose product is −1, then the lines are perpendicular.

Figure 42

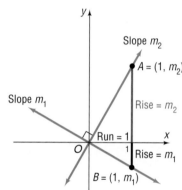

Proof   Let m_1 and m_2 denote the slopes of the two lines. There is no loss in generality (that is, neither the angle nor the slopes are affected) if we situate the lines so that they meet at the origin. See Figure 42. The point $A = (1, m_2)$ is on the line having slope m_2, and the point $B = (1, m_1)$ is on the line having slope m_1. (Do you see why this must be true?)

Suppose that the lines are perpendicular. Then triangle OAB is a right triangle. As a result of the Pythagorean Theorem, it follows that

$$[d(O, A)]^2 + [d(O, B)]^2 = [d(A, B)]^2 \qquad \textbf{(5)}$$

By the distance formula, we can write each of these distances as

$$[d(O, A)]^2 = (1 - 0)^2 + (m_2 - 0)^2 = 1 + m_2^2$$
$$[d(O, B)]^2 = (1 - 0)^2 + (m_1 - 0)^2 = 1 + m_1^2$$
$$[d(A, B)]^2 = (1 - 1)^2 + (m_2 - m_1)^2 = m_2^2 - 2m_1 m_2 + m_1^2$$

Using these facts in equation (5), we get

$$(1 + m_2^2) + (1 + m_1^2) = m_2^2 - 2m_1 m_2 + m_1^2$$

which, upon simplification, can be written as

$$m_1 m_2 = -1$$

If the lines are perpendicular, the product of their slopes is −1.   ∎

You may find it easier to remember the condition for two nonvertical lines to be perpendicular by observing that the equality $m_1 m_2 = -1$ means that m_1 and m_2 are negative reciprocals of each other; that is, $m_1 = -\dfrac{1}{m_2}$ and $m_2 = -\dfrac{1}{m_1}$.

EXAMPLE 11   **Finding the Slope of a Line Perpendicular to a Given Line**

If a line has slope $\dfrac{3}{2}$, any line having slope $-\dfrac{2}{3}$ is perpendicular to it.   ∎

EXAMPLE 12	Finding the Equation of a Line Perpendicular to a Given Line

Find an equation of the line containing the point $(1, -2)$ that is perpendicular to the line $x + 3y = 6$. Graph the two lines.

Solution We first write the equation of the given line in slope–intercept form to find its slope.

$$x + 3y = 6$$
$$3y = -x + 6 \qquad \text{Proceed to solve for } y.$$
$$y = -\frac{1}{3}x + 2 \qquad \text{Place in the form } y = mx + b.$$

Figure 43

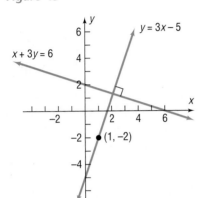

The given line has slope $-\dfrac{1}{3}$. Any line perpendicular to this line will have slope 3. Because we require the point $(1, -2)$ to be on this line with slope 3, we use the point–slope form of the equation of a line.

$$y - (-2) = 3(x - 1) \qquad \text{Point–slope form}$$

To obtain other forms of the equation, we proceed as follows:

$$y + 2 = 3(x - 1)$$
$$y + 2 = 3x - 3$$
$$y = 3x - 5 \qquad \text{Slope–intercept form}$$
$$3x - y = 5 \qquad \text{General form}$$

Figure 43 shows the graphs.

 NOW WORK PROBLEM 47.

WARNING: Be sure to use a square screen when you graph perpendicular lines. Otherwise, the angle between the two lines will appear distorted.

A.7 EXERCISES

In Problems 1–4, (a) find the slope of the line and (b) interpret the slope.

1.

2.

3.

4.

In Problems 5–12, plot each pair of points and determine the slope of the line containing them. Graph the line.

5. $(2, 3)$; $(4, 0)$ **6.** $(4, 2)$; $(3, 4)$ **7.** $(-2, 3)$; $(2, 1)$ **8.** $(-1, 1)$; $(2, 3)$

9. $(-3, -1)$; $(2, -1)$ **10.** $(4, 2)$; $(-5, 2)$ **11.** $(-1, 2)$; $(-1, -2)$ **12.** $(2, 0)$; $(2, 2)$

In Problems 13–20, graph the line containing the point P and having slope m.

13. $P = (1, 2)$; $m = 3$ **14.** $P = (2, 1)$; $m = 4$ **15.** $P = (2, 4)$; $m = -\dfrac{3}{4}$

16. $P = (1, 3)$; $m = -\dfrac{2}{5}$ **17.** $P = (-1, 3)$; $m = 0$ **18.** $P = (2, -4)$; $m = 0$

19. $P = (0, 3)$; slope undefined   **20.** $P = (-2, 0)$; slope undefined

In Problems 21–28, find an equation of the line L. Express your answer using either the general form or the slope–intercept form of the equation of a line, whichever you prefer.

21.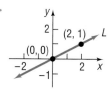

22.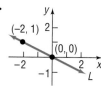

23.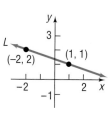

24.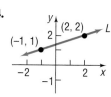

25.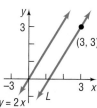

L is parallel to y = 2x

26.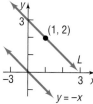

L is parallel to y = −x

27.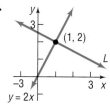

L is perpendicular to y = 2x

28.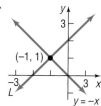

L is perpendicular to y = −x

In Problems 29–52, find an equation for the line with the given properties. Express your answer using either the general form or the slope–intercept form of the equation of a line, whichever you prefer.

29. Slope $= 3$; containing the point $(-2, 3)$

30. Slope $= 2$; containing the point $(4, -3)$

31. Slope $= -\dfrac{2}{3}$; containing the point $(1, -1)$

32. Slope $= \dfrac{1}{2}$; containing the point $(3, 1)$

33. Containing the points $(1, 3)$ and $(-1, 2)$

34. Containing the points $(-3, 4)$ and $(2, 5)$

35. Slope $= -3$; y-intercept $= 3$

36. Slope $= -2$; y-intercept $= -2$

37. x-intercept $= 2$; y-intercept $= -1$

38. x-intercept $= -4$; y-intercept $= 4$

39. Slope undefined; containing the point $(2, 4)$

40. Slope undefined; containing the point $(3, 8)$

41. Parallel to the line $y = 2x$;
containing the point $(-1, 2)$

42. Parallel to the line $y = -3x$;
containing the point $(-1, 2)$

43. Parallel to the line $2x - y = -2$;
containing the point $(0, 0)$

44. Parallel to the line $x - 2y = -5$;
containing the point $(0, 0)$

45. Parallel to the line $x = 5$;
containing the point $(4, 2)$

46. Parallel to the line $y = 5$;
containing the point $(4, 2)$

47. Perpendicular to the line $y = \dfrac{1}{2}x + 4$;

containing the point $(1, -2)$

48. Perpendicular to the line $y = 2x - 3$;

containing the point $(1, -2)$

49. Perpendicular to the line $2x + y = 2$;
containing the point $(-3, 0)$

50. Perpendicular to the line $x - 2y = -5$;
containing the point $(0, 4)$

51. Perpendicular to the line $x = 8$;
containing the point $(3, 4)$

52. Perpendicular to the line $y = 8$;
containing the point $(3, 4)$

In Problems 53–72, find the slope and y-intercept of each line. Graph the line.

53. $y = 2x + 3$

54. $y = -3x + 4$

55. $\dfrac{1}{2}y = x - 1$

56. $\dfrac{1}{3}x + y = 2$

57. $y = \dfrac{1}{2}x + 2$

58. $y = 2x + \dfrac{1}{2}$

59. $x + 2y = 4$

60. $-x + 3y = 6$

61. $2x - 3y = 6$

62. $3x + 2y = 6$

63. $x + y = 1$

64. $x - y = 2$

65. $x = -4$

66. $y = -1$

67. $y = 5$

68. $x = 2$

69. $y - x = 0$

70. $x + y = 0$

71. $2y - 3x = 0$

72. $3x + 2y = 0$

73. Find an equation of the x-axis.

74. Find an equation of the y-axis.

75. **Geometry** Use slopes to show that the triangle whose vertices are $(-2, 5)$, $(1, 3)$, and $(-1, 0)$ is a right triangle.

76. **Geometry** Use slopes to show that the quadrilateral whose vertices are $(1, -1)$, $(4, 1)$, $(2, 2)$, and $(5, 4)$ is a parallelogram.

77. **Geometry** Use slopes to show that the quadrilateral whose vertices are $(-1, 0)$, $(2, 3)$, $(1, -2)$, and $(4, 1)$ is a rectangle.

78. **Geometry** Use slopes and the distance formula to show that the quadrilateral whose vertices are $(0, 0)$, $(1, 3)$, $(4, 2)$, and $(3, -1)$ is a square.

79. **Measuring Temperature** The relationship between Celsius (°C) and Fahrenheit (°F) degrees for measuring temperature is linear. Find an equation relating °C and °F if 0°C corresponds to 32°F and 100°C corresponds to 212°F. Use the equation to find the Celsius measure of 70°F.

80. **Measuring Temperature** The Kelvin (K) scale for measuring temperature is obtained by adding 273 to the Celsius temperature.
    (a) Write an equation relating K and °C.
    (b) Write an equation relating K and °F (see Problem 79).

81. **Business: Computing Profit** Each Sunday a newspaper agency sells x copies of a newspaper for $1.00 per copy. The cost to the agency of each newspaper is $0.50. The agency pays a fixed cost for storage, delivery, and so on, of $100 per Sunday.
    (a) Write an equation that relates the profit P, in dollars, to the number x of copies sold. Graph this equation.
    (b) What is the profit to the agency if 1000 copies are sold?
    (c) What is the profit to the agency if 5000 copies are sold?

82. **Business: Computing Profit** Repeat Problem 81 if the cost to the agency is $0.45 per copy and the fixed cost is $125 per Sunday.

83. **Cost of Electricity** In 2000, Florida Power and Light Company supplied electricity to residential customers for a monthly customer charge of $5.65 plus 6.543¢ per kilowatt-hour supplied in the month for the first 750 kilowatt-hours used.* Write an equation that relates the monthly charge C, in dollars, to the number x of kilowatt-hours used in the month. Graph this equation. What is the monthly charge for using 300 kilowatt-hours? For using 750 kilowatt-hours?

84. Prove that if two nonvertical lines have slopes whose product is -1 then the lines are perpendicular.
    [**Hint:** Refer to Figure 42, and use the converse of the Pythagorean Theorem.]

*Source: Florida Power and Light Co., Miami, Florida, 2000.

In Problems 85–88, match each graph with the correct equation:

(a) $y = x$ (b) $y = 2x$ (c) $y = \dfrac{x}{2}$ (d) $y = 4x$

85.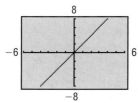

86.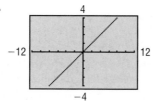

87.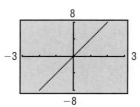

88.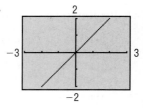

In Problems 89–92, write an equation of each line. Express your answer using either the general form or the slope–intercept form of the equation of a line, whichever you prefer.

89.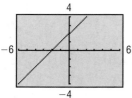

90.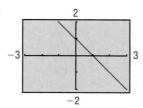

91.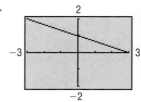

92.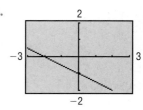

93. Which of the following equations might have the graph shown? (More than one answer is possible.)

(a) $2x + 3y = 6$

(b) $-2x + 3y = 6$

(c) $3x - 4y = -12$

(d) $x - y = 1$

(e) $x - y = -1$

(f) $y = 3x - 5$

(g) $y = 2x + 3$

(h) $y = -3x + 3$

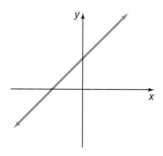

94. Which of the following equations might have the graph shown? (More than one answer is possible.)

(a) $2x + 3y = 6$

(b) $2x - 3y = 6$

(c) $3x + 4y = 12$

(d) $x - y = 1$

(e) $x - y = -1$

(f) $y = -2x + 1$

(g) $y = -\dfrac{1}{2}x + 10$

(h) $y = x + 4$

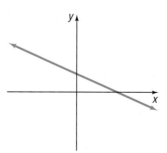

95. The figure below shows the graph of two parallel lines. Which of the following pairs of equations might have such a graph?

(a) $x - 2y = 3$
$x + 2y = 7$

(b) $x + y = 2$
$x + y = -1$

(c) $x - y = -2$
$x - y = 1$

(d) $x - y = -2$
$2x - 2y = -4$

(e) $x + 2y = 2$
$x + 2y = -1$

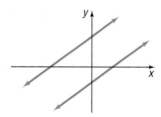

96. The figure below shows the graph of two perpendicular lines. Which of the following pairs of equations might have such a graph?

(a) $y - 2x = 2$
$y + 2x = -1$

(b) $y - 2x = 0$
$2y + x = 0$

(c) $2y - x = 2$
$2y + x = -2$

(d) $y - 2x = 2$
$x + 2y = -1$

(e) $2x + y = -2$
$2y + x = -2$

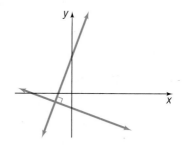

97. **Geometry**   The **tangent line** to a circle may be defined as the line that intersects the circle in a single point, called the **point of tangency** (see the figure). If the equation of the circle is $x^2 + y^2 = r^2$ and the equation of the tangent line is $y = mx + b$, show that:

(a) $r^2(1 + m^2) = b^2$

[**Hint:** The quadratic equation $x^2 + (mx + b)^2 = r^2$ has exactly one solution.]

(b) The point of tangency is $\left(-\dfrac{r^2 m}{b}, \dfrac{r^2}{b}\right)$.

(c) The tangent line is perpendicular to the line containing the center of the circle and the point of tangency.

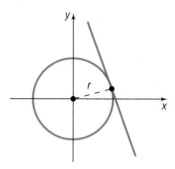

98. The **Greek method** for finding the equation of the tangent line to a circle used the fact that at any point on a circle the line containing the radius and the tangent line are perpendicular (see Problem 97). Use this method to find an equation of the tangent line to the circle $x^2 + y^2 = 9$ at the point $(1, 2\sqrt{2})$.

99. Use the Greek method described in Problem 98 to find an equation of the tangent line to the circle $x^2 + y^2 - 4x + 6y + 4 = 0$ at the point $(3, 2\sqrt{2} - 3)$.

100. Refer to Problem 97. The line $x - 2y = -4$ is tangent to a circle at $(0, 2)$. The line $y = 2x - 7$ is tangent to the same circle at $(3, -1)$. Find the center of the circle.

101. Find an equation of the line containing the centers of the two circles

$$x^2 + y^2 - 4x + 6y + 4 = 0$$

and $$x^2 + y^2 + 6x + 4y + 9 = 0$$

102. Show that the line containing the points (a, b) and (b, a), $a \neq b$, is perpendicular to the line $y = x$. Also show that the midpoint of (a, b) and (b, a) lies on the line $y = x$.

103. The equation $2x - y + C = 0$ defines a **family of lines,** one line for each value of C. On one set of coordinate axes, graph the members of the family when $C = -4$, $C = 0$, and $C = 2$. Can you draw a conclusion from the graph about each member of the family?

104. Rework Problem 103 for the family of lines $Cx + y + 4 = 0$.

105. If a circle of radius 2 is made to roll along the x-axis, what is an equation for the path of the center of the circle?

106. Which form of the equation of a line do you prefer to use? Justify your position with an example that shows that your choice is better than another. Have reasons.

107. Can every line be written in slope–intercept form? Explain.

108. Does every line have two distinct intercepts? Explain. Are there lines that have no intercepts? Explain.

109. What can you say about two lines that have equal slopes and equal y-intercepts?

110. What can you say about two lines with the same x-intercept and the same y-intercept? Assume that the x-intercept is not 0.

111. If two lines have the same slope, but different x-intercepts, can they have the same y-intercept?

112. If two lines have the same y-intercept, but different slopes, can they have the same x-intercept? What is the only way that this can happen?

113. The accepted symbol used to denote the slope of a line is the letter m. Investigate the origin of this symbolism. Begin by consulting a French dictionary and looking up the French word *monter*. Write a brief essay on your findings.

114. Grade of a Road The term *grade* is used to describe the inclination of a road. How does this term relate to the notion of slope of a line? Is a 4% grade very steep? Investigate the grades of some mountainous roads and determine their slopes. Write a brief essay on your findings.

115. Carpentry Carpenters use the term *pitch* to describe the steepness of staircases and roofs. How does pitch relate to slope? Investigate typical pitches used for stairs and for roofs. Write a brief essay on your findings.

A.8 SCATTER DIAGRAMS; LINEAR CURVE FITTING

OBJECTIVES
1. Draw and Interpret Scatter Diagrams
2. Distinguish between Linear and Nonlinear Relations
3. Use a Graphing Utility to Find the Line of Best Fit

SCATTER DIAGRAMS

1. A **relation** is a correspondence between two sets. If x and y are two elements in these sets and if a relation exists between x and y, then we say that x **corresponds to** y or that y **depends on** x and write $x \rightarrow y$. We may also write $x \rightarrow y$ as the ordered pair (x, y). Here, y is referred to as the **dependent** variable and x is called the **independent** variable.

Often we are interested in specifying the type of relation (such as an equation) that might exist between two variables. The first step in finding this relation is to plot the ordered pairs using rectangular coordinates. The resulting graph is called a **scatter diagram.**

EXAMPLE 1	Drawing a Scatter Diagram

The data listed in Table 2 represent the apparent temperature versus the relative humidity in a room whose actual temperature is 72° Fahrenheit.

	TABLE	2					
Relative Humidity (%), x	Apparent Temperature (°F) y	(x, y)		Relative Humidity (%), x	Apparent Temperature (°F) y	(x, y)	
0	64	(0, 64)		60	72	(60, 72)	
10	65	(10, 65)		70	73	(70, 73)	
20	67	(20, 67)		80	74	(80, 74)	
30	68	(30, 68)		90	75	(90, 75)	
40	70	(40, 70)		100	76	(100, 76)	
50	71	(50, 71)					

(a) Draw a scatter diagram by hand.

(b) Use a graphing utility to draw a scatter diagram.*

(c) Describe what happens to the apparent temperature as the relative humidity increases.

Solution   (a) To draw a scatter diagram by hand, we plot the ordered pairs listed in Table 2, with the relative humidity as the x-coordinate and the apparent temperature as the y-coordinate. See Figure 44(a). Notice that the points in a scatter diagram are not connected.

(b) Figure 44(b) shows a scatter diagram using a graphing utility.

(c) We see from the scatter diagrams that, as the relative humidity increases, the apparent temperature increases.

Figure 44

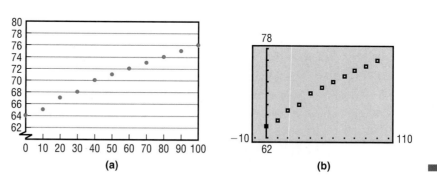

(a)   (b)

NOW WORK PROBLEM 7(a).

CURVE FITTING

2 Scatter diagrams are used to help us to see the type of relation that may exist between two variables. In this text, we will discuss a variety of different relations that may exist between two variables. For now, we concentrate on distinguishing between linear and nonlinear relations. See Figure 45.

---

* Consult your owner's manual for the proper keystrokes.

Figure 45

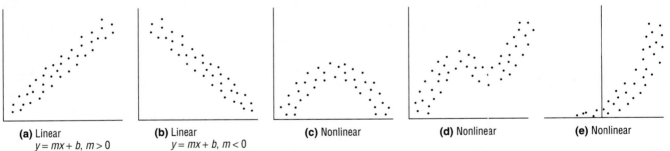

(a) Linear
$y = mx + b$, $m > 0$

(b) Linear
$y = mx + b$, $m < 0$

(c) Nonlinear

(d) Nonlinear

(e) Nonlinear

EXAMPLE 2 | **Distinguishing between Linear and Nonlinear Relations**

Determine whether the relation between the two variables in Figure 46 is linear or nonlinear.

Figure 46

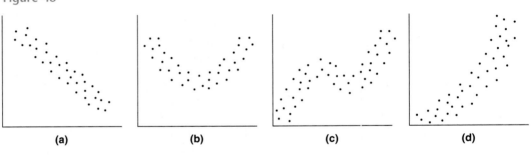

(a)

(b)

(c)

(d)

Solution (a) Linear   (b) Nonlinear   (c) Nonlinear   (d) Nonlinear ▬

NOW WORK PROBLEM **1.**

Suppose that the scatter diagram of a set of data appears to be linearly related as in Figure 45(a) or (b). We might wish to find an equation of a line that relates the two variables. One way to obtain an equation for such data is to draw a line through two points on the scatter diagram and estimate the equation of the line.

EXAMPLE 3 | **Find an Equation for Linearly Related Data**

Using the data in Table 2 from Example 1, select two points from the data and find an equation of the line containing the points.

(a) Graph the line on the scatter diagram obtained in Example 1(a).

(b) Graph the line on the scatter diagram obtained in Example 1(b).

Solution Select two points, say $(10, 65)$ and $(70, 73)$. (You should select your own two points and complete the solution.) The slope of the line joining the points $(10, 65)$ and $(70, 73)$ is

$$m = \frac{73 - 65}{70 - 10} = \frac{8}{60} = \frac{2}{15}$$

The equation of the line with slope $\dfrac{2}{15}$ and passing through $(10, 65)$ is found using the point–slope form with $m = \dfrac{2}{15}$, $x_1 = 10$, and $y_1 = 65$.

$$y - y_1 = m(x - x_1)$$

$$y - 65 = \frac{2}{15}(x - 10)$$

$$y = \frac{2}{15}x + \frac{191}{3}$$

(a) Figure 47(a) shows the scatter diagram with the graph of the line drawn by hand.

 (b) Figure 47(b) shows the scatter diagram with the graph of the line using a graphing utility.

Figure 47

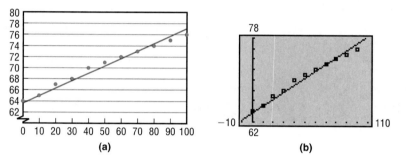

(a)   (b)

NOW WORK PROBLEMS 7(b) AND (c).

 ## LINE OF BEST FIT

3 The line obtained in Example 3 depends on the selection of points, which will vary from person to person. So the line that we found might be different from the line that you found. Although the line that we found in Example 3 appears to "fit" the data well, there may be a line that "fits it better." Do you think your line fits the data better? Is there a line of *best fit*? As it turns out, there is a method for finding the line that best fits linearly related data (called the *line of best fit*).*

EXAMPLE 4   Finding the Line of Best Fit

Using the data in Table 2 from Example 1:

(a) Find the line of best fit using a graphing utility.
(b) Graph the line of best fit on the scatter diagram obtained in Example 1(b).
(c) Interpret the slope of the line of best fit.
(d) Use the line of best fit to predict the apparent temperature of a room whose actual temperature is 72°F and relative humidity is 45%.

Solution   (a) Graphing utilities contain built-in programs that find the line of best fit for a collection of points in a scatter diagram. (Look in your owner's manual under Linear Regression or Line of Best Fit for details on how to execute

---

* We shall not discuss in this book the underlying mathematics of lines of best fit. Most books in statistics and many in linear algebra discuss this topic.

Figure 48

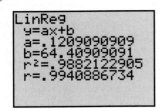

Figure 49

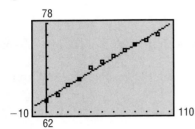

the program.) Upon executing the LINear REGression program, we obtain the results shown in Figure 48. The output that the utility provides shows us the equation $y = ax + b$, where a is the slope of the line and b is the y-intercept. The line of best fit that relates relative humidity to apparent temperature may be expressed as the line $y = 0.121x + 64.409$.

(b) Figure 49 shows the graph of the line of best fit, along with the scatter diagram.

(c) The slope of the line of best fit is approximately 0.121, which means that, for every 1% increase in the relative humidity, apparent room temperature increases 0.121°F.

(d) Letting $x = 45$ in the equation of the line of best fit, we obtain $y = 0.121(45) + 64.409 \approx 70°F$, which is the apparent temperature in the room.

NOW WORK PROBLEMS 7(d) AND (e).

Does the line of best fit appear to be a good fit? In other words, does the line appear to accurately describe the relation between temperature and relative humidity?

And just how "good" is this line of best fit? The answers are given by what is called the *correlation coefficient*. Look again at Figure 48. The last line of output is $r = 0.994$. This number, called the **correlation coefficient, r,** $-1 \leq r \leq 1$, is a measure of the strength of the *linear relation* that exists between two variables. The closer that $|r|$ is to 1, the more perfect the linear relationship is. If r is close to 0, there is little or no *linear* relationship between the variables. A negative value of r, $r < 0$, indicates that as x increases y decreases; a positive value of r, $r > 0$, indicates that as x increases y does also. The data given in Example 1, having a correlation coefficient of 0.994, are indicative of a strong linear relationship with positive slope.

A.8 EXERCISES

In Problems 1–6, examine the scatter diagram and determine whether the type of relation, if any, that may exist is linear or nonlinear.

1.

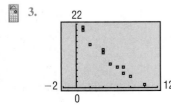

2.

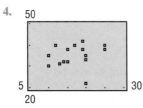

3.

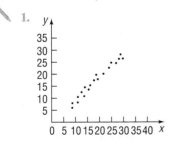

4.

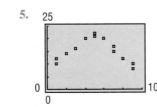

5.

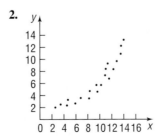

6.

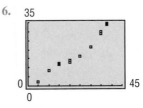

In Problems 7–14:
  (a) *Draw a scatter diagram by hand.*
  (b) *Select two points from the scatter diagram and find the equation of the line containing the points selected.**
  (c) *Graph the line found in part (b) on the scatter diagram.*
  (d) *Use a graphing utility to find the line of best fit.*
  (e) *Use a graphing utility to graph the line of best fit on the scatter diagram.*

7.
x	3	4	5	6	7	8	9
y	4	6	7	10	12	14	16

8.
x	3	5	7	9	11	13
y	0	2	3	6	9	11

9.
x	-2	-1	0	1	2
y	-4	0	1	4	5

10.
x	-2	-1	0	1	2
y	7	6	3	2	0

11.
x	20	30	40	50	60
y	100	95	91	83	70

12.
x	5	10	15	20	25
y	2	4	7	11	18

13.
x	-20	-17	-15	-14	-10
y	100	120	118	130	140

14.
x	-30	-27	-25	-20	-14
y	10	12	13	13	18

15. Consumption and Disposable Income   An economist wants to estimate a line that relates personal consumption expenditures C and disposable income I. Both C and I are in thousands of dollars. She interviews eight heads of households for families of size 3 and obtains the data below. Let I represent the independent variable and C the dependent variable.
  (a) Draw a scatter diagram by hand.
  (b) Find a line that fits the data.*
  (c) Interpret the slope. The slope of this line is called the **marginal propensity to consume.**
  (d) Predict the consumption of a family whose disposable income is $42,000.
  (e) Use a graphing utility to find the line of best fit to the data.

I (000)	C (000)
20	16
20	18
18	13
27	21
36	27
37	26
45	36
50	39

16. Marginal Propensity to Save   The same economist as in Problem 15 wants to estimate a line that relates savings S and disposable income I. Let $S = I - C$ be the dependent variable and I the independent variable.
  (a) Draw a scatter diagram by hand.
  (b) Find a line that fits the data.
  (c) Interpret the slope. The slope of this line is called the **marginal propensity to save.**
  (d) Predict the savings of a family whose income is $42,000.
  (e) Use a graphing utility to find the line of best fit.

17. Mortgage Qualification   The amount of money that a lending institution will allow you to borrow mainly depends on the interest rate and your annual income. The following data represent the annual income I required by a bank in order to lend L dollars at an interest rate of 7.5% for 30 years.

Annual Income, I ($)	Loan Amount, L ($)
15,000	44,600
20,000	59,500
25,000	74,500
30,000	89,400
35,000	104,300
40,000	119,200
45,000	134,100
50,000	149,000
55,000	163,900
60,000	178,800
65,000	193,700
70,000	208,600

Source: *Information Please Almanac,* 1999

Let I represent the independent variable and L the dependent variable.
  (a) Use a graphing utility to draw a scatter diagram of the data.
  (b) Use a graphing utility to find the line of best fit to the data.
  (c) Graph the line of best fit on the scatter diagram drawn in part (a).
  (d) Interpret the slope of the line of best fit.
  (e) Determine the loan amount that an individual will qualify for if her income is $42,000.

* Answers will vary. We will use the first and last data points in the answer section.

18. **Mortgage Qualification** The amount of money that a lending institution will allow you to borrow mainly depends on the interest rate and your annual income. The following data represent the annual income I required by a bank in order to lend L dollars at an interest rate of 8.5% for 30 years.

Annual Income, I ($)	Loan Amount, L ($)
15,000	40,600
20,000	54,100
25,000	67,700
30,000	81,200
35,000	94,800
40,000	108,300
45,000	121,900
50,000	135,400
55,000	149,000
60,000	162,500
65,000	176,100
70,000	189,600

Source: Information Please Almanac, 1999

Let I represent the independent variable and L the dependent variable.
(a) Use a graphing utility to draw a scatter diagram of the data.
(b) Use a graphing utility to find the line of best fit to the data.
(c) Graph the line of best fit on the scatter diagram drawn in part (a).
(d) Interpret the slope of the line of best fit.
(e) Determine the loan amount that an individual will qualify for if her income is $42,000.

19. **Apparent Room Temperature** The following data represent the apparent temperature versus the relative humidity in a room whose actual temperature is 65° Fahrenheit.

Relative Humidity, h (%)	Apparent Temperature, T (°F)
0	59
10	60
20	61
30	61
40	62
50	63
60	64
70	65
80	65
90	66
100	67

Source: National Oceanic and Atmospheric Administration

Let h represent the independent variable and T the dependent variable.

(a) Use a graphing utility to draw a scatter diagram of the data.

(b) Use a graphing utility to find the line of best fit to the data.

(c) Graph the line of best fit on the scatter diagram drawn in part (a).

(d) Interpret the slope of the line of best fit.

(e) Determine the apparent temperature of a room whose actual temperature is 65°F if the relative humidity is 75%.

20. **Apparent Room Temperature** The following data represent the apparent temperature versus the relative humidity in a room whose actual temperature is 75° Fahrenheit.

Relative Humidity, h (%)	Apparent Temperature, T (°F)
0	68
10	69
20	71
30	72
40	74
50	75
60	76
70	76
80	77
90	78
100	79

Source: National Oceanic and Atmospheric Administration

Let h represent the independent variable and let T be the dependent variable.

(a) Use a graphing utility to draw a scatter diagram of the data.

(b) Use a graphing utility to find the line of best fit to the data.

(c) Graph the line of best fit on the scatter diagram drawn in part (a).

(d) Interpret the slope of the line of best fit.

(e) Determine the apparent temperature of a room whose actual temperature is 75°F if the relative humidity is 75%.

21. Average Miles per Car   The following data represent the average miles driven per car (in thousands) in the United States for the years 1985 to 1996. Let the year, x, represent the independent variable and average miles per car, M, represent the dependent variable.

(a) Use a graphing utility to draw a scatter diagram of the data.

(b) Use a graphing utility to find the line of best fit to the data.

(c) Graph the line of best fit on the scatter diagram drawn in part (a).

(d) Interpret the slope of the line of best fit.

(e) Predict the average number of miles driven per car in 1997.

Year, x	Average Miles per Car, M
1985	9.4
1986	9.5
1987	9.7
1988	10.0
1989	10.2
1990	10.3
1991	10.3
1992	10.6
1993	10.5
1994	10.8
1995	11.1
1996	11.3

Source: U.S. Federal Highway Administration

Graphing Utilities

B.1   THE VIEWING RECTANGLE

All graphing utilities, that is, all graphing calculators and all computer software graphing packages, graph equations by plotting points on a screen. The screen itself actually consists of small rectangles, called **pixels.** The more pixels the screen has, the better the resolution. Most graphing calculators have 48 pixels per square inch; most computer screens have 32 to 108 pixels per square inch. When a point to be plotted lies inside a pixel, the pixel is turned on (lights up). The graph of an equation is a collection of pixels. Figure 1 shows how the graph of $y = 2x$ looks on a TI-83 graphing calculator.

The screen of a graphing utility will display the coordinate axes of a rectangular coordinate system. However, you must set the scale on each axis. You must also include the smallest and largest values of x and y that you want included in the graph. This is called **setting the viewing rectangle** or **viewing window.** Figure 2 illustrates a typical viewing window.

To select the viewing window, we must give values to the following expressions:

Figure 1 $y = 2x$

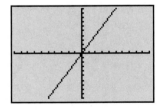

Figure 2

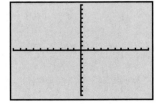

Xmin:	the smallest value of x
Xmax:	the largest value of x
Xscl:	the number of units per tick mark on the x-axis
Ymin:	the smallest value of y
Ymax:	the largest value of y
Yscl:	the number of units per tick mark on the y-axis

Figure 3 illustrates these settings and their relation to the Cartesian coordinate system.

Figure 3

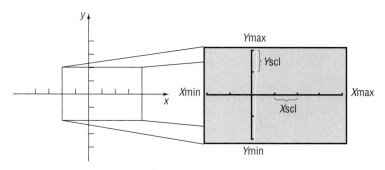

If the scale used on each axis is known, we can determine the minimum and maximum values of x and y shown on the screen by counting the tick marks. Look again at Figure 2. For a scale of 1 on each axis, the minimum and maximum values of x are -10 and 10, respectively; the minimum and maximum values of y are also -10 and 10. If the scale is 2 on each axis, then the minimum and maximum values of x are -20 and 20, respectively; and the minimum and maximum values of y are -20 and 20, respectively.

Conversely, if we know the minimum and maximum values of x and y, we can determine the scales being used by counting the tick marks displayed. We shall follow the practice of showing the minimum and maximum values of x and y in our illustrations so that you will know how the viewing window was set. See Figure 4.

Figure 4

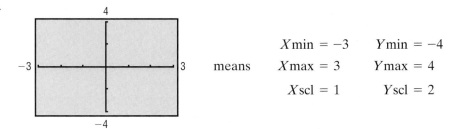

means

$$X\text{min} = -3 \qquad Y\text{min} = -4$$
$$X\text{max} = 3 \qquad Y\text{max} = 4$$
$$X\text{scl} = 1 \qquad Y\text{scl} = 2$$

EXAMPLE 1

Finding the Coordinates of a Point Shown on a Graphing Utility Screen

Find the coordinates of the point shown in Figure 5. Assume that the coordinates are integers.

Figure 5

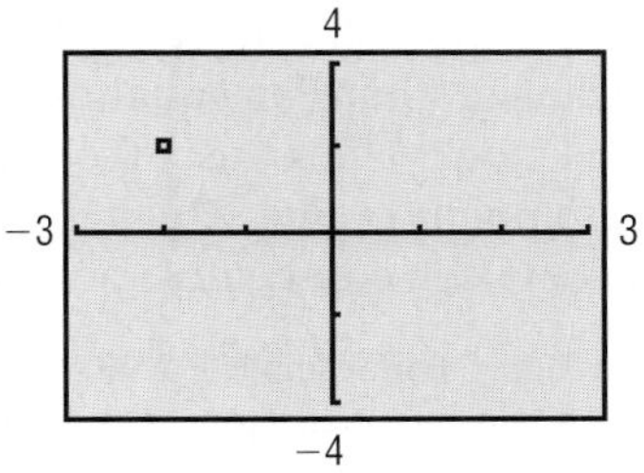

Solution   First we note that the viewing window used in Figure 5 is

$$X\text{min} = -3 \qquad Y\text{min} = -4$$
$$X\text{max} = 3 \qquad Y\text{max} = 4$$
$$X\text{scl} = 1 \qquad Y\text{scl} = 2$$

The point shown is 2 tick units to the left on the horizontal axis (scale = 1) and 1 tick up on the vertical axis (scale = 2). The coordinates of the point shown are $(-2, 2)$.   ■

In Problems 1–4, determine the coordinates of the points shown. Tell in which quadrant each point lies. Assume that the coordinates are integers.

1.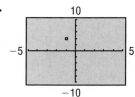

2.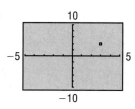

3.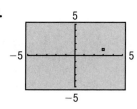

4.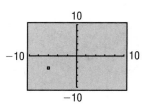

In Problems 5–10, determine the viewing window used.

5.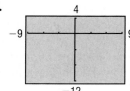

6.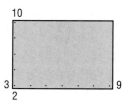

7.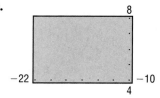

8.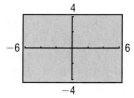

9.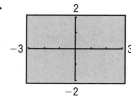

10.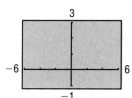

In Problems 11–16, select a setting so that each of the given points will lie within the viewing rectangle.

11. $(-10, 5), (3, -2), (4, -1)$ **12.** $(5, 0), (6, 8), (-2, -3)$ **13.** $(40, 20), (-20, -80), (10, 40)$

14. $(-80, 60), (20, -30), (-20, -40)$ **15.** $(0, 0), (100, 5), (5, 150)$ **16.** $(0, -1), (100, 50), (-10, 30)$

In Problems 17–20, find the length of the line segment. Assume that the endpoints of each line segment have integer coordinates.

17.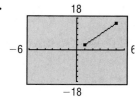

18.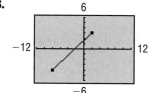

19.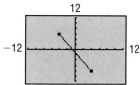

20.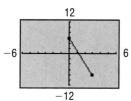

B.2 **USING A GRAPHING UTILITY TO GRAPH EQUATIONS**

From Examples 2 and 3 of Section 1.2, we see that a graph can be obtained by plotting points in a rectangular coordinate system and connecting them. Graphing utilities perform these same steps when graphing an equation. For example, the TI-83 determines 95 evenly spaced input values,* uses the equation to determine the output values, plots these points on the screen, and finally (if in the connected mode) draws a line between consecutive points.

*These input values depend on the values of Xmin and Xmax. For example, if Xmin $= -10$ and Xmax $= 10$, then the first input value will be -10 and the next input value will be $-10 + (10 - (-10))/94 = -9.7872$, and so on.

To graph an equation in two variables x and y using a graphing utility requires that the equation be written in the form $y = \{\text{expression in } x\}$. If the original equation is not in this form, replace it by equivalent equations until the form $y = \{\text{expression in } x\}$ is obtained. In general, there are four ways to obtain equivalent equations.

PROCEDURES THAT RESULT IN EQUIVALENT EQUATIONS

1. Interchange the two sides of the equation:
   Replace $3x + 5 = y$ by $y = 3x + 5$

2. Simplify the sides of the equation by combining like terms, eliminating parentheses, and so on:
   Replace $(2y + 2) + 6 = 2x + 5(x + 1)$
   by $2y + 8 = 7x + 5$

3. Add or subtract the same expression on both sides of the equation:
   Replace $y + 3x - 5 = 4$
   by $y + 3x - 5 + 5 = 4 + 5$

4. Multiply or divide both sides of the equation by the same nonzero expression:
   Replace $3y = 6 - 2x$

   by $\dfrac{1}{3} \cdot 3y = \dfrac{1}{3}(6 - 2x)$

| EXAMPLE 1 | **Expressing an Equation in the Form $y = \{\text{expression in } x\}$** |

Solve for y: $2y + 3x - 5 = 4$

Solution   We replace the original equation by a succession of equivalent equations.

$$2y + 3x - 5 = 4$$

$$2y + 3x - 5 + 5 = 4 + 5 \qquad \text{Add 5 to both sides.}$$

$$2y + 3x = 9 \qquad \text{Simplify.}$$

$$2y + 3x - 3x = 9 - 3x \qquad \text{Subtract } 3x \text{ from both sides.}$$

$$2y = 9 - 3x \qquad \text{Simplify.}$$

$$\frac{2y}{2} = \frac{9 - 3x}{2} \qquad \text{Divide both sides by 2.}$$

$$y = \frac{9 - 3x}{2} \qquad \text{Simplify.} \qquad ■$$

Now we are ready to graph equations using a graphing utility. Most graphing utilities require the following steps:

STEPS FOR GRAPHING AN EQUATION USING A GRAPHING UTILITY

STEP 1: Solve the equation for y in terms of x.

STEP 2: Get into the graphing mode of your graphing utility. The screen will usually display $y =$, prompting you to enter the expression involving x that you found in Step 1. (Consult your manual for the correct way to enter the expression; for example, $y = x^2$ might be entered as $x^\wedge 2$ or as $x*x$ or as $x \; x^Y \; 2$).

STEP 3: Select the viewing window. Without prior knowledge about the behavior of the graph of the equation, it is common to select the **standard viewing window*** initially. The viewing window is then adjusted based on the graph that appears. In this text the standard viewing window will be

$$X\text{min} = -10 \qquad Y\text{min} = -10$$
$$X\text{max} = 10 \qquad Y\text{max} = 10$$
$$X\text{scl} = 1 \qquad Y\text{scl} = 1$$

STEP 4: Execute.

STEP 5: Adjust the viewing window until a complete graph is obtained.

EXAMPLE 2 | Graphing an Equation on a Graphing Utility

Graph the equation: $6x^2 + 3y = 36$

Solution **STEP 1:** We solve for y in terms of x.

$$6x^2 + 3y = 36$$
$$3y = -6x^2 + 36 \qquad \text{Subtract } 6x^2 \text{ from both sides of the equation.}$$
$$y = -2x^2 + 12 \qquad \text{Divide both sides of the equation by 3 and simplify.}$$

STEP 2: From the graphing mode, enter the expression $-2x^2 + 12$ after the prompt $y =$.

STEP 3: Set the viewing window to the standard viewing window.

STEP 4: Execute. The screen should look like Figure 6.

STEP 5: The graph of $y = -2x^2 + 12$ is not complete. The value of Ymax must be increased so that the top portion of the graph is visible. After increasing the value of Ymax to 12, we obtain the graph in Figure 7. The graph is now complete.

Figure 6

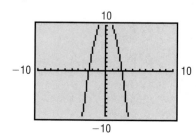

Figure 7

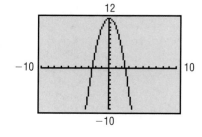

* Some graphing utilities have a ZOOM-STANDARD feature that automatically sets the viewing window to the standard viewing window and graphs the equation.

Look again at Figure 7. Although a complete graph is shown, the graph might be improved by adjusting the values of Xmin and Xmax. Figure 8 shows the graph of $y = -2x^2 + 12$ using Xmin $= -4$ and Xmax $= 4$. Do you think this is a better choice for the viewing window?

Figure 8

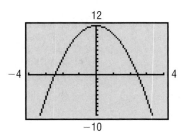

| EXAMPLE 3 | Creating a Table and Graphing an Equation |

Create a table and graph the equation: $y = x^3$

Solution   Most graphing utilities have the capability of creating a table of values for an equation. (Check your manual to see if your graphing utility has this capability.) Table 1 illustrates a table of values for $y = x^3$ on a TI-83. See Figure 9 for the graph.

Figure 9

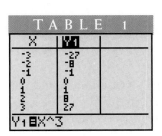

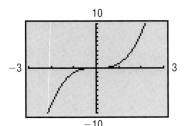

B.2 EXERCISES

In Problems 1–16, graph each equation using the following viewing windows:

(a) Xmin $= -5$	(b) Xmin $= -10$	(c) Xmin $= -10$	(d) Xmin $= -5$
Xmax $= 5$	Xmax $= 10$	Xmax $= 10$	Xmax $= 5$
Xscl $= 1$	Xscl $= 1$	Xscl $= 2$	Xscl $= 1$
Ymin $= -4$	Ymin $= -8$	Ymin $= -8$	Ymin $= -20$
Ymax $= 4$	Ymax $= 8$	Ymax $= 8$	Ymax $= 20$
Yscl $= 1$	Yscl $= 1$	Yscl $= 2$	Yscl $= 5$

1. $y = x + 2$ **2.** $y = x - 2$ **3.** $y = -x + 2$ **4.** $y = -x - 2$

5. $y = 2x + 2$ **6.** $y = 2x - 2$ **7.** $y = -2x + 2$ **8.** $y = -2x - 2$

9. $y = x^2 + 2$ **10.** $y = x^2 - 2$ **11.** $y = -x^2 + 2$ **12.** $y = -x^2 - 2$

13. $3x + 2y = 6$ **14.** $3x - 2y = 6$ **15.** $-3x + 2y = 6$ **16.** $-3x - 2y = 6$

17.–32. *For each of the above equations, create a table, $-3 \le x \le 3$, and list points on the graph.*

B.3 USING A GRAPHING UTILITY TO LOCATE INTERCEPTS AND CHECK FOR SYMMETRY

VALUE AND ZERO (OR ROOT)

Most graphing utilities have an eVALUEate feature that, given a value of x, determines the value of y for an equation. We can use this feature to evaluate an equation at $x = 0$ to determine the y-intercept. Most graphing utilities also have a ZERO (or ROOT) feature that can be used to determine the x-intercept(s) of an equation.

EXAMPLE 1 | **Finding Intercepts Using a Graphing Utility**

Use a graphing utility to find the intercepts of the equation $y = x^3 - 8$.

Solution Figure 10(a) shows the graph of $y = x^3 - 8$.

Figure 10

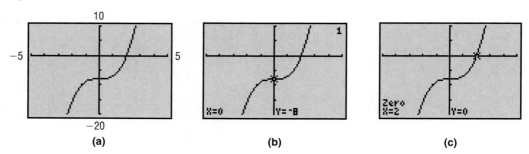

(a)　　　　　　　(b)　　　　　　　(c)

The eVALUEate feature of a TI-83 graphing calculator accepts as input a value of x and determines the value of y. If we let $x = 0$, we find that the y-intercept is -8. See Figure 10(b).

The ZERO feature of a TI-83 is used to find the x-intercept(s). See Figure 10(c). The x-intercept is 2. ∎

TRACE

Most graphing utilities allow you to move from point to point along the graph, displaying on the screen the coordinates of each point. This feature is called TRACE.

EXAMPLE 2 | **Using TRACE to Locate Intercepts**

Graph the equation $y = x^3 - 8$. Use TRACE to locate the intercepts.

Solution Figure 11 shows the graph of $y = x^3 - 8$.

Figure 11

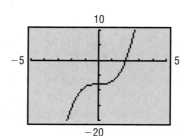

Activate the TRACE feature. As you move the cursor along the graph, you will see the coordinates of each point displayed. When the cursor is on the y-axis, we find that the y-intercepts is −8. See Figure 12.

Figure 12

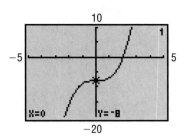

Continue moving the cursor along the graph. Just before you get to the x-axis, the display will look like the one in Figure 13(a). (Due to differences in graphing utilities, your display may be slightly different from the one shown here.)

Figure 13

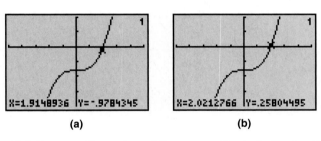

(a)   (b)

In Figure 13(a), the negative value of the y-coordinate indicates that we are still below the x-axis. The next position of the cursor is shown in Figure 13(b). The positive value of the y-coordinate indicates that we are now above the x-axis. This means that between these two points the x-axis was crossed. The x-intercept lies between 1.9148936 and 2.0212766.   ■

EXAMPLE 3

Graphing the Equation $y = \dfrac{1}{x}$

Figure 14

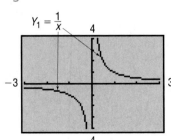

Graph the equation: $y = \dfrac{1}{x}$

With the viewing window set as

$$X\min = -3 \qquad Y\min = -4$$
$$X\max = 3 \qquad Y\max = 4$$
$$X\text{scl} = 1 \qquad Y\text{scl} = 1$$

Use TRACE to infer information about the intercepts and symmetry.

Solution   Figure 14 illustrates the graph. We infer from the graph that there are no intercepts; we may also infer that symmetry with respect to the origin is a possibility. The TRACE feature on a graphing utility can provide further evidence of symmetry with respect to the origin. Using TRACE, we observe that for any ordered pair (x, y) the ordered pair $(-x, -y)$ is also a point on the graph. For example, the points $(0.95744681, 1.0444444)$ and $(-0.95744681, -1.0444444)$ both lie on the graph.   ■

B.3   EXERCISES

In Problems 1–6, use ZERO (or ROOT) to approximate the smaller of the two x-intercepts of each equation. Express the answer rounded to two decimal places.

1. $y = x^2 + 4x + 2$ **2.** $y = x^2 + 4x - 3$ **3.** $y = 2x^2 + 4x + 1$

4. $y = 3x^2 + 5x + 1$ **5.** $y = 2x^2 - 3x - 1$ **6.** $y = 2x^2 - 4x - 1$

*In Problems 7–14, use ZERO (or ROOT) to approximate the **positive** x-intercepts of each equation. Express each answer rounded to two decimal places.*

7. $y = x^3 + 3.2x^2 - 16.83x - 5.31$ **8.** $y = x^3 + 3.2x^2 - 7.25x - 6.3$

9. $y = x^4 - 1.4x^3 - 33.71x^2 + 23.94x + 292.41$ **10.** $y = x^4 + 1.2x^3 - 7.46x^2 - 4.692x + 15.2881$

11. $y = \pi x^3 - (8.88\pi + 1)x^2 - (42.066\pi - 8.88)x + 42.066$

12. $y = \pi x^3 - (5.63\pi + 2)x^2 - (108.392\pi - 11.26)x + 216.784$

13. $y = x^3 + 19.5x^2 - 1021x + 1000.5$

14. $y = x^3 + 14.2x^2 - 4.8x - 12.4$

In Problems 15–18, the graph of an equation is given.
  (a) *List the intercepts of the graph.*
  (b) *Based on the graph, tell whether the graph is symmetric with respect to the x-axis, y-axis, and/or origin.*

15. 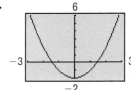   **16.** 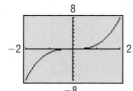   **17.** 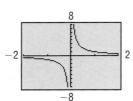   **18.**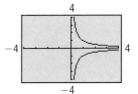

B.4   USING A GRAPHING UTILITY TO SOLVE EQUATIONS

For many equations, there are no algebraic techniques that lead to a solution. For such equations, a graphing utility can often be used to investigate possible solutions. When a graphing utility is used to solve an equation, usually *approximate* solutions are obtained. Unless otherwise stated, we shall follow the practice of giving approximate solutions *rounded to two decimal places*.

The ZERO (or ROOT) feature of a graphing utility can be used to find the solutions of an equation when one side of the equation is 0. In using this feature to solve equations, we make use of the fact that the x-intercepts (or zeros) of the graph of an equation are found by letting $y = 0$ and solving the equation for x. Solving an equation for x when one side of the equation is 0 is equivalent to finding where the graph of the corresponding equation crosses or touches the x-axis.

EXAMPLE 1

Using ZERO (or ROOT) to Approximate Solutions of an Equation

Find the solution(s) of the equation $x^2 - 6x + 7 = 0$. Round answers to two decimal places.

Solution   The solutions of the equation $x^2 - 6x + 7 = 0$ are the same as the x-intercepts of the graph of $Y_1 = x^2 - 6x + 7$. We begin by graphing the equation. See Figure 15(a).

Figure 15

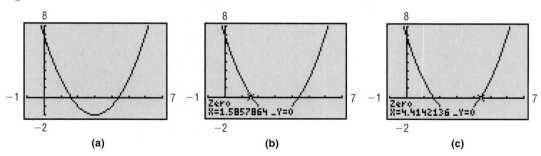

(a)   (b)   (c)

From the graph there appear to be two x-intercepts (solutions to the equation): one between 1 and 2, the other between 4 and 5.

Using the ZERO (or ROOT) feature of our graphing utility, we determine that the x-intercepts, and so the solutions to the equation, are $x = 1.59$ and $x = 4.41$, rounded to two decimal places. See Figures 15(b) and (c). ■

A second method for solving equations using a graphing utility involves the INTERSECT feature of the graphing utility. This feature is used most effectively when one side of the equation is not 0.

EXAMPLE 2

Using INTERSECT to Approximate Solutions of an Equation

Find the solution(s) to the equation $3(x - 2) = 5(x - 1)$. Round answers to two decimal places.

Solution   We begin by graphing each side of the equation as follows: graph $Y_1 = 3(x - 2)$ and $Y_2 = 5(x - 1)$. See Figure 16(a).

Figure 16

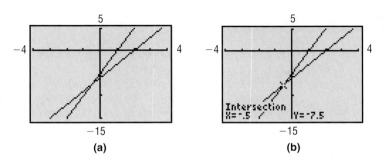

(a)   (b)

At the point of intersection of the graphs, the value of the y-coordinate is the same. We conclude that the x-coordinate of the point of intersection represents the solution to the equation. Do you see why? The INTERSECT feature on a graphing utility determines the point of intersection of the graphs. Using this feature, we find that the graphs intersect at $(-0.5, -7.5)$. See Figure 16(b). The solution of the equation is therefore $x = -0.5$. ■

Check: We can verify our solution by evaluating each side of the equation with −0.5 STOred in x. See Figure 17. Since the left side of the equation equals the right side of the equation, the solution checks.

Figure 17

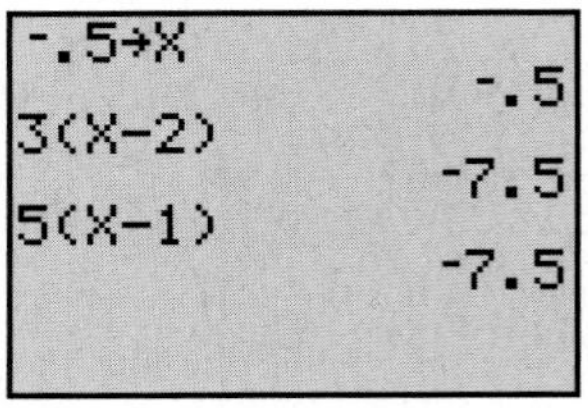

```
-.5→X
           -.5
3(X-2)
           -7.5
5(X-1)
           -7.5
```

SUMMARY

The steps to follow for approximating solutions of equations are given next.

STEPS FOR APPROXIMATING SOLUTIONS OF EQUATIONS USING ZERO (OR ROOT)

STEP 1: Write the equation in the form {expression in x} = 0.

STEP 2: Graph Y_1 = {expression in x}.

STEP 3: Use ZERO (or ROOT) to determine each x-intercept of the graph.

STEPS FOR APPROXIMATING SOLUTIONS OF EQUATIONS USING INTERSECT

STEP 1: Graph Y_1 = {expression in x on left side of equation}.

Graph Y_2 = {expression in x on right side of equation}.

STEP 2: Use INTERSECT to determine each x-coordinate of the point(s) of intersection, if any.

EXAMPLE 3   Solving a Radical Equation

Find the real solutions of the equation $\sqrt[3]{2x - 4} - 2 = 0$.

Solution Figure 18 shows the graph of the equation $Y_1 = \sqrt[3]{2x - 4} - 2$. From the graph, we see one x-intercept near 6. Using ZERO (or ROOT), we find that the x-intercept is 6. The only solution is $x = 6$.

Figure 18

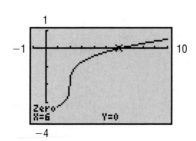

B.5 SQUARE SCREENS

Figure 19

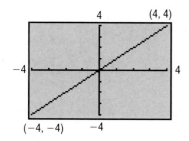

Most graphing utilities have a rectangular screen. Because of this, using the same settings for both x and y will result in a distorted view. For example, Figure 19 shows the graph of the line $y = x$ connecting the points $(-4, -4)$ and $(4, 4)$.

We expect the line to bisect the first and third quadrants, but it doesn't. We need to adjust the selections for Xmin, Xmax, Ymin, and Ymax so that a **square screen** results. On most graphing utilities, this is accomplished by setting the ratio of x to y at $3:2$.* In other words,

$$2(X\text{max} - X\text{min}) = 3(Y\text{max} - Y\text{min})$$

EXAMPLE 1

Examples of Viewing Rectangles That Result in Square Screens

Figure 20

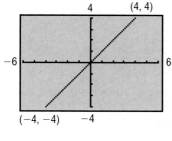

(a) Xmin $= -3$
Xmax $= 3$
Xscl $= 1$
Ymin $= -2$
Ymax $= 2$
Yscl $= 1$

(b) Xmin $= -6$
Xmax $= 6$
Xscl $= 1$
Ymin $= -4$
Ymax $= 4$
Yscl $= 1$

(c) Xmin $= -6$
Xmax $= 6$
Xscl $= 2$
Ymin $= -4$
Ymax $= 4$
Yscl $= 1$ ■

Figure 20 shows the graph of the line $y = x$ on a square screen using the viewing rectangle given in Example 1(b). Notice that the line now bisects the first and third quadrants. Compare this illustration to Figure 19.

*Some graphing utilities have a built-in function that automatically squares the screen. For example, the TI-85 has a ZSQR function that does this. Some graphing utilities require a ratio other than $3:2$ to square the screen. For example, the HP 48G requires the ratio of x to y to be $2:1$ for a square screen. Consult your manual.

B.5 EXERCISES

In Problems 1–8, determine which of the given viewing rectangles result in a square screen.

1. Xmin $= -3$
Xmax $= 3$
Xscl $= 2$
Ymin $= -2$
Ymax $= 2$
Yscl $= 2$

2. Xmin $= -5$
Xmax $= 5$
Xscl $= 1$
Ymin $= -4$
Ymax $= 4$
Yscl $= 1$

3. Xmin $= 0$
Xmax $= 9$
Xscl $= 3$
Ymin $= -2$
Ymax $= 4$
Yscl $= 2$

4. Xmin $= -6$
Xmax $= 6$
Xscl $= 1$
Ymin $= -4$
Ymax $= 4$
Yscl $= 2$

5. Xmin $= -6$
Xmax $= 6$
Xscl $= 1$
Ymin $= -2$
Ymax $= 2$
Yscl $= 0.5$

6. Xmin $= -6$
Xmax $= 6$
Xscl $= 2$
Ymin $= -4$
Ymax $= 4$
Yscl $= 1$

7. Xmin $= 0$
Xmax $= 9$
Xscl $= 1$
Ymin $= -2$
Ymax $= 4$
Yscl $= 1$

8. Xmin $= -6$
Xmax $= 6$
Xscl $= 2$
Ymin $= -4$
Ymax $= 4$
Yscl $= 2$

9. If Xmin $= -4$, Xmax $= 8$, and Xscl $= 1$, how should Ymin, Ymax, and Yscl be selected so that the viewing rectangle contains the point $(4, 8)$ and the screen is square?

10. If Xmin $= -6$, Xmax $= 12$, and Xscl $= 2$, how should Ymin, Ymax, and Yscl be selected so that the viewing rectangle contains the point $(4, 8)$ and the screen is square?

B.6   USING A GRAPHING UTILITY TO GRAPH A POLAR EQUATION

Most graphing utilities require the following steps in order to obtain the graph of a polar equation. Be sure to be in POLar mode.

> **GRAPHING A POLAR EQUATION USING A GRAPHING UTILITY**
>
> **STEP 1:** Set the mode to POLar. Solve the equation for r in terms of θ.
>
> **STEP 2:** Select the viewing rectangle in polar mode. Besides setting Xmin, Xmax, Xscl, and so forth, the viewing rectangle in polar mode requires setting the minimum and maximum values for θ and an increment setting for θ (θstep). In addition, a square screen and radian measure should be used.
>
> **STEP 3:** Enter the expression involving θ that you found in Step 1. (Consult your manual for the correct way to enter the expression.)
>
> **STEP 4:** Execute.

EXAMPLE 1   **Graphing a Polar Equation Using a Graphing Utility**

Use a graphing utility to graph the polar equation $r \sin \theta = 2$.

Solution   **STEP 1:** We solve the equation for r in terms of θ.

$$r \sin \theta = 2$$

$$r = \frac{2}{\sin \theta}$$

STEP 2: From the POLar mode, select the viewing rectangle. We will use the one given next.

θmin $= 0$	Xmin $= -9$	Ymin $= -6$
θmax $= 2\pi$	Xmax $= 9$	Ymax $= 6$
θstep $= \pi/24$	Xscl $= 1$	Yscl $= 1$

θstep determines the number of points that the graphing utility will plot. For example, if θstep is $\dfrac{\pi}{24}$, then the graphing utility will evalu-

Figure 21

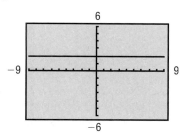

ate r at $\theta = 0(\theta\min), \dfrac{\pi}{24}, \dfrac{2\pi}{24}, \dfrac{3\pi}{24}$, and so forth, up to $2\pi(\theta\max)$. The smaller θstep is, the more points that the graphing utility will plot. The student is encouraged to experiment with different values for $\theta\min$, $\theta\max$, and θstep to see how the graph is affected.

STEP 3: Enter the expression $\dfrac{2}{\sin\theta}$ after the prompt $r_1 = \quad$.

STEP 4: Execute.

The graph is shown in Figure 15. ∎

B.7 USING A GRAPHING UTILITY TO GRAPH PARAMETRIC EQUATIONS

Most graphing utilities have the capability of graphing parametric equations. The following steps are usually required in order to obtain the graph of parametric equations. Check your owner's manual to see how yours works.

> **GRAPHING PARAMETRIC EQUATIONS USING A GRAPHING UTILITY**
>
> STEP 1: Set the mode to PARametric. Enter $x(t)$ and $y(t)$.
> STEP 2: Select the viewing window. In addition to setting $X\min$, $X\max$, Xscl, and so on, the viewing window in parametric mode requires setting minimum and maximum values for the parameter t and an increment setting for t (Tstep).
> STEP 3: Execute.

EXAMPLE 1   Graphing a Curve Defined by Parametric Equations Using a Graphing Utility

Graph the curve defined by the parametric equations

$$x = 3t^2, \qquad y = 2t, \qquad -2 \le t \le 2$$

Solution   STEP 1: Enter the equations $x(t) = 3t^2$, $y(t) = 2t$ with the graphing utility in PARametric mode.

STEP 2: Select the viewing window. The interval is $-2 \le t \le 2$, so we select the following square viewing window:

$$
\begin{array}{lll}
T\min = -2 & X\min = 0 & Y\min = -5 \\
T\max = 2 & X\max = 15 & Y\max = 5 \\
T\text{step} = 0.1 & X\text{scl} = 1 & Y\text{scl} = 1
\end{array}
$$

We choose $T\min = -2$ and $T\max = 2$ because $-2 \le t \le 2$. Finally, the choice for Tstep will determine the number of points that the graphing utility will plot. For example, with Tstep at 0.1, the graphing utility will evaluate x and y at $t = -2, -1.9, -1.8$, and so on. The smaller the Tstep, the more points the graphing utility will plot. The reader is encouraged to experiment with different values of Tstep to see how the graph is affected.

Figure 22

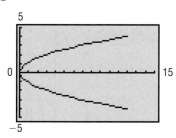

STEP 3: Execute. Notice the direction in which the graph is drawn. This direction shows the orientation of the curve.

The graph shown in Figure 22 is complete. ∎

EXPLORATION   Graph the following parametric equations using a graphing utility with Xmin $= 0$, Xmax $= 15$, Ymin $= -5$, Ymax $= 5$, and Tstep $= 0.1$.

1. $x = \dfrac{3t^2}{4}$, $y = t$, $-4 \le t \le 4$

2. $x = 3t^2 + 12t + 12$, $y = 2t + 4$, $-4 \le t \le 0$

3. $x = 3t^{2/3}$, $y = 2\sqrt[3]{t}$, $-8 \le t \le 8$

Compare these graphs to the graph in Figure 26. Conclude that parametric equations defining a curve are not unique; that is, different parametric equations can represent the same graph.   ▬

EXPLORATION   In FUNCtion mode, graph $x = \dfrac{3y^2}{4}$ $\left(Y_1 = \sqrt{\dfrac{4x}{3}} \text{ and } Y_2 = -\sqrt{\dfrac{4x}{3}} \right)$ with Xmin $= 0$, Xmax $= 15$, Ymin $= -5$, Ymax $= 5$. Compare this graph with Figure 26. Why do the graphs differ?   ▬

CHAPTER 1   Functions and Their Graphs

1.1 Exercises *(page 7)*

1. (a) Quadrant II   **(b)** Positive *x*-axis   **(c)** Quadrant III
(d) Quadrant I   **(e)** Negative *y*-axis   **(f)** Quadrant IV

3. The points will be on a vertical line that
is 2 units to the right of the *y*-axis

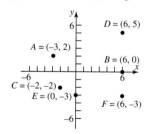

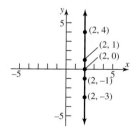

5. $\sqrt{5}$ **7.** $\sqrt{10}$ **9.** $2\sqrt{17}$ **11.** $\sqrt{85}$ **13.** $\sqrt{53}$ **15.** $\sqrt{6.89} \approx 2.625$ **17.** $\sqrt{a^2 + b^2}$

19. $d(A, B) = \sqrt{13}$
$d(B, C) = \sqrt{13}$
$d(A, C) = \sqrt{26}$
$(\sqrt{13})^2 + (\sqrt{13})^2 = (\sqrt{26})^2$
Area $= \dfrac{13}{2}$ square units

21. $d(A, B) = \sqrt{130}$
$d(B, C) = \sqrt{26}$
$d(A, C) = 2\sqrt{26}$
$(\sqrt{26})^2 + (2\sqrt{26})^2 = (\sqrt{130})^2$
Area $= 26$ square units

23. $d(A, B) = 4$
$d(B, C) = \sqrt{41}$
$d(A, C) = 5$
$4^2 + 5^2 = (\sqrt{41})^2$
Area $= 10$ square units

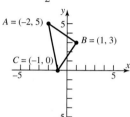

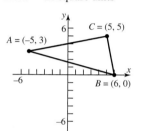

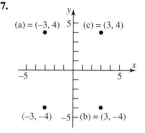

25. $(2, 2); (2, -4)$ **27.** $(0, 0); (8, 0)$ **29.** $(4, -1)$ **31.** $\left(\dfrac{3}{2}, 1\right)$ **33.** $(5, -1)$ **35.** $(1.05, 0.7)$ **37.** $\left(\dfrac{a}{2}, \dfrac{b}{2}\right)$ **39.** $\sqrt{17}; 2\sqrt{5}; \sqrt{29}$

41. $d(P_1, P_2) = 6; d(P_2, P_3) = 4; d(P_1, P_3) = 2\sqrt{13}$; right triangle

43. $d(P_1, P_2) = 2\sqrt{17}; d(P_2, P_3) = \sqrt{34}; d(P_1, P_3) = \sqrt{34}$; isosceles right triangle   **45.** $4\sqrt{10}$ **47.** $2\sqrt{65}$ **49.** $\left(\dfrac{s}{2}, \dfrac{s}{2}\right)$

51. $90\sqrt{2} \approx 127.28$ ft   **53. (a)** $(90, 0), (90, 90), (0, 90)$ **(b)** $5\sqrt{2161}$ ft or ≈ 232.4 ft   **(c)** $30\sqrt{149}$ ft or ≈ 366.2 ft   **55.** $d = 50t$

1.2 Exercises *(page 18)*

1.

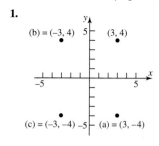

3.

5.

7.

9.

11. (a) $(-1, 0), (1, 0)$ **(b)** *x*-axis, *y*-axis, origin   **13. (a)** $\left(-\dfrac{\pi}{2}, 0\right), (0, 1), \left(\dfrac{\pi}{2}, 0\right)$ **(b)** *y*-axis

15. (a) $(0, 0)$ **(b)** *x*-axis   **17. (a)** $(1, 0)$ **(b)** none   **19. (a)** $(-1.5, 0), (0, -2), (1.5, 0)$ **(b)** *y*-axis

21. (a) none   **(b)** origin   **23.** $(0, 0)$ is on the graph.   **25.** $(0, 3)$ is on the graph.

27. $(0, 2)$ and $(\sqrt{2}, \sqrt{2})$ are on the graph.   **29.** $(0, 0)$; symmetric with respect to the *y*-axis

31. $(0, 0)$; symmetric with respect to the origin

33. $(0, 9), (3, 0), (-3, 0)$; symmetric with respect to the *y*-axis

35. $(-2, 0), (2, 0), (0, -3), (0, 3)$; symmetric with respect to the *x*-axis, *y*-axis, and origin

37. $(0, -27), (3, 0)$; no symmetry   **39.** $(0, -4), (4, 0), (-1, 0)$; no symmetry

41. $(0, 0)$; symmetric with respect to the origin   **43.** $(0, 0)$; symmetric with respect to the origin

45.
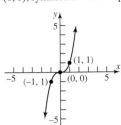

47.
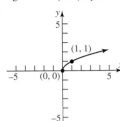

49. $a = -1$
51. $2a + 3b = 6$
53. Center $(2, 1)$; Radius 2; $(x - 2)^2 + (y - 1)^2 = 4$
55. Center $\left(\dfrac{5}{2}, 2\right)$; Radius $\dfrac{3}{2}$; $\left(x - \dfrac{5}{2}\right)^2 + (y - 2)^2 = \dfrac{9}{4}$

57. $x^2 + y^2 = 4$;
$x^2 + y^2 - 4 = 0$
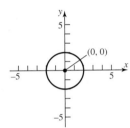

59. $(x - 1)^2 + (y + 1)^2 = 1$;
$x^2 + y^2 - 2x + 2y + 1 = 0$
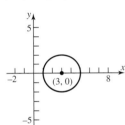

61. $x^2 + (y - 2)^2 = 4$;
$x^2 + y^2 - 4y = 0$
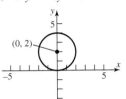

63. $(x - 4)^2 + (y + 3)^2 = 25$;
$x^2 + y^2 - 8x + 6y = 0$
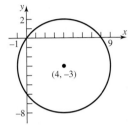

65. $(h, k) = (0, 0)$; $r = 2$
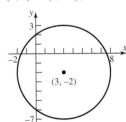

67. $(h, k) = (3, 0)$; $r = 2$

69. $(h, k) = (-2, 2)$; $r = 3$
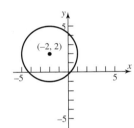

71. $(h, k) = \left(\dfrac{1}{2}, -1\right)$; $r = \dfrac{1}{2}$
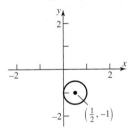

73. $(h, k) = (3, -2)$; $r = 5$
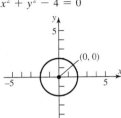

75. $x^2 + y^2 - 13 = 0$
77. $x^2 + y^2 - 4x - 6y + 4 = 0$
79. $x^2 + y^2 + 2x - 6y + 5 = 0$
81. c
83. b
85. b, c, e, g
87. $x^2 + y^2 + 2x + 4y - 4168.16 = 0$

1.3 Exercises *(page 33)*

1. Function; Domain: {Dad, Colleen, Kaleigh, Marissa}, Range: {January 8, March 15, September 17}   **3.** Not a function   **5.** Not a function
7. Function; Domain: $\{1, 2, 3, 4\}$; Range: $\{3\}$ **9.** Not a function   **11.** Function; Domain: $\{-2, -1, 0, 1\}$, Range: $\{0, 1, 4\}$
13. **(a)** -4 **(b)** 1 **(c)** -3 **(d)** $3x^2 - 2x - 4$ **(e)** $-3x^2 - 2x + 4$ **(f)** $3x^2 + 8x + 1$ **(g)** $12x^2 + 4x - 4$
(h) $3x^2 + 6xh + 3h^2 + 2x + 2h - 4$ **15.** **(a)** 0 **(b)** $\dfrac{1}{2}$ **(c)** $-\dfrac{1}{2}$ **(d)** $\dfrac{-x}{x^2 + 1}$ **(e)** $\dfrac{-x}{x^2 + 1}$ **(f)** $\dfrac{x + 1}{x^2 + 2x + 2}$ **(g)** $\dfrac{2x}{4x^2 + 1}$
(h) $\dfrac{x + h}{x^2 + 2xh + h^2 + 1}$ **17.** **(a)** 4 **(b)** 5 **(c)** 5 **(d)** $|x| + 4$ **(e)** $-|x| - 4$ **(f)** $|x + 1| + 4$ **(g)** $2|x| + 4$ **(h)** $|x + h| + 4$
19. **(a)** $-\dfrac{1}{5}$ **(b)** $-\dfrac{3}{2}$ **(c)** $\dfrac{1}{8}$ **(d)** $\dfrac{-2x + 1}{-3x - 5}$ **(e)** $\dfrac{-2x - 1}{3x - 5}$ **(f)** $\dfrac{2x + 3}{3x - 2}$ **(g)** $\dfrac{4x + 1}{6x - 5}$ **(h)** $\dfrac{2x + 2h + 1}{3x + 3h - 5}$ **21.** Function   **23.** Function
25. Not a function   **27.** Not a function   **29.** Function   **31.** Not a function   **33.** All real numbers   **35.** All real numbers
37. $\{x | x \neq -4, x \neq 4\}$ **39.** $\{x | x \neq 0\}$ **41.** $\{x | x \geq 4\}$ **43.** $\{x | x > 9\}$ **45.** $\{x | x > 1\}$ **47.** **(a)** $f(0) = 3$; $f(-6) = -3$
(b) $f(6) = 0$; $f(11) = 1$ **(c)** Positive   **(d)** Negative   **(e)** $-3, 6$, and 10 **(f)** $-3 < x < 6$; $10 < x \leq 11$ **(g)** $\{x | -6 \leq x \leq 11\}$
(h) $\{y | -3 \leq y \leq 4\}$ **(i)** $-3, 6, 10$ **(j)** 3 **(k)** 3 times   **(l)** once   **(m)** $0, 4$ **(n)** $-5, 8$ **49.** Not a function

51. Function **(a)** Domain: $\{x|-\pi \le x \le \pi\}$; Range: $\{y|-1 \le y \le 1\}$ **(b)** $\left(-\dfrac{\pi}{2}, 0\right), \left(\dfrac{\pi}{2}, 0\right), (0, 1)$ **(c)** y-axis **53.** Not a function

55. Function **(a)** Domain: $\{x|x > 0\}$; Range: all real numbers **(b)** $(1, 0)$ **(c)** None

57. Function **(a)** Domain: all real numbers; Range: $\{y|y \le 2\}$ **(b)** $(-3, 0), (3, 0), (0, 2)$ **(c)** y-axis

59. Function **(a)** Domain: all real numbers; Range: $\{y|y \ge -3\}$ **(b)** $(1, 0), (3, 0), (0, 9)$ **(c)** None

61. (a) Yes **(b)** $f(-2) = 9$; $(-2, 9)$ **(c)** $0, \dfrac{1}{2}$; $(0, -1), \left(\dfrac{1}{2}, -1\right)$ **(d)** All real numbers **(e)** $-\dfrac{1}{2}, 1$ **(f)** -1

63. (a) No **(b)** $f(4) = -3$; $(4, -3)$ **(c)** 14; $(14, 2)$ **(d)** $\{x|x \ne 6\}$ **(e)** -2 **(f)** $-\dfrac{1}{3}$ **65. (a)** Yes **(b)** $f(2) = \dfrac{8}{17}$; $\left(2, \dfrac{8}{17}\right)$

(c) $-1, 1$; $(-1, 1), (1, 1)$ **(d)** All real numbers **(e)** 0 **(f)** 0 **67.** $A = -\dfrac{7}{2}$ **69.** $A = -4$ **71.** $A = 8$; undefined at $x = 3$

73. (a) III **(b)** IV **(c)** I **(d)** V **(e)** II

75.

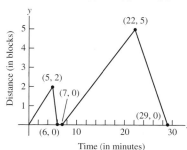

77. (a) 2 hr ellapsed during which Kevin was between 0 and 3 mi from home.
(b) 0.5 hr ellapsed during which Kevin was 3 mi from home.
(c) 0.3 hr ellapsed during which Kevin was between 0 and 3 mi from home.
(d) 0.2 hr ellapsed during which Kevin was 0 mi from home.
(e) 0.9 hr ellapsed during which Kevin was between 0 and 2.8 mi from home.
(f) 0.3 hr ellapsed during which Kevin was 2.8 mi from home.
(g) 1.1 hr ellapsed during which Kevin was between 0 and 2.8 mi from home.
(h) 3 mi **(i)** 2 times

79. (a) 15.1 m, 14.07 m, 12.94 m, 11.72 m **(b)** 1.01 sec, 1.43 sec, 1.75 sec **(c)** 2.02 sec
81. (a) 81.1 ft **(b)** 129.6 ft **(c)** 26.6 ft **(d)** 528.125 ft

83. (a) $222 **(b)** $225 **(c)** $220 **(d)** $230

85. $A(x) = \dfrac{1}{2}x^2$

(e)

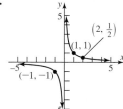

(f) The cost varies from $220–$230.

87. $G(x) = 10x$
89. Only $h(x) = 2x$
91. No; f has a domain of all real numbers, while g has a domain of $\{x|x \ne -1\}$.

1.4 Exercises *(page 44)*

1. C **3.** E **5.** B **7.** F

9.
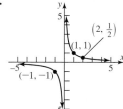

11.

13.

15. Yes
17. No
19. $(-8, -2)$; $(0, 2)$; $(5, \infty)$
21. Yes; 10
23. $-2, 2$; 6, 10

25. (a) $(-2, 0), (0, 3), (2, 0)$ **(b)** Domain: $\{x|-4 \le x \le 4\}$ or $[-4, 4]$; Range: $\{y|0 \le y \le 3\}$ or $[0, 3]$
(c) Increasing on $(-2, 0)$ and $(2, 4)$; Decreasing on $(-4, -2)$ and $(0, 2)$ **(d)** Even **27. (a)** $(0, 1)$ **(b)** Domain: all real numbers;
Range: $\{y|y > 0\}$ or $(0, \infty)$. **(c)** Increasing on $(-\infty, \infty)$ **(d)** Neither **29. (a)** $(-\pi, 0), (0, 0), (\pi, 0)$ **(b)** Domain: $\{x|-\pi \le x \le \pi\}$ or

$[-\pi, \pi]$; Range: $\{y|-1 \le y \le 1\}$ or $[-1, 1]$ **(c)** Increasing on $\left(-\dfrac{\pi}{2}, \dfrac{\pi}{2}\right)$; Decreasing on $\left(-\pi, -\dfrac{\pi}{2}\right)$ and $\left(\dfrac{\pi}{2}, \pi\right)$ **(d)** Odd

31. (a) $\left(0, \dfrac{1}{2}\right), \left(\dfrac{1}{2}, 0\right), \left(\dfrac{5}{2}, 0\right)$ **(b)** Domain: $\{x|-3 \le x \le 3\}$ or $[-3, 3]$; Range: $\{y|-1 \le y \le 2\}$ or $[-1, 2]$

(c) Increasing on $(2, 3)$; Decreasing on $(-1, 1)$; Constant on $(-3, -1)$ and $(1, 2)$ **(d)** Neither **33. (a)** $0; 3$ **(b)** $-2, 2; 0, 0$

35. (a) $\dfrac{\pi}{2}; 1$ **(b)** $-\dfrac{\pi}{2}; -1$ **37.** Odd **39.** Even **41.** Odd **43.** Neither **45.** Even **47.** Odd **49.** Each graph is that of $y = x^2$,

but shifted vertically. If $y = x^2 + k, k > 0$, the shift is up k units; if $y = x^2 + k, k < 0$, the shift is down $|k|$ units.
51. Each graph is that of $y = |x|$, but either compressed or stretched. If $y = k|x|$ and $k > 1$, the graph is stretched vertically; if $y = k|x|$,
$0 < k < 1$, the graph is compressed vertically. **53.** The graph of $y = f(-x)$ is the reflection about the y-axis of the graph of $y = f(x)$.

55. They are all ∪-shaped and open upward. All three go through the points $(-1, 1)$, $(0, 0)$ and $(1, 1)$. As the exponent increases, the steepness of the curve increases (except near $x = 0$).

57. Yes; all real numbers; $\{0, 1\}$; $(0, 1)$; points of the form $(x, 0)$, where x is an irrational number; even   **59.** at most one

1.5 Exercises (page 56)

1. B **3.** H **5.** I **7.** L **9.** F **11.** G **13.** $y = (x - 4)^3$ **15.** $y = x^3 + 4$ **17.** $y = -x^3$ **19.** $y = 4x^3$
21. (1) $y = \sqrt{x} + 2$; (2) $y = -(\sqrt{x} + 2)$; (3) $y = -(\sqrt{-x} + 2)$ **23.** (1) $y = -\sqrt{x}$; (2) $y = -\sqrt{x} + 2$; (3) $y = -\sqrt{x + 3} + 2$
25. (c) **27.** (c) **29.**

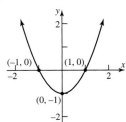

31.

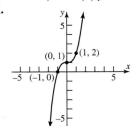

33.

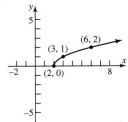

35.

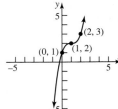

37.

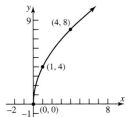

39.

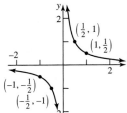

41.

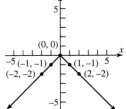

43.

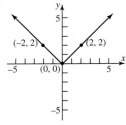

45.

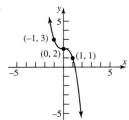

47.

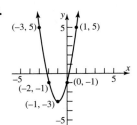

49.

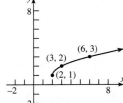

51.

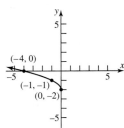

53.

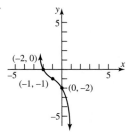

55.

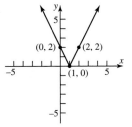

57. (a) $F(x) = f(x) + 3$ **(b)** $G(x) = f(x + 2)$ **(c)** $P(x) = -f(x)$ **(d)** $H(x) = f(x + 1) - 2$

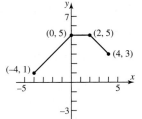

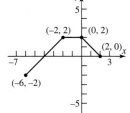

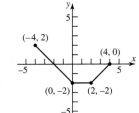

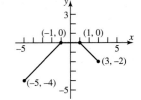

(e) $Q(x) = \frac{1}{2}f(x)$

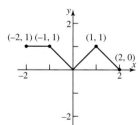

(f) $g(x) = f(-x)$

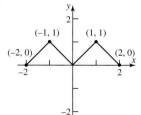

(g) $h(x) = f(2x)$

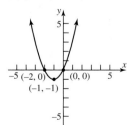

59. (a) $F(x) = f(x) + 3$

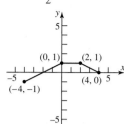

(b) $G(x) = f(x + 2)$

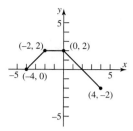

(c) $P(x) = -f(x)$

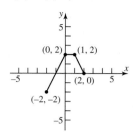

(d) $H(x) = f(x + 1) - 2$

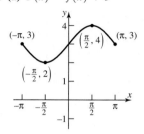

(e) $Q(x) = \frac{1}{2}f(x)$

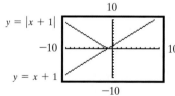

(f) $g(x) = f(-x)$

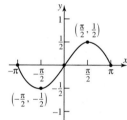

(g) $h(x) = f(2x)$

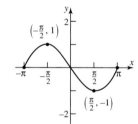

61. (a)

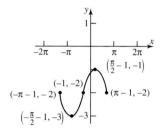

$y = |x + 1|$

$y = x + 1$

(b) $y = |4 - x^2|$

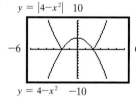

$y = 4 - x^2$

(c) $y = |x^3 + x|$

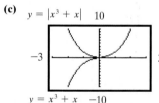

$y = x^3 + x$

(d) Any part of the graph of $y = f(x)$ that lies below the x-axis is reflected about the x-axis to obtain the graph of $y = |f(x)|$.

63. (a)

(b)

65. $f(x) = (x + 1)^2 - 1$

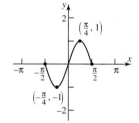

67. $f(x) = (x - 4)^2 - 15$

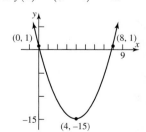

69. $f(x) = \left(x + \dfrac{1}{2}\right)^2 + \dfrac{3}{4}$

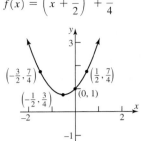

71.

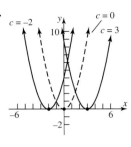

73.

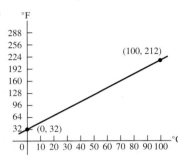

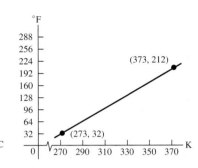

75. (a)

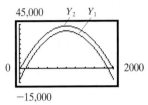

(b) 10% tax

(c) Y_1 is the graph of $p(x)$ shifted down vertically 10,000 units. Y_2 is the graph of $p(x)$ vertically compressed by a factor of 0.9.

(d) 10% tax

1.6 Exercises *(page 69)*

1. (a)

Domain	Range
$200 →	20 hours
$300 →	25 hours
$350 →	30 hours
$425 →	40 hours

(b) Inverse is a function

3. (a)

Domain	Range
$200	20 hours
	25 hours
$350	30 hours
$425	40 hours

(b) Inverse is not a function

5. (a) $\{(6, 2), (6, -3), (9, 4), (10, 1)\}$

(b) Inverse is not a function

7. (a) $\{(0, 0), (1, 1), (16, 2), (81, 3)\}$

(b) Inverse is a function

9. One-to-one   **11.** Not one-to-one   **13.** One-to-one

15.

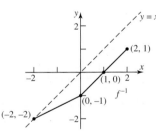

17.

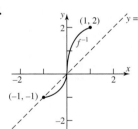

19.

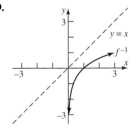

21. $f(g(x)) = f\left(\dfrac{1}{3}(x - 4)\right) = 3\left[\dfrac{1}{3}(x - 4)\right] + 4$
$= (x - 4) + 4 = x;$
$g(f(x)) = g(3x + 4) = \dfrac{1}{3}[(3x + 4) - 4] = \dfrac{1}{3}3x = x$

23. $f(g(x)) = 4\left[\dfrac{x}{4} + 2\right] - 8 = (x + 8) - 8 = x;$
$g(f(x)) = \dfrac{4x - 8}{4} + 2 = (x - 2) + 2 = x$

25. $f(g(x)) = (\sqrt[3]{x + 8})^3 - 8 = (x + 8) - 8 = x;$
$g(f(x)) = \sqrt[3]{(x^3 - 8) + 8} = \sqrt[3]{x^3} = x$

27. $f(g(x)) = \dfrac{1}{\left(\dfrac{1}{x}\right)} = x; g(f(x)) = \dfrac{1}{\left(\dfrac{1}{x}\right)} = x$

29. $f(g(x)) = \dfrac{2\left(\dfrac{4x-3}{2-x}\right)+3}{\dfrac{4x-3}{2-x}+4} = \dfrac{2(4x-3)+3(2-x)}{4x-3+4(2-x)}$

$\qquad = \dfrac{5x}{5} = x;$

$g(f(x)) = \dfrac{4\left(\dfrac{2x+3}{x+4}\right)-3}{2-\dfrac{2x+3}{x+4}} = \dfrac{4(2x+3)-3(x+4)}{2(x+4)-(2x+3)}$

$\qquad = \dfrac{5x}{5} = x$

31. $f^{-1}(x) = \dfrac{1}{3}x$

$\qquad f(f^{-1}(x)) = 3\left(\dfrac{1}{3}x\right) = x$

$\qquad f^{-1}(f(x)) = \dfrac{1}{3}(3x) = x$

Domain f = Range f^{-1} = All real numbers
Range f = Domain f^{-1} = All real numbers

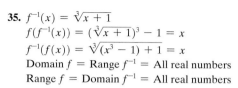

33. $f^{-1}(x) = \dfrac{x}{4} - \dfrac{1}{2}$

$\qquad f(f^{-1}(x)) = 4\left(\dfrac{x}{4} - \dfrac{1}{2}\right) + 2 = (x-2) + 2 = x$

$\qquad f^{-1}(f(x)) = \dfrac{4x+2}{4} - \dfrac{1}{2} = \left(x + \dfrac{1}{2}\right) - \dfrac{1}{2} = x$

Domain f = Range f^{-1} = All real numbers
Range f = Domain f^{-1} = All real numbers

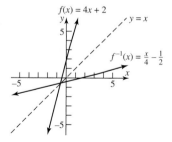

35. $f^{-1}(x) = \sqrt[3]{x+1}$

$\qquad f(f^{-1}(x)) = (\sqrt[3]{x+1})^3 - 1 = x$

$\qquad f^{-1}(f(x)) = \sqrt[3]{(x^3-1)+1} = x$

Domain f = Range f^{-1} = All real numbers
Range f = Domain f^{-1} = All real numbers

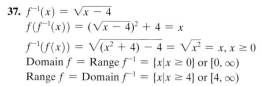

37. $f^{-1}(x) = \sqrt{x-4}$

$\qquad f(f^{-1}(x)) = (\sqrt{x-4})^2 + 4 = x$

$\qquad f^{-1}(f(x)) = \sqrt{(x^2+4)-4} = \sqrt{x^2} = x, x \geq 0$

Domain f = Range f^{-1} = $\{x \,|\, x \geq 0\}$ or $[0, \infty)$
Range f = Domain f^{-1} = $\{x \,|\, x \geq 4\}$ or $[4, \infty)$

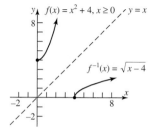

39. $f^{-1}(x) = \dfrac{4}{x}$

$\qquad f(f^{-1}(x)) = \dfrac{4}{\left(\dfrac{4}{x}\right)} = x$

$\qquad f^{-1}(f(x)) = \dfrac{4}{\left(\dfrac{4}{x}\right)} = x$

Domain f = Range f^{-1} = All real numbers except 0
Range f = Domain f^{-1} = All real numbers except 0

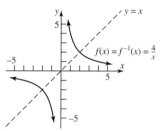

41. $f^{-1}(x) = \dfrac{2x + 1}{x}$

$$f(f^{-1}(x)) = \dfrac{1}{\dfrac{2x+1}{x} - 2} = \dfrac{x}{(2x+1) - 2x} = x$$

$$f^{-1}(f(x)) = \dfrac{2\left(\dfrac{1}{x-2}\right) + 1}{\dfrac{1}{x-2}} = \dfrac{2 + (x-2)}{1} = x$$

Domain f = Range f^{-1} = All real numbers except 2
Range f = Domain f^{-1} = All real numbers except 0

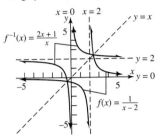

$f^{-1}(x) = \dfrac{2x+1}{x}$

43. $f^{-1}(x) = \dfrac{2 - 3x}{x}$

$$f(f^{-1}(x)) = \dfrac{2}{3 + \dfrac{2-3x}{x}} = \dfrac{2x}{3x + 2 - 3x} = \dfrac{2x}{2} = x$$

$$f^{-1}(f(x)) = \dfrac{2 - 3\left(\dfrac{2}{3+x}\right)}{\dfrac{2}{3+x}} = \dfrac{2(3+x) - 3 \cdot 2}{2} = \dfrac{2x}{2} = x$$

Domain f = All real numbers except -3
Range f = Domain f^{-1} = All real numbers except 0

45. $f^{-1}(x) = \dfrac{-2x}{x - 3}$

$$f(f^{-1}(x)) = \dfrac{3\left(\dfrac{-2x}{x-3}\right)}{\dfrac{-2x}{x-3} + 2} = \dfrac{3(-2x)}{-2x + 2(x-3)} = \dfrac{-6x}{-6} = x$$

$$f^{-1}(f(x)) = \dfrac{-2\left(\dfrac{3x}{x+2}\right)}{\dfrac{3x}{x+2} - 3} = \dfrac{-2(3x)}{3x - 3(x+2)} = \dfrac{-6x}{-6} = x$$

Domain f = All real numbers except -2
Range f = Domain f^{-1} = All real numbers except 3

47. $f^{-1}(x) = \dfrac{x}{3x - 2}$

$$f(f^{-1}(x)) = \dfrac{2\left(\dfrac{x}{3x-2}\right)}{3\left(\dfrac{x}{3x-2}\right) - 1} = \dfrac{2x}{3x - (3x-2)} = \dfrac{2x}{2} = x$$

$$f^{-1}(f(x)) = \dfrac{\dfrac{2x}{3x-1}}{3\left(\dfrac{2x}{3x-1}\right) - 2} = \dfrac{2x}{6x - 2(3x-1)} = \dfrac{2x}{2} = x$$

Domain f = All real numbers except $\dfrac{1}{3}$

Range f = Domain f^{-1} = All real numbers except $\dfrac{2}{3}$

49. $f^{-1}(x) = \dfrac{3x + 4}{2x - 3}$

$$f(f^{-1}(x)) = \dfrac{3\left(\dfrac{3x+4}{2x-3}\right) + 4}{2\left(\dfrac{3x+4}{2x-3}\right) - 3} = \dfrac{3(3x+4) + 4(2x-3)}{2(3x+4) - 3(2x-3)} = \dfrac{17x}{17} = x$$

$$f^{-1}(f(x)) = \dfrac{3\left(\dfrac{3x+4}{2x-3}\right) + 4}{2\left(\dfrac{3x+4}{2x-3}\right) - 3} = \dfrac{3(3x+4) + 4(2x-3)}{2(3x+4) - 3(2x-3)} = \dfrac{17x}{17} = x$$

Domain f = All real numbers except $\dfrac{3}{2}$

Range f = Domain f^{-1} = All real numbers except $\dfrac{3}{2}$

51. $f^{-1}(x) = \dfrac{-2x + 3}{x - 2}$

$$f(f^{-1}(x)) = \dfrac{2\left(\dfrac{-2x + 3}{x - 2}\right) + 3}{\dfrac{-2x + 3}{x - 2} + 2} = \dfrac{2(-2x + 3) + 3(x - 2)}{-2x + 3 + 2(x - 2)} = \dfrac{-x}{-1} = x$$

$$f^{-1}(f(x)) = \dfrac{-2\left(\dfrac{2x + 3}{x + 2}\right) + 3}{\dfrac{2x + 3}{x + 2} - 2} = \dfrac{-2(2x + 3) + 3(x + 2)}{2x + 3 - 2(x + 2)} = \dfrac{-x}{-1} = x$$

Domain f = All real numbers except -2

Range f = Domain f^{-1} = All real numbers except 2

53. $f^{-1}(x) = \dfrac{2}{\sqrt{1 - 2x}}$

$$f(f^{-1}(x)) = \dfrac{\dfrac{4}{1 - 2x} - 4}{2 \cdot \dfrac{4}{1 - 2x}} = \dfrac{4 - 4(1 - 2x)}{2 \cdot 4} = \dfrac{8x}{8} = x$$

$$f^{-1}(f(x)) = \dfrac{2}{\sqrt{1 - 2\left(\dfrac{x^2 - 4}{2x^2}\right)}} = \dfrac{2}{\sqrt{\dfrac{4}{x^2}}} = \sqrt{x^2} = x, \text{ since } x > 0.$$

Domain f = $\{x | x > 0\}$ or $(0, \infty)$

Range f = Domain f^{-1} = $\left\{x \middle| x < \dfrac{1}{2}\right\}$ or $\left(-\infty, \dfrac{1}{2}\right)$

55. $f^{-1}(x) = \dfrac{1}{m}(x - b), m \neq 0$ **57.** Quadrant I   **59.** $f(x) = |x|, x \geq 0,$ is one-to-one; $f^{-1}(x) = x, x \geq 0$

61. $f(g(x)) = \dfrac{9}{5}\left[\dfrac{5}{9}(x - 32)\right] + 32 = x; g(f(x)) = \dfrac{5}{9}\left[\left(\dfrac{9}{5}x + 32\right) - 32\right] = x$ **63.** $l(T) = \dfrac{gT^2}{4\pi^2}, T > 0$

Fill-in-the-Blank Items *(page 73)*

1. x-coordinate or abscissa; y-coordinate or ordinate   **3.** y-axis   **5.** independent; dependent   **7.** even; odd   **9.** One-to-one

True/False Items *(page 74)*

1. F   **3.** F   **5.** T   **7.** F   **9.** F

Review Exercises *(page 74)*

1. $(0, 0)$; Symmetric with respect to the x-axis   **3.** $(-4, 0), (0, 2), (0, -2), (4, 0)$; Symmetric with respect to the x-axis, y-axis, and origin
5. $(0, 1)$; Symmetric with respect to the y-axis   **7.** $(-1, 0), (0, 0), (0, -2)$; No symmetry
9. Center $(0, 1)$; Radius = 2 **11.** Center $(1, -2)$; Radius = 3 **13.** Center $(1, -2)$; Radius = $\sqrt{5}$ **15.** $A = 11$

17. b, c, d

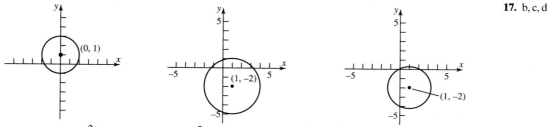

19. (a) $f(-x) = \dfrac{-3x}{x^2 - 4}$ (b) $-f(x) = \dfrac{-3x}{x^2 - 4}$ (c) $f(x + 2) = \dfrac{3x + 6}{x^2 + 4x}$ (d) $f(x - 2) = \dfrac{3x - 6}{x^2 - 4x}$ (e) $f(2x) = \dfrac{6x}{4x^2 - 4}$
21. (a) $f(-x) = \sqrt{x^2 - 4}$ (b) $-f(x) = -\sqrt{x^2 - 4}$ (c) $f(x + 2) = \sqrt{x^2 + 4x}$ (d) $f(x - 2) = \sqrt{x^2 - 4x}$ (e) $f(2x) = 2\sqrt{x^2 - 1}$
23. (a) $f(-x) = \dfrac{x^2 - 4}{x^2}$ (b) $-f(x) = -\dfrac{x^2 - 4}{x^2}$ (c) $f(x + 2) = \dfrac{x^2 + 4x}{x^2 + 4x + 4}$ (d) $f(x - 2) = \dfrac{x^2 - 4x}{x^2 - 4x + 4}$ (e) $f(2x) = \dfrac{x^2 - 1}{x^2}$
25. $\{x | x \neq -3, x \neq 3\}$ **27.** $\{x | x \leq 2\}$ **29.** $\{x | x > 0\}$ **31.** $\{x | x \neq -3, x \neq 1\}$ **33.** Odd   **35.** Even   **37.** Neither   **39.** Odd

41.

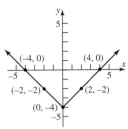

Intercepts: $(-4, 0), (4, 0), (0, -4)$
Domain: all real numbers
Range: $\{y | y \geq -4\}$ or $[-4, \infty)$

43.

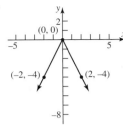

Intercept: $(0, 0)$
Domain: all real numbers
Range: $\{y | y \leq 0\}$ or $(-\infty, 0]$

45.

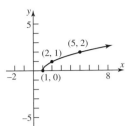

Intercept: $(1, 0)$
Domain: $\{x | x \geq 1\}$ or $[1, \infty)$
Range: $\{y | y \geq 0\}$ or $[0, \infty)$

47.

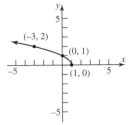

Intercepts: $(0, 1), (1, 0)$
Domain: $\{x | x \leq 1\}$ or $(-\infty, 1]$
Range: $\{y | y \geq 0\}$ or $[0, \infty)$

49.

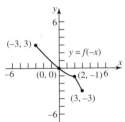

51.

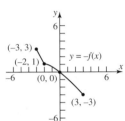

53. (a)

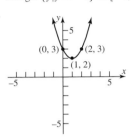

(b)

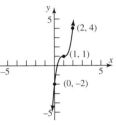

(c)

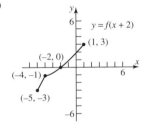

(d)

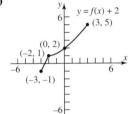

(e)

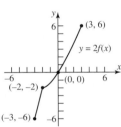

(f)

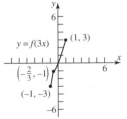

55. $f^{-1}(x) = \dfrac{2x + 3}{5x - 2}$; $f(f^{-1}(x)) = \dfrac{2\left(\dfrac{2x + 3}{5x - 2}\right) + 3}{5\left(\dfrac{2x + 3}{5x - 2}\right) - 2}$

$= \dfrac{2(2x + 3) + 3(5x - 2)}{5(2x + 3) - 2(5x - 2)} = \dfrac{19x}{19} = x;$

$f^{-1}(f(x)) = \dfrac{2\left(\dfrac{2x + 3}{5x - 2}\right) + 3}{5\left(\dfrac{2x + 3}{5x - 2}\right) - 2} = \dfrac{2(2x + 3) + 3(5x - 2)}{5(2x + 3) - 2(5x - 2)} = \dfrac{19x}{19} = x;$

Domain f = Range f^{-1} = All real numbers except $\dfrac{2}{5}$;

Range f = Domain f^{-1} = All real numbers except $\dfrac{2}{5}$

57. $f^{-1}(x) = \dfrac{x+1}{x}; f(f^{-1}(x)) = \dfrac{1}{\dfrac{x+1}{x} - 1} = \dfrac{x}{x+1-x} = x;$

$f^{-1}(f(x)) = \dfrac{\dfrac{1}{x-1} + 1}{\dfrac{1}{x-1}} = \dfrac{1 + x - 1}{1} = x;$

Domain f = Range f^{-1} = All real numbers except 1;
Range f = Domain f^{-1} = All real numbers except 0

59. $f^{-1}(x) = \dfrac{27}{x^3}; f(f^{-1}(x)) = \dfrac{3}{\left(\dfrac{27}{x^3}\right)^{1/3}} = \dfrac{3}{\left(\dfrac{3}{x}\right)} = x;$

$f^{-1}(f(x)) = \dfrac{27}{\left(\dfrac{3}{x^{1/3}}\right)^3} = \dfrac{27}{\left(\dfrac{27}{x}\right)} = x;$

Domain f = Range f^{-1} = All real numbers except 0;
Range f = Domain f^{-1} = All real numbers except 0

61. Center = $(1, -2)$; Radius = $4\sqrt{2}; x^2 + y^2 - 2x + 4y - 27 = 0$ **63.** $T(h) = -0.0025h + 30, 0 \le x \le 10{,}000$

65. (a) $C(r) = 0.12\pi r^2 + \dfrac{40}{r}$ **(b)** \$16.03   **(c)** \$29.13

(d) The cost is least for $r \approx 3.76$ cm.

67. (a) y-axis   **(b)** x-axis
(c) line through origin, slope -1
(d) the x-axis and y-axis   **(e)** the origin

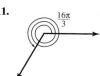

C H A P T E R   2   Trigonometric Functions

2.1 Exercises *(page 88)*

1. **3.** **5.** **7.** **9.** **11.**

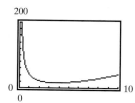

13. $\dfrac{\pi}{6}$ **15.** $\dfrac{4\pi}{3}$ **17.** $-\dfrac{\pi}{3}$ **19.** π **21.** $-\dfrac{3\pi}{4}$ **23.** $-\dfrac{\pi}{2}$ **25.** $60°$ **27.** $-225°$ **29.** $90°$ **31.** $15°$ **33.** $-90°$ **35.** $-30°$
37. 5 m   **39.** 6 ft

41. 0.6 radian   **43.** $\dfrac{\pi}{3} \approx 1.047$ in.   **45.** 25 m^2 **47.** $2\sqrt{3} \approx 3.464$ ft   **49.** 0.24 radian   **51.** $\dfrac{\pi}{3} \approx 1.047$ in^2 **53.** $s = 2.094$ ft; $A = 2.094$ ft^2
55. $s = 14.661$ yd; $A = 87.965$ yd^2 **57.** 0.30   **59.** -0.70 **61.** 2.18   **63.** 179.91°   **65.** 114.59°   **67.** 362.11°   **69.** 40.17°   **71.** 1.03°
73. 9.15°   **75.** 40°19′12″   **77.** 18°15′18″   **79.** 19°59′24″   **81.** $3\pi \approx 9.4248$ in.; $5\pi \approx 15.7080$ in.   **83.** $2\pi \approx 6.28$ m^2
85. $\dfrac{675\pi}{2} \approx 1060.29$ ft^2 **87.** $\omega = \dfrac{1}{60}$ radian/sec; $v = \dfrac{1}{12}$ cm/sec   **89.** Approximately 452.5 rpm   **91.** Approximately 359 mi

93. Approximately 898 mi/hr   **95.** Approximately 2292 mi/hr   **97.** $\dfrac{3}{4}$ rpm   **99.** Approximately 2.86 mi/hr   **101.** Approximately 31.47 rpm

103. Approximately 1037 mi/hr   **105.** $v_1 = r_1\omega_1, v_2 = r_2\omega_2$, and $v_1 = v_2$ so $r_1\omega_1 = r_2\omega_2 \Rightarrow \dfrac{r_1}{r_2} = \dfrac{\omega_2}{\omega_1}$.

2.2 Exercises *(page 105)*

1. $\sin t = \dfrac{1}{2}; \cos t = \dfrac{\sqrt{3}}{2}; \tan t = \dfrac{\sqrt{3}}{3}; \csc t = 2; \sec t = \dfrac{2\sqrt{3}}{3}; \cot t = \sqrt{3}$ **3.** $\sin t = \dfrac{\sqrt{21}}{5}; \cos t = -\dfrac{2}{5}; \tan t = -\dfrac{\sqrt{21}}{2}; \csc t = \dfrac{5\sqrt{21}}{21};$

$\sec t = -\dfrac{5}{2}; \cot t = -\dfrac{2\sqrt{21}}{21}$ **5.** $\sin t = \dfrac{\sqrt{2}}{2}; \cos t = -\dfrac{\sqrt{2}}{2}; \tan t = -1; \csc t = \sqrt{2}; \sec t = -\sqrt{2}; \cot t = -1$ **7.** $\sin t = -\dfrac{1}{3}; \cos t = \dfrac{2\sqrt{2}}{3};$

$\tan t = -\dfrac{\sqrt{2}}{4}; \csc t = -3; \sec t = \dfrac{3\sqrt{2}}{4}; \cot t = -2\sqrt{2}$ **9.** -1 **11.** 0   **13.** -1 **15.** 0   **17.** -1 **19.** $\dfrac{1}{2}(\sqrt{2} + 1)$ **21.** 2   **23.** $\dfrac{1}{2}$ **25.** $\sqrt{6}$

27. 4   **29.** 0   **31.** 0   **33.** $2\sqrt{2} + \dfrac{4\sqrt{3}}{3}$ **35.** -1 **37.** 1   **39.** $\sin \dfrac{2\pi}{3} = \dfrac{\sqrt{3}}{2}; \cos \dfrac{2\pi}{3} = -\dfrac{1}{2}; \tan \dfrac{2\pi}{3} = -\sqrt{3}; \csc \dfrac{2\pi}{3} = \dfrac{2\sqrt{3}}{3}; \sec \dfrac{2\pi}{3} = -2;$

$\cot \dfrac{2\pi}{3} = -\dfrac{\sqrt{3}}{3}$ **41.** $\sin 210° = -\dfrac{1}{2}; \cos 210° = -\dfrac{\sqrt{3}}{2}; \tan 210° = \dfrac{\sqrt{3}}{3}; \csc 210° = -2; \sec 210° = -\dfrac{2\sqrt{3}}{3}; \cot 210° = \sqrt{3}$

43. $\sin\dfrac{3\pi}{4}=\dfrac{\sqrt{2}}{2}$; $\cos\dfrac{3\pi}{4}=-\dfrac{\sqrt{2}}{2}$; $\tan\dfrac{3\pi}{4}=-1$; $\csc\dfrac{3\pi}{4}=\sqrt{2}$; $\sec\dfrac{3\pi}{4}=-\sqrt{2}$; $\cot\dfrac{3\pi}{4}=-1$ **45.** $\sin\dfrac{8\pi}{3}=\dfrac{\sqrt{3}}{2}$; $\cos\dfrac{8\pi}{3}=-\dfrac{1}{2}$;

$\tan\dfrac{8\pi}{3}=-\sqrt{3}$; $\csc\dfrac{8\pi}{3}=\dfrac{2\sqrt{3}}{3}$; $\sec\dfrac{8\pi}{3}=-2$; $\cot\dfrac{8\pi}{3}=-\dfrac{\sqrt{3}}{3}$ **47.** $\sin405°=\dfrac{\sqrt{2}}{2}$; $\cos405°=\dfrac{\sqrt{2}}{2}$; $\tan405°=1$; $\csc405°=\sqrt{2}$;

$\sec405°=\sqrt{2}$; $\cot405°=1$ **49.** $\sin\left(-\dfrac{\pi}{6}\right)=-\dfrac{1}{2}$; $\cos\left(-\dfrac{\pi}{6}\right)=\dfrac{\sqrt{3}}{2}$; $\tan\left(-\dfrac{\pi}{6}\right)=-\dfrac{\sqrt{3}}{3}$; $\csc\left(-\dfrac{\pi}{6}\right)=-2$; $\sec\left(-\dfrac{\pi}{6}\right)=\dfrac{2\sqrt{3}}{3}$;

$\cot\left(-\dfrac{\pi}{6}\right)=-\sqrt{3}$ **51.** $\sin(-45°)=-\dfrac{\sqrt{2}}{2}$; $\cos(-45°)=\dfrac{\sqrt{2}}{2}$; $\tan(-45°)=-1$; $\csc(-45°)=-\sqrt{2}$; $\sec(-45°)=\sqrt{2}$; $\cot(-45°)=-1$

53. $\sin\dfrac{5\pi}{2}=1$; $\cos\dfrac{5\pi}{2}=0$; $\tan\dfrac{5\pi}{2}$ is not defined; $\csc\dfrac{5\pi}{2}=1$; $\sec\dfrac{5\pi}{2}$ is not defined; $\cot\dfrac{5\pi}{2}=0$ **55.** $\sin720°=0$; $\cos720°=1$;

$\tan720°=0$; $\csc720°$ is undefined; $\sec720°=1$; $\cot720°$ is undefined **57.** 0.47 **59.** 0.38 **61.** 1.33 **63.** 0.31 **65.** 3.73 **67.** 1.04

69. 0.84 **71.** 0.02 **73.** $\sin\theta=\dfrac{4}{5}$; $\cos\theta=-\dfrac{3}{5}$; $\tan\theta=-\dfrac{4}{3}$; $\csc\theta=\dfrac{5}{4}$; $\sec\theta=-\dfrac{5}{3}$; $\cot\theta=-\dfrac{3}{4}$ **75.** $\sin\theta=-\dfrac{3\sqrt{13}}{13}$; $\cos\theta=\dfrac{2\sqrt{13}}{13}$;

$\tan\theta=-\dfrac{3}{2}$; $\csc\theta=-\dfrac{\sqrt{13}}{3}$; $\sec\theta=\dfrac{\sqrt{13}}{2}$; $\cot\theta=-\dfrac{2}{3}$ **77.** $\sin\theta=-\dfrac{\sqrt{2}}{2}$; $\cos\theta=-\dfrac{\sqrt{2}}{2}$; $\tan\theta=1$; $\csc\theta=-\sqrt{2}$; $\sec\theta=-\sqrt{2}$; $\cot\theta=1$

79. $\sin\theta=-\dfrac{2\sqrt{13}}{13}$; $\cos\theta=-\dfrac{3\sqrt{13}}{13}$; $\tan\theta=\dfrac{2}{3}$; $\csc\theta=-\dfrac{\sqrt{13}}{2}$; $\sec\theta=-\dfrac{\sqrt{13}}{3}$; $\cot\theta=\dfrac{3}{2}$ **81.** $\sin\theta=-\dfrac{3}{5}$; $\cos\theta=\dfrac{4}{5}$; $\tan\theta=-\dfrac{3}{4}$;

$\csc\theta=-\dfrac{5}{3}$; $\sec\theta=\dfrac{5}{4}$; $\cot\theta=-\dfrac{4}{3}$ **83.** 0 **85.** -0.1 **87.** 3 **89.** 5 **91.** $\dfrac{\sqrt{3}}{2}$ **93.** $\dfrac{1}{2}$ **95.** $\dfrac{3}{4}$ **97.** $\dfrac{\sqrt{3}}{2}$ **99.** $\sqrt{3}$ **101.** $-\dfrac{\sqrt{3}}{2}$

105. $R\approx310.56$ ft; $H\approx77.64$ ft

107. $R\approx19,542$ m; $H\approx2278$ m

103.

θ	0.5	0.4	0.2	0.1	0.01	0.001	0.0001	0.00001
$\sin\theta$	0.4794	0.3894	0.1987	0.0998	0.0100	0.0010	0.0001	0.00001
$\dfrac{\sin\theta}{\theta}$	0.9589	0.9735	0.9933	0.9983	1.0000	1.0000	1.0000	1.0000

109. (a) 1.20 sec
(b) 1.11 sec
(c) 1.20 sec

$\dfrac{\sin\theta}{\theta}$ approaches 1 as θ approaches 0.

111. (a) 1.9 hr; 0.57 hr (b) 1.69 hr; 0.75 hr (c) 1.63 hr; 0.86 hr (d) 1.67 hr **113.** (a) 16.6 ft

(b)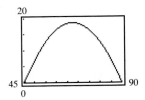

(c) 67.5°

2.3 Exercises (page 120)

1. $\dfrac{\sqrt{2}}{2}$ **3.** 1 **5.** 1 **7.** $\sqrt{3}$ **9.** $\dfrac{\sqrt{2}}{2}$ **11.** 0 **13.** $\sqrt{2}$ **15.** $\dfrac{\sqrt{3}}{3}$ **17.** II **19.** IV **21.** IV **23.** II **25.** $\tan\theta=-\dfrac{3}{4}$; $\cot\theta=-\dfrac{4}{3}$; $\sec\theta=\dfrac{5}{4}$;

$\csc\theta=-\dfrac{5}{3}$ **27.** $\tan\theta=2$; $\cot\theta=\dfrac{1}{2}$; $\sec\theta=\sqrt{5}$; $\csc\theta=\dfrac{\sqrt{5}}{2}$ **29.** $\tan\theta=\dfrac{\sqrt{3}}{3}$; $\cot\theta=\sqrt{3}$; $\sec\theta=\dfrac{2\sqrt{3}}{3}$; $\csc\theta=2$ **31.** $\tan\theta=-\dfrac{\sqrt{2}}{4}$;

$\cot\theta=-2\sqrt{2}$; $\sec\theta=\dfrac{3\sqrt{2}}{4}$; $\csc\theta=-3$ **33.** $\cos\theta=-\dfrac{5}{13}$; $\tan\theta=-\dfrac{12}{5}$; $\csc\theta=\dfrac{13}{12}$; $\sec\theta=-\dfrac{13}{5}$; $\cot\theta=-\dfrac{5}{12}$ **35.** $\sin\theta=-\dfrac{3}{5}$;

$\tan\theta=\dfrac{3}{4}$; $\csc\theta=-\dfrac{5}{3}$; $\sec\theta=-\dfrac{5}{4}$; $\cot\theta=\dfrac{4}{3}$ **37.** $\cos\theta=-\dfrac{12}{13}$; $\tan\theta=-\dfrac{5}{12}$; $\csc\theta=\dfrac{13}{5}$; $\sec\theta=-\dfrac{13}{12}$; $\cot\theta=-\dfrac{12}{5}$ **39.** $\sin\theta=\dfrac{2\sqrt{2}}{3}$;

$\tan\theta=-2\sqrt{2}$; $\csc\theta=\dfrac{3\sqrt{2}}{4}$; $\sec\theta=-3$; $\cot\theta=-\dfrac{\sqrt{2}}{4}$ **41.** $\cos\theta=-\dfrac{\sqrt{5}}{3}$; $\tan\theta=-\dfrac{2\sqrt{5}}{5}$; $\csc\theta=\dfrac{3}{2}$; $\sec\theta=-\dfrac{3\sqrt{5}}{5}$; $\cot\theta=-\dfrac{\sqrt{5}}{2}$

43. $\sin\theta=-\dfrac{\sqrt{3}}{2}$; $\cos\theta=\dfrac{1}{2}$; $\tan\theta=-\sqrt{3}$; $\csc\theta=-\dfrac{2\sqrt{3}}{3}$; $\cot\theta=-\dfrac{\sqrt{3}}{3}$ **45.** $\sin\theta=-\dfrac{3}{5}$; $\cos\theta=-\dfrac{4}{5}$; $\csc\theta=-\dfrac{5}{3}$; $\sec\theta=-\dfrac{5}{4}$;

$\cot\theta=\dfrac{4}{3}$ **47.** $\sin\theta=\dfrac{\sqrt{10}}{10}$; $\cos\theta=-\dfrac{3\sqrt{10}}{10}$; $\csc\theta=\sqrt{10}$; $\sec\theta=-\dfrac{\sqrt{10}}{3}$; $\cot\theta=-3$ **49.** $-\dfrac{\sqrt{3}}{2}$ **51.** $-\dfrac{\sqrt{3}}{3}$ **53.** 2 **55.** -1 **57.** -1

59. $\dfrac{\sqrt{2}}{2}$ **61.** 0 **63.** $-\sqrt{2}$ **65.** $\dfrac{2\sqrt{3}}{3}$ **67.** 1 **69.** 1 **71.** 0 **73.** 1 **75.** -1 **77.** 0 **79.** 0.9 **81.** 9 **83.** 0 **85.** All real numbers

87. Odd multiples of $\dfrac{\pi}{2}$ **89.** Odd multiples of $\dfrac{\pi}{2}$ **91.** $-1\le y\le1$ **93.** All real numbers **95.** $|y|\ge1$ **97.** Odd; yes; origin

99. Odd; yes; origin **101.** Even; yes; y-axis **103.** (a) $-\dfrac{1}{3}$ (b) 1 **105.** (a) -2 (b) 6 **107.** (a) -4 (b) -12 **109.** 15.8 min

111. Let a be a real number and $P = (x, y)$ be the point on the unit circle that corresponds to t. Consider the equation $\tan t = \dfrac{y}{x} = a$. Then
$y = ax$. But $x^2 + y^2 = 1$ so that $x^2 + a^2 x^2 = 1$. Thus, $x = \pm\dfrac{1}{\sqrt{1 + a^2}}$ and $y = \pm\dfrac{a}{\sqrt{1 + a^2}}$; that is, for any real number a, there is a
point $P = (x, y)$ on the unit circle for which $\tan t = a$. In other words, the range of the tangent function is the set of all real numbers.

113. Suppose there is a number p, $0 < p < 2\pi$, for which $\sin(\theta + p) = \sin \theta$ for all θ. If $\theta = 0$, then $\sin(0 + p) = \sin p = \sin 0 = 0$; --
so that $p = \pi$. If $\theta = \dfrac{\pi}{2}$, then $\sin\left(\dfrac{\pi}{2} + p\right) = \sin\left(\dfrac{\pi}{2}\right)$. But $p = \pi$. Thus, $\sin\left(\dfrac{3\pi}{2}\right) = -1 = \sin\left(\dfrac{\pi}{2}\right) = 1$. This is impossible.
Therefore, the smallest positive number p for which $\sin(\theta + p) = \sin \theta$ for all θ is 2π.

115. $\sec \theta = \dfrac{1}{\cos \theta}$; since $\cos \theta$ has period 2π, so does $\sec \theta$.

117. If $P = (a, b)$ is the point on the unit circle corresponding to θ, then $Q = (-a, -b)$ is the point on the unit circle corresponding to
$\theta + \pi$. Thus, $\tan(\theta + \pi) = \dfrac{-b}{-a} = \dfrac{b}{a} = \tan \theta$. Suppose there exists a number p, $0 < p < \pi$, for which $\tan(\theta + p) = \tan \theta$ for all θ.
Then if $\theta = 0$, then $\tan p = \tan 0 = 0$. But this means that p is a multiple of π. Since no multiple of π exists in the interval $(0, \pi)$,
this is a contradiction. Therefore, the period of $f(\theta) = \tan \theta$ is π.

119. Let $P = (a, b)$ be the point on the unit circle corresponding to θ. Then $\csc \theta = \dfrac{1}{b} = \dfrac{1}{\sin \theta}$; $\sec \theta = \dfrac{1}{a} = \dfrac{1}{\cos \theta}$; $\cot \theta = \dfrac{a}{b} = \dfrac{1}{b/a} = \dfrac{1}{\tan \theta}$.

121. $(\sin \theta \cos \phi)^2 + (\sin \theta \sin \phi)^2 + \cos^2 \theta = \sin^2 \theta \cos^2 \phi + \sin^2 \theta \sin^2 \phi + \cos^2 \theta$
$$= \sin^2 \theta(\cos^2 \phi + \sin^2 \phi) + \cos^2 \theta = \sin^2 \theta + \cos^2 \theta = 1$$

2.4 Exercises (page 133)

1. 0   **3.** $-\dfrac{\pi}{2} < x < \dfrac{\pi}{2}$ **5.** 1   **7.** $0, \pi, 2\pi$ **9.** $\sin x = 1$ for $x = -\dfrac{3\pi}{2}, \dfrac{\pi}{2}$; $\sin x = -1$ for $x = -\dfrac{\pi}{2}, \dfrac{3\pi}{2}$ **11.** B, C, F

13. 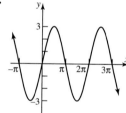   **15.** 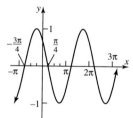   **17.** 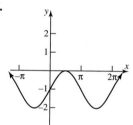   **19.**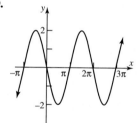

21. 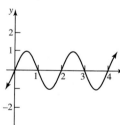   **23.** 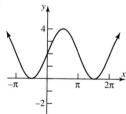   **25.** 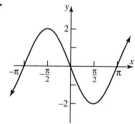   **27.**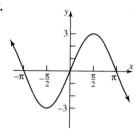

29. Amplitude $= 2$; Period $= 2\pi$ **31.** Amplitude $= 4$; Period $= \pi$ **33.** Amplitude $= 6$; Period $= 2$

35. Amplitude $= \dfrac{1}{2}$; Period $= \dfrac{4\pi}{3}$ **37.** Amplitude $= \dfrac{5}{3}$; Period $= 3$ **39.** F   **41.** A   **43.** H   **45.** C   **47.** J   **49.** A   **51.** B

53. 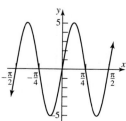   **55.** 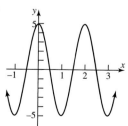   **57.** 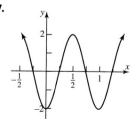   **59.**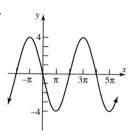

61.

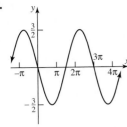

63. $y = 3 \sin(2x)$ **65.** $y = 3 \sin(\pi x)$ **67.** $y = 5 \cos\left(\dfrac{\pi}{4}x\right)$

69. $y = -3 \cos\left(\dfrac{1}{2}x\right)$ **71.** $y = \dfrac{3}{4} \sin(2\pi x)$ **73.** $y = -\sin\left(\dfrac{3}{2}x\right)$

75. $y = -2 \cos\left(\dfrac{3\pi}{2}x\right)$ **77.** $y = 3 \sin\left(\dfrac{\pi}{2}x\right)$ **79.** $y = -4 \cos(3x)$

81. Period $= \dfrac{1}{30}$; Amplitude $= 220$

83. (a) Amplitude $= 220$; Period $= \dfrac{1}{60}$

(b), (e)

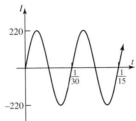

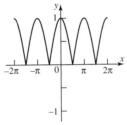

(c) $I = 22 \sin(120\pi t)$

(d) Amplitude $= 22$; Period $= \dfrac{1}{60}$

85. (a) $P = \dfrac{[V_0 \sin(2\pi ft)]^2}{R} = \dfrac{V_0^2}{R} \sin^2(2\pi ft)$

(b) Since the graph of P has amplitude $\dfrac{V_0^2}{2R}$, period $\dfrac{1}{2f}$, and is

of the form $y = A \cos(\omega t) + B$, then $A = -\dfrac{V_0^2}{2R}$ and $B = \dfrac{V_0^2}{2R}$.

Since $\dfrac{1}{2f} = \dfrac{2\pi}{\omega}$, then $\omega = 4\pi f$. Therefore,

$P = -\dfrac{V_0^2}{2R} \cos(4\pi ft) + \dfrac{V_0^2}{2R} = \dfrac{V_0^2}{2R}[1 - \cos(4\pi ft)]$.

87.

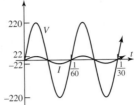

2.5 Exercises *(page 142)*

1. 0   **3.** 1   **5.** $\sec x = 1$ for $x = -2\pi, 0, 2\pi$; $\sec x = -1$ for $x = -\pi, \pi$ **7.** $-\dfrac{3\pi}{2}, -\dfrac{\pi}{2}, \dfrac{\pi}{2}, \dfrac{3\pi}{2}$ **9.** $-\dfrac{3\pi}{2}, -\dfrac{\pi}{2}, \dfrac{\pi}{2}, \dfrac{3\pi}{2}$ **11.** D   **13.** B

15.

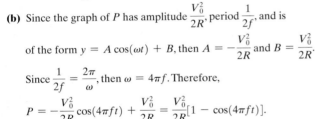

17.

19.

21.

23.

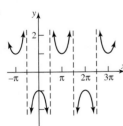

25.

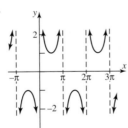

27.

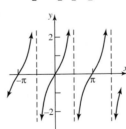

29.

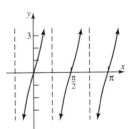

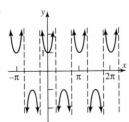

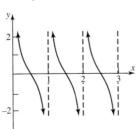

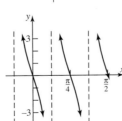

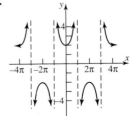

31.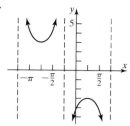

33.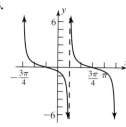

35. (a) $L(\theta) = \dfrac{3}{\cos \theta} + \dfrac{4}{\sin \theta} = 3 \sec \theta + 4 \csc \theta$

(b)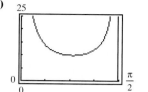

(c) 0.83
(d) 9.86 ft

2.6 Exercises *(page 152)*

1. Amplitude = 4

Period = π

Phase shift = $\dfrac{\pi}{2}$

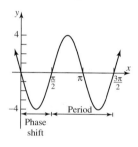

3. Amplitude = 2

Period = $\dfrac{2\pi}{3}$

Phase shift = $-\dfrac{\pi}{6}$

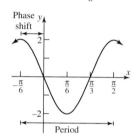

5. Amplitude = 3

Period = π

Phase shift = $-\dfrac{\pi}{4}$

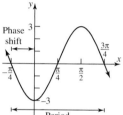

7. Amplitude = 4

Period = 2

Phase shift = $-\dfrac{2}{\pi}$

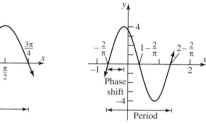

9. Amplitude = 3

Period = 2

Phase shift = $\dfrac{2}{\pi}$

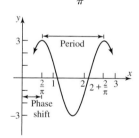

11. Amplitude = 3

Period = π

Phase shift = $\dfrac{\pi}{4}$

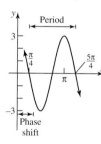

13. $y = 2 \sin\left[2\left(x - \dfrac{1}{2}\right)\right]$ or $y = 2 \sin(2x - 1)$

15. $y = 3 \sin\left[\dfrac{2}{3}\left(x + \dfrac{1}{3}\right)\right]$ or $y = 3 \sin\left(\dfrac{2}{3}x + \dfrac{2}{9}\right)$

17. Period = $\dfrac{1}{15}$; Amplitude = 120; Phase shift = $\dfrac{1}{90}$

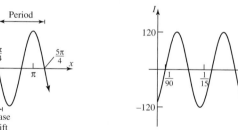

19. (a)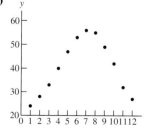

(b) $y = 15.9 \sin\left(\dfrac{\pi}{6}x - \dfrac{2\pi}{3}\right) + 40.1$

(c)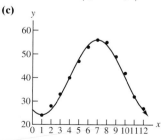

(d) $y = 15.62 \sin(0.517x - 2.096) + 40.377$

(e)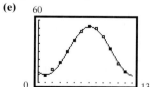

21. (a)
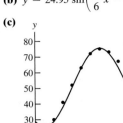

(b) $y = 24.95 \sin\left(\dfrac{\pi}{6}x - \dfrac{2\pi}{3}\right) + 50.45$

(c)
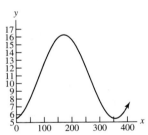

(d) $y = 25.693 \sin(0.476x - 1.814) + 49.854$

(e)
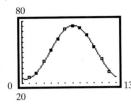

23. (a) 4:08 PM

(b) $y = 4.4 \sin\left(\dfrac{4\pi}{25}x - 6.6643\right) + 3.8$

(c)
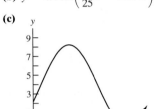

(d) 8.2 ft

25. (a) $y = 1.0835 \sin\left(\dfrac{2\pi}{365}x - \dfrac{357\pi}{146}\right) + 11.6665$

(b)
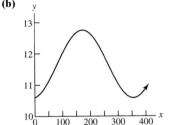

(c) 11.85 hr

27. (a) $y = 5.3915 \sin\left(\dfrac{2\pi}{365}x - \dfrac{357\pi}{146}\right) + 10.8415$ **(b)**

(c) 11.74 hr

Fill-in-the-Blank Items *(page 157)*

1. angle; initial side; terminal side   **3.** π **5.** 2π; π **7.** sine, tangent, cosecant, cotangent   **9.** $3; \dfrac{\pi}{3}$

True/False Items *(page 157)*

1. F   **3.** F   **5.** T   **7.** F

Review Exercises *(page 157)*

1. $\dfrac{3\pi}{4}$ **3.** $\dfrac{\pi}{10}$ **5.** $135°$ **7.** $-450°$ **9.** $\dfrac{1}{2}$ **11.** $\dfrac{3\sqrt{2}}{2} - \dfrac{4\sqrt{3}}{3}$ **13.** $-3\sqrt{2} - 2\sqrt{3}$ **15.** 3   **17.** 0   **19.** 0   **21.** 1   **23.** 1   **25.** 1

27. $\cos\theta = \dfrac{3}{5}; \tan\theta = -\dfrac{4}{3}; \csc\theta = -\dfrac{5}{4}; \sec\theta = \dfrac{5}{3}; \cot\theta = -\dfrac{3}{4}$ **29.** $\sin\theta = -\dfrac{12}{13}; \cos\theta = -\dfrac{5}{13}; \csc\theta = -\dfrac{13}{12}; \sec\theta = -\dfrac{13}{5}; \cot\theta = \dfrac{5}{12}$

31. $\sin\theta = \dfrac{3}{5}; \cos\theta = -\dfrac{4}{5}; \tan\theta = -\dfrac{3}{4}; \csc\theta = \dfrac{5}{3}; \cot\theta = -\dfrac{4}{3}$ **33.** $\cos\theta = -\dfrac{5}{13}; \tan\theta = -\dfrac{12}{5}; \csc\theta = \dfrac{13}{12}; \sec\theta = -\dfrac{13}{5}; \cot\theta = -\dfrac{5}{12}$

35. $\cos\theta = \dfrac{12}{13}; \tan\theta = -\dfrac{5}{12}; \csc\theta = -\dfrac{13}{5}; \sec\theta = \dfrac{13}{12}; \cot\theta = -\dfrac{12}{5}$ **37.** $\sin\theta = -\dfrac{\sqrt{10}}{10}; \cos\theta = -\dfrac{3\sqrt{10}}{10}; \csc\theta = -\sqrt{10};$

$\sec\theta = -\dfrac{\sqrt{10}}{3}; \cot\theta = 3$ **39.** $\sin\theta = -\dfrac{2\sqrt{2}}{3}; \cos\theta = \dfrac{1}{3}; \tan\theta = -2\sqrt{2}; \csc\theta = -\dfrac{3\sqrt{2}}{4}; \cot\theta = -\dfrac{\sqrt{2}}{4}$

41. $\sin\theta = \dfrac{\sqrt{5}}{5}; \cos\theta = -\dfrac{2\sqrt{5}}{5}; \tan\theta = -\dfrac{1}{2}; \csc\theta = \sqrt{5}; \sec\theta = -\dfrac{\sqrt{5}}{2}$

43.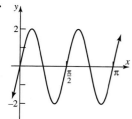

45.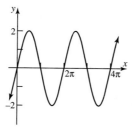

47.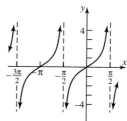

49.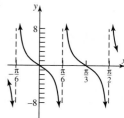

51.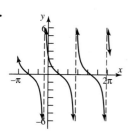

53.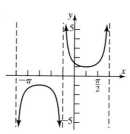

55. Amplitude $= 4$; Period $= 2\pi$

57. Amplitude $= 8$; Period $= 4$

59. Amplitude $= 4$

Period $= \dfrac{2\pi}{3}$

Phase shift $= 0$

61. Amplitude $= 2$

Period $= 4$

Phase shift $= -\dfrac{1}{\pi}$

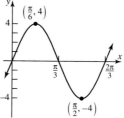

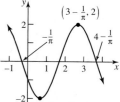

63. Amplitude $= \dfrac{1}{2}$

Period $= \dfrac{4\pi}{3}$

Phase shift $= \dfrac{2\pi}{3}$

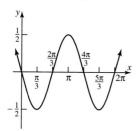

65. Amplitude $= \dfrac{2}{3}$

Period $= 2$

Phase shift $= \dfrac{6}{\pi}$

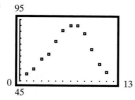

67. $y = 5 \cos\left(\dfrac{1}{4}x\right)$ **69.** $y = -6 \cos\left(\dfrac{\pi}{4}x\right)$

71. $\dfrac{\pi}{3} \approx 1.047$ ft; $\dfrac{\pi}{3} \approx 1.047$ ft^2

73. Approximately 114.59 revolutions/hr

75. 0.1 revolution/sec $= \dfrac{\pi}{5}$ radian/sec

77. **(a)** 120   **(b)** $\dfrac{1}{60}$ **(c)**

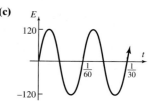

79. (a)

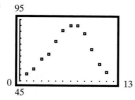

(c)

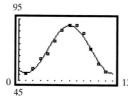

(e)

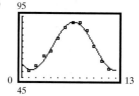

(b) $y = 19.5 \sin\left(\dfrac{\pi}{6}x - \dfrac{2\pi}{3}\right) + 70.5$ **(d)** $y = 19.52 \sin(0.54x - 2.28) + 71.01$

C H A P T E R 3   Analytic Trigonometry

3.1 Exercises (page 170)

1. 0   **3.** $-\dfrac{\pi}{2}$ **5.** 0   **7.** $\dfrac{\pi}{4}$ **9.** $\dfrac{\pi}{3}$ **11.** $\dfrac{5\pi}{6}$ **13.** 0.10   **15.** 1.37   **17.** 0.51   **19.** -0.38 **21.** -0.12 **23.** 1.08   **25.** 0.54   **27.** $\dfrac{4\pi}{5}$ **29.** -3.5

31. $-\dfrac{3\pi}{7}$ **33.** Yes; $-\dfrac{\pi}{6}$ lies in the interval $\left[-\dfrac{\pi}{2}, \dfrac{\pi}{2}\right]$.   **35.** No; 2 is not in the domain of $\sin^{-1} x$.   **37.** No; $-\dfrac{\pi}{6}$ does not lie in the interval $[0, \pi]$.

39. Yes, $-\dfrac{1}{2}$ is in the domain of $\cos^{-1} x$.   **41.** Yes; $-\dfrac{\pi}{3}$ lies in the interval $\left(-\dfrac{\pi}{2}, \dfrac{\pi}{2}\right)$.   **43.** Yes; 2 is in the domain of $\tan^{-1} x$.

45. (a) 13.92 hr or 13 hr, 55 min   **(b)** 12 hr   **(c)** 13.85 hr or 13 hr, 51 min   **47. (a)** 13.3 hr of 13 hr, 18 min   **(b)** 12 hr
(c) 13.26 hr or 13 hr, 15 min   **49. (a)** 12 hr   **(b)** 12 hr   **(c)** 12 hr   **(d)** It's 12 hr.   **51.** 3.35 min

3.2 Exercises (page 176)

1. $\dfrac{\sqrt{2}}{2}$ **3.** $-\dfrac{\sqrt{3}}{3}$ **5.** 2   **7.** $\sqrt{2}$ **9.** $-\dfrac{\sqrt{2}}{2}$ **11.** $\dfrac{2\sqrt{3}}{3}$ **13.** $\dfrac{3\pi}{4}$ **15.** $\dfrac{\pi}{6}$ **17.** $\dfrac{\sqrt{2}}{4}$ **19.** $\dfrac{\sqrt{5}}{2}$ **21.** $-\dfrac{\sqrt{14}}{2}$ **23.** $-\dfrac{3\sqrt{10}}{10}$ **25.** $\sqrt{5}$ **27.** $-\dfrac{\pi}{4}$

29. $\dfrac{\pi}{6}$ **31.** $-\dfrac{\pi}{2}$ **33.** $\dfrac{\pi}{6}$ **35.** $\dfrac{2\pi}{3}$ **37.** 1.32   **39.** 0.46   **41.** -0.34 **43.** 2.72   **45.** -0.73 **47.** 2.55

49.

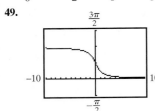

51.

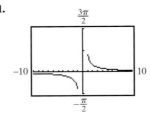

3.3 Exercises (page 181)

1. $\csc \theta \cdot \cos \theta = \dfrac{1}{\sin \theta} \cdot \cos \theta = \dfrac{\cos \theta}{\sin \theta} = \cot \theta$ **3.** $1 + \tan^{2}(-\theta) = 1 + (-\tan \theta)^{2} = 1 + \tan^{2} \theta = \sec^{2} \theta$

5. $\cos \theta(\tan \theta + \cot \theta) = \cos \theta\left(\dfrac{\sin \theta}{\cos \theta} + \dfrac{\cos \theta}{\sin \theta}\right) = \cos \theta\left(\dfrac{\sin^{2} \theta + \cos^{2} \theta}{\cos \theta \sin \theta}\right) = \cos \theta\left(\dfrac{1}{\cos \theta \sin \theta}\right) = \dfrac{1}{\sin \theta} = \csc \theta$

7. $\tan \theta \cot \theta - \cos^{2} \theta = \dfrac{\sin \theta}{\cos \theta} \cdot \dfrac{\cos \theta}{\sin \theta} - \cos^{2} \theta = 1 - \cos^{2} \theta = \sin^{2} \theta$ **9.** $(\sec \theta - 1)(\sec \theta + 1) = \sec^{2} \theta - 1 = \tan^{2} \theta$

11. $(\sec \theta + \tan \theta)(\sec \theta - \tan \theta) = \sec^{2} \theta - \tan^{2} \theta = 1$

13. $\cos^{2} \theta(1 + \tan^{2} \theta) = \cos^{2} \theta + \cos^{2} \theta \tan^{2} \theta = \cos^{2} \theta + \cos^{2} \theta \cdot \dfrac{\sin^{2} \theta}{\cos^{2} \theta} = \cos^{2} \theta + \sin^{2} \theta = 1$

15. $(\sin \theta + \cos \theta)^{2} + (\sin \theta - \cos \theta)^{2} = \sin^{2} \theta + 2 \sin \theta \cos \theta + \cos^{2} \theta + \sin^{2} \theta - 2 \sin \theta \cos \theta + \cos^{2} \theta$
$= \sin^{2} \theta + \cos^{2} \theta + \sin^{2} \theta + \cos^{2} \theta = 1 + 1 = 2$

17. $\sec^{4} \theta - \sec^{2} \theta = \sec^{2} \theta(\sec^{2} \theta - 1) = (1 + \tan^{2} \theta)\tan^{2} \theta = \tan^{4} \theta + \tan^{2} \theta$

19. $\sec \theta - \tan \theta = \dfrac{1}{\cos \theta} - \dfrac{\sin \theta}{\cos \theta} = \dfrac{1 - \sin \theta}{\cos \theta} \cdot \dfrac{1 + \sin \theta}{1 + \sin \theta} = \dfrac{1 - \sin^{2} \theta}{\cos \theta(1 + \sin \theta)} = \dfrac{\cos^{2} \theta}{\cos \theta(1 + \sin \theta)} = \dfrac{\cos \theta}{1 + \sin \theta}$

21. $3 \sin^{2} \theta + 4 \cos^{2} \theta = 3 \sin^{2} \theta + 3 \cos^{2} \theta + \cos^{2} \theta = 3(\sin^{2} \theta + \cos^{2} \theta) + \cos^{2} \theta = 3 + \cos^{2} \theta$

23. $1 - \dfrac{\cos^{2} \theta}{1 + \sin \theta} = 1 - \dfrac{1 - \sin^{2} \theta}{1 + \sin \theta} = 1 - \dfrac{(1 + \sin \theta)(1 - \sin \theta)}{1 + \sin \theta} = 1 - (1 - \sin \theta) = \sin \theta$

25. $\dfrac{1 + \tan \theta}{1 - \tan \theta} = \dfrac{1 + \dfrac{1}{\cot \theta}}{1 - \dfrac{1}{\cot \theta}} = \dfrac{\dfrac{\cot \theta + 1}{\cot \theta}}{\dfrac{\cot \theta - 1}{\cot \theta}} = \dfrac{\cot \theta + 1}{\cot \theta - 1}$ **27.** $\dfrac{\sec \theta}{\csc \theta} + \dfrac{\sin \theta}{\cos \theta} = \dfrac{\dfrac{1}{\cos \theta}}{\dfrac{1}{\sin \theta}} + \tan \theta = \dfrac{\sin \theta}{\cos \theta} + \tan \theta = \tan \theta + \tan \theta = 2 \tan \theta$

29. $\dfrac{1 + \sin \theta}{1 - \sin \theta} = \dfrac{1 + \dfrac{1}{\csc \theta}}{1 - \dfrac{1}{\csc \theta}} = \dfrac{\dfrac{\csc \theta + 1}{\csc \theta}}{\dfrac{\csc \theta - 1}{\csc \theta}} = \dfrac{\csc \theta + 1}{\csc \theta - 1}$

31. $\dfrac{1 - \sin \theta}{\cos \theta} + \dfrac{\cos \theta}{1 - \sin \theta} = \dfrac{(1 - \sin \theta)^{2} + \cos^{2} \theta}{\cos \theta(1 - \sin \theta)} = \dfrac{1 - 2 \sin \theta + \sin^{2} \theta + \cos^{2} \theta}{\cos \theta(1 - \sin \theta)} = \dfrac{2 - 2 \sin \theta}{\cos \theta(1 - \sin \theta)} = \dfrac{2(1 - \sin \theta)}{\cos \theta(1 - \sin \theta)} = \dfrac{2}{\cos \theta}$
$= 2 \sec \theta$

33. $\dfrac{\sin \theta}{\sin \theta - \cos \theta} = \dfrac{1}{\dfrac{\sin \theta - \cos \theta}{\sin \theta}} = \dfrac{1}{1 - \dfrac{\cos \theta}{\sin \theta}} = \dfrac{1}{1 - \cot \theta}$

35. $(\sec \theta - \tan \theta)^2 = \sec^2 \theta - 2 \sec \theta \tan \theta + \tan^2 \theta = \dfrac{1}{\cos^2 \theta} - \dfrac{2 \sin \theta}{\cos^2 \theta} + \dfrac{\sin^2 \theta}{\cos^2 \theta} = \dfrac{1 - 2 \sin \theta + \sin^2 \theta}{\cos^2 \theta} = \dfrac{(1 - \sin \theta)^2}{1 - \sin^2 \theta}$

$= \dfrac{(1 - \sin \theta)^2}{(1 - \sin \theta)(1 + \sin \theta)} = \dfrac{1 - \sin \theta}{1 + \sin \theta}$

37. $\dfrac{\cos \theta}{1 - \tan \theta} + \dfrac{\sin \theta}{1 - \cot \theta} = \dfrac{\cos \theta}{1 - \dfrac{\sin \theta}{\cos \theta}} + \dfrac{\sin \theta}{1 - \dfrac{\cos \theta}{\sin \theta}} = \dfrac{\cos \theta}{\dfrac{\cos \theta - \sin \theta}{\cos \theta}} + \dfrac{\sin \theta}{\dfrac{\sin \theta - \cos \theta}{\sin \theta}} = \dfrac{\cos^2 \theta}{\cos \theta - \sin \theta} + \dfrac{\sin^2 \theta}{\sin \theta - \cos \theta}$

$= \dfrac{\cos^2 \theta - \sin^2 \theta}{\cos \theta - \sin \theta} = \dfrac{(\cos \theta - \sin \theta)(\cos \theta + \sin \theta)}{\cos \theta - \sin \theta} = \sin \theta + \cos \theta$

39. $\tan \theta + \dfrac{\cos \theta}{1 + \sin \theta} = \dfrac{\sin \theta}{\cos \theta} + \dfrac{\cos \theta}{1 + \sin \theta} = \dfrac{\sin \theta(1 + \sin \theta) + \cos^2 \theta}{\cos \theta(1 + \sin \theta)} = \dfrac{\sin \theta + \sin^2 \theta + \cos^2 \theta}{\cos \theta(1 + \sin \theta)} = \dfrac{\sin \theta + 1}{\cos \theta(1 + \sin \theta)} = \dfrac{1}{\cos \theta} = \sec \theta$

41. $\dfrac{\tan \theta + \sec \theta - 1}{\tan \theta - \sec \theta + 1} = \dfrac{\tan \theta + (\sec \theta - 1)}{\tan \theta - (\sec \theta - 1)} \cdot \dfrac{\tan \theta + (\sec \theta - 1)}{\tan \theta + (\sec \theta - 1)} = \dfrac{\tan^2 \theta + 2 \tan \theta(\sec \theta - 1) + \sec^2 \theta - 2 \sec \theta + 1}{\tan^2 \theta - (\sec^2 \theta - 2 \sec \theta + 1)}$

$= \dfrac{\sec^2 \theta - 1 + 2 \tan \theta(\sec \theta - 1) + \sec^2 \theta - 2 \sec \theta + 1}{\sec^2 \theta - 1 - \sec^2 \theta + 2 \sec \theta - 1} = \dfrac{2 \sec^2 \theta - 2 \sec \theta + 2 \tan \theta(\sec \theta - 1)}{-2 + 2 \sec \theta}$

$= \dfrac{2 \sec \theta(\sec \theta - 1) + 2 \tan \theta(\sec \theta - 1)}{2(\sec \theta - 1)} = \dfrac{2(\sec \theta - 1)(\sec \theta + \tan \theta)}{2(\sec \theta - 1)} = \tan \theta + \sec \theta$

43. $\dfrac{\tan \theta - \cot \theta}{\tan \theta + \cot \theta} = \dfrac{\dfrac{\sin \theta}{\cos \theta} - \dfrac{\cos \theta}{\sin \theta}}{\dfrac{\sin \theta}{\cos \theta} + \dfrac{\cos \theta}{\sin \theta}} = \dfrac{\dfrac{\sin^2 \theta - \cos^2 \theta}{\cos \theta \sin \theta}}{\dfrac{\sin^2 \theta + \cos^2 \theta}{\cos \theta \sin \theta}} = \dfrac{\sin^2 \theta - \cos^2 \theta}{1} = \sin^2 \theta - \cos^2 \theta$

45. $\dfrac{\tan \theta - \cot \theta}{\tan \theta + \cot \theta} + 1 = \dfrac{\dfrac{\sin \theta}{\cos \theta} - \dfrac{\cos \theta}{\sin \theta}}{\dfrac{\sin \theta}{\cos \theta} + \dfrac{\cos \theta}{\sin \theta}} + 1 = \dfrac{\dfrac{\sin^2 \theta - \cos^2 \theta}{\cos \theta \sin \theta}}{\dfrac{\sin^2 \theta + \cos^2 \theta}{\cos \theta \sin \theta}} + 1 = \sin^2\theta - \cos^2 \theta + 1 = \sin^2 \theta + (1 - \cos^2 \theta) = 2 \sin^2 \theta$

47. $\dfrac{\sec \theta + \tan \theta}{\cot \theta + \cos \theta} = \dfrac{\dfrac{1}{\cos \theta} + \dfrac{\sin \theta}{\cos \theta}}{\dfrac{\cos \theta}{\sin \theta} + \cos \theta} = \dfrac{\dfrac{1 + \sin \theta}{\cos \theta}}{\dfrac{\cos \theta + \cos \theta \sin \theta}{\sin \theta}} = \dfrac{1 + \sin \theta}{\cos \theta} \cdot \dfrac{\sin \theta}{\cos \theta(1 + \sin \theta)} = \dfrac{\sin \theta}{\cos \theta} \cdot \dfrac{1}{\cos \theta} = \tan \theta \sec \theta$

49. $\dfrac{1 - \tan^2 \theta}{1 + \tan^2 \theta} + 1 = \dfrac{1 - \tan^2 \theta}{\sec^2 \theta} + 1 = \dfrac{1}{\sec^2 \theta} - \dfrac{\tan^2 \theta}{\sec^2 \theta} + 1 = \cos^2 \theta - \dfrac{\dfrac{\sin^2 \theta}{\cos^2 \theta}}{\dfrac{1}{\cos^2 \theta}} + 1 = \cos^2 \theta - \sin^2 \theta + 1$

$= \cos^2 \theta + (1 - \sin^2 \theta) = 2 \cos^2 \theta$

51. $\dfrac{\sec \theta - \csc \theta}{\sec \theta \csc \theta} = \dfrac{\sec \theta}{\sec \theta \csc \theta} - \dfrac{\csc \theta}{\sec \theta \csc \theta} = \dfrac{1}{\csc \theta} - \dfrac{1}{\sec \theta} = \sin \theta - \cos \theta$

53. $\sec \theta - \cos \theta - \sin \theta \tan \theta = \left(\dfrac{1}{\cos \theta} - \cos \theta \right) - \sin \theta \cdot \dfrac{\sin \theta}{\cos \theta} = \dfrac{1 - \cos^2 \theta}{\cos \theta} - \dfrac{\sin^2 \theta}{\cos \theta} = \dfrac{\sin^2 \theta}{\cos \theta} - \dfrac{\sin^2 \theta}{\cos \theta} = 0$

55. $\dfrac{1}{1 - \sin \theta} + \dfrac{1}{1 + \sin \theta} = \dfrac{1 + \sin \theta + 1 - \sin \theta}{(1 + \sin \theta)(1 - \sin \theta)} = \dfrac{2}{1 - \sin^2 \theta} = \dfrac{2}{\cos^2 \theta} = 2 \sec^2 \theta$

57. $\dfrac{\sec \theta}{1 - \sin \theta} = \dfrac{\sec \theta}{1 - \sin \theta} \cdot \dfrac{1 + \sin \theta}{1 + \sin \theta} = \dfrac{\sec \theta(1 + \sin \theta)}{1 - \sin^2 \theta} = \dfrac{\sec \theta(1 + \sin \theta)}{\cos^2 \theta} = \dfrac{1 + \sin \theta}{\cos^3 \theta}$

59. $\dfrac{(\sec \theta - \tan \theta)^2 + 1}{\csc \theta(\sec \theta - \tan \theta)} = \dfrac{\sec^2 \theta - 2 \sec \theta \tan \theta + \tan^2 \theta + 1}{\dfrac{1}{\sin \theta}\left(\dfrac{1}{\cos \theta} - \dfrac{\sin \theta}{\cos \theta} \right)} = \dfrac{2 \sec^2 \theta - 2 \sec \theta \tan \theta}{\dfrac{1}{\sin \theta}\left(\dfrac{1 - \sin \theta}{\cos \theta} \right)} = \dfrac{\dfrac{2}{\cos^2 \theta} - \dfrac{2 \sin \theta}{\cos^2 \theta}}{\dfrac{1 - \sin \theta}{\sin \theta \cos \theta}} = \dfrac{2 - 2 \sin \theta}{\cos^2 \theta} \cdot \dfrac{\sin \theta \cos \theta}{1 - \sin \theta}$

$= \dfrac{2(1 - \sin \theta)}{\cos \theta} \cdot \dfrac{\sin \theta}{1 - \sin \theta} = \dfrac{2 \sin \theta}{\cos \theta} = 2 \tan \theta$

61. $\dfrac{\sin \theta + \cos \theta}{\cos \theta} - \dfrac{\sin \theta - \cos \theta}{\sin \theta} = \dfrac{\sin \theta}{\cos \theta} + 1 - 1 + \dfrac{\cos \theta}{\sin \theta} = \dfrac{\sin^2 \theta + \cos^2 \theta}{\cos \theta \sin \theta} = \dfrac{1}{\cos \theta \sin \theta} = \sec \theta \csc \theta$

63. $\dfrac{\sin^3\theta + \cos^3\theta}{\sin\theta + \cos\theta} = \dfrac{(\sin\theta + \cos\theta)(\sin^2\theta - \sin\theta\cos\theta + \cos^2\theta)}{\sin\theta + \cos\theta} = \sin^2\theta + \cos^2\theta - \sin\theta\cos\theta = 1 - \sin\theta\cos\theta$

65. $\dfrac{\cos^2\theta - \sin^2\theta}{1 - \tan^2\theta} = \dfrac{\cos^2\theta - \sin^2\theta}{1 - \dfrac{\sin^2\theta}{\cos^2\theta}} = \dfrac{\cos^2\theta - \sin^2\theta}{\dfrac{\cos^2\theta - \sin^2\theta}{\cos^2\theta}} = \cos^2\theta$

67. $\dfrac{(2\cos^2\theta - 1)^2}{\cos^4\theta - \sin^4\theta} = \dfrac{[2\cos^2\theta - (\sin^2\theta + \cos^2\theta)]^2}{(\cos^2\theta - \sin^2\theta)(\cos^2\theta + \sin^2\theta)} = \dfrac{(\cos^2\theta - \sin^2\theta)^2}{\cos^2\theta - \sin^2\theta} = \cos^2\theta - \sin^2\theta = (1 - \sin^2\theta) - \sin^2\theta = 1 - 2\sin^2\theta$

69. $\dfrac{1 + \sin\theta + \cos\theta}{1 + \sin\theta - \cos\theta} = \dfrac{(1 + \sin\theta) + \cos\theta}{(1 + \sin\theta) - \cos\theta} \cdot \dfrac{(1 + \sin\theta) + \cos\theta}{(1 + \sin\theta) + \cos\theta} = \dfrac{1 + 2\sin\theta + \sin^2\theta + 2(1 + \sin\theta)(\cos\theta) + \cos^2\theta}{1 + 2\sin\theta + \sin^2\theta - \cos^2\theta}$

$= \dfrac{1 + 2\sin\theta + \sin^2\theta + 2(1 + \sin\theta)(\cos\theta) + (1 - \sin^2\theta)}{1 + 2\sin\theta + \sin^2\theta - (1 - \sin^2\theta)} = \dfrac{2 + 2\sin\theta + 2(1 + \sin\theta)(\cos\theta)}{2\sin\theta + 2\sin^2\theta}$

$= \dfrac{2(1 + \sin\theta) + 2(1 + \sin\theta)(\cos\theta)}{2\sin\theta(1 + \sin\theta)} = \dfrac{2(1 + \sin\theta)(1 + \cos\theta)}{2\sin\theta(1 + \sin\theta)} = \dfrac{1 + \cos\theta}{\sin\theta}$

71. $(a\sin\theta + b\cos\theta)^2 + (a\cos\theta - b\sin\theta)^2 = a^2\sin^2\theta + 2ab\sin\theta\cos\theta + b^2\cos^2\theta + a^2\cos^2\theta - 2ab\sin\theta\cos\theta + b^2\sin^2\theta$
$= a^2(\sin^2\theta + \cos^2\theta) + b^2(\cos^2\theta + \sin^2\theta) = a^2 + b^2$

73. $\dfrac{\tan\alpha + \tan\beta}{\cot\alpha + \cot\beta} = \dfrac{\tan\alpha + \tan\beta}{\dfrac{1}{\tan\alpha} + \dfrac{1}{\tan\beta}} = \dfrac{\tan\alpha + \tan\beta}{\dfrac{\tan\beta + \tan\alpha}{\tan\alpha\,\tan\beta}} = (\tan\alpha + \tan\beta) \cdot \dfrac{\tan\alpha\,\tan\beta}{\tan\alpha + \tan\beta} = \tan\alpha\,\tan\beta$

75. $(\sin\alpha + \cos\beta)^2 + (\cos\beta + \sin\alpha)(\cos\beta - \sin\alpha) = (\sin^2\alpha + 2\sin\alpha\cos\beta + \cos^2\beta) 1 + (\cos^2\beta - \sin^2\alpha)$
$= 2\cos^2\beta + 2\sin\alpha\cos\beta = 2\cos\beta(\cos\beta + \sin\alpha)$

77. $\ln|\sec\theta| = \ln|\cos\theta|^{-1} = -\ln|\cos\theta|$

79. $\ln|1 + \cos\theta| + \ln|1 - \cos\theta| = \ln(|1 + \cos\theta||1 - \cos\theta|) = \ln|1 - \cos^2\theta| = \ln|\sin^2\theta| = 2\ln|\sin\theta|$

81. Let $\theta = \tan^{-1} v$. Then $\tan\theta = v, -\dfrac{\pi}{2} < \theta < \dfrac{\pi}{2}$. Now, $\sec\theta > 0$ and $\tan^2\theta + 1 = \sec^2\theta$. Thus $\sec(\tan^{-1} v) = \sec\theta = \sqrt{1 + v^2}$.

83. Let $\theta = \cos^{-1} v$. Then $\cos\theta = v, 0 \le \theta \le \pi$, and $\tan(\cos^{-1} v) = \tan\theta = \dfrac{\sin\theta}{\cos\theta} = \dfrac{\sqrt{1 - \cos^2\theta}}{\cos\theta} = \dfrac{\sqrt{1 - v^2}}{v}$.

85. Let $\theta = \sin^{-1} v$. Then $\sin\theta = v, -\dfrac{\pi}{2} \le \theta \le \dfrac{\pi}{2}$, and $\cos(\sin^{-1} v) = \cos\theta = \sqrt{1 - \sin^2\theta} = \sqrt{1 - v^2}$.

3.4 Exercises *(page 190)*

1. $\dfrac{1}{4}(\sqrt{6} + \sqrt{2})$ **3.** $\dfrac{1}{4}(\sqrt{2} - \sqrt{6})$ **5.** $-\dfrac{1}{4}(\sqrt{2} + \sqrt{6})$ **7.** $2 - \sqrt{3}$ **9.** $-\dfrac{1}{4}(\sqrt{6} + \sqrt{2})$ **11.** $\sqrt{6} - \sqrt{2}$ **13.** $\dfrac{1}{2}$ **15.** 0 **17.** 1

19. -1 **21.** $\dfrac{1}{2}$ **23. (a)** $\dfrac{2\sqrt{5}}{25}$ **(b)** $\dfrac{11\sqrt{5}}{25}$ **(c)** $\dfrac{2\sqrt{5}}{5}$ **(d)** 2 **25. (a)** $\dfrac{4 - 3\sqrt{3}}{10}$ **(b)** $\dfrac{-3 - 4\sqrt{3}}{10}$ **(c)** $\dfrac{4 + 3\sqrt{3}}{10}$ **(d)** $\dfrac{25\sqrt{3} + 48}{39}$

27. (a) $-\dfrac{5 + 12\sqrt{3}}{26}$ **(b)** $\dfrac{12 - 5\sqrt{3}}{26}$ **(c)** $-\dfrac{5 - 12\sqrt{3}}{26}$ **(d)** $\dfrac{-240 + 169\sqrt{3}}{69}$ **29. (a)** $-\dfrac{2\sqrt{2}}{3}$ **(b)** $\dfrac{-2\sqrt{2} + \sqrt{3}}{6}$ **(c)** $\dfrac{-2\sqrt{2} + \sqrt{3}}{6}$

(d) $\dfrac{9 - 4\sqrt{2}}{7}$ **31.** $\sin\left(\dfrac{\pi}{2} + \theta\right) = \sin\dfrac{\pi}{2}\cos\theta + \cos\dfrac{\pi}{2}\sin\theta = 1 \cdot \cos\theta + 0 \cdot \sin\theta = \cos\theta$

33. $\sin(\pi - \theta) = \sin\pi\cos\theta - \cos\pi\sin\theta = 0 \cdot \cos\theta - (-1)\sin\theta = \sin\theta$

35. $\sin(\pi + \theta) = \sin\pi\cos\theta + \cos\pi\sin\theta = 0 \cdot \cos\theta + (-1)\sin\theta = -\sin\theta$

37. $\tan(\pi - \theta) = \dfrac{\tan\pi - \tan\theta}{1 + \tan\pi\tan\theta} = \dfrac{0 - \tan\theta}{1 + 0 \cdot \tan\theta} = -\tan\theta$

39. $\sin\left(\dfrac{3\pi}{2} + \theta\right) = \sin\dfrac{3\pi}{2}\cos\theta + \cos\dfrac{3\pi}{2}\sin\theta = (-1)\cos\theta + 0 \cdot \sin\theta = -\cos\theta$

41. $\sin(\alpha + \beta) + \sin(\alpha - \beta) = \sin\alpha\cos\beta + \cos\alpha\sin\beta + \sin\alpha\cos\beta - \cos\alpha\sin\beta = 2\sin\alpha\cos\beta$

43. $\dfrac{\sin(\alpha + \beta)}{\sin\alpha\cos\beta} = \dfrac{\sin\alpha\cos\beta + \cos\alpha\sin\beta}{\sin\alpha\cos\beta} = \dfrac{\sin\alpha\cos\beta}{\sin\alpha\cos\beta} + \dfrac{\cos\alpha\sin\beta}{\sin\alpha\cos\beta} = 1 + \cot\alpha\tan\beta$

45. $\dfrac{\cos(\alpha + \beta)}{\cos\alpha\cos\beta} = \dfrac{\cos\alpha\cos\beta - \sin\alpha\sin\beta}{\cos\alpha\cos\beta} = \dfrac{\cos\alpha\cos\beta}{\cos\alpha\cos\beta} - \dfrac{\sin\alpha\sin\beta}{\cos\alpha\cos\beta} = 1 - \tan\alpha\tan\beta$

47. $\dfrac{\sin(\alpha + \beta)}{\sin(\alpha - \beta)} = \dfrac{\sin\alpha\cos\beta + \cos\alpha\sin\beta}{\sin\alpha\cos\beta - \cos\alpha\sin\beta} = \dfrac{\dfrac{\sin\alpha\cos\beta + \cos\alpha\sin\beta}{\cos\alpha\cos\beta}}{\dfrac{\sin\alpha\cos\beta - \cos\alpha\sin\beta}{\cos\alpha\cos\beta}} = \dfrac{\dfrac{\sin\alpha\cos\beta}{\cos\alpha\cos\beta} + \dfrac{\cos\alpha\sin\beta}{\cos\alpha\cos\beta}}{\dfrac{\sin\alpha\cos\beta}{\cos\alpha\cos\beta} - \dfrac{\cos\alpha\sin\beta}{\cos\alpha\cos\beta}} = \dfrac{\tan\alpha + \tan\beta}{\tan\alpha - \tan\beta}$

49. $\cot(\alpha + \beta) = \dfrac{\cos(\alpha + \beta)}{\sin(\alpha + \beta)} = \dfrac{\cos\alpha\cos\beta - \sin\alpha\sin\beta}{\sin\alpha\cos\beta + \cos\alpha\sin\beta} = \dfrac{\frac{\cos\alpha\cos\beta - \sin\alpha\sin\beta}{\sin\alpha\sin\beta}}{\frac{\sin\alpha\cos\beta + \cos\alpha\sin\beta}{\sin\alpha\sin\beta}} = \dfrac{\frac{\cos\alpha\cos\beta}{\sin\alpha\sin\beta} - \frac{\sin\alpha\sin\beta}{\sin\alpha\sin\beta}}{\frac{\sin\alpha\cos\beta}{\sin\alpha\sin\beta} + \frac{\cos\alpha\sin\beta}{\sin\alpha\sin\beta}} = \dfrac{\cot\alpha\cot\beta - 1}{\cot\beta + \cot\alpha}$

51. $\sec(\alpha + \beta) = \dfrac{1}{\cos(\alpha + \beta)} = \dfrac{1}{\cos\alpha\cos\beta - \sin\alpha\sin\beta} = \dfrac{\frac{1}{\sin\alpha\sin\beta}}{\frac{\cos\alpha\cos\beta - \sin\alpha\sin\beta}{\sin\alpha\sin\beta}} = \dfrac{\frac{1}{\sin\alpha}\cdot\frac{1}{\sin\beta}}{\frac{\cos\alpha\cos\beta}{\sin\alpha\sin\beta} - \frac{\sin\alpha\sin\beta}{\sin\alpha\sin\beta}} = \dfrac{\csc\alpha\csc\beta}{\cot\alpha\cot\beta - 1}$

53. $\sin(\alpha - \beta)\sin(\alpha + \beta) = (\sin\alpha\cos\beta - \cos\alpha\sin\beta)(\sin\alpha\cos\beta + \cos\alpha\sin\beta) = \sin^2\alpha\cos^2\beta - \cos^2\alpha\sin^2\beta$
$= (\sin^2\alpha)(1 - \sin^2\beta) - (1 - \sin^2\alpha)(\sin^2\beta) = \sin^2\alpha - \sin^2\beta$

55. $\sin(\theta + k\pi) = \sin\theta\cos k\pi + \cos\theta\sin k\pi = (\sin\theta)(-1)^k + (\cos\theta)(0) = (-1)^k\sin\theta, k$ any integer **57.** $\dfrac{\sqrt3}{2}$ **59.** $-\dfrac{24}{25}$ **61.** $-\dfrac{33}{65}$

63. $\dfrac{63}{65}$ **65.** $\dfrac{48 + 25\sqrt3}{39}$ **67.** $\dfrac{4}{3}$ **69.** $u\sqrt{1 - v^2} - v\sqrt{1 - u^2}$ **71.** $\dfrac{u\sqrt{1 - v^2} - v}{\sqrt{1 + u^2}}$ **73.** $\dfrac{uv - \sqrt{1 - u^2}\sqrt{1 - v^2}}{v\sqrt{1 - u^2} + u\sqrt{1 - v^2}}$

75. Let $\alpha = \sin^{-1}v$ and $\beta = \cos^{-1}v$. Then $\sin\alpha = \cos\beta = v$, and since $\sin\alpha = \cos\left(\dfrac{\pi}{2} - \alpha\right)$, $\cos\left(\dfrac{\pi}{2} - \alpha\right) = \cos\beta$.

If $v \ge 0$, then $0 \le \alpha \le \dfrac{\pi}{2}$, so that $\left(\dfrac{\pi}{2} - \alpha\right)$ and β both lie on $\left[0, \dfrac{\pi}{2}\right]$. If $v < 0$, then $-\dfrac{\pi}{2} \le \alpha < 0$, so that $\left(\dfrac{\pi}{2} - \alpha\right)$ and β both lie on

$\left(\dfrac{\pi}{2}, \pi\right]$. Either way, $\cos\left(\dfrac{\pi}{2} - \alpha\right) = \cos\beta$ implies $\dfrac{\pi}{2} - \alpha = \beta$, or $\alpha + \beta = \dfrac{\pi}{2}$.

77. Let $\alpha = \tan^{-1}\dfrac{1}{v}$, and $\beta = \tan^{-1}v$. Because $v \ne 0$, $\alpha, \beta \ne 0$. Then $\tan\alpha = \dfrac{1}{v} = \dfrac{1}{\tan\beta} = \cot\beta$, and since

$\tan\alpha = \cot\left(\dfrac{\pi}{2} - \alpha\right)$, $\cot\left(\dfrac{\pi}{2} - \alpha\right) = \cot\beta$. Because $v > 0, 0 < \alpha < \dfrac{\pi}{2}$ and so $\left(\dfrac{\pi}{2} - \alpha\right)$ and β both lie on $\left(0, \dfrac{\pi}{2}\right)$.

Then $\cot\left(\dfrac{\pi}{2} - \alpha\right) = \cot\beta$ implies $\dfrac{\pi}{2} - \alpha = \beta$, or $\alpha + \beta = \dfrac{\pi}{2}$.

79. $\sin(\sin^{-1}v + \cos^{-1}v) = \sin(\sin^{-1}v)\cos(\cos^{-1}v) + \cos(\sin^{-1}v)\sin(\cos^{-1}v) = (v)(v) + \sqrt{1 - v^2}\sqrt{1 - v^2} = v^2 + 1 - v^2 = 1$

81. $\dfrac{\sin(x + h) - \sin x}{h} = \dfrac{\sin x\cos h + \cos x\sin h - \sin x}{h} = \dfrac{\cos x\sin h - \sin x(1 - \cos h)}{h} = \cos x\cdot\dfrac{\sin h}{h} - \sin x\cdot\dfrac{1 - \cos h}{h}$

83. $\tan\dfrac{\pi}{2}$ is not defined; $\tan\left(\dfrac{\pi}{2} - \theta\right) = \dfrac{\sin\left(\dfrac{\pi}{2} - \theta\right)}{\cos\left(\dfrac{\pi}{2} - \theta\right)} = \dfrac{\cos\theta}{\sin\theta} = \cot\theta$ **85.** $\tan\theta = \tan(\theta_2 - \theta_1) = \dfrac{\tan\theta_2 - \tan\theta_1}{1 + \tan\theta_1\tan\theta_2} = \dfrac{m_2 - m_1}{1 + m_1m_2}$

87. No; $\tan\left(\dfrac{\pi}{2}\right)$ is undefined.

3.5 Exercises *(page 199)*

1. (a) $\dfrac{24}{25}$ **(b)** $\dfrac{7}{25}$ **(c)** $\dfrac{\sqrt{10}}{10}$ **(d)** $\dfrac{3\sqrt{10}}{10}$ **3. (a)** $\dfrac{24}{25}$ **(b)** $-\dfrac{7}{25}$ **(c)** $\dfrac{2\sqrt5}{5}$ **(d)** $-\dfrac{\sqrt5}{5}$

5. (a) $-\dfrac{2\sqrt2}{3}$ **(b)** $\dfrac{1}{3}$ **(c)** $\sqrt{\dfrac{3 + \sqrt6}{6}}$ **(d)** $\sqrt{\dfrac{3 - \sqrt6}{6}}$ **7. (a)** $\dfrac{4\sqrt2}{9}$ **(b)** $-\dfrac{7}{9}$ **(c)** $\dfrac{\sqrt3}{3}$ **(d)** $\dfrac{\sqrt6}{3}$

9. (a) $-\dfrac{4}{5}$ **(b)** $\dfrac{3}{5}$ **(c)** $\sqrt{\dfrac{5 + 2\sqrt5}{10}}$ **(d)** $\sqrt{\dfrac{5 - 2\sqrt5}{10}}$ **11. (a)** $-\dfrac{3}{5}$ **(b)** $-\dfrac{4}{5}$ **(c)** $\dfrac{1}{2}\sqrt{\dfrac{10 - \sqrt{10}}{5}}$ **(d)** $-\dfrac{1}{2}\sqrt{\dfrac{10 + \sqrt{10}}{5}}$

13. $\dfrac{\sqrt{2 - \sqrt2}}{2}$ **15.** $1 - \sqrt2$ **17.** $-\dfrac{\sqrt{2 + \sqrt3}}{2}$ **19.** $\dfrac{2}{\sqrt{2 + \sqrt2}} = (2 - \sqrt2)\sqrt{2 + \sqrt2}$ **21.** $-\dfrac{\sqrt{2 - \sqrt2}}{2}$

23. $\sin^4\theta = (\sin^2\theta)^2 = \left(\dfrac{1 - \cos(2\theta)}{2}\right)^2 = \dfrac{1}{4}(1 - 2\cos(2\theta) + \cos^2(2\theta)) = \dfrac{1}{4} - \dfrac{1}{2}\cos(2\theta) + \dfrac{1}{4}\cos^2(2\theta)$

$= \dfrac{1}{4} - \dfrac{1}{2}\cos(2\theta) + \dfrac{1}{4}\left(\dfrac{1 + \cos(4\theta)}{2}\right) = \dfrac{1}{4} - \dfrac{1}{2}\cos(2\theta) + \dfrac{1}{8} + \dfrac{1}{8}\cos(4\theta) = \dfrac{3}{8} - \dfrac{1}{2}\cos(2\theta) + \dfrac{1}{8}\cos(4\theta)$

25. $\sin(4\theta) = \sin[2(2\theta)] = 2\sin(2\theta)\cos(2\theta) = (4\sin\theta\cos\theta)(1 - 2\sin^2\theta) = 4\sin\theta\cos\theta - 8\sin^3\theta\cos\theta = (\cos\theta)(4\sin\theta - 8\sin^3\theta)$
27. $\sin(5\theta) = 16\sin^5\theta - 20\sin^3\theta + 5\sin\theta$
29. $\cos^4\theta - \sin^4\theta = (\cos^2\theta + \sin^2\theta)(\cos^2\theta - \sin^2\theta) = \cos(2\theta)$

31. $\cot(2\theta) = \dfrac{1}{\tan(2\theta)} = \dfrac{1}{\dfrac{2\tan\theta}{1-\tan^2\theta}} = \dfrac{1-\tan^2\theta}{2\tan\theta} = \dfrac{1-\dfrac{1}{\cot^2\theta}}{2\left(\dfrac{1}{\cot\theta}\right)} = \dfrac{\dfrac{\cot^2\theta-1}{\cot^2\theta}}{\dfrac{2}{\cot\theta}} = \dfrac{\cot^2\theta-1}{\cot^2\theta}\cdot\dfrac{\cot\theta}{2} = \dfrac{\cot^2\theta-1}{2\cot\theta}$

33. $\sec(2\theta) = \dfrac{1}{\cos(2\theta)} = \dfrac{1}{2\cos^2\theta-1} = \dfrac{1}{\dfrac{2}{\sec^2\theta}-1} = \dfrac{1}{\dfrac{2-\sec^2\theta}{\sec^2\theta}} = \dfrac{\sec^2\theta}{2-\sec^2\theta}$ **35.** $\cos^2(2\theta) - \sin^2(2\theta) = \cos[2(2\theta)] = \cos(4\theta)$

37. $\dfrac{\cos(2\theta)}{1+\sin(2\theta)} = \dfrac{\cos^2\theta-\sin^2\theta}{1+2\sin\theta\cos\theta} = \dfrac{(\cos\theta-\sin\theta)(\cos\theta+\sin\theta)}{\sin^2\theta+\cos^2\theta+2\sin\theta\cos\theta} = \dfrac{(\cos\theta-\sin\theta)(\cos\theta+\sin\theta)}{(\sin\theta+\cos\theta)(\sin\theta+\cos\theta)} = \dfrac{\cos\theta-\sin\theta}{\cos\theta+\sin\theta}$

$= \dfrac{\dfrac{\cos\theta-\sin\theta}{\sin\theta}}{\dfrac{\cos\theta+\sin\theta}{\sin\theta}} = \dfrac{\dfrac{\cos\theta}{\sin\theta}-\dfrac{\sin\theta}{\sin\theta}}{\dfrac{\cos\theta}{\sin\theta}+\dfrac{\sin\theta}{\sin\theta}} = \dfrac{\cot\theta-1}{\cot\theta+1}$ **39.** $\sec^2\dfrac{\theta}{2} = \dfrac{1}{\cos^2\left(\dfrac{\theta}{2}\right)} = \dfrac{1}{\dfrac{1+\cos\theta}{2}} = \dfrac{2}{1+\cos\theta}$

41. $\cot^2\dfrac{\theta}{2} = \dfrac{1}{\tan^2\left(\dfrac{\theta}{2}\right)} = \dfrac{1}{\dfrac{1-\cos\theta}{1+\cos\theta}} = \dfrac{1+\cos\theta}{1-\cos\theta} = \dfrac{1+\dfrac{1}{\sec\theta}}{1-\dfrac{1}{\sec\theta}} = \dfrac{\dfrac{\sec\theta+1}{\sec\theta}}{\dfrac{\sec\theta-1}{\sec\theta}} = \dfrac{\sec\theta+1}{\sec\theta}\cdot\dfrac{\sec\theta}{\sec\theta-1} = \dfrac{\sec\theta+1}{\sec\theta-1}$

43. $\dfrac{1-\tan^2\left(\dfrac{\theta}{2}\right)}{1+\tan^2\left(\dfrac{\theta}{2}\right)} = \dfrac{1-\dfrac{1-\cos\theta}{1+\cos\theta}}{1+\dfrac{1-\cos\theta}{1+\cos\theta}} = \dfrac{\dfrac{1+\cos\theta-(1-\cos\theta)}{1+\cos\theta}}{\dfrac{1+\cos\theta+1-\cos\theta}{1+\cos\theta}} = \dfrac{2\cos\theta}{1+\cos\theta}\cdot\dfrac{1+\cos\theta}{2} = \cos\theta$

45. $\dfrac{\sin(3\theta)}{\sin\theta} - \dfrac{\cos(3\theta)}{\cos\theta} = \dfrac{\sin(3\theta)\cos\theta-\cos(3\theta)\sin\theta}{\sin\theta\cos\theta} = \dfrac{\sin(3\theta-\theta)}{\dfrac{1}{2}(2\sin\theta\cos\theta)} = \dfrac{2\sin(2\theta)}{\sin(2\theta)} = 2$

47. $\tan(3\theta) = \tan(\theta+2\theta) = \dfrac{\tan\theta+\tan(2\theta)}{1-\tan\theta\tan(2\theta)} = \dfrac{\tan\theta+\dfrac{2\tan\theta}{1-\tan^2\theta}}{1-\dfrac{\tan\theta(2\tan\theta)}{1-\tan^2\theta}} = \dfrac{\tan\theta-\tan^3\theta+2\tan\theta}{1-\tan^2\theta-2\tan^2\theta} = \dfrac{3\tan\theta-\tan^3\theta}{1-3\tan^2\theta}$

49. $\dfrac{1}{2}(\ln|1-\cos(2\theta)| - \ln 2) = \ln\left(\dfrac{|1-\cos(2\theta)|}{2}\right)^{1/2} = \ln|\sin^2\theta|^{1/2} = \ln|\sin\theta|$

51. $\dfrac{\sqrt{3}}{2}$ **53.** $\dfrac{7}{25}$ **55.** $\dfrac{24}{7}$ **57.** $\dfrac{24}{25}$ **59.** $\dfrac{1}{5}$ **61.** $\dfrac{25}{7}$ **63.** $\sin(2\theta) = \dfrac{4x}{4+x^2}$ **65.** $-\dfrac{1}{4}$

67. $\dfrac{2z}{1+z^2} = \dfrac{2\tan\left(\dfrac{\alpha}{2}\right)}{1+\tan^2\left(\dfrac{\alpha}{2}\right)} = \dfrac{2\tan\left(\dfrac{\alpha}{2}\right)}{\sec^2\left(\dfrac{\alpha}{2}\right)} = \dfrac{\dfrac{2\sin\left(\dfrac{\alpha}{2}\right)}{\cos\left(\dfrac{\alpha}{2}\right)}}{\dfrac{1}{\cos^2\left(\dfrac{\alpha}{2}\right)}} = 2\sin\left(\dfrac{\alpha}{2}\right)\cos\left(\dfrac{\alpha}{2}\right) = \sin\left(2\cdot\dfrac{\alpha}{2}\right) = \sin\alpha$

69. $A = \dfrac{1}{2}h(\text{base}) = h\left(\dfrac{1}{2}\text{ base}\right) = s\cos\dfrac{\theta}{2}\cdot s\sin\dfrac{\theta}{2} = \dfrac{1}{2}s^2\sin\theta$ **71.**

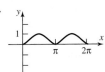

73. $\sin\dfrac{\pi}{24} = \dfrac{\sqrt{2}}{4}\sqrt{4-\sqrt{6}-\sqrt{2}}; \cos\dfrac{\pi}{24} = \dfrac{\sqrt{2}}{4}\sqrt{4+\sqrt{6}+\sqrt{2}}$

75. $\sin^3\theta + \sin^3(\theta+120°) + \sin^3(\theta+240°) = \sin^3\theta + (\sin\theta\cos 120° + \cos\theta\sin 120°)^3 + (\sin\theta\cos 240° + \cos\theta\sin 240°)^3$

$= \sin^3\theta + \left(-\dfrac{1}{2}\sin\theta + \dfrac{\sqrt{3}}{2}\cos\theta\right)^3 + \left(-\dfrac{1}{2}\sin\theta - \dfrac{\sqrt{3}}{2}\cos\theta\right)^3$

$= \sin^3\theta + \dfrac{1}{8}(3\sqrt{3}\cos^3\theta - 9\cos^2\theta\sin\theta + 3\sqrt{3}\cos\theta\sin^2\theta - \sin^3\theta) - \dfrac{1}{8}(\sin^3\theta + 3\sqrt{3}\sin^2\theta\cos\theta + 9\sin\theta\cos^2\theta + 3\sqrt{3}\cos^3\theta)$

$= \dfrac{3}{4}\sin^3\theta - \dfrac{9}{4}\cos^2\theta\sin\theta = \dfrac{3}{4}[\sin^3\theta - 3\sin\theta(1-\sin^2\theta)] = \dfrac{3}{4}(4\sin^3\theta - 3\sin\theta) = -\dfrac{3}{4}\sin(3\theta)$ (from Example 2)

77. (a) $R = \dfrac{v_0^2\sqrt{2}}{16}(\sin\theta\cos\theta - \cos^2\theta)$ **(b)** 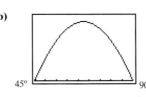 **(c)** $\theta = 67.5°$ makes R largest

$= \dfrac{v_0^2\sqrt{2}}{16}\left(\dfrac{1}{2}\sin(2\theta) - \dfrac{1+\cos(2\theta)}{2}\right)$

$= \dfrac{v_0^2\sqrt{2}}{32}[\sin(2\theta) - \cos(2\theta) - 1]$

45° 90°

3.6 Exercises *(page 203)*

1. $\dfrac{1}{2}[\cos(2\theta) - \cos(6\theta)]$ **3.** $\dfrac{1}{2}[\sin(6\theta) + \sin(2\theta)]$ **5.** $\dfrac{1}{2}[\cos(2\theta) + \cos(8\theta)]$ **7.** $\dfrac{1}{2}[\cos\theta - \cos(3\theta)]$ **9.** $\dfrac{1}{2}[\sin(2\theta) + \sin\theta]$

11. $2\sin\theta\cos(3\theta)$ **13.** $2\cos(3\theta)\cos\theta$ **15.** $2\sin(2\theta)\cos\theta$ **17.** $2\sin\theta\sin\dfrac{\theta}{2}$ **19.** $\dfrac{\sin\theta + \sin(3\theta)}{2\sin(2\theta)} = \dfrac{2\sin(2\theta)\cos\theta}{2\sin(2\theta)} = \cos\theta$

21. $\dfrac{\sin(4\theta) + \sin(2\theta)}{\cos(4\theta) + \cos(2\theta)} = \dfrac{2\sin(3\theta)\cos\theta}{2\cos(3\theta)\cos\theta} = \dfrac{\sin(3\theta)}{\cos(3\theta)} = \tan(3\theta)$ **23.** $\dfrac{\cos\theta - \cos(3\theta)}{\sin\theta + \sin(3\theta)} = \dfrac{2\sin(2\theta)\sin\theta}{2\sin(2\theta)\cos\theta} = \dfrac{\sin\theta}{\cos\theta} = \tan\theta$

25. $\sin\theta[\sin\theta + \sin(3\theta)] = \sin\theta[2\sin(2\theta)\cos\theta] = \cos\theta[2\sin(2\theta)\sin\theta] = \cos\theta\left[2\cdot\dfrac{1}{2}[\cos\theta - \cos(3\theta)]\right] = \cos\theta[\cos\theta - \cos(3\theta)]$

27. $\dfrac{\sin(4\theta) + \sin(8\theta)}{\cos(4\theta) + \cos(8\theta)} = \dfrac{2\sin(6\theta)\cos(2\theta)}{2\cos(6\theta)\cos(2\theta)} = \dfrac{\sin(6\theta)}{\cos(6\theta)} = \tan(6\theta)$

29. $\dfrac{\sin(4\theta) + \sin(8\theta)}{\sin(4\theta) - \sin(8\theta)} = \dfrac{2\sin(6\theta)\cos(-2\theta)}{2\sin(-2\theta)\cos(6\theta)} = \dfrac{\sin(6\theta)}{\cos(6\theta)}\cdot\dfrac{\cos(2\theta)}{-\sin(2\theta)} = \tan(6\theta)[-\cot(2\theta)] = -\dfrac{\tan(6\theta)}{\tan(2\theta)}$

31. $\dfrac{\sin\alpha + \sin\beta}{\sin\alpha - \sin\beta} = \dfrac{2\sin\dfrac{\alpha+\beta}{2}\cos\dfrac{\alpha-\beta}{2}}{2\sin\dfrac{\alpha-\beta}{2}\cos\dfrac{\alpha+\beta}{2}} = \dfrac{\sin\dfrac{\alpha+\beta}{2}}{\cos\dfrac{\alpha+\beta}{2}}\cdot\dfrac{\cos\dfrac{\alpha-\beta}{2}}{\sin\dfrac{\alpha-\beta}{2}} = \tan\dfrac{\alpha+\beta}{2}\cot\dfrac{\alpha-\beta}{2}$

33. $\dfrac{\sin\alpha + \sin\beta}{\cos\alpha + \cos\beta} = \dfrac{2\sin\dfrac{\alpha+\beta}{2}\cos\dfrac{\alpha-\beta}{2}}{2\cos\dfrac{\alpha+\beta}{2}\cos\dfrac{\alpha-\beta}{2}} = \dfrac{\sin\dfrac{\alpha+\beta}{2}}{\cos\dfrac{\alpha+\beta}{2}} = \tan\dfrac{\alpha+\beta}{2}$

35. $1 + \cos(2\theta) + \cos(4\theta) + \cos(6\theta) = [1 + \cos(6\theta)] + [\cos(2\theta) + \cos(4\theta)] = 2\cos^2(3\theta) + 2\cos(3\theta)\cos(-\theta)$
$= 2\cos(3\theta)[\cos(3\theta) + \cos\theta] = 2\cos(3\theta)[2\cos(2\theta)\cos\theta] = 4\cos\theta\cos(2\theta)\cos(3\theta)$

37. (a) $y = 2\sin(2061\pi t)\cos(357\pi t)$ **(b)** $y_{max} = 2$ **(c)**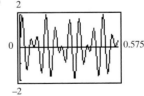

2

0   0.575

−2

39. $\sin(2\alpha) + \sin(2\beta) + \sin(2\gamma) = 2\sin(\alpha+\beta)\cos(\alpha-\beta) + \sin(2\gamma) = 2\sin(\alpha+\beta)\cos(\alpha-\beta) + 2\sin\gamma\cos\gamma$
$= 2\sin(\pi-\gamma)\cos(\alpha-\beta) + 2\sin\gamma\cos\gamma = 2\sin\gamma\cos(\alpha-\beta) + 2\sin\gamma\cos\gamma = 2\sin\gamma[\cos(\alpha-\beta) + \cos\gamma]$
$= 2\sin\gamma\left(2\cos\dfrac{\alpha-\beta+\gamma}{2}\cos\dfrac{\alpha-\beta-\gamma}{2}\right) = 4\sin\gamma\cos\dfrac{\pi-2\beta}{2}\cos\dfrac{2\alpha-\pi}{2} = 4\sin\gamma\cos\left(\dfrac{\pi}{2} - \beta\right)\cos\left(\alpha - \dfrac{\pi}{2}\right)$
$= 4\sin\gamma\sin\beta\sin\alpha$

41.
$\sin(\alpha - \beta) = \sin\alpha\cos\beta - \cos\alpha\sin\beta$
$\sin(\alpha + \beta) = \sin\alpha\cos\beta + \cos\alpha\sin\beta$
$\sin(\alpha - \beta) + \sin(\alpha + \beta) = 2\sin\alpha\cos\beta$

$\sin\alpha\cos\beta = \dfrac{1}{2}[\sin(\alpha+\beta) + \sin(\alpha-\beta)]$

43. $2\cos\dfrac{\alpha+\beta}{2}\cos\dfrac{\alpha-\beta}{2} = 2\cdot\dfrac{1}{2}\left[\cos\left(\dfrac{\alpha+\beta}{2} + \dfrac{\alpha-\beta}{2}\right) + \cos\left(\dfrac{\alpha+\beta}{2} - \dfrac{\alpha-\beta}{2}\right)\right] = \cos\dfrac{2\alpha}{2} + \cos\dfrac{2\beta}{2} = \cos\alpha + \cos\beta$

3.7 Exercises *(page 208)*

1. $\theta = \dfrac{\pi}{6} + 2k\pi, \theta = \dfrac{5\pi}{6} + 2k\pi; \dfrac{\pi}{6}, \dfrac{5\pi}{6}, \dfrac{13\pi}{6}, \dfrac{17\pi}{6}, \dfrac{25\pi}{6}, \dfrac{29\pi}{6}$ **3.** $\theta = \dfrac{5\pi}{6} + k\pi; \dfrac{5\pi}{6}, \dfrac{11\pi}{6}, \dfrac{17\pi}{6}, \dfrac{23\pi}{6}, \dfrac{29\pi}{6}, \dfrac{35\pi}{6}$

5. $\theta = \dfrac{\pi}{2} + 2k\pi, \theta = \dfrac{3\pi}{2} + 2k\pi; \dfrac{\pi}{2}, \dfrac{3\pi}{2}, \dfrac{5\pi}{2}, \dfrac{7\pi}{2}, \dfrac{9\pi}{2}, \dfrac{11\pi}{2}$ **7.** $\theta = \dfrac{\pi}{3} + k\pi, \theta = \dfrac{2\pi}{3} + k\pi; \dfrac{\pi}{3}, \dfrac{2\pi}{3}, \dfrac{4\pi}{3}, \dfrac{5\pi}{3}, \dfrac{7\pi}{3}, \dfrac{8\pi}{3}$

9. $\theta = \dfrac{8\pi}{3} + 4k\pi, \theta = \dfrac{10\pi}{3} + 4k\pi; \dfrac{8\pi}{3}, \dfrac{10\pi}{3}, \dfrac{20\pi}{3}, \dfrac{22\pi}{3}, \dfrac{32\pi}{3}, \dfrac{34\pi}{3}$ **11.** $\dfrac{7\pi}{6}, \dfrac{11\pi}{6}$ **13.** $\dfrac{\pi}{3}, \dfrac{2\pi}{3}, \dfrac{4\pi}{3}, \dfrac{5\pi}{3}$ **15.** $\dfrac{\pi}{4}, \dfrac{3\pi}{4}, \dfrac{5\pi}{4}, \dfrac{7\pi}{4}$

17. $\dfrac{\pi}{2}, \dfrac{7\pi}{6}, \dfrac{11\pi}{6}$ **19.** $\dfrac{\pi}{3}, \dfrac{2\pi}{3}, \dfrac{4\pi}{3}, \dfrac{5\pi}{3}$ **21.** $\dfrac{4\pi}{9}, \dfrac{8\pi}{9}, \dfrac{16\pi}{9}$ **23.** $\dfrac{3\pi}{4}, \dfrac{7\pi}{4}$ **25.** $\dfrac{11\pi}{6}$ **27.** $\dfrac{7\pi}{6}, \dfrac{11\pi}{6}$ **29.** $\dfrac{3\pi}{4}, \dfrac{7\pi}{4}$ **31.** $\dfrac{2\pi}{3}, \dfrac{4\pi}{3}$ **33.** $\dfrac{3\pi}{4}, \dfrac{5\pi}{4}$

35. 0.41, 2.73   **37.** 1.37, 4.51   **39.** 2.69, 3.59   **41.** 1.82, 4.46   **43.** 28.9°   **45.** Yes; it varies from 1.28 to 1.34   **47.** 1.47

49. If θ is the original angle of incidence and ϕ is the angle of refraction, then $\dfrac{\sin\theta}{\sin\phi} = n_2$. The angle of incidence of the emerging

beam is also ϕ, and the index of refraction is $\dfrac{1}{n_2}$. Thus, θ is the angle of refraction of the emerging beam.

3.8 Exercises *(page 214)*

1. $\dfrac{\pi}{2}, \dfrac{2\pi}{3}, \dfrac{4\pi}{3}, \dfrac{3\pi}{2}$ **3.** $\dfrac{\pi}{2}, \dfrac{7\pi}{6}, \dfrac{11\pi}{6}$ **5.** $0, \dfrac{\pi}{4}, \dfrac{5\pi}{4}$ **7.** $\dfrac{\pi}{2}, \dfrac{2\pi}{3}, \dfrac{4\pi}{3}, \dfrac{3\pi}{2}$ **9.** π **11.** $\dfrac{\pi}{3}, \dfrac{2\pi}{3}, \dfrac{4\pi}{3}, \dfrac{5\pi}{3}$ **13.** $\dfrac{\pi}{4}, \dfrac{5\pi}{4}$ **15.** $0, \dfrac{\pi}{3}, \pi, \dfrac{5\pi}{3}$ **17.** $\dfrac{\pi}{2}, \dfrac{3\pi}{2}$

19. $0, \dfrac{2\pi}{3}, \dfrac{4\pi}{3}$ **21.** $0, \dfrac{\pi}{3}, \dfrac{\pi}{2}, \dfrac{2\pi}{3}, \pi, \dfrac{4\pi}{3}, \dfrac{3\pi}{2}, \dfrac{5\pi}{3}$ **23.** $0, \dfrac{\pi}{5}, \dfrac{2\pi}{5}, \dfrac{3\pi}{5}, \dfrac{4\pi}{5}, \pi, \dfrac{6\pi}{5}, \dfrac{7\pi}{5}, \dfrac{8\pi}{5}, \dfrac{9\pi}{5}$ **25.** $\dfrac{\pi}{6}, \dfrac{5\pi}{6}, \dfrac{3\pi}{2}$ **27.** $\dfrac{\pi}{3}, \dfrac{5\pi}{3}$

29. No real solutions   **31.** No real solutions   **33.** $\dfrac{\pi}{2}, \dfrac{7\pi}{6}$ **35.** $0, \dfrac{\pi}{3}, \pi, \dfrac{5\pi}{3}$ **37.** $\dfrac{\pi}{4}$

39. $-1.29, 0$

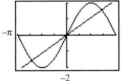

41. $-2.24, 0, 2.24$

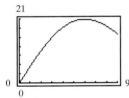

43. $-0.82, 0.82$

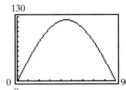

45. $-1.31, 1.98, 3.84$
47. 0.52
49. 1.26
51. $-1.02, 1.02$
53. 0, 2.15
55. 0.76, 1.35

57. (a) 60° **(b)** 60°
(c) $A(60°) = 12\sqrt{3}$ sq in.
(d)

$\theta_{max} = 60°$
Maximum Area $= 20.78$ sq in.

59. 2.03, 4.91
61. (a) 29.99° or 60.01° **(b)** 123.6 m
(c)

Fill-in-the-Blank Items *(page 217)*

1. identity; conditional   **3.** $+$ **5.** $1 - \cos\alpha$ **7.** 0

True/False Items *(page 218)*

1. T   **3.** T   **5.** F   **7.** F

Review Exercises *(page 218)*

1. $\dfrac{\pi}{2}$ **3.** $\dfrac{\pi}{4}$ **5.** $\dfrac{5\pi}{6}$ **7.** $\dfrac{\sqrt{2}}{2}$ **9.** $-\sqrt{3}$ **11.** $\dfrac{2\sqrt{3}}{3}$ **13.** $\dfrac{3}{5}$ **15.** $-\dfrac{4}{3}$ **17.** $-\dfrac{\pi}{6}$ **19.** $-\dfrac{\pi}{4}$ **21.** $\tan\theta\cot\theta - \sin^2\theta = 1 - \sin^2\theta = \cos^2\theta$

23. $\cos^2\theta(1 + \tan^2\theta) = \cos^2\theta\sec^2\theta = 1$ **25.** $4\cos^2\theta + 3\sin^2\theta = \cos^2\theta + 3(\cos^2\theta + \sin^2\theta) = 3 + \cos^2\theta$

27. $\dfrac{1 - \cos\theta}{\sin\theta} + \dfrac{\sin\theta}{1 - \cos\theta} = \dfrac{(1 - \cos\theta)^2 + \sin^2\theta}{\sin\theta(1 - \cos\theta)} = \dfrac{1 - 2\cos\theta + \cos^2\theta + \sin^2\theta}{\sin\theta(1 - \cos\theta)} = \dfrac{2(1 - \cos\theta)}{\sin\theta(1 - \cos\theta)} = 2\csc\theta$

29. $\dfrac{\cos\theta}{\cos\theta - \sin\theta} = \dfrac{\dfrac{\cos\theta}{\cos\theta}}{\dfrac{\cos\theta - \sin\theta}{\cos\theta}} = \dfrac{1}{1 - \dfrac{\sin\theta}{\cos\theta}} = \dfrac{1}{1 - \tan\theta}$

31. $\dfrac{\csc\theta}{1 + \csc\theta} = \dfrac{\dfrac{1}{\sin\theta}}{1 + \dfrac{1}{\sin\theta}} = \dfrac{1}{1 + \sin\theta} = \dfrac{1}{1 + \sin\theta} \cdot \dfrac{1 - \sin\theta}{1 - \sin\theta} = \dfrac{1 - \sin\theta}{1 - \sin^2\theta} = \dfrac{1 - \sin\theta}{\cos^2\theta}$

33. $\csc\theta - \sin\theta = \dfrac{1}{\sin\theta} - \sin\theta = \dfrac{1-\sin^2\theta}{\sin\theta} = \dfrac{\cos^2\theta}{\sin\theta} = \cos\theta \cdot \dfrac{\cos\theta}{\sin\theta} = \cos\theta\cot\theta$

35. $\dfrac{1-\sin\theta}{\sec\theta} = \cos\theta(1-\sin\theta)\cdot\dfrac{1+\sin\theta}{1+\sin\theta} = \dfrac{\cos\theta(1-\sin^2\theta)}{1+\sin\theta} = \dfrac{\cos^3\theta}{1+\sin\theta}$

37. $\cot\theta - \tan\theta = \dfrac{\cos\theta}{\sin\theta} - \dfrac{\sin\theta}{\cos\theta} = \dfrac{\cos^2\theta - \sin^2\theta}{\sin\theta\cos\theta} = \dfrac{1-2\sin^2\theta}{\sin\theta\cos\theta}$

39. $\dfrac{\cos(\alpha+\beta)}{\cos\alpha\sin\beta} = \dfrac{\cos\alpha\cos\beta - \sin\alpha\sin\beta}{\cos\alpha\sin\beta} = \dfrac{\cos\alpha\cos\beta}{\cos\alpha\sin\beta} - \dfrac{\sin\alpha\sin\beta}{\cos\alpha\sin\beta} = \cot\beta - \tan\alpha$

41. $\dfrac{\cos(\alpha-\beta)}{\cos\alpha\cos\beta} = \dfrac{\cos\alpha\cos\beta + \sin\alpha\sin\beta}{\cos\alpha\cos\beta} = \dfrac{\cos\alpha\cos\beta}{\cos\alpha\cos\beta} + \dfrac{\sin\alpha\sin\beta}{\cos\alpha\cos\beta} = 1 + \tan\alpha\tan\beta$

43. $(1+\cos\theta)\left(\tan\dfrac{\theta}{2}\right) = \left(2\cos^2\dfrac{\theta}{2}\right)\dfrac{\sin\left(\dfrac{\theta}{2}\right)}{\cos\left(\dfrac{\theta}{2}\right)} = 2\sin\dfrac{\theta}{2}\cos\dfrac{\theta}{2} = \sin\theta$

45. $2\cot\theta\cot2\theta = 2\left(\dfrac{\cos\theta}{\sin\theta}\right)\left(\dfrac{\cos2\theta}{\sin2\theta}\right) = \dfrac{2\cos\theta(\cos^2\theta-\sin^2\theta)}{2\sin^2\theta\cos\theta} = \dfrac{\cos^2\theta-\sin^2\theta}{\sin^2\theta} = \cot^2\theta - 1$

47. $1 - 8\sin^2\theta\cos^2\theta = 1 - 2(2\sin\theta\cos\theta)^2 = 1 - 2\sin^2(2\theta) = \cos(4\theta)$ **49.** $\dfrac{\sin(2\theta)+\sin(4\theta)}{\cos(2\theta)+\cos(4\theta)} = \dfrac{2\sin(3\theta)\cos(-\theta)}{2\cos(3\theta)\cos(-\theta)} = \tan(3\theta)$

51. $\dfrac{\cos(2\theta)-\cos(4\theta)}{\cos(2\theta)+\cos(4\theta)} - \tan\theta\tan(3\theta) = \dfrac{-2\sin(3\theta)\sin(-\theta)}{2\cos(3\theta)\cos(-\theta)} - \tan\theta\tan(3\theta) = \tan(3\theta)\tan\theta - \tan\theta\tan(3\theta) = 0$ **53.** $\dfrac{1}{4}(\sqrt{6}-\sqrt{2})$

55. $\dfrac{1}{4}(\sqrt{6}-\sqrt{2})$ **57.** $\dfrac{1}{2}$ **59.** $\sqrt{2}-1$ **61.** (a) $-\dfrac{33}{65}$ (b) $-\dfrac{56}{65}$ (c) $-\dfrac{63}{65}$ (d) $\dfrac{33}{56}$ (e) $\dfrac{24}{25}$ (f) $\dfrac{119}{169}$ (g) $\dfrac{5\sqrt{26}}{26}$ (h) $\dfrac{2\sqrt{5}}{5}$

63. (a) $-\dfrac{16}{65}$ (b) $-\dfrac{63}{65}$ (c) $-\dfrac{56}{65}$ (d) $\dfrac{16}{63}$ (e) $\dfrac{24}{25}$ (f) $\dfrac{119}{169}$ (g) $\dfrac{\sqrt{26}}{26}$ (h) $-\dfrac{\sqrt{10}}{10}$ **65.** (a) $-\dfrac{63}{65}$ (b) $\dfrac{16}{65}$ (c) $\dfrac{33}{65}$ (d) $-\dfrac{63}{16}$

(e) $\dfrac{24}{25}$ (f) $-\dfrac{119}{169}$ (g) $\dfrac{2\sqrt{13}}{13}$ (h) $-\dfrac{\sqrt{10}}{10}$ **67.** (a) $\dfrac{-\sqrt{3}-2\sqrt{2}}{6}$ (b) $\dfrac{1-2\sqrt{6}}{6}$ (c) $\dfrac{-\sqrt{3}+2\sqrt{2}}{6}$ (d) $\dfrac{8\sqrt{2}+9\sqrt{3}}{23}$ (e) $-\dfrac{\sqrt{3}}{2}$

(f) $-\dfrac{7}{9}$ (g) $\dfrac{\sqrt{3}}{3}$ (h) $\dfrac{\sqrt{3}}{2}$ **69.** (a) 1 (b) 0 (c) $-\dfrac{1}{9}$ (d) Not defined (e) $\dfrac{4\sqrt{5}}{9}$ (f) $-\dfrac{1}{9}$ (g) $\dfrac{\sqrt{30}}{6}$ (h) $-\dfrac{\sqrt{6}\sqrt{3}-\sqrt{5}}{6}$

71. $\dfrac{4+3\sqrt{3}}{10}$ **73.** $\dfrac{48+25\sqrt{3}}{-39}$ **75.** $-\dfrac{24}{25}$ **77.** $\dfrac{\pi}{3},\dfrac{5\pi}{3}$ **79.** $\dfrac{3\pi}{4},\dfrac{5\pi}{4}$ **81.** $\dfrac{3\pi}{4},\dfrac{7\pi}{4}$ **83.** $0,\dfrac{\pi}{2},\pi,\dfrac{3\pi}{2}$ **85.** $\dfrac{\pi}{3},\dfrac{2\pi}{3},\dfrac{4\pi}{3},\dfrac{5\pi}{3}$ **87.** $0,\pi$

89. $0,\dfrac{2\pi}{3},\pi,\dfrac{4\pi}{3}$ **91.** $0,\dfrac{\pi}{6},\dfrac{5\pi}{6}$ **93.** $\dfrac{\pi}{6},\dfrac{\pi}{2},\dfrac{5\pi}{6}$ **95.** $\dfrac{\pi}{3},\dfrac{5\pi}{3}$ **97.** $\dfrac{\pi}{4},\dfrac{\pi}{2},\dfrac{3\pi}{4},\dfrac{3\pi}{2}$ **99.** $\dfrac{\pi}{2},\pi$ **101.** 1.11   **103.** 0.87   **105.** 2.22

C H A P T E R   4   Applications of Trigonometric Functions

4.1 Exercises *(page 230)*

1. $\sin\theta = \dfrac{5}{13}; \cos\theta = \dfrac{12}{13}; \tan\theta = \dfrac{5}{12}; \csc\theta = \dfrac{13}{5}; \sec\theta = \dfrac{13}{12}; \cot\theta = \dfrac{12}{5}$ **3.** $\sin\theta = \dfrac{2\sqrt{13}}{13}; \cos\theta = \dfrac{3\sqrt{13}}{13}; \tan\theta = \dfrac{2}{3}; \csc\theta = \dfrac{\sqrt{13}}{2};$

$\sec\theta = \dfrac{\sqrt{13}}{3}; \cot\theta = \dfrac{3}{2}$ **5.** $\sin\theta = \dfrac{\sqrt{3}}{2}; \cos\theta = \dfrac{1}{2}; \tan\theta = \sqrt{3}; \csc\theta = \dfrac{2\sqrt{3}}{3}; \sec\theta = 2; \cot\theta = \dfrac{\sqrt{3}}{3}$ **7.** $\sin\theta = \dfrac{\sqrt{6}}{3}; \cos\theta = \dfrac{\sqrt{3}}{3};$

$\tan\theta = \sqrt{2}; \csc\theta = \dfrac{\sqrt{6}}{2}; \sec\theta = \sqrt{3}; \cot\theta = \dfrac{\sqrt{2}}{2}$ **9.** $\sin\theta = \dfrac{\sqrt{5}}{5}; \cos\theta = \dfrac{2\sqrt{5}}{5}; \tan\theta = \dfrac{1}{2}; \csc\theta = \sqrt{5}; \sec\theta = \dfrac{\sqrt{5}}{2}; \cot\theta = 2$

11. 0   **13.** 1   **15.** 0   **17.** 0   **19.** 1   **21.** $a \approx 13.74, c \approx 14.62, \alpha = 70°$ **23.** $b \approx 5.03, c \approx 7.83, \alpha = 50°$
25. $a \approx 0.71, c \approx 4.06, \beta = 80°$ **27.** $b \approx 10.72, c \approx 11.83, \beta = 65°$ **29.** $b \approx 3.08, a \approx 8.46, \alpha = 70°$ **31.** $c \approx 5.83, \alpha \approx 59.0°, \beta = 31.0°$
33. $b \approx 4.58, \alpha \approx 23.6°, \beta = 66.4°$ **35.** 4.59 in., 6.55 in.   **37.** 5.52 in. or 11.83 in.   **39.** 23.6° and 66.4°   **41.** 70.02 ft   **43.** 985.91 ft
45. 137.37 m   **47.** 20.67 ft   **49.** 15.9°   **51.** 60.27 ft   **53.** 530.18 ft   **55.** 554.52 ft   **57.** (a) 111.96 ft/sec or 76.3 mi/hr
(b) 82.42 ft/sec or 56.2 mi/hr   (c) Under 18.8°   **59.** S76.6°E   **61.** 14.9°   **63.** 3.83 mi   **65.** No; Move the tripod back about 1 ft.
67. (a) $A(\theta) = 2\sin\theta\cos\theta$ (b) From double-angle formula, since $2\sin\theta\cos\theta = \sin(2\theta)$ (c) $\theta = 45°$ (d) $\dfrac{\sqrt{2}}{2}$ by $\sqrt{2}$

69. (a) $T(\theta) = \dfrac{2}{3 \sin \theta} - \dfrac{1}{4 \tan \theta} + 1$

(b)

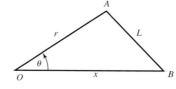

(graph from $0°$ to $90°$, range 0 to 4) ; $68°$; 1.62 hr; 0.90 hr

71. (a) 10 min   **(b)** 20 min

(c) $T(\theta) = 5 - \dfrac{5}{3 \tan \theta} + \dfrac{5}{\sin \theta}$ **(d)** 10.4 min

(e) (graph from $0°$ to $90°$, range 0 to 20) ; $70.5°$; 9.7 min; approximately 177 ft

73. Drop $\perp h$ to base b. Then $h = a \sin \theta$ and $\dfrac{1}{2}b = a \cos \theta$. Area $= \dfrac{1}{2}bh = \dfrac{1}{2}(2a \cos \theta)(a \sin \theta) = a^2 \sin \theta \cos \theta$.

75. Define h to be the length of the hypotenuse of the smaller triangle. Then $\sin \theta = \dfrac{a}{h}$, so $h = \dfrac{a}{\sin \theta}$. Since the two right triangles are similar,

then $\dfrac{b}{a} = \dfrac{h + a + b}{h} = 1 + \dfrac{a + b}{h} = 1 + \dfrac{(a + b) \sin \theta}{a}$. Solving for $\sin \theta$ in the equation $\dfrac{b}{a} = 1 + \dfrac{(a + b) \sin \theta}{a}$ yields $\sin \theta = \dfrac{b - a}{b + a}$.

Since θ is acute, then $\cos \theta = \sqrt{1 - \sin^2 \theta} = \sqrt{1 - \left(\dfrac{b - a}{b + a}\right)^2} = \sqrt{\dfrac{4ab}{(a + b)^2}} = \dfrac{2\sqrt{ab}}{a + b}$, or $\cos \theta = \dfrac{\sqrt{ab}}{\left(\dfrac{a + b}{2}\right)}$.

77. Since $\cos \alpha = \tan \beta = \dfrac{\sin \beta}{\cos \beta}$ and $\cos \beta = \tan \alpha = \dfrac{\sin \alpha}{\cos \alpha}$, then $\cos \alpha \cos \beta = \sin \alpha = \sin \beta$. Since sine is a one-to-one function for

angles between 0 and $\dfrac{\pi}{2}$ radians, then $\alpha = \beta$. Therefore, $\sin \alpha = \cos^2 \alpha = 1 - \sin^2 \alpha$. Solving the quadratic equation $\sin^2 \alpha + \sin \alpha - 1 = 0$

for the variable $\sin \alpha$ yields $\sin \alpha = \dfrac{-1 \pm \sqrt{5}}{2}$. Since $\sin \alpha$ is positive when α is acute, then $\sin \alpha = \sin \beta = \dfrac{-1 + \sqrt{5}}{2}$, which equals $\sqrt{\dfrac{3 - \sqrt{5}}{2}}$.

4.2 Exercises *(page 243)*

1. $a \approx 3.23, b \approx 3.55, \alpha = 40°$ **3.** $a \approx 3.25, c \approx 4.23, \beta = 45°$ **5.** $\gamma = 95°, c \approx 9.86, a \approx 6.36$ **7.** $\alpha = 40°, a = 2, c \approx 3.06$
9. $\gamma = 120°, b \approx 1.06, c \approx 2.69$ **11.** $\alpha = 100°, a \approx 5.24, c \approx 0.92$ **13.** $\beta = 40°, a \approx 5.64, b \approx 3.86$ **15.** $\gamma = 100°, a \approx 1.31, b \approx 1.31$
17. One triangle; $\beta \approx 30.7°, \gamma \approx 99.3°, c \approx 3.86$ **19.** One triangle; $\gamma \approx 36.2°, \alpha \approx 43.8°, a \approx 3.51$ **21.** No triangle
23. Two triangles; $\gamma_1 \approx 30.9°, \alpha_1 \approx 129.1°, a_1 \approx 9.07$ or $\gamma_2 \approx 149.1°, \alpha_2 \approx 10.9°, a_2 \approx 2.20$ **25.** No triangle   **27.** Two triangles; $a_1 \approx 57.7°$,
$\beta_1 \approx 97.3°, b_1 \approx 2.35$ or $\alpha_2 \approx 122.3°, \beta_2 \approx 32.7°, b_2 \approx 1.28$ **29. (a)** Station Able is about 143.33 mi from the ship; Station Baker is about
135.58 mi from the ship.   **(b)** Approximately 41 min   **31.** 1490.48 ft   **33.** 381.69 ft   **35. (a)** 169.18 mi   **(b)** 161.3°   **37.** 84.7°; 183.72 ft
39. 2.64 mi   **41.** 1.88 mi   **43.** 449.36 ft   **45.** 39.39 ft   **47.** 29.97 ft

49. $\dfrac{a - b}{c} = \dfrac{a}{c} - \dfrac{b}{c} = \dfrac{\sin \alpha}{\sin \gamma} - \dfrac{\sin \beta}{\sin \gamma} = \dfrac{\sin \alpha - \sin \beta}{\sin \gamma} = \dfrac{2 \sin \left(\dfrac{\alpha - \beta}{2}\right) \cos \left(\dfrac{\alpha + \beta}{2}\right)}{2 \sin \dfrac{\gamma}{2} \cos \dfrac{\gamma}{2}} = \dfrac{\sin \left(\dfrac{\alpha - \beta}{2}\right) \cos \left(\dfrac{\pi}{2} - \dfrac{\gamma}{2}\right)}{\sin \dfrac{\gamma}{2} \cos \dfrac{\gamma}{2}} = \dfrac{\sin \left(\dfrac{\alpha - \beta}{2}\right)}{\cos \dfrac{\gamma}{2}}$

51. $\dfrac{a - b}{a + b} = \dfrac{\dfrac{a - b}{c}}{\dfrac{a + b}{c}} = \dfrac{\dfrac{\sin\left[\frac{1}{2}(\alpha - \beta)\right]}{\cos \dfrac{\gamma}{2}}}{\dfrac{\cos\left[\frac{1}{2}(\alpha - \beta)\right]}{\sin \dfrac{\gamma}{2}}} = \dfrac{\tan\left[\frac{1}{2}(\alpha - \beta)\right]}{\cot \dfrac{\gamma}{2}} = \dfrac{\tan\left[\frac{1}{2}(\alpha - \beta)\right]}{\tan\left(\dfrac{\pi}{2} - \dfrac{\gamma}{2}\right)} = \dfrac{\tan\left[\frac{1}{2}(\alpha - \beta)\right]}{\tan\left[\frac{1}{2}(\alpha + \beta)\right]}$

4.3 Exercises *(page 250)*

1. $b \approx 2.95, \alpha \approx 28.7°, \gamma \approx 106.3°$ **3.** $c \approx 3.75, \alpha \approx 32.1°, \beta \approx 52.9°$ **5.** $\alpha \approx 48.5°, \beta \approx 38.6°, \gamma \approx 92.9°$ **7.** $\alpha \approx 127.2°, \beta \approx 32.1°, \gamma \approx 20.7°$
9. $c \approx 2.57, \alpha \approx 48.6°, \beta \approx 91.4°$ **11.** $a \approx 2.99, \beta \approx 19.2°, \gamma \approx 80.8°$ **13.** $b \approx 4.14, \alpha \approx 43.0°, \gamma \approx 27.0°$ **15.** $c \approx 1.69, \alpha \approx 65.0°, \beta \approx 65.0°$
17. $\alpha \approx 67.4°, \beta = 90°, \gamma \approx 22.6°$ **19.** $\alpha = 60°, \beta = 60°, \gamma = 60°$ **21.** $\alpha \approx 33.6°, \beta \approx 62.2°, \gamma \approx 84.3°$ **23.** $\alpha \approx 97.9°, \beta \approx 52.4°, \gamma \approx 29.7°$
25. 70.75 ft   **27. (a)** 12.0°   **(b)** 220.8 mph   **29. (a)** 63.7 ft   **(b)** 66.8 ft   **(c)** 92.8°   **31. (a)** 492.6 ft   **(b)** 269.3 ft   **33.** 342.3 ft
35. Using the Law of Cosines:
$L^2 = x^2 + r^2 - 2rx \cos \theta$
$x^2 - 2rx \cos \theta + r^2 - L^2 = 0$
Then, using the quadratic formula:
$x = r \cos \theta + \sqrt{r^2 \cos^2 \theta + L^2 - r^2}$

37. $\cos\dfrac{\gamma}{2} = \sqrt{\dfrac{1+\cos\gamma}{2}} = \sqrt{\dfrac{1+\dfrac{a^2+b^2-c^2}{2ab}}{2}} = \sqrt{\dfrac{2ab+a^2+b^2-c^2}{4ab}} = \sqrt{\dfrac{(a+b)^2-c^2}{4ab}} = \sqrt{\dfrac{(a+b+c)(a+b-c)}{4ab}}$

$= \sqrt{\dfrac{2s(2s-2c)}{4ab}} = \sqrt{\dfrac{s(s-c)}{ab}}$

39. $\dfrac{\cos\alpha}{a} + \dfrac{\cos\beta}{b} + \dfrac{\cos\gamma}{c} = \dfrac{b^2+c^2-a^2}{2abc} + \dfrac{a^2+c^2-b^2}{2abc} + \dfrac{a^2+b^2-c^2}{2abc} = \dfrac{b^2+c^2-a^2+a^2+c^2-b^2+a^2+b^2-c^2}{2abc}$

$= \dfrac{a^2+b^2+c^2}{2abc}$

4.4 Exercises *(page 255)*

1. 2.83   **3.** 2.99   **5.** 14.98   **7.** 9.56   **9.** 3.86   **11.** 1.48   **13.** 2.82   **15.** 1.53   **17.** 30   **19.** 1.73   **21.** 19.90   **23.** 19.81   **25.** 9.03 sq ft

27. \$5446.38   **29.** 9.26 sq cm   **31.** $A = \dfrac{1}{2}ab\sin\gamma = \dfrac{1}{2}a\sin\gamma\left(\dfrac{a\sin\beta}{\sin\alpha}\right) = \dfrac{a^2\sin\beta\sin\gamma}{2\sin\alpha}$ **33.** 0.92   **35.** 2.27   **37.** 5.44

39. $A = \dfrac{1}{2}r^2(\theta+\sin\theta)$ **41.** 31,145.15 sq ft

43. $h_1 = 2\dfrac{K}{a}, h_2 = 2\dfrac{K}{b}, h_3 = 2\dfrac{K}{c}$. Then $\dfrac{1}{h_1}+\dfrac{1}{h_2}+\dfrac{1}{h_3} = \dfrac{a}{2K}+\dfrac{b}{2K}+\dfrac{c}{2K} = \dfrac{a+b+c}{2K} = \dfrac{2s}{2K} = \dfrac{s}{K}$.

45. Angle AOB measures $180° - \left(\dfrac{\alpha}{2}+\dfrac{\beta}{2}\right) = 180° - \dfrac{1}{2}(180°-\gamma) = 90° + \dfrac{\gamma}{2}$, and $\sin\left(90°+\dfrac{\gamma}{2}\right) = \cos\left(-\dfrac{\gamma}{2}\right) = \cos\dfrac{\gamma}{2}$

since cosine is an even function. Therefore, $r = \dfrac{c\sin\dfrac{\alpha}{2}\sin\dfrac{\beta}{2}}{\sin\left(90°+\dfrac{\gamma}{2}\right)} = \dfrac{c\sin\dfrac{\alpha}{2}\sin\dfrac{\beta}{2}}{\cos\dfrac{\gamma}{2}}$.

47. $\cot\dfrac{\alpha}{2}+\cot\dfrac{\beta}{2}+\cot\dfrac{\gamma}{2} = \dfrac{s-a}{r}+\dfrac{s-b}{r}+\dfrac{s-c}{r} = \dfrac{3s-(a+b+c)}{r} = \dfrac{3s-2s}{r} = \dfrac{s}{r}$

49. (a) Area $\triangle OAC = \dfrac{1}{2}|AC|\,|OC| = \dfrac{1}{2}\cdot\dfrac{|AC|}{1}\cdot\dfrac{|OC|}{1} = \dfrac{1}{2}\cos\alpha\sin\alpha$

(b) Area $\triangle OCB = \dfrac{1}{2}|BC|\,|OC| = \dfrac{1}{2}|OB|^2\dfrac{|BC|}{|OB|}\cdot\dfrac{|OC|}{|OB|} = \dfrac{1}{2}|OB|^2\sin\beta\cos\beta$

(c) Area $\triangle OAB = \dfrac{1}{2}|BD|\,|OA| = \dfrac{1}{2}|OB|\dfrac{|BD|}{|OB|} = \dfrac{1}{2}|OB|\sin(\alpha+\beta)$

(d) $\dfrac{\cos\alpha}{\cos\beta} = \dfrac{\dfrac{|OC|}{1}}{\dfrac{|OC|}{|OB|}} = |OB|$

(e) Area $\triangle OAB = $ Area $\triangle OAC + $ Area $\triangle OCB$

$\dfrac{1}{2}|OB|\sin(\alpha+\beta) = \dfrac{1}{2}\sin\alpha\cos\alpha + \dfrac{1}{2}|OB|^2\sin\beta\cos\beta$

$\sin(\alpha+\beta) = \dfrac{\sin\alpha\cos\alpha + |OB|^2\sin\beta\cos\beta}{|OB|}$

$\sin(\alpha+\beta) = \dfrac{\sin\alpha(|OB|\cos\beta) + |OB|^2\sin\beta\left(\dfrac{\cos\alpha}{|OB|}\right)}{|OB|}$

$\sin(\alpha+\beta) = \sin\alpha\cos\beta + \cos\alpha\sin\beta$

4.5 Exercises *(page 265)*

1. $d = -5\cos(\pi t)$ **3.** $d = -6\cos(2t)$ **5.** $d = -5\sin(\pi t)$ **7.** $d = -6\sin(2t)$ **9. (a)** Simple harmonic   **(b)** 5 m   **(c)** $\dfrac{2\pi}{3}$ sec

(d) $\dfrac{3}{2\pi}$ oscillation/sec   **11. (a)** Simple harmonic   **(b)** 6 m   **(c)** 2 sec   **(d)** $\dfrac{1}{2}$ oscillation/sec   **13. (a)** Simple harmonic   **(b)** 3 m

(c) 4π sec   **(d)** $\dfrac{1}{4\pi}$ oscillation/sec   **15. (a)** Simple harmonic   **(b)** 2 m   **(c)** 1 sec   **(d)** 1 oscillation/sec

17.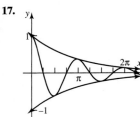

19.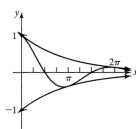

21.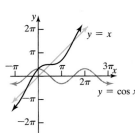

23.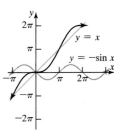

25.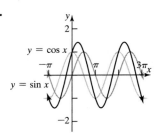

27.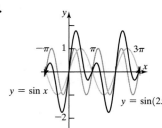

29. (a)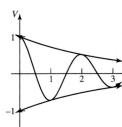

(b) At $t = 0, 2$; at $t = 1, t = 3$

(c) $\{t | 0.25 \le t \le 0.67$ or $1.29 \le t \le 1.75$ or $2.19 \le t \le 3\}$

31.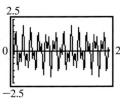

33.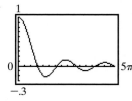

35. $y = \dfrac{1}{x} \sin x$ 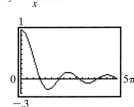 $y = \dfrac{1}{x^2} \sin x$ 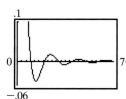 $y = \dfrac{1}{x^3} \sin x$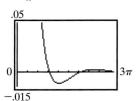

Fill-in-the-Blank Items *(page 267)*

1. Complementary **3.** Sines **5.** Heron's

True/False Items *(page 267)*

1. F **3.** F **5.** T

Review Exercises *(page 267)*

1. $\alpha = 70°$, $b \approx 3.42$, $a \approx 9.4$ **3.** $a \approx 4.58$, $\alpha = 66.4°$, $\beta \approx 23.6°$ **5.** $\gamma = 100°$, $b \approx 0.65$, $c \approx 1.29$ **7.** $\beta \approx 56.8°$, $\gamma \approx 23.2°$, $b \approx 4.25$
9. No triangle **11.** $b \approx 3.32$, $\alpha \approx 62.8°$, $\gamma \approx 17.2°$ **13.** No triangle **15.** $c \approx 2.32$, $\alpha \approx 16.1°$, $\beta \approx 123.9°$ **17.** $\beta = 36.2°$, $\gamma = 63.8°$, $c = 4.55$
19. $\alpha = 39.6°$, $\beta = 18.6°$, $\gamma = 121.9°$ **21.** Two triangles: $\beta_1 \approx 13.4°$, $\gamma_1 \approx 156.6°$, $c_1 \approx 6.86$ or $\beta_2 \approx 166.6°$, $\gamma_2 \approx 3.4°$, $c_2 \approx 1.02$
23. $a = 5.23$, $\beta = 46.0°$, $\gamma = 64.0°$ **25.** 1.93 **27.** 18.79 **29.** 6 **31.** 3.80 **33.** 0.32 **35.** 839.10 ft **37.** 23.32 ft **39.** 2.15 mi **41.** 204.07 mi
43. (a) 2.59 mi **(b)** 2.92 mi **(c)** 2.53 mi **45. (a)** 131.8 mi **(b)** 23.1° **(c)** 0.21 hr **47.** 8798.67 sq ft **49.** 1.92 sq in. **51.** 76.94 in.

53. (a) Simple harmonic **(b)** 6 ft **(c)** π sec **(d)** $\dfrac{1}{\pi}$ oscillation/sec **55. (a)** Simple harmonic **(b)** 2 ft **(c)** 2 sec **(d)** $\dfrac{1}{2}$ oscillation/sec

57.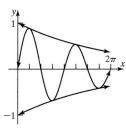

59.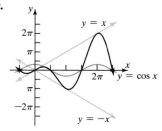

61.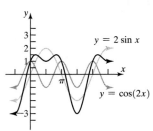

CHAPTER 5   Polar Coordinates; Vectors

5.1 Exercises *(page 281)*

1. A **3.** C **5.** B **7.** A

9.

11.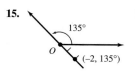

13.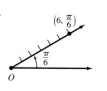

15.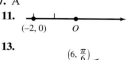

17.

19.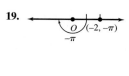

21.

(a) $\left(5, -\dfrac{4\pi}{3}\right)$

(b) $\left(-5, \dfrac{5\pi}{3}\right)$

(c) $\left(5, \dfrac{8\pi}{3}\right)$

23.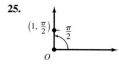

(a) $(2, -2\pi)$

(b) $(-2, \pi)$

(c) $(2, 2\pi)$

25.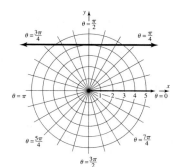

(a) $\left(1, -\dfrac{3\pi}{2}\right)$

(b) $\left(-1, \dfrac{3\pi}{2}\right)$

(c) $\left(1, \dfrac{5\pi}{2}\right)$

27.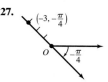

(a) $\left(3, -\dfrac{5\pi}{4}\right)$

(b) $\left(-3, \dfrac{7\pi}{4}\right)$

(c) $\left(3, \dfrac{11\pi}{4}\right)$

29. $(0, 3)$ **31.** $(-2, 0)$ **33.** $(-3\sqrt{3}, 3)$ **35.** $(\sqrt{2}, -\sqrt{2})$ **37.** $\left(-\dfrac{1}{2}, \dfrac{\sqrt{3}}{2}\right)$ **39.** $(2, 0)$ **41.** $(-2.57, 7.05)$ **43.** $(-4.98, -3.85)$ **45.** $(3, 0)$

47. $(1, \pi)$ **49.** $\left(\sqrt{2}, -\dfrac{\pi}{4}\right)$ **51.** $\left(2, \dfrac{\pi}{6}\right)$ **53.** $(2.47, -1.02)$ **55.** $(9.30, 0.47)$ **57.** $r^2 = \dfrac{3}{2}$ **59.** $r^2 \cos^2 \theta - 4r \sin \theta = 0$ **61.** $r^2 \sin 2\theta = 1$

63. $r \cos \theta = 4$ **65.** $x^2 + y^2 - x = 0$ or $\left(x - \dfrac{1}{2}\right)^2 + y^2 = \dfrac{1}{4}$ **67.** $(x^2 + y^2)^{3/2} - x = 0$ **69.** $x^2 + y^2 = 4$ **71.** $y^2 = 8(x + 2)$

73. $d = \sqrt{(r_2 \cos \theta_2 - r_1 \cos \theta_1)^2 + (r_2 \sin \theta_2 - r_1 \sin \theta_1)^2}$

$\quad = \sqrt{(r_2^2 \cos^2 \theta_2 - 2r_2 \cos \theta_2\, r_1 \cos \theta_1 + r_1^2 \cos^2 \theta_1) + (r_2^2 \sin^2 \theta_2 - 2r_2 \sin \theta_2\, r_1 \sin \theta_1 + r_1^2 \sin^2 \theta_1)}$

$\quad = \sqrt{r_1^2 + r_2^2 - 2r_1 r_2(\cos \theta_2 \cos \theta_1 + \sin \theta_2 \sin \theta_1)}$

$\quad = \sqrt{r_1^2 + r_2^2 - 2r_1 r_2 \cos(\theta_2 - \theta_1)}$

5.2 Exercises *(page 296)*

1. $x^2 + y^2 = 16$;

Circle, radius 4, center at pole

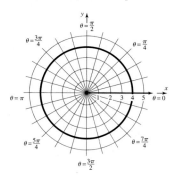

3. $y = x\sqrt{3}$; Line through pole,

making an angle of $\dfrac{\pi}{3}$ with polar axis

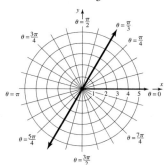

5. $y = 4$; Horizontal line 4 units

above the pole

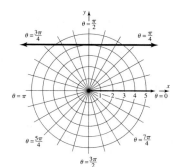

7. $x = -2$; Vertical line 2 units to the left of the pole

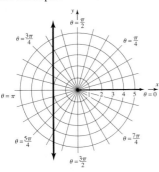

9. $(x - 1)^2 + y^2 = 1$; Circle, radius 1, center $(1, 0)$ in rectangular coordinates

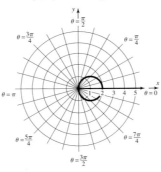

11. $x^2 + (y + 2)^2 = 4$; Circle, radius 2, center at $(0, -2)$ in rectangular coordinates

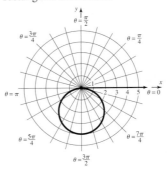

13. $(x - 2)^2 + y^2 = 4$; Circle, radius 2, center at $(2, 0)$ in rectangular coordinates

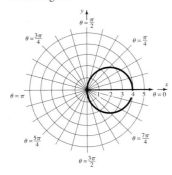

15. $x^2 + (y + 1)^2 = 1$; Circle, radius 1, center at $(0, -1)$ in rectangular coordinates

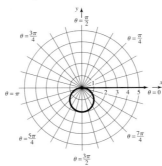

17. E   **19.** F   **21.** H   **23.** D
25. Cardioid

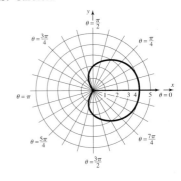

27. Cardioid

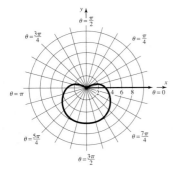

29. Limaçon without inner loop

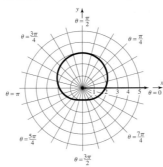

31. Limaçon without inner loop

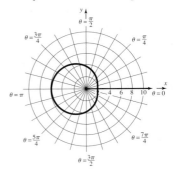

33. Limaçon with inner loop

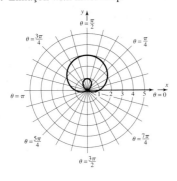

35. Limaçon with inner loop

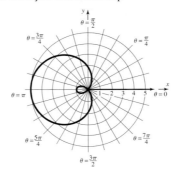

37. Rose

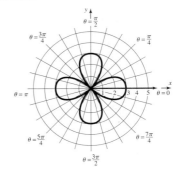

39. Rose

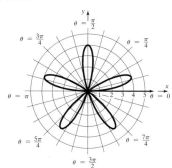

41. Lemniscate

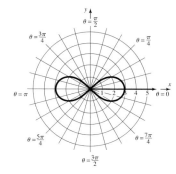

43. Spiral

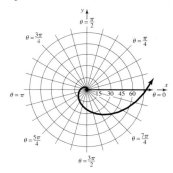

45. Cardioid

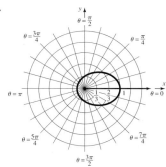

47. Limaçon with inner loop

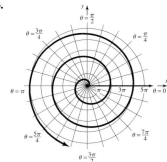

49.

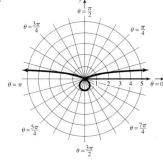

51.

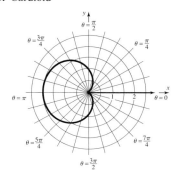

53.

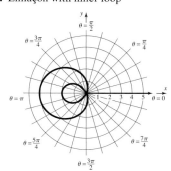

55.

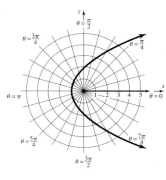

57.

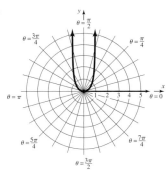

59. $r \sin \theta = a$
$y = a$

61. $r = 2a \sin \theta$
$r^2 = 2ar \sin \theta$
$x^2 + y^2 = 2ay$
$x^2 + y^2 - 2ay = 0$
$x^2 + (y - a)^2 = a^2$
Circle, radius a, center at $(0, a)$
in rectangular coordinates

63.
$r = 2a \cos \theta$
$r^2 = 2ar \cos \theta$
$x^2 + y^2 = 2ax$
$x^2 - 2ax + y^2 = 0$
$(x - a)^2 + y^2 = a^2$
Circle, radius a, center at $(a, 0)$
in rectangular coordinates

65. (a) $r^2 = \cos \theta$
Test for symmetry to y-axis by replacing θ by $\pi - \theta$:
$r^2 = \cos(\pi - \theta)$
$r^2 = -\cos \theta$
Not equivalent; test fails
Test for symmetry to y-axis by replacing r by $-r$ and θ by $-\theta$:
$(-r)^2 = \cos(-\theta)$
$r^2 = \cos \theta$
New test works

(b) $r^2 = \sin \theta$
Test for symmetry to y-axis by replacing θ by $\pi - \theta$:
$r^2 = \sin(\pi - \theta)$
$r^2 = \sin \theta$
Test works
Test for symmetry to y-axis by replacing r by $-r$ and θ by $-\theta$:
$(-r)^2 = \sin(-\theta)$
$r^2 = -\sin \theta$
Not equivalent; new test fails

Historical Problems *(page 304)*

1. (a) $1 + 4i, 1 + i$ **(b)** $-1, 2 + i$

5.3 Exercises *(page 305)*

1.

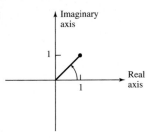

$$\sqrt{2}(\cos 45° + i \sin 45°)$$

3.

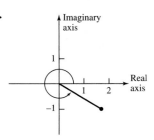

$$2(\cos 330° + i \sin 330°)$$

5.

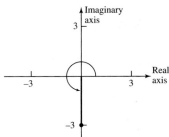

$$3(\cos 270° + i \sin 270°)$$

7.

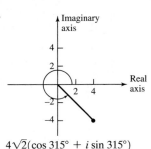

$$4\sqrt{2}(\cos 315° + i \sin 315°)$$

9.

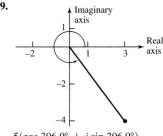

$$5(\cos 306.9° + i \sin 306.9°)$$

11.

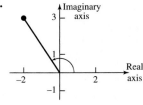

$$\sqrt{13}(\cos 123.7° + i \sin 123.7°)$$

13. $-1 + \sqrt{3}i$ **15.** $2\sqrt{2} - 2\sqrt{2}i$ **17.** $-3i$ **19.** $-0.035 + 0.197i$ **21.** $1.970 + 0.347i$

23. $zw = 8(\cos 60° + i \sin 60°); \dfrac{z}{w} = \dfrac{1}{2}(\cos 20° + i \sin 20°)$ **25.** $zw = 12(\cos 40° + i \sin 40°); \dfrac{z}{w} = \dfrac{3}{4}(\cos 220° + i \sin 220°)$

27. $zw = 4\left(\cos \dfrac{9\pi}{40} + i \sin \dfrac{9\pi}{40}\right); \dfrac{z}{w} = \cos \dfrac{\pi}{40} + i \sin \dfrac{\pi}{40}$ **29.** $zw = 4\sqrt{2}(\cos 15° + i \sin 15°); \dfrac{z}{w} = \sqrt{2}(\cos 75° + i \sin 75°)$

31. $-32 + 32\sqrt{3}i$ **33.** $32i$ **35.** $\dfrac{27}{2} + \dfrac{27\sqrt{3}}{2}i$ **37.** $-\dfrac{25\sqrt{2}}{2} + \dfrac{25\sqrt{2}}{2}i$

39. $-4 + 4i$ **41.** $-23 + 14.142i$ **43.** $\sqrt[6]{2}(\cos 15° + i \sin 15°), \sqrt[6]{2}(\cos 135° + i \sin 135°), \sqrt[6]{2}(\cos 255° + i \sin 255°)$

45. $\sqrt[4]{8}(\cos 75° + i \sin 75°), \sqrt[4]{8}(\cos 165° + i \sin 165°), \sqrt[4]{8}(\cos 255° + i \sin 255°), \sqrt[4]{8}(\cos 345° + i \sin 345°)$

47. $2(\cos 67.5° + i \sin 67.5°), 2(\cos 157.5° + i \sin 157.5°), 2(\cos 247.5° + i \sin 247.5°), 2(\cos 337.5° + i \sin 337.5°)$

49. $\cos 18° + i \sin 18°, \cos 90° + i \sin 90°, \cos 162° + i \sin 162°, \cos 234° + i \sin 234°, \cos 306° + i \sin 306°$

51. $1, i, -1, -i$

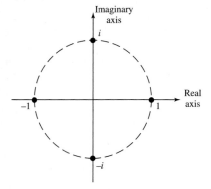

53. Look at formula (8); $|z_k| = \sqrt[n]{r}$ for all k.

55. Look at formula (8). The z_k are spaced apart by an angle of $\dfrac{2\pi}{n}$.

5.4 Exercises *(page 316)*

1.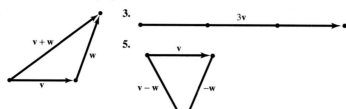

3.

5.

7.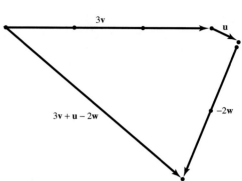

9. T   **11.** F   **13.** F   **15.** T   **17.** 12   **19.** $\mathbf{v} = 3\mathbf{i} + 4\mathbf{j}$ **21.** $\mathbf{v} = 2\mathbf{i} + 4\mathbf{j}$ **23.** $\mathbf{v} = 8\mathbf{i} - \mathbf{j}$ **25.** $\mathbf{v} = -\mathbf{i} + \mathbf{j}$ **27.** 5   **29.** $\sqrt{2}$ **31.** $\sqrt{13}$

33. $-\mathbf{j}$ **35.** $\sqrt{89}$ **37.** $\sqrt{34} - \sqrt{13}$ **39.** $\mathbf{i}$ **41.** $\dfrac{3}{5}\mathbf{i} - \dfrac{4}{5}\mathbf{j}$ **43.** $\dfrac{\sqrt{2}}{2}\mathbf{i} - \dfrac{\sqrt{2}}{2}\mathbf{j}$ **45.** $\mathbf{v} = \dfrac{8\sqrt{5}}{5}\mathbf{i} + \dfrac{4\sqrt{5}}{5}\mathbf{j}$ or $\mathbf{v} = -\dfrac{8\sqrt{5}}{5}\mathbf{i} - \dfrac{4\sqrt{5}}{5}\mathbf{j}$

47. $\{-2 + \sqrt{21}, -2 - \sqrt{21}\}$ **49.** $\mathbf{v} = \dfrac{5}{2}(\mathbf{i} + \sqrt{3}\mathbf{j})$ **51.** $\mathbf{v} = 7(-\mathbf{i} + \sqrt{3}\mathbf{j})$ **53.** $\mathbf{v} = \dfrac{25}{2}(\sqrt{3}\mathbf{i} - \mathbf{j})$ **55.** $\mathbf{F} = 20(\sqrt{3}\mathbf{i} + \mathbf{j})$

57. $\mathbf{F} = (20\sqrt{3} + 30\sqrt{2})\mathbf{i} + (20 - 30\sqrt{2})\mathbf{j}$ **59.** Tension in right cable: 1000 lb; Tension in left cable: 845.2 lb
61. Tension in right part: 1088.4 lb; Tension in left part: 1089.1 lb   **63.**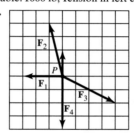

Historical Problems *(page 324)*

1. $(a\mathbf{i} + b\mathbf{j}) \cdot (c\mathbf{i} + d\mathbf{j}) = ac + bd$
real part $[(\overline{a + bi})(c + di)] = $ real part$[(a - bi)(c + di)] = $ real part$[ac + adi - bci - bdi^2] = ac + bd$

5.5 Exercises *(page 324)*

1. (a) 0   **(b)** 90°   **(c)** orthogonal   **3. (a)** 4   **(b)** 36.9°   **(c)** neither   **5. (a)** $\sqrt{3} - 1$ **(b)** 75°   **(c)** neither   **7. (a)** 24   **(b)** 16.3°
(c) neither   **9. (a)** 0   **(b)** 90°   **(c)** orthogonal   **11.** $\dfrac{2}{3}$ **13.** $\mathbf{v}_1 = \dfrac{5}{2}\mathbf{i} - \dfrac{5}{2}\mathbf{j}, \mathbf{v}_2 = -\dfrac{1}{2}\mathbf{i} - \dfrac{1}{2}\mathbf{j}$ **15.** $\mathbf{v}_1 = -\dfrac{1}{5}\mathbf{i} - \dfrac{2}{5}\mathbf{j}, \mathbf{v}_2 = \dfrac{6}{5}\mathbf{i} - \dfrac{3}{5}\mathbf{j}$

17. $\mathbf{v}_1 = \dfrac{14}{5}\mathbf{i} + \dfrac{7}{5}\mathbf{j}, \mathbf{v}_2 = \dfrac{1}{5}\mathbf{i} - \dfrac{2}{5}\mathbf{j}$ **19.** 496.7 mi/hr; 38.5° west of south   **21.** 8.6° off direct heading across the current, upstream; 1.52 min
23. Force required to keep Sienna from rolling down the hill: 737.6 lb; Force perpendicular to the hill: 5248.4 lb.
25. $\mathbf{v} = (250\sqrt{2} - 30)\mathbf{i} + (250\sqrt{2} + 30\sqrt{3})\mathbf{j}$; 518.8 km/hr; N38.6°E
27. $\mathbf{v} = 3\mathbf{i} + 20\mathbf{j}$; 20.2 mi/hr; N8.5°E (Assuming boat traveling north and current traveling east.)   **29.** 3 ft-lb   **31.** $1000\sqrt{3}$ ft-lb ≈ 1732 ft-lb
33. Let $\mathbf{u} = a_1\mathbf{i} + b_1\mathbf{j}, \mathbf{v} = a_2\mathbf{i} + b_2\mathbf{j}, \mathbf{w} = a_3\mathbf{i} + b_3\mathbf{j}$. Compute $\mathbf{u} \cdot (\mathbf{v} + \mathbf{w})$ and $\mathbf{u} \cdot \mathbf{v} + \mathbf{u} \cdot \mathbf{w}$.
35. $\cos \alpha = \dfrac{\mathbf{v} \cdot \mathbf{i}}{\|\mathbf{v}\| \|\mathbf{i}\|} = \mathbf{v} \cdot \mathbf{i}$; if $\mathbf{v} = x\mathbf{i} + y\mathbf{j}$, then $\mathbf{v} \cdot \mathbf{i} = x = \cos \alpha$ and $\mathbf{v} \cdot \mathbf{j} = y = \cos\left(\dfrac{\pi}{2} - \alpha\right) = \sin \alpha$.
37. $\mathbf{v} = a\mathbf{i} + b\mathbf{j}$; the vector projection of $\mathbf{v}$ onto $\mathbf{i}$ is $\dfrac{\mathbf{v} \cdot \mathbf{i}}{\|\mathbf{i}\|^2}\mathbf{i} = (\mathbf{v} \cdot \mathbf{i})\mathbf{i}$; $\mathbf{v} \cdot \mathbf{i} = a, \mathbf{v} \cdot \mathbf{j} = b$, so $\mathbf{v} = (\mathbf{v} \cdot \mathbf{i})\mathbf{i} + (\mathbf{v} \cdot \mathbf{j})\mathbf{j}$.
39. $(\mathbf{v} - \alpha\mathbf{w}) \cdot \mathbf{w} = \mathbf{v} \cdot \mathbf{w} - \alpha\mathbf{w} \cdot \mathbf{w} = \alpha\|\mathbf{w}\|^2 - \alpha\|\mathbf{w}\|^2 = 0$ since the dot product of any vector with itself equals to the square of its magnitude.
41. $W = \mathbf{F} \cdot \overrightarrow{AB} = 0$ when $\mathbf{F}$ is orthogonal to $\overrightarrow{AB}$

5.6 Exercises *(page 335)*

1. All points of the form $(x, 0, z)$ **3.** All points of the form $(x, y, 2)$ **5.** All points of the form $(-4, y, z)$ **7.** All points of the form $(1, 2, z)$
9. $\sqrt{21}$ **11.** $\sqrt{33}$ **13.** $\sqrt{26}$ **15.** $(2, 0, 0); (2, 1, 0); (0, 1, 0); (2, 0, 3); (0, 1, 3); (0, 0, 3)$ **17.** $(1, 4, 3); (3, 2, 3); (3, 4, 3); (3, 2, 5); (1, 4, 5); (1, 2, 5)$
19. $(-1, 2, 2); (4, 0, 2); (4, 2, 2); (-1, 2, 5); (4, 0, 5); (-1, 0, 5)$ **21.** $\mathbf{v} = 3\mathbf{i} + 4\mathbf{j} - \mathbf{k}$ **23.** $\mathbf{v} = 2\mathbf{i} + 4\mathbf{j} + \mathbf{k}$ **25.** $\mathbf{v} = 8\mathbf{i} - \mathbf{j}$ **27.** 7

29. $\sqrt{3}$ **31.** $\sqrt{22}$ **33.** $-\mathbf{j} - 2\mathbf{k}$ **35.** $\sqrt{105}$ **37.** $\sqrt{38} - \sqrt{17}$ **39.** $\dfrac{\mathbf{v}}{\|\mathbf{v}\|} = \mathbf{i}$ **41.** $\dfrac{\mathbf{v}}{\|\mathbf{v}\|} = \dfrac{3}{7}\mathbf{i} - \dfrac{6}{7}\mathbf{j} - \dfrac{2}{7}\mathbf{k}$

43. $\dfrac{\mathbf{v}}{\|\mathbf{v}\|} = \dfrac{\sqrt{3}}{3}\mathbf{i} + \dfrac{\sqrt{3}}{3}\mathbf{j} + \dfrac{\sqrt{3}}{3}\mathbf{k}$ **45.** $\mathbf{v} \cdot \mathbf{w} = 0$; $\theta = 90°$ **47.** $\mathbf{v} \cdot \mathbf{w} = -2, \theta \approx 100.3°$ **49.** $\mathbf{v} \cdot \mathbf{w} = 0$; $\theta = 90°$

51. $\mathbf{v} \cdot \mathbf{w} = 52$; $\theta = 0°$ **53.** $\alpha \approx 64.6°$; $\beta \approx 149.0°$; $\gamma \approx 106.6°$; $\mathbf{v} = 7(\cos 64.6°\mathbf{i} + \cos 149.0°\mathbf{j} + \cos 106.6°\mathbf{k})$

55. $\alpha = \beta = \gamma \approx 54.7°$; $\mathbf{v} = \sqrt{3}(\cos 54.7°\mathbf{i} + \cos 54.7°\mathbf{j} + \cos 54.7°\mathbf{k})$ **57.** $\alpha = \beta = 45°$; $\gamma = 90°$; $\mathbf{v} = \sqrt{2}(\cos 45°\mathbf{i} + \cos 45°\mathbf{j} + \cos 90°\mathbf{k})$

59. $\alpha \approx 60.9°$; $\beta \approx 144.2°$; $\gamma \approx 71.1°$; $\mathbf{v} = \sqrt{38}(\cos 60.9°\mathbf{i} + \cos 144.2°\mathbf{j} + \cos 71.1°\mathbf{k})$

61. If the point $P = (x, y, z)$ is on the sphere with center $C = (x_0, y_0, z_0)$ and radius r, then the distance between P and C is $r = \sqrt{(x - x_0)^2 + (y - y_0)^2 + (z - z_0)^2}$. Therefore, the equation for a sphere is $(x - x_0)^2 + (y - y_0)^2 + (z - z_0)^2 = r^2$.

63. $(x - 1)^2 + (y - 2)^2 + (z - 2)^2 = 4$ **65.** radius $= 2$, center $(-1, 1, 0)$ **67.** radius $= 3$, center $(2, -2, -1)$

69. radius $= \dfrac{3\sqrt{2}}{2}$, center $(2, 0, -1)$ **71.** 2 joules **73.** 9

5.7 Exercises *(page 341)*

1. 2 **3.** 4 **5.** $-11A + 2B + 5C$ **7.** $-6A + 23B - 15C$ **9.** (a) $5\mathbf{i} + 5\mathbf{j} + 5\mathbf{k}$ (b) $-5\mathbf{i} - 5\mathbf{j} - 5\mathbf{k}$ (c) 0 (d) 0

11. (a) $\mathbf{i} - \mathbf{j} - \mathbf{k}$ (b) $-\mathbf{i} + \mathbf{j} + \mathbf{k}$ (c) 0 (d) 0 **13.** (a) $-\mathbf{i} + 2\mathbf{j} + 2\mathbf{k}$ (b) $\mathbf{i} - 2\mathbf{j} - 2\mathbf{k}$ (c) 0 (d) 0

15. (a) $3\mathbf{i} - \mathbf{j} + 4\mathbf{k}$ (b) $-3\mathbf{i} + \mathbf{j} - 4\mathbf{k}$ (c) 0 (d) 0 **17.** $-9\mathbf{i} - 7\mathbf{j} - 3\mathbf{k}$ **19.** $9\mathbf{i} + 7\mathbf{j} + 3\mathbf{k}$ **21.** 0 **23.** $-27\mathbf{i} - 21\mathbf{j} - 9\mathbf{k}$

25. $-18\mathbf{i} - 14\mathbf{j} - 6\mathbf{k}$ **27.** 0 **29.** -25 **31.** 25 **33.** 0 **35.** Any vector of the form $c(-9\mathbf{i} - 7\mathbf{j} - 3\mathbf{k})$, where c is a nonzero scalar

37. Any vector of the form $c(-\mathbf{i} + \mathbf{j} + 5\mathbf{k})$, where c is a nonzero scalar **39.** $\sqrt{166}$ **41.** $\sqrt{555}$ **43.** $\sqrt{34}$ **45.** $\sqrt{998}$

47. $\dfrac{11\sqrt{19}}{57}\mathbf{i} + \dfrac{\sqrt{19}}{57}\mathbf{j} + \dfrac{7\sqrt{19}}{57}\mathbf{k}$ or $-\dfrac{11\sqrt{19}}{57}\mathbf{i} - \dfrac{\sqrt{19}}{57}\mathbf{j} - \dfrac{7\sqrt{19}}{57}\mathbf{k}$

49. $\mathbf{u} \times \mathbf{v} = \begin{vmatrix} \mathbf{i} & \mathbf{j} & \mathbf{k} \\ a_1 & b_1 & c_1 \\ a_2 & b_2 & c_2 \end{vmatrix}$

$= (b_1c_2 - b_2c_1)\mathbf{i} - (a_1c_2 - a_2c_1)\mathbf{j} + (a_1b_2 - a_2b_1)\mathbf{k}$

$= -[(b_2c_1 - b_1c_2)\mathbf{i} - (a_2c_1 - a_1c_2)\mathbf{j} + (a_2b_1 - a_1b_2)\mathbf{k}]$

$= -\begin{vmatrix} \mathbf{i} & \mathbf{j} & \mathbf{k} \\ a_2 & b_2 & c_2 \\ a_1 & b_1 & c_1 \end{vmatrix}$

$= -(\mathbf{v} \times \mathbf{u})$

51. $\mathbf{u} \times \mathbf{v} = \begin{vmatrix} \mathbf{i} & \mathbf{j} & \mathbf{k} \\ a_1 & b_1 & c_1 \\ a_2 & b_2 & c_2 \end{vmatrix} = (b_1c_2 - b_2c_1)\mathbf{i} - (a_1c_2 - a_2c_1)\mathbf{j} + (a_1b_2 - a_2b_1)\mathbf{k}$

$\|\mathbf{u} \times \mathbf{v}\|^2 = (\sqrt{(b_1c_2 - b_2c_1)^2 + (a_1c_2 - a_2c_1)^2 + (a_1b_2 - a_2b_1)^2})^2$

$\qquad = b_1^2c_2^2 - 2b_1b_2c_1c_2 + b_2^2c_1^2 + a_1^2c_2^2 - 2a_1a_2c_1c_2 + a_2^2c_1^2 + a_1^2b_2^2 - 2a_1a_2b_1b_2 + a_2^2b_1^2$

$\|\mathbf{u}\|^2 = a_1^2 + b_1^2 + c_1^2, \|\mathbf{v}\|^2 = a_2^2 + b_2^2 + c_2^2$

$\|\mathbf{u}\|^2\|\mathbf{v}\|^2 = (a_1^2 + b_1^2 + c_1^2)(a_2^2 + b_2^2 + c_2^2) = a_1^2a_2^2 + a_1^2b_2^2 + a_1^2c_2^2 + b_1^2a_2^2 + b_1^2b_2^2 + b_1^2c_2^2 + a_2^2c_1^2 + b_2^2c_1^2 + c_1^2c_2^2$

$(\mathbf{u} \cdot \mathbf{v})^2 = (a_1a_2 + b_1b_2 + c_1c_2)^2 = (a_1a_2 + b_1b_2 + c_1c_2)(a_1a_2 + b_1b_2 + c_1c_2)$

$\qquad = a_1^2a_2^2 + a_1a_2b_1b_2 + a_1a_2c_1c_2 + b_1b_2c_1c_2 + b_1b_2a_1a_2 + b_1^2b_2^2 + b_1b_2c_1c_2 + a_1a_2c_1c_2 + c_1^2c_2^2$

$\qquad = a_1^2a_2^2 + b_1^2b_2^2 + c_1^2c_2^2 + 2a_1a_2b_1b_2 + 2b_1b_2c_1c_2 + 2a_1a_2c_1c_2$

$\|\mathbf{u}\|^2\|\mathbf{v}\|^2 - (\mathbf{u} \cdot \mathbf{v})^2 = a_1^2b_2^2 + a_1^2c_2^2 + b_1^2a_2^2 + b_1^2c_2^2 + a_2^2c_1^2 + b_2^2c_1^2 - 2a_1a_2b_1b_2 - 2b_1b_2c_1c_2 - 2a_1a_2c_1c_2$, which equals $\|\mathbf{u} \times \mathbf{v}\|^2$.

53. We know for any two vectors that $\|\mathbf{u} \times \mathbf{v}\| = \|\mathbf{u}\|\|\mathbf{v}\| \sin \theta$, where θ is the angle between $\mathbf{u}$ and $\mathbf{v}$, so that if $\mathbf{u}$ and $\mathbf{v}$ are orthogonal, then $\theta = 90°$, and so the result follows.

Fill-in-the-Blank Items *(page 343)*

1. pole; polar axis **3.** $r = 2 \cos \theta$ **5.** magnitude; modulus; argument **7.** 0

True/False Items *(page 343)*

1. F **3.** F **5.** T **7.** T

Review Exercises (page 343)

1. $\left(\dfrac{3\sqrt{3}}{2}, \dfrac{3}{2}\right)$ **3.** $(1, \sqrt{3})$ **5.** $(0, 3)$ **7.** $\left(3\sqrt{2}, \dfrac{3\pi}{4}\right), \left(-3\sqrt{2}, -\dfrac{\pi}{4}\right)$ **9.** $\left(2, -\dfrac{\pi}{2}\right), \left(-2, \dfrac{\pi}{2}\right)$

11. $(5, 0.93), (-5, 4.07)$ **13.** $r - 2\sin\theta = 0$
15. $r^2(2\cos^2\theta - \sin^2\theta) - \tan\theta = 0$
17. $r^3\cos\theta = 4$ **19.** $x^2 + y^2 - 2y = 0$
21. $x^2 + y^2 = 25$ **23.** $x + 3y = 6$

25. Circle: radius 2, center at $(2, 0)$ in rectangular coordinates; symmetric with respect to the polar axis

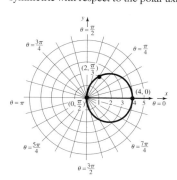

27. Cardioid; symmetric with respect to the line $\theta = \dfrac{\pi}{2}$

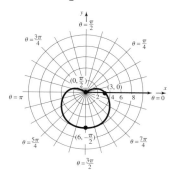

29. Limaçon without inner loop; symmetric with respect to the polar axis

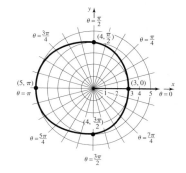

31. $\sqrt{2}(\cos 225° + i\sin 225°)$ **33.** $5(\cos 323.1° + i\sin 323.1°)$ **35.** $-\sqrt{3} + i$ **37.** $-\dfrac{3}{2} + \left(\dfrac{3\sqrt{3}}{2}\right)i$ **39.** $0.098 - 0.017i$

41. $zw = \cos 130° + i\sin 130°; \dfrac{z}{w} = \cos 30° + i\sin 30°$ **43.** $zw = 6(\cos 0 + i\sin 0) = 6; \dfrac{z}{w} = \dfrac{3}{2}\left(\cos\dfrac{8\pi}{5} + i\sin\dfrac{8\pi}{5}\right)$

45. $zw = 5(\cos 5° + i\sin 5°); \dfrac{z}{w} = 5(\cos 15° + i\sin 15°)$ **47.** $\dfrac{27}{2} + \dfrac{27\sqrt{3}}{2}i$ **49.** $4i$ **51.** 64 **53.** $-527 - 336i$

55. $3, 3(\cos 120° + i\sin 120°), 3(\cos 240° + i\sin 240°)$ **57.** $\mathbf{v} = 2\mathbf{i} - 4\mathbf{j}; \|\mathbf{v}\| = 2\sqrt{5}$ **59.** $\mathbf{v} = -\mathbf{i} + 3\mathbf{j}; \|\mathbf{v}\| = \sqrt{10}$

61. $\mathbf{v} = -3\mathbf{i} - 2\mathbf{j} + \mathbf{k}; \|\mathbf{v}\| = \sqrt{14}$ **63.** $\mathbf{v} = 3\mathbf{i} - \mathbf{k}; \|\mathbf{v}\| = \sqrt{10}$ **65.** $-20\mathbf{i} + 13\mathbf{j}$ **67.** $\sqrt{5}$ **69.** $\sqrt{5} + 5$ **71.** $-\dfrac{2\sqrt{5}}{5}\mathbf{i} + \dfrac{\sqrt{5}}{5}\mathbf{j}$

73. $21\mathbf{i} - 2\mathbf{j} - 5\mathbf{k}$ **75.** $\sqrt{38}$ **77.** 0 **79.** $3\mathbf{i} + 9\mathbf{j} + 9\mathbf{k}$ **81.** $\dfrac{3\sqrt{14}}{14}\mathbf{i} + \dfrac{\sqrt{14}}{14}\mathbf{j} - \dfrac{\sqrt{14}}{7}\mathbf{k}; -\dfrac{3\sqrt{14}}{14}\mathbf{i} - \dfrac{\sqrt{14}}{14}\mathbf{j} + \dfrac{\sqrt{14}}{7}\mathbf{k}$

83. $\mathbf{v}\cdot\mathbf{w} = -11; \theta \approx 169.7°$ **85.** $\mathbf{v}\cdot\mathbf{w} = -4; \theta \approx 153.4°$ **87.** $\mathbf{v}\cdot\mathbf{w} = 1; \theta \approx 70.5°$ **89.** $\mathbf{v}\cdot\mathbf{w} = 0; \theta = 90°$ **91.** $\mathbf{v}_1 = \dfrac{9}{10}(3\mathbf{i} + \mathbf{j})$

93. $\alpha \approx 56.1°; \beta \approx 138°; \gamma \approx 68.2°$ **95.** $2\sqrt{83}$ **97.** $\sqrt{29} \approx 5.39$ mph; 0.4 mi   **99.** Left cable: 1843.2 lb; right cable: 1630.4 lb

CHAPTER 6   Analytic Geometry

6.2 Exercises (page 356)

1. B   **3.** E   **5.** H   **7.** C

9. $y^2 = 16x$ **11.** $x^2 = -12y$ **13.** $y^2 = -8x$ **15.** $x^2 = 2y$

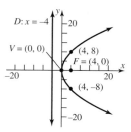

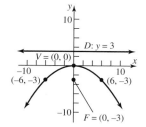

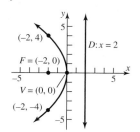

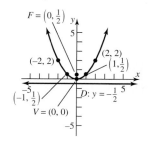

17. $(x - 2)^2 = -8(y + 3)$ **19.** $x^2 = \dfrac{4}{3}y$ **21.** $(x + 3)^2 = 4(y - 3)$ **23.** $(y + 2)^2 = -8(x + 1)$

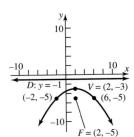
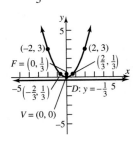
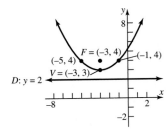
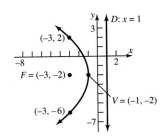

25. Vertex: $(0, 0)$; Focus: $(0, 1)$; Directrix: $y = -1$

27. Vertex: $(0, 0)$; Focus: $(-4, 0)$; Directrix: $x = 4$

29. Vertex: $(-1, 2)$; Focus: $(1, 2)$; Directrix: $x = -3$

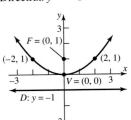
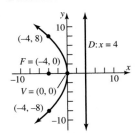
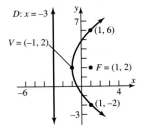

31. Vertex: $(3, -1)$; Focus: $\left(3, -\dfrac{5}{4}\right)$; Directrix: $y = -\dfrac{3}{4}$

33. Vertex: $(2, -3)$; Focus: $(4, -3)$; Directrix: $x = 0$

35. Vertex: $(0, 2)$; Focus: $(-1, 2)$; Directrix: $x = 1$

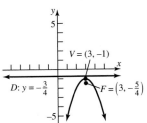

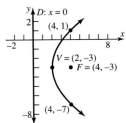
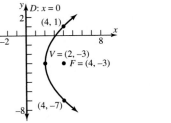

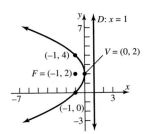

37. Vertex: $(-4, -2)$; Focus: $(-4, -1)$; Directrix: $y = -3$

39. Vertex: $(-1, -1)$; Focus: $\left(-\dfrac{3}{4}, -1\right)$; Directrix: $x = -\dfrac{5}{4}$

41. Vertex: $(2, -8)$; Focus: $\left(2, -\dfrac{31}{4}\right)$; Directrix: $y = -\dfrac{33}{4}$

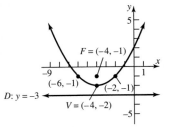
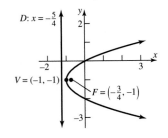
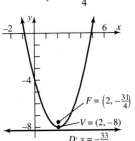

43. $(y - 1)^2 = x$ **45.** $(y - 1)^2 = -(x - 2)$ **47.** $x^2 = 4(y - 1)$ **49.** $y^2 = \dfrac{1}{2}(x + 2)$

51. 1.5625 ft from the base of the dish, along the axis of symmetry   **53.** 1 in. from the vertex   **55.** 20 ft   **57.** 0.78125 ft
59. 4.17 ft from the base along the axis of symmetry   **61.** 24.31 ft, 18.75 ft, 7.64 ft

63. $Ax^2 + Ey = 0, A \neq 0, E \neq 0$

$$Ax^2 = -Ey$$

$$x^2 = -\frac{E}{A}y$$

This is the equation of a parabola with vertex at $(0, 0)$ and axis of symmetry the y-axis.

The focus is $\left(0, -\frac{E}{4A}\right)$; the directrix is the line $y = \frac{E}{4A}$. The parabola opens up if $-\frac{E}{A} > 0$ and down if $-\frac{E}{A} < 0$.

65. $Ax^2 + Dx + Ey + F = 0, A \neq 0$

$$Ax^2 + Dx = -Ey - F$$

$$x^2 + \frac{D}{A}x = -\frac{E}{A}y - \frac{F}{A}$$

$$\left(x + \frac{D}{2A}\right)^2 = -\frac{E}{A}y - \frac{F}{A} + \frac{D^2}{4A^2}$$

$$\left(x + \frac{D}{2A}\right)^2 = -\frac{E}{A}y + \frac{D^2 - 4AF}{4A^2}$$

(a) If $E \neq 0$, then the equation may be written as

$$\left(x + \frac{D}{2A}\right)^2 = -\frac{E}{A}\left(y - \frac{D^2 - 4AF}{4AE}\right)$$

This is the equation of a parabola with vertex at

$$\left(-\frac{D}{2A}, \frac{D^2 - 4AF}{4AE}\right)$$ and axis of symmetry parallel to the y-axis.

(b)–(d) If $E = 0$, the graph of the equation contains no points if

$D^2 - 4AF < 0$, is a single vertical line if $D^2 - 4AF = 0$, and is two vertical lines if $D^2 - 4AF > 0$.

6.3 Exercises *(page 367)*

1. C   **3.** B

5. Vertices: $(-5, 0), (5, 0)$

Foci: $(-\sqrt{21}, 0), (\sqrt{21}, 0)$

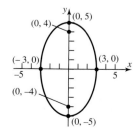

7. Vertices: $(0, -5), (0, 5)$

Foci: $(0, -4), (0, 4)$

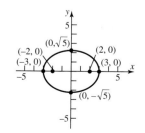

9. $\dfrac{x^2}{4} + \dfrac{y^2}{16} = 1$

Vertices: $(0, -4), (0, 4)$

Foci: $(0, -2\sqrt{3}), (0, 2\sqrt{3})$

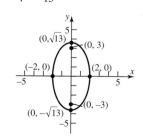

11. $\dfrac{x^2}{8} + \dfrac{y^2}{2} = 1$

Vertices: $(-2\sqrt{2}, 0), (2\sqrt{2}, 0)$

Foci: $(-\sqrt{6}, 0), (\sqrt{6}, 0)$

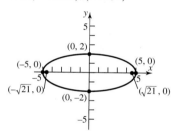

13. $\dfrac{x^2}{16} + \dfrac{y^2}{16} = 1$

Vertices: $(-4, 0), (4, 0), (0, -4), (0, 4)$

Focus: $(0, 0)$

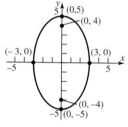

15. $\dfrac{x^2}{25} + \dfrac{y^2}{16} = 1$

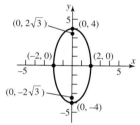

17. $\dfrac{x^2}{9} + \dfrac{y^2}{25} = 1$

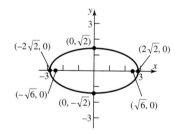

19. $\dfrac{x^2}{9} + \dfrac{y^2}{5} = 1$

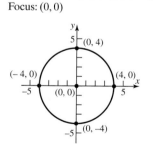

21. $\dfrac{x^2}{4} + \dfrac{y^2}{13} = 1$

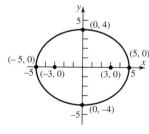

23. $x^2 + \dfrac{y^2}{16} = 1$

25. $\dfrac{(x+1)^2}{4} + (y-1)^2 = 1$

27. $(x-1)^2 + \dfrac{y^2}{4} = 1$

29. Center: $(3, -1)$; Vertices: $(3, -4)$, $(3, 2)$

Foci: $(3, -1 - \sqrt{5})$, $(3, -1 + \sqrt{5})$

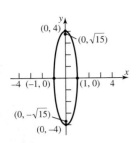

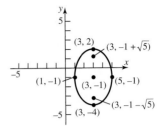

31. $\dfrac{(x+5)^2}{16} + \dfrac{(y-4)^2}{4} = 1$

Center: $(-5, 4)$; Vertices: $(-9, 4)$, $(-1, 4)$
Foci: $(-5 - 2\sqrt{3}, 4)$, $(-5 + 2\sqrt{3}, 4)$

33. $\dfrac{(x+2)^2}{4} + (y-1)^2 = 1$

Center: $(-2, 1)$; Vertices: $(-4, 1)$, $(0, 1)$
Foci: $(-2 - \sqrt{3}, 1)$, $(-2 + \sqrt{3}, 1)$

35. $\dfrac{(x-2)^2}{3} + \dfrac{(y+1)^2}{2} = 1$

Center: $(2, -1)$; Vertices: $(2 - \sqrt{3}, -1)$,
$(2 + \sqrt{3}, -1)$; Foci: $(1, -1)$, $(3, -1)$

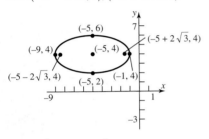

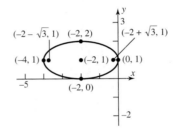

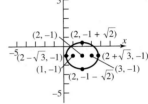

37. $\dfrac{(x-1)^2}{4} + \dfrac{(y+2)^2}{9} = 1$

Center: $(1, -2)$; Vertices: $(1, -5)$, $(1, 1)$
Foci: $(1, -2 - \sqrt{5})$, $(1, -2 + \sqrt{5})$

39. $x^2 + \dfrac{(y+2)^2}{4} = 1$

Center: $(0, -2)$; Vertices: $(0, -4)$, $(0, 0)$
Foci: $(0, -2 - \sqrt{3})$, $(0, -2 + \sqrt{3})$

41. $\dfrac{(x-2)^2}{25} + \dfrac{(y+2)^2}{21} = 1$

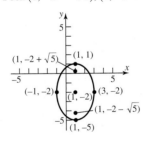

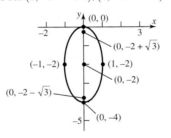

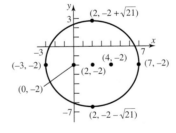

43. $\dfrac{(x-4)^2}{5} + \dfrac{(y-6)^2}{9} = 1$

45. $\dfrac{(x-2)^2}{16} + \dfrac{(y-1)^2}{7} = 1$

47. $\dfrac{(x-1)^2}{10} + (y-2)^2 = 1$

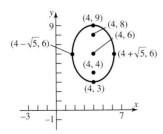

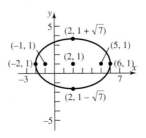

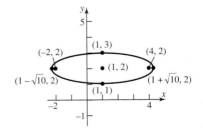

49. $\dfrac{(x-1)^2}{9} + (y-2)^2 = 1$

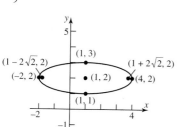

51.

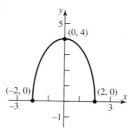

53.

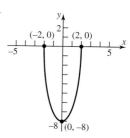

55. $\dfrac{x^2}{100} + \dfrac{y^2}{36} = 1$ **57.** 43.3 ft   **59.** 24.65 ft, 21.65 ft, 13.82 ft   **61.** 0 ft, 12.99 ft, 15 ft, 12.99 ft, 0 ft   **63.** 91.5 million mi; $\dfrac{x^2}{(93)^2} + \dfrac{y^2}{8646.75} = 1$

65. perihelion: 460.6 million mi; mean distance: 483.8 million mi; $\dfrac{x^2}{(483.8)^2} + \dfrac{y^2}{233,524.2} = 1$ **67.** 30 ft

69. (a) $Ax^2 + Cy^2 + F = 0$

$Ax^2 + Cy^2 = -F$

If A and C are of the same and F is of opposite sign, then the equation takes the form

$\dfrac{x^2}{\left(-\dfrac{F}{A}\right)} + \dfrac{y^2}{\left(-\dfrac{F}{C}\right)} = 1$, where $-\dfrac{F}{A}$ and $-\dfrac{F}{C}$ are positive. This is the equation of an ellipse

with center at $(0,0)$.

(b) If $A = C$, the equation may be written as $x^2 + y^2 = -\dfrac{F}{A}$.

This is the equation of a circle with center at $(0,0)$ and radius equal to $\sqrt{-\dfrac{F}{A}}$.

6.4 Exercises (page 381)

1. B   **3.** A

5. $x^2 - \dfrac{y^2}{8} = 1$

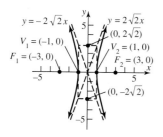

7. $\dfrac{y^2}{16} - \dfrac{x^2}{20} = 1$

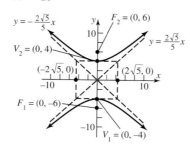

9. $\dfrac{x^2}{9} - \dfrac{y^2}{16} = 1$

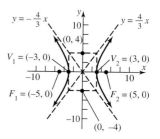

11. $\dfrac{y^2}{36} - \dfrac{x^2}{9} = 1$

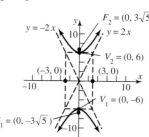

13. $\dfrac{x^2}{8} - \dfrac{y^2}{8} = 1$

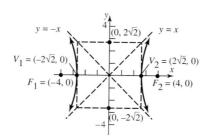

15. $\dfrac{x^2}{25} - \dfrac{y^2}{9} = 1$

Center: $(0,0)$
Transverse axis: x-axis
Vertices: $(-5,0)$, $(5,0)$
Foci: $(-\sqrt{34},0)$, $(\sqrt{34},0)$
Asymptotes: $y = \pm\dfrac{3}{5}x$

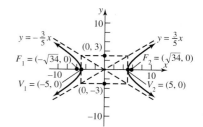

17. $\dfrac{x^2}{4} - \dfrac{y^2}{16} = 1$

Center: $(0,0)$

Transverse axis: x-axis

Vertices: $(-2, 0), (2, 0)$

Foci: $(-2\sqrt{5}, 0), (2\sqrt{5}, 0)$

Asymptotes: $y = \pm 2x$

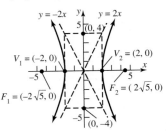

19. $\dfrac{y^2}{9} - x^2 = 1$

Center: $(0,0)$

Transverse axis: y-axis

Vertices: $(0, -3), (0, 3)$

Foci: $(0, -\sqrt{10}), (0, \sqrt{10})$

Asymptotes: $y = \pm 3x$

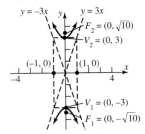

21. $\dfrac{y^2}{25} - \dfrac{x^2}{25} = 1$

Center: $(0,0)$

Transverse axis: y-axis

Vertices: $(0, -5), (0, 5)$

Foci: $(0, -5\sqrt{2}), (0, 5\sqrt{2})$

Asymptotes: $y = \pm x$

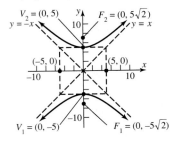

23. $x^2 - y^2 = 1$ **25.** $\dfrac{y^2}{36} - \dfrac{x^2}{9} = 1$

27. $\dfrac{(x-4)^2}{4} - \dfrac{(y+1)^2}{5} = 1$

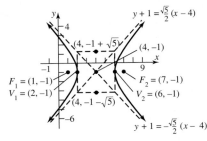

29. $\dfrac{(y+4)^2}{4} - \dfrac{(x+3)^2}{12} = 1$

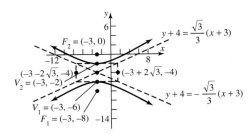

31. $(x-5)^2 - \dfrac{(y-7)^2}{3} = 1$

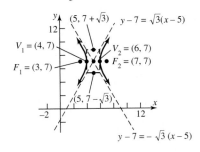

33. $\dfrac{(x-1)^2}{4} - \dfrac{(y+1)^2}{9} = 1$

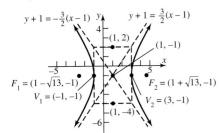

35. $\dfrac{(x-2)^2}{4} - \dfrac{(y+3)^2}{9} = 1$

Center: $(2, -3)$

Transverse axis: Parallel to x-axis

Vertices: $(0, -3)$, $(4, -3)$

Foci: $(2 - \sqrt{13}, -3)$, $(2 + \sqrt{13}, -3)$

Asymptotes: $y + 3 = \pm\dfrac{3}{2}(x - 2)$

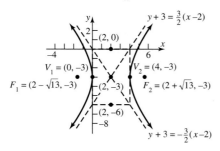

37. $\dfrac{(y-2)^2}{4} - (x+2)^2 = 1$

Center: $(-2, 2)$

Transverse axis: Parallel to y-axis

Vertices: $(-2, 0)$, $(-2, 4)$

Foci: $(-2, 2 - \sqrt{5})$, $(-2, 2 + \sqrt{5})$

Asymptotes: $y - 2 = \pm 2(x + 2)$

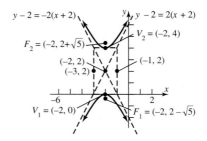

39. $\dfrac{(x+1)^2}{4} - \dfrac{(y+2)^2}{4} = 1$

Center: $(-1, -2)$

Transverse axis: Parallel to x-axis

Vertices: $(-3, -2)$, $(1, -2)$

Foci: $(-1 - 2\sqrt{2}, -2)$, $(-1 + 2\sqrt{2}, -2)$

Asymptotes: $y + 2 = \pm(x + 1)$

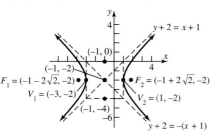

41. $(x-1)^2 - (y+1)^2 = 1$

Center: $(1, -1)$

Transverse axis: Parallel to x-axis

Vertices: $(0, -1)$, $(2, -1)$

Foci: $(1 - \sqrt{2}, -1)$, $(1 + \sqrt{2}, -1)$

Asymptotes: $y + 1 = \pm(x - 1)$

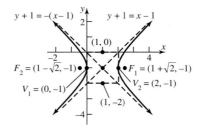

43. $\dfrac{(y-2)^2}{4} - (x+1)^2 = 1$

Center: $(-1, 2)$

Transverse axis: Parallel to y-axis

Vertices: $(-1, 0)$, $(-1, 4)$

Foci: $(-1, 2 - \sqrt{5})$, $(-1, 2 + \sqrt{5})$

Asymptotes: $y - 2 = \pm 2(x + 1)$

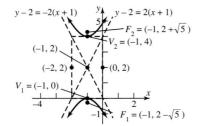

45. $\dfrac{(x-3)^2}{4} - \dfrac{(y+2)^2}{16} = 1$

Center: $(3, -2)$

Transverse axis: Parallel to x-axis

Vertices: $(1, -2)$, $(5, -2)$

Foci: $(3 - 2\sqrt{5}, -2)$, $(3 + 2\sqrt{5}, -2)$

Asymptotes: $y + 2 = \pm 2(x - 3)$

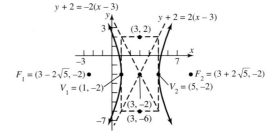

47. $\dfrac{(y-1)^2}{4} - (x+2)^2 = 1$

Center: $(-2, 1)$
Transverse axis: Parallel to y-axis
Vertices: $(-2, -1), (-2, 3)$
Foci: $(-2, 1 - \sqrt{5}), (-2, 1 + \sqrt{5})$

Asymptotes: $y - 1 = \pm 2(x + 2)$

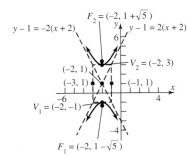

49.

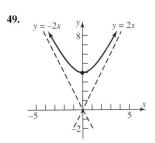

51.

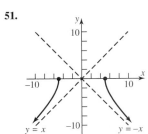

53. (a) The ship will reach shore at a point 64.66 mi from the master station.   **(b)** 0.00086 sec   **(c)** $(104, 50)$
55. (a) 450 ft   **57.** If e is close to 1, narrow hyperbola; if e is very large, wide hyperbola
59. $\dfrac{x^2}{4} - y^2 = 1$; asymptotes $y = \pm\dfrac{1}{2}x$, $y^2 - \dfrac{x^2}{4} = 1$; asymptotes $y = \pm\dfrac{1}{2}x$

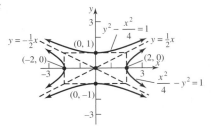

61. $Ax^2 + Cy^2 + F = 0$ If A and C are of opposite sign and $F \neq 0$, this equation may be written as $\dfrac{x^2}{\left(-\dfrac{F}{A}\right)} + \dfrac{y^2}{\left(-\dfrac{F}{C}\right)} = 1$,

$\qquad Ax^2 + Cy^2 = -F$ where $-\dfrac{F}{A}$ and $-\dfrac{F}{C}$ are opposite in sign. This is the equation of a hyperbola with center $(0, 0)$.

$\qquad\qquad$ The transverse axis is the x-axis if $-\dfrac{F}{A} > 0$; the transverse axis is the y-axis if $-\dfrac{F}{A} < 0$.

6.5 Exercises *(page 390)*

1. Parabola   **3.** Ellipse   **5.** Hyperbola   **7.** Hyperbola   **9.** Circle   **11.** $x = \dfrac{\sqrt{2}}{2}(x' - y'), y = \dfrac{\sqrt{2}}{2}(x' + y')$

13. $x = \dfrac{\sqrt{2}}{2}(x' - y'), y = \dfrac{\sqrt{2}}{2}(x' + y')$ **15.** $x = \dfrac{1}{2}(x' - \sqrt{3}y'), y = \dfrac{1}{2}(\sqrt{3}x' + y')$ **17.** $x = \dfrac{\sqrt{5}}{5}(x' - 2y'), y = \dfrac{\sqrt{5}}{5}(2x' + y')$

19. $x = \dfrac{\sqrt{13}}{13}(3x' - 2y'), y = \dfrac{\sqrt{13}}{13}(2x' + 3y')$

21. $\theta = 45°$ (see Problem 11)

$$x'^2 - \frac{y'^2}{3} = 1$$

Hyperbola
Center at origin
Transverse axis is the x'-axis.
Vertices at $(\pm 1, 0)$

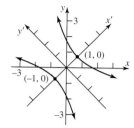

23. $\theta = 45°$ (see Problem 13)

$$x'^2 + \frac{y'^2}{4} = 1$$

Ellipse
Center at $(0, 0)$
Major axis is the y'-axis.
Vertices at $(0, \pm 2)$

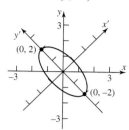

25. $\theta = 60°$ (see Problem 15)

$$\frac{x'^2}{4} + y'^2 = 1$$

Ellipse
Center at $(0, 0)$
Major axis is the x'-axis.
Vertices at $(\pm 2, 0)$

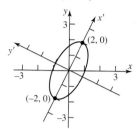

27. $\theta \approx 63°$ (see Problem 17)

$$y'^2 = 8x'$$

Parabola

Vertex at $(0, 0)$

Focus at $(2, 0)$

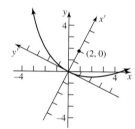

29. $\theta \approx 34°$ (see Problem 19)

$$\frac{(x' - 2)^2}{4} + y'^2 = 1$$

Ellipse

Center at $(2, 0)$

Major axis is the x'-axis.

Vertices at $(4, 0)$ and $(0, 0)$

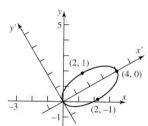

31. $\cot(2\theta) = \frac{7}{24}$;

$$\theta = \sin^{-1}\left(\frac{3}{5}\right) \approx 37°$$

$$(x' - 1)^2 = -6\left(y' - \frac{1}{6}\right)$$

Parabola

Vertex at $\left(1, \frac{1}{6}\right)$

Focus at $\left(1, -\frac{4}{3}\right)$

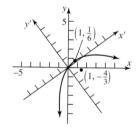

33. Hyperbola **35.** Hyperbola **37.** Parabola **39.** Ellipse **41.** Ellipse

43. Refer to equation (6): $A' = A \cos^2 \theta + B \sin \theta \cos \theta + C \sin^2 \theta$
$B' = B(\cos^2 \theta - \sin^2 \theta) + 2(C - A)(\sin \theta \cos \theta)$
$C' = A \sin^2 \theta - B \sin \theta \cos \theta + C \cos^2 \theta$
$D' = D \cos \theta + E \sin \theta$
$E' = -D \sin \theta + E \cos \theta$
$F' = F$

45. Use Problem 43 to find $B'^2 - 4A'C'$.
After much cancellation, $B'^2 - 4A'C' = B^2 - 4AC$.

47. The distance between P_1 and P_2 in the $x'y'$-plane equals $\sqrt{(x_2' - x_1')^2 + (y_2' - y_1')^2}$.
Assuming $x' = x \cos \theta - y \sin \theta$ and $y' = x \sin \theta + y \cos \theta$, then $(x_2' - x_1')^2 = (x_2 \cos \theta - y_2 \sin \theta - x_1 \cos \theta + y_1 \sin \theta)^2$
$= \cos^2 \theta (x_2 - x_1)^2 - 2 \sin \theta \cos \theta (x_2 - x_1)(y_2 - y_1) + \sin^2 \theta (y_2 - y_1)^2$, and
$(y_2' - y_1')^2 = (x_2 \sin \theta + y_2 \cos \theta - x_1 \sin \theta - y_1 \cos \theta)^2 = \sin^2 \theta (x_2 - x_1)^2 + 2 \sin \theta \cos \theta (x_2 - x_1)(y_2 - y_1) + \cos^2 \theta (y_2 - y_1)^2$.
Therefore, $(x_2' - x_1')^2 + (y_2' - y_1')^2 = \cos^2 \theta (x_2 - x_1)^2 + \sin^2 \theta (x_2 - x_1)^2 + \sin^2 \theta (y_2 - y_1)^2 + \cos^2 \theta (y_2 - y_1)^2$
$= (x_2 - x_1)^2 (\cos^2 \theta + \sin^2 \theta) + (y_2 - y_1)^2(\sin^2 \theta + \cos^2 \theta) = (x_2 - x_1)^2 + (y_2 - y_1)^2$.

6.6 Exercises (page 396)

1. Parabola; directrix is perpendicular to the polar axis 1 unit to the right of the pole. **3.** Hyperbola; directrix is parallel to the polar axis $\frac{4}{3}$ units below the pole. **5.** Ellipse; directrix is perpendicular to the polar axis $\frac{3}{2}$ units to the left of the pole.

7. Parabola; directrix is perpendicular to the polar axis 1 unit to the right of the pole; vertex is at $\left(\dfrac{1}{2}, 0\right)$.

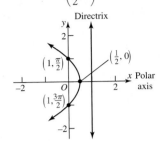

9. Ellipse; directrix is parallel to the polar axis $\dfrac{8}{3}$ units above the pole; vertices are at $\left(\dfrac{8}{7}, \dfrac{\pi}{2}\right)$ and $\left(8, \dfrac{3\pi}{2}\right)$.

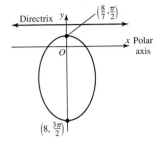

11. Hyperbola; directrix is perpendicular to the polar axis $\dfrac{3}{2}$ units to the left of the pole; vertices are at $(-3, 0)$ and $(1, \pi)$.

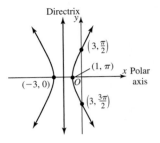

13. Ellipse; directrix is parallel to the polar axis 8 units below the pole; vertices are at $\left(8, \dfrac{\pi}{2}\right)$ and $\left(\dfrac{8}{3}, \dfrac{3\pi}{2}\right)$.

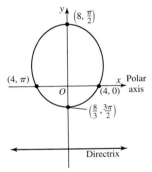

15. Ellipse; directrix is parallel to the polar axis 3 units below the pole; vertices are at $\left(6, \dfrac{\pi}{2}\right)$ and $\left(\dfrac{6}{5}, \dfrac{3\pi}{2}\right)$.

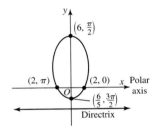

17. Ellipse; directrix is perpendicular to the polar axis 6 units to the left of the pole; vertices are at $(6, 0)$ and $(2, \pi)$.

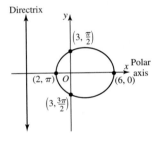

19. $y^2 + 2x - 1 = 0$ **21.** $16x^2 + 7y^2 + 48y - 64 = 0$ **23.** $3x^2 - y^2 + 12x + 9 = 0$ **25.** $4x^2 + 3y^2 - 16y - 64 = 0$

27. $9x^2 + 5y^2 - 24y - 36 = 0$ **29.** $3x^2 + 4y^2 - 12x - 36 = 0$ **31.** $r = \dfrac{1}{1 + \sin \theta}$ **33.** $r = \dfrac{12}{5 - 4 \cos \theta}$ **35.** $r = \dfrac{12}{1 - 6 \sin \theta}$

37. Use $d(D, P) = p - r \cos \theta$ in the derivation of equation (a) in Table 5.
39. Use $d(D, P) = p + r \sin \theta$ in the derivation of equation (a) in Table 5.

6.7 Exercises (page 407)

1.

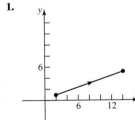

$x - 3y + 1 = 0$

3.

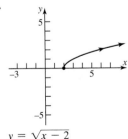

$y = \sqrt{x - 2}$

5.

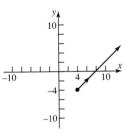

$x = y + 8$

7.

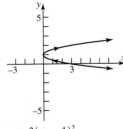

$x = 3(y - 1)^2$

9.

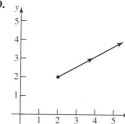

$2y = 2 + x$

11.

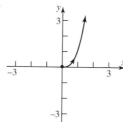

$y = x^3$

13.

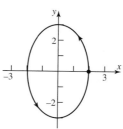

$\dfrac{x^2}{4} + \dfrac{y^2}{9} = 1$

15.

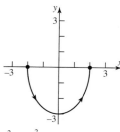

$\dfrac{x^2}{4} + \dfrac{y^2}{9} = 1$

17.

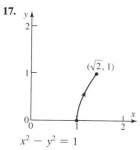

$x^2 - y^2 = 1$

19.

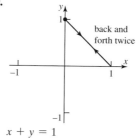

back and forth twice

$x + y = 1$

21. (a) $x = 3$

$y = -16t^2 + 50t + 6$

(b) 3.24 sec

(c) 1.5625 sec; 45.0625 ft

(d)

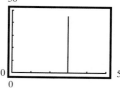

23. (a) Train: $x_1 = t^2, y_1 = 1$;
Bill: $x_2 = 5(t - 5), y_2 = 3$

(b) Bill won't catch the train.

(c)

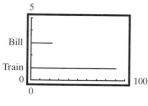

25. (a) $x = (145 \cos 20°)t$

$y = -16t^2 + (145 \sin 20°)t + 5$

(b) 3.197 sec

(c) 1.55 sec; 43.43 ft

(d) 435.61 ft

(e)

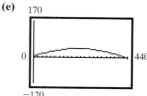

27. (a) $x = (40 \cos 45°)t$

$y = -4.9t^2 + (40 \sin 45°)t + 300$

(b) 11.226 sec

(c) 2.886 sec; 340.8 m

(d) 317.5 m

(e)

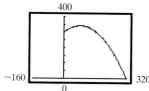

29. (a) Paseo: $x = 40t - 5, y = 0$; Bonneville: $x = 0, y = 30t - 4$ **(b)** $d = \sqrt{(40t - 5)^2 + (30t - 4)^2}$

(c)

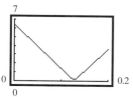

(d) 0.2 mi; 7.68 min   **(e)** Turn axes off to see the graph:

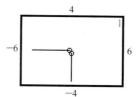

31. $x = t$ $x = \dfrac{t + 1}{4}$

             or

$y = 4t - 1$ $y = t$

33. $x = t$ $x = t^3$

        or

$y = t^2 + 1$ $y = t^6 + 1$

35. $x = t$ $x = \sqrt[3]{t}$

       or

$y = t^3$ $y = t$

37. $x = t$ $x = t^3$

     or

$y = t^{2/3}$ $y = t^2$

39. $x = t + 2, y = t, 0 \le t \le 5$

41. $x = 3 \cos t, y = 2 \sin t, 0 \le t \le 2\pi$

43. $x = 2 \cos(\pi t), y = -3 \sin(\pi t), 0 \le t \le 2$ **45.** $x = 2 \sin(2\pi t), y = 3 \cos(2\pi t), 0 \le t \le 1$

47.

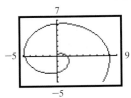 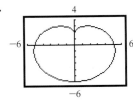 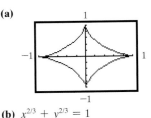 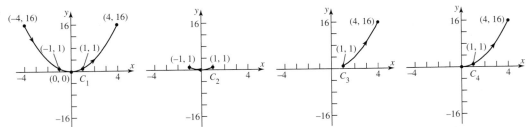

49. The orientation is from (x_1, y_1) to (x_2, y_2).

51.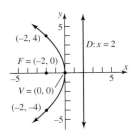

53.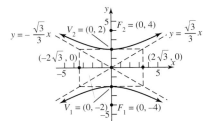

55. (a)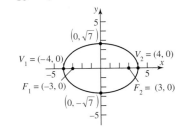

(b) $x^{2/3} + y^{2/3} = 1$

Fill-in-the-Blank Items *(page 411)*

1. Parabola **3.** Hyperbola **5.** y-axis **7.** $\cot(2\theta) = \dfrac{A - C}{B}$ **9.** ellipse

True/False Items *(page 411)*

1. T **3.** T **5.** T **7.** T **9.** T

Review Exercises *(page 412)*

1. Parabola; vertex $(0, 0)$, focus $(-4, 0)$, directrix $x = 4$ **3.** Hyperbola; center $(0, 0)$, vertices $(5, 0)$ and $(-5, 0)$, foci $(\sqrt{26}, 0)$ and $(-\sqrt{26}, 0)$, asymptotes $y = \dfrac{1}{5}x$ and $y = -\dfrac{1}{5}x$ **5.** Ellipse; center $(0, 0)$, vertices $(0, 5)$ and $(0, -5)$, foci $(0, 3)$ and $(0, -3)$

7. $x^2 = -4(y - 1)$: Parabola; vertex $(0, 1)$, focus $(0, 0)$, directrix $y = 2$ **9.** $\dfrac{x^2}{2} - \dfrac{y^2}{8} = 1$: Hyperbola; center $(0, 0)$, vertices $(\sqrt{2}, 0)$ and $(-\sqrt{2}, 0)$, foci $(\sqrt{10}, 0)$ and $(-\sqrt{10}, 0)$, asymptotes $y = 2x$ and $y = -2x$ **11.** $(x - 2)^2 = 2(y + 2)$: Parabola; vertex $(2, -2)$, focus $\left(2, -\dfrac{3}{2}\right)$, directrix $y = -\dfrac{5}{2}$ **13.** $\dfrac{(y - 2)^2}{4} - (x - 1)^2 = 1$: Hyperbola; center $(1, 2)$, vertices $(1, 4)$ and $(1, 0)$, foci $(1, 2 + \sqrt{5})$ and $(1, 2 - \sqrt{5})$, asymptotes $y - 2 = \pm 2(x - 1)$ **15.** $\dfrac{(x - 2)^2}{9} + \dfrac{(y - 1)^2}{4} = 1$: Ellipse; center $(2, 1)$, vertices $(5, 1)$ and $(-1, 1)$, foci $(2 + \sqrt{5}, 1)$ and $(2 - \sqrt{5}, 1)$ **17.** $(x - 2)^2 = -4(y + 1)$: Parabola; vertex $(2, -1)$, focus $(2, -2)$, directrix $y = 0$

19. $\dfrac{(x - 1)^2}{4} + \dfrac{(y + 1)^2}{9} = 1$: Ellipse; center $(1, -1)$, vertices $(1, 2)$ and $(1, -4)$, foci $(1, -1 + \sqrt{5})$ and $(1, -1 - \sqrt{5})$

21. $y^2 = -8x$

23. $\dfrac{y^2}{4} - \dfrac{x^2}{12} = 1$

25. $\dfrac{x^2}{16} + \dfrac{y^2}{7} = 1$

27. $(x - 2)^2 = -4(y + 3)$

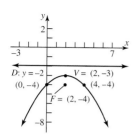

29. $(x + 2)^2 - \dfrac{(y + 3)^2}{3} = 1$

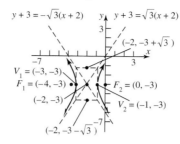

31. $\dfrac{(x + 4)^2}{16} + \dfrac{(y - 5)^2}{25} = 1$

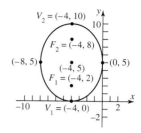

33. $\dfrac{(x + 1)^2}{9} - \dfrac{(y - 2)^2}{7} = 1$

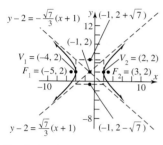

35. $\dfrac{(x - 3)^2}{9} - \dfrac{(y - 1)^2}{4} = 1$

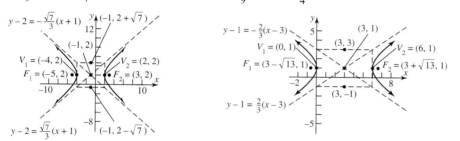

37. Parabola   **39.** Ellipse   **41.** Parabola   **43.** Hyperbola   **45.** Ellipse

47. $x'^2 - \dfrac{y'^2}{9} = 1$

Hyperbola

Center at the origin

Transverse axis the x'-axis

Vertices at $(\pm 1, 0)$

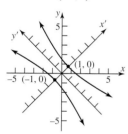

49. $\dfrac{x'^2}{2} + \dfrac{y'^2}{4} = 1$

Ellipse

Center at origin

Major axis the y'-axis

Vertices at $(0, \pm 2)$

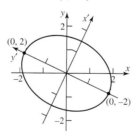

51. $y'^2 = -\dfrac{4\sqrt{13}}{13}x'$

Parabola

Vertex at the origin

Focus on the x'-axis at $\left(-\dfrac{\sqrt{13}}{13}, 0\right)$

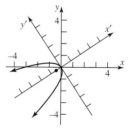

53. Parabola; directrix is perpendicular to the polar axis 4 units to the left of the pole; vertex is $(2, \pi)$.

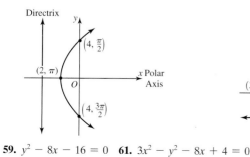

55. Ellipse; directrix is parallel to the polar axis 6 units below the pole; vertices are $\left(6, \dfrac{\pi}{2}\right)$ and $\left(2, \dfrac{3\pi}{2}\right)$.

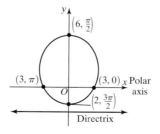

57. Hyperbola; directrix is perpendicular to the polar axis 1 unit to the right of the pole; vertices are $\left(\dfrac{2}{3}, 0\right)$ and $(-2, \pi)$.

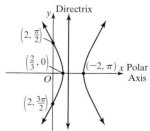

59. $y^2 - 8x - 16 = 0$ **61.** $3x^2 - y^2 - 8x + 4 = 0$

63.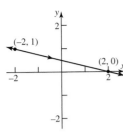

$x + 4y = 2$

65.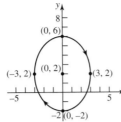

$$\frac{x^2}{9} + \frac{(y-2)^2}{16} = 1$$

67.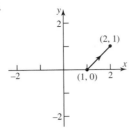

$1 + y = x$

69. $\dfrac{x^2}{5} - \dfrac{y^2}{4} = 1$

71. The ellipse $\dfrac{x^2}{16} + \dfrac{y^2}{7} = 1$

73. $\dfrac{1}{4}$ ft or 3 in.

75. 19.72 ft, 18.86 ft, 14.91 ft

77. **(a)** 45.24 mi from the Master Station
(b) 0.000645 sec
(c) $(66, 20)$

79. **(a)** $x = (100 \cos 35°)t$
$y = -16t^2 + (100 \sin 35°)t + 6$
(b) 3.6866 sec   **(c)** 1.7924 sec; 57.4 ft
(d) 302 ft   **(e)**

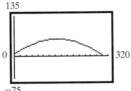

C H A P T E R   7   Exponential and Logarithmic Functions

7.1 Exercises *(page 426)*

1. (a) 11.212   **(b)** 11.587   **(c)** 11.664   **(d)** 11.665   **3. (a)** 8.815   **(b)** 8.821   **(c)** 8.824   **(d)** 8.825
5. (a) 21.217   **(b)** 22.217   **(c)** 22.440   **(d)** 22.459   **7.** 3.320   **9.** 0.427   **11.** B   **13.** D   **15.** A   **17.** E

19.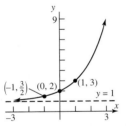

Domain: All real numbers
Range: $\{y \mid y > 1\}$ or $(1, \infty)$
Horizontal asymptote: $y = 1$

21.

Domain: All real numbers
Range: $\{y \mid y > -2\}$ or $(-2, \infty)$
Horizontal asymptote: $y = -2$

23.

Domain: All real numbers
Range: $\{y \mid y > 2\}$ or $(2, \infty)$
Horizontal asymptote: $y = 2$

25.

Domain: All real numbers
Range: $\{y \mid y > 2\}$ or $(2, \infty)$
Horizontal asymptote: $y = 2$

27.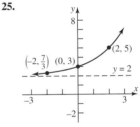

Domain: All real numbers
Range: $\{y \mid y > 0\}$ or $(0, \infty)$
Horizontal asymptote: $y = 0$

29.

Domain: All real numbers
Range: $\{y \mid y > 0\}$ or $(0, \infty)$
Horizontal asymptote: $y = 0$

31.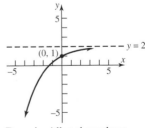

Domain: All real numbers
Range: $\{y \mid y < 5\}$ or $(-\infty, 5)$
Horizontal asymptote: $y = 5$

33.

Domain: All real numbers
Range: $\{y \mid y < 2\}$ or $(-\infty, 2)$
Horizontal asymptote: $y = 2$

35. $\dfrac{1}{2}$ **37.** $\{-\sqrt{2}, 0, \sqrt{2}\}$ **39.** $\left\{1 - \dfrac{\sqrt{6}}{3}, 1 + \dfrac{\sqrt{6}}{3}\right\}$ **41.** 0   **43.** 4   **45.** $\dfrac{3}{2}$ **47.** $\{1, 2\}$ **49.** $\dfrac{1}{49}$ **51.** $\dfrac{1}{4}$

53.

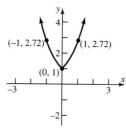

Domain: All real numbers
Range: $\{y|y \geq 1\}$ or $[1, \infty)$
Intercept: $(0, 1)$

55.

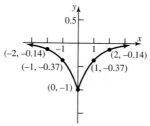

Domain: All real numbers
Range: $\{y|-1 \leq y < 0\}$ or $[-1, 0)$
Intercept: $(0, -1)$

57. (a) 74% **(b)** 47% **59. (a)** 44.3 watts **(b)** 11.6 watts
61. 3.35 mg; 0.45 mg
63. (a) 0.63 **(b)** 0.98 **(c)** 1
(d)

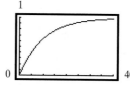

(e) About 7 min

65. (a) 5.16% **(b)** 8.88% **67. (a)** 71% **(b)** 73% **(c)** 100%
69. (a) 5.414 amp, 7.585 amp, 10.376 amp **(b)** 12 amp **(d)** 3.343 amp, 5.309 amp, 9.443 amp **(e)** 24 amp
(c), (f)

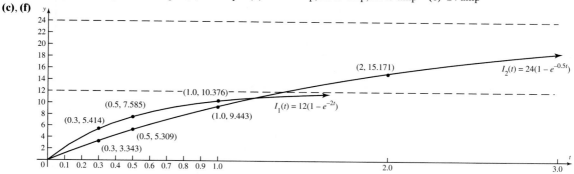

71. $n = 4: 2.7083; n = 6: 2.7181; n = 8: 2.7182788; n = 10: 2.7182818$

73. $\dfrac{f(x + h) - f(x)}{h} = \dfrac{a^{x+h} - a^x}{h} = \dfrac{a^x a^h - a^x}{h} = \dfrac{a^x(a^h - 1)}{h} = a^x \left(\dfrac{a^h - 1}{h} \right)$ **75.** $f(-x) = a^{-x} = \dfrac{1}{a^x} = \dfrac{1}{f(x)}$

77. (a) 9.23×10^{-3}, or about 0 **(b)** 0.81, or about 1   **79.** 59 min
(c) 5.01, or about 5 **(d)** 57.91°, 43.99°, 30.07°

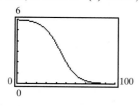

7.2 Exercises *(page 437)*

1. $2 = \log_3 9$ **3.** $2 = \log_a 1.6$ **5.** $2 = \log_{1.1} M$ **7.** $x = \log_2 7.2$ **9.** $\sqrt{2} = \log_x \pi$ **11.** $x = \ln 8$ **13.** $2^3 = 8$ **15.** $a^6 = 3$ **17.** $3^x = 2$

19. $2^{1.3} = M$ **21.** $(\sqrt{2})^x = \pi$ **23.** $e^x = 4$ **25.** 0 **27.** 2 **29.** -4 **31.** $\dfrac{1}{2}$ **33.** 4 **35.** $\dfrac{1}{2}$ **37.** $\{x|x > 3\}; (3, \infty)$

39. All real numbers except 0; $\{x|x \neq 0\}$ **41.** All real numbers except 1; $\{x|x \neq 1\}$ **43.** $\{x|x > -1\}; (-1, \infty)$
45. $\{x|x < -1 \text{ or } x > 0\}; (-\infty, -1) \text{ or } (0, \infty)$ **47.** 0.511 **49.** 30.099 **51.** $\sqrt{2}$ **53.** B **55.** D **57.** A **59.** E

61.

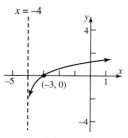

Domain: $\{x|x > -4\}$ or $(-4, \infty)$
Range: All real numbers
Vertical asymptote: $x = -4$

63.

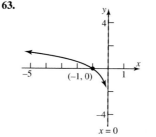

Domain: $\{x|x < 0\}$ or $(-\infty, 0)$
Range: All real numbers
Vertical asymptote: $x = 0$

65.

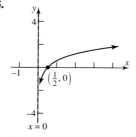

Domain: $\{x|x > 0\}$ or $(0, \infty)$
Range: All real numbers
Vertical asymptote: $x = 0$

67.

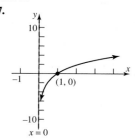

Domain: $\{x|x > 0\}$ or $(0, \infty)$
Range: All real numbers
Vertical asymptote: $x = 0$

69.

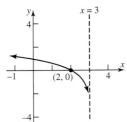

Domain: $\{x \mid x < 3\}$ or $(-\infty, 3)$
Range: All real numbers
Vertical asymptote: $x = 3$

71.

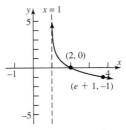

Domain: $\{x \mid x > 1\}$ or $(1, \infty)$
Range: All real numbers
Vertical asymptote: $x = 1$

73. 9 **75.** $\dfrac{7}{2}$ **77.** 2 **79.** 5 **81.** 3 **83.** 2 **85.** $\dfrac{\ln 10}{3}$ **87.** $\dfrac{\ln 8 - 5}{2}$

89. $\{-2\sqrt{2}, 2\sqrt{2}\}$ **91.** -1

93.

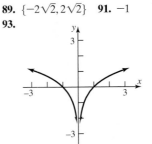

Domain: All real numbers except 0 or $\{x \mid x \neq 0\}$
Range: All real numbers
Intercepts: $(-1, 0), (1, 0)$

95.

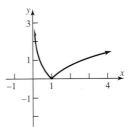

Domain: $\{x \mid x > 0\}$ or $(0, \infty)$
Range: $\{y \mid y \geq 0\}$ or $[0, \infty)$
Intercept: $(1, 0)$

97. (a) $n \approx 6.93$ so 7 panes are necessary (b) $n \approx 13.86$ so 14 panes are necessary
99. (a) $d \approx 127.7$ so it takes about 128 days (b) $d \approx 575.6$ so it takes about 576 days
101. (a) 6.93 min (b) 16.09 min (c) No, since $F(t)$ can never equal one.
103. $h \approx 2.29$ so the time between injections is about 2 hr, 17 min **105.** 0.2695 sec
0.8959 sec

107. (a) $k = 20.07$
(b) 91% (c) 0.175
(d) 0.08

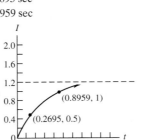

7.3 Exercises (page 450)

1. 71 **3.** -4 **5.** 7 **7.** 1 **9.** 1 **11.** 2 **13.** $\dfrac{5}{4}$ **15.** 4 **17.** $a + b$ **19.** $b - a$ **21.** $3a$ **23.** $\dfrac{1}{5}(a + b)$ **25.** $2 \log_a u + 3 \log_a v$

27. $-3 \log M$ **29.** $\dfrac{1}{2}[3 \log_5 a - \log_5 b]$ **31.** $2 \ln x + \dfrac{1}{2} \ln(1 - x)$ **33.** $3 \log_2 x - \log_2(x - 3)$ **35.** $\log x + \log(x + 2) - 2 \log(x + 3)$

37. $\dfrac{1}{3} \ln(x - 2) + \dfrac{1}{3} \ln(x + 1) - \dfrac{2}{3} \ln(x + 4)$ **39.** $\ln 5 + \ln x + \dfrac{1}{2} \ln(1 - 3x) - 3 \ln(x - 4)$ **41.** $\log_5 u^3 v^4$ **43.** $-\dfrac{5}{2} \log_{1/2} x$

45. $-2 \ln(x - 1)$ **47.** $\log_2[x(3x - 2)^4]$ **49.** $\log_a\left(\dfrac{25x^6}{\sqrt{2x + 3}}\right)$ **51.** 2.771 **53.** -3.880 **55.** 5.615 **57.** 0.874 **59.** 3 **61.** 1

63. $y = \left(\dfrac{\log x}{\log 4}\right) = \left(\dfrac{\ln x}{\ln 4}\right)$ **65.** $y = \left(\dfrac{\log(x + 2)}{\log 2}\right) = \left(\dfrac{\ln(x + 2)}{\ln 2}\right)$ **67.** $y = \dfrac{\log(x + 1)}{\log(x - 1)} = \dfrac{\ln(x + 1)}{\ln(x - 1)}$

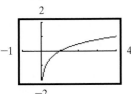

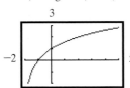

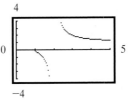

69. $y = Cx$ **71.** $y = Cx(x + 1)$ **73.** $y = Ce^{3x}$ **75.** $y = Ce^{-4x} + 3$ **77.** $y = \dfrac{\sqrt[3]{C}(2x + 1)^{1/6}}{(x + 4)^{1/9}}$

79. $\log_a(x + \sqrt{x^2 - 1}) + \log_a(x - \sqrt{x^2 - 1}) = \log_a[(x + \sqrt{x^2 - 1})(x - \sqrt{x^2 - 1})] = \log_a[x^2 - (x^2 - 1)] = \log_a 1 = 0$
81. $\ln(1 + e^{2x}) = \ln(e^{2x}(e^{-2x} + 1)) = \ln e^{2x} + \ln(e^{-2x} + 1) = 2x + \ln(1 + e^{-2x})$ **83.** $y = f(x) = \log_a x$;
$a^y = x$ implies $a^y = \left(\dfrac{1}{a}\right)^{-y} = x$, so $-y = \log_{1/a} x = -f(x)$. **85.** $f(x) = \log_a x$; $f\left(\dfrac{1}{x}\right) = \log_a \dfrac{1}{x} = \log_a 1 - \log_a x = -f(x)$

87. If $A = \log_a M$ and $B = \log_a N$, then $a^A = M$ and $a^B = N$.

Then $\log_a\left(\dfrac{M}{N}\right) = \log_a\left(\dfrac{a^A}{a^B}\right) = \log_a a^{A-B} = A - B = \log_a M - \log_a N.$

89. (a)

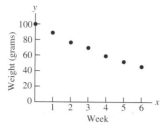

91. (a)

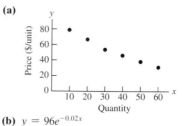

93. (a)

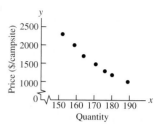

(b) $A = 100e^{-0.128t}$

(c) 5.4 weeks

(d) 0.17 g

(f)

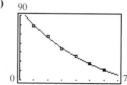

(b) $y = 96e^{-0.02x}$

(c) 23.5 units

(e)

(b) 168 computers

(d)

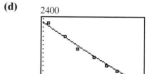

7.4 Exercises *(page 457)*

1. 6 **3.** 16 **5.** 8 **7.** 3 **9.** 5 **11.** $\{-1 + \sqrt{1 + e^4}, -1 - \sqrt{1 + e^4}\} \approx \{6.456, -8.456\}$ **13.** $\dfrac{\ln 3}{\ln 2} \approx 1.585$ **15.** 0 **17.** $\dfrac{\ln 10}{\ln 2} \approx 3.322$

19. $-\dfrac{\ln 1.2}{\ln 8} \approx -0.088$ **21.** $\dfrac{\ln 3}{2 \ln 3 + \ln 4} \approx 0.307$ **23.** $\dfrac{\ln 7}{\ln 0.6 + \ln 7} \approx 1.356$ **25.** 0 **27.** $\dfrac{\ln \pi}{1 + \ln \pi} \approx 0.534$ **29.** $\dfrac{\ln 1.6}{3 \ln 2} \approx 0.226$ **31.** $\dfrac{9}{2}$

33. 2 **35.** 1 **37.** 16 **39.** $\left\{-1, \dfrac{2}{3}\right\}$ **41.** 0 **43.** $\ln(2 + \sqrt{5}) \approx 1.444$ **45.** 1.92 **47.** 2.79 **49.** -0.57 **51.** -0.70 **53.** 0.57

55. $\{0.39, 1.00\}$ **57.** 1.32 **59.** 1.31

7.5 Exercises *(page 465)*

1. \$108.29 **3.** \$609.50 **5.** \$697.09 **7.** \$12.46 **9.** \$125.23 **11.** \$88.72 **13.** \$860.72 **15.** \$554.09 **17.** \$59.71 **19.** \$361.93 **21.** 5.35%

23. 26% **25.** $6\frac{1}{4}$% compounded annually **27.** 9% compounded monthly **29.** 104.32 months (about 8.7 yr); 103.97 mo (about 8.66 yr)

31. 61.02 mo; 60.82 mo **33.** 15.27 yr or 15 yr, 4 mo **35.** \$104,335 **37.** \$12,910.62 **39.** About \$30.17 per share or \$3017 **41.** 9.35%

43. Not quite. Jim will have \$1057.60. The second bank gives a better deal, since Jim will have \$1060.62 after 1 yr.

45. Will has \$11,632.73; Henry has \$10,947.89. **47. (a)** Interest is \$30,000 **(b)** Interest is \$38,613.59

(c) Interest is \$37,752.73. Simple interest at 12% is best. **49. (a)** \$1364.62 **(b)** \$1353.35 **51.** \$4631.93

53. (a) 6.1 yr **(b)** 18.45 yr **(c)** $mP = P\left(1 + \dfrac{r}{n}\right)^{nt}$

$$m = \left(1 + \dfrac{r}{n}\right)^{nt}$$

$$\ln m = \ln\left(1 + \dfrac{r}{n}\right)^{nt}$$

$$\ln m = nt \ln\left(1 + \dfrac{r}{n}\right)$$

$$t = \dfrac{\ln m}{n \ln\left(1 + \dfrac{r}{n}\right)}$$

7.6 Exercises *(page 474)*

1. 34.7 days; 69.3 days **3. (a)** 28.4 yr **(b)** 94.4 yr **5.** 5832; 3.9 days **7.** 25,198 **9.** 9.797 g **11.** 9727 yr ago

13. (a) 5:18 P.M. **(b)** After 14.3 min, the pizza will be 160°F. **(c)** As time passes, the temperature of the pizza gets closer to 70°F.

15. 18.63°C; 25.1°C **17.** 7.34 kg; 76.6 hr **19.** 26.6 days **21. (a)** 0.1286 **(b)** 0.9 **(c)** 1996 **23. (a)** 1000 **(b)** 30 **(c)** 11.076 hr

7.7 Exercises *(page 479)*

1. 70 decibels   **3.** 111.76 decibels   **5.** 10 watt/m²   **7.** 4.0 on the Richter scale   **9.** 125,892.54 mm; the Mexico City earthquake was 15.85 times as intense as the one in San Francisco.   **11.** Crowd noise was 31.6 times as intense as guidelines allow.

Fill-in-the Blank Items *(page 481)*

1. $(0, 1), (1, a),$ and $\left(-1, \dfrac{1}{a}\right)$ **3.** 4   **5.** 1   **7.** All real numbers greater than 0   **9.** 1

True/False Items *(page 482)*

1. T   **3.** F   **5.** T   **7.** F

Review Exercises *(page 482)*

1. -3 **3.** $\sqrt{2}$ **5.** 0.4   **7.** $\log_3 u + 2 \log_3 v - \log_3 w$ **9.** $2 \log x + \dfrac{1}{2}\log(x^3 + 1)$ **11.** $\ln x + \dfrac{1}{3}\ln(x^2 + 1) - \ln(x - 3)$

13. $\dfrac{25}{4}\log_4 x$ **15.** $-2\ln(x + 1)$ **17.** $\log\left(\dfrac{4x^3}{[(x + 3)(x - 2)]^{1/2}}\right)$ **19.** 2.124   **21.** $y = Ce^{2x^2}$ **23.** $y = \sqrt{e^{x+C} + 9}$

25. $y = \ln(x^2 + 4) - C$

27.

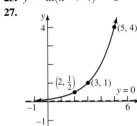

Domain: All real numbers
Range: $\{y|y > 0\}$ or $(0, \infty)$
Horizontal asymptote: $y = 0$

29.

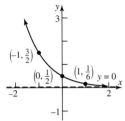

Domain: All real numbers
Range: $\{y|y > 0\}$ or $(0, \infty)$
Horizontal asymptote: $y = 0$

31.

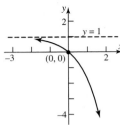

Domain: All real numbers
Range: $\{y|y < 1\}$ or $(-\infty, 1)$
Horizontal asymptote: $y = 1$

33.

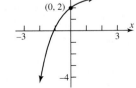

Domain: $\{x|x > 0\}$ or $(0, \infty)$
Range: All real numbers
Vertical asymptote: $x = 0$

35.

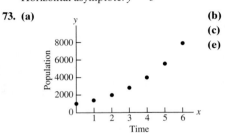

Domain: All real numbers
Range: $\{y|y < 3\}$ or $(-\infty, 3)$
Horizontal asymptote: $y = 3$

37. $\dfrac{1}{4}$ **39.** $\left\{\dfrac{-1 - \sqrt{3}}{2}, \dfrac{-1 + \sqrt{3}}{2}\right\}$ **41.** $\dfrac{1}{4}$ **43.** $\dfrac{2\ln 3}{\ln 5 - \ln 3} \approx 4.301$ **45.** $\dfrac{12}{5}$

47. 83   **49.** $\left\{\dfrac{1}{2}, -3\right\}$ **51.** -1 **53.** $1 - \ln 5 \approx -0.609$ **55.** $\dfrac{\ln 3}{3\ln 2 - 2\ln 3} \approx -9.327$

57. 3229.5 m   **59.** 7.6 mm of mercury   **61. (a)** 37.3 watt   **(b)** 6.9 decibels
63. (a) 71%   **(b)** 85.5%   **(c)** 90%   **(d)** About 1.6 mo   **(e)** About 4.8 mo
65. (a) 9.85 yr   **(b)** 4.27 yr   **67.** $41,669   **69.** 80 decibels   **71.** 24,203 yr ago

73. (a)

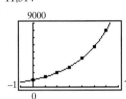

(b) $N = 1000e^{0.346574t}$
(c) 11,314
(e)

75. 6,078,190,457
77. (a) 0.3   **(b)** 0.8
(c)

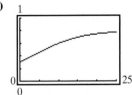

(d) When $t \approx 20.1$ or about 2008.

APPENDIX A   Review

A.1 Exercises (page 495)

1. 4   **3.** -28 **5.** $\dfrac{4}{5}$ **7.** 0   **9.** $x = 0$ **11.** $x = 3$ **13.** None   **15.** $x = 1, x = 0, x = -1$ **17.** $\{x \mid x \neq 5\}$ **19.** $\{x \mid x \neq -4\}$

21.   **23.** $>$ **25.** $>$ **27.** $>$ **29.** $=$ **31.** $<$ **33.** $x > 0$ **35.** $x < 2$ **37.** $x \leq 1$

39.   **41.**   **43.** 1   **45.** 5   **47.** 1   **49.** 22   **51.** 2   **53.** 1

55. 2   **57.** 6   **59.** 16   **61.** $\dfrac{1}{16}$ **63.** $\dfrac{1}{9}$ **65.** 9   **67.** 5   **69.** 4   **71.** $\dfrac{1}{64x^6}$ **73.** $\dfrac{x^4}{y^2}$ **75.** $\dfrac{1}{x^3 y}$ **77.** $-\dfrac{8x^3 z}{9y}$ **79.** $\dfrac{16x^2}{9y^2}$

81. $A = lw$; all the variables are positive real numbers   **83.** $C = \pi d$; d and C are positive real numbers

85. $A = \dfrac{\sqrt{3}}{4} x^2$; x and A are positive real numbers   **87.** $V = \dfrac{4}{3} \pi r^3$; r and V are positive real numbers

89. $V = x^3$; x and V are positive real numbers   **91. (a)** $2 \leq 5$ **(b)** $6 > 5$ **93. (a)** Acceptable   **(b)** Not acceptable

95. No; $\dfrac{1}{3}$ is larger; $0.000333\ldots$ **97.** No

A.2 Exercises (page 500)

1. 13   **3.** 26   **5.** 25   **7.** Right triangles; 5   **9.** Not a right triangle   **11.** Right triangle; 25   **13.** Not a right triangle   **15.** 8 in^2 **17.** 4 in^2

19. $A = 25\pi$ m^2; $C = 10\pi$ m   **21.** 224 ft^3 **23.** $V = \dfrac{256}{3} \pi$ cm^3; $S = 64\pi$ cm^2 **25.** 648π in^3 **27.** π square units   **29.** 2π square units

31. About 16.8 ft   **33.** 64 ft^2 **35.** $24 + 2\pi \approx 30.28$ ft^2; $16 + 2\pi \approx 22.28$ ft^2 **37.** About 5.477 mi   **39.** From 100 ft: 12.247 mi;
From 150 ft: 15.000 mi   **41.** The area of a rectangular pool with a 1000-ft perimeter has an area that is between 0 sq ft and 62,500 sq ft;
square; about 79,577 sq ft; answers may vary; answers may vary; circle

A.3 Exercises (page 511)

1. 7   **3.** -3 **5.** 4   **7.** $\dfrac{5}{4}$ **9.** -1 **11.** -18 **13.** -3 **15.** -16 **17.** 0.5   **19.** 2   **21.** 2   **23.** 3   **25.** $\{0, 9\}$ **27.** $\{0, 9\}$ **29.** 21

31. $\{-2, 2\}$ **33.** 6   **35.** $\{-3, 3\}$ **37.** $\{-4, 1\}$ **39.** $\left\{-1, \dfrac{3}{2}\right\}$ **41.** $\{-4, 4\}$ **43.** 2   **45.** No real solution   **47.** $\{-2, 2\}$ **49.** $\{-1, 3\}$

51. $\{-2, -1, 0, 1\}$ **53.** $\{0, 4\}$ **55.** $\{-6, 2\}$ **57.** $\left\{-\dfrac{1}{2}, 3\right\}$ **59.** $\{3, 4\}$ **61.** $\dfrac{3}{2}$ **63.** $\left\{-\dfrac{2}{3}, \dfrac{3}{2}\right\}$ **65.** $\left\{-\dfrac{3}{4}, 2\right\}$ **67.** $\{-5, 5\}$

69. $\{-1, 3\}$ **71.** $\{-3, 0\}$ **73.** 16   **75.** $\dfrac{1}{16}$ **77.** $\dfrac{1}{9}$ **79.** $\{-7, 3\}$ **81.** $\left\{-\dfrac{1}{4}, \dfrac{3}{4}\right\}$ **83.** $\left\{\dfrac{-1 - \sqrt{7}}{6}, \dfrac{-1 + \sqrt{7}}{6}\right\}$ **85.** $\{2 - \sqrt{2}, 2 + \sqrt{2}\}$

87. $\left\{\dfrac{5 - \sqrt{29}}{2}, \dfrac{5 + \sqrt{29}}{2}\right\}$ **89.** $\left\{1, \dfrac{3}{2}\right\}$ **91.** No real solution   **93.** $\left\{\dfrac{-1 - \sqrt{5}}{4}, \dfrac{-1 + \sqrt{5}}{4}\right\}$ **95.** $\left\{\dfrac{-\sqrt{3} - \sqrt{15}}{2}, \dfrac{-\sqrt{3} + \sqrt{15}}{2}\right\}$

97. No real solution   **99.** Repeated real solution   **101.** Two unequal real solutions   **103.** $x = \dfrac{b + c}{a}$ **105.** $x = \dfrac{abc}{a + b}$ **107.** $x = a^2$

109. $R = \dfrac{R_1 R_2}{R_1 + R_2}$ **111.** $R = \dfrac{mv^2}{F}$ **113.** $r = \dfrac{S - a}{S}$ **115.** $\dfrac{-b + \sqrt{b^2 - 4ac}}{2a} + \dfrac{-b - \sqrt{b^2 - 4ac}}{2a} = \dfrac{-2b}{2a} = \dfrac{-b}{a}$

117. $k = -\dfrac{1}{2}$ or $\dfrac{1}{2}$ **119.** The solutions of $ax^2 - bx + c = 0$ are $\dfrac{b + \sqrt{b^2 - 4ac}}{2a}$ and $\dfrac{b - \sqrt{b^2 - 4ac}}{2a}$.   **121. (b)**

A.4 Exercises (page 520)

1. $8 + 5i$ **3.** $-7 + 6i$ **5.** $-6 - 11i$ **7.** $6 - 18i$ **9.** $6 + 4i$ **11.** $10 - 5i$ **13.** 37   **15.** $\dfrac{6}{5} + \dfrac{8}{5}i$ **17.** $1 - 2i$ **19.** $\dfrac{5}{2} - \dfrac{7}{2}i$ **21.** $-\dfrac{1}{2} + \dfrac{\sqrt{3}}{2}i$

23. $2i$ **25.** $-i$ **27.** i **29.** -6 **31.** $-10i$ **33.** $-2 + 2i$ **35.** 0   **37.** 0   **39.** $2i$ **41.** $5i$ **43.** $5i$ **45.** $\{-2i, 2i\}$ **47.** $\{-4, 4\}$

49. $\{3 - 2i, 3 + 2i\}$ **51.** $\{3 - i, 3 + i\}$ **53.** $\left\{\dfrac{1}{4} - \dfrac{1}{4}i, \dfrac{1}{4} + \dfrac{1}{4}i\right\}$ **55.** $\left\{\dfrac{1}{5} - \dfrac{2}{5}i, \dfrac{1}{5} + \dfrac{2}{5}i\right\}$ **57.** $\left\{-\dfrac{1}{2} - \dfrac{\sqrt{3}}{2}i, -\dfrac{1}{2} + \dfrac{\sqrt{3}}{2}i\right\}$

59. $\{2, -1 - \sqrt{3}i, -1 + \sqrt{3}i\}$ **61.** $\{-2, 2, -2i, 2i\}$ **63.** $\{-3i, -2i, 2i, 3i\}$ **65.** Two complex solutions that are conjugates of each other.
67. Two unequal real solutions.   **69.** A repeated real solution.   **71.** $2 - 3i$ **73.** 6   **75.** 25
77. $z + \bar{z} = (a + bi) + (a - bi) = 2a$; $z - \bar{z} = (a + bi) - (a - bi) = 2bi$
79. $\overline{z + w} = \overline{(a + bi) + (c + di)} = \overline{(a + c) + (b + d)i} = (a + c) - (b + d)i = (a - bi) + (c - di) = \bar{z} + \bar{w}$

A.5 Exercises *(page 528)*

1. $[0, 2]; 0 \le x \le 2$ **3.** $(-1, 2); -1 < x < 2$ **5.** $[0, 3); 0 \le x < 3$

7. $[0, 4]$

9. $[4, 6)$

11. $[4, \infty)$

13. $(-\infty, -4)$

15. $2 \le x \le 5$

17. $-3 < x < -2$

19. $x \ge 4$

21. $x < -3$

23. (a) $6 < 8$ **(b)** $-2 < 0$ **(c)** $9 < 15$ **(d)** $-6 > -10$ **25. (a)** $7 > 0$ **(b)** $-1 > -8$ **(c)** $12 > -9$ **(d)** $-8 < 6$
27. (a) $2x + 4 < 5$ **(b)** $2x - 4 < -3$ **(c)** $6x + 3 < 6$ **(d)** $-4x - 2 > -4$
29. $<$ **31.** $>$ **33.** $\ge$ **35.** $<$ **37.** $\le$ **39.** $>$ **41.** $\ge$

43. $\{x | x < 4\}; (-\infty, 4)$

45. $\{x | x \ge -1\}; [-1, \infty)$

47. $\{x | x > 3\}; (3, \infty)$

49. $\{x | x \ge 2\}; [2, \infty)$

51. $\{x | x > -7\}; (-7, \infty)$

53. $\left\{x \middle| x \le \dfrac{2}{3}\right\}; \left(-\infty, \dfrac{2}{3}\right]$

55. $\{x | x < -20\}; (-\infty, -20)$

57. $\left\{x \middle| x \ge \dfrac{4}{3}\right\}; \left[\dfrac{4}{3}, \infty\right)$

59. $\{x | 3 \le x \le 5\}; [3, 5]$

61. $\left\{x \middle| \dfrac{2}{3} \le x \le 3\right\}; \left[\dfrac{2}{3}, 3\right]$

63. $\left\{x \middle| -\dfrac{11}{2} < x < \dfrac{1}{2}\right\}; \left(-\dfrac{11}{2}, \dfrac{1}{2}\right)$

65. $\{x | -6 < x < 0\}; (-6, 0)$

67. $\left\{x \middle| x < -\dfrac{1}{2}\right\}; \left(-\infty, -\dfrac{1}{2}\right)$

69. $\left\{x \middle| x > \dfrac{10}{3}\right\}; \left(\dfrac{10}{3}, \infty\right)$

71. $\{x | x > 3\}; (3, \infty)$

73. $\{x | -4 < x < 4\}; (-4, 4)$

75. $\{x | x < -4 \text{ or } x > 4\}; (-\infty, -4) \text{ or } (4, \infty)$

77. $\{x | 0 \le x \le 1\}; [0, 1]$

79. $\{x | x < -1 \text{ or } x > 2\}; (-\infty, -1) \text{ or } (2, \infty)$

81. $\{x | -1 \le x \le 1\}; [-1, 1]$

83. $\{x | x \le -2 \text{ or } x \ge 2\}; (-\infty, -2] \text{ or } [2, \infty)$

85. $|x - 2| < \dfrac{1}{2}; \left\{x \middle| \dfrac{3}{2} < x < \dfrac{5}{2}\right\}$ **87.** $|x + 3| > 2; \{x | x < -5 \text{ or } x > 1\}$ **89.** $21 < \text{age} < 30$ **91.** $|x - 98.6| \ge 1.5$;
$\{x | x \le 97.1 \,°\text{F} \text{ or } x \ge 100.1 \,°\text{F}\}$ **93. (a)** Male ≥ 73.4 **(b)** Female ≥ 79.7 **(c)** A female can expect to live 6.3 yr longer.
95. The agent's commission ranges from \$45,000 to \$95,000, inclusive. As a percent of selling price, the commission ranges from 5% to 8.6%,
inclusive. **97.** The amount withheld varies from \$72.14 to \$93.14, inclusive. **99.** The usage varies from 675.43 kW · hr to 2500.86 kW · hr,
inclusive. **101.** The dealer's cost varies from \$7457.63 to \$7857.14, inclusive. **103.** You need at least a 74 on the fifth test.

105. $\dfrac{a + b}{2} - a = \dfrac{a + b - 2a}{2} = \dfrac{b - a}{2} > 0;$ therefore, $a < \dfrac{a + b}{2}$

$b - \dfrac{a + b}{2} = \dfrac{2b - a - b}{2} = \dfrac{b - a}{2} > 0;$ therefore, $b > \dfrac{a + b}{2}$

107. $(\sqrt{ab})^2 - a^2 = ab - a^2 = a(b - a) > 0;$ thus, $(\sqrt{ab})^2 > a^2$ and $\sqrt{ab} > a$
$b^2 - (\sqrt{ab})^2 = b^2 - ab = b(b - a) > 0;$ thus $b^2 > (\sqrt{ab})^2$ and $b > \sqrt{ab}$

109. $h - a = \dfrac{2ab}{a + b} - a = \dfrac{ab - a^2}{a + b} = \dfrac{a(b - a)}{a + b} > 0;$ thus, $h > a$

$b - h = b - \dfrac{2ab}{a + b} = \dfrac{b^2 - ab}{a + b} = \dfrac{b(b - a)}{a + b} > 0;$ thus $h < b$

A.6 Exercises *(page 536)*

1. 3 **3.** -2 **5.** $2\sqrt{2}$ **7.** $-2x\sqrt[3]{x}$ **9.** $x^3 y^2$ **11.** $x^2 y$ **13.** $6\sqrt{x}$ **15.** $6x\sqrt{x}$ **17.** $15\sqrt[3]{3}$ **19.** $12\sqrt{3}$ **21.** $2\sqrt{3}$ **23.** $x - 2\sqrt{x} + 1$
25. $-5\sqrt{2}$ **27.** $(2x - 1)\sqrt[3]{2x}$ **29.** $\dfrac{\sqrt{2}}{2}$ **31.** $-\dfrac{\sqrt{15}}{5}$ **33.** $\dfrac{\sqrt{3}(5 + \sqrt{2})}{23}$ **35.** $\dfrac{-19 + 8\sqrt{5}}{41}$ **37.** $\dfrac{5\sqrt[3]{4}}{2}$ **39.** $\dfrac{2x + h - 2\sqrt{x(x + h)}}{h}$

41. $\dfrac{9}{2}$ **43.** 3 **45.** 4 **47.** -3 **49.** 64 **51.** $\dfrac{1}{27}$ **53.** $\dfrac{27\sqrt{2}}{32}$ **55.** $\dfrac{27\sqrt{2}}{32}$ **57.** $x^{7/12}$ **59.** xy^2 **61.** $x^{4/3}y^{5/3}$ **63.** $\dfrac{8x^{3/2}}{y^{1/4}}$ **65.** $\dfrac{3x+2}{(1+x)^{1/2}}$

67. $\dfrac{2+x}{2(1+x)^{3/2}}$ **69.** $\dfrac{4-x}{(x+4)^{3/2}}$ **71.** $\dfrac{1}{2}(5x+2)(x+1)^{1/2}$ **73.** $2x^{1/2}(3x-4)(x+1)$ **75.** $\dfrac{2(2-x)(2+x)}{(8-x^2)^{1/2}}$

A.7 Exercises *(page 548)*

1. (a) $\dfrac{1}{2}$ **(b)** For every 2 unit increase in x, y will increase by 1 unit. **3. (a)** $-\dfrac{1}{3}$ **(b)** For every 3 unit increase in x, y will decrease by 1 unit.

5. Slope $= -\dfrac{3}{2}$ **7.** Slope $= -\dfrac{1}{2}$ **9.** Slope $= 0$ **11.** Slope undefined

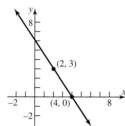 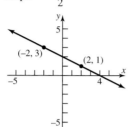 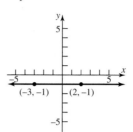 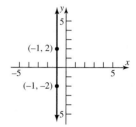

13. **15.** **17.** **19.**

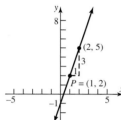 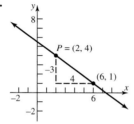 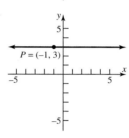 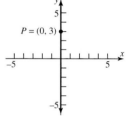

21. $x - 2y = 0$ or $y = \dfrac{1}{2}x$ **23.** $x + 3y = 4$ or $y = -\dfrac{1}{3}x + \dfrac{4}{3}$ **25.** $2x - y = 3$ or $y = 2x - 3$ **27.** $x + 2y = 5$ or $y = -\dfrac{1}{2}x + \dfrac{5}{2}$

29. $3x - y = -9$ or $y = 3x + 9$ **31.** $2x + 3y = -1$ or $y = -\dfrac{2}{3}x - \dfrac{1}{3}$ **33.** $x - 2y = -5$ or $y = \dfrac{1}{2}x + \dfrac{5}{2}$

35. $3x + y = 3$ or $y = -3x + 3$ **37.** $x - 2y = 2$ or $y = \dfrac{1}{2}x - 1$ **39.** $x = 2$; no slope–intercept form **41.** $2x - y = -4$ or $y = 2x + 4$

43. $2x - y = 0$ or $y = 2x$ **45.** $x = 4$; no slope–intercept form **47.** $2x + y = 0$ or $y = -2x$ **49.** $x - 2y = -3$ or $y = \dfrac{1}{2}x + \dfrac{3}{2}$

51. $y = 4$ or $y = 4$

53. Slope $= 2$; y-intercept $= 3$ **55.** $y = 2x - 2$; Slope $= 2$; y-intercept $= -2$ **57.** Slope $= \dfrac{1}{2}$; y-intercept $= 2$

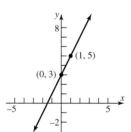 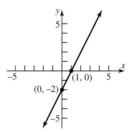 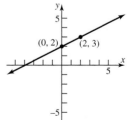

59. $y = -\frac{1}{2}x + 2$; Slope $= -\frac{1}{2}$; y-intercept $= 2$

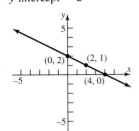

61. $y = \frac{2}{3}x - 2$; Slope $= \frac{2}{3}$; y-intercept $= -2$

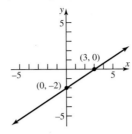

63. $y = -x + 1$; Slope $= -1$; y-intercept $= 1$

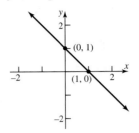

65. Slope undefined; no y-intercept

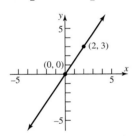

67. Slope $= 0$; y-intercept $= 5$

69. $y = x$; Slope $= 1$; y-intercept $= 0$

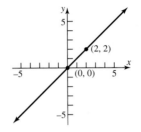

71. $y = \frac{3}{2}x$; Slope $= \frac{3}{2}$; y-intercept $= 0$

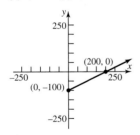

73. $y = 0$

75. $P_1 = (-2, 5), P_2 = (1, 3), m_1 = -\frac{2}{3}; P_2 = (1, 3), P_3 = (-1, 0), m_2 = \frac{3}{2}$; because $m_1 m_2 = -1$, the lines are perpendicular; thus, the points P_1, P_2, and P_3 are the vertices of a right triangle.

77. $P_1 = (-1, 0), P_2 = (2, 3), m = 1; P_3 = (1, -2), P_4 = (4, 1), m = 1; P_1 = (-1, 0), P_3 = (1, -2), m = -1; P_2 = (2, 3), P_4 = (4, 1), m = -1$; opposite sides are parallel, and adjacent sides are perpendicular; the points are the vertices of a rectangle.

79. $°C = \frac{5}{9}(°F - 32)$; approximately 21 °C

81. (a) $P = 0.5x - 100$
(b) $400 (c) $2400

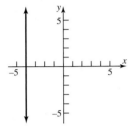

83. $C = 0.06543x + 5.65; C = 25.28; $C = 54.72

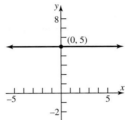

85. (b)
87. (d)
89. $x - y = -2$ or $y = x + 2$
91. $x + 3y = 3$ or $y = -\frac{1}{3}x + 1$
93. b, c, e, g
95. c

97. (a) $x^2 + (mx + b)^2 = r^2$
$(1 + m^2)x^2 + 2mbx + b^2 - r^2 = 0$
One solution if and only if discriminant $= 0$
$(2mb)^2 - 4(1 + m^2)(b^2 - r^2) = 0$
$-4b^2 + 4r^2 + 4m^2r^2 = 0$
$r^2(1 + m^2) = b^2$

(b) $x = \dfrac{-2mb}{2(1 + m^2)} = \dfrac{-2mb}{2b^2/r^2} = -\dfrac{r^2m}{b}$

$y = m\left(-\dfrac{r^2m}{b}\right) + b = -\dfrac{r^2m^2}{b} + b = \dfrac{-r^2m^2 + b^2}{b} = \dfrac{r^2}{b}$

(c) Slope of tangent line $= m$
Slope of line joining center to point of tangency $= \dfrac{r^2/b}{-r^2m/b} = -\dfrac{1}{m}$

99. $\sqrt{2}x + 4y = 11\sqrt{2} - 12$ **101.** $x + 5y = -13$

103. All have the same slope, 2; the lines are parallel.

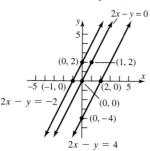

105. $y = 2$
107. No, if the line is horizontal.
109. They are the same line.
111. No

A.8 Exercises *(page 556)*

1. Linear   **3.** Linear   **5.** Nonlinear

7. (a), (c)

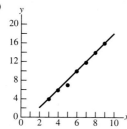

(b) Using $(3, 4)$ and $(9, 16)$, $y = 2x - 2$.
(d) $y = 2.0357x - 2.3571$ **(e)**

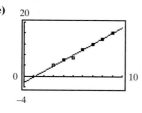

9. (a), (c)

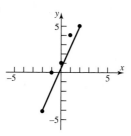

(b) Using $(-2, -4)$ and $(2, 5)$, $y = \dfrac{9}{4}x + \dfrac{1}{2}$.
(d) $y = 2.2x + 1.2$ **(e)**

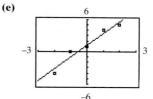

11. (a), (c)

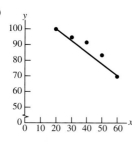

(b) Using $(20, 100)$ and $(60, 70)$, $y = -\dfrac{3}{4}x + 115$.
(d) $y = -0.72x + 116.6$ **(e)**

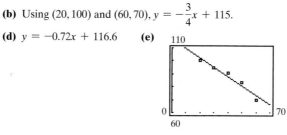

13. (a), (c)

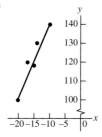

(b) Using $(-20, 100)$ and $(-10, 140)$, $y = 4x + 180$.
(d) $y = 3.8613x + 180.292$ **(e)**

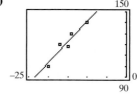

15. (a)

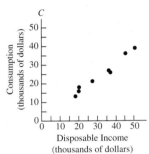

(b) Using $(20, 16)$ and $(50, 39)$, $C = \dfrac{23}{30}I + \dfrac{2}{3}$.

(c) As disposable income increases by \$1, consumption increases by about \$0.77.

(d) \$32,867

(e) $C = 0.755I + 0.6266$

17. (a) 225,000

(b) $L = 2.98I - 76.11$

(c) 225,000

(d) For each additional dollar of income, the amount that the institution will loan you increases by \$2.98.

(e) \$125,143

19. (a) 70

(b) $T = 0.0782h + 59.091$

(c) 70

(d) If relative humidity increases by 1%, the apparent temperature increases by 0.0782 degrees Fahrenheit.

(e) about 65°F

21. (a) 12

(b) $M = 0.1633x - 314.7139$

(c) 12

(d) For each year that passes, the average miles driven per car increases by 0.1633 thousand miles (163.3 miles).

(e) about 11.4 thousand miles

A P P E N D I X   B   **Graphing Utilities**

B.1 Exercises *(page 563)*

1. $(-1, 4)$; II   **3.** $(3, 1)$; I   **5.** $X\min = -6, X\max = 6, X\mathrm{scl} = 2, Y\min = -4, Y\max = 4, Y\mathrm{scl} = 2$ **7.** $X\min = -6, X\max = 6,$
$X\mathrm{scl} = 2, Y\min = -1, Y\max = 3, Y\mathrm{scl} = 1$ **9.** $X\min = 3, X\max = 9, X\mathrm{scl} = 1, Y\min = 2, Y\max = 10, Y\mathrm{scl} = 2$
11. $X\min = -11, X\max = 5, X\mathrm{scl} = 1, Y\min = -3, Y\max = 6, Y\mathrm{scl} = 1$ **13.** $X\min = -30, X\max = 50, X\mathrm{scl} = 10, Y\min = -90,$
$Y\max = 50, Y\mathrm{scl} = 10$ **15.** $X\min = -10, X\max = 110, X\mathrm{scl} = 10, Y\min = -10, Y\max = 160, Y\mathrm{scl} = 10$ **17.** $4\sqrt{10}$ **19.** $2\sqrt{65}$

B.2 Exercises *(page 566)*

1. (a)

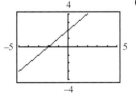

(b)

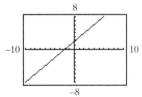

(c)

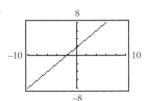

(d)

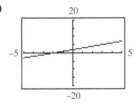

3. (a) 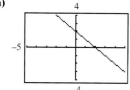 **(b)** 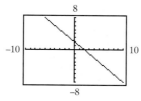 **(c)** 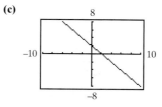 **(d)**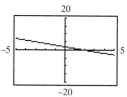

5. (a) 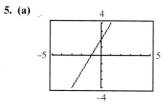 **(b)** 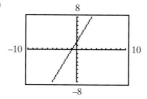 **(c)** 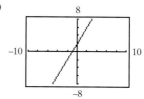 **(d)**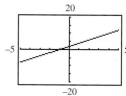

7. (a) 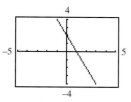 **(b)** 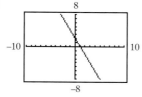 **(c)** 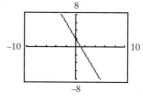 **(d)**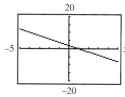

9. (a) 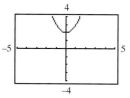 **(b)** 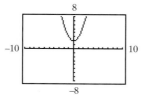 **(c)** 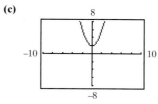 **(d)**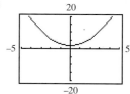

11. (a) 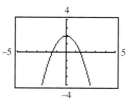 **(b)** 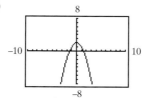 **(c)** 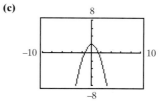 **(d)**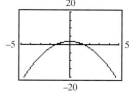

13. (a) 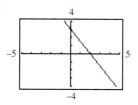 **(b)** 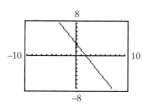 **(c)** 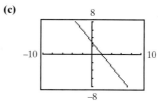 **(d)**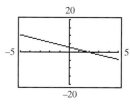

15. (a) 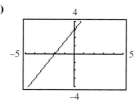 **(b)** 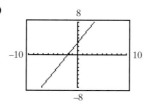 **(c)** 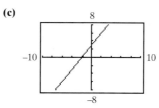 **(d)**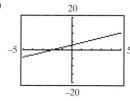

17.

X	Y1
-3	-1
-2	0
-1	1
0	2
1	3
2	4
3	5

Y1=X+2

19.

X	Y1
-3	5
-2	4
-1	3
0	2
1	1
2	0
3	-1

Y1=-X+2

21.

X	Y1
-3	-4
-2	-2
-1	0
0	2
1	4
2	6
3	8

Y1=2X+2

23.

X	Y1
-3	8
-2	6
-1	4
0	2
1	0
2	-2
3	-4

Y1=-2X+2

25.

X	Y1
-3	11
-2	6
-1	3
0	2
1	3
2	6
3	11

Y1=X²+2

27.

X	Y1
-3	-7
-2	-2
-1	1
0	2
1	1
2	-2
3	-7

Y1=-X²+2

29.

X	Y1
-3	7.5
-2	6
-1	4.5
0	3
1	1.5
2	0
3	-1.5

Y1=-(3/2)X+3

31.

X	Y1
-3	-1.5
-2	0
-1	1.5
0	3
1	4.5
2	6
3	7.5

Y1=(3/2)X+3

B.3 Exercises *(page 569)*

1. -3.41 **3.** -1.71 **5.** -0.28 **7.** 3.00 **9.** 4.50 **11.** $0.32, 12.30$ **13.** $1.00, 23.00$ **15. (a)** $(-1.5, 0), (0, -2), (1.5, 0)$ **(b)** y-axis
17. (a) none   **(b)** origin

B.5 Exercises *(page 572)*

1. Yes   **3.** Yes   **5.** No   **7.** Yes   **9.** $Y\min = 4$ Other answers are possible.
$Y\max = 12$
$Y\text{scl} = 1$

Index